4th International Conference on Computing and Solutions in Manufacturing Engineering 2016 (CoSME'16)

MATEC Web of Conferences Volume 94 (2017)

Brasnov, Romania
3 - 4 November 2016

Editors:

G. Oancea
M.V. Dragoi

ISBN: 978-1-5108-3511-5

Printed from e-media with permission by:

Curran Associates, Inc.
57 Morehouse Lane
Red Hook, NY 12571

Some format issues inherent in the e-media version may also appear in this print version.

Printed by Curran Associates, Inc. (2017)

For additional information, please contact EDP Sciences – Web of Conferences at the address below.

EDP Sciences – Web of Conferences
17, Avenue du Hoggar
Parc d'Activité de Courtabœuf
BP 112
F-91944 Les Ulis Cedex A
France

Phone: +33 (0) 1 69 18 75 75
Fax: +33 (0) 1 69 28 84 91

contact-edps@webofconferences.org

Additional copies of this publication are available from:

Curran Associates, Inc.
57 Morehouse Lane
Red Hook, NY 12571 USA
Phone: 845-758-0400
Fax: 845-758-2633
Email: curran@proceedings.com
Web: www.proceedings.com

TABLE OF CONTENTS

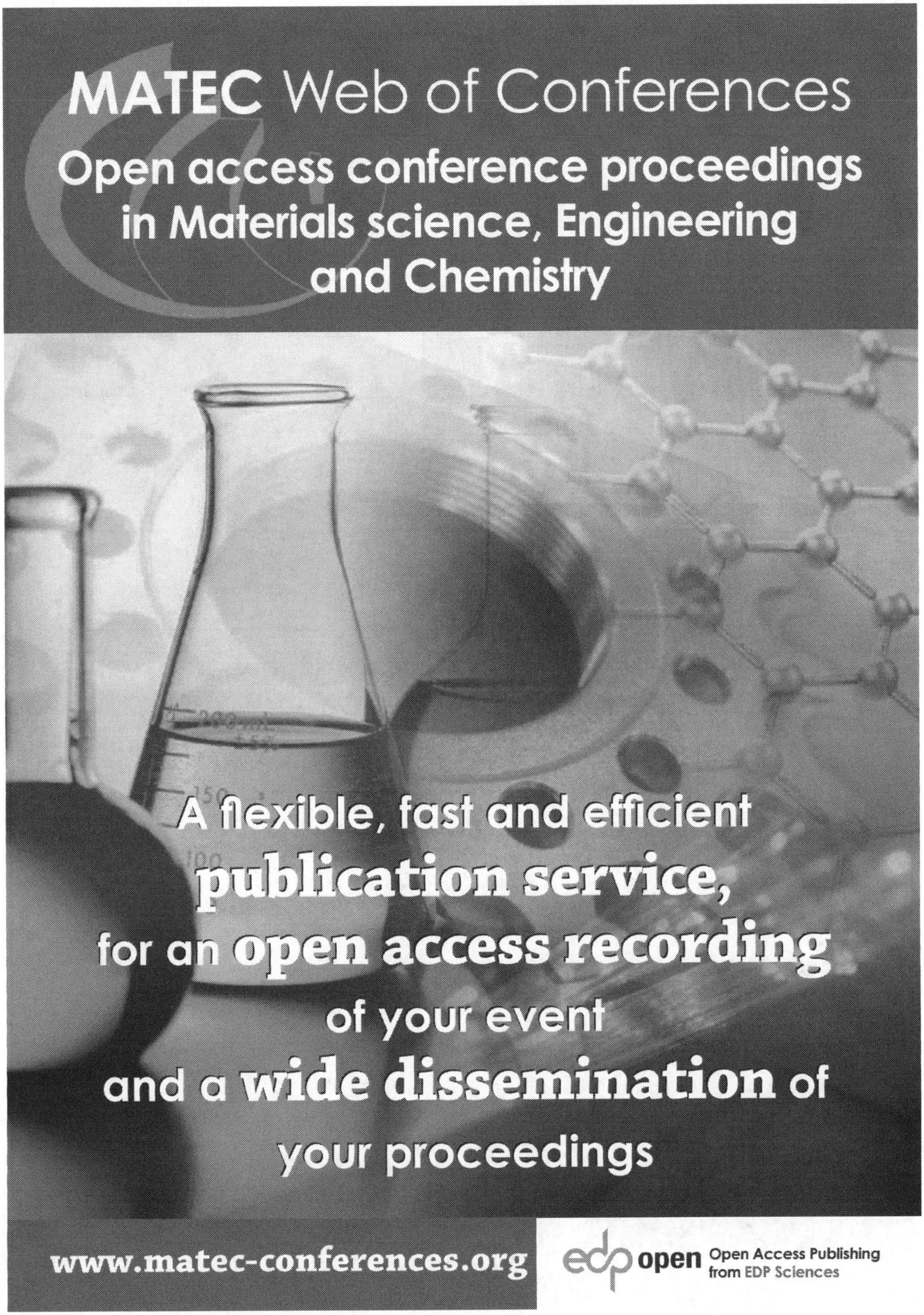
MATEC Web of Conferences
Open access conference proceedings in Materials science, Engineering and Chemistry
A flexible, fast and efficient publication service, for an open access recording of your event and a wide dissemination of your proceedings
www.matec-conferences.org
edp open Open Access Publishing from EDP Sciences

PREFACE

The 4th International Conference on **Co**mputing and **S**olutions in **M**anufacturing **E**ngineering **2016 - CoSME'16** organised by Department of Manufacturing Engineering, Faculty of Technological Engineering and Industrial Management, *Transilvania* University of Braşov, during 3 - 4 November 2016, took place in the "probably the most beautiful city of Romania", Braşov, which is located in the heart of Romania, in the region of Transilvania.

The objective of the CoSME Conference is to bring together professionals from universities and industry, to present and discuss recent advances in manufacturing engineering, to stimulate technical and scientific innovations on sustainable development of designing and manufacturing products, and to provide a focus for the exchange of latest ideas on products engineering.

All papers submitted to CoSME'16 Conference were subjected to a rigorous peer reviews process. Each paper was reviewed by at least two independent expert referees selected from the Reviewing Committee, which mainly consists of Members of International Scientific Committee. After the reviewing process, 95 papers written by authors from 12 countries and 41 independent academic organisations have been accepted to be presented in the conference. A few information after the CoSME'16 Conference are as follows: 95 accepted papers, 90 presented papers, 6 invited keynote speakers, 1 plenary presentation, 7 sections (in the conference program the papers were distributed in 7 sections), 134 participants from universities/research institutes and industry, and 8 sponsors.

Extended abstracts of all accepted papers were published in a *Book of Abstracts* and the presented papers in the conference are published, on-line with open-access, in this volume of *MATEC Web of Conferences*.

We express our deepest gratitude to members in the International Scientific Committee, invited keynote speakers, reviewers, sponsors, all participants in the conference, and members in the Organising Committee for their valuable contribution to the successful organisation of the CoSME'16 Conference. We would like to give special thanks to MATEC team for its support in online publishing of CoSME'16 proceedings.

Professor Gheorghe OANCEA, PhD
Chairman of CoSME'16 Conference
Dean of Technological Engineering and Industrial Management Faculty
Transilvania University of Braşov

Braşov, November 2016

INTERNATIONAL SCIENTIFIC COMMITTEE

CHAIR PERSONS

ORGANISING COMMITTEE

MEMBERS

INVITED KEYNOTE SPEAKERS

Cécile LANGLADE,
Université de Technologie Belfort-Montbéliard, France

HYBRID SURFACE TREATMENTS: NEW PROCESSES COMBINING SEVERE PLASTIC DEFORMATION AND THERMAL SPRAY PROCESSES TO ACHIEVE IMPROVED MECHANICAL PROPERTIES

Andreas GEBHARDT,
University of Applied Sciences Aachen, Germany

LEAVING THE LABS BEHIND. AM IS HEADING FOR SERIES PRODUCTION

Michael BUJARA,
Regional Sales Manager – Sales EMEA, Carl Zeiss Industrielle Messtechnik GMBH, Germany

THE ZEISS METROLOGY - FACING NEW CHALLENGES

Panagiotis KYRATSIS,
Western Macedonia University of Applied Sciences, Greece

CAD BASED MANUFACTURING

Tineke EGYEDI,
Delft Institute for Research on Standardization, Netherlands

ON TECHNOLOGY CHOICE IN STANDARDIZATION: IMPLICATIONS FOR PATENT VALUATION

Angelos ZACHARIADIS,
Union of Hellenes Scientists for Standardization and Protypation (ENEPROT), Greece

DIFFERENTIATIONS OF THE INDUSTRIAL FINANCIAL CREDIT SYSTEM FROM THE PREINDUSTRIAL MODELS

PLENARY SPEAKER

Gheorghe OANCEA,
Transilvania University of Brasov, Romania

FACULTY OF TECHNOLOGICAL ENGINEERING AND INDUSTRIAL MANAGEMENT: A HISTORY OF OVER 50 YEARS

edp sciences

MATEC Web of Conferences

Statement of Peer review

In submitting conference proceedings to MATEC Web of Conferences, I certify to the Publisher that I adhere to the **Policy on Publishing Integrity** of the journal in order to safeguard good scientific practice in publishing.

1. All articles have been subjected to peer review administered by the proceedings editors.
2. Reviews have been conducted by expert referees, who have been requested to provide unbiased and constructive comments aimed, whenever possible, at improving the work.
3. Proceedings editors have taken all reasonable steps to ensure the quality of the materials they publish and their decision to accept or reject a paper for publication has been based only on the merits of the work and the relevance to the journal.

Title, date and place of the conference

The 4th International Conference on Computing and Solutions in Manufacturing Engineering – CoSME'16. November, 3-4, Brașov, Romania.

Proceedings editor(s):

Professor Gheorghe CANCEA, PhD.
Professor Mircea Viorel DRAGOI, PhD.
Transilvania University of Brasov, Romania.

Date and editor's signature

November 29, 2016

Correlative analysis of shoe insoles acting in moist and cold/warm environment

Barbu Braun[1,*], and *Mihaela* Baritz[1]

[1]Transilvania University of Brasov, Product of Design Mechatronics and Environment Department, Brasov, Romania

Abstract. Insoles represent a mean by which humans can improve postural comfort or existence and manifestation of posture static and dynamic. The variety of materials and sizes for different kinds of foot insoles are for manufacturers a challenge in terms of getting products flexible, versatile and easily adaptable to a larger range of subjects. Biomechanical foot dysfunction influences plantar load distribution and mechanical tissue stress. Greater influence over plantar loads and tissue stress may lead to optimize the insole design to improve foot biomechanics. Some investigations founded that the use of cushioned insoles reduces the risk of stress fractures and injuries; other research has shown no protective effect. The use of cushioned or shock-absorbing insoles invoked reducing the impact forces associated with running, protecting against injuries. The paper emphasizes a new method for analysis of the insoles materials behavior in different alkalis and acids, similar to those encountered in real situations when walking and standing. The behavior study of composite structure insoles started to establish constructive variants and the modality of exposure of these samples at different degrees of softening in simple water, in water with salinity and under negative and positive temperature, also to mechanical stress.

1 Introduction

Insoles represent a multitude of means by which humans can improve comfort and postural dysfunctions existence and manifestation of a static posture and/or dynamic. The variety of materials and sizes as appropriate to different kinds of foot insoles are for manufacturers a challenge in terms of getting those products flexible, versatile and easily adaptable to a range of subjects as large [1].

Ensuring postural comfort through the use of well-made insoles and performance characteristics is the most important requirement in the construction category of footwear, either daily wear or for sports activities. As established in international standards there are a number of categories of shoes that protect human loco-motor system. "Safety shoes or boots with compression protection are required for work where a heavy object or item may roll over or a heavy animal may step on the toe (e.g., operating skid trucks and pallet jacks, working around bulk paper rolls or heavy pipes, etc.). In accordance with ASTM

*Corresponding author: braun@unitbv.ro

performance standards - USA, compression resistant footwear must also meet impact resistance requirements. Again, there are two classes for compression resistance: class 50 which provides protection against exposure to a compressive force of 794 kg and class 75 which provides protection against exposure to a compressive force of 1134 kg" [2]. There are also other categories of standardized limits for planting surface protection of human loco-motor system, like: metatarsal protection, penetration of footwear bottoms, chainsaw use, electric shock, electric conduction, static dissipation, dielectric insulation, ankle protection, water resistant, chemical resistant, thermal insulated (cold).

Another important issue, with applications in rehabilitation of patients with diseases on the plantar surface is the choice of materials "when the physical and mechanical properties - durability, resilience, compressive stiffness, and coefficient of friction - are generally taken as the key requirements for the evaluation and selection of suitable fabrication materials for orthotic insoles" [3].

As shown by the author in his paper, there are still a number of "traditional tests which also neglect the importance of comfort perception in foot orthotic treatment, which affects the rate of compliance. A new in-shoe microclimate, which is relatively higher in temperature and humidity than the outer environment, is created during gait. Factors presented in previous experience, presence of pain or injury, neurophysiology and psychological issues, as well as the design, contour, and hardness of insole manufacturing may contribute to the overall comfort perception of foot orthoses". Many companies or clinics specialized in performance athletic footwear made a series of crash tests for soles with or without insoles to assess resistance to jumping, falling or shares with high energy loads. "Therefore, the energy absorbed (ΔE between with and without insole conditions as having the same peak deceleration) by the insole could be calculated from the regression equations as follows:

$$a_s = \alpha \times E + \beta = \alpha_i \times E_i + \beta_i \tag{1}$$

where α, β, α_i, and β_i, are the coefficients of the linear regression equations for without insole and with insole conditions, respectively. The term a_s indicates the same peak deceleration under each condition, and E and E_i are the impact energies resulting in the same peak deceleration for without insole and with insole conditions, respectively; E_i is larger than E because of the cushioning ability of the insole, and:

$$\Delta E = E_i - E = [(\alpha - \alpha_i) \times E_i + (\beta - \beta_i)]\alpha \tag{2}$$

For another testing condition, the shock absorption foam was interposed below the insole. With the same way, the energy absorbed by the shock absorption foam could be calculated from the regression equations with insole only and with insole plus shock absorption foam conditions" [4].

Other researches on the insoles were focused on analyzing the methods of materials selection used to achieve them, focusing on their proper selection after simulation and modeling activities. As materials, producers have chosen those, based on polyethylene due to the configuration stability and possess adequate flexibility and strength. "Foot patterns are studied and tested to identify the different pressure at the insole to produce results in stress, displacement, and deformation" [5].

In most analysis of insoles behavior and after choosing materials, research has focused on determining the parameters of these mounting positions through their use in orthopedic studies on the forms of walking, running or jumping. Thus the concept of senzorimotor insoles was developed and built for applications in musculoskeletal rehabilitation. With this concept many patients could improve their gait without an operation of their musculoskeletal system [6].

"In an attempt to gain a greater understanding of the characteristics of orthotics materials and how they affect normal gait, Healy (2012) performed a repeated measures study analyzing kinematics and plantar pressures of 10 healthy participants while walking on a treadmill in various conditions, including shoes only, shoes with 4 types of flat insoles and 4 types of custom orthoses. The material properties for both custom and non-custom insert conditions included one of each of:

1. Low density polyurethane (PU) (Shore A hardness 20-25),
2. Medium density PU (Shore A hardness 55 ±3),
3. Low density ethyl vinyl acetate (EVA) (Shore A hardness 25) and
4. Medium density EVA (Shore hardness 50) [7].

These materials were selected based that these were commonly prescribed in orthoses for individuals with diabetes. "Results for the flat insoles indicated that medium density EVA produced the greatest peak pressure. Low and medium density PU were most effective at increasing average contact area and reducing the pressure time interval. In the custom orthoses category there were no significant differences when compared to the shoe only condition, but medium density PU increased the average contact area by a greater percentage than the other materials. The authors concluded that medium density PU may have a different loading characteristic than the others, indicating it to be most suitable for patient with diminished plantar pressure sensitivities" [7].

2 Method to evaluate the behavior of insoles

The study consisted of two steps: the first step was related to analysis of water influence on insoles materials and the second step was related to analysis of cold and warm environment influences on the same insoles materials. First the first step of studies 7 samples of insoles was choosing (Fig.1.) to be analyzed in different environmental conditions (immersion in water and water solution, cold/warm environment and mechanical pressure).

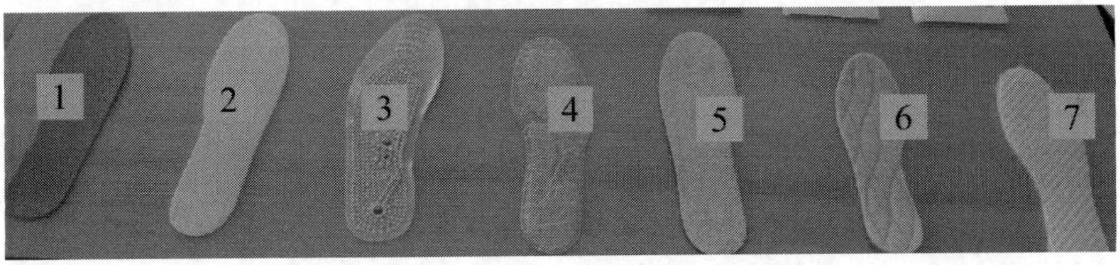

Fig. 1. The used materials for plantar shoe insole.

In the first step, two samples of each insole category were weighed, in dry conditions, at ambient temperature of 19^0 C. After this step two samples of each insoles were then placed completely in water, in 5% saline solution (two samples of each insoles), other two samples of each insoles insert into refrigerator at minus 18^0C (during 12 hours), other two samples of each insoles placed in a heating chamber (dry autoclave) up at 50^0C temperature (during 1 hour) and other two samples of each insoles placed under a mechanical force during 24 hours.

Entire duration of immersion in the prepared solutions was 7 days and weighing was performed after 1 hour, 2 hours, 24 hours, 48 hours and after 7 days for each sample. Each sample was monitored for 7 days from a visual standpoint and also the environmental parameters that were these samples (neutered containers, monitored temperature to remain constant throughout the experiment, natural and normal lighting - 200 Lx).

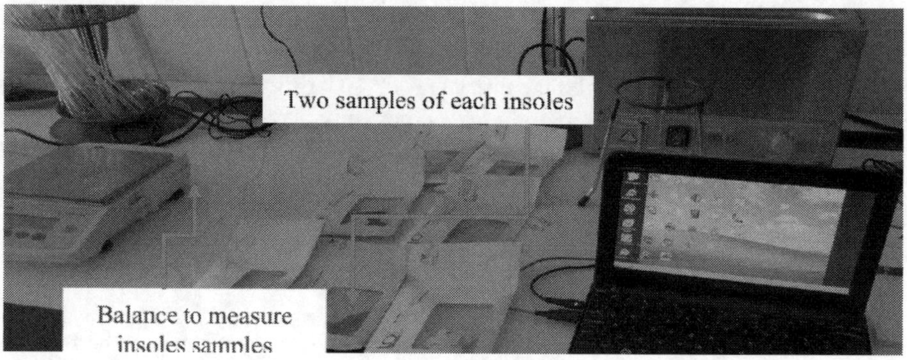

Fig. 2. The procedure of weighing the insoles samples before immersion in water, saline, dry autoclave and refrigerator and also after immersion process.

The duration of immersion was determined according to the determination of the samples estimated average exposure period after which to emphasize the substantial changes at the level of structure in the insoles construction.

Fig. 3. The procedure of insoles immersion in water and saline.

3 Results of experiments

In the first phase of experiments was conducted to the weight's variation determination - two samples for each type of insoles - immersed in water (Fig. 4.), and in saline (Fig. 5.), both at constant temperature of 21^0 C.

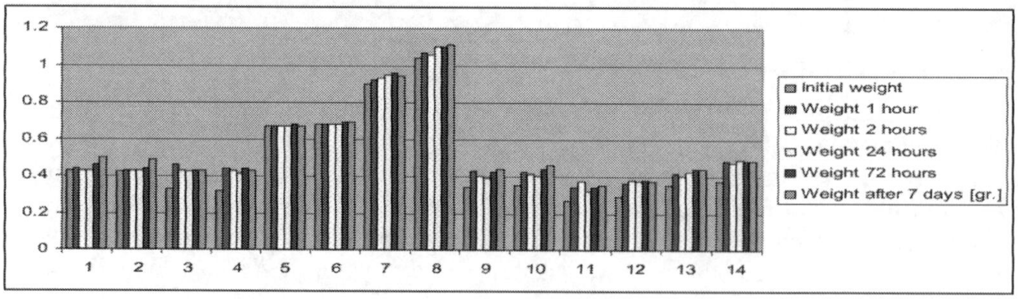

Fig. 4. The weight variation for insole samples immersed in simple water.

For samples immersed in saline, weight variation undergoes a series of changes from situation of immersion only in simple water which indicates a particular behavior of insoles samples.

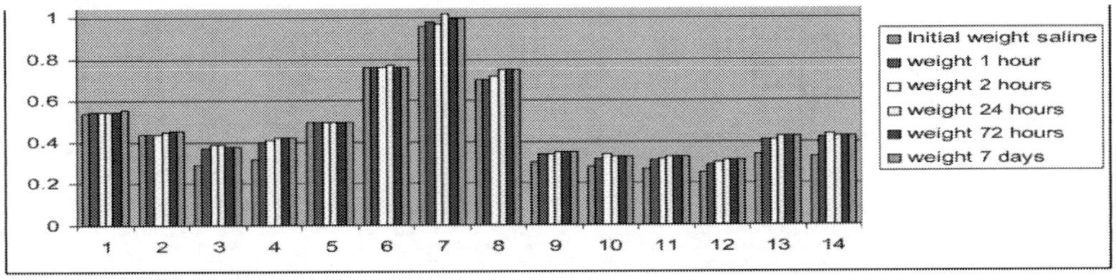

Fig. 5. The weight variation for insole samples immersed in saline.

However, there are a set of insoles that almost did not react at all when immersed in both water and saline, namely insoles samples made by silicon with magnetic pills. Also, procedures for sample introduction of insoles in refrigerator at a temperature of -18^0 C and others under the action of mechanical forces did not lead to weight variation during the experiment. The only procedure that changed in less insoles weight was that they were placed in dry autoclave at temperature of + 50^0 C. The weights of all samples decreased by an average of 2.5% which may be considered as not substantially alter their weight.

In a next step the samples were analyzed by a digital microscope, before and after immersion in water, saline, chilled and placed in dry autoclave. The images of the insoles were captured with a digital microscope type Keyence VHX-600 having magnifications between 20x and 5000x. Insoles material behaved differently especially the absorption of water and saline. In case of heat stress they have behavioral changes (stiffness/flexibility, brittleness/softening, color changed) immediately after removing from refrigerator and respective dry autoclave, they still returning to initial quality about 30 minutes.

Initial state Water immersion *Saline immersion*

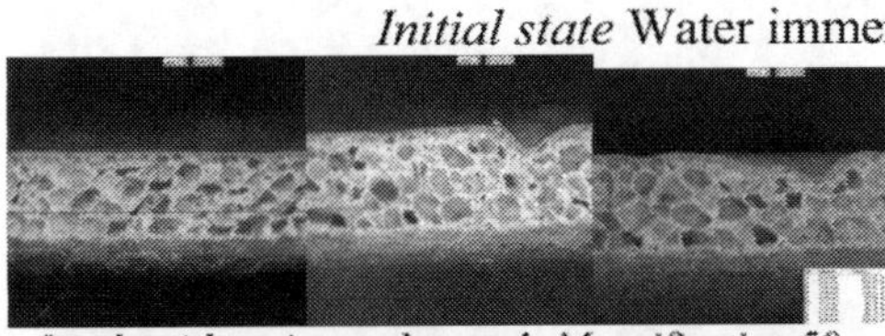

Insole with active carbon no1. Magnification 50x

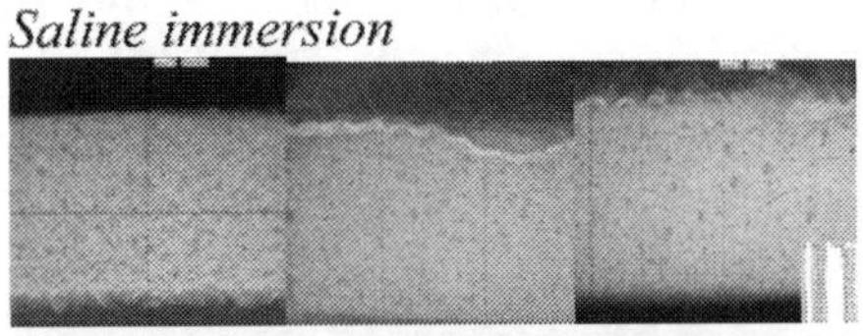

Insole with active carbon no2. Magnification 50x

Insole with magnetic plate no3. Magnification 50x (20x)

Insole with gel no4. Magnification 50x

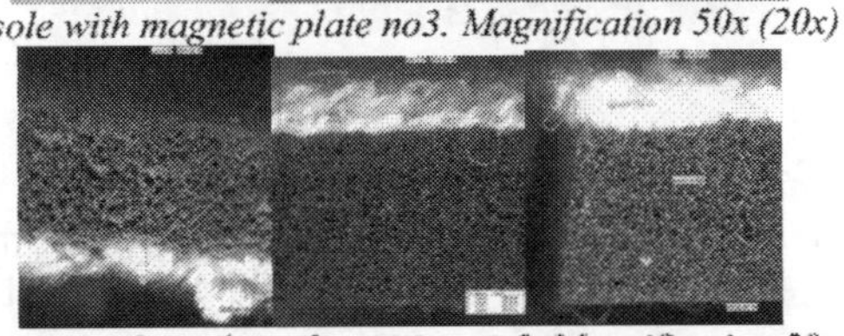

Foam insole and textile coating no5. Magnification 50x

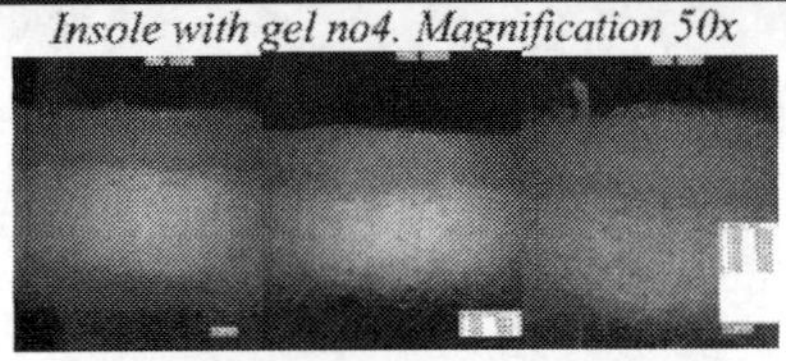

Foam insole, wool and Al coating no6. Magnification 50x

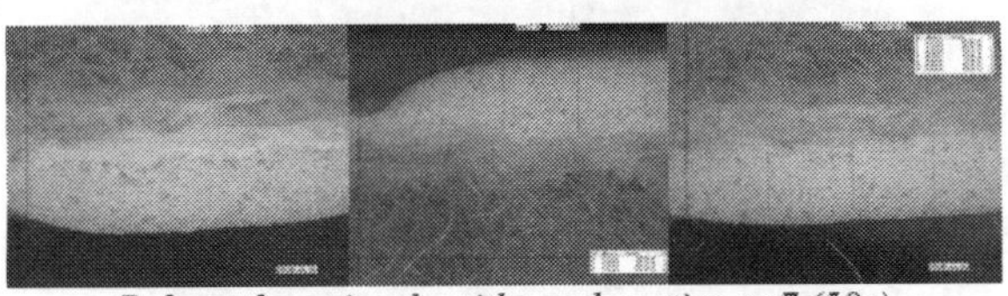

Polyurethane insole with wool coating no7.(50x)

Keyence digital microscope (20x-5000x) to insoles analysis

Fig. 6. Microscope images of insole samples in initial state, water immersion and saline immersion.

 DOI: 10.1051/matecconf/20179401001

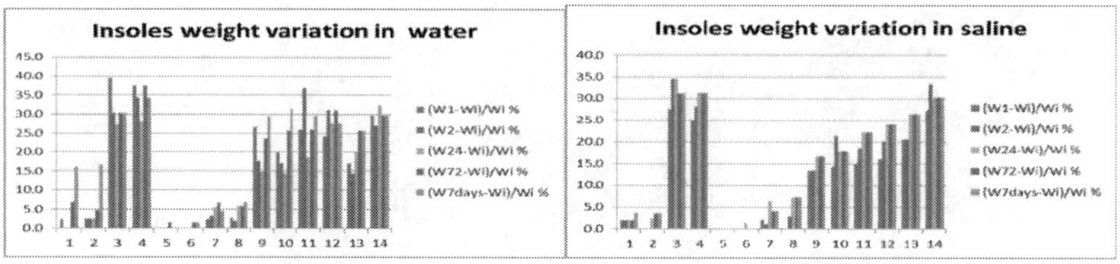

Fig. 7. Insoles weight variation in simple water (left) and saline (1%) (right).

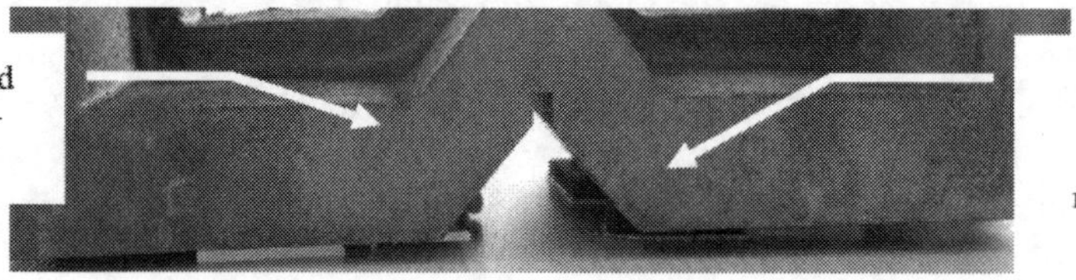

Fig. 8. Insoles samples under mechanical stress during 7 days.

In the case of insoles subjected to mechanical stress it has been observed a structure deformation, but returned to initial state in 30 min. Insoles were under a weight equal to 10 kg for 7 days in the identical conditions to those that were analyzed in immersion (21^{0}C ambient temperature, relative humidity 40%, constant atmospheric pressure 750 mm Hg).

The size of these insoles samples evaluated before and after exposure to mechanical stress (assessments conducted with digital microscope) was their thickness in the pressure area with calibrated weight of 10 kg. (Fig. 8.). Thus it is found that the most important structural changes of the analyzed insoles material samples were occurred at the stage of immersion in water and saline.

4 Conclusions

After analyzing the data obtained from measuring the weight of insoles sample we can say that these materials react most to the absorption of liquid – water, and forms a compact structure especially those that are made of foam and polyurethane. In case of foam insoles samples, polyurethane or those reached with active carbon, immersed in saline, the salt crystals enter in spaces of structure and when water evaporates, the crystals change the structure stiffen of insoles.

Correlating the actions of stress on the effects on insoles material structure we're able to identify the following aspects: Insoles made from silicone and gel-like material enriched with active carbon or magnetic tablets (insoles no.1, 3 and 4) partial change their color, do not absorb water or saline, not brittle after exposure to low or high temperatures and the main dimension (thickness) does not change after exposure to mechanical stress; Foam or polyurethane insoles (insoles no. 2, 5, 6 and 7) absorb much more quantity of water or saline and contact materials (wool, fabric protective, Al foil) with foot or the shoe surface undergoes changes of appearance, color or even come off from the insole base.

Therefore the behavior analysis of insoles selected types highlighted the importance of keeping the initial characteristics (flexibility, constant size, elasticity and toughness) to give comfort and ease of use on the plantar surface of the human foot. Any physical or chemical aggression on these insoles structures, their use in inappropriate conditions may cause discomfort, dysfunction in the gait cycle or may develop over time even local damages.

The next step in our research will mean to capture, to process and to interpret the images in terms of longitudinal and transversal sections to the tested insoles, due to the used digital microscope, having the possibility to measure the degree of material deformation under mechanical, thermal and chemical conditions. In this way it will be possible to obtain a statistical conclusion related to all categories of foot insoles behavior, for all condition.

This paper is part of the current researches in Applied Optometric Laboratory and Advanced Mechatronic Researches Center from University Transilvania Brasov.

References

1. M. Baritz et al., *Microscopic analyze of insoles composite structures behavior under water immersion, temperature variation and saline*, Conference *POLCOM* Bucharest (2014)
2. University of Nebraska Lincoln-Environmental Health and Safety, Personal Protective Equipment (PPE) - Foot Protection Evaluation And Checklist, http://ehs.unl.edu (2015)
3. W. L Ting et al., JRRD **51**, 311-324 (2014)
4. H. Chiu et al., J. of Appl. Biomech. **23**, 119-127 (2007)
5. J. Whaley et al., *EURECA*, 41 – 42, (2013)
6. S. Kolling et al., Material modeling of orthopedic insoles, 7th *European LS-DYNA* Conf. (2009)
7. Pedorthics Quatrterly, *Pedorthic Association of Canada* (2012)

FEM modelling of shoes insoles components for standing and walking simulation

Barbu Braun[1,*] , and *Mihaela* Baritz[1]

[1]Transilvania University of Brasov, Department of Product Design, Mechatronics and Environment, Brasov, Romania

Abstract. The paper presents a research stage in which a new method applied for foot insoles components rapid prototyping was successfully tested in case of two young persons with small orthopaedic diseases. The research in this stage is focused on the FEA model analysis before and after prototyping, the model consisting in two sets of specific items to be inserted into the plantar supporters, with orthopaedic correction role. The simulation and testing was performed in condition of wearing shoes containing such of plantar supporters, when standing and walking, these situations being the most common. The main studied problem was to verify if the CAD modelled orthotic items to be prototyped should resist in case of static and dynamic loads, similar to those found in case of standing and walking. It was demonstrated a good correlation in terms of testing results before and after items prototyping, especially for the second person. Besides, it was demonstrated that the FEA analysis applied method could be successfully used to verify the prototyped orthotic items endurance and resistance.

1 Role of shoes insoles for health and daily activities

Walking and standing are activities that occur daily in your current life and as a result, wearing comfortable shoes and suitable is crucial. For this reason, the insoles play a key-role. For this reason in recent years, research on foot insoles optimising in terms of design, especially the comfort, is in a continuous increase. Concretely, research has developed in three directions: to increase comfort when wearing different type of shoes, for daily activities, to increase athletic performance, to correct some foot malformation (flatfoot, Hallux, Valgus, pronounced arch) (especially for children) [1, 2], or to monitor and improve the health status.

The first direction addresses researches referring to new comfortable and biocompatible materials, to increase the degree of comfort when standing or walking [3, 4].

Referring to the athletic performances, different sensory micro-systems encapsulated were developed to measure any body behaviour parameters (e.g. measuring the intensity of physical activity and connecting a piezoelectric film (inserted into the sole of a shoe) to a Blue tooth or Smart phone devices) [5, 6].

* Correspondng author: braun@unitbv.ro

 DOI: 10.1051/matecconf/20179401002

On the last aspect, in the last years there were developed different types of foot insoles, having the possibility to control and even to monitor the human body parameters (e.g. blood pressure, pulse), via pressure / force / acceleration sensors encapsulated [7].

2 Insoles components rapid prototyping for any foot diseases

The paper presents a stage of research on the modelling of shoe insoles components, for dedicated orthopaedic footwear in case of two persons with plantar diseases. The first one was identified with pronounced arch, and the second person was identified with flatfeet, low degree. For both persons it was took the problem to improve the shoes, in order to correct more efficiently and easily the feet deformities. Thus means that for each one, a pair of foot insoles including orthotic elements was manufactured. The research is about the stage in which the orthotic items, corrective role were modelled and prototyped. For this reason, five steps were established: CAD modelling of the plantar corrective items [8], simulation using finite element analysis (FEA) of the CAD modelled items (virtual model), items rapid prototyping, foot insoles manufacturing, including the prototyped corrective items [9] and FEA of the prototyped items (as demonstrative model) to demonstrate their endurance and strength during walking and standing.

2.1 CAD modelling of the orthotic corrective items

For both persons, for CAD modelling, it was started from the scanning procedure applied for two pair of mouldings, representing the feet conformation [8]. For this reason, the prototyping items were modelled having the foot sole CAD model as reference surface, the final model being generated by applying successive specific operations, using CATIA V5 software environment. In Fig. 1 there are presented some CAD models specific to prototyping items:

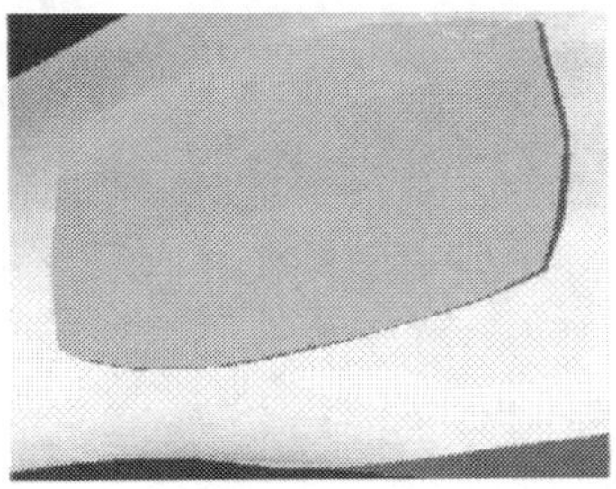

Fig. 1. Example of CAD modelling for a prototyping item specific to insoles for flatfeet correction

Respecting the same procedure, there were generated the CAD models specific to insoles for pronounced arch correction (the second case).

The next step necessary for foot insoles manufacturing was to prototype the CAD modelled items, specific to both cases. To ensure that these would resist to static and dynamic loads (corresponding to daily activities – standing, walking) when manufactured foot insoles wearing, a simulation test, using FEA, for each CAD modelled item was performed. It was taken into account that the material for items prototyping is considered to be the ABS plastic. In order to simulate the behaviour, the Poisson coefficient specific to the material was known ($\nu = 0.33$), and the acting loads were determined, taking into account the body weight. The CAD models allowed measuring the prototyping models dimension, thus allowing determining the items surfaces reported to the foot sole area. Knowing the investigated person's body weight, the assigned load to the items could be estimated, using the relations (1) and (2):

$$F_s = \frac{m \cdot g}{2} \cdot \frac{A_i}{A_{f\_s}} \tag{1}$$

where: F_s [N] – acting load when standing (in static mode); m [kg] – body weight; g [m/s²] - gravitational acceleration; A_i – surface area for the modeled item; A_{f_s} – surface area for the entire foot sole;

$$F_d = \frac{m \cdot g}{2} \cdot \frac{A_i}{A_{f\_s}} \cdot \Psi \tag{2}$$

where: F_d [N] - load when walking (in dynamic mode); Ψ- dynamic coefficient correlation (experimentally estimated to be about 2.2 ÷ 2.4).

For the first person, having plantar pronounced arch, in terms of foot insoles, it was necessary to prototype two items to be encapsulated into the insoles, to correct the disease: the first one being specific to the heel area and the second corresponding to the metatarsal area. Knowing that its body weight is about 71 kg, it was proceeded to determine the estimative compression loads for both prototyping items, using the relations (1) and (2).

In case of the second person, with flatfeet, low degree, in terms of foot insoles, it was necessary to prototype one item, specific to the median foot area, for plantar correction. Using the relations (1) and (2) and knowing that his body weight is about 59 kg, it was proceeded to determine the estimative compression load for the prototyping item, also.

Once obtained the CAD models for the prototyping items, for both tested persons, these were converted into .IGS files and imported into ANSYS software environment for FEA analysis (Fig. 2).

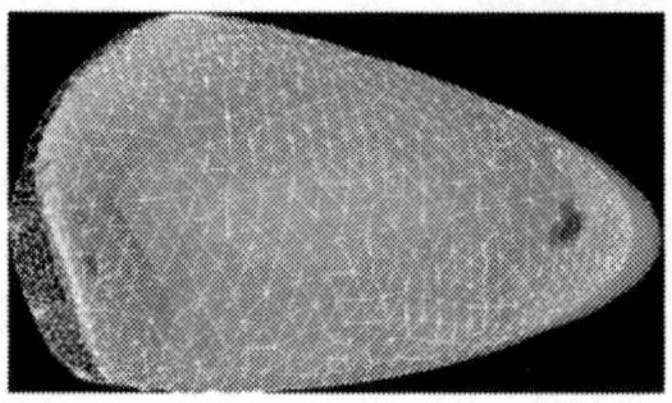

Fig. 2. Example of FEA in ANSYS for a CAD model, specific to the foot sole median area, to correct the flatfoot in case of the second person.

Fig. 2 presents an example of a prototyping item deformation, by FEA simulation. It can be observed also the stress distribution stress due to the compression load. The FEA results in terms of loads, strengths and deformations for prototyping items, in static and dynamic modes are presented in tables 1÷4.

Table 1. FEA results for prototyping items, due to simulation in ANSYS of the CAD models for the first person (with plantar pronounced arch), in case of standing (static mode).

	Prototyping element specific to the plantar area	Calculated load acting on the prototyping item (F) [N]	Determined strength (σ) [MPa]	Maximum admissible strength (σ_a) [MPa]	Determined deformation (f) [mm]	Maximum admissible deformation (f_{max}) [mm]
for left foot	for heel area	38.59	0.0257	60	0.74 10⁻⁵	0.0016
	for metatarsal area	51.46			0.55 10⁻⁵	
for right foot	for heel area	38.21	0.03		0.8 10⁻⁵	
	for metatarsal area	50.95			1.1 10⁻⁵	

 DOI: 10.1051/matecconf/20179401002

Table 2. FEA results for prototyping items, due to simulation in ANSYS of the CAD models for the first person (with plantar pronounced arch), in case of walking (dynamic mode).

Prototyping element specific to the plantar area	Calculated load acting on the prototyping item (F) [N]	Determined strength (σ) [MPa]	Maximum admissible strength (σ_a) [MPa]	Determined deformation (f) [mm]	Maximum admissible deformation (f_{max}) [mm]
for left foot					
for heel area	88.76	0.059	60	$1.71\ 10^{-5}$	0.0016
for metatarsal area	72.04	0.036		$1.25\ 10^{-5}$	
for right foot					
for heel area	87.89	0.06	60	$1.7\ 10^{-5}$	0.0016
for metatarsal area	71.33	0.0357		$1.24\ 10^{-5}$	

Table 3. FEA results for prototyping items, due to simulation in ANSYS of the CAD models for the second person (with flatfeet, low degree), in case of standing (static mode).

Prototyping element specific to the plantar area	Calculated load acting on the prototyping item (F) [N]	Determined strength (σ) [MPa]	Maximum admissible strength (σ_a) [MPa]	Determined deformation (f) [mm]	Maximum admissible deformation (f_{max}) [mm]
for left foot					
for arch sole	120.83	0.02	60	$2.4\ 10^{-5}$	0.00161
for right foot					
for arch sole	116.18	0.02	60	$2.3\ 10^{-5}$	0.0016

Table 4. FEA results for prototyping items, due to simulation in ANSYS of the CAD models for the second person (with flatfeet, low degree), in case of walking (dynamic mode).

Prototyping element specific to the plantar area	Calculated load acting on the prototyping item (F) [N]	Determined strength (σ) [MPa]	Maximum admissible strength (σ_a) [MPa]	Determined deformation (f) [mm]	Maximum admissible deformation (f_{max}) [mm]
for left foot					
for arch sole	205.41	0.04	60	$4\ 10^{-5}$	0.00163
for right foot					
for arch sole	199.47	0.04	60	$3.9\ 10^{-5}$	0.00162

2.2 Foot insoles manufacturing

The results on FEA analysis of the items CAD models leaded to the information that their behaviour in case in which these will be prototyped and inserted into the orthopaedic foot insoles would be appropriate in terms of endurance and resistance when standing and walking. As a result, the next step was to perform the rapid prototyping of the concerned items, by converting the CAD models as .IGS files into .STL files. For this operation there it was used a 3D printing, in ABS plastic and special printing support, by importing the

 DOI: 10.1051/matecconf/20179401002

.STL files into the CatalystEX software environment associated to the 3D Printer. In Fig. 3 there are presented some examples of prototyped orthotic items to be inserted into the foot insoles.

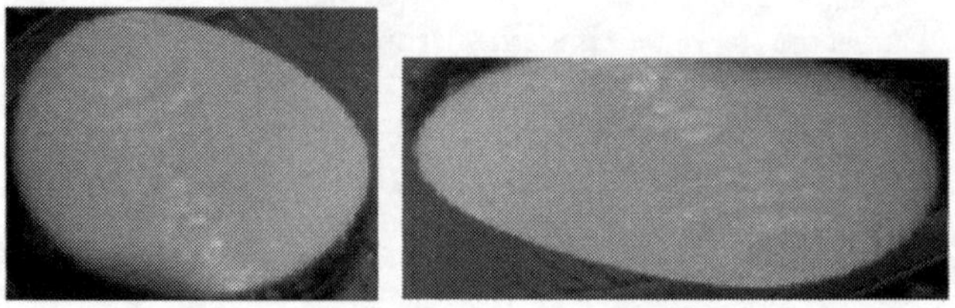

Fig. 3. Example of prototyped orthotic items for heel area (a) and for metatarsal area (b), for inserting into the foot insoles to correct the pronounced arch in case of the first person.

The next step was to insert the prototyped items into the foot soles, using different insoles covers made of biocompatible and conformable materials, to ensure the comfort when wearing (Fig. 4).

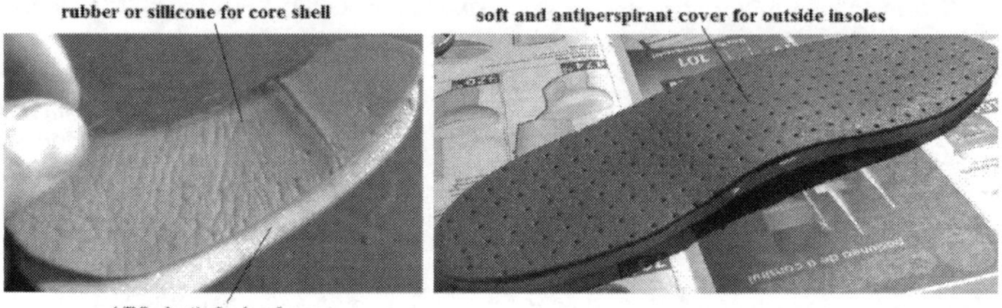

Fig. 4 The used materials for manufacturing the plantar insole for flatfoot progressive correction.

3 FEA of the prototyped insoles items

To complete the study on the prototyped items endurance and resistance, these were tested for the second time, this time in real condition, respectively when wearing the foot insoles during standing and walking. In this order, it was proceeded to measure in real time (by wearing the foot insoles) the plantar pressures, using a Foot Scan plate (equipped with 65,536 piezoelectric sensors), for both persons. The procedure was applied both in case of standing and walking on the force plate. Endurance and resistance of the inserted prototyped items invoked several steps: first of all the pressure distribution charts (Fig. 5 and 6) were done for both tested persons, left and right feet, when standing and when walking. Based on pressure distribution charts, the next step was to determine the averaged plantar pressures for areas of interest. It is about the heel and metatarsal area (addressed by the prototyping items) in case of the first person (with pronounced arch sole) and also the median area in case of the second person (with flatfeet, low degree). Then, knowing the mean pressure values and also the dimensions of the prototyped items, it was simple to determine the real loads acting on the orthotic items, when standing (F_s) and when walking (F_d) (relations 3).

$$F_s = p_i \cdot A_i \text{ and } F_d = p_i \cdot A_i \cdot \Psi \tag{3}$$

where p_i represents the mean pressure acting on the prototyped item (determined due to the plantar pressure charts), A_i – the surface area of the prototyped items and Ψ - multiplier impact coefficient ($\Psi > 1$), due to the dynamic shock, given by the energy conservation law:

$$\Psi = \sqrt{\frac{H}{f_s}} \quad (4)$$

where H is the height from which the person stepping while walking and f_s is the prototyping item deformation if standing (in static mode). Experimentally it could be determined the height H in case of both tested persons.

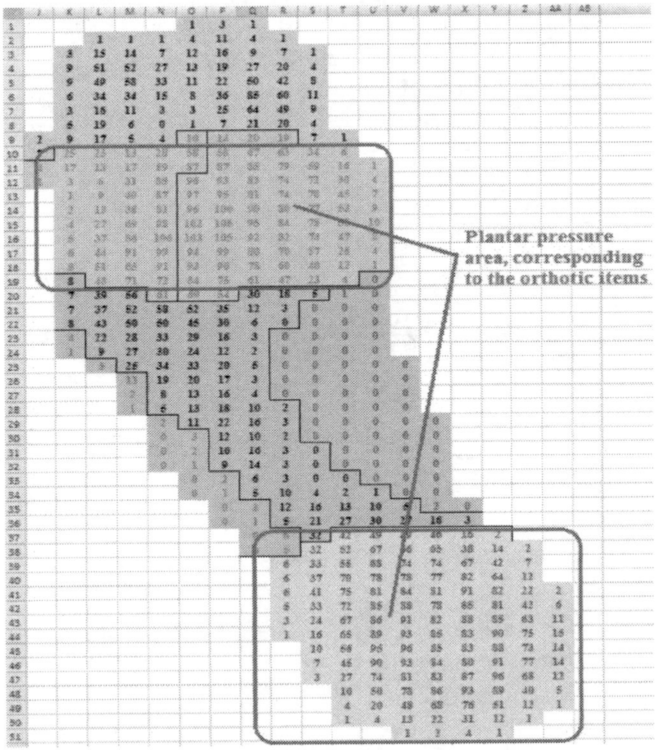

Fig. 5. Plantar pressures chart distribution for the left foot in the case of the first person (pronounced arch), tested when standing.

Table 5. FEA results, by using ANSYS, for the first tested person (with plantar pronounced arch), in case of standing (static mode) and wearing foot insoles including the prototyped orthotic items.

Prototyping element specific to the plantar area	Calculated load acting on the prototyping item (F) [N]	Determined strength (σ) [MPa]	Maximum admissible strength (σ_a) [MPa]	Determined deformation (f) [mm]	Maximum admissible deformation (f_{max}) [mm]
for left foot					
for heel area	59.04	0.04	60	$0.9 \; 10^{-5}$	0.0016
for metatarsal area	83.84			$1.3 \; 10^{-5}$	
for right foot					
for heel area	51.16	0.03	60	$0.8 \; 10^{-5}$	0.0016
for metatarsal area	88.62	0.05		$1.4 \; 10^{-5}$	

DOI: 10.1051/matecconf/20179401002

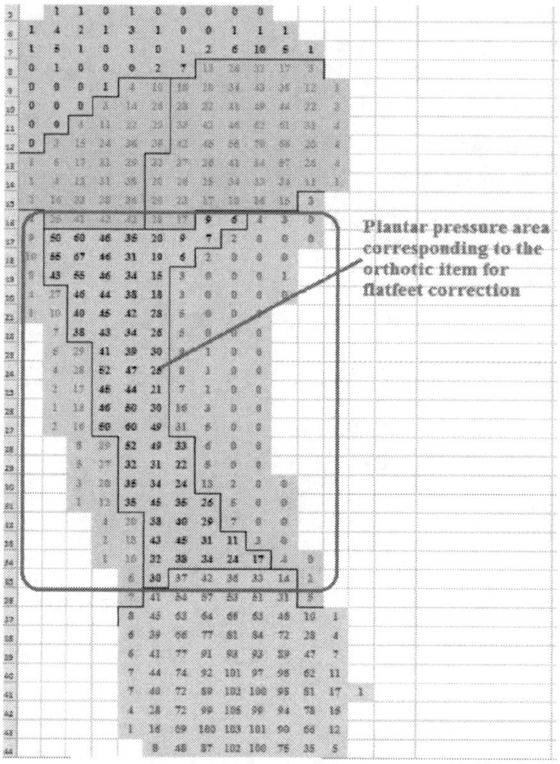

Fig. 6. Plantar pressures chart distribution for the left foot in the case of the second person (flat feet), tested when standing.

In this context, the FEA for the prototyped orthotic items was performed similarly to the FEA for the CAD modelled orthotic items, the results in terms of loads, strength and deformation being presented in the tables 5 ÷ 8.

Table 6. FEA results, by using ANSYS, for the first tested person (with plantar pronounced arch), in case of walking (dynamic mode) and wearing foot insoles including the prototyped orthotic items.

Prototyping element specific to the plantar area	Calculated load acting on the prototyping item (F) [N]	Determined strength (σ) [MPa]	Maximum admissible strength (σ_a) [MPa]	Determined deformation (f) [mm]	Maximum admissible deformation (f_{max}) [mm]
for left foot					
for heel area	112.14	0.07	60	$1.7\ 10^{-5}$	0.0016
for metatarsal area	116.92	0.06		$1.8\ 10^{-5}$	
for right foot					
for heel area	120.76	0.08	60	$1.9\ 10^{-5}$	0.0016
for metatarsal area	102.96	0.05		$1.6\ 10^{-5}$	

Table 7. FEA results, by using ANSYS, for the second tested person (with flat feet, low degree), in case of standing (static mode) and wearing foot insoles including the prototyped orthotic items.

Prototyping element specific to the plantar area	Calculated load acting on the prototyping item (F) [N]	Determined strength (σ) [MPa]	Maximum admissible strength (σ_a) [MPa]	Determined deformation (f) [mm]	Maximum admissible deformation (f_{max}) [mm]
for left foot					
for arch sole	148.85	0.03	60	2.9 10^{-5}	0.00163
for right foot					
for arch sole	132.53	0.02	60	2.6 10^{-5}	0.00164

Table 8. FEA results, by using ANSYS, for the second tested person (with flat feet, low degree), in case of walking (dynamic mode) and wearing foot insoles including the prototyped orthotic items.

Prototyping element specific to the plantar area	Calculated load acting on the prototyping item (F) [N]	Determined strength (σ) [MPa]	Maximum admissible strength (σ_a) [MPa]	Determined deformation (f) [mm]	Maximum admissible deformation (f_{max}) [mm]
for left foot					
for arch sole	237.15	0.04	60	4.7 10^{-5}	0.00164
for right foot					
for arch sole	228.58	0.04	60	4.5 10^{-5}	0.00163

4 Results and discussion

A very important aspect of the research related to the orthotic items prototyping was to verify the FEA results by CAD models simulation with the FEA results based on real information when wearing the foot insoles. For this reason it was considered to compare results obtained in ANSYS before prototyping (by estimating the loads distribution) with the results obtained in ANSYS after prototyping (due to the information related to the plantar pressure charts).

Fig. 7 and 8 present the correlation diagrams in terms of prototyped orthotic items deformation, before and after manufacturing:

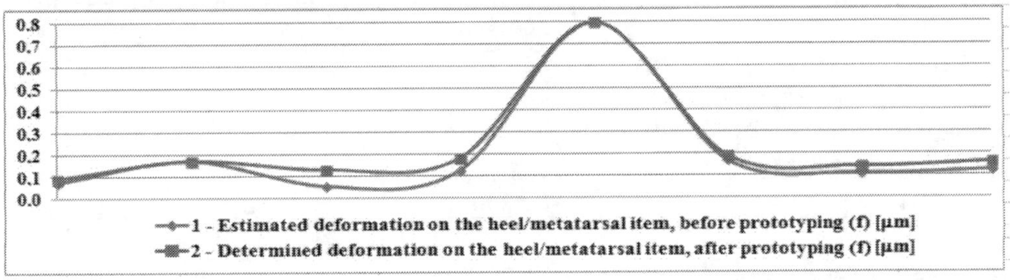

Fig. 7. Correlation diagrams on the results in terms of prototyped orthotic items for the foot insoles in case of the first person (with pronounced arch).

DOI: 10.1051/matecconf/20179401002

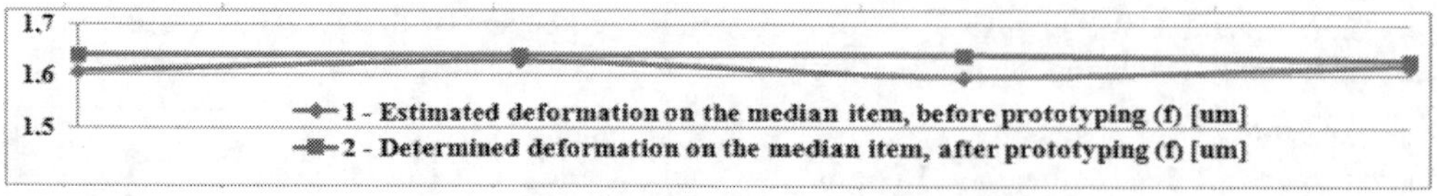

Fig. 8. Correlation diagrams on the results in terms of prototyped orthotic items for the foot insoles in case of the second person (with flatfeet, low degree).

Table 9. Results correlation in terms of prototyped orthotic items in case of the 1st person.

Testing step		Before prototyping: Estimated deformation on the item, before prototyping (f) [μm]	After prototyping: Determined deformation on the item, after prototyping (f) [μm]	Correlation in terms of items deformation [%]
Left foot				
for heel area	when standing	0.074	0.09	17.78
	when walking	0.171	0.17	0.59
for metatarsal area	when standing	0.055	0.13	57.69
	when walking	0.124	0.18	31.11
Right foot				
for heel area	when standing	0.8	0.8	0.00
	when walking	0.17	0.19	10.53
for metatarsal area	when standing	0.11	0.14	21.43
	when walking	0.124	0.16	22.5

Table 10. Results correlation in terms of prototyped orthotic items in case of the 2nd person.

Testing step		Before prototyping: Estimated deformation on the item, before prototyping (f) [μm]	After prototyping: Determined deformation on the item,after prototyping (f) [μm]	Correlation in terms of items deformation [%]
Left foot				
for arch sole	when standing	1.61	1.64	1.83
	when walking	1.63	1.64	0.61
Right foot				
for arch sole	when standing	1.6	1.64	2.44
	when walking	1.62	1.63	0.61

In the tables 9 and 10 there are presented all information on correlation between estimated results and obtained results in real conditions, in terms of prototyped orthotic items deformation:

It was observed that for the first person, the correlation deviation was greater than in case of the second one. This could be explained through the fact that in case of the second person the flatfeet degree is less than the pronounced arch in case of the first person. In both cases the correlation deviation indicated that the determined loads acting on the orthotic items due to the foot insoles testing by their wearing were slightly higher than the estimative determined loads in the stage of their CAD modelling. However, even in this case it was found that the prototyped items should resist without any problem in any situation regarding the plantar supporters wearing when walking or standing, for both persons. Moreover, the orthotic items should resist even in case of their wearing (inside plantar supporters) by persons weighing up to about 120 kg.

The presented method applied for plantar supporters manufacturing for the two presented cases proved to be very useful and efficient. Wearing the foot insoles by the two persons helped them to improve their walking and postural position reducing the problem regarding the hip, knees or ankle joints overloading. As a result, the undertaken researches in case of the two persons could be successfully applied in the future different persons (different age groups) with other plantar diseases (Hallux, Valgus etc.) or other orthopaedic diseases.

The research was supported by the Sector/Operational Program Human Resources Development (SOP HRD), ID 59323, financed from the European Social Fund and by the Romanian Government under the project number POSTDRU/89/1.5/S/59323.

References

1. A. Panjan, N. Sarabon, Methods for the Evaluation of Human Balance Body, Sport Sci Rev, **XIX**, 5-6 (2010)
2. V. Papilian, *Anatomia omului* (Pub H BIC ALL, Bucharest, 1998)
3. J. W. Brodsky, J. Brajtbord, S. C. Coleman, S. Raut, F. E. Polo, Effect of heating on the mechanical properties of insole materials, Foot Ankle Int **33**, 772-778 (2012)
4. D. Mihov, B. Katerska, Some biocompatible materials used in medical practice, Trakia J of Sci **8**, 119-125 (2010)
5. L. W. Thomas, Smart Shoes, United States Patent, number: 5373651 (1994)
6. C. A. Vock, Y. Perry, Shoe wear-out sensor, body-bar sensing system, United States Patent, number: 7911339 (2008)
7. D. R. Carmo, M. Soares, F. A. Rocha, J.L. Carvalho, S.S Rodrigues, Insole with pressure control and tissue neoformation induction systems for diabetic foot, Conf Proc IEEE Eng Med Biol Soc , **51** (2010)
8. B. C. Braun, I. C. Roşca, I. Şerban, C. Cobliş, CAD Methods for Orthopedic Orthosis Prototyping, MediTECH **2**, 388 – 391 (2011)
9. B. C. Braun, M. I. Baritz, I. C. Roşca, Research on the influence of progressive plantar supporters on the stability in biped position, OPTIROB **8**, 465-470 (2013)

Design of equipment for training machine tools

Miroslav Císar[1,*], *Nadežda* Čuboňová[1], and *Ivan* Kuric[1]

[1]Department of Automation and Production Systems, Faculty of Mechanical Engineering, University of Zilina, Univerzitna 8215/1, 01026 Zilina, Slovak Republic

Abstract. The article proposes multiple devices designed to be used with training machine tools EMCO Concept series. Design was focused to enchant educational potential of training machine tools located in the laboratory of CNC programing, department of automation and production systems, Faculty of mechanical engineering, University of Zilina. The described device allows monitoring of machine tool, to measure tool offset and dimension, to use alternative ways of clamping, and to create video of machining.

1 Introduction

Constant improvement development and application of results of research affects our lives, whether it takes place in industry, training, or education. Implementation of progressive modern technologies to education process is essential for sustaining competitiveness of educational institutions and graduates. Even implementation of seemingly trivial solution can help students to learn various techniques that can improve their creativity and effectiveness in real life situation in industry. The article deals with the application of such resources to support the teaching which deals with CNC manufacturing technology [1, 2].

The article deals with training machine tool Emco Concept Mill (CM) 105 using other production resources available at Department of Automation and Production Systems, Faculty of Mechanical Engineering, University of Zilina. The proposed solution include design of camera mount for GoPro Hero camera in interior of machine tool, design of technological palate, and design of device for measurement of tool offsets. Necessity of designing and implementing of these devices resulted from specificity of used machine tool. The machine tool described in next chapter slightly deviates from machine tools commonly used in industry as it is designed specifically for industrial training with focus on versatility, easiness of operation and maintenance.

2 Emco Concept Mill

Training with real machine tools is important part of educational process in field of mechanical engineering. However common machine tools have limited options in terms of control system and it usually requires significant space. In terms of training and education it is good to have one machine that can be controlled with multiple control systems.

*Corresponding author: miroslav.cisar@fstroj.uniza.sk

DOI: 10.1051/matecconf/20179401003

EMCO CM 105 (Figure 1) is compact machine tool designed for industrial training and educational purposes. Such machine tools are commonly used at technical high schools and universities. Tools are stored in ten station turret. Slides and load-bearing elements are made of gray cast iron. This machine is equipped with infinitely variable main drive and 10-station turret tool changer [3, 4].

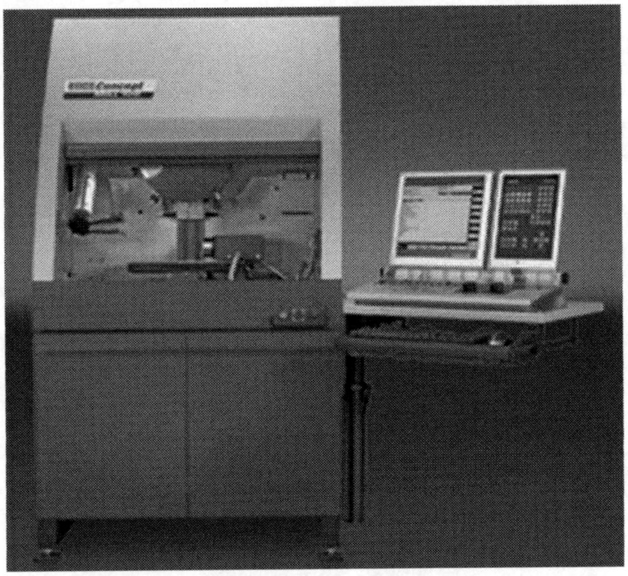

Fig. 1. EMCO Concept MILL 105 [3].

Compact table format is suitable for teaching of common manufacturing technologies. Safety glass window provides high level of protection against chips and prevents coolant leakage, but in same time it render recording of high quality video, from outside of the machine, virtually impossible because of reflections. The control for the CM 105 is connected via PC with interchangeable WinNC control from EMCO. The uncommon arrangement of control system where machine tool is controlled by simulated control system what is actually an computer program brings some problems with compatibility of control system with real production machines but it also allow to use another software to interact with simulated control system [3, 4].

The worktable have T-slots, but workpiece is usually mounted to with mechanical vice permanently mounted to worktable. That reduces necessity to repeatedly measure and set workpiece offset as it can be aligned with corner of vice. Machine tool workspace provides capability of clamping workpiece with dimensions 200×150mm and its maximal size in Z axis depends on length of used tool. Effective length of Z axis is 150mm. Inner space of machine tool is significantly larger than effective workspace what allows placement of small action camera and other equipment inside of machine tool [3, 4].

Sadly, the control system of CM 105 lacks ability to connect automatic device for measuring of tool length and diameter and therefor the investment to such device commonly used in industry would be waste of money. The machine is relatively limited in terms of peripherals accessories, for example it is not possible to use it with 3D measuring probe commonly used in industry. But again, these shortcomings are more than enough compensated by ability to be used with various control systems, what saves space, finances. Another benefit of this machine tool is that it uses rather small tools which are more affordable and thus more suitable for training and education.

3 Camera mount

Capturing of video of machining in machine tool is seemingly trivial task but it brings several problem such as camera size, its strength and resistance to specific environmental factors (swarf, coolant, vibrations, etc.). That led us to simple choice of so called action camera - GoPro HD HERO 3 which was highest available model in time when this problem was solved. The camera features video resolutions up to 1080p30, 5MP photos up to 3 frames per second, an ultra-wide angle lens and built-in Wi-Fi.

The camera can be equipped with wide scale of stands and mounts usually connected together with forks that forms an angle joints. One of these forks is fitted with M5 closed cap nut and joints are fastened with knob screw. Combination of specific selected camera and machine tool provides several possible places for mounting, shown on Figure 2, each of which provides different type of view:

1. Mounting on turret - Tool POV.
2. Mounting on worktable - Workpiece POV.
3. Mounting on machine tool frame - Static video.

Fig. 2. Places for camera placement at workspace of EMCO CM 105 [7].

Standard set of mounts for this camera can be used to be mount it to worktable or to machine tool frame but it does not offer the angle of view that we desire – the view with tool in center of picture. It would require sticking ten mountpads around the turret one for each tool position what would require manual realignment of camera after each tool change during machining. Therefor we designed special pendulum-like mount mounted to center of rotation of turret. As turret spins during tool exchange, camera keeps facing downwards because it is self-oriented by gravity. This way camera is constantly facing to the tool which is currently active. Such placement also eliminate possible problems with changing focusing distance that could rise when camera is mounted on machine tool frame or on worktable [1, 5].

The design was based on fact that the turret have ten positions, it rotates during tool exchange,and mount had to allow recording individual tools in their specific positions. Camera and mount components should not affect production capability of machine tool in any negative way and operation of machine should not affect camera and its stability.

The design of prototype was also focused to use as much standardized components as possible in order to verify its functionality and reduce costs of production and still keep relative freedom in terms of adjustability of camera position relative to active tool. Fluent rotation of pendulum during tool exchange is ensured by using of ball bearing mounted to the center of turret by center pivot.

The final construction (Figure 3) consists of rotary joint, angle joint connecting rotary section with carrier rod with attached clamp with standard fork that allows connection of camera or its equipment. Carrier rod is made of long Allen screw screwed into nut glued in hexagonal hole in one part of angle joint. Using standard size thread allows us to easily replace carrier rod by another screw with different length if necessary. Screw head also serve as stopper that prevents slip off and fall of the camera [1].

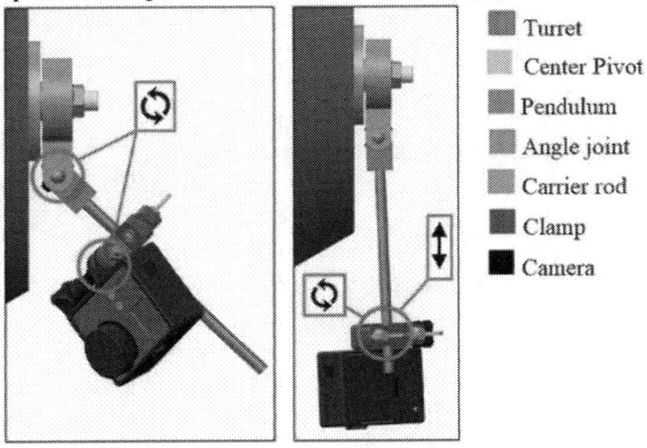

Fig. 3. Model of camera mount construction and possibilities of adjusting of camera position [5].

The design and concepts were verified by producing prototype (Figure 4 – left) using standardized and 3D printed parts. Testing of prototype showed problems with focusing to tool placed too close to camera lens. This was solved with 58mm close up filter ×2 and step up ring 37 to 58 mm. Rectangle of same size as front part of camera case was milled into the step up ring which was glued on 3D printed adaptor designed to fit on front element of camera lens. Inner surface of 3D printed part was covered by rubber tape in order to increase friction and secure proper mounting of filter to camera case [5].

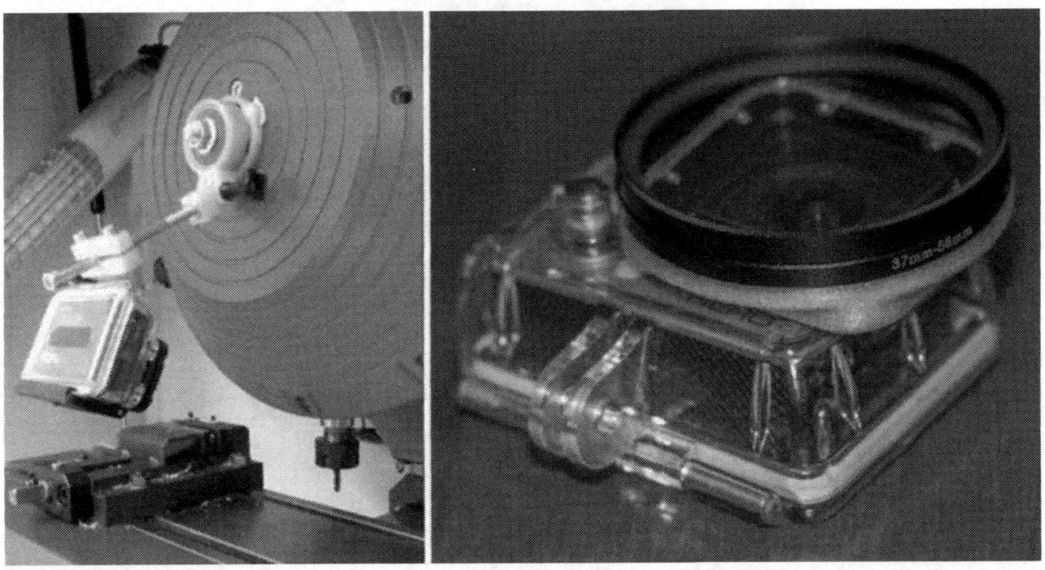

Fig. 4. Camera on prototype of mount fitted in machine tool (left) and detail of mounting close up filter to camera case (right).

The diameter of close up filter significantly larger than diameter of lens was selected in order to eliminate picture vignetting. Test confirmed that with close up filter attached to camera it is possible to focus on tool properly and thus dramatically increase quality of recorded video [1].

Camera placed in machine tool workspace can be used thru wireless network what allow implementing remote observation of machining process.

DOI: 10.1051/matecconf/20179401003

4 Device for measurement of tools

Measurement of tool length and diameter is common task in standard operation machine tool that is necessary to ensure accuracy of dimensions of produced parts. Commonly it is automatized and even added to programs for production used together with conditional tool exchange based on tool state.

However, the CM 105 does not allow input from device and thus we can implement only device that can be used by operator and not by automatic cycle in program. Therefor we designed simple battery operated tool setter optical gateway with acoustic output that can be used to measure tool offsets in manual mode of machine tool.

Proposed design consist of two separated parts – the emitter and the receiver (Figure 5). The functional part of emitter is simple battery laser aligned with base with pushbutton to switch on or off. The receiver unit is also battery operated, it contains receiver for laser beam, buzzer that indicates laser beam interruption, and pushbutton for switching device on or off. Both parts have its own batteries and switches as they are not connected by wires, each unit is controlled by simple microcontrollers Digispark ATtiny85. In the emitter the microcontroller serve just as timed switch in order to save battery. The receiver uses microcontroller in order to analyze frequency of signal from laser emitter resulting from rotating flutes of tool crossing the beam [6].

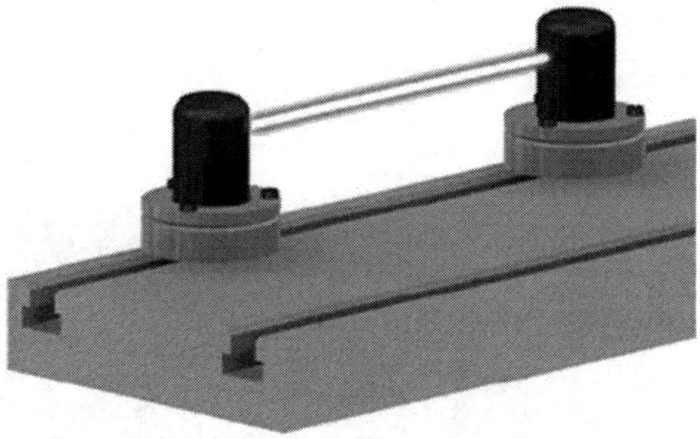

Fig. 5. Placement of units of custom tool setter on worktable.

In order to ensure proper functionality of device it is firstly necessary to properly align laser beam of emitter with X axis of machine tool. It can be easily done by moving worktable along the axis and adjusting horizontal angle of emitter until the red dot (laser spot) stops moving to the sides with moving of worktable. Second step is placing of receiver in path of the beam in way that allows reception. The strength of signal from laser can be measured by long press of pushbutton (at least ten seconds), in such setup mode tone of acoustic signal represent strength of signal [6].

Main mechanical construction of units is made of machined aluminum grinded and fitted together with Allen screws. All electronics are covered with 3D printed plastic covers that protect construction from swarf and still allow to pushbutton to be pressed by applying force in Z axis to the cover. Base construction is divided to two parts bottom of which is fixed to worktable using T-slot and top part allows precise adjustment with three Allen screws that allows leveling of unit in order to make laser beam perpendicular with table movement. Programing of microcontroller in receiver allows to filter signals from interruption of laser beam and thus remove unnecessary acoustic signals during machine tool operation. It is possible because of measurement of tool dimensions is done with tool rotating on low speed. The tool setter design described in this chapter is not very precise but

 DOI: 10.1051/matecconf/20179401003

it help us to explain and clarify principles, commonly used in industry, on machine that in normal circumstances would not allow that. The main benefit of described construction is that it allows to measure tool dimensions near the vice what is necessary property for CM 105 as its effective workspace is too small to place such device as it is usual on industrial machines.

5 Technological pallets

The technological pallets are commonly used in mass production but it is not so common in smaller scale productions. The training and educational machines such as CM 105 are typically equipped with very limited workspace what renders common technological pallets virtually unusable. Therefor we decided to design small quick releasable technological pallet compatible with our machine tool that would provide practical example of usage of such aids in industry.

Basic design was based on requirement of hand manipulation of pallet, it should be easily and automatically positioned and aligned in machine tool workspace and it should also allow easy fixation and release of parts. It should provide secure clamping and it should reduce workspace as little as possible.

The final design of exchangeable technological pallet system (Figure 6) consists of base plate mounted on worktable, pallet itself clamped by eccentric levers. Position of pallet on base plate is ensured by taper pins fitted with housing drilled into pallet [7].

Fig. 6. Model of exchangeable technological pallet system.

Mounting of the pallet is done by friction between pallet and base plate, the clamping force is produced by turning of lever – there are helical surfaces on bottom of the levers and on the cavities at the sides of palette. The sizing and optimization of construction was done thru simulations. Complete system should contain at least two pallets in order to present benefits of implementation of such system especially reduction of the time necessary to clamping workpieces in workspace [7, 8].

Pallets can be modified to another system for clamping such as system with matrix of threated holes or it can be equipped with various clamping devices such as magnetic table, mechanical vice, etc.

6 Monitoring of machine tool usage

Monitoring of machine tool Emco CM 105 is possible with relatively simple computer program as it is not controlled by official control system of common producer. It have system of interchangeable control systems which is implemented as computer program with user interface and behavior simulating official control systems used on common industrial machines. Computer that runs WinNC software that mediates user interface, communicate with PLC in machine tool via Ethernet [3, 4, 9].

If we want to monitor some technological parameters in real time it can relatively easily be done by one of two ways. Either by monitoring network communication between machine tool and controlling computer, or by reading memory allocated by control software. We decided to use second method.

The reverse engineering methods were used to find addresses of variables storing values we desired to monitor. The Cheat Engine,an open-source software designed to read memory allocated by programs in order to cheat in computer games, was used as tool for scanning memory allocated by process Sie840Du.exe for addresses of variables [9, 10].

Using addresses found with the Cheat Engine software we created custom software for monitoring technological parameters. This software is not meant to be used by machine tool operator so it is designed to start with system as hidden. User interface is accessible thru icon in notification area. Graphical user interface is therefore relatively simple and it serves primarily for initial setup and debugging. Also in order to reduce allocation of system resources, the instance of the form with user interface is not loaded until it is manually requested by double clicking icon in notification area (system tray) [9].

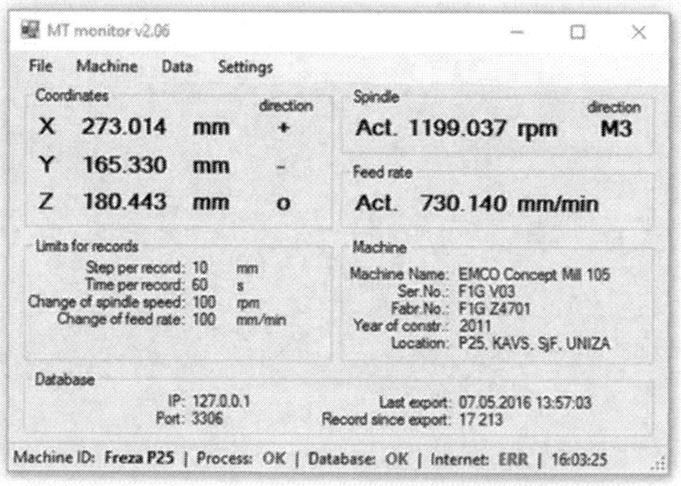

Fig. 7. Custom software "MT monitor" for monitoring of Emco Concept Machine tools.

Current version of MT monitor is capable of real time reading of tool position in machine coordinate system, spindle speed, and feed rate. It also displays current direction per axis and direction of spindle rotation. User interface also shows state of process that simulates control system, state of connection to database and to internet (if it is possible). There are also basic information about machine and count of records since last data export.

The data captured thru monitoring software will be analyzed together with data gained by technical diagnostic such as results of measurements of tool positioning precision in order to get relevant data for establishing trend and correlation between machine tool usage and its condition. For example monitored data can be used to compare with measured wear of feeds, to check most used parts of workspace, and more. Functionalities of the MT monitor will have to be widened to support other control systems supported in WinNC software used for controlling Emco Concept machines and to collect data independently on control system used for machine tool operation. This is necessary because same machine tool can be used with multiple control system, but effects to hardware remains regardless of system used [9].

7 Conclusion

Constant development and improvement should be essential part of educational processes. Implementation of machine tools to educational process offers lots of opportunities to enhance education by developing new and improving old accessories. This article contains small sample of work in this field. The process of improvement is never finished and it requires constant effort and care. All described solutions was designed with focus on increasing educational potential of training machine tools located in the laboratory of CNC programing, department of automation and production systems, Faculty of mechanical engineering, University of Zilina.

This article was made under support projects KEGA 037ŽU-4/2014 and KEGA 024ŽU-4/2016

References

1. N. Čuboňová, M. Císar, Adv. Sci. Technol. Res. J. **9**, 34, (2015)
2. J. Jozwik, I. Kuric, 14th Intern. Conference Automation in Production Planning and Manufacturing, **1**, (2013)
3. Didactic Martin, s.r.o. *Technickýpopis Concept MILL 105: Stolné CNC obrábacie centrum pre frézovanie vŕtanie a rezanie závitov s vymeniteľným riadiacim systémom [Technical description of Concept MILL 105: Table CNC machine tool for miling, drilling and threading with interchangeabke control system]*, Online:www.didactic.sk/Data/802/UserFiles/emco_tech_popisy/spec_cm105-sk.pdf (2007)
4. EMCO Maier Ges. M.b. H., *Machine Description EMCO CM 105*, Edition A2003-02, Online: http://www.emco-world.com/en/products/industrialtraining/machines/milling /cat/26/d/2/p/1000045%2C26/pr/concept-mill-105/view/11.html, (2013)
5. L. Martinka, *Proposal and Production of Sensing Device Clamping* (in Slovak), MSc. thesis, University of Zilina, (2014)
6. D. Gonda, *Making device for determining corrections of tool on the machine Emco Concept Mill 105* (in Slovak), MSc. thesis, University of Zilina, (2015)
7. M. Gálusová, *Design of the exchange of technological pallets for Emco Concept Mill 105,* (in Slovak) MSc. thesis, University of Zilina, (2015)
8. A. Sapietova, M. Saga, P. Novak, Communication - Scientific Letters **14**, 432, (2012)
9. Š. Vlček, *Monitoring of the operation parameters on machine EMCO Concept Mill 105* (in Slovak), MSc. thesis, University of Zilina, (2016)
10. V. Poppeová, J. Uríček, V. Bulej, P. Šindler, Joint 41st International Symposium on Robotics and 6th German Conference on Robotics, **1**, (2010)

Designing parts for a robust behaviour against vibrations. A new concept

Florin Corciovă[1,*], and *Mircea Viorel* Drăgoi[1]

[1]Transilvania University of Braşov, Department of Manufacturing Engineering, Mihai Viteazu No.5, Braşov, Romania

Abstract. The paper presents a new concept on how to design parts as they have a robust behavior against vibrations, that is to say they can remain stable, even if they were subject to vibrations. The new idea is that the value of the own-frequency must be kept away from the vibration frequency the part or assembly it belongs to is subjected. A way to control the own-frequency of a part is to slightly modify geometrical or material features of the part. The paper displays the way the modification of the shape of the part can modify its own-frequency.

1 Introduction

Vibrations are a wide spread phenomenon in nature and in technique. In both areas, they are either welcome, or on the contrary undesirable. In technique/engineering, in most cases vibrations are to be eliminated. Vibrations are not always dangerous, but often they are annoying or even irritating. Note, that vibration are mostly dangerous if their frequency equals or is near to the own-frequency of the body that is subjected to vibrations.

The most used technique to avoid or eliminate vibrations of a certain part or ensemble is to insulate from the source of vibrations by means of dumpers. Literature displays many scientific papers that deal with the subject in various approaches, as it is briefly described below.

In [1] is presented an "electro-hydraulic actuator that attenuates and isolates ground motion to keep dynamic excitations transmitted to machine tools below permissible levels"; [2] describes an absorber, which has "a linear viscous damper connecting the absorber mass directly to the ground instead of the main mass"; in [3] is presented the "optimum design of damped vibration absorber for viscoelastic bladed disk assemblies"; in [4-7] different type of non-conventional types of damping systems ("with material-inlaid plates", "a class of piecewise smooth vibration isolation system", other).

Other scientists propose different means to reduce/avoid vibrations of mechanical systems. E.g. selecting optimally the nut parameters in case of ball screw driven machine tools [8].

The present paper proposes a new approach of the problem: preventing, or at least decreasing effect of vibrations of slim parts, by means of adequate adjusting of geometrical features. None of the cited works refer to such a manner to solve the problem.

*Corresponding author: corciova.florin@unitbv.ro

DOI: 10.1051/matecconf/20179401004

2 The concept

It is known that thin parts have a specific behaviour under the stress of vibration. One of the particularities is that they display big displacements in specific points. This happens only at some certain frequencies of the vibrations. The key point of the concept here proposed is to simulate by means of specific software the behaviour of the part subjected to a large range of frequencies vibrations. In this way the "bad" frequencies, those which cause big displacements of part's sides, are identified. If any of these frequencies fits to the own-frequency of the part, it becomes a dangerous one.

Adding to the part some specific geometrical features, the "bad" values of the frequencies, as they were defined above, can be pushed to upper or lower values, which do not overlap to the dangerous ones. Of course, the geometry alterations have to be slight, and they may not affect the correct work of the part. In the next sections of the paper will be demonstrated how different specific geometrical features can (sometimes significantly) move the bad frequencies away.

3 Carrying on the researches

Simulation of behaviour under stress of vibration was carried on using COMSOL software [9]. A special designed sample part and some variants of it were used for simulations.

3.1 The sample part and its variants

A thin part was used as standard and some variants of it were designed to study the modifications in terms of frequencies that affect it significantly (figures 1 and 2). Apart from standard, the variants were obtained by adding some geometric features: a holes pattern, a nervure, and holes inside the part. The holes inside the part are an axial one in the median horizontal plane, and a bended one in the same plane. For the last two, in figure 2 a detail and a section through median plane are displayed. These models have no modification in terms of external geometry compared to the standard.

The bended hole in the part aims to emphasize how the dissymmetry of the part affects the way it distorts under the stress of vibrations. It is important to note that even if the external geometry of the part is symmetrical, in terms of mass properties, it is not.

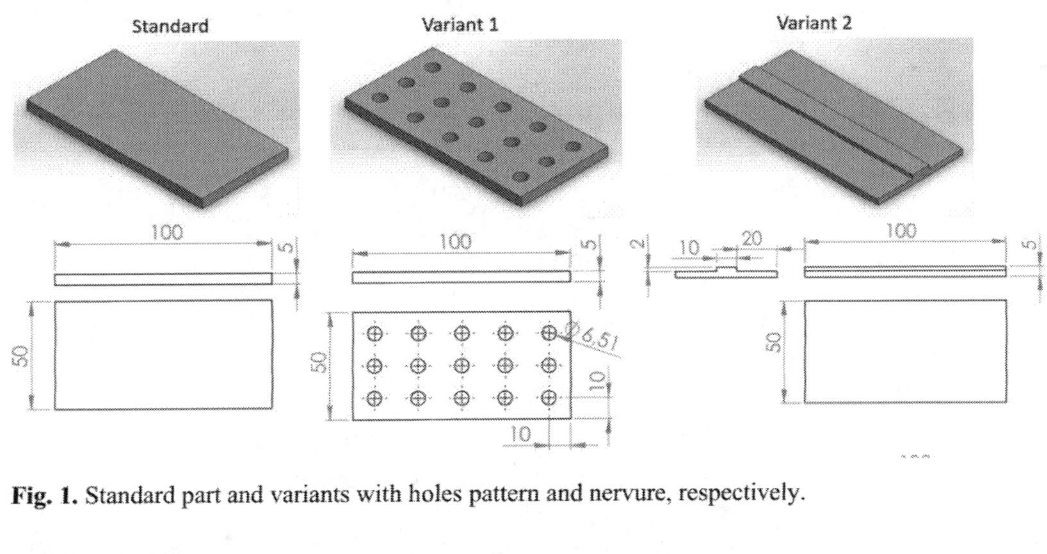

Fig. 1. Standard part and variants with holes pattern and nervure, respectively.

DOI: 10.1051/matecconf/20179401004

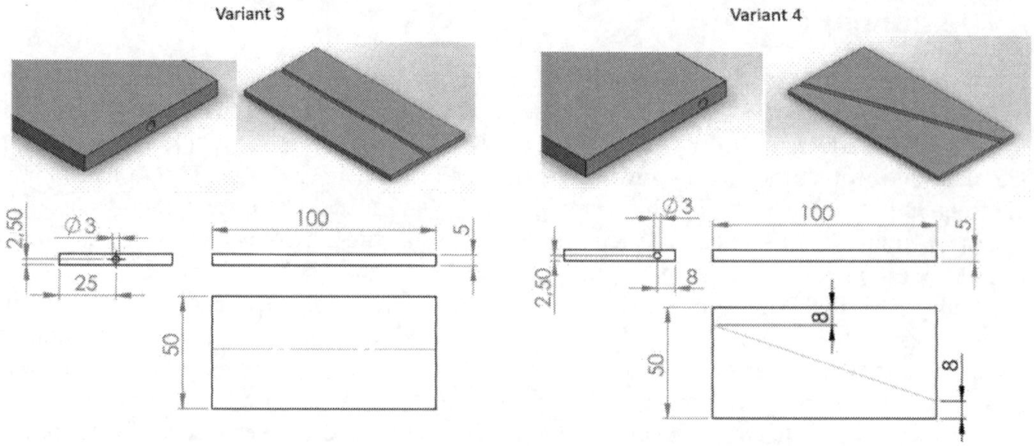

Fig.2. Variants with internal holes, having the same external geometry as the standard.

3.2 Finding the frequencies that affect significantly the parts

Research was carried on by means of simulating the behaviour of the parts with COMSOL software. The mathematical model is described below. All the notations and formula above are according to [9].

$$s^{ij} = s_0^{ij} + C^{ijkl}\left(\varepsilon_{kl} - \alpha_{kl}\left(T - T_{ref}\right) - \varepsilon_{0kl}\right) \tag{1}$$

Formula for elasticity tensor component:

$$C^{ijkl} = \lambda g^{ij} g^{kl} + \mu(g^{ik} g^{il} + g^{il} g^{ik}) \tag{2}$$

For the research model equation (3) was used:

$$s = S_0 + C : (\epsilon - \epsilon_0 - \epsilon_{inel}) \tag{3}$$

Where, s - total stress, S_0 - initial stress, C- elasticity tensor
The total strain tensor is written in terms of the displacement gradient (4).

$$\epsilon = \frac{1}{2}(\nabla u + (\nabla u)^T) \tag{4}$$

The gradient of the displacement, which occurs frequently in the following theory, is always computed with respect to material coordinates in 3D:

$$\nabla u = \begin{bmatrix} \frac{\partial u}{\partial X} & \frac{\partial u}{\partial Y} & \frac{\partial u}{\partial Z} \\ \frac{\partial v}{\partial X} & \frac{\partial v}{\partial Y} & \frac{\partial v}{\partial Z} \\ \frac{\partial w}{\partial X} & \frac{\partial w}{\partial Y} & \frac{\partial w}{\partial Z} \end{bmatrix} \tag{5}$$

$$\sigma \cdot n = F_A \tag{6}$$

$$F_A = \frac{F_{tot}}{A} \tag{7}$$

F_{tot} - the force applied on the Y axis, A - surface area, n - coefficient of deformation, σ - stress.

To perform simulation, some parameters related to part's material had to be set in the software. They are according to those presented in Table 1.

Table 1. General properties of the used material.

Property	Symbol	Value	Unit
Density	rho	1020	Kg/m^3
Young's module	E	2300	MPa
Poisson's ratio	nu	0.24	1

The standard and its four variants were subjected to a continuous range of frequencies from 0 to 120 Hz. For each part, some frequencies caused significant distortions (1 mm) in certain zones of the sample. The values of those frequencies are displayed in Table 2. They are graphically represented in figure 3. It shows that always Variant 2 (that reinforced with a nervure) reacts at lower values of frequency, and Variant 4 (that having a bended hole inside) is affected rather by bigger values of frequency. The way a variant behaves compared to the standard is the same for each "bad" frequency.

Table 2. Frequencies that caused significant distortions to standard and variant parts.

Part name	Standard	Variant 1	Variant 2	Variant 3	Variant 4
Frequencies	22.74	21.42	17.49	23.22	24.42
	28.87	26.52	21.64	28.14	29.88
	61.82	57.38	45.26	60.74	64.36
	88.36	82.23	55.72	92.03	94.86
	103.82	96.69	65.36	102.96	109.67

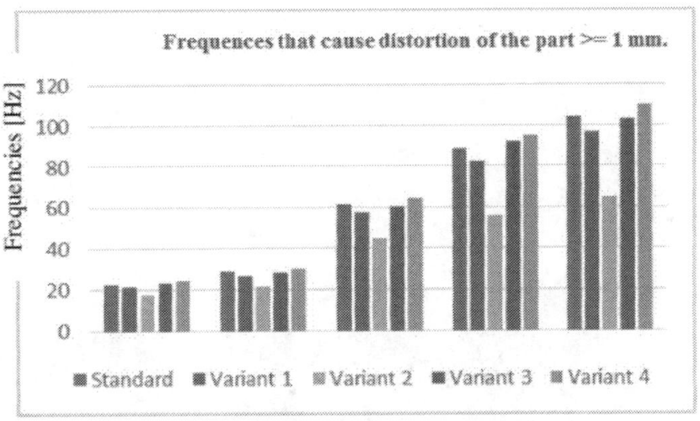

Fig.3. Groups of "bad" frequencies, and the way they affect different samples.

3.3 Discussion on results of simulation

A very good view of the influence of different frequencies on the samples can be gathered from figures 4 and 5. Figure 4 shows clearly how a certain group of frequencies (those displayed on the third line of Table 3) causes similar distortions on the samples. On contrary in figure 5 one can see how another group of frequencies (displayed on the second line of Table 3) cause different types of distortions on the samples, depending on their particularities/geometrical features. These observations reveal that in design stage, if a part needs to modify its features to be free of danger of vibrations, the modifications must be carefully chosen, taking into account the frequency of the vibration the part will be subjected to.

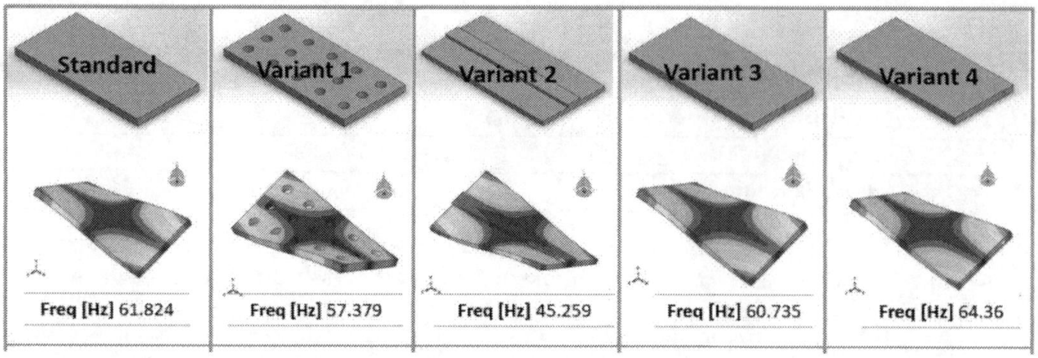

Fig.4. Group of frequencies that produce similar distortions on parts, regardless to their features.

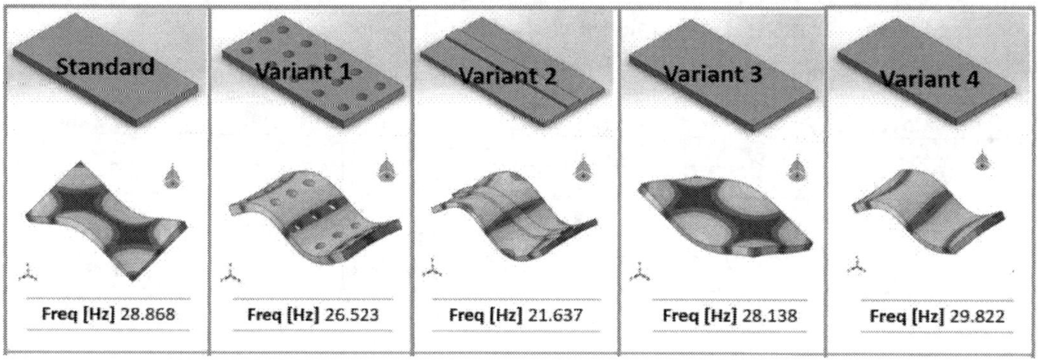

Fig. 5. Group of frequencies that produce different type of distortion on parts, depending on their features.

If the geometry of the part is dissymmetric, the distortion is as well, dissymmetric as one can see in figure 6. This means, that if the specific features which modify the shape of the part are clever chosen, a desired effect can be obtained. That is to say that if designer is skilled, he can insulate the side that distorts maximum in an area of the part where it should not affect the correct work of the whole.

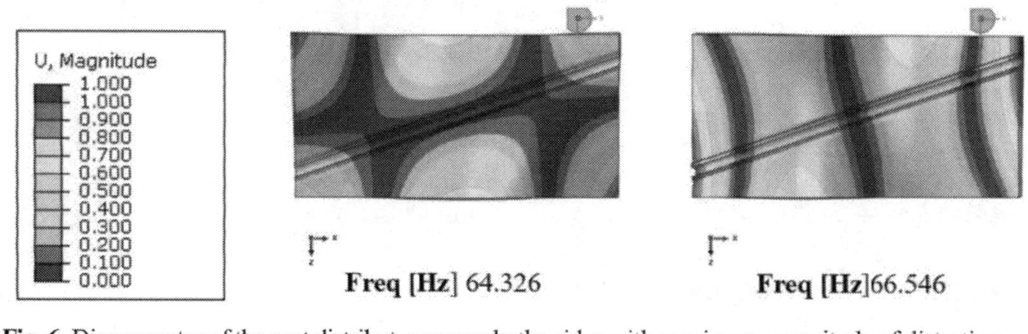

Fig. 6. Dissymmetry of the part distributes unevenly the sides with maximum magnitude of distortion.

Note that even a small difference of frequency pushes the areas with maximum magnitude of distortion to opposite sides. This shows that any particular part and its modifications for avoiding the bad influence of vibrations must be judged in a strict correlation with the vibration frequency analysed.

Series of the "bad" frequencies values follow similar patterns of increasing, regardless to the shape of the sample, as one can see in figure 7.

 DOI: 10.1051/matecconf/20179401004

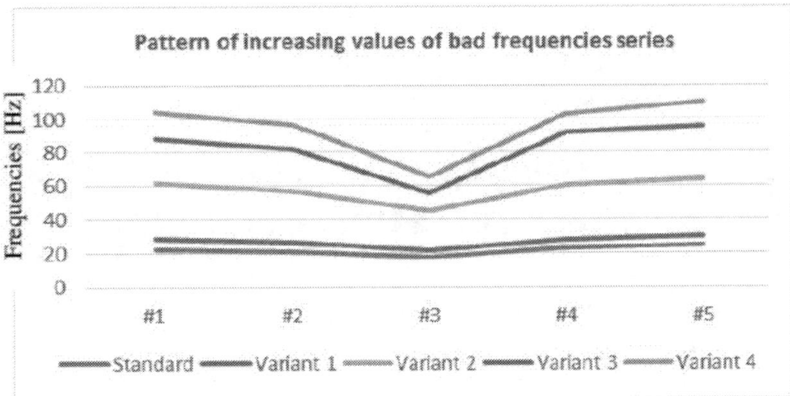

Fig. 7. Patterns of increasing "bad" frequencies.

4 Conclusions

Parts are different stressed by vibrations depending on the frequency they are subjected to. Some frequencies make the part to distort significantly (with big magnitude) – those presented in Table 1 – they were called "bad" frequencies. Other frequencies affect only slightly, or not at all the parts.

The geometry of the part may be modified by adding or removing some features. These modifications should have as effect a better behaviour of the part under the stress of vibrations. Such modifications might not change the external shape of the part, but only the mass distribution of the part. In this way, either the value of a "bad" frequency is pushed to another value that does not affect negatively the work of the part, or moves the side with big magnitude of distortion to another area of the part, as it no more has a negative impact on the way the part works. However, any modification in terms of geometry should be applied, it may not affect the functionality of the part or of the assembly it belongs to.

We hereby acknowledge the structural funds project PRO-DD (POS-CCE, O.2.2.1., ID 123, SMIS 2637, ctr. No 11/2009) for providing the infrastructure used in this work.

References

1. N. D. Anh and N. Nguyen, J. Mec. Sci. Tec. **30**, 593 (2016)
2. M. Law, M. Wabner, A. Colditz, et al. CIRP J. Man. Sci. Tec **10**, 36 (2015)
3. N. D. Anh, and N Nguyen, J. Mec. Eng. Sci **228**, 45 (2014)
4. A. A. Ghaderi, A Mohammadzaleh, M. N. Mansour, Mechanica **6**, 465 (2105)
5. J. X. Shi, R. Hirano, and M. Shimoda, Comp Struct. **148**, 50 (2016)
6. X. Gao, Q. Chen, X. B. Liu, NonlDyn **84**, 1715 (2016)
7. R. F., Harik, and J. S. Issa, J Vib. Cont **21**, 1810 (2015)
8. C. Okwudire and P. Zhao, *Proceedings of the ASME International Manufacturing Science and Engineering Conference,*597-604 (Univ Notre Dame, 2012)
9. https://www.comsol.com/model/download/215621/models.mph.micromixer.pdf, (accessed 17.05.2016)

Finite-element analysis of some pneumatically-actuated wrist-rehabilitation equipment

Ovidiu Filip[1,*] , and *Tudor* Deaconescu[1]

[1]Department of Industrial Engineering and Management, Transilvania University of Brasov, B-dul Eroilor nr. 29, Romania

Abstract. The paper presents some wrist-rehabilitation equipment, actuated by a pneumatic muscle. This one drives, through a rack-pinion mechanism, a Fin Ray-type mechanism, with a view to mobilizing the injured palm. Unlike other rehabilitation equipments, the one presented in this paper, ensures the movement of both palm and fingers.

1 Introduction

Passive motion is a part of the dynamic kinetic techniques for rehabilitation [1]. If the motion is made by means of muscle contraction, it is called active; and if the motion is made with external help, without muscle contraction, then it is called passive. Conscious motion, by muscle contraction, of the bodily segments, divides the kinetic techniques into active and passive techniques.

Disputes upon passive mobilization, have existed for a very long time; there being currently specialists opposing this technique, as well some others supporting it. Although these disputes date far back, passive kinetotherapy has been developing for centuries. This purpose is served by the wrist-rehabilitation equipment presented in this paper.

2 Wrist-rehabilitation equipments

On the medical-equipment market, there are several constructive types destined for hand rehabilitation [2]. According to the type of drive or actuation, they may be:

- with electric drive;
- with pneumatic actuation.

At this time, on the market, only electric-drive equipments are commercialized, as pneumatic actuation has not been yet sufficiently well studied and implemented in this system of applications.

According to their mobility, rehabilitation equipments are fixed and portable. In order to understand the operation of these rehabilitation systems, a few types of wrist-rehabilitation equipments will be presented in the following.

In figure 1, a portable equipment, with electric drive, produced by QAL Medical (former Otto Bock) [3] is presented. Because the equipment is placed on the patient's

* Corresponding author : ovidiu.filip@unitbv.ro

 DOI: 10.1051/matecconf/20179401005

forearm, it belongs to the group of portable equipments. It is driven by means of an electric motor.

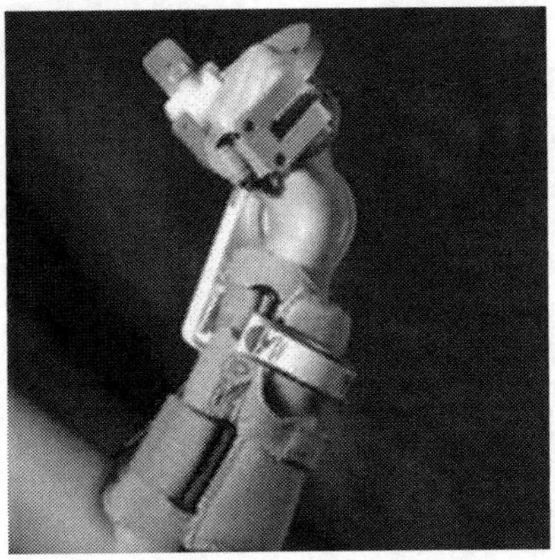

Fig. 1. W2 Wrist CPM.

Another portable equipment, produced by Kinetic Muscles Inc, is shown in figure 2. Paper [4] contains the description and operation of this equipment, called The Hand Mentor, actuated by a pneumatic muscle.

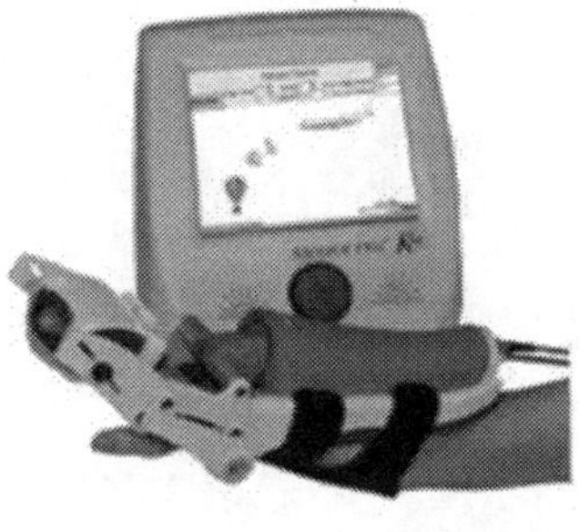

Fig. 2. The Hand Mentor.

Figure 3 presents a patent of the company Kinetic Muscle Inc [5], which is similar to the above equipment; but which, for actuation, uses two pneumatic muscles operating out of time.

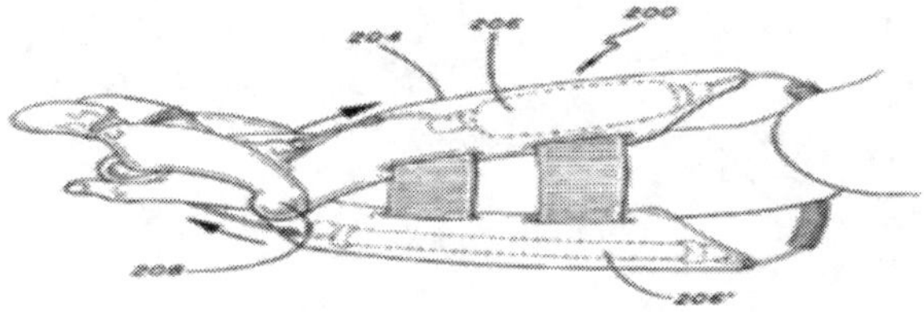

Fig. 3. Patent US2010198115.

The rehabilitation equipment in figure 4 is fixed, with electric drive [6]. This is a more complex equipment, by which rehabilitation exercises for both wrists and finger joints, can be made. In order to perform these movements, the equipment is provided with several accessories, assembled as needed. This is a fixed equipment, driven by an electric motor.

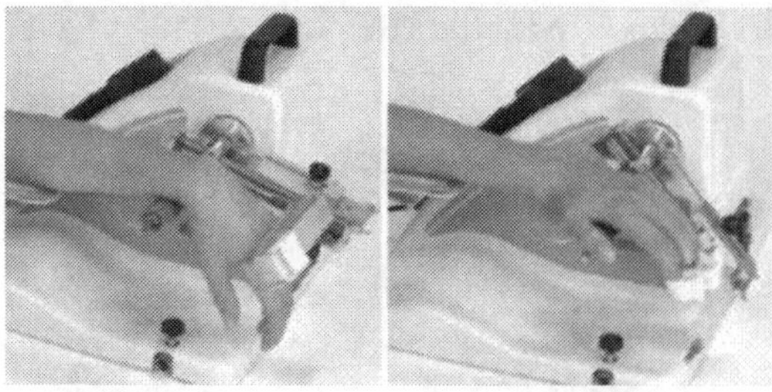

Fig. 4. Kinetec Maestra.

A study conducted by the authors of this paper upon the motion of some fish species, has led to the conclusion there are similarities between the palm movement and, for instance the swinging in a horizontal plane of the caudal fin of the perch. The flexion-extension motion of the palm is similar to the swinging motion of the fish fin. In both cases, the bone structure is set in motion by muscle groups situated sideways from the skeleton and operating out of time.

Based on these studies, in paper [7] the authors proposed a new wrist-rehabilitation equipment (fig. 5), of a fixed type, actuated by a pneumatic muscle. The motion imprinted to the palm is similar to the one of the fish fin, and it is based on the effect known as Fin Ray.

Fig. 5. Wrist-rehabilitation equipment.

The equipment consists of a framework on which the forearm support, the palm support, the pneumatic muscle and the rack-pinion mechanism are assembled. The linear

 DOI: 10.1051/matecconf/20179401005

advancement of the loose end of the pneumatic muscle is transformed into a rotary motion of the palm-supporting mechanism.

This is feasible by means of a rack-pinion mechanism. Unlike the other equipments presented above, this one enables the passive mobilization, at the same time, of both palm and fingers.

As follows, in this paper, the results obtained by a finite-element analysis upon the resistance of the whole equipment to the mechanical stresses undergone during operation, will be presented.

3 Verification at mechanical stresses of the wrist-rehabilitation equipment

For the verification at mechanical stresses, the forearm mass will be considered equal to 8 kg (approximately 80 N), and the palm mass 0,5 kg (approximately 5 N).

3.1 Calculation of the force necessary for palm-lifting

In order to determine the necessary force developed by the pneumatic muscle, with a view to lifting a force of 5 N, the support of the mechanism and the palm-support mechanism are imported from the software Pro/Engineer Wildfire 5.0 into Adams.

The palm-support mechanism is remade again, by means of the constraints in Adams, obtaining a new mechanism, shown in figure 6. In order to determine the value of the force, at rack level, two cases will be considered:

- in the former case, the torsional springs will not be considered;
- in the latter case, the torsional springs will be considered.

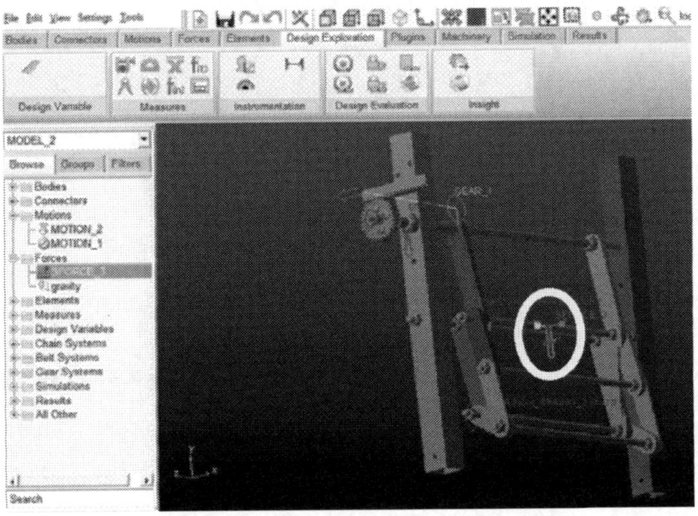

Fig. 6. Ensemble imported in Adams.

Palm weight was introduced in the calculations as a concentrated force, which acts in the middle of the bar supporting the palm (the circled area in figure 6). Following the simulations, the following force variations were obtained:

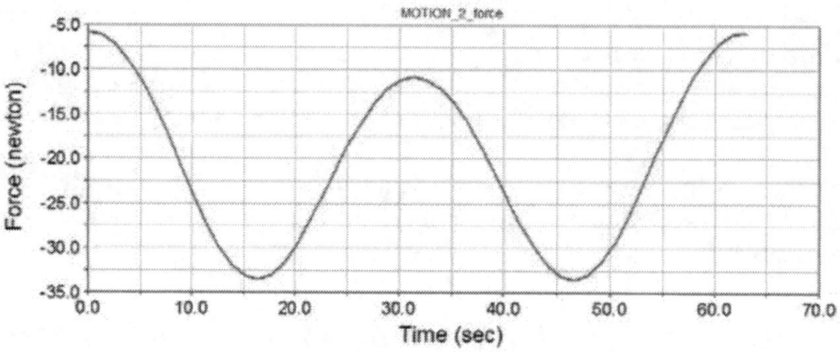

Fig. 7. Force variation at the rack (annulled springs; weight is considered in calculations).

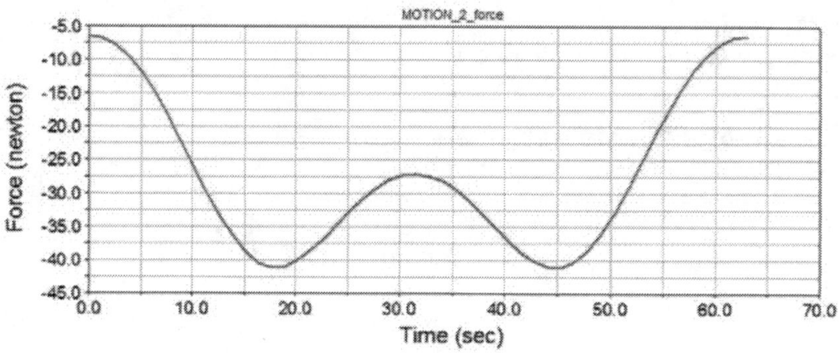

Fig. 8. Force variation at the rack (springs and mechanism weight are considered in calculations).

The necessary force at the rack, with a view to lifting a body of 5 N in weight, can be briefly calculated, applying the law of levers. Noting with:

- Fc – force at rack
- rc – pinion radius
- F – weight to be lifted
- rf – distance between axes

This relation can be written:

$$Fc \cdot rc = F \cdot rf \tag{1}$$

It follows that:

$$Fc = \frac{F \cdot rf}{rc} \tag{2}$$

As rc = 15 mm, rf = 70 mm and F = 5 N, then Fc = 23.3 N is obtained, in case the gravitation is not taken into account.

If 200 g (mass of the ensemble in motion) are added to the force F, then the value of the force Fc calculated with the relation (2) is of 32.67 N.

In the case of the dynamic analysis made in Adams, if we consider the weight of the ensemble in motion and if we annul the two torsional springs, then we will obtain a maximal force value of approximately 33 N (figure 7). It is noticeable that the value theoretically obtained is confirmed by the simulation in Adams.

From figure 8, the force of 45N will be kept as a force for the future calculations.

3.2 Verification at strains of the equipment

The material of which the rehabilitation-equipment components have been made, is aluminium. According to [8], the mechanical properties of a few aluminium alloys can be found in Table 1.

Table 1. Mechanical properties of the aluminium alloys

Alloy	Temper	Tensile Strength (MPa)	Shear Strength (MPa)
AA1050A	H2	100	60
AA1050A	H4	115	70
AA1050A	H6	130	80

The verification at mechanical stresses was made with the finite-element analysis (FEA). In order to perform this analysis, the meshing and processing were achieved with the software Pro/Engineer Wildfire 5.0.

A first analysis was made considering the mechanical stress of the equipment by the weight of the forearm. Reckoning the forearm weight of 80 N, the force being uniformly distributed on the four legs, this means a leg is subjected to an axial load of 20 N. Following this load, the results in figures 9 and 10 are obtained.

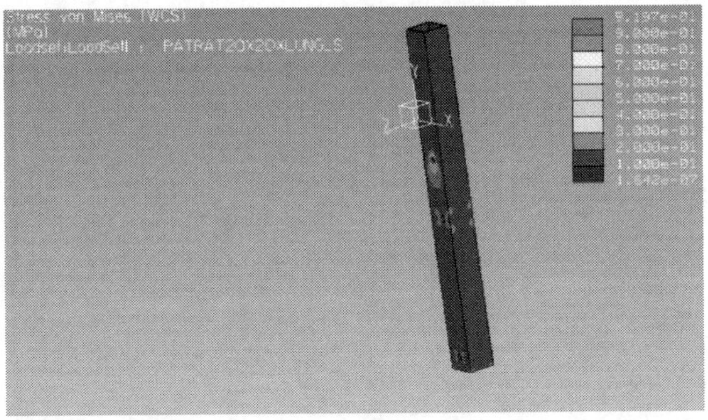

Fig. 9. Tension variation in the leg.

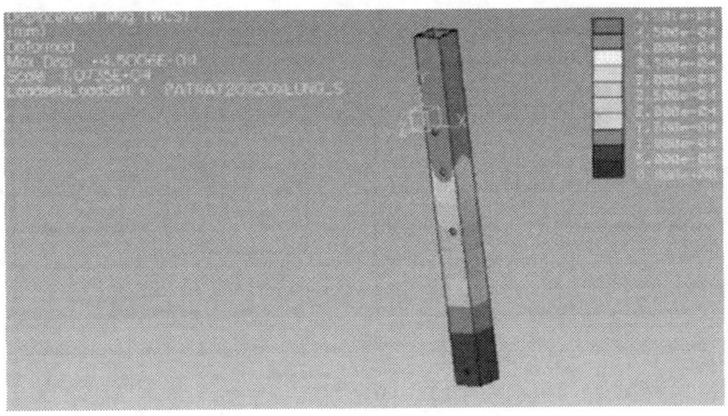

Fig. 10. Leg deformation.

It is noticeable that the deformation in the leg of the equipment is very small, of only 0,0004 mm.

The palm-movement process mechanically stresses several components of the equipment. In figure 11, the components contributing to the motion of the palm, can be seen.

DOI: 10.1051/matecconf/20179401005

Fig. 11. Simplified ensemble.

The stressed elements are the legs of the equipment, the support in the fixed end of the pneumatic muscle, the support in the loose end of the muscle, the rack and the pinion, the shaft transmitting the rotary motion, the metallic plates which support the bar on which the palm rests.

The finite-element analysis was achieved for all components of the mechanical ensemble; all of which present loads below the yield point, and small deformations. In the following, for exemplification, the analyses afferent to the support in fixed end of the pneumatic muscle, to the rack and to the bar on which the palm rests, are presented.

In figures 12 and 13, the tensions and the deformations of the bar on which the palm rests, are noticeable.

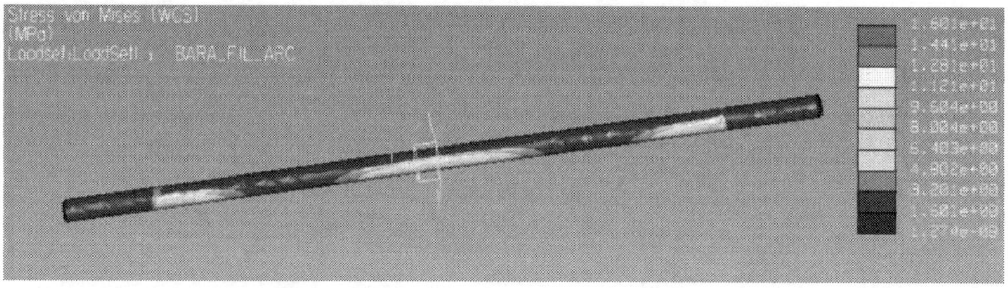

Fig. 12. Tensions in the palm-supporting bar.

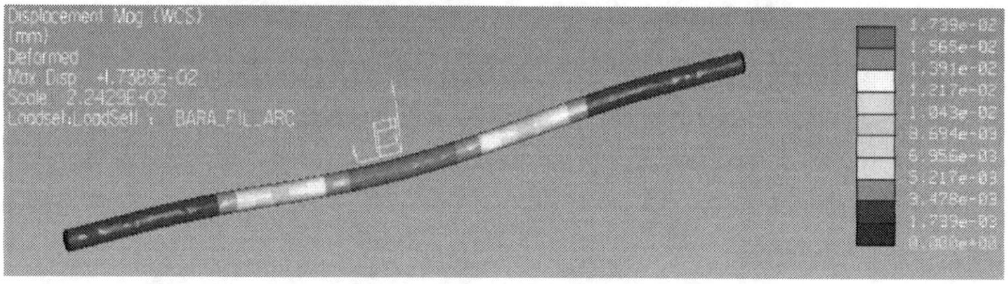

Fig. 13. Deformation of the palm-supporting bar.

It is noticeable that the maximal deformation in the case of the bar is of approximately 0,02 mm.

In the case of the rack, the maximal stress occurs when the contact between rack and cogwheel is made on one tooth. The maximal force which can stress the rack, according to figure 8, is of approximately 45 N. Figures 14 and 15 show the tensions and deformations appearing in the rack.

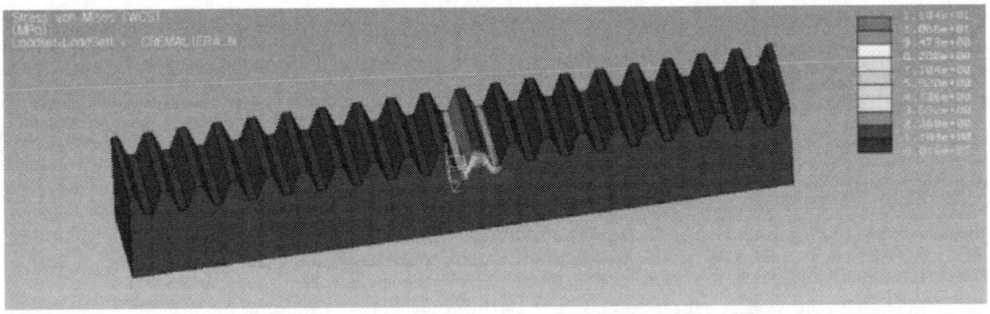

Fig. 14. Tension variation in a rack tooth.

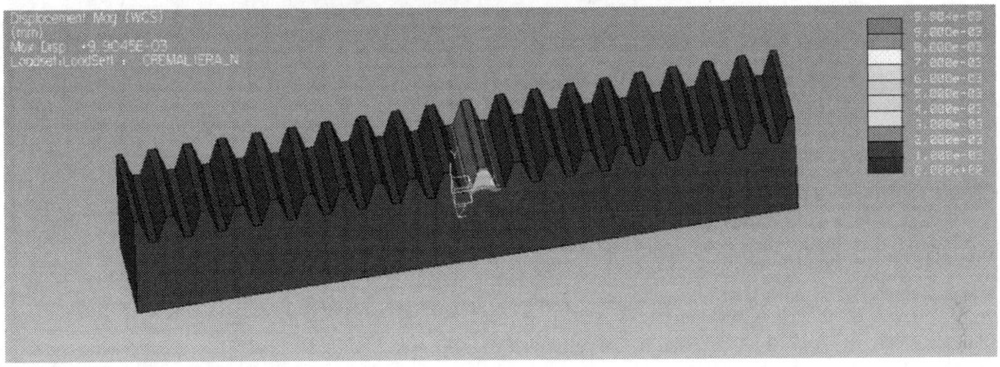

Fig. 15. Deformation of the rack tooth.

Also in this case, the shift at the tooth tip is small, of approximately 0,01 mm.

The tensions and the deformations in the pinion are shown in the figure 16 and 17

Fig. 16. Tension variation in the pinion.

Fig. 17. Deformation variation in the in the pinion.

The maximal deformation of the pinion is 0.04 mm.

The support in the fixed end of the pneumatic muscle, because of the great distance between the embedment plane and the force-application point, presents higher stresses. Figures 18 and 19 show the variation of the tensions and the deformations in the support.

Fig. 18. Tension variation in the fixed support of the pneumatic muscle.

Fig. 19. Deformation variation in the fixed support of the pneumatic muscle.

All of the above point to the fact that the maximal deformation of the components does not exceed 0,2 mm, the tensions therein not exceeding the yield point.

4 Conclusions

The wrist-rehabilitation equipment is original, resisting the stresses and strains for which it was designed. The deformations in the components of the equipment are small, not influencing its operation. Achieving the three-dimensional model enabled verifying the equipment at mechanical stresses and making thereby the physical model.

References

1. T. Sbenghe, *Prophylactical, therapeutical and recovering kinesiotherapy.* (Editura Medicala, Bucureşti, 1987)
2. O. Filip, T. Deaconescu, *Advances in Production, Automation and Transportation System, Proceedings of the 6th International Conference on Manufacturing Engineering, Quality and Production Systems (MEQAPS '13), Proceedings of the 4th International Conference on Automotive and Transportation Systems (ICAT '13),* 55-59, (2013)
3. *** http://qalmedical.com/w2-wrist-cpm-device/ Accessed: 10.04.2016
4. N.G. Kutner, et al. Phys Ther, **90**, 4, 493-504, (2010)
5. E. J. Koeneman, et al, *System and Method for Neuromuscular Reeducation,* US2010198115 (2010)
6. *** http://www.kinetec.fr/en/kinetec-selection/cpm-continuous-passive-motion/hand-wrist.html Accessed: 10.04.2016
7. O. Filip, T. Deaconescu, *ICEEMS 2013, 7th International Conference on Economic Engineering and Manufacturing System, Recent,* **14**, 4, 263-266 (2013)
8. *** http://www.azom.com/article.aspx?ArticleID=2863#13, Accessed: 3.10.2016

DOI: 10.1051/matecconf/20179401006

Mathematical modelling of a Fin Ray type mechanism, used in the case of the wrist rehabilitation equipment

Ovidiu Filip[1,*], and *Tudor* Deaconescu[1]

[1]Department of Industrial Engineering and Management, Transilvania University of Brasov, B-dul Eroilor nr. 29, Romania

Abstract. The paper presents the mathematical modelling of some Fin Ray-type mechanism, destined for the wrist-rehabilitation equipment, its driving being made by a pneumatic muscle. The validity of the mathematical modelling of Fin Ray mechanism is verified by three-dimensional mechanism made in Pro/Engineer.

1 Introduction

In modern times, the successful evolution, on the international market, of a company, depends on its ability to achieve products of a high qualitative level, at low costs, which should satisfy to the highest degree, the consumers' demands [1]. This is the case of the companies manufacturing medical-rehabilitation equipment, which should keep pace with the new driving systems appeared on the market. For this purpose, a new concept is presented in this paper, of some wrist-rehabilitation equipment, based on the Fin Ray effect.

Following the accidents they suffered, many times, the patients are required to go through a period of limb immobilization. Prolonged immobilization leads to muscle wasting (hypotrophy), to dysfunctions of the circulatory system and to bone demineralization [2, 3]. Surgical treatment of hand must be always followed by a physical-kinetic rehabilitation program [3, 4]. The mobilizations shall be gently made and they aim at the patient's consciously regaining motion.

In the event the patient cannot consciously achieve the imposed movement, passive mobilization shall be resorted to. This mobilization is recommended even in the case of irreversible lesions [2].

Passive mobilization can be made in two ways:

- by the kinotherapist, through direct action on the patient;
- by means of rehabilitation equipment.

The use of the rehabilitation equipment must allow adjusting the extent of the course and the speed by which the motion of the immobile limb is achieved. Various exercises can be thereby made, at different speeds and movements, adapted to each patient's needs. These requirements are necessary in the case of the wrist-rehabilitation equipments.

* Corresponding author : ovidiu.filip@unitbv.ro

In paper [5] several wrist-rehabilitation equipments are presented, which ensure palm motion by means of electrical or pneumatic drives. Most equipments marketed nowadays resort to electric motors for driving purposes, and ensure palm rotation in relation to the forearm. In this paper, a new pneumatically-driven rehabilitation-equipment is presented, which ensures both palm-rotation in relation to the forearm, and finger (proximal phalanges)-rotation in relation to the palm.

The palm-motion limits can be found in the literature; as follows, the values afferent to the flexion and extension of the two articulations, being presented [6]. In figure 1, the fist flexion or palmar flexion is presented, which has the maximal value of 80° [6].

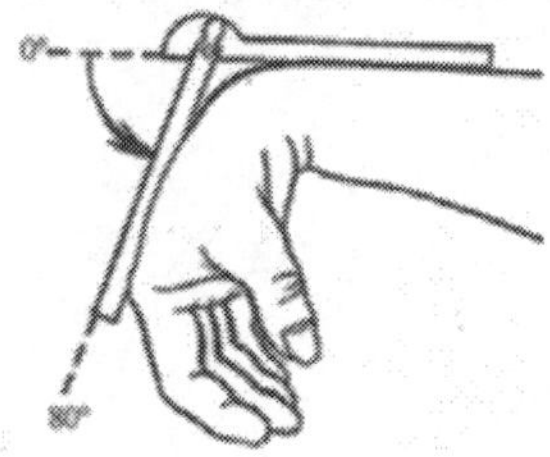

Fig. 1. Palmar flexion.

The maximal value of fist extension can be seen in figure 2, having the value of 70° [6].

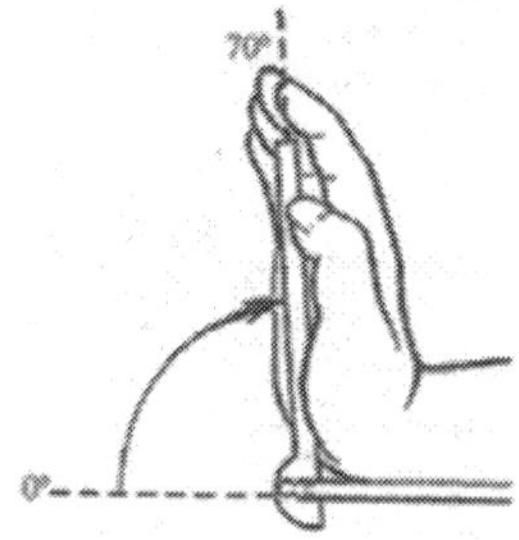

Fig. 2. Fist extension.

The following figures illustrate the finger flexion and extension. In figure 3, it is noticeable that, in the case of flexing, the common value to all fingers is of 90° [6].

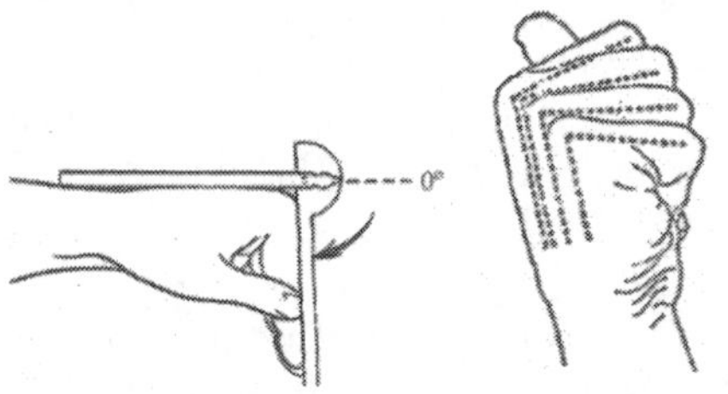

Fig. 3. Finger flexion.

Finger extension (figure 4) varies from one individual to another; and it has values ranging between 0° and 45° [6]. The high values of the finger extension are called hyperextension.

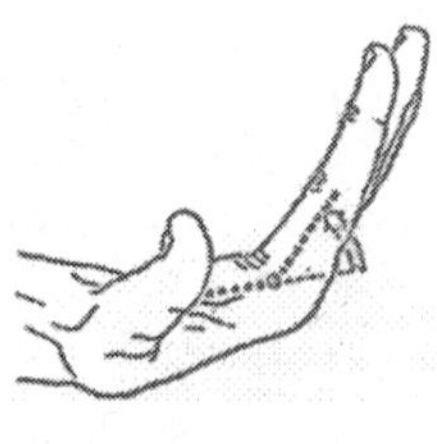

Fig. 4. Finger extension.

 DOI: 10.1051/matecconf/20179401006

Given the above presented facts, a mechanism has been designed, with a view to ensuring palm and finger motion, according to the areas indicated in the literature.

2 Structure of the mechanism

The wrist-rehabilitation equipment proposed in this paper, is based on a mechanism imitating the movement of fish, by means of the swim-fins. These are elastic structures crossed by thin bony fibres, which have a different behaviour from an elastic lamella, moving on the basis of the Fin Ray effect. The Fin Ray-type mechanism allowing palm movement, is shown in figure 5.

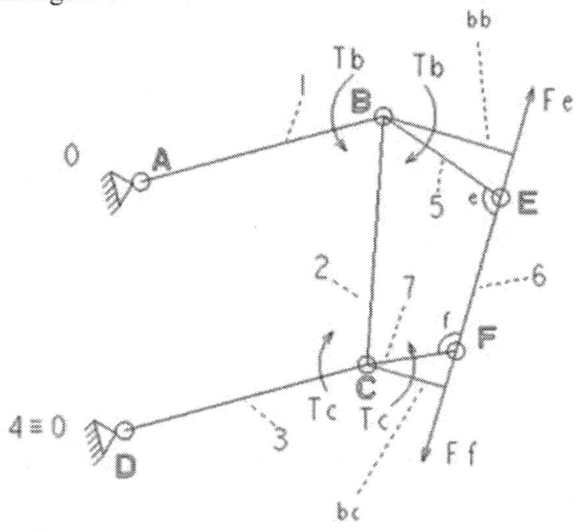

Fig. 5. Support mechanism for the palm.

The above assembly consists of two quadrilateral mechanisms. The quadrilateral mechanism ABCD is driven in a rotatory motion by a pneumatic muscle, by means of a cog rack-pinion mechanism. The quadrilateral mechanism BEFC is coupled to the mechanism ABCD by means of two torsional springs placed in the torques B and C. The spring in speed torque B connects the arm AB in the first mechanism, and the arm BE in the second mechanism; and the spring in the torque C makes the connection between the arm DC and the arm CF.

In order to determine the mathematical model of the mechanism, the following assumptions are taken as a starting point:

- the masses of the elements will be neglected,
- the frictional forces are null,
- the elastic constant of the spring moment is the same for both torsional springs.

With a view to sizing the mechanism, the following constraints are taken as a starting point:

- the crank AB must rotate upwards by 70° to the horizontal and by 80° downwards to the horizontal;
- the lengths of the balancers AB and BE depend on the conformation of the injured palm. In this case, AB = 70 mm and BE = 40 mm.

The other sizes of the quadrilateral mechanism were determined by simulating the operation of the mechanism, resorting to the software Pro/Engineer Wildfire 5.0. Following these simulations, the following sizes were obtained: BC = 70 mm, CD = 70 mm, CF = 25 mm, FE = 45 mm, the distance AD measured vertically, ADv = 70 mm, and horizontally ADh = 5 mm.

2.1 The mathematical model

In the simulation made in Pro/Engineer Wildfire 5.0, as in reality, the position of the mechanism ABCD determines the position of the mechanism BEFC, the value of the angle $\widehat{DAB}$ influencing the value of the angle $\widehat{CBE}$. In order to simplify the calculations, in the mathematical model, the angle $\widehat{CBE}$ will be considered as input data, whereas the angle $\widehat{DAB}$ as output data.

Using the law of cosines in the triangles ABD and BCD (figure 5), the following relations can be written for the mechanism ABCD:

$$BD^2 = AB^2 + AD^2 - 2AB \cdot AD \cdot \cos\widehat{DAB} \quad (1)$$

$$AD^2 = AB^2 + BD^2 - 2AB \cdot BD \cdot \cos\widehat{ABD} \quad (2)$$

$$CD^2 = BC^2 + BD^2 - 2BC \cdot BD \cdot \cos\widehat{DBC} \quad (3)$$

$$BD^2 = DC^2 + BC^2 - 2DC \cdot BC \cdot \cos\widehat{BCD} \quad (4)$$

$$AB^2 = AD^2 + BD^2 - 2AD \cdot BD \cdot \cos\widehat{ADB} \quad (5)$$

$$BC^2 = DC^2 + BD^2 - 2DC \cdot BD \cdot \cos\widehat{BDC} \quad (6)$$

Identically, in the quadrilateral mechanism BEFC, using the law of cosines in the triangles BCE and CEF, the following relations are obtained:

$$CE^2 = BC^2 + BE^2 - 2BC \cdot BE \cdot \cos\widehat{CBE} \quad (7)$$

$$CE^2 = EF^2 + CF^2 - 2EF \cdot CF \cdot \cos\widehat{CFE} \quad (8)$$

$$BC^2 = BE^2 + CE^2 - 2BE \cdot CE \cdot \cos\widehat{BEC} \quad (9)$$

$$CF^2 = EF^2 + CE^2 - 2EF \cdot CE \cdot \cos\widehat{CEF} \quad (10)$$

$$BE^2 = BC^2 + CE^2 - 2BC \cdot CE \cdot \cos\widehat{BCE} \quad (11)$$

$$EF^2 = CF^2 + CE^2 - 2CF \cdot CE \cdot \cos\widehat{ECF} \quad (12)$$

So that the quadrilateral mechanism BCFE should operate properly, it is necessary that the torsional springs assembled in the torques B and C be strained over the entire working area of this mechanism (90°). Therefore, springs with arms disposed at 270° (figure 6) will be used.

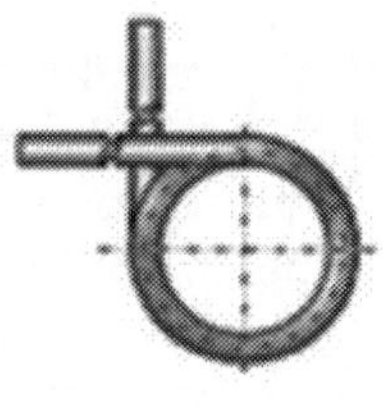

Fig. 6. Torsional spring.

The torsion moment in the torque B has the value:

$$Tb = Fe \cdot bb = K \cdot (270° - \widehat{ABE}) \quad (13)$$

where Fe – force applied in the point E, bb – arm of the force Fe, K – constant afferent to the spring moment.

The torsion moment in the torque C has the value:

$$Tc = Ff \cdot bc = K \cdot (270° - \widehat{DCF}) \quad (14)$$

where Fe is the force applied in the point E, bb – arm of the force Fe.

So that the crank EF should be in equilibrium, the forces Fe and Ff must be equal:

$$\frac{K \cdot (270° - \widehat{ABE})}{bb} = \frac{K \cdot (270° - \widehat{DCF})}{bc} \quad (15)$$

where:

$$bb = BE \cdot \sin\widehat{BEF} \quad (16)$$

$$bc = CF \cdot \sin\widehat{CFE} \quad (17)$$

From the above relations, it follows:

$$270° \cdot (bc - bb) + bb \cdot \widehat{DCF} - bc \cdot \widehat{ABE} = 0 \quad (18)$$

The angles $\widehat{DCF}$ and $\widehat{ABE}$ can be expressed as a sum of angles which may be determined from the two quadrilateral mechanisms:

$$\widehat{ABE} = \widehat{ABC} + \widehat{CBE} = \widehat{ABD} + \widehat{DBC} + \widehat{CBE} \quad (19)$$

$$\widehat{DCF} = \widehat{BCD} + \widehat{BCF} = \widehat{DCB} + \widehat{BCE} + \widehat{ECF} \quad (20)$$

Given that the movement of entry into the mechanism for the support of the palm, in the case of this analysis, is the rotation of the arm BE it follows that the angle $\widehat{CBE}$ is known. By attributing different values to the angle $\widehat{CBE}$ the positions of the mechanism BCEF can be determined.

The connection between the mechanism BCEF and the mechanism ABCD is achieved by the equilibrium of the forces, at the level of the crank EF. In this respect, the angles $\widehat{ABD}$, $\widehat{DBC}$ and $\widehat{BCD}$ become:

$$\widehat{ABD} = \cos^{-1}\left(\frac{AB^2-AD^2+AB^2+AD^2-2AB\cdot AD\cdot\cos\widehat{DAB}}{2AB\cdot\sqrt{AB^2+AD^2-2AB\cdot AD\cdot\cos\widehat{DAB}}}\right) \quad (21)$$

$$\widehat{DBC} = \cos^{-1}\left(\frac{BC^2-CD^2+AB^2+AD^2-2AB\cdot AD\cdot\cos\widehat{DAB}}{2BC\cdot\sqrt{AB^2+AD^2-2AB\cdot AD\cdot\cos\widehat{DAB}}}\right) \quad (22)$$

$$\widehat{BCD} = \cos^{-1}\left(\frac{DC^2+BC^2-AB^2-AD^2+2AB\cdot AD\cdot\cos\widehat{DAB}}{2DC\cdot BC}\right) \quad (23)$$

By introducing the relations 19 and 20 in the relation 18, there is obtained:

$$270^\circ\cdot(bc-bb)+bb\cdot\left(\widehat{BCD}+\widehat{BCF}\right)-bc\cdot\left(\widehat{ABD}+\widehat{DBC}+\widehat{CBE}\right)=0 \quad (24)$$

By introducing in the relation 24, the relations 21, 22, 23, an equation is obtained, having cos$\widehat{DAB}$ as unknown. The equation is solved with Matlab software.

2.2 Matlab application used to determine the angles of the mechanism

As follows, a Matlab application is presented, by which $\widehat{DAB}$ is calculated, resorting to different values of the angle $\widehat{CBE}$ as input data.

The calculation software for the angle $\widehat{DAB}$

```
%arms mec1
l1=70;
l2=70;
l3=70;
l4=sqrt(70^2+5^2);
%arms mec2
l2=70;
l5=40;
l6=45;
l7=25;
disp ('type the angle ub=')
disp ('then the secondary quadrilateral mechanism will be calculated')
disp ('type "calc"')
b=degtorad(ub)
d2=sqrt((l2)^2+(l5)^2-2*l2*l5*cos(b));
e1=acos(((l5)^2+(d2)^2-(l2)^2)/(2*l5*d2));
e1g=radtodeg(e1);
e2=acos(((l6)^2+(d2)^2-(l7)^2)/(2*l6*d2));
e2g=radtodeg(e2);
c1=acos(((l2)^2+(d2)^2-(l5)^2)/(2*l2*d2));
c1g=radtodeg(c1);
c2=acos(((l7)^2+(d2)^2-(l6)^2)/(2*l7*d2));
c2g=radtodeg(c2);
```

```
f=acos(((l6)^2+(l7)^2-(d2)^2)/(2*l6*l7));
fg=radtodeg(f);
eg=e1g+e2g;
cg=c1g+c2g;
e=e1+e2;
c=c1+c2;
%calculation of the arms of the torsional moments
arm_b=l5*sin(e);
arm_c=l7*sin(f);
rap_arms=arm_b/armc;
disp (cosine of the angle A <input>' is calculated)
disp (type "ec"')
z=270;
zr=degtorad(270);
syms x
ab1=acos(((l1)^2+(l1)^2+(l4)^2-2*l1*l4*x-(l4)^2)/(2*l1*sqrt((l1)^2+(l4)^2-
2*l1*l4*x)));
ab2=acos(((l2)^2+(l1)^2+(l4)^2-2*l1*l4*x-(l3)^2)/(2*l2*sqrt((l1)^2+(l4)^2-
2*l1*l4*x)));
solve(zr*(arm_c-arm_b)-(ab1+ab2)*arm_c-b*arm_c+c*arm_b+acos(((l3)^2+(l2)^2-
((l1)^2+(l4)^2-2*l1*l4*x))/(2*l3*l2))*arm_b == 0, x)
disp ('the angle A <input> is calculated')
disp ('type "val_a"')
```

The mathematical model was verified by using the obtained data in the simulation of the mechanism achieved in Pro/Engineer.

Following the simulation of the mechanism, several positions of this one were obtained. 16 positions were retained, which describe the entire course of 150° to be covered by the mechanism, so as to ensure the movement of the palm between the maximal values of the fist flexion and extension. In these positions, the values of the angles $\widehat{DAB}$ and $\widehat{CBE}$ were measured.

Introducing the values of the angle $\widehat{CBE}$ (variable *ub* in the application) the values of the angle $\widehat{DAB}$ (variable *val_a* in the application) will be obtained.

The obtained values in Pro/Engineer and in Matlab are presented in figure 7.

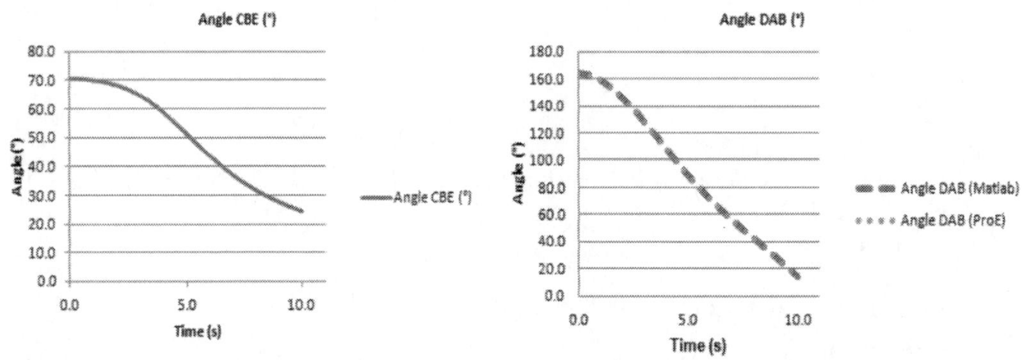

Fig. 7. Values of the angles $\widehat{CBE}$ and $\widehat{DAB}$.

Small differences (at the level of 0.000x degrees) can be noticed between the two values of the angle $\widehat{DAB}$ obtained in Pro/Engineer and in Matlab.

The values obtained in Pro/Engineer are presented in figure 8.

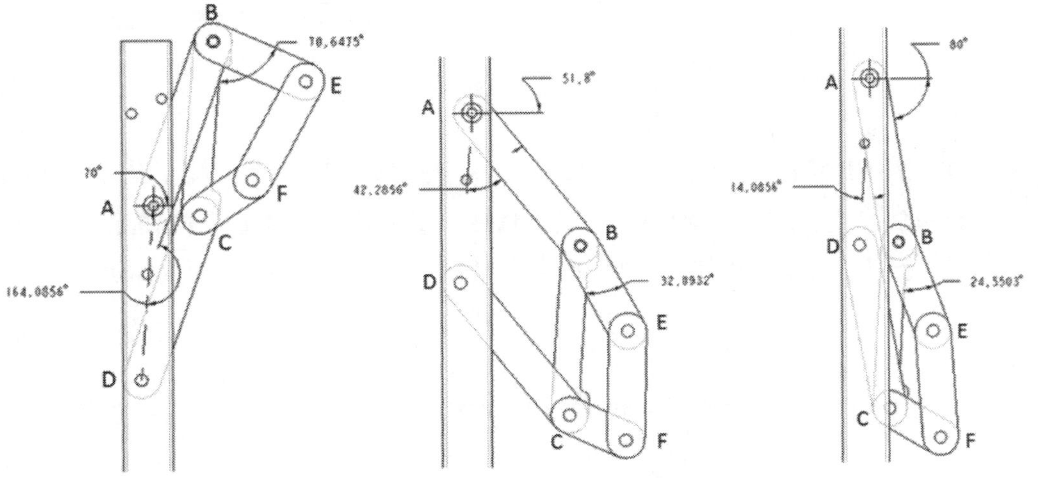

Fig. 8. Values of the angles $\widehat{DAB}$ and $\widehat{CBE}$ obtained in Pro/Engineer.

Figure 8 also shows that this mechanism can be used for mobilizing the fingers over their active area, between the maximal-flexion (90°) and the maximal-extension areas, ranging between (0°- 45°).

3 Conclusions

As a result of the facts presented above, the following ideas are worth keeping in mind:
- the presented mechanism ensures palm motion over its entire active area;
- this mechanism can be used also for the finger (proximal phalanges) motion, over their active area;
- a mathematical model was designed, which is validated by the simulation in Pro/Engineer;
- a Matlab application was designed, in order to determine the angle $\widehat{DAB}$.
- in the future, this mechanism can be used for the wrist-rehabilitation equipment.

References

1. A. Deaconescu, RMEE, **14**, 1, 29-38, (2015)
2. C. Albu, T.L. Vlad, A. Albu, *Passive kinesiotherapy* (Editura Polirom, Iaşi, 2004)
3. O. Filip, T. Deaconescu, *Advances in Production, Automation and Transportation System, Proceedings of the 6th International Conference on Manufacturing Engineering, Quality and Production Systems (MEQAPS '13), Proceedings of the 4th International Conference on Automotive and Transportation Systems (ICAT '13)*, 55-59, (2013)
4. I. Kiss, *Physio-kinesiotherapy and medical recovery* (Editura Medicala, Bucureşti, 2012)
5. O. Filip, T. Deaconescu, *ICEEMS, Recent*, **14**, 4, 263-266, (2013)
6. E.L. Sidenco, *Practical guide for joint and muscle evaluation,* (Editura Fundaţiei „Romania de Maine", Bucureşti, 2009)

Application of the method based on technological knowledge and expertise for adding the design of a technological process

Piotr Ociepka [1,*]

[1]Institute of Engineering Processes Automation and Integrated Manufacturing Systems, Faculty of Mechanical Engineering, Silesian University of Technology, ul. Konarskiego 18a, 44-100 Gliwice, Poland

Abstract. The manufacturing process is an important step in creating new products. Market trends cause that this process must be very flexible to facilitate fast and cost-effective adaptation to changing market conditions. Commonly used CNC machine tools, grouped in flexible manufacturing systems, greatly facilitate the rapid adjustment of production to these dynamic changes. The programs for CNC machine tools could be easily generated using software of the CAM class. It is also easy to change or modify them. However, the technological process itself still must be planned and continuously optimized with respect to each, new produced detail. The paper presents a method, based on the engineering knowledge and experience, designated to assist the technological process preparation. The method, developed by authors, is based on the CBR method. It is the method of problems solving, basing on seeking an analogy (similarity) between the currently being solved task and the previously solved tasks. In the paper is presented the algorithm of this approach and the formalized description of the developed method.

1 Introduction

The changing market demand and the increasingly shorter "product life cycle" are forcing manufacturers to accelerate the processes of designing and manufacturing. It results that engineers need, carrying out their routine activities, to use more powerful tools that are able to significantly accelerate activities carried out by them. It is important that, at such operations one should keep the acceptable cost of production as well as the required quality and durability of products. With regard to the processes of designing and constructing, there are many advanced and specialized tools of the CAD/CAE class, which allow significantly accelerating the realization of these activities [1-13]. In the case of the manufacturing process are commonly used the flexible manufacturing systems, which in combination with the multi-purpose CNC machines and computer programs of the CAM class, allows quickly and efficiently adapting production systems to the changeable productions schedules. The important and key element, at the stage of new products manufacturing, are: an adequately

* Corresponding author: piotr.ociepka@polsl.pl

planned technological process, correctly chosen machine tools and correctly determined machining parameters [14-15]. These activities require from a technologist an extensive knowledge and experience of the implemented machining process. The experience is particularly important in these engineering activities, which is the result of practical application by the technologist the knowledge possessed by him. That is why more and more attempts are being made to automate and facilitate the task performed by a technologist. There are created informatics tools, often based on artificial intelligence methods that support the technological process preparation. In the paper are presented the results of works related with the development of the method based on knowledge and experience for supporting the technological process preparation.

2 CBR method

Case Based Reasoning (CBR) is a method of problems solving, basing on engineering experience processing [16-21]. It involves the search for an analogy between the currently solved task and previously solved ones that have been saved in the "case base" of the CBR system. The main advantage of the CBR method is facilitating the process of knowledge and experience acquiring. This process is realized simultaneously with the given solution creating that is when the knowledge of this solution is the most complete.

Another advantage of this method is that the new solution (new case), stored in the CBR database, could be immediately available during solving the following tasks. It is considered with a very simple, yet effective learning process. With the continuous use of such system it gives the opportunity to gather a wide range of technological knowledge and experience in the certain field of work. The other advantage of this method is the possibility of saving also negative cases or solutions that have been proven to be incorrect. This allows avoiding errors during conducting similar tasks in the future.

The process of problems solving, using the CBR method, could be described as a series of actions, which in the literature is called the loop of four R (Retrieval, Reuse, Revision, Retainment) or shortly the R^4 model. This cycle is presented in Fig. 1.

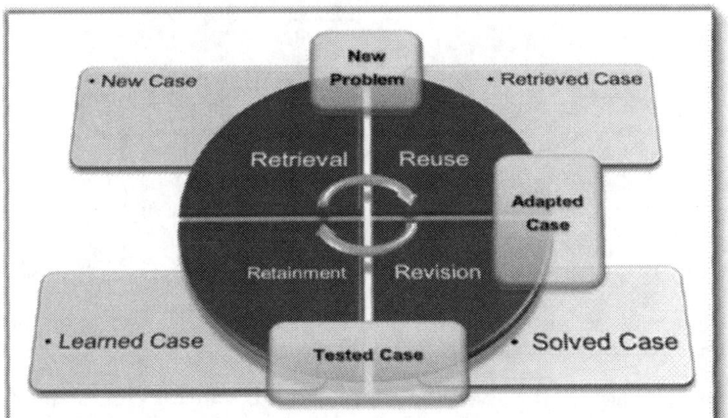

Fig. 1. R^4 model of the CBR cycle.

3 Algorithm of functioning of the elaborated method

The developed method of computer-aided technological process preparation could be included to the group of variants methods. In these methods is used the principle of similarity of elements, according to which one could assume that, in the case of elements, characterized by a high degree of technological and constructional similarities one could use, with a high probability, the similar technological processes. The important element of such approach is the adequate description of the technological conditions and the description of the element construction in order to determine the degree of similarity of the analyzed case. In Fig. 2 is shown the algorithm of the system, based on the proposed method, for the manufacturing process.

In the first step is created the description of the technological task. It contains the assumptions for the implemented process. It includes the technological conditions and the description of the construction (geometry, dimensions, tolerances, material) of the analyzed element. On the basis of such description, the CBR reasoning mechanism looks for the most similar case and proposes it as the solution to the problem. In the next stage the technologist could make modifications to the proposed technological process in order to adapt it to the currently analyzed task. Afterwards the adapted technological process could be stored in the CBR database as a new case.

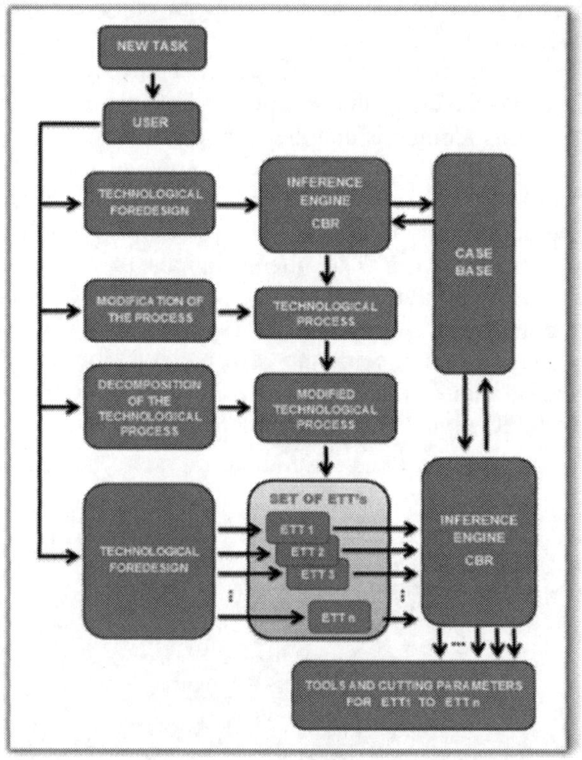

Fig. 2. Algorithm of the system functioning.

The next important step is the selection of machining (cutting) tools and parameters. For this purpose, the adapted technological process (the set of technological activities) is decomposed into so called Elementary Technological Tasks (ETT), for which one must choose the proper cutting tools and parameters. Then, for particular ETT is created the task description *T_case*. In a subsequent step, with the developed system, are sought analogous cases (ETT), implemented in the past. Basing on these cases the cutting tools and parameters are selected for the currently realized task (ETT). The method, presented in this paper, could be used for supporting the preparation of technological processes for rotationally-symmetrical elements (like shafts, bushes, rings, etc.). Of course, there is the possibility to modify the developed method taking into account other types of elements.

3.1 Formalized description of technological process selection

For the process of selection the cutting tools and machining parameters the formalized description of functioning of this method is as follows. Given is the description of the task, *T_case*, in which the user defines the attributes and weights describing the analyzed technological task.

$$T\_case = \{ ZT, PM, PK\} \tag{1}$$

where:
ZT – technological foredesign,
PM – description of the initial material,
PK – constructional form,

$$ZT = \{AT_1, WT_1, w_1, AT_2, WT_2, w_2, \ ..., \ AT_j, WT_j, w_j,\} \tag{2}$$

where:
AT_j – j-th attribute describing the technological task,
WT_j – value of the j-th technological attribute describing the technological task,
w_j – weight of the j-th technological attribute describing the technological task.

$$PK = \{ PKz, PKw \} \tag{3}$$

where:
PKz – outer constructional form,
PKw – inner constructional form.

$$PKz=\{ Sz_1, Sz_2, \ ..., Sz_n \} \tag{4}$$

$$PKw=\{ Sw_1, Sw_2, \ ..., Sw_n \} \tag{5}$$

where:
Sz_n – n -th step on the element, outer,
Sw_n – n -th step on the element, inner.

$$Sz =\{EOKz_1, EOKz_2, \ ..., EOKz_m \} \tag{6}$$

$$Sw =\{EOKw_1, EOKw_2, \ ..., EOKw_m \} \tag{7}$$

where:
$EOKz_m$ – m -th outer constructional feature,
$EOKw_m$ – m -th inner constructional feature.

$EOKz$ and $EOKw$ are finite sets of typical features of constructional objects which may occur as logical components rotational-symmetric elements.

$$EOKz = \{A_{EOKz1}, WA_{EOKz1}, A_{EOKz2}, WA_{EOKz2}, \ ..., A_{EOKzz}, WA_{EOKzz} \} \tag{8}$$

$$EOKw = \{A_{EOKw1}, WA_{EOKw1}, A_{EOKw2}, WA_{EOKw2}, \ ..., A_{EOKwz}, WA_{EOKwz} \} \tag{9}$$

where:
A_{EOKzz} – z -th attribute describing outer constructional feature,
WA_{EOKzz} – z -th value of the attribute describing outer constructional feature,
A_{EOKwz} – z -th attribute describing inner constructional feature,
WA_{EOKwz} – z -th value of the attribute describing inner constructional feature.

During comparing the geometric form of the analyzed element it is analyzed both the incidence of constructional features as well as their order of occurrence in the analyzed stage (part) of an object [22-27]. It was assumed that the analysis and decomposition of the element runs from the left side of this element. In Fig. 3 is shown the description of an constructional form of an exemplary shaft.

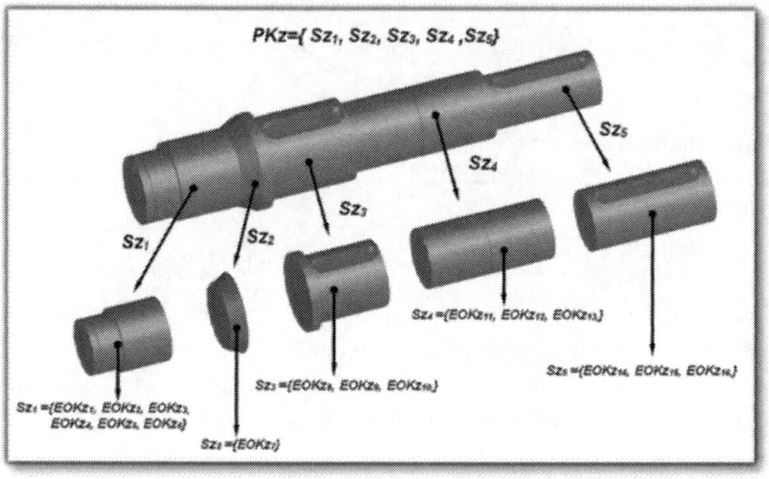

Fig. 3. Description of the constructional form of an exemplary shaft.

After this step the following procedures of the CBR are performed. In the CBR case base is stored a set of cases, $Cases_s$, which includes the technological solutions (processes) implemented in the past.

$$Cases_s = \{ case^1 \ [T\_case_1, PT_1], \ldots \ case^i \ [T\_case_i, PT_i]\} \tag{10}$$

where:
$case^i \ [T\_case_1, PT_1]$ – i-th technological case,
T_case_i – description of the i-th technological task,
PT_i – structure of the technological process for the i-th case.

Basing on the values of particular attributes and weights, stored in the description T_case the calculation mechanism of CBR, according to the formulas (11) and (12), determines the degree of similarity between the task description T_case and cases $case_i$ stored in the database.

$$Sim\ (T\_case,\ case^i)) = 1 - Dist\ (T\_case,\ case^i) \tag{11}$$

$$Dist\ (T\_case\ ,\ case^i) = \left(\frac{1}{k} \cdot \sum_{j=1}^{k} w_j^2 \cdot \left[T\_case_j - case_j^i \right]^2 \right)^{\frac{1}{2}} \tag{12}$$

where:
$case^i_j$ – value of the j-th attribute in the i-th case,
k – number of corresponding attributes,
w_j – weight coefficient of the j-th attribute in the analyzed case.

Then is generated the set of cases with the required (acceptable) degree of similarity $Cases^s$.

$$Cases^s = \{case^1 \ [T\_case_1, PT_1], \ldots . \ case^k \ [T\_case_k, PT_k]\} \tag{13}$$

In the next step, the system sorts the chosen solutions, according to the degree of similarity, and creates the ordered set of cases $Cases^s_u$.

$$Cases^s_u = < case^1\ [T\_case_1, PT_1], \ldots.\ case^k\ [T\_case_k, PT_k] > \tag{14}$$

The ordered set $Cases^s_u$ = <....> is the set of the most similar technological processes that could be used in the task currently being under consideration.

3.2 Formalized description of cutting tools and parameters selection

For the process of selection the cutting tools and machining parameters the formalized description of functioning of this method is as follows. Given is the description of the task, *T_case*, in which the user defines the attributes and weights describing the analyzed technological task.

$$T\_case = \{\ WA_1, w_1,\ WA_2, w_2,\ \ldots.\ WA_j, w_j\} \tag{15}$$

where:
WA_j – the value of the *j*-th attribute describing the technological task,
w_j – the weight of the *j*-th attribute describing the technological task.

In the CBR database is stored the set of cases $Case_s$, which includes technological solutions (used tools and machining parameters) implemented in the past.

$$Case_s = \{\ case^1\ [T\_case_1, T_1, P_1, H_1, M_1], \ldots\ case^i\ [T\_case_i, T_i, P_i, H_i, M_i]\} \tag{16}$$

where:
$case^i\ [T\_case_1, T_1, P_1, H_1, M_1]$ – the *i*- th technological case,
T_case_i – description of the *i*-th technological case,
T_i – the tools description for the *i*- th technological case,
P_i – parameters description for the *i*- th technological case,
H_i – fixing method description for the *i*- th technological case,
M_i – machine tool description for the *i*- th technological case.

Table 1. Attributes and exemplar values with respect to the *T_case* task description.

Data set	Attribute	Exemplar values
T_case	Material	P, M, K, N, S, H
	Production	(piece, series, mass)
	Machining type	Rough, Medium, Finish
	Operation type	(outer/inner)
	Outer diameter	(min/-/max)
	Inner diameter	(min/-/max)
	Feed type	transverse, longitudinal, mixed
	Pass	R, L, BB_L, BB_R
	Allowance value	(min/max)
	Accuracy class	IT 7
	Surface roughness	Ra=1.25
	Goal function	Output, costs

Table 2. Register of attributes and exemplar values in relations to the data sets T_i, P_i, H_i, M_i.

Data set	Attribute	Exemplar values
T_i	Tip form	R, S, C, D, V
	Tip size	R (06), S(12), C(25)
	Nose radius	04, 08, 1.2, 1.6
	Clearance angle Kr	75°, 90 °, 135 °
	Type	Positive, negative
	For the type of a material	P, M, K, N, S, H
	Tool body type	A90 °, B75 °, D45°
	Type of tip fixing	C, D, M, W, P, S
	Tip version	R, N, L
	Tip marking	SNMG 120408-R2
P_i	Cutting speed	V_c=...
	Cutting depth	a_p=...
	Rotational speed	N =...
	Feed	f_n=...
H_i	Fixing type	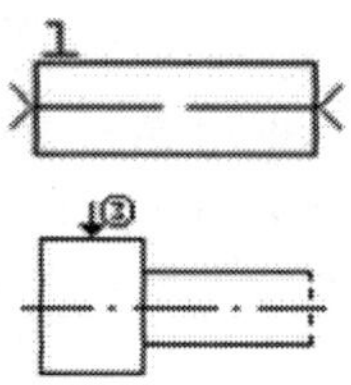
M_i	Machine tool type	CNC lathe
	Machine tool model	AVIAturn 35SM

On the basis of values of particular attributes and weights, recorded in the description *T_case*, the CBR calculation mechanism, according to the formulas (11) and (12), determines the degree of similarity between the task description *T_case*, and cases recorded in the database. Then is generated the set of cases with the desired degree of similarity $Cases^s$.

$$Cases^s=\{case_1\ [T\_case_1,\ T_1,\ P_1,\ H_1,\ M_1],....\ case_k\ [T\_case_k,\ T_k,\ P_k,\ H_k,\ M_k]\} \quad (17)$$

In the next step, the system sorts the selected solutions according to the degree of similarity and creates the ordered set of cases $Cases^s_u$.

$$Cases^s_u= <\ case_1\ [T\_case_1,\ T_1,\ P_1,\ H_1,\ M_1],....\ case_k\ [T\_case_k,\ T_k,\ P_k,\ H_k,\ M_k]\ > \quad (18)$$

The ordered set $Cases^s_u$ = <....> is the set of tools and cutting parameters recommended for the use in the currently considered technological task (ETT).

4 Conclusions

The informatics methods and tools, supporting the manufacturing process, are being increasingly used by contemporary engineers. They accelerate the preparation stage of the technological process and the same the manufacturing process itself. They facilitate "young" technologist's effective and realization of routine tasks. Moreover they allow reducing making errors resulting from monotonous work for the often repeated routine tasks. At the same time they allow the engineer implementing the more creative solutions. Creating the software, basing on the CBR method, is time consuming and absorbing its users, but in the long run it lets to save a lot of time and to eliminate unnecessary errors. Developing such software could be profitable, only if it will be used by a larger number of people for a longer period of time. The proposed method only supports the technologist in its activities. At the current level of development of the informatics methods and systems it cannot be completely automated the production preparation process. It is however possible to effectively support the activities performed at each stage of this process. The man, in contrast to the machine, has the ability to abstract thinking. He is able to combine multiple disciplines of science, as well as to think creatively, not only adapting the earlier cases to current needs.

References

1. A. Sękala, A. Gwiazda, A Dobrzańska-Danikiewicz, Applied Mechanics and Materials, **657** 854-858 (2014)
2. A. Sękala, G. Ćwikła, G. Kost, IOP Conf. Series: Materials Science and Engineering, **95**, 012097 (2015)
3. K. Herbuś, P. Ociepka, IOP Conf. Series: Materials Science and Engineering, **95**, 012096 (2015)
4. A. Gwiazda, K. Herbuś, G. Kost, P. Ociepka, Solid State Phenomena, **220/221**, 479-484 (2015)
5. A. Gwiazda, Adv. Mat. Res. **837** 393-98 (2014)
6. P. Ociepka, W. Banaś, K. Herbuś, G. Kost, Advanced Materials Research, **1036,** 817-822 (2014)
7. W. Banaś, K. Herbuś, G. Kost, A. Nierychlok, P. Ociepka, D. Reclik, Advanced Materials Research, **837**, 537-542 (2014)
8. K. Herbuś, G. Kost, D. Reclik, J. Świder, Advanced Materials Research, **837**, 582-587 (2014)
9. P. Ociepka, K. Herbuś, IOP Conf. Series: Materials Science and Engineering, **95**, 012101 (2015)
10. A. Dymarek, T. Dzitkowski, K. Herbuś, G. Kost, P. Ociepka, Solid State Phenomena, **198**, 59-64 (2013)
11. A. Dymarek, T. Dzitkowski, K. Herbuś, G. Kost, P. Ociepka, Advanced Materials Research, **837**, 351-356 (2014)
12. K. Herbuś, P. Ociepka, A. Gwiazda, Advanced Materials Research, **1036**, 732-736 (2014)
13. K. Herbuś, P. Ociepka, IOP Conf. Series: Materials Science and Engineering, **95**, 012084 (2015)
14. G. Ćwikła, C. Grabowik, W. Janik, Applied Mechanics and Materials, **657,** 808-812 (2014)
15. A. Gwiazda, R. Knosala, J. of Mat. Pr. Tech., **64/1-3** 181-88 (1997)
16. A. Aamodt, E. Plaza; AI Communications. IOS Press, **7**, 39-59 (1994)

17. P. Ociepka, J. Świder, Journal of materials processing technology, **157-158**, 221-227 (2004)
18. P. Ociepka, K. Herbuś, A. Gwiazda, Advanced Materials Research, **1036**, 1005-1010 (2014)
19. P. Ociepka, K. Herbuś, A. Gwiazda, Advanced Materials Research, **1036**, 1011-1016 (2014)
20. P. Ociepka, K. Herbuś, IOP Conf. Series: Materials Science and Engineering, **95**, 012100 (2015)
21. P. Ociepka, K. Herbuś, Applied Mechanics and Materials **809-810**, 1243-1248 (2015)
22. K. Herbuś, P. Ociepka, A. Gwiazda, Advanced Materials Research, **1036**, 1001-1004 (2014)
23. C. Grabowik, G. Ćwikła, W. Janik, Applied Mechanics and Materials, **657**, 750-754 (2014)
24. P. Ociepka, K. Herbuś, IOP Conf. Ser.: Mater. Sci. Eng., **145** 022030 (2016)
25. K. Herbuś, P. Ociepka, IOP Conf. Ser.: Mater. Sci. Eng., **145** 042018 (2016)
26. K. Herbuś, P. Ociepka, IOP Conf. Ser.: Mater. Sci. Eng., **145** 052010 (2016)
27. P. Ociepka, K. Herbuś, IOP Conf. Ser.: Mater. Sci. Eng., **145** 022029 (2016)

MATEC Web of Conferences **94**, 01008 (2017) DOI: 10.1051/matecconf/20179401008
CoSME'16

Graphic interface for numerical commands on the USB port of PC compatible computers

Elena Popa [1,*], and *Viorel* Popa[2]

[1]Transilvania University of Brasov, Department of Manufacturing Engineering, Mihai Viteazul No.5, Brasov, Romania
[2]Transilvania University of Brasov, Department of Wood processing and wood products design, University street nr.1, Brasov, Romania

Abstract. Computers are increasingly used in the present technological processes. Several numerical input/ output electronic modules were designed and made to support computer automated technological process in the wood industry. This paper presents the software for these modules, built in the Delphi language and aimed to obtain numerical commands by using the USB port of a computer. As modern computers are no longer provided with parallel ports, a K8055 USB Experiment Interface Board manufactured by VELLEMAN was used. The board includes a PIC16C745-IP microcontroller, which enables communication via specific software.

1 Introduction

The present technological processes in the wood industry are increasingly computer automated. An example in this respect is the use of numerical input/output electronic modules. These modules can be used both in production and research to verify the correct functioning of the electric, pneumatic and hydraulic actuators as well as to test controls and connectors (buttons, limiters, proximity sensors, pressostats etc.).

According to their version, the electronic modules can command 8 electromagnetic relays or 8 thyristors with the aid of a computer, by means of an electronic interface connected to the USB port. The K8055 interface is manufactured by Velleman and includes the PIC16C745-IP microcontroller, which is programmed to ensure communication with a PC compatible computer in the following way:

- 5 Digital inputs (0= ground, 1= open). On board test buttons provided;
- 8 Digital open collector output switches (max.50V/100mA). On board LED indication;
- General conversion time: 20mS per control,
- Power supply through USB approx.70mA,
- Diagnostic software and communication DLL included.

A USB cable connects the command electronic interface to the computer.

The command electronic module is outside the scope of this paper.

Focus is directed to an original graphic interface built in Delphi [1], whose purpose is to

* Corresponding author: elenapopa@unitbv.ro

work with the above mentioned electronic modules and interface. Specifically, the graphic interface aims to take over numerical signals from the process and/ or obtain (8-bit) numerical commands with the aid of a PC compatible computer [2, 3].

2 The graphic interface for numerical Input / Output

The graphic interface (Fig. 1) includes a main window with 11 buttons and a BMP figure. The first 10 buttons are used to open secondary windows, while the last one closes the interface. The scope of the secondary windows is to carry out different numerical commands or to take over exterior numerical signals. The included figure presents a few command and connection elements that can be used by the electronic interface (on and off buttons, limiters, photoelectric or inductive detectors, pressostats, contactors etc.). All the secondary windows are essentially specialized command panels featuring virtual on and off buttons.

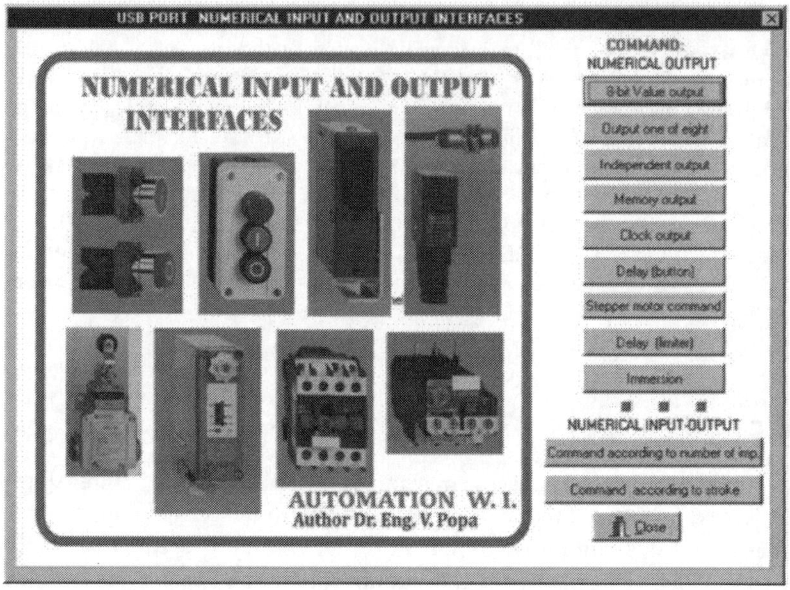

Fig.1. The main window of the graphic interface.

On panels there are virtual rectangular electroluminescent diodes, whose light is turned on or off depending on the given commands.

To use the secondary graphic interfaces, which are appealable from the main window, initially check if the electronic command interface is connected to a USB port, then click the CONNECT virtual button. If the connection is realized, the numerical commands can later be carried out.

Figure 2 presents the secondary window for commands carried out by entering into the port values ranging between 0 and 255. The output data on the microcontroller port is signalled through 8 electroluminescent diodes, which correspond to the B0...B7 bits.

Another possibility of direct command is shown in Figure 3. The window allows commands that can be carried out independently at bit level. Electric, pneumatic, or hydraulic motors can be operated this way. Lit LEDS and the display of the decimal value signal the state of the commanded bits. The on and off commands are carried out by clicking the button corresponding to the bit.

DOI: 10.1051/matecconf/20179401008

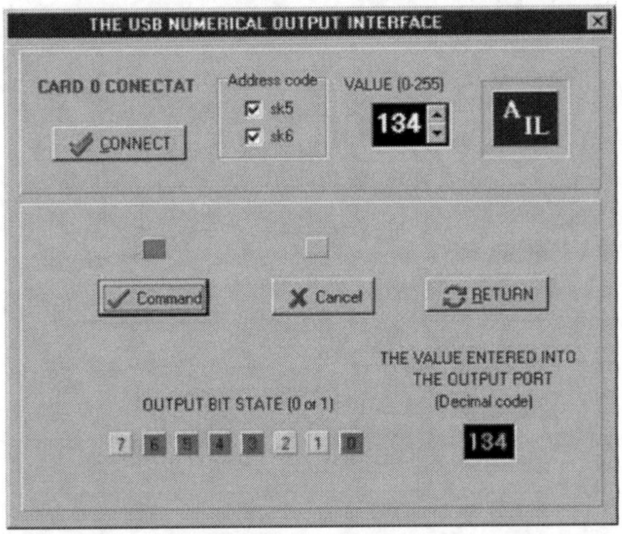

Fig.2. The window commanding the numerical output port.

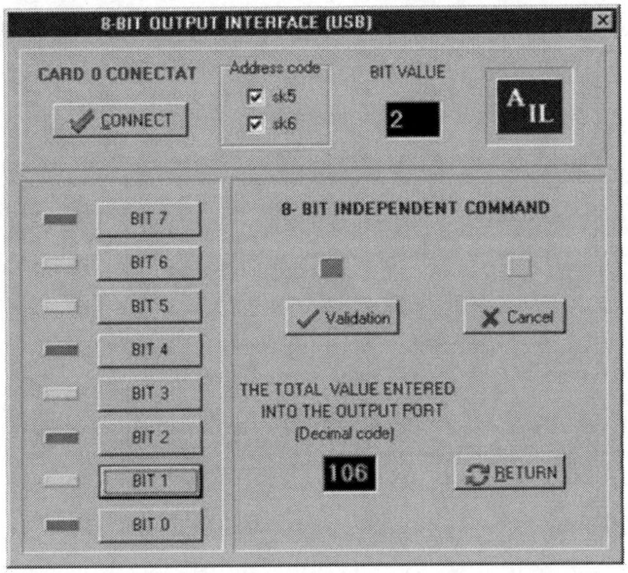

Fig.3. The window for the independent command of the output port at bit level.

The commands can be carried out only after the validation button has been hit. To cancel the commands on the whole port entails activating the cancel button.

The window presented in Figure 4 allows the command of a stepper motor through the use of an electronic interface built for the automated distribution of impulses. The command panel shown in the window allows the input of such values as the rotation direction, the number of impulses (therefore the motion that has to be performed), and the speed of the stepper motor. The speed and the rotation direction can be modified dynamically. The panel continuously displays the entered number of impulses as well as the active state of the output bits of the commanded port.

Figure 5 demonstrates the command window through which one or several execution elements is turned on or off at a certain time of the day.

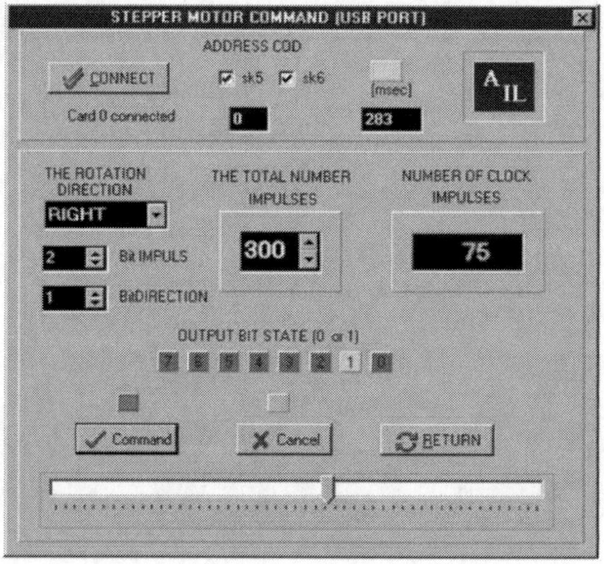

Fig.4. The window for a stepper motor command.

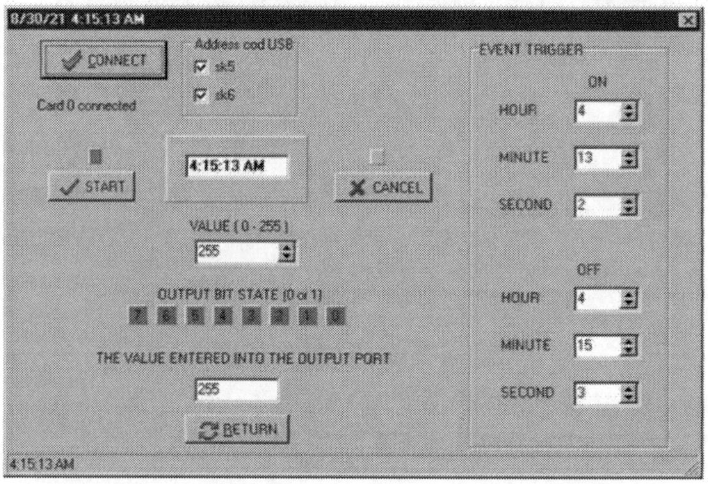

Fig.5. The command window at a certain time of the day.

The automated command according to time is carried out with the aid of a computer without the presence of a time delay relay. The start delay command can be carried out either by clicking a button (Fig. 6), by means of an electric limiter (Fig. 7) or through an inductive, photoelectric etc. proximity sensor.

The time (in seconds) is to be entered into the window's panel, the seconds being counted by the internal clock of the computer. Once the delay is started, the passing of the seconds is continuously displayed until the execution elements are turned on or off. The state of the 8 bits of the output port is presented by means of 8 electroluminescent diodes.

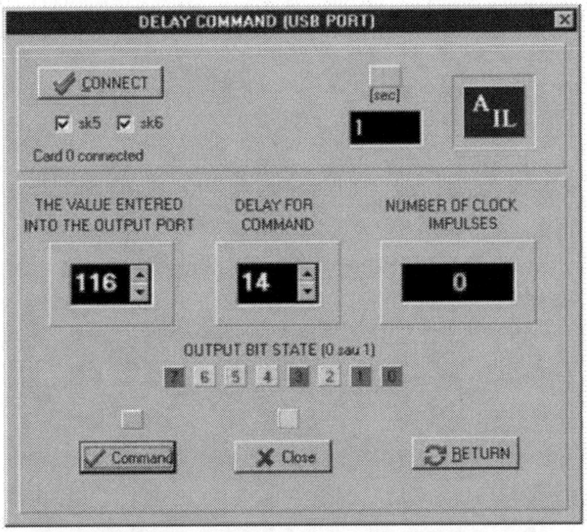

Fig.6. The window commanding the delay through a start delay button

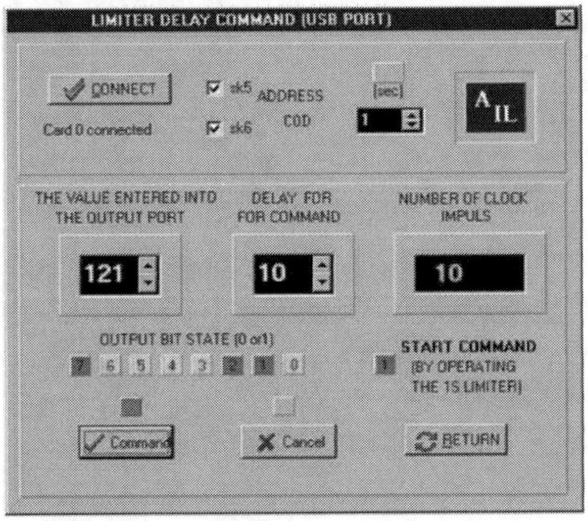

Fig.7. The window commanding the delay through a limiter start.

The command according to a certain number of impulses (Fig. 8) is carried out after the computer has read the state of the input bits of the microcontroller. The state of bit sk1, to which an electric limiter is connected, is tested and counted. The panel displays the number of impulses given by the limiter. When the programmed impulses correspond to the ones given by the limiter, the commands that turn on or off some electric, pneumatic or hydraulic motors are carried out.

The automated command according to stroke (Fig. 9) entails using both the input port (to which two limiters sk1 and sk2 are connected) and the output port of the microcontroller so that the driving motor is activated in one direction or another.

The following commands can easily be carried out by means of a computer:

- by clicking a start button, the advance stroke is executed with automated return to the initial position, and stop;
- advance stroke and repeated withdrawal with a final stop in withdrawn position.

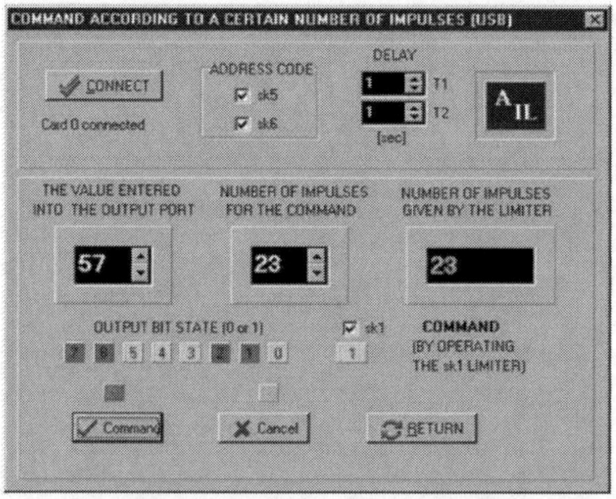

Fig. 8. The window for the command according to a certain number of impulses given by the limiter.

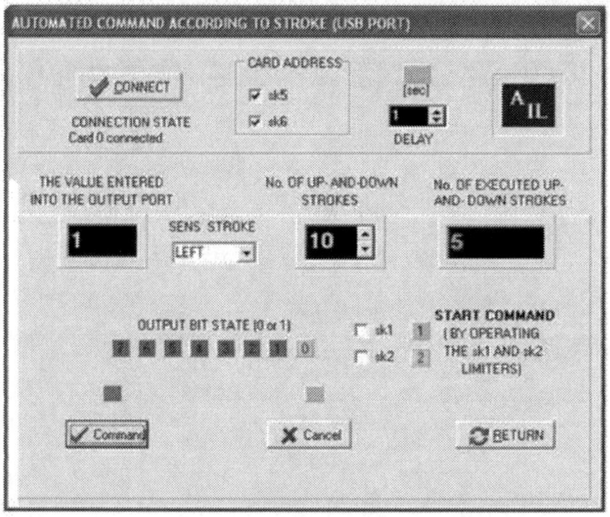

Fig. 9. The window for the automated command according to stroke.

In the case of commands carried out via a computer, the number of up-and-down strokes can be imposed, with the functioning coming to a stop in withdrawn position.

The numerical input/ output interface program can be installed on a PC laptop and used only in conjunction with the electronic interface that includes an optocoupler galvanic isolation module.

The presented graphic interfaces can be utilized in the wood industry to simulate, verify and debug automation equipment. Examples include the command of stepper motors on the mini numerical control machines (Fig. 4), and the command of adhesive heating before adhesive use (Fig. 5). Likewise, the interfaces are effective in the case of 90-degree transfer conveyors to adjust the delay (Fig. 6 and Fig. 7), or in strip counting to form the panel core (Fig. 8).

3 Conclusion

The demonstrated graphic interface for numerical input/ output is authentic. It was built for applications using the USB port and aims to be employed in the wood industry technological processes. The use of the presented graphic interface entails the existence of the electronic modules that take over exterior numerical signals and command electric, pneumatic and hydraulic actuators.

The electronic modules are connected through an electronic interface fitted with a microcontroller, which ensures the bidirectional communication (input/ output) on the USB port of modern computers.

The commands are usually carried out by clicking virtual buttons on the panel of the graphic interface windows. These buttons can command up to 8 actuators. The commands can also be carried out by 5 classical limit stops or binary proximity sensors.

References

1. S. Kovács, *Delphi 3.0. User's Guide* (Albatros Publishing House, Cluj-Napoca, 2000)
2. V. Popa, *Flow Automation in Wood Industry. Automation and Control Devices* (Transilvania University of Braşov, 2001)
3. V. Popa, *Automation in Wood Industry* (Transilvania University of Braşov, 2012)

Dynamic analysis of scraper conveyor operation with external loads

Jerzy Świder[1], *Krzysztof* Herbuś[1] , and *Kamil* Szewerda[2,*]

[1]Institute of Engineering Processes Automation and Integrated Manufacturing Systems, Faculty of Mechanical Engineering, Silesian University of Technology, ul. Konarskiego 18a, 44-100 Gliwice, Poland
[2]The Laboratory of Virtual Prototyping Methods, The Institute of Mining Technology KOMAG ul. Pszczyńska 37, 44-100 Gliwice, Poland

Abstract. A load to an armoured face conveyor (AFC) during coal mining is changeable and very difficult or even impossible to be predicted. Changes of the load to the upper scraper chain affect the load of the driving motor and generate changes in a scraper chain tension. Impact of increasing the external load to the upper scraper chain on the operation of electric motors and on the scraper chain tension is presented. The developed numerical model of the Rybnik 850 conveyor enabled identifying the places of the scraper chain high tension or places of its loosening. An impact of changing frequency of driving motor voltage on AFC's operational conditions was tested and analysed using the AFC's numerical model. During tests, tension of the scraper chain on the discharge end and the return end was recorded. High tension of the scraper chain and its loosening during the changeable load were also recorded on upward and downward transportation of run-of-mine material.

1 Introduction

Armoured face conveyors are the basic machines of a longwall system. The length of conveyors and the method of their installation (horizontal or inclined) are imposed by mining-and-geological conditions including a position of a coal seam. They operate in longwall panels in conditions of the changeable load, which is difficult to be predicted. The main component of the load results from the volume of the transported material. Chain tension and cooperation of conveyor's driving motors also have an impact on dynamic overloads. Majority of conveyors in Polish mines operate with constant (nominal) speed after a start-up phase. However, state-of-the-art conveyors have a possibility to be equipped with a frequency transducer. The use of the frequency transducer enables controlling the speed of conveyor movement, i.e. controlling its output [1 - 3].

Results of tests and numerical analyses aiming at identification of changes in load to conveyor driving motors in the case of the changeable load to the upper conveyor's chain, are presented. Numerical simulations were carried out on the developed computational model of RYBNIK 850 AFC. The impact of changes in the mains frequency of the electric

* Corresponding author : kszewerda@komag.eu

motor (changes in settings of a frequency transducer) on dynamics of conveyor operation was also analysed.

2 Computational Model

The computational model consists of the physical model of RYBNIK 850 AFC and of the models of driving motors (Fig. 1). The physical model was built of rigid solids, geometric constraints, elastic-attenuating elements and contacts between the selected solids. The model was developed in MSC. Adams computer program, designed for analyses of kinematics and dynamics of multi-body systems. The model of electric motors was developed in MatLab/Simulink software environment. Frequency transducers, used to control speed of conveyor movement were also included.

After defining the output/input signals from both models, the physical model of the conveyor was connected with the model of electric motors. The operation of the conveyor was then simulated using the co–simulation technology [4 – 9].

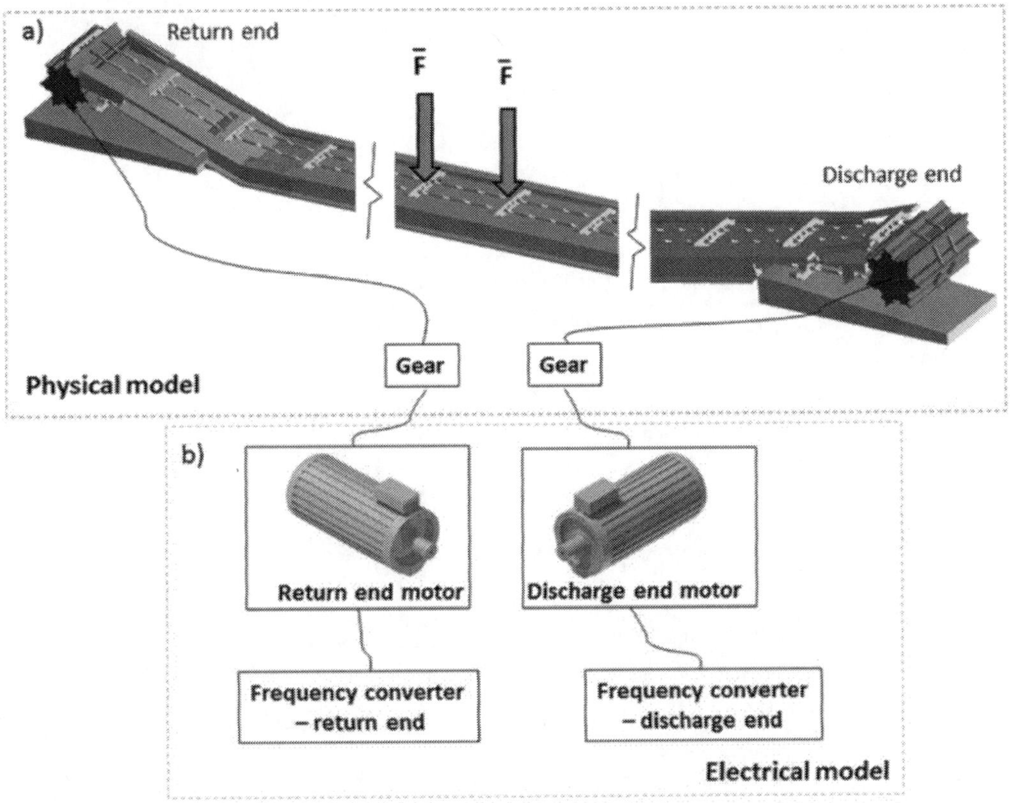

Fig. 1. Computational model of AFC a) physical model, b) electrical model [6].

Estimation of the load to the upper scraper chain is a difficult task and even impossible. Such load is of stochastic character. For numerical tests, vectors of forces $\bar{F}$ connected to centres of the selected scrapers (Fig. 1) were defined. Load to the upper part of the chain was then simplified to the friction force induced by the weight of transported run-of-mine. By changing the value of the load vector, movement resistance, which should be overcome by driving motors to move run-of-mine was increased or decreased.

DOI: 10.1051/matecconf/20179401009

3 Analysis of changes of current in motors under changes of a load value and voltage frequency

Numerical simulations were carried out at the constant length of the conveyor (the distance between centres of driving drums at the discharge end and the return end) was not changed. The external load was changed by increase of the vector of the loading force from 0 to 400 kN by 50 kN. Voltage frequency was 50 Hz. During simulations, currents in electric motors of driving drums and torques were recorded. In Fig. 2 diagrams of minimum, maximum and average currents in the motor at the discharge end (a) and the returned end (b) of the conveyor in a function of load to the upper part of the conveyor's chain are presented.

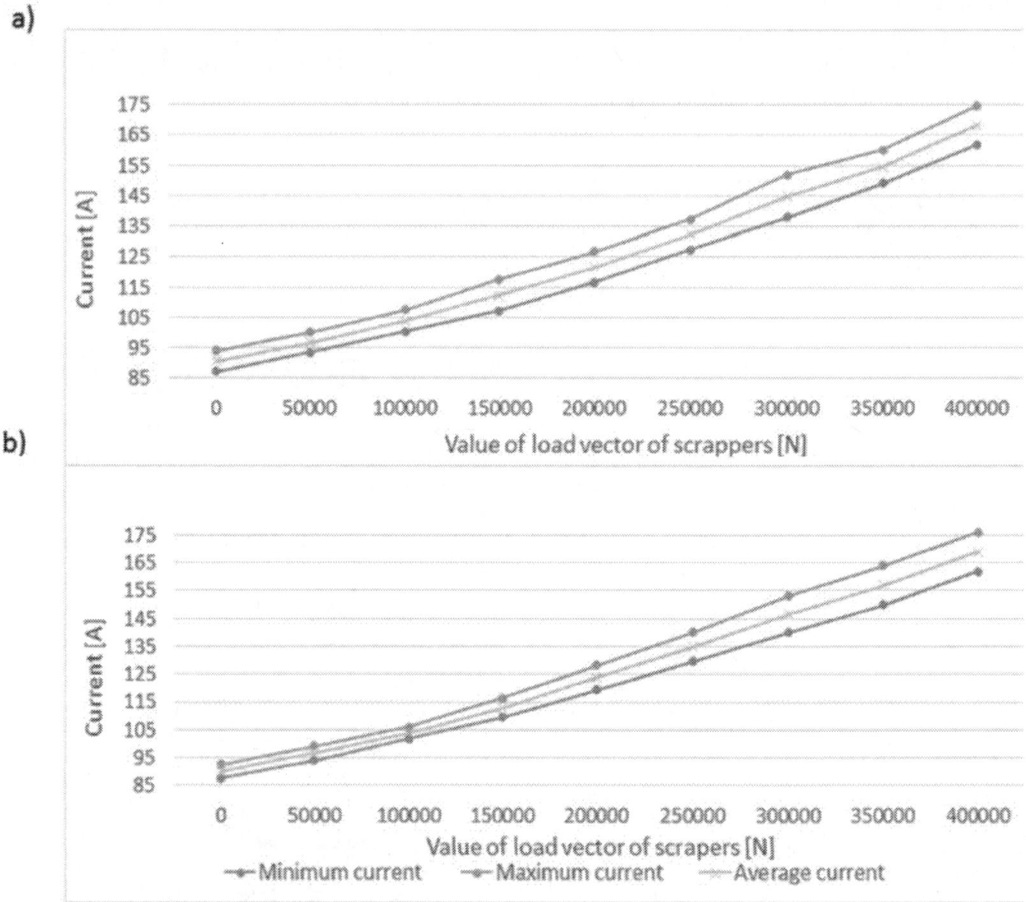

Fig. 2. Current in the motor a) at the discharge end b) at the return end in a function of the load to the conveyor [6].

Then changes in mains frequency of the electric motor by changing the setting of the frequency transducer in the motor's model were simulated. Simulations were carried out for the following frequencies 25Hz, 35Hz and 50Hz. At each frequency the conveyor was loaded by the force vector changed from 0 to 400 kN with a skip 50 kN. Other boundary conditions in the computational model as well as the method of results recording were the same as in the previous case. In Fig. 3, average current recorded in the loaded (a) electric motor at the discharge end, (b) electric motor at the return end in a function of the conveyor load and mains frequency is given.

DOI: 10.1051/matecconf/20179401009

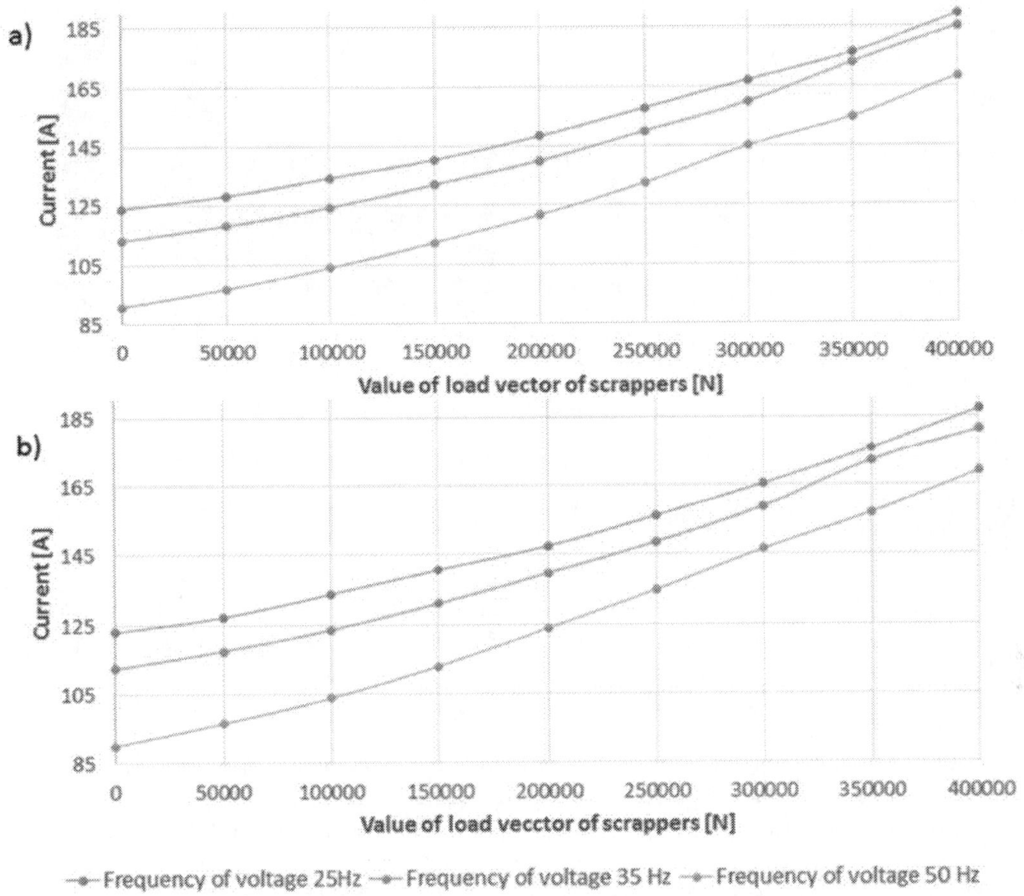

Fig. 3. Average current in electric motors in a function of the conveyor's load and mains frequency a) the motor located at the discharge end, b) the motor located at the return end [6].

Current in the electric motor was proportional to its load, what means that the motor generated a torque indispensable to balance the set load. When analysing the operation of the conveyor with two driving motors and assuming their correct cooperation, we found that the sum of torques generated by motors, including reducer's gear ratio, is equal to the sum of conveyor movement resistances. Considering that fact and the rule of current proportionality to the torque value, the current recorded in both motors was summed up. The sum of currents in the conveyor's motors in a function of the load, considering mains frequency, is presented in Fig. 4.

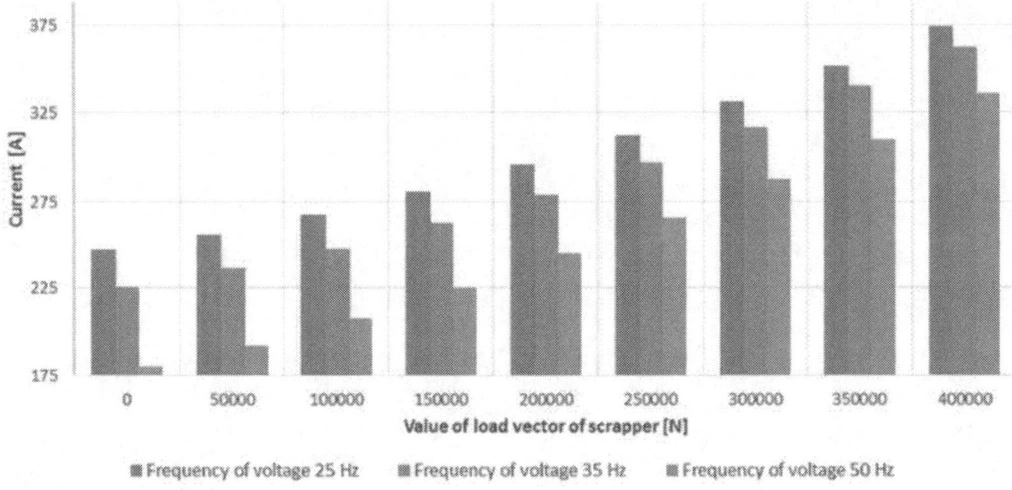

Fig. 4. Average sums of currents in electric motors in a function of the load to the conveyor, considering mains frequency [6].

It was also found that the increase of the load to the upper part of the chain resulted in the increase of current in electric motors. Reduction of motor's voltage frequency (reduction of its rotational speed) resulted in the increase of current at the same load to the conveyor. For example, the sum of currents in motors equal to 265 A was recorded:

- at mains frequency equal to 25 Hz with the load vector equal to 100 kN,
- at mains frequency equal to 35 Hz with the load vector equal to 150 kN,
- at mains frequency equal to 50 Hz with the load vector equal to 250 kN.

4 Conclusions

AFCs operating in hard coal mines as a part of a longwall system are exposed to a stochastic load, which results from a changeable volume of the transported run-of-mine. During simulation tests and analyses it was found that:

- the change in the load to the upper chain causes the change of torques generated by the motors; the generated torque is proportional to the current in electric motors,
- the sum of currents recorded in motors is proportional to the conveyor load at correct cooperation of both driving motors,
- the change in mains frequency of electric motors enables controlling the speed of conveyor movement, controlling its output at the same time; the change of mains frequency of electric motors causes change in motor's current, the reduction of mains frequency reduces a speed of the conveyor and increase of current in the electric motor at the same load to the conveyor,
- determination of the relationship between the change in mains frequency and current in motors will enable estimating the load to driving motors by measuring the current during the conveyor operation; then it will be possible to develop an algorithm for controlling operational parameters of the conveyor and at the same time to control its output. Development of this algorithm, which will control operational parameters of the conveyor (mains frequency and extension of return end of conveyor) will be the objective of future research work.

The article is a result of the project realized within the doctoral grant financed by KOMAG Institute of Mining Technology.

Calculations were made on the computers of Centrum Informatyczne Trójmiejskiej Akademickiej Sieci Komputerowej (Academic Computer Center in Gdansk -TASK).

References

1. J. Przybyłka: *Electric drives for belt conveyors with controlled rotational speed manufactured by DFME DAMEL S.A.* Transport Przemysłowy i maszyny robocze **1(27)**/2015, 37 -39 (2015)
2. J. Suchoń: *Mining scraper conveyors. The construction and use* (KOMAG Institute of Mining Technology, Gliwice, 2012)
3. U. Paschedag, World Coal **18/6**, 14-20 (2009)
4. J. Świder, K. Herbuś, K. Szewerda, J. Achiev. Mater. Manuf. Eng. **73/2**, 165 – 175 (2015)
5. K. Herbuś, K. Szewerda, J. Świder, Model. Inż. **55,** 34-41 (2015)
6. K. Szewerda: *Method of parametrization and selection of algorithms for AFC's control.* KOMAG Institute of Mining Technology (not published)
7. K. Herbuś, G. Kost, D. Reclik, J. Świder, Advanced Materials Research **837**, 582-587 (2014)
8. K. Herbuś, P. Ociepka, IOP Conf. Series: Materials Science and Engineering **95** 012084 (2015)
9. P. Ociepka, K. Herbuś, IOP Conf. Series: Materials Science and Engineering **95** 012101 (2015)

Analysis of impact of longitudinal inclination of a chain conveyor on dynamical phenomena during operation

Kamil Szewerda[2,*], *Jerzy* Świder[1], and *Krzysztof* Herbuś[1]

[1]Institute of Engineering Processes Automation and Integrated Manufacturing Systems, Faculty of Mechanical Engineering, Silesian University of Technology, ul. Konarskiego 18a, 44-100 Gliwice, Poland
[2]The Laboratory of Virtual Prototyping Methods, The Institute of Mining Technology KOMAG ul. Pszczyńska 37, 44-100 Gliwice, Poland

Abstract. Armoured face conveyors (AFCs) of length even up to 500 meters are used in coal mines for transportation of run-of-mine. The method of their installation depends on the geological and mining conditions and on a coal seam position. Operation of AFC in a horizontal or inclined position is allowed. The maximum accepted inclination is determined by the machine manufacturer. Two directions of longitudinal inclination of the conveyor have to be considered: upward and downward transportation of excavated material. Both the angle and the direction of inclination of the conveyor have a significant impact on the dynamics of its operation. Change in inclination of AFC results in tension of the chain. Loosening the scraper chain can lead to blockage, what can be a reason of damage. Excessive chain tension results in speeding up frictional wear of conveyor components. Determination of an impact of longitudinal inclination of the conveyor on a degree and a place of loosening or excessive tension of the chain was the objective of the numerical analyses, the results of which are presented. Analysis were performed using the developed numerical model of the Rybnik 850 conveyor. These analyses are a part of the preparatory work aiming at development of the control algorithm of AFC, to prevent the formation of unfavorable dynamic effects during the conveyor operation.

1 Introduction

AFCs are the main machines used in longwall complexes in hard coalmines. Transportation of run-of-mine from the longwall face to beam stage loaders (BSL) and cooperation with other machines of the complex, i.e. powered roof supports and a mining machine are their main tasks. Geological conditions, including the seam position, force the required supporting system and AFC operation. AFCs can operate in horizontal position, but they can also be inclined. Inclination of AFC changes its operational dynamics. Series of tests of AFC with use of numerical simulations were carried out to identify changes in the

* Corresponding author : kszewerda@komag.eu

conveyor's dynamics in the case of its longitudinal inclination. Two cases of the conveyor's longitudinal inclination were considered. The first of them is a situation, when the conveyor's return end is placed above the discharge end and run-of-mine is transported downwards. The second case is a situation, when discharge end is placed above the return end and run-of mine is transported upwards. Amount and direction of conveyor's inclination forced the method of chain load and decided about the places in which the chain was loosened or tensioned [1, 2]. Results of analyses of simulation tests of the conveyor's operation under changeable angle and direction of longitudinal inclination are presented.

Such analysis is one of the phases of development of an algorithm for control of conveyors operation aiming at improvement of its operational parameters, which minimise unfavourable phenomena during transportation of run-of-mine upwards and downwards.

2 Computational model of the conveyor

Analysis of impact of longitudinal inclination of AFC route on dynamic phenomena that occur during its operation was made on the basis of simulation tests with use of the computational model of RYBNIK 850 conveyor with E260 conveyor's pan profile with two scraper chains of size 34x126 placed in the middle. The computational model (Fig.1) consisted of the following components: 773 rigid solids, 182 geometric nodes, 3213 defined contacts between selected rigid solids, 676 elastic-attenuating components and of two electric motors supplied from frequency transducers. The model had 1779 degrees of freedom. Numerical simulations were carried out with use of co–simulation technology, which enables integrating the software used in analyses of kinematics and dynamic of a multi-body system with Matlab\Simulink software, in which the electric motors were modelled [3-5].

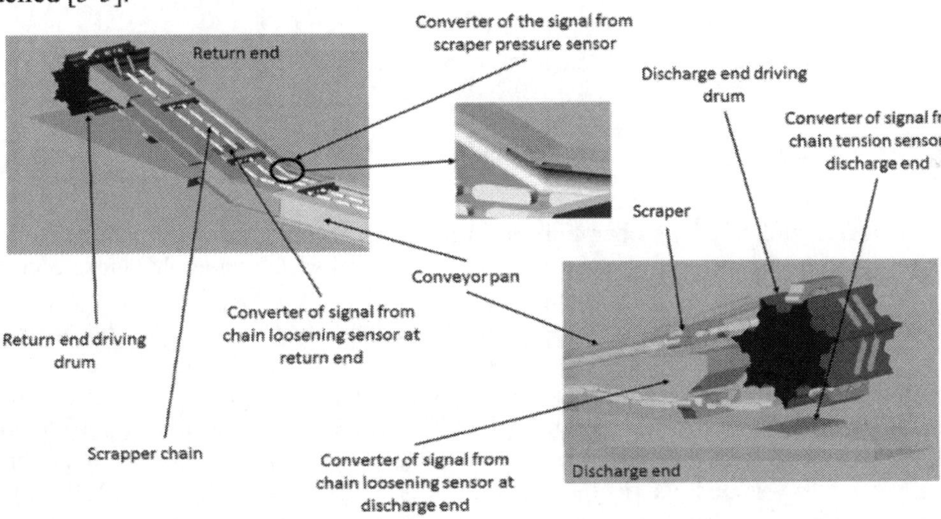

Fig. 1. Computational model of AFC [5].

Two – gear electric motors of 2SGS 400S-12/4 type manufactured by Damel S.A. were used in the computational model. The torque generated by motors was transferred to the driving drums through a gear. Vectors of load to upper line of the scraper chain, simulating the transported material, were defined. Vectors were applied to centres of solid of the selected scrapers, were attached to them and moved with them during chain motion.

Direction of load vectors was vertical. The force was defined as the parameters, so it was possible to control it during simulation. Movement resistance resulting from transported material on upper line of the scraper chain was recreated and controlled by the vectors [6, 7]. The vector of acceleration of gravity $\bar{g}$ during operation of a horizontal conveyor (not inclined) was defined in the computational model. Direction of the vector action was vertical (according to Y axis of the global coordination system). Sense of the gravity vector was accepted towards negative values of Y axis and its value was equal to 9.81 m/s^2. The vector defined the action of the field of gravity on all solids describing the conveyor's geometry. Change in the conveyor inclination resulted in appearance of two components of the gravitation vector. Knowing the inclination angle and direction of inclination (upward, downward), it was possible to modify the gravitation vector's components. The diagram of modification of the gravitation vector according to direction and inclination of the conveyor is presented in Fig. 2.

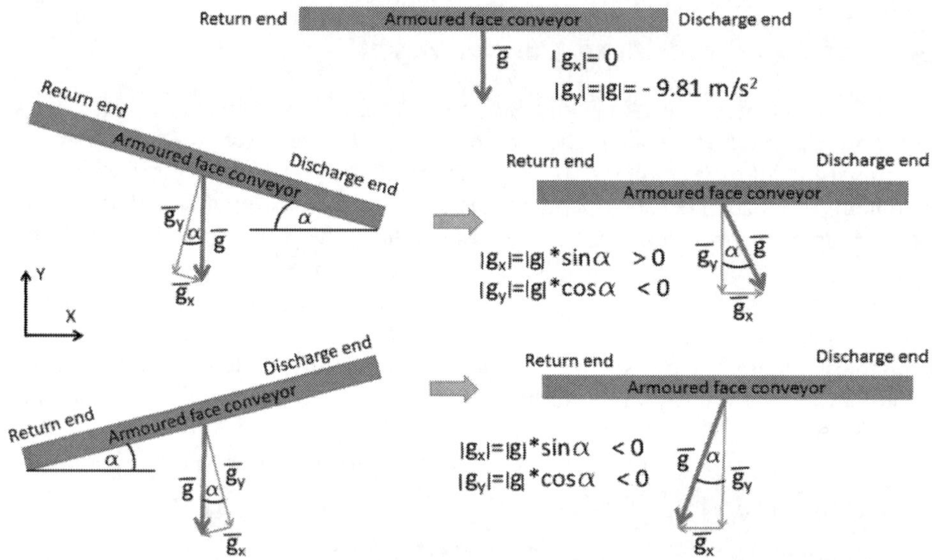

Fig. 2. Components of the vector of gravitation in the computational model [5].

Numerical simulations of operation of the inclined conveyor was possible due to implementation of the gravitation vector. A change of the inclination angle causes a change of components of the gravitation vector – the vertical component (Y axis) was equal to acceleration of gravity multiplied by cosine of the inclination angle and the horizontal component (X axis) was equal to acceleration of gravity multiplied by sine of the inclination angle. Direction of the conveyor inclination is described by a sense of the horizontal component of the gravitation angle. In the case of downward transportation (the discharge station situated below the conveyor's return end), the horizontal component of the gravity vector was positive. In the case of upward transportation (the discharge station situated above the conveyor's return end), the horizontal component of the gravity vector was negative.

3 Analysis of operation of horizontal conveyors and those inclined upwards or downwards

Simulations of conveyor operation were carried out at the following boundary conditions:
•Electric motors were supplied through frequency transducers (frequency 50 Hz).
•The conveyor had a constant length.

•At 7th second of simulation, two vectors recreating the load to the upper line of the scraper chain with the transported material were activated. The force pressing the scraper to the conveyor pan was 200kN.

To analyse the impact of longitudinal inclination of a conveyor's route on the phenomena occurring during its operation, first simulation of the operation of the not inclined conveyor was made and then operation of the conveyor of longitudinal inclination equal to 10°, 20°and 30° was simulated and analysed, both in the case of upward and downward transportation. Current of electric motors and tension of the chain at the place of leaving the driving drums were recorded during simulation.

In Table 1, minimal, maximal and average current in the motor, recorded during simulation of stable operation with the load to the upper chain line at different longitudinal inclinations is given.

Table 1. Current in electric motors during stable operation of the conveyor with load [5].

Inclination	Min. current at discharge [A]	Max. current at discharge [A]	Average current at discharge [A]	Min. current at return end [A]	Max. current at return end [A]	Average current at return end [A]
Stable operation with load						
Downward 30°	129.9	179.8	139.4	124.7	168	136.7
Downward 20°	134.2	144.9	139.3	127.1	148.3	136.8
Downward 10°	136.7	142.9	139.8	131.4	145.7	137.5
No inclination	**133.5**	**149.3**	**140.6**	**135.7**	**144.5**	**140.4**
Upward 10°	138.7	154.4	145.3	136.8	151.8	144.2
Upward 20°	119.2	188.9	144.3	137.9	151.1	143.8
Upward 30°	98.5	262.8	147.4	138	175.3	147.3

In the same way the results of simulations regarding other phases of the conveyor operation i.e. stable operation without the load, start-up phase and the uneven load to the conveyor were collected. After analysis it was found that:

•regarding the conveyor operation on inclination, the amplitude between minimum and maximum current in electric motors increases in comparison with horizontal operation, what is especially visible in transportation of run-of-mine upwards with inclination 20° and 30°

•there is a significant increase of dynamics of the electric motor at the discharge end during transportation of run-of-mine with high inclination; loosening the chain observed in the case of stable operation, without the load to the upper line of the chain in its upper line during leaving the driving drum at the return end, causes high oscillations of current in the motor at the discharge end

•at the moment of applying the load to the upper line of the chain, the current increases in near the same way, when the conveyor is moving horizontally or transporting run-of-mine up at high inclination (30°); in the case of transportation of run-of-mine downward at high inclination, current of the electric motor at the discharge end increased more gently than in the case, when the conveyor was not inclined; and in the case of the motor at return end, current at the moment of applying the load to the upper line of the chain increased faster, when the conveyor operated downwardly than in the case of the not inclined conveyor, what results from the fact that a sliding force of transported material reduces movement resistance that has to be overcome by the electric motor at the discharge end. More rapid increase of current in a the motor at the return end at the moment of loading the upper chain in the case of the inclined conveyor resulted in necessity of overcoming the component force of chain gravity.

Together with a change of longitudinal inclination and its direction, tension of the chain and a place of the chain loosening change as well. Chain tension was recorded by the sensors installed in the computational model [5,7]. The sensors enabled detecting the chain loosening, when leaving the driving drum both at the discharge and return ends as well as detecting the excessive chain tension at the discharge end. Besides, the pressing force of scrapers on the conveyor's pan surface at the return end was also measured. Loosening of the chain, what is disadvantageous regarding the conveyor's operational dynamics, as well as possibilities of a scraper jamming or loose chain link leading to a chain break, were identified. Excessive chain tension, which negatively affects the conveyor operation, speeding up frictional wear of conveyor's subassemblies and causing heat emission around the machine, was also measured. In Table 2 chain tensions recorded during numerical simulations of conveyor operation at different longitudinal inclinations are given. In the third column of Table 2, values of signals from the sensor about chain condition, recorded during simulation are presented. In the case of chain loosening and chain tension sensors at discharge end the signals have value 0 or 1, what means that the given state is present or not. For the sensor of the scraper's pressing force on the conveyor's pan surface, range of its values recorded during simulations are given.

Table 2. Chain tension at the selected angle and direction of the conveyor inclination [5].

Downward 30°	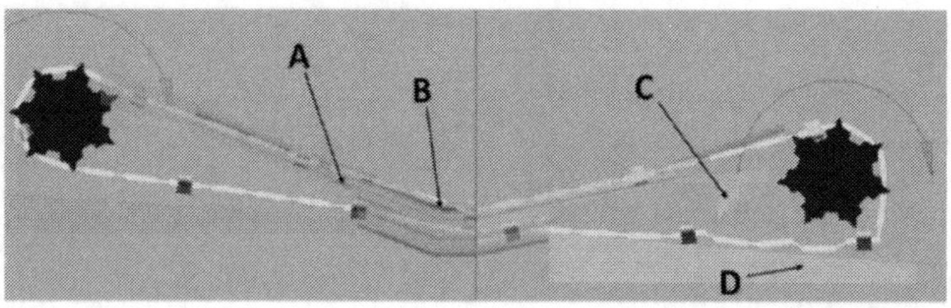	Loosening discharge (D) = **1** Tension discharge (C) = **0** Loosening return end (A) = **0** Tension return end (B)=**15÷25kN**
Inclination 0°	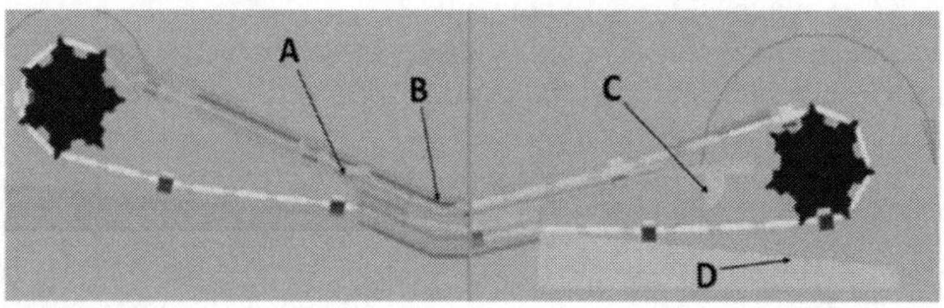	Loosening discharge (D) = **0** Tension discharge (C) = **0** Loosening return end (A) = **1** Tension return end (B)= **0.5÷2kN**
Upward 30°	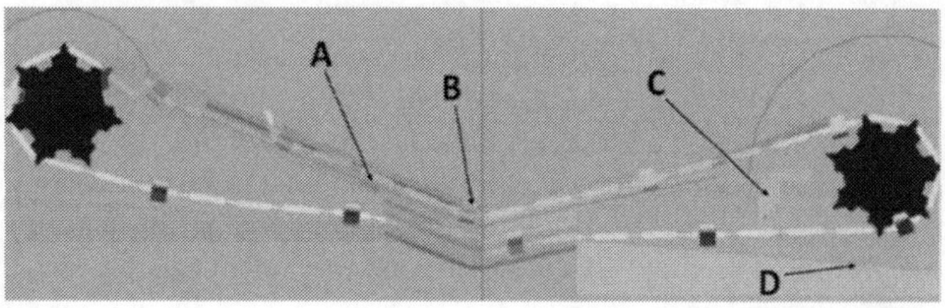	Loosening discharge (D) = **0** Tension discharge (C)= **1** Loosening return end (A) = **1** Tension return end (B)=**0÷2 kN**

During operation of the conveyor in downward inclination, the chain was loosened at leaving the driving drum located at the discharge end. With an increase of the downward inclination angle, chain tension at leaving the driving drum located at the return end increased (there was a high pressing force of the scraper on the pan surface and on the sensor recording that pressure). At inclinations 20° and 30°, the chain loosening sensor at the return end reached a value "0". In this case chain tension was high. At inclination 10° signal recorded by the loosening sensor at the return end reached a value "1", what meant that scrapers were in a position within the sensor's range of detection. At the same time, the signal from chain tension sensor at the return end informed about occurrence of the small pressing force of the scraper, so in this situation the chain tensioning condition at leaving the driving drum at return was interpreted as the proper one (the chain was tensioned enough to avoid its loosening and jamming the scraper, without excessive pressure on the conveyor's pan).

During operation of the not inclined conveyor, there was neither excessive chain tension nor its loosening at leaving the discharge end. Such chain condition was advantageous. At this chain condition, minimal, periodical slight chain loosening, when leaving the driving drum at the return end, was observed.

At the other side, during upward operation of the conveyor, the chain was loosened, when leaving the driving drum at the return end (Fig. 3). In addition, chain tension was recorded, when leaving the driving drum at the discharge end. Significant loosening of the chain, when leaving the driving drum at the return end led to dangerous scraper blockade, causing high dynamic overloads and even to chain break.

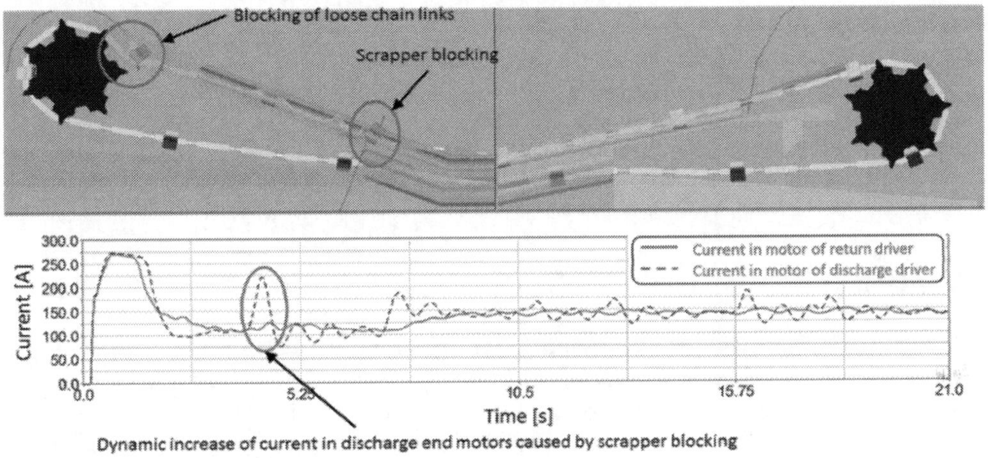

Fig. 3. Results of chain loosening at the conveyor's return end [5].

Rapid increase of current in the motor at the discharge end in the result of scrapers blockade at the return end was observed during operation of the conveyor inclined upwardly under angle 20°, at 4th second of simulation (Fig. 3).

4 Conclusions

During conveyor operation, its driving motors generate the torque, which is multiplied and transferred to driving drums. Cooperation of driving drums with a chain results in a transformation of torque into the pulling force moving the chain with scrapers and transported run-of-mine. Resistance of conveyor movement and the transported materials has to be overcome to maintain constant transportation speed. During correct operation of the conveyor, the sum of torques generated by motors is proportional to movement resistance. Basing on numerical analyses we can say that change in a longitudinal inclination results in change of movement resistances of the upper and bottom chain. The changes result mainly from action of the sliding force generated by mass of conveyor

 DOI: 10.1051/matecconf/20179401010

components and transported run-of-mine. The sliding force increases with increase of the conveyor inclination angle. Depending on a direction of inclination, the force differently affects the load to the conveyor. During downward transportation, the rolling force reduces movement resistance in the upper chain. In turn, pulling the bottom part of the chain up causes increased movement resistance of the bottom chain. The reverse situation is during transportation of run-of-mine in the upward direction.

Differentiation in movement resistance of the upper and bottom part of the chain causes significant difference in loads to the conveyor's chain. These differences result in a chain loosening at different places of the conveyor. During upward transportation, the chain loosens, when leaving the driving drum at the return end. During downward transportation, the chain loosens, when leaving the driving drum at the discharge end. Increase of conveyor's longitudinal inclination in downward operation causes increase of the scrapers pressing force on the conveyor pan at the return end. That results in an increase of movement resistance, temperature as well as speeded up frictional wear of conveyor components.

Change in longitudinal inclination of the conveyor does not result in significant increase of the load to the driving motors, but causes big differences in chain tensions at the return end and the discharge end. Excessive tension and loosening of the chain are very disadvantageous regarding the dynamics of conveyor's operation and they may lead to failure or damage to the conveyor.

Results of analyses justify taking into account the conveyor's longitudinal inclination in a control system to minimise the risk of failure.

The article is a result of the project realized within doctoral grant financed by KOMAG Institute of Mining Technology. Calculations were made on the computers of Centrum Informatyczne Trójmiejskiej Akademickiej Sieci Komputerowej (Academic Computer Center in Gdansk -TASK).

References

1. J. Antoniak, *Improvement of operational parameters of run-of-mine transportation to increase coal production from longwall panels. Kierunki Mechanizacji Górnictwa do 2010 roku*. KOMTECH 2002 (Komdruk-Komag, Gliwice, 2002)
2. M. Dolipski, *The dynamics of the chain conveyors* (Silesian University of Technology, Gliwice, 1997)
3. K. Herbuś P. Ociepka, IOP Conf. Series: Materials Science and Engineering 95 012084 (2015)
4. K. Herbuś, G. Kost, D. Reclik, J. Świder, Advanced Materials Research **837**, 582-587 (2014)
5. K. Szewerda: Method of parametrization and selection of algorithms for AFC's control. KOMAG Institute of Mining Technology (to be published)
6. J. Świder, K. Herbuś, K. Szewerda, J. Achiev. Mater. Manuf. Eng. **73/2**, 165 – 175 (2015)
7. K. Herbuś, K. Szewerda, J. Świder, Model. Inż. **55**, 34-41 (2015)

MATEC Web of Conferences **94**, 01011 (2017) DOI: 10.1051/matecconf/20179401011
CoSME'16

Engineering applications using CAD based application programming interface

Anastasios Tzotzis[1], *Cesar* Garcia-Hernandez[2], *Jose-Luis* Huertas-Talon[2], *Dimitrios* Tzetzis[3], and *Panagiotis* Kyratsis[1,*]

[1]Western Macedonia University of Applied Sciences, Department of Mechanical Engineering and Industrial Design, Kila Kozani, GR50100, Greece
[2]University of Zaragoza, Department of Design and Manufacturing Engineering, Campus Rio Ebro, C/Maria de Luna, 3-50018, Zaragoza, Spain
[3]International Hellenic University, School of Economics and Business Administration, 14th km Thessaloniki-Moudania, GR57001 Thermi, Greece

Abstract. Automating the design process of a product or a system can provide engineers and designers with many benefits. As such, repeatable tasks that are time consuming can be handled automatically and with minimal human attention. This is achieved by using the API (Application Programmable Interface) of CAD systems in order to create macros or even develop software applications. The present paper deals with an application that has been developed with the API of a general purposes CAD system. This application automates the design process of a standard pneumatic double acting cylinder based on the appropriate inserted parameters (ISO 15552).The design process begins with the creation of a series of components developed as solids, and extends to the extraction of basic attributes and properties from the complete mechanical assembly. Finally, the assembled models and the extracted data can be used to further study the design of the pneumatic double acting cylinder. It is expected in the future to expand the features of the presented application in order to automate the design process of other related mechanical systems.

1 Introduction

Almost every feature that is available in a CAD system can be automated. Thus, the use of the API is required when developing engineering applications. Applications like these can aid the productiveness of many workshops that produce a limited range of products or that are specialized in a specific range of mechanical systems (i.e. machinery).

The most common advantages from the use of the API in the design process are easily noticed. First of all the development time of a product decreases drastically. Moreover, the applications that can be created with the use of the API can combine various functionalities of a general purposes CAD system, in a sole and easy to use interface. In addition, the interconnection between different software is applicable, which supplements the functionality of an all-around software tool.

*Corresponding author: pkyratsis@teiwm.gr

2 Literature survey

Many researchers have used the API of various CAD systems in order to develop applications and tools, and always the conclusions dictate that the benefits and the advantages of the API are vast. According to Kyratsis et al., up to date CAD software environments are filled with powerful modelling capabilities and by using the API of a modern CAD system, the DRILL3D software routine was developed. The application creates parametrically controlled tool geometries and by using different cutting conditions, achieves the generation of solid models for all the relevant data involved during the drilling manufacturing process (drilling tool, cut workpiece, undeformed chip etc.)[1].

Neto et al. developed a CAD based system in order to program a robot from a 3D CAD model, allowing users with basic CAD skills to generate robot programs off-line [2].Another use of the API can be found in the study of Haba and Oancea on the manufacturing technologies that are designed for the assembly of a single cylinder, four strokes, air cooled engine block [3]. Similarly, Badan and Oancea developed a software tool that allows user to generate different types of holes, on parallelepipedic solids, starting from different references as edges, vertexes, known points on the solids, or different types of entities which are placed on the solids [4].

The present paper acknowledges that with the use of the automatic programming interface of a general purposes CAD system, it is possible to develop a collection of CAD based applications that can be easily accessed with the help of a simple user interface. The presented applications were developed with the VBA (Visual Basic for Applications) programming language that is embedded in the selected CAD system. One such application can be used to build the solid models of the pneumatic cylinder's components complying with the ISO 15552:2004 [5]. Then, the built models are combined to form the assembly of the pneumatic cylinder. Subsequently, with the complete assembly model, more functions become available. Such as the automatic mechanical drawings creation for each component and the extraction of the assembly's properties that can be stored to a Microsoft Excel™ file for further study. With this collection of applications, the pneumatic cylinder application can become a workstation for routine CAD based processes.

3 Pneumatic cylinder application presentation

All tools that are available in this application are related to an important mechanical element, a pneumatic double acting cylinder which complies with the ISO 15552 standardization. This standardization refers to pneumatic cylinders with a maximum rated pressure of 10 bars. Pneumatic cylinder scan be found in many industrial processes as parts of various machines, whereas the double acting type is the most used cylinder. This is due to the fact that this type is very reliable, can be repaired and have increased durability.

3.1 Pneumatic cylinder tool flow

In figure 1, the flowchart of the pneumatic cylinder application is presented. From left to right, the flow is straightforward. First of all, the necessary data are imported; the required force of the piston, the number of duty cycles that the piston must perform and finally the pressure of the compressed air. With these parameters defined, the application can calculate the theoretical diameter of the piston. Next, the materials are set depending on the cylinder use. At this point, the modeling process begins. At first, the cylinder's components are built, in order to form the assembly of the pneumatic cylinder at any time.

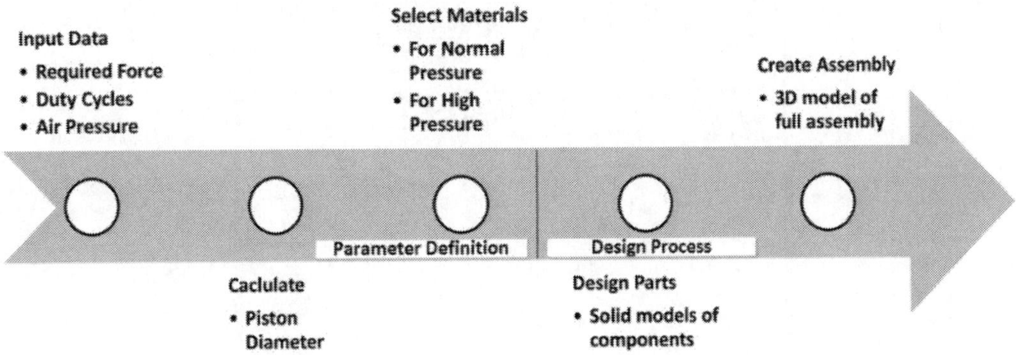

Fig. 1. The flow chart of the presented application.

Figure 2 presents the functions that become available once the assembly model is completed. It also presents the interconnection between SolidWorks™ and other software like Microsoft Office™, which is necessary for the application to perform certain functions. These functions are the creation of the mechanical drawings for every component, the creation of the required process plan for the cylinder's support (sheet metal part), as well as the creation of the bill of materials (BOM) of the assembly. In addition, the application can produce rendered models of the 3D solid models. These functions do not interfere with each other, which mean that they can be performed at any order.

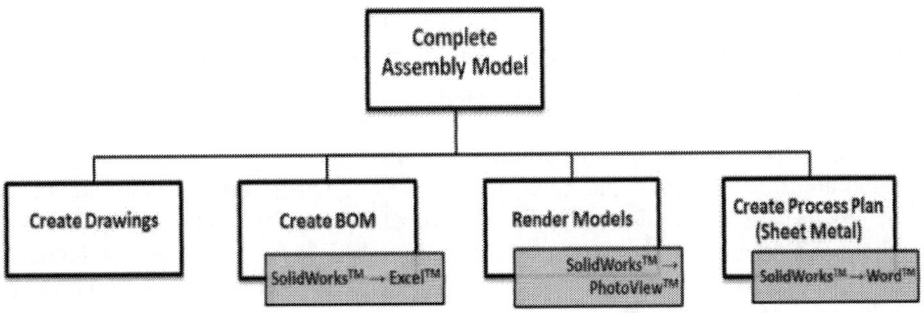

Fig. 2. Assembly-dependent functions of the application.

3.2 Basic calculations

In order for the application to perform the calculation of the piston, the below stated equations are used. These equations derive from Pascal's law and depend on the volume of the air that occupies the inside of the cylinder.

$$F_1=[(P*\pi*d^2)/4]*\eta \tag{1}$$
$$F_2=\left[\left(P*\pi*(d^2-d_1^2)\right)/4\right]*\eta \tag{2}$$

Equation 1 represents the force that the piston can exercise when extracting, and equation 2 represents the force that the piston can exercise when retracting. Where P is the pressure of the air in bars, d is the outside diameter in mm of the piston and η is a theoretical coefficient that determines the loss of force due to friction.

Based on these equations, the application determines the recommended diameter of the piston with the use of the inserted data. Figure 3 presents the main dialogue window of the pneumatic cylinder tool, which is used to perform the calculations and build the parts.

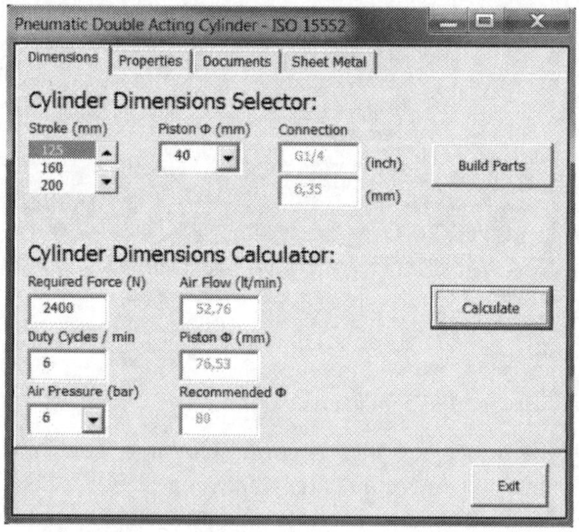

Fig. 3. Main dialogue window of the application.

3.3 Part design process

The design process of the cylinder begins with the design of its components. As seen in figure 3, with only two parameters inserted, all of the components can be built. Because most mechanical elements comply with specific standardization, their secondary dimensions can be derived from the basic dimensions. So in the case of the ISO 15552 pneumatic cylinder, the diameter and the stroke of the piston act as reference to all other dimensions.

Before using the application to design the cylinder's components, it is necessary to select a template, a saving path and a set of materials. These parameters are set by default, but can be changed as required with the use of the tab 'Properties' (figure 4). The template path is where the application loads the appropriate template, depending on whether a part, an assembly or a drawing is about to be created. The save path is where all the documents created are stored. And the selected materials label indicates the set of materials that is going to be used for the components creation.

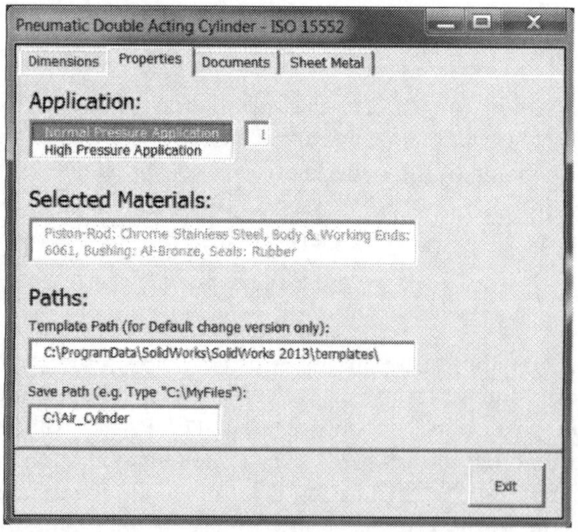

Fig. 4. The Properties tab of the application.

DOI: 10.1051/matecconf/20179401011

As far as it concerns the structure of the design process code, it is divided in six steps. These steps are always executed in order to build the 3D solid model of every component. Figure 5 depicts the steps used in the case of the cylinder piston. The application uses the inserted data in order to initialize the necessary variables. The values of the primary dimensions (piston diameter and stroke) are given by the user, while the secondary dimensions derive from the primary dimensions and the mathematical equations (1) and (2). For example the width of the piston is known to be a specific percentage of its diameter.

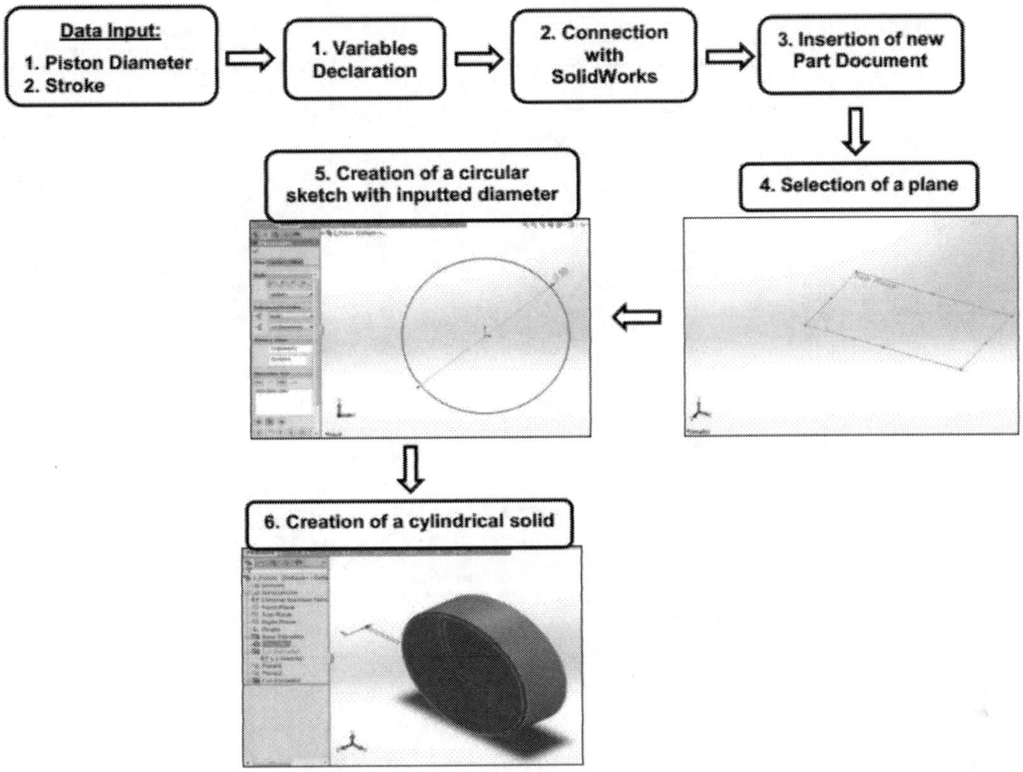

Fig. 5. Flow chart for every part designed.

3.4 Assembly creation process

The final stage of the design process is the creation of the cylinder's assembly. The required parts that comprise the assembly are loaded from the save folder, so it is obvious that prerequisite for the assembly creation is the set of the component models. Each part is assigned with specific mechanical properties, which are predetermined as described in the ISO 15552 standardization. The values of the mechanical properties are taken from a library that contains a collection of information about various materials available with the CAD software. The materials of the components and their properties, change accordingly based on the application that the cylinder is scheduled to be used. This means that the material selection is depended on the compressed air pressure.

The used design method for the assembly is the bottom-up method, where parts are modeled first and then inserted into an assembly for positioning. The structure of the assembly and the dimensions are almost identical to a physical pneumatic cylinder model; some design deviations may exist as different manufacturers make use of different design approaches.

The design process of the parts and the design process of the assembly both share a similar flow. It is divided in six steps and begins with the connection of the application with SolidWorks™ and the declaration of a group of variables. The creation process flow of the assembly is presented in figure 6.

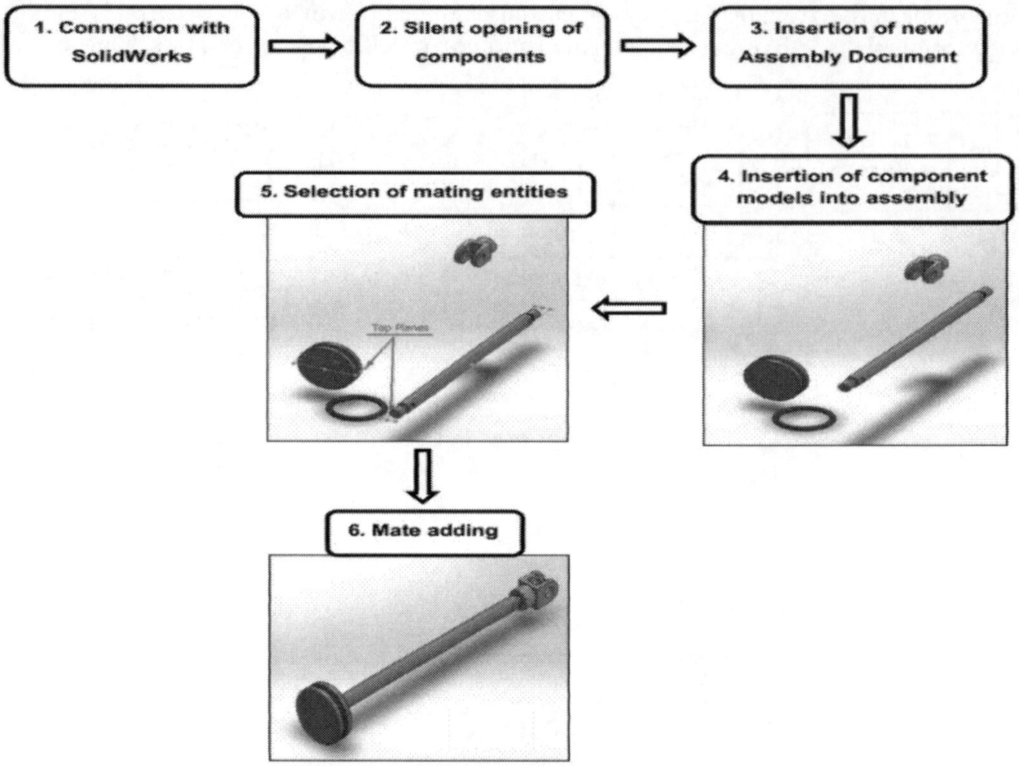

Fig. 6. Flow of the assembly design process.

3.5 Other applications

Besides the part and the assembly design process, the application provides an extra set of four tools that require a completed assembly model in order to function properly. These tools can be used at any order.

The first one regards the creation of mechanical drawings with basic dimensioning, for each part and assembly. The dimensions and the annotations that are used in the mechanical drawings are those that were used during the design process of the parts and the assemblies. The second tool is responsible for the rendering process of any solid model that is on the screen. The rendering is carried out at a fixed resolution. The third tool can be used to export the basic attributes of any solid model that is on the screen. The extracted attributes can then be stored at an Excel™ file as columns of a bill of materials or printed on screen in a dialog box. The last tool creates a Word™ document that acts as a process plan; it contains all the necessary information for the manufacturing of the sheet metal parts included. These sheet metal parts are used as supports for the final assembly. All required information and data are derived from the solid model of the parts, which then is exported to Microsoft Excel™ and Microsoft Word™ accordingly. Figure 7 depicts the final assembly of the pneumatic double acting cylinder.

 DOI: 10.1051/matecconf/20179401011

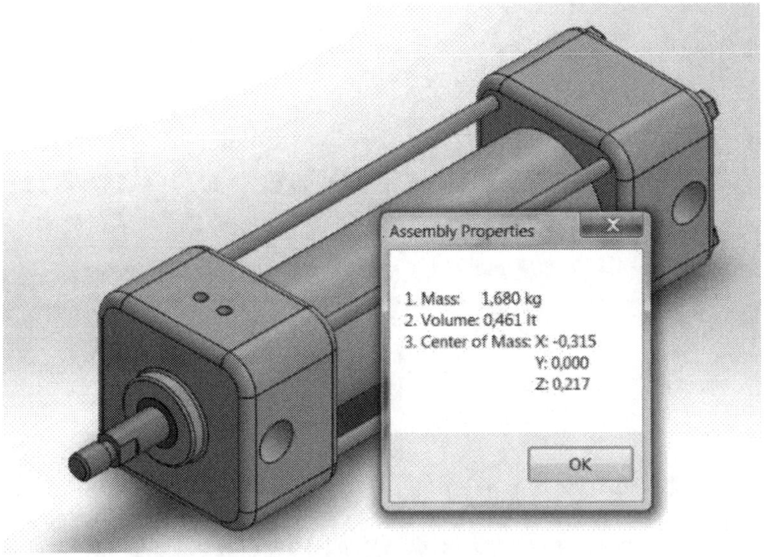

Fig. 7. The final assembly of the pneumatic double acting cylinder.

4 Conclusions

With the use of the API of a general purposes CAD system it is possible to develop applications in short time that can be used to analyze various engineering situations. The main advantage of such applications is that there is no need of any physical model in order to simulate and analyze any engineering processes, or even to extract data. It is also clear that the programming of the API can speed up many standard CAD processes and save time that can be better allocated. The development of human to machine interfaces can also be achieved with the use of the API and can drastically increase the manufacturing productivity in today's modern industries.

The publication was financed by the 'Teaching and research activities sponsored by HAAS Automation Ltd' project, Special Account of Education and Research, Western Macedonia University of Applied Sciences, Greece.

References

1. P. Kyratsis, N. Bilalis, A. Antoniadis, Int. J. Mech. Syst. Sci. &Eng. **1**,129 (2009)
2. P. Neto, J. N. Pires, A. P. Moreira, *IEEE International Conference*, **1**, (2010)
3. S. A. Haba, G. Oancea, Int. J. Adv. Manuf. Technol. **80**,747 (2015)
4. I. Badan, G. Oancea, *EMESEG International Conference*, **1**, (2010)
5. ISO 15552 Pneumatic fluid power - Cylinders with detachable mountings, 1000 kPa (10 bar) series, bores from 32 mm to 320 mm - Basic, mounting and accessories dimensions (2004)

MATEC Web of Conferences **94**, 01012 (2017) DOI: 10.1051/matecconf/20179401012
CoSME'16

Research regarding reverse engineering for aircraft components

Razvan Udroiu[1,*]

[1]Transilvania University of Brasov, Department of Manufacturing Engineering, Mihai Viteazu No.5, Brasov, Romania

Abstract. Reverse engineering is a useful technique used in manufacturing and design process of new components. In aerospace industry new components can be developed, based on existing components without technical Computer Aided Design (CAD) data, in order to reduce the development cycle of new products. This paper proposes a methodology wherein the CAD model of turbine blade can be build using computer aided reverse engineering technique utilising a 5 axis Coordinate Measuring Machine (CMM). The proposed methodology uses a scanning strategy by features, followed by a design methodology for 3D modelling of complex shapes.

1 Introduction

In present, most companies face a huge demand for diversified products from customers. They must adapt to the different requirements, through experimentation methods, techniques, and innovative actions on the market.

Traditionally a technical drawing or Computer Aided Design (CAD) model is created before the manufacturing of the part. But in several cases, where the CAD model does not exist, or the design of the product should be amended because of poor performance or manufacturing difficulties, Reverse Engineering (RE) technology becomes indispensable. The purpose of reverse engineering technology [1-3] is to capture the geometry of an existing physical model, through digitization, and to create a 3D virtual model for use in CAD, Computer Aided Engineering (CAE), Computational Fluid Dynamics (CFD) or Computer Aided Manufacturing (CAM) software. Liang & Grier [4] have defined the reverse engineering as a process of obtaining a geometric CAD model from 3D measurements acquired by contact or non-contact scanning technique of an existing physical model. The purpose of RE results from the following reasons:

- Manufacturing of products for which there is not a 3D CAD model,
- Optimize the shape of the products (redesign) for which 3D date is not available [5],
- Development of new products based on existing one with no CAD data,
- Development of customized products (medical prostheses and sports equipment),
- Inspect object dimension, deformations, etc. to confirm the quality and performance by computer-aided engineering analysis and inspection,

* Corresponding author: udroiu.r@unitbv.ro

• The repair and maintenance of complex products [6].

The main methods of capturing 3D geometry information on a product are classified into contact methods and non-contact methods. Which method is most attractive in terms better, faster, cheaper? Fast scanners will not be as accurate, more accurate scanners will cost more money. This means that you will need to choose up to two out of the three available factors: accuracy, speed or cost [7]. Computer Aided Inspection (CAI) and Computer Aided Reverse Engineering (CARE) are two techniques which use contact or non-contact methods and computer aided tools.

Contact methods comes in two main varieties: stationary CMMs and portable CMMs. Stationary coordinate measuring machines (CMMs) are large installation/ systems that are highly accurate, expensive and much slower compared to non-contact methods [7].

The noncontact solutions are a relatively new approach which consists in laser scanners, structured light scanners and industrial CT scanners (computed tomography). The advantages of non contact scanners over contact devices include fast data collection, and more comprehensive collection of points means a clearer view of the entire part. There are many parameters that affect the optical methods and hence the quality of scanning data, such as: reflective surfaces, narrow openings and sharp edges that can be difficult to scan. Thus, translucent, reflective or black / near-black parts require preprocessing operations, consisting in spraying with matte white powder of the surface, which affect the accuracy of the measurement in a range between 0.003-0.005 mm [8].

This paper proposes a 3D contact scanning methodology and a design reconstruction of a component, with applications for turbine blades. The case study regarding the 3D scanning of a turbine blade from Tumansky R-13 turbojet engine is a result of research from Aircraft engineering specialization from Transilvania University of Brasov. The Tumansky R-13 is a turbojet engine used by some variants of MiG-21 aircraft and Sukhoi Su-15M.

2 3D scanning of a turbine blade using CMM machine

The modern aerospace industry uses reverse engineering method as a key to the future of aerospace manufacturing as CAD tool [8-10]. Following the main purpose of RE, capturing and converting a discrete data set into a 3D model, one of the most used scanning techniques is contact scanning using a coordinate measuring machine (CMM).

2.1 Preparing the data for CMM scanning

The main steps before starting the scanning process of the blade consist in the following operations: define the scanning tools, tools calibration, define the device used for positioning of the blade on the CMM table and the alignment of the device.

The turbine blade was 3D scanning using a DEA Global Brown & Sharpe 5 axis CMM machine having a Renishaw SP 600 M scanning probe with stylus tools attached on it (Fig.1.a.). The stylus is the first link between CMM and workpiece its choice being determined by the geometry of the workpiece [11].

Two stylus tip type was chosen, the first one a ball tip, used for scanning the surface of the blade, and the other a cylinder tip, used for scanning the sharp edges of the blade. The ball tip stylus is the most frequently used type of stylus, suitable for the majority of applications. The ball tip stylus tool has the following main characteristics: 8 mm diameter ruby ball and 100 mm in length of a ceramic stem. The cylinder stylus is used for thin workpieces with which proper contact cannot be guaranteed with ball tip stylus. An 8 mm diameter ruby cylinder stylus was chosen in order to scanning the sharp edges of the turbine blade. Renishaw's SP600 M analogue scanning probes having a high speed scanning up to

300 mm/s and high frequency response offers high-performance digitising and profile scanning capability [11].

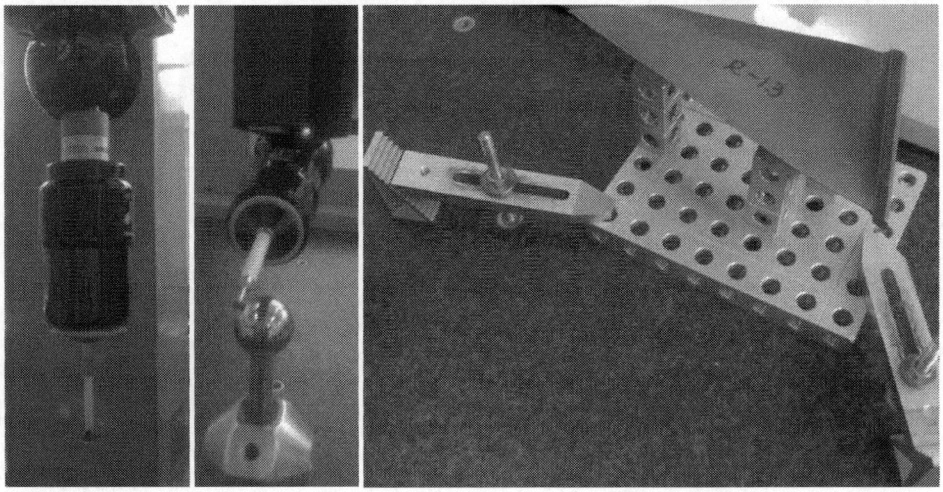

a) b) c)

Fig. 1. a) Renishaw SP 600 M scanning probe, b) calibration procedure and c) the device used for positioning the turbine blade.

The probe calibration procedure (Fig.1.b.) consists in measuring of minimum 25 points on a ceramic sphere having 20.002 mm in diameter, establishing the stylus tip's effective diameters when measuring, and their positions relative to machine's coordinate system. The measurement speed in calibration procedure is recommended to be the same with the scanning speed of the workpiece. The results of the calibration procedure are saved in PC-DMIS software and show a 0.003 mm standard deviation. PC-DMIS CMM is coordinate measurement machine software used to measure everything from simple prismatic parts to the most complex aerospace and automotive components. The device used for positioning of the blade on the CMM table is shown in the Fig.1.c.

The next preparing step consists in defining of the workpiece coordinate system. The origin of the workpiece coordinate system was chosen in the centre of the hole of the foot blade (Fig.2.).

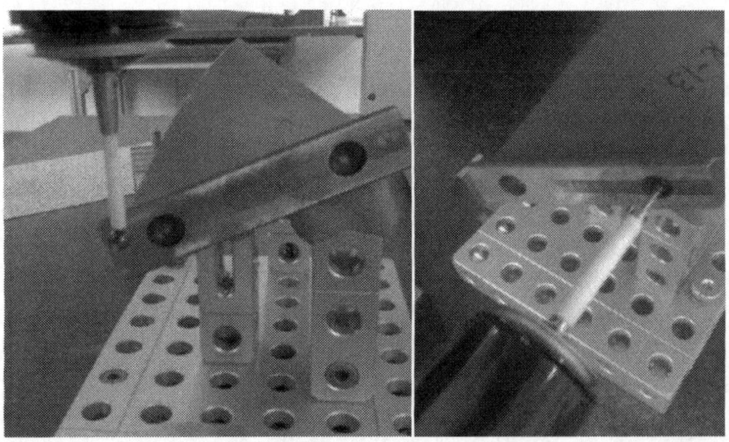

Fig. 2. Defining of the workpiece coordinate system.

 DOI: 10.1051/matecconf/20179401012

2.2 The blade scanning using the CMM machine

Based on the complex geometry of the turbine blade, the geometry of the blade was decomposed in the following main features:

- Blade root feature,
- The main surface feature consisting in upper and lower surfaces of the blade,
- The boundary feature of the blade,
- The link surface feature between the root and the main surface of the blade.

The scanning methodology of the blade was made taking into accord these features.

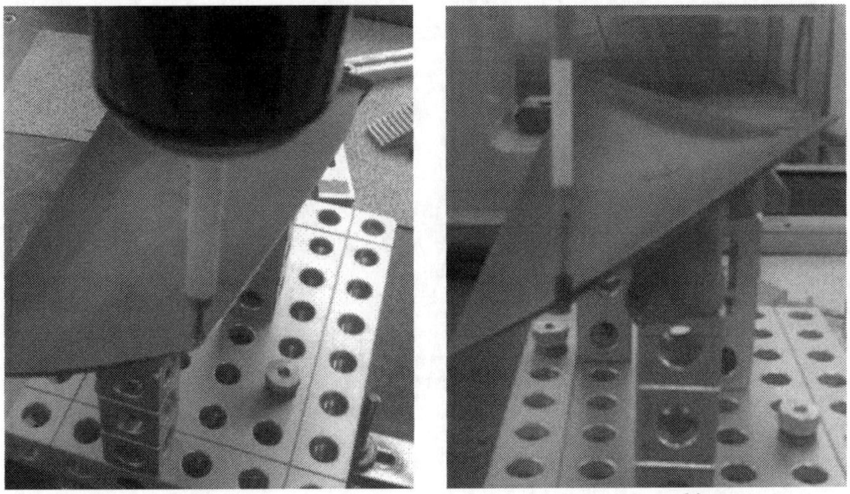

a) b)

Fig. 3. Scanning process of the blade features for a) the upper and lower surface, b) the sharp boundary feature.

The upper and lower surfaces of the blade and the link surface were scanned using a ball tip stylus, following a patch scan strategy (Fig.3.a.). An open linear strategy was used for scanning the boundary feature of the blade using a cylinder tip stylus (Fig.3.b.). The boundary feature consists of sharp edges. The resulting scanning paths are shown in Fig.4. The programming of the scanning paths was made using PC-DMIS software, based on the blade features. The point cloud data resulting from 3D scanning were exported in IGES file format.

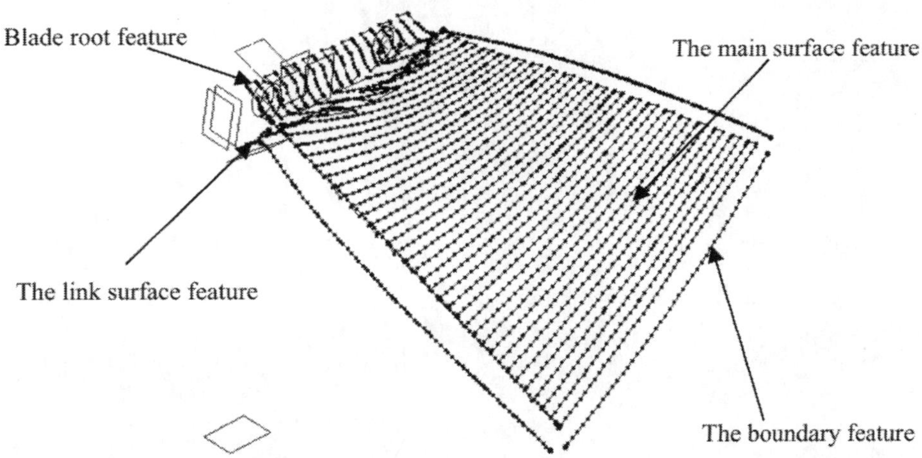

Fig. 4. The scanning paths of the blade features.

DOI: 10.1051/matecconf/20179401012

3 3D modelling of the scanning date using CATIA software

There are several inspection / reverse engineering software products on the market, bridges the gap between the point collection and the design requirements. Thus can be mentioned standalone software products such as GEOMAGIC, RAPIDFORM, GOM Inspect or some software modules of the complex software systems, e.g. Catia, Creo and SolidWorks.

The design methodology (Fig.5.) proposed for 3D modelling of the turbine blade, is made using Catia version 5 release 21 by Dassault Systemes.

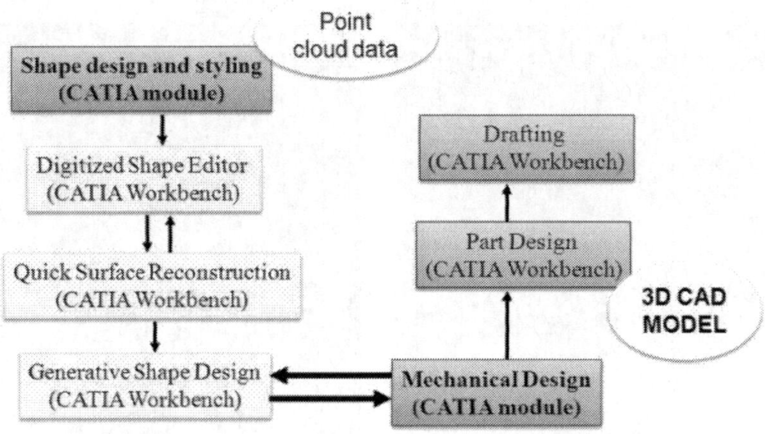

Fig. 5. Design methodology of the blade.

The point cloud data obtained from scanning process of the blade were imported in Digitized Shape Editor workbench of CATIA. Digitized Shape Editor is a CATIA's workbench that allows importing and processing of clouds of points [12]. These clouds of points are used in Quick Surface Reconstruction and then in Generative Shape Design, in order to obtain the blade surface. Following the methodology from Fig.5 was obtained the 3D blade model of the turbine blade (Fig.6.).

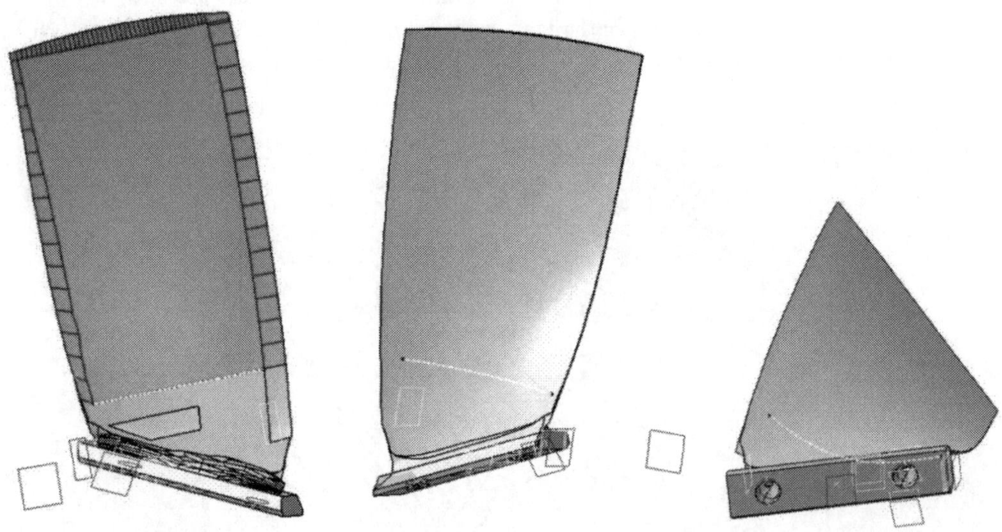

Fig. 6. 3D model of the scanned blade.

The design methodology of the blade using CATIA consists in the following main steps:

- Import of scanning data (clouds of points),
- Alignment and union of the clouds of points in one cloud of point,
- Processing of cloud of point,
- Generating of curves and surfaces,
- Check the quality of the surfaces,
- Save the 3D model of the blade in IGES or STL file format.

4 Conclusions

Reverse engineering is an important method utilized in manufacturing and design process. When there is no technical data of a product or it is unusable, the CAD model is obtained with this method. Computer Aided Reverse Engineering (CARE) integrated into CAD/ CAE/ CAM systems allows rapid product development by redesigning and optimizing the product, in a short period and decreasing the manufacturing cost.

The turbine blade has a complex shape and required a scanning process by features. The proposed scanning methodology using features, presented in this paper can be used also for other complex parts from aerospace industry.

The errors from scan programming could produce some damage of the stylus, scanning probe or scanning part. The CMM operator must be specialized and he should pay attention to the scanning process. Compared with non-contact scanning, contact scanning presents higher risk factor regarding the influence of the scanning programming relative to the damage of some CMM components, but the scanning accuracy is better.

A future research will be focused on comparing of contact and non-contact scanning procedures of the turbine blade in terms of accuracy, speed and cost.

References

1. V. Raja, K.J. Fernandes, *Reverse Engineering: An Industrial Perspective* (Springer-Verlag London Limited, 2008)
2. M. Sokovic, J. Kopac, J. of Mat. Proc. Techn., **175**, 398 (2006)
3. R. Vinesh, J.F. Kiran, *Reverse Engineering – An Industrial Perspective* (Springer, 2008)
4. L.C. Chen, G.C.I Lin, Robot. Comput.-Integr. Manuf., **16**,161 (2000)
5. V. Tut, A. Tulcan, C. Cosma, I. Serban, Int. J. of Mech., **4**, 79 (2010)
6. D. Tsakatikas, G. Kaisarlis, *4th International Conference on NDT*, **4** (2007)
7. ***, Contact vs. Noncontact Measurement for Computer-Aided Inspection, Quality Magazine,http://www.qualitymag.com/articles/91150-contact-vs-noncontact-measurement-for-computer-aided-inspection (2013)
8. A.A.G. Madrigal, H.R.S. Carrillo, J.O.F. Cavazos, *Memorias del XIX Congreso internacional anual de la SOMIM*, **897** (2013)
9. K. Mohaghegh, M.H. Sadeghi, A. Abdullah, Int. J. Adv. Manuf. Tech., **32**, 1009 (2007)
10. E. Savio, L. De Chiffre, R. Scmitt, Annals of the CIRP, **56**, 810 (2007)
11. ***, *Renishaw*, http://www.renishaw.com/en/products--32083, (2016)
12. ***, *CATIA documentation*, http://www.3ds.com/support/documentation/overview/ (2015)

MATEC Web of Conferences **94**, 01013 (2017) DOI: 10.1051/matecconf/20179401013
CoSME'16

Studies regarding the processing of small thickness containers

Mihaela Urdea[1,*]

[1]Transilvania University of Braşov, Department of Manufacturing Engineering, Mihai Viteazul No.5, Braşov, Romania

Abstract. This paper studies the possibility of modeling containers made of small thickness sheet metal for the final goal to process them. It's about large external buffer containers for products that come or leave by truck. Modeling is done in Sheet Metal module and it is based on the possibility of obtaining the containers' unfolded components. It studies the issue of the efficient use of sheet metal with standard dimensions in the context of an optimal processing. These containers, completely built of sheet metal, are fabricated in a variety of sizes and are made up of welded parts. In this paper different varieties of components are defined in order to develop a design strategy for a rapid modeling which will result in a rapid construction. This study was conducted with the participation of Raptronic Process Engineering Company.

1 The use and appearance of containers

Boxes and containers for transport and storage goods, for example the containers manufactured at Raptronic Process Engineering Company, can be modeled in SolidWorks, using the Sheet Metal tools. Figure 1 presents a general type for these containers [1].

Fig. 1. Containers for goods.

Raptronic Process Engineering is specialized in the development and manufacturing of equipment and process technologies for food, chemical and pharmaceutical industries. RPE

* Corresponding author : urdeam@unitbv.ro

DOI: 10.1051/matecconf/20179401013

produces customized containers or tanks, bunkers, agitators, kneading and blending machines of different capacities, made out of stainless steel and manufactured based on the customers' needs. RPE delivers automatic carbonation and pre-mix equipment for any types of liquid products, functioning at certain pressures or adjustable pressures. In the transport activity large containers are used [1].

These containers are completely built of sheet metal and can be fabricated in a variety of sizes and are made up of welded parts. It is adopted the general shape of the containers with weldable components according to Figure 2 in which 4 major sub-assemblies can be notice: the cover with the locking system, the body, the base and 3 legs. There are observed as useful components a spigot outlet, 2 sensor levels, an elbow for ventilation, a manhole, etc.

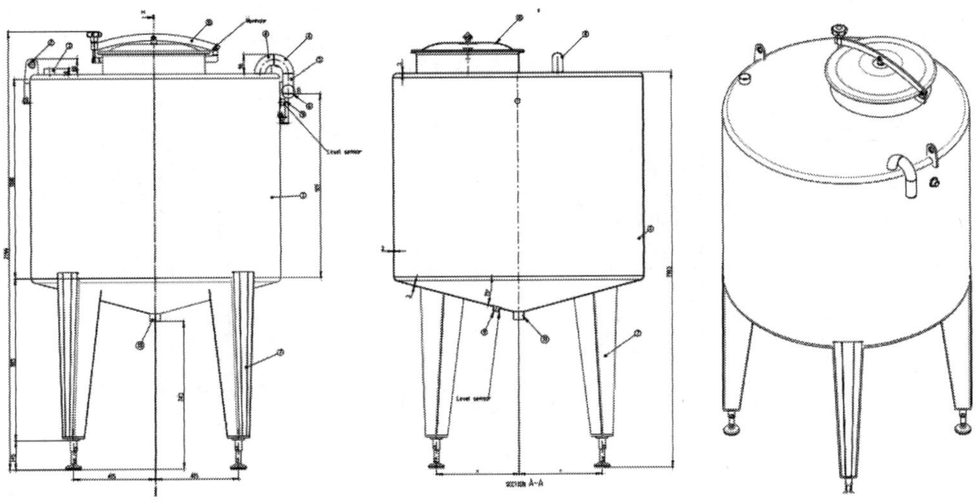

Fig. 2. Container represented in AutoCAD.

The sheet metal bodies can be easily processed and are also easy to be modeled in SolidWorks. Sheet Metal tools contain specialized functions for modeling sheet metal parts. The modeling process starts from a basic pattern, similar to the conventional solid modeling, which is developed through the addition of new entities. The studied containers are large exterior buffer containers for products that are transported by truck and have different shapes and dimensions.

The constant thickness of the walls allows a special strategy for the modeling selection and similarly a design strategy depending on the blank material available on the market. Also the finished product must be studied and processed from the point of view of internal cleaning systems (CIP S).

2 The container

In the sheets metal modeling activity, Sheet Metal tools can be used. There are some exceptions depending on the components shape. In this case the sheet parts can be modeled from a solid wall thickness, without using the specific advanced functions of Sheet Metal tools. Finally SolidWorks offers the possibility of transforming the parts in Sheet Metal tools. The alloy steel, symbolized AISI 304 in SolidWorks, is the used material for the studied containers.

Studying in this context the welded components of the containers, it was noticed that the transformation in Sheet Metal tools is not always possible. This paper offers some steps in

 DOI: 10.1051/matecconf/20179401013

this transformation, and the study is made permanently from the viewpoint of the blank material.

As a sheet blank it can be used sheet with the following dimensions: 1000x2000, 1250x2500, 1500x3000, but often it is preferable role sheet metal, due to uncommon sizes [1].

In case of the fluid container, its body can be built from more overlapping welded cylinders, with different thicknesses for every segment, increasing towards the base. This paper focuses on designing the body components with constant thickness.

2.1 The cylinder

In the situation of constant thickness, the cylindrical part of the container can be modeled easily, like a sheet model, in two ways; using only the Sheet Metal tools (Fig. 3 a) or using first the Part tools and after it the Sheet Metal tools (Fig. 3 b).

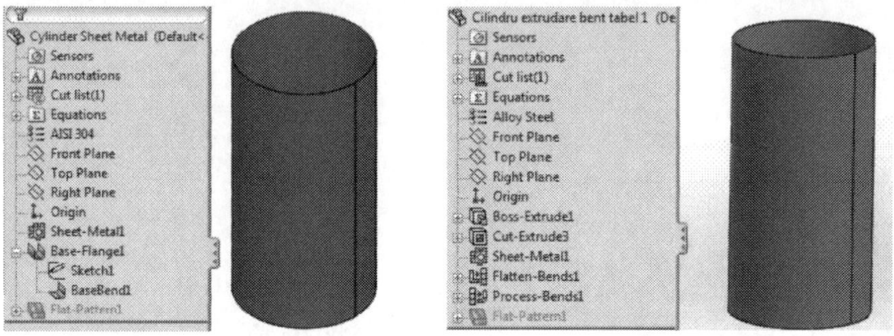

a b

Fig. 3. The cylindrical part of the container.

The final goal is to obtain the flattened cylinder, cut on the height and defining the developed dimensions in order to choose the blank. With **Flatten** function one can individually flatten any component in a multibody sheet metal part (Fig. 4). Another purpose is to measure the blank dimensions, necessary for obtaining this component.

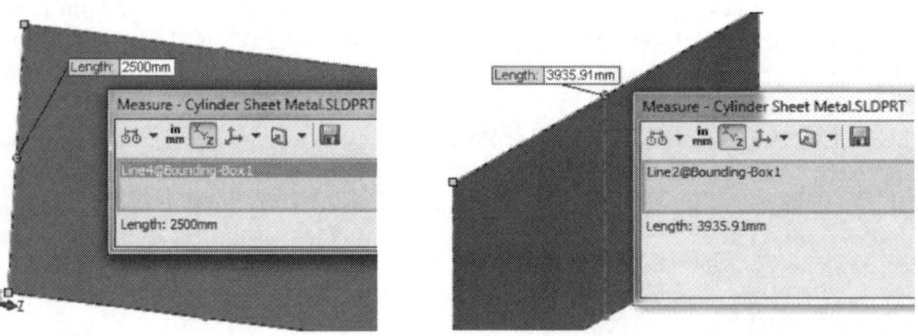

Fig. 4. The flat pattern for the cylinder, measured with Measure function.

For the body both versions are preferred considering several modeling options for a cylindrical shape using the **Design Table** tools. Due to the fact that the designer uses several sizes of the containers a table of sizes will be generated. For this purpose, all sizes are defined as variable using **Link-value**. The Design Table tools will develop an Excel table with the linked values (Fig. 5 a). Also considering the possible width of the sheet blank it will assign a relationship related on the cylinder height.

DOI: 10.1051/matecconf/20179401013

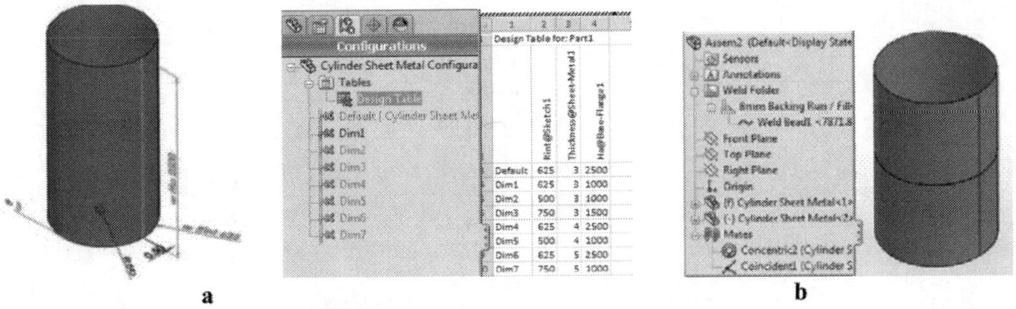

Fig. 5. Design Table for the cylinder.

In the same way all variable components of the container will be processed. So, these flattened parts will be defined and measured. For cylinders with great heights, without sheet blank dimensions, welded sheet metal is used (Fig. 5 b).

2.2 The container bottom

The container bottom is a tapered surface, rounded at the edges for joining by welding with the cylinder. It can be noticed that the only way for obtaining the base with the proposed shape, is the classic Part way. The Sheet Metal functions cannot process a profile like this, but in this situation the flattened shape will not be obtained. To solve this problem a perfect conical shape which can be modeled by Sheet Metal, for the base, was proposed. It will then be performed a bending for joining by welding with cylinder.

Figure 6 presents the base sketch with the calculated dimensions before the bending for the variant Dim 625. In the graphic construction the sketch length is 633.8 mm.

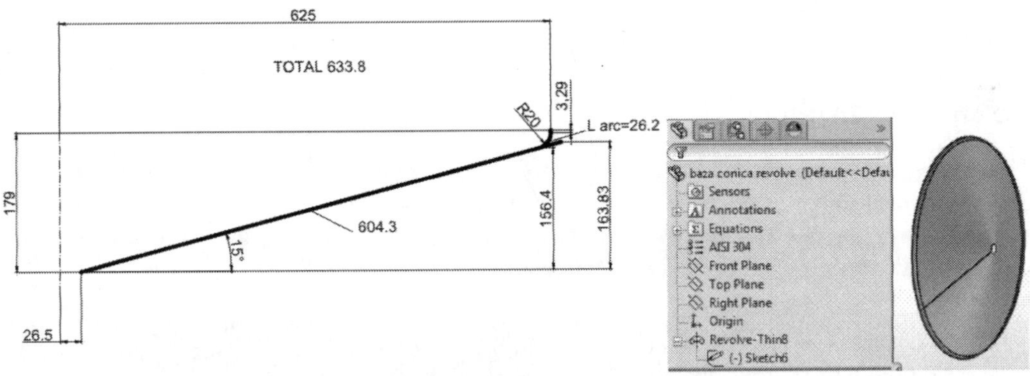

Fig. 6. The container bottom before and after bending.

K-Factor is a ratio that represents the location of the neutral sheet with respect to the thickness of the sheet metal part. In case of using sheet metal bending K-Factor, the following calculations result (Fig. 7) [2, 3]:

$$\textit{Bend Allowance} = \pi(R + KT)*A/180 = \pi\,(20+1.5)*75/180=4.581 \tag{1}$$

$$\textit{Total length} = 604.3+3.29+\textit{Bend Allowance} = 612.171 \tag{2}$$

It can be noticed a difference between the two studied situations, due to stretching of the material during bending.

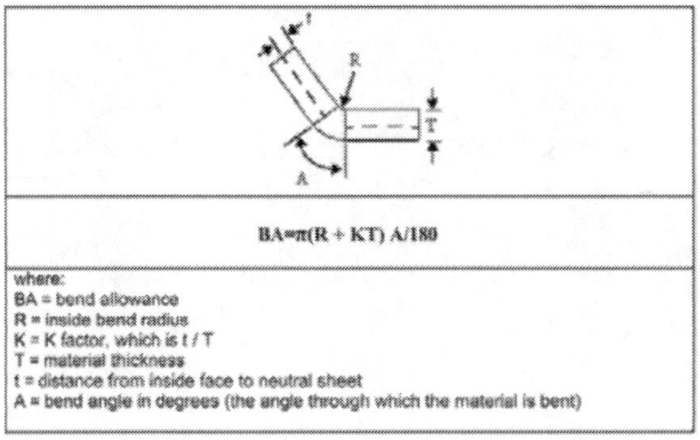

Fig. 7. Information about the K-Factor.

It is measured the dimensions of the cone before and after flattening. **Measure** function makes all the necessary calculations for the cone and blank (Fig. 8).

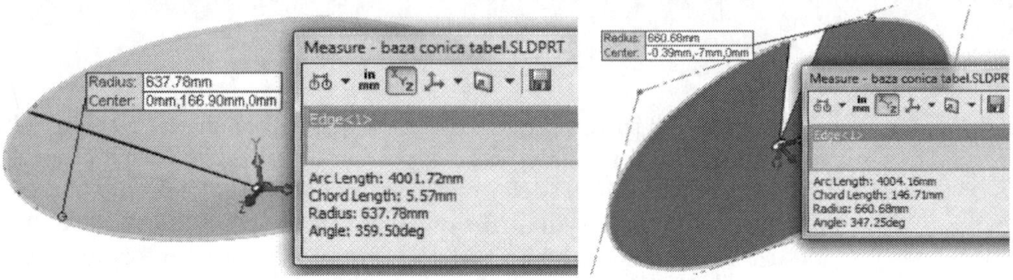

Fig. 8. Measure function used for measuring the size of the cone and of the blank.

The shown dimensions are for the studied shape, the inner diameter for cylinder 1250 mm. Also in this case it has been developed a table of dimensions related to the cylinder table with 7 dimensions.

Similarly, the cover was modeled and measured.

2.3 The container foot

Figure 9 a, is an example of a sheet metal foot used for small capacity containers. Also, to this component a Design Table, according to the possible dimensions can be associated. For this, links are generated, with link value function (Fig. 9 b). The flat pattern for the foot can be measured starting from the overall dimensions (Fig. 9 c).

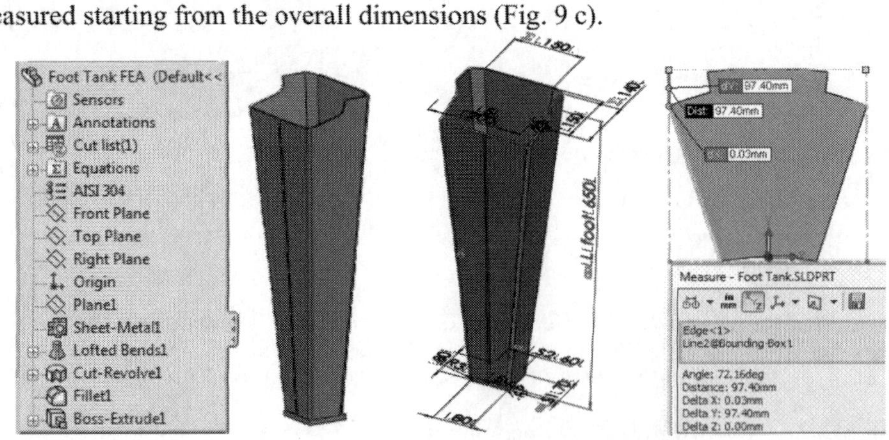

a b c

Fig. 9. The container foot.

An important step is to check the feet resistance and possibly there resizing. In this case a finite element analysis is performed on one foot. In order to analyze the response of a component to an applied load, one uses the SolidWorks Simulation finite element analysis (FEA) program. For preparing the analysis model it is very important to define the feet material (alloy steel AISI 304 with Young's Modulus 1.9 e^{+001} and Poisson's Ratio 0.9 N/A) and the external loads.

The containers can have three or four feet. For security reasons at a container with 3 feet one will consider only two. They support the entire container load and in the studied case, the oil with 1.2 kg / l (1200kg/ $m^{3)}$ density. The total weight, for the studied case, is estimated at ~4955kg. This corresponds to a calculated pressing force ~ 48559N. The force is distributed on the body surface support. It was calculated the stress in the two situations, namely for sheet thickness of 4 mm (pressure =32N/mm^2) with a 5 mm distance in the area of joining sheet foot and similarly for feet with 6 mm sheet thickness (pressure = 21 N/mm^2) with 1 mm distance in the area of joining sheet foot.

It was applied **Fixed Geometry** and **Pressure** on the two ends of the foot. The result is presented in the Figures 10 and 11.

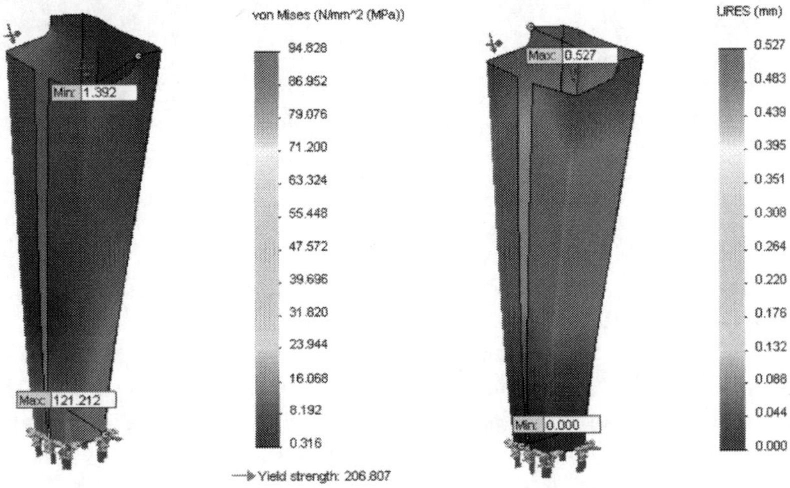

a b

Fig. 10. Von Mises (a) and Displacement (b) for 4 mm sheet thickness.

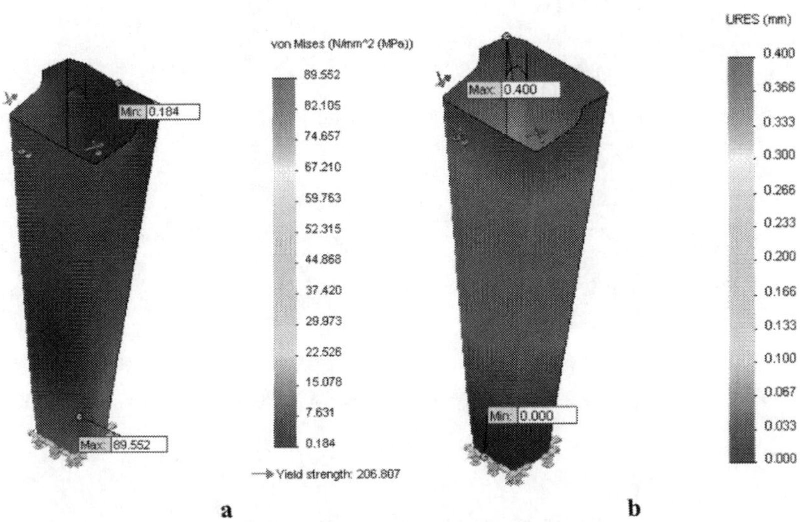

a b

Fig. 11. Von Mises (a) and Displacement (b) for 6 mm sheet thickness.

Von Mises equivalent stresses, is a scalar quantity obtained from the principal stress values and it is used to measure the state of stress. It is seen that equivalent stress **Von Mises** is below the maximum allowable flow of the material. **Displacement** measures the maximum deflection [4]. To avoid all the possible problems, in practical situation, it is preferred the container with 6 mm sheet thickness.

In order to attach the container pipes, **Pipes** module is used. SolidWorks provides a library of products required for rapid generation of pipes. From the library named **Piping** one can attach the necessary components: flanges (Fig. 12 a), valves, elbows, gaskets, equipment. Finally the customized container looks like the Figure below (Fig. 12 b).

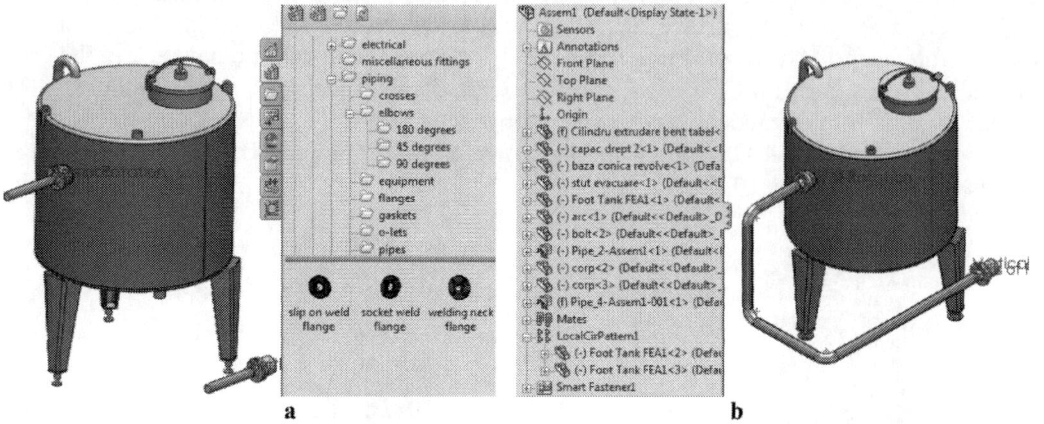

Fig. 12. The container with pipes components.

3 Other container shapes

Figure 13 presents an example of a double jacketed (for heating) mixing container, modeled in SolidWorks. This container is equipped with an impeller for mixing. The temperature is controlled with a precision digital thermostat and these containers have a heated stainless steel drain valve with independent temperature control [1].

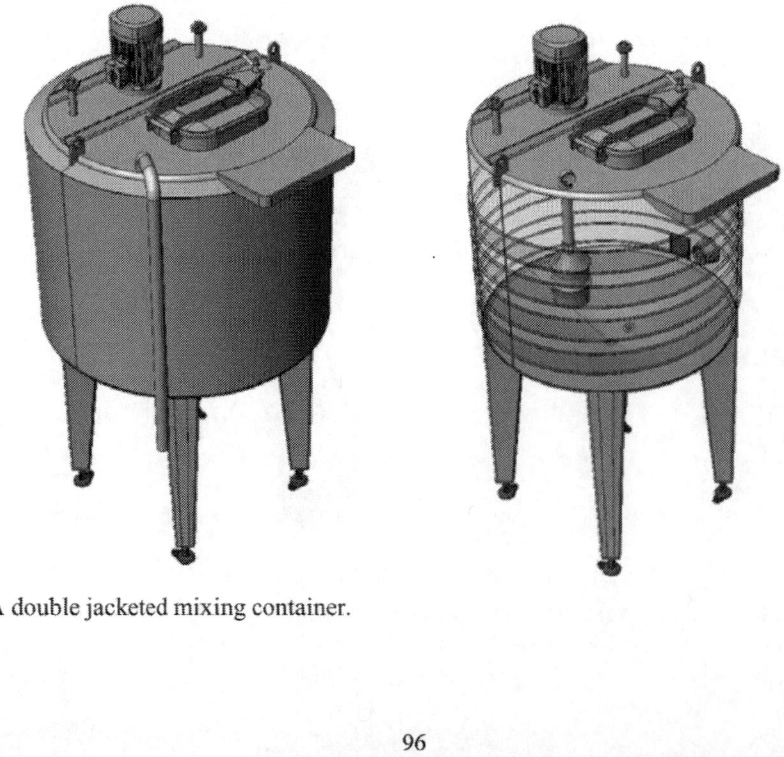

Fig. 13. A double jacketed mixing container.

Figure 14 a presents an example of a melting container for melting wax for making candles [5]. Melting containers have a hot grid that melts fat blocks and the molten mass is stored inside the container, which has insulation and a hot water jacket [1]. Figure 14 b presents a melting container modeled in SolidWorks.

a b

Fig. 14. Melting container.

4 Conclusions

The small thickness containers for transport and storage goods, liquids, powder and different raw materials are very used in the food industry. These containers are made of alloy steel sheet in various sizes. The blank material must be properly cut without losses. In this purpose, 3D models have been created in SolidWorks, easily to be flattened and measured.

For the fast modeling of the containers with variable dimensions, the Design Table function has been used. In this activity it was taken into account the size of the blank and the possibility of welding the components.

Certain parts of the container cannot be classically modeled with Sheet Metal functions, namely the container bottom and the cover. The bending area creates this problem. In this case these parts have been modeled before the bending. The length of the sketch, necessary to obtain the desired profile before bending, was calculated. The resulted dimensions were compared with the results from the modeling of the sheets, through Sheet Metal functions.

The resistance of the foot was checked with finite element analysis. In this way it was confirmed the designed foot resistance in the sheet metal area. The containers can be used, after washing, for the storage of other fluids (for example chocolate). Due to this the feet are oversized.

References

1. http://www.rap-group.ro/raptronic-sisteme-de-proces_en.htm
2. http://help.solidworks.com/2015/English/SolidWorks/sldworks/r_welcome_sw_online_help.htm
3. E., Maican, *SolidWorks – 3D modeling for engineers*, (Printech Bucureşti, 2006)
4. Gh., L,. Mogan, S., L., Butnariu, *Finite element analysis in engineering. Practical applications in CATIA* (Transilvania University, 2007)
5. http://www.satyasaifabrications.com/melting-containers.html

Analysis of bearings behaviour with cylindrical rollers with variable center of gravity

Sorin Barabas[1,*] and *Adriana* Florescu[1]

[1]Transilvania University of Brasov, Faculty of Technological Engineering and Industrial Management, Department of Engineering and Industrial Management, Romania

Abstract. This paper consists in analysis of the contact stress and deformations appeared to the cylindrical rollers with variable center of gravity who equips large bearings, with finite elements method. Also it proposed an innovative model of lubrication system mounts in the hollow cylindrical rollers of the large bearings where it is stored the lubricant who is driven to the raceways through holes fitted with closing/opening systems made in the cover, controlled by a command system, monitoring and control, equipped with temperature sensors and wireless command. Elimination of lubricant, in stages, leads to changes in position of center of gravity of roller, thus to change its behavior in functioning. Comparison of deformations occurred in cylindrical rollers, solid, with hollowness, and with variable center of gravity, allow favorable conclusions, on the implementation of the proposed lubrication system.

1 Introduction

Suitable bearings lubrication minimize wear, increase their lifetime significantly, and makes possible rolling processes. Using large bearings in wind power assemblies, which have permanent operation in difficult conditions with high variability of both, the load and speed, is a problem widely studied because of difficulties arising. The importance of developing green energy, increasing the energy efficiency of wind power plants and their wide distribution, turn maintenance process, therefore, the lubrication process, a matter of great importance.

The possibility of using hollow rollers in large bearings construction, allows the development of a new lubrication system, automated, controlled, and managed by computer, placed in the cavity of rollers.

To improve efficiency of lubrication, it is necessary to maintain the lubricant film thickness to constant values, the amount of lubricant introduced to the raceways being strictly controlled and in accordance with the lubricant consumed or evacuated. Lubrication process is influenced by temperature, rotation speed, geometry of raceway, static and dynamic loads, type and quantity of lubricant. All these factors determine the quality of process with consequences on lifetime and efficiency of energetic assembly.

* Corresponding author: sorin.barabas@unitbv.ro

DOI: 10.1051/matecconf/20179402001

Contacts of rolling elements has the following possibilities for interaction [1]: contact elastohydrodynamic, mixed frictional contact, wear contact and contact with presence of particles between the surfaces.

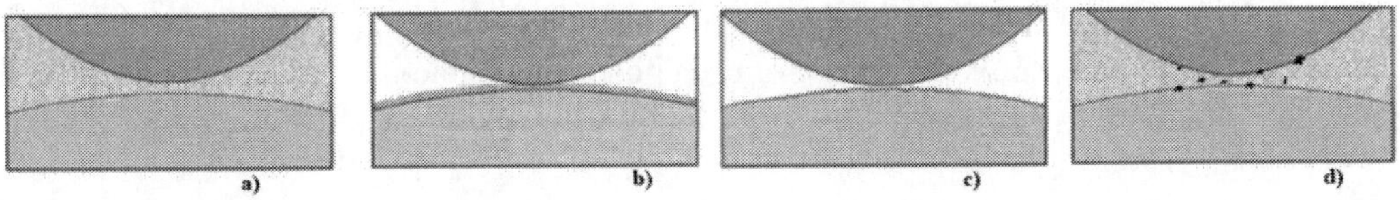

Fig. 1. Contacts of rolling elements a) elastohydrodynamic, b) mixed, c) wear contact, d) contact with particles between the surfaces.

The amount of lubricant can be calculated by the formula:

$$Q_l = f_v\left[\rho_b \frac{10^{-3}\pi T(D^2-d^2)}{4} - \frac{m_b}{7,8*10^{-3}}\right] \quad (1)$$

where: Q_l is lubricant quantity[N];
f_v is a parameter who depends by rotation speed (approx. 0,3…0,5);
ρ_b is bearing steel density;
T is bearing width [mm];
D is exterior diameter [mm];
d is interior diameter [mm];
m_b is bearing mass [kg].

The contact stress and the deformations occurred in the contact zone, between roller and raceway, strongly influence the lubrication film. The elastohydrodynamic theory [2,3,4] uses, for calculation of thickness of lubrication film, the point contact equation (Hamrock and Dowson, equation (2)) and line contact equation (Dowson, equation (3)). (Figure 2.).

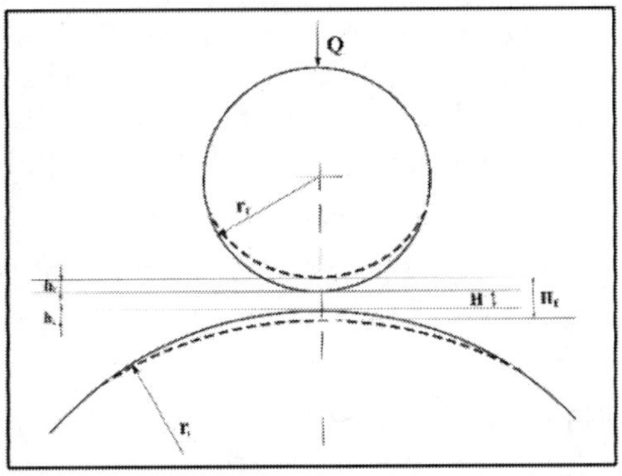

Fig. 2. The influence of deformations on thickness of lubricant film

$$H = 3.63U^{0.68}G^{0.49}W^{-0.073}(1-e^{-0.68k})R_r \quad (2)$$

$$H = 2.65U^{0.68}G^{0.54}W'^{-0.13}R_r \quad (3)$$

where: H is minimum lubricant thickness [mm];
U is a speed parameter who depends by rotation speed and lubricant viscosity;
G is material parameter;
W, W' are load parameters for point contact and line contact;
k is a parameter who depends by contact area;
R_r is curvature between roller and raceway [mm];
h_r, h_i are deformations in roller and inner ring [mm]

The elastohydrodynamic theory and the theory of linear elasticity [5,6] was used for modelling the process of lubrication

$$Hr = \frac{1}{\pi}\left(\frac{1-v_I^2}{E_I} + \frac{1-v_{II}^2}{E_{II}}\right) \iint_x^y \frac{p(\xi.\eta)}{\sqrt{(x-\xi)^2+(y-\eta)^2}} d\xi d\eta \tag{4}$$

where: H_r is distance between surfaces ;
E_I, E_{II} ,are Young modulus for each bodies in contact;
v_I and v_{II} are Poisson's ratio for each bodies in contact;
$p(x, y)$ is the pressure of the common area.

$$\int_x^y p(\xi.\eta)\ d\xi d\eta = \frac{dQ_d}{dt} \tag{5}$$

$$\frac{dQ_d}{dt} = \frac{d(Q + Q_s)}{dt} \tag{6}$$

where: Q_d is dynamic load [N] ;
Q is normal load of bearing [N] ;
Q_s is load due to the weight of lubricant stored in hollow roller.

Application of equilibrium equation (5) and and variability equation of gravity center of roller in relation to time (6) in integral equation of elastic contact, Boussinesq - Flamant (4) allows the analysis of stresses and strains in relation to time. The condition for elastohydrodynamic contact is $H_r > H$.

2 Proposed concept for automated lubrication system mounted in hollow rollers of large bearings

The concept of automated lubrication implemented for hollow rollers equipping large bearings installed in wind power plants presents the latest fundamental elements of innovation consisting of positioning inside the hollow roller of an automated system equipped with control unit, temperature sensors, opening / closing module of the area with lubricant, wireless transmitter to the computer, power supply and monitoring software, command and control of the functional parameters of the bearing [7].

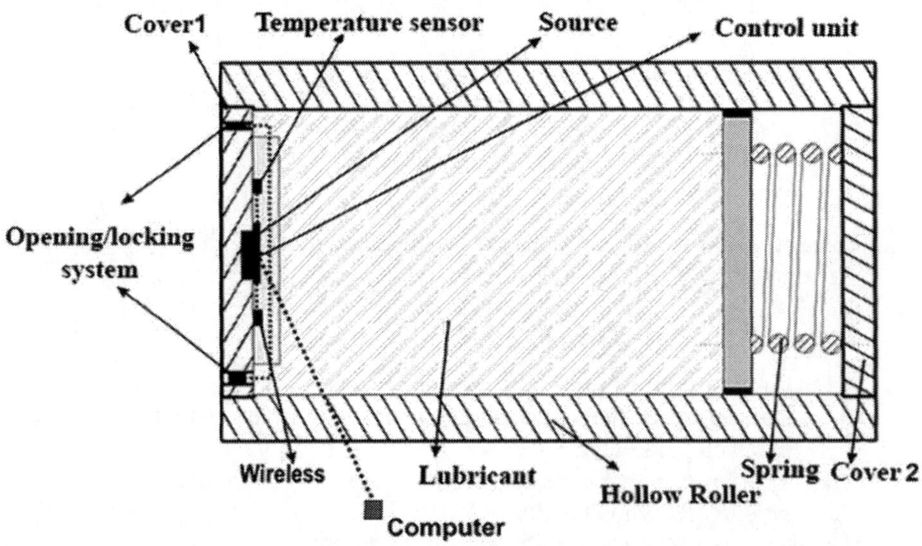

Fig. 3. New concept for automated lubrication system

In Figure 3 it can be seen the proposed lubrication system scheme, following that research conducted during the project to establish the exact design and operation of each module. Thus the locking - unlocking system may comprise electrically operated valve or thermal or magnetic operated closing valves.

The source can be a sustainable battery or carbon brushes. The control unit can be programmable or programmed. The assembly may be inside or outside of the roller.

When opening the valves, the lubricant liquid is pushed toward the runways, driven by the piston spring mounted under pressure. In the roller a shift of the centre of gravity occurs which can lead to the appearance of further tensions in the contact areas.

3 Stress and deformations analysis with Finite Element Method for bearings equipped with proposed lubrication system

Stress analysis by finite element for bearings with hollow roller wit variable center of gravity, due to lubricant evacuation consists in application of equations (4) and (6) in Nastran software.

Were chosen rollers with D_{ext} = 120 mm and L = 240mm. Material of roller is 15NiCr13 carburized at 63HRC. The diameter of bearing is D_{rul} = 2100 mm. For the inside diameter (hole diameter) were chosen four cases according to the following values: solid roller, hollow roller Di_3 = 90 mm with cavity filled with lubricant, with cavity without lubricant and with half of the quantity of lubricant. For bearing with solid rollers and for bearings with hollow rollers with different quantity of lubricant was made the finite element analysis.

When applied finite element analysis method to rollers, was used iteration cinematic method, that assumes that the nodal forces are known by evenly distributing, remaining to calculate the contact forces. The value of these forces is determined by successive iterations resulting when the speed of the node is null.

The results obtained by applying the finite element method were calculated in the position of maximum stress when the loading force is perpendicular to roller. Redistributed components of normal loading force is considered lower, in the angular positions, without influencing the results [8-10].

It was considered the case of normal loading with value Q=280 kN. For analyzed bearing the loading per hollow roller is 8 kN and for the study cases total loads are Q_1 = 308 kN, Q_2 = 285.25 kN, Q_3 = 282.45 kN, Q_4 = 280 kN

Table 1: The values of contact elements

Type	Loads (kN)	Contact pressure (N/mm²)	Contact lenght (mm)	Von Mises stress (N/mm²)	Deformation (mm)
1	308	2402.4	0.5934	1039.5	0.0199
2	285.25	2406.5	0.6581	1041.9	0.0226
3	282.45	2410.6	0.7512	1046.2	0.0231
4	280	2412.8	0.7612	1049.7	0.0252

1-solid roller; 2-hollow roller (D_i=90mm, filled with lubricant); 3-hollow roller (D_i=90mm, half filled); 4- hollow roller (D_i=90mm, empty);

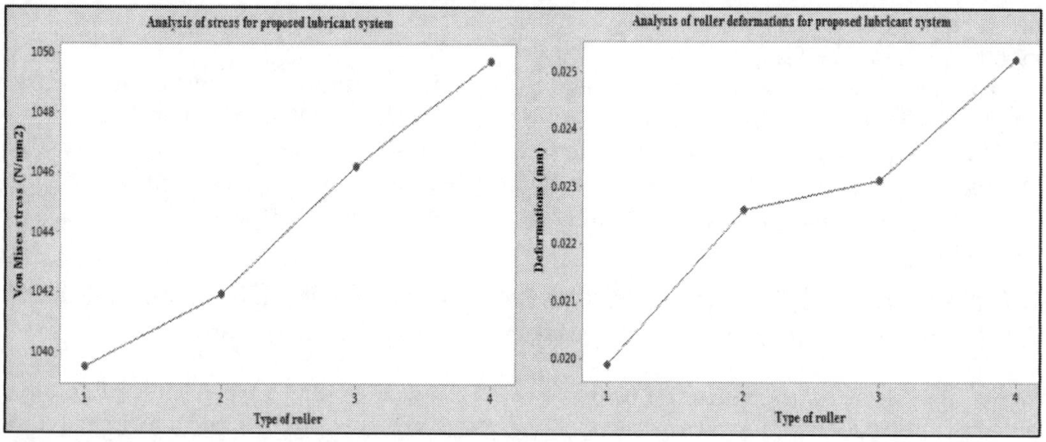

Fig. 4. Graphical analysis for stress and deformations in case of lubricant system for large bearings with hollow rollers

An analysis made on the results (Figure 4), clearly shows slight increase of stresses and deformations with evacuation of lubricant to raceways. The results are interesting from another point of view, namely that advantages of using of hollow rollers, completely reduces the disadvantage of increasing Von Mises stress and deformations.

4 Conclusions

The paper also proposed a hollow roller for large bearing having in cavity an automated lubricant evacuation system, to the raceways of the bearing. With the evacuation of lubricant is shown a displacement of the center of gravity rollers, which can cause in system additional wear, who requires a dynamic computing for new stresses occurred on the runways. Finite element analysis performed with Nastran program revealed a slight increase in tensions and strains, insignificant for the modification the thickness of film of lubricant, between roller and raceway. From the research done, lubrication system proposed shows the complete novelty, such as placing it in hollow roller, the lubricant injection system, monitoring, control and its command. The behavior of hollow rollers with lubrication system implemented is identical, differences in stress and strain being negligible, with the maximum deformation of the solid roller 0.025 mm, and the change in center of gravity causes deformations increase only, with 2 microns. Results that implementation of the proposed lubrication system bring significant benefits.

References

1. F. J. Ebert, Chi. J. Aer., **23,** 122 (2010)
2. E.V. Zaretsky NASA Tec.Mem., 102575, (1990)
3. K. Bobzin, T. Brogelmann, Sur. & Coat. Tec., **290**, 100 (2016)
4. B. J. Hamrock, D. Dowson, *ASME JOT,* **98,** 375 (1976)
5. G. Nijebanning, C. H. Venner, H. Moes, Wear J., **176,** 217 (1994)
6. M.Y. Panovko, Prob. Mash. i Mash. **5**, 54 (2009)
7. S. A. Barabas, A. Florescu, Rev. Rec. **43,** 138 (2014)
8. E. Mermoz, D. Fages, L. Zamponi, J. M. Linares, J. M. Sprauel, CIRP. Man. Tec, **65,** 157 (2016)
9. V. Liqin, C. Li, Z. Dezhi, G. Le, Chi. J. Aer., **21,** 86, (2008)
10. Y. Wei, R. Balendra, J. Mat. Proc. Tec., **145,** 233, (2004)

MATEC Web of Conferences **94**, 02002 (2017) DOI: 10.1051/matecconf/20179402002
CoSME'16

CADOProSys Software. Logistics

Cătălin-Iulian Chivu[1], *Catrina* Chivu[1,*], and *Cristina* Găvruş[1]

[1]Engineering and Industrial Management Department, Transilvania University of Brasov, 500174 Mihai Viteazu, Brasov, Romania

Abstract. CADOProSys Software was developed by a group of teachers from Engineering and Industrial Management Department of Transilvania University of Brasov, Romania. The main purpose of software is to help the beginners and graduates to design production systems in the field of automotive industry. CADOProSys Software has many modules. One of these modules is focused on logistics, from suppliers and customers point of view. Starting a production system implies the selection of suppliers from many points of view. The user of the module is able to introduce the criteria and the importance of each criterion. Based on these the modules will provide the appropriate supplier and the inquiry documents.

1 Introduction

1.1 Logistics: meaning and evolution

Logistics is being recognized as a very important factor for major corporations to significantly increase shareholder and customer value. There were lots of definitions for logistic, but one of the most simple and complex in the same time is that given by Edward Frazelle, according to whom "*logistics is the flow of material, information, and money between consumers and suppliers*" [1].

Starting to '50s logistics expanded its meaning (figure 1). Thus, from workplace logistics to global logistics, the concept developed from material handling between workstations to flow of material, information and money between countries.

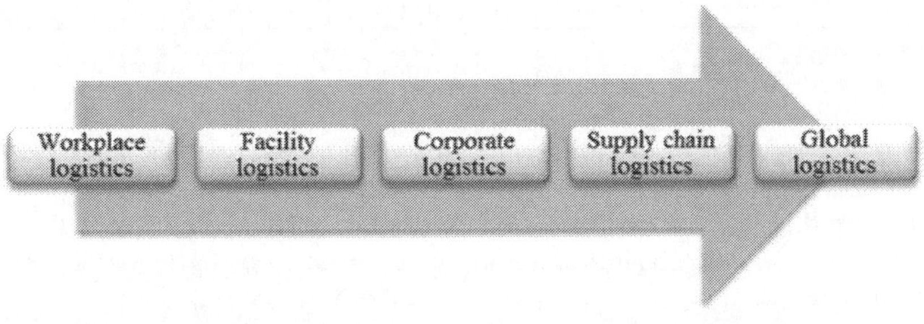

Fig. 1. Logistics evolution.

* Corresponding author: catrina.c@unitbv.ro

There are five independent activities that define logistics: customer satisfaction, inventory planning and management, supply, transportation and warehousing (figure 2).

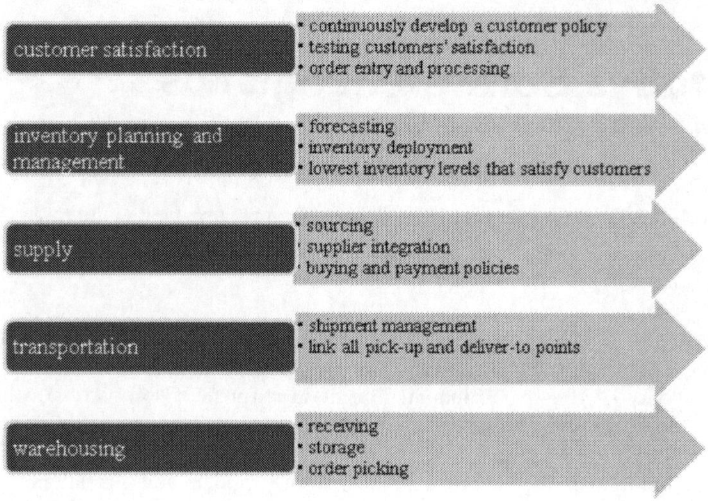

Fig. 2. Logistics activities.

1.2 Logistics software: market solutions

There are on the market lots of software oriented on different activities of logistics. According to site *Software advice* [2], the best rated logistics software are presented in Table 1.

Table 1. Logistics software [2].

No.	Software	Firm type[1]	Platform	Price[2]
1.	Freightview	S, M, L	Mac OS, Windows, Linux	*
2.	U Route	S, M	Mac OS, Windows, Linux	*
3.	SAP - SCM	L	Mac OS, Windows, Linux	*****
4.	UltraShipTMS	S, M, L	Mac OS, Windows, Linux	***
5.	S2K Warehouse Management Software	S, M, L	Mac OS, Windows, Linux	****
6.	ShippersEdge	S, M	Mac OS, Windows, Linux	*
7.	LogFire	S, M, L	Mac OS, Windows, Linux	*
8.	Cadence WMS	S, M, L	Mac OS, Windows, Linux	*****
9.	WISE	S, M	Mac OS, Windows, Linux	****
10.	LeanTMS	S, M, L	Mac OS, Windows, Linux	*****

[1] S=small; M=medium; L=large
[2] * = low, 5*=very expensive

Analysing de properties of the top ten logistics (figure 3) software it can be underlined that none of this software has a module for new companies, to select the appropriate supplier based on the technical properties of the products. There are no modules oriented to supplier offer, to automatically introduce the offer (based on particularised inquiry) and only two of them are designed for automotive industry.

Selecting the right suppliers becomes even more critical where the organizational strategy is to reduce its supply base and thereby deal with a smaller number of selected vendors. It is vital that the reasons for sourcing and supplier selection should follow agreed vendor appraisal criteria [3].

DOI: 10.1051/matecconf/20179402002

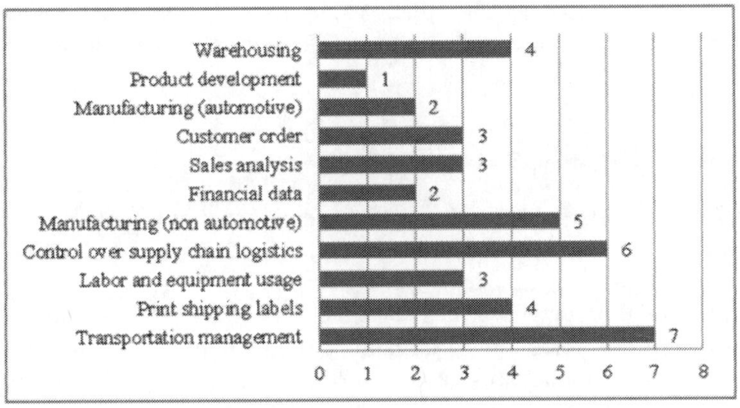

Fig. 3. Characteristics of market logistics software.

2 CADOProSys logistics module

Logistics module was developed [4-6] based on the market research of existing software and curriculum of undergraduates. Thus, the authors believed that the logistics module should have the structure presented in figure 4.

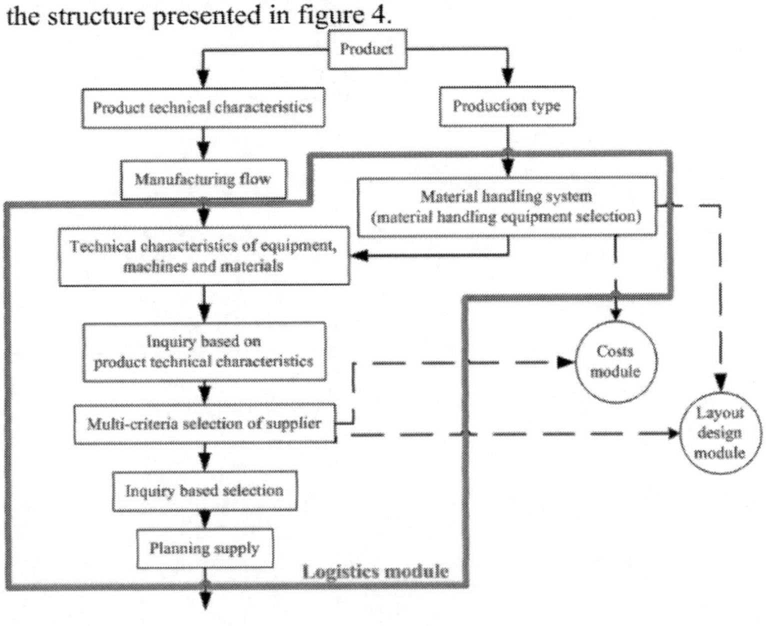

Fig. 4. Logistics module of CADOProSyst.

Designer of the production system is able to introduce the technical requirements for each component: raw material, materials, machine-tools, machines, material handling equipment, etc. Based on these data it is generated the target specifications matrix.

The selection of the supplier is done based on multi-attribute utility theory (MAUT) [7-10]. This theory is one of the most applied due to its ability to take into account both qualitative and quantitative factors, influencing supplier selection in the presence of risk and uncertainty. One of its advantages is the large number of potential suppliers (up to 500 prospective suppliers) and supplier selection attributes (up to 100 different attributes) that can be handled. The method rigorously measures value by identifying what is important (hierarchy), identifying relative importance (weights), identifying how well each alternative does on each criterion (score). The final value may be obtained either linearly or nonlinearly. CADOProSys allows the user to define hierarchy, weights and score, and the final value is determined automatically.

3 CADOProSys logistics application

Even if CADOProSys Software was developed especially for automotive industry, some of its modules may be used for any industry (there are module where some coefficients are those specific to automotive industry). Such a module is one focused on selection of suppliers. Thus, in the following, are presented the steps and results of using CADOProSys – Logistics Software for selection of organic cotton supplier, raw material used in a production system of ergonomic baby carriers.

First is generated the data sheet with product technical specification (figure 5a). This data sheet is send electronically to the suppliers, but it can be also generated as Excel sheet (figure 5b).

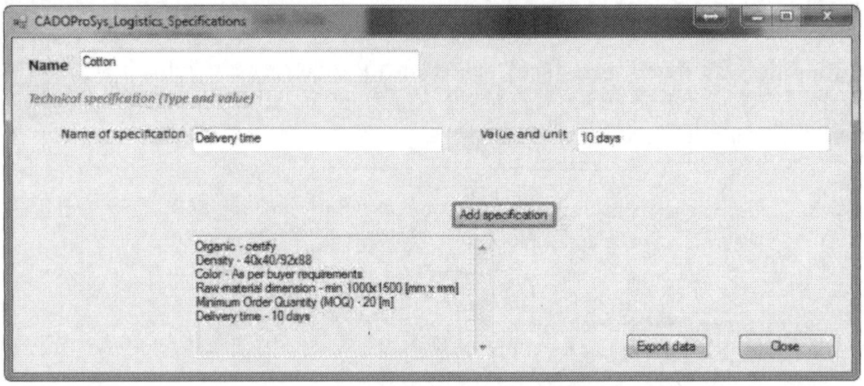

a)

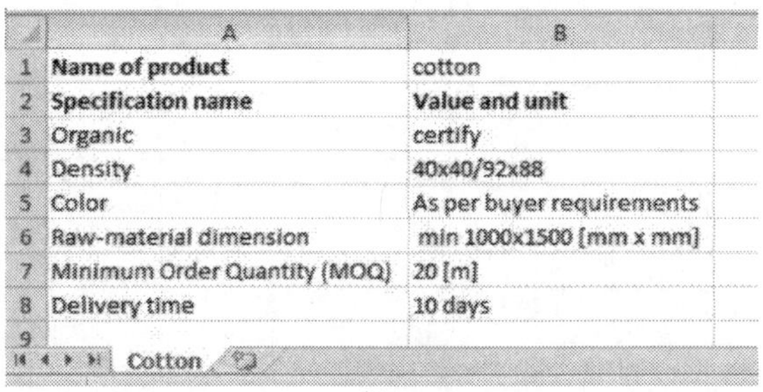

	A	B
1	**Name of product**	cotton
2	**Specification name**	**Value and unit**
3	Organic	certify
4	Density	40x40/92x88
5	Color	As per buyer requirements
6	Raw-material dimension	min 1000x1500 [mm x mm]
7	Minimum Order Quantity (MOQ)	20 [m]
8	Delivery time	10 days
9		

Cotton

b)

Fig. 5. Logistics module of CADOProSyst applied for organic cotton suppliers' selection: a) electronic data sheet; b) Excel data sheet.

Based on the above sheet (either on-line filled or Excel sheet) the next step is to upload data from each supplier in the selection module (a windows as that in figure 6 for each supplier).

 DOI: 10.1051/matecconf/20179402002

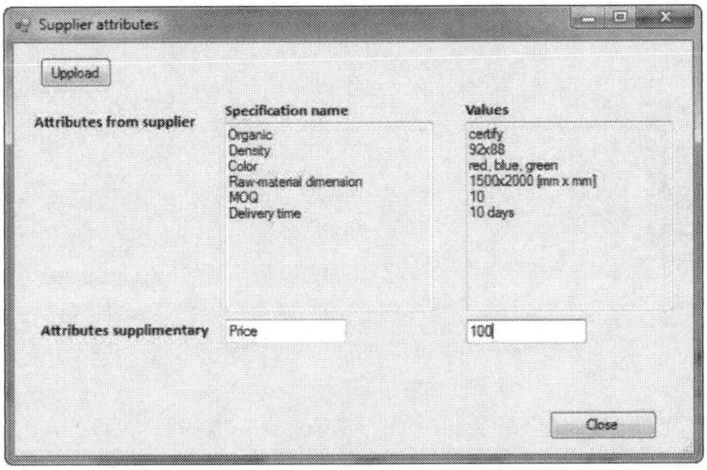

Fig. 6. Supplier attributes.

After all suppliers are defined from attributes point of view is generated a synthesis with the attributes and values (Figure 7).

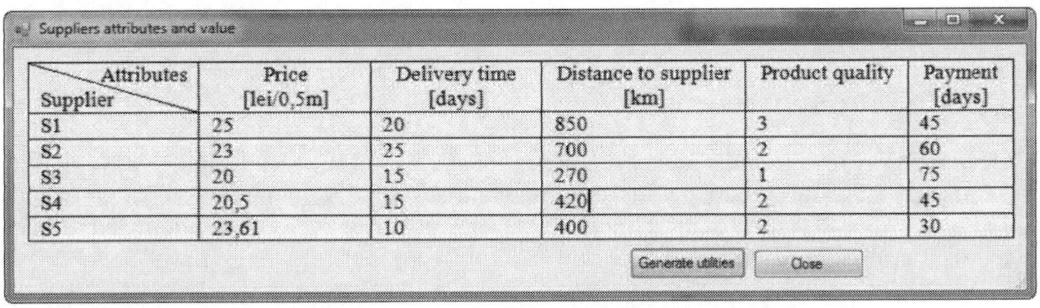
Suppliers attributes and value

Attributes / Supplier	Price [lei/0,5m]	Delivery time [days]	Distance to supplier [km]	Product quality	Payment [days]
S1	25	20	850	3	45
S2	23	25	700	2	60
S3	20	15	270	1	75
S4	20,5	15	420	2	45
S5	23,61	10	400	2	30

Generate utilities   Close

Fig. 7. Supplier attributes and values.

User is requested to define the weight on each attribute and the software will compute the score based on the values of the attribute with the highest weight (Figure 8).

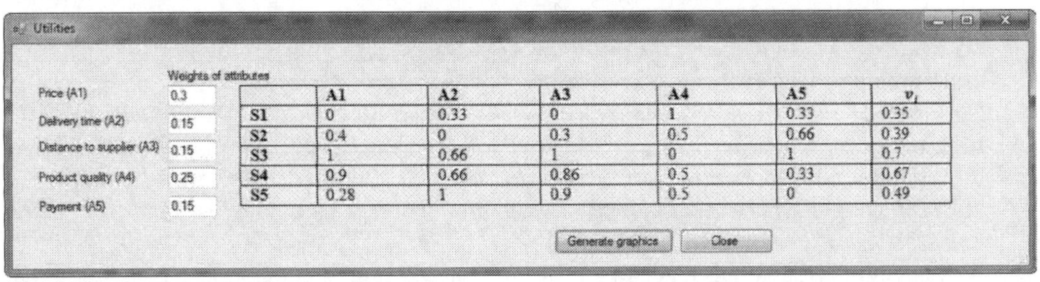
Utilities

Weights of attributes

Price (A1)	0.3
Delivery time (A2)	0.15
Distance to supplier (A3)	0.15
Product quality (A4)	0.25
Payment (A5)	0.15

	A1	A2	A3	A4	A5	v_i
S1	0	0.33	0	1	0.33	0.35
S2	0.4	0	0.3	0.5	0.66	0.39
S3	1	0.66	1	0	1	0.7
S4	0.9	0.66	0.86	0.5	0.33	0.67
S5	0.28	1	0.9	0.5	0	0.49

Generate graphics   Close

Fig. 8. Utilities.

CADOProSys – Logistics offers the possibility to obtain graphic representation of the results for the total value of utility (scores) but also charts per each attributes (Figure 9). Thus, based on the results shown in Figure 9, the optimum supplier, with higher utility score is S3, but this supplier did not obtained the highest score on each attribute. Thus, based on importance, on management policy, the firm may choose other supplier that satisfies better one attributes.

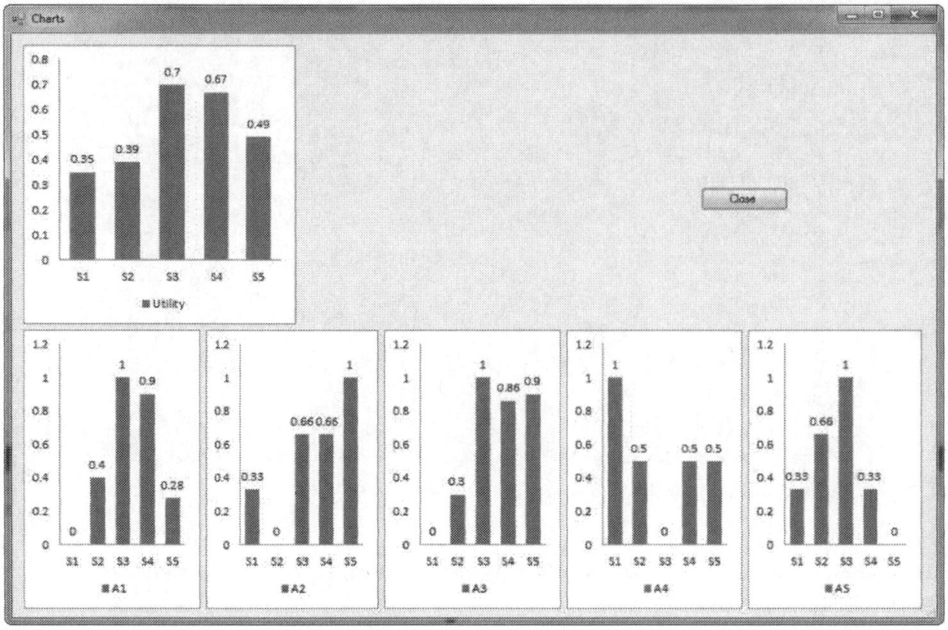

Fig. 9. Charts for total score and those for each attribute.

4 Conclusions

CADOProSys is user friendly and, from educational point of view is designed according to the curricula. Logistics module, with its component designed for suppliers selection reduce the time of this selection with almost 40%, has no error of computation and it can dynamically adapt to requirements.

References

1. E. Frazelle. *Supply chain strategy. The Logistics of Supply Chain Management.* (McGraw-Hill PH., 2002)
2. Software Advice. http://www.softwareadvice.com/scm/logistics-comparison (automotive industry).
3. F. Harrison. *Supply chain management workbook.* (Butterworth Heinemann P.H, 2001)
4. E. Klimczak. *Design for Software: A Playbook for Developers.* (Wiley P.H., 2013)
5. P. Tchounikine. *Computer science and educational software design* (Springer, 2011)
6. H. Min. *The Essentials of supply chain management new business concepts and applications.* (Pearson P.H., 2015)
7. A. Langevin, D. Riopel. *Logistics systems: design and optimisation.* (Springer P.H., 2005)
8. J. Baron. *Thinking and deciding.* (Cambridge P.H., 2008)
9. S. Budidedu. *Supplier selection by using multi criteria decision making methods.* (LAP, 2015)
10. R. M. Garfamy. *Supplier selection and business process improvement: an exploratory multiple case study.* (LAP, 2010)

CADOProSys Software. General structure

Catrina Chivu[1,*], *Cătălin-Iulian* Chivu[1], and *Magdalena* Barbu[1]

[1]Engineering and Industrial Management Department, Transilvania University of Brasov, 500174 MihaiViteazu, Brasov, Romania

Abstract. Design of flexible manufacturing systems is a challenge based on the field, structure, medium and long term management strategies of the production system. On the market there are many ERP software designed for integrated management. Generally, this type of software is not focused on manufacturing lines, managing only the documents inside the manufacturing systems, being not appropriate for design. Anyway, all the applications from the market assumed that the user knows how to design PLM software. This assumption was the motivation for the authors of the present paper to design a computer aided application that will help the beginners, the graduates, to design a manufacturing line. The software includes many modules. Thus, there are modules for designing: space, number of machine-tools and/ or equipment, human resource, layout, production planning, logistics (supply-chain and material handling), automation. The software was designed especially for automotive industry, industry that dominates the Brasov area. Present paper is focused on the general structure of the CADOProSys (Computer Aided Design and Optimisation of Production Systems) Software, including the logic diagram and some opportunities provided by the application.

1 Introduction

1.1 Market solutions

For any company, to be competitive via market implies to make difficult decisions, to be as much as it can, innovative. Obviously, innovation requires investment (time and capital): it is risky, and return on investment can span several years. Thus, the majority of products introduced into the market are only marginally different from their competitors and rarely survive beyond an initial period.

In a business plan, designing and planning the new production system is one of the most important parts of the start-up. Is a challenge process and requires experience and sometimes intuition because is all about productivity, quality, e-business, local/& global competition and customer service, or in other words is about operations management.

Any business organizations, but especially an automotive one, have three basic functional areas: finance, marketing and operations (figure 1). "Finance" is responsible for securing financial resources, allocating those resources, budgeting, analysing investment

*Corresponding author: catrina.c@unitbv.ro

proposals, and providing funds for operations. "Marketing" is responsible for assessing consumer wants and needs, selling and promoting the organization's goods. "Operations" is responsible for producing the goods [1].

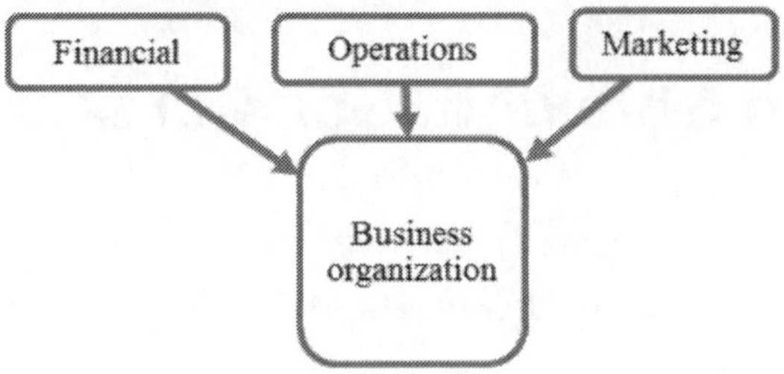

Fig. 1. Functional areas of a business organization.

Based on these areas, on the market, there is a lot of ERP (Enterprise Resource Planning) software, focused on company's management. Thus, these systems gather, under the umbrella of a single IT platform, human resources, know-how, software and hardware. So whether we are talking about sales, purchasing, production, finance or accounting department, all these departments within a firm can be integrated into a single system.

According to a specialised website [2], first ten most used software, in the field of manufacturing, are listed in Table 1.

Table 1. Software used for management in the field of manufacturing [2].

No.	Software	Firm type[1]	Platform	Price[2]
1.	E2 Shop System Software	S, M	Windows	*
2.	MIE Trak PRO	S, M	Mac OS, Windows, Linux	*
3.	Global Shop Solutions	S, M	Mac OS, Windows, Linux	**
4.	EnterpriseIQ	S, M	Mac OS, Windows, Linux	**
5.	QT9 Quality Management	S, M, L	Mac OS, Windows, Linux	*
6.	SAP Business One	S, M, L	Mac OS, Windows, Linux	***
7.	Oracle JD Edwards - Manufacturing	S, M, L	Mac OS, Windows, Linux	****
8.	Synergy SPC	S, M, L	Mac OS, Windows, Linux	*
9.	OmegaCube ERP	A	Mac OS, Windows, Linux	*
10.	Infor VISUAL ERP	S, M	Mac OS, Windows, Linux	***

[1]S=small; M=medium; L=large
[2]* = low, 5*=very expensive

All these package offer management for both discrete and repetitive manufactures in the automotive. There are managed accounting, business analytics human resources, inventory control, invoicing, quality control & inspection, supply chain, repair & maintenance, etc. Each of this software may be used for complete enterprise resource planning and manufacturing execution functionality including generating shipping labels and packing slips and then tracks shipments online to multiple shipping addresses.

All these software are focused in managing existing manufacturing systems, with fewer instruments for designing new systems. From digital manufacturing solutions point of view, there is very few number of software, most of the being "object" oriented. Thus, there is software for assembly process simulation (e.g. DELMIA) [3]. One of the most powerful and complex instrument is offered by Siemens, through Tecnomatix software [4]. This software allows designing innovative systems based on product engineering, manufacturing engineering and production. It is a very complex instrument that may be used to increase productivity through planning and to manage manufacturing by optimising subsystems. The most disadvantageous aspect of this product is the complexity, which make it usable only

 DOI: 10.1051/matecconf/20179402003

by specialists, not beginners in the field of manufacturing system design. Thus, if we are talking about undergraduates, they do not have all information and knowledge to be able to accurately and realistically design a manufacturing system with Tecnomatics because they do not control the technical parameters involved in design. This is the main reason that led to the development of software, enabling the undergraduates to design a manufacturing system.

1.2 Product design concept

Product development or product design, regardless of the type of product, even software, implies a set of activities beginning with the market opportunity and ending in the production, sale and delivery of the product.

There is no ideal recipe to design software. Basically, almost half of the methods are focused on how the template of a program is derived from the description of the input data [5-9]. The methods used in developing software may be grouped as shown in Figure 2 (as difficulty increases).

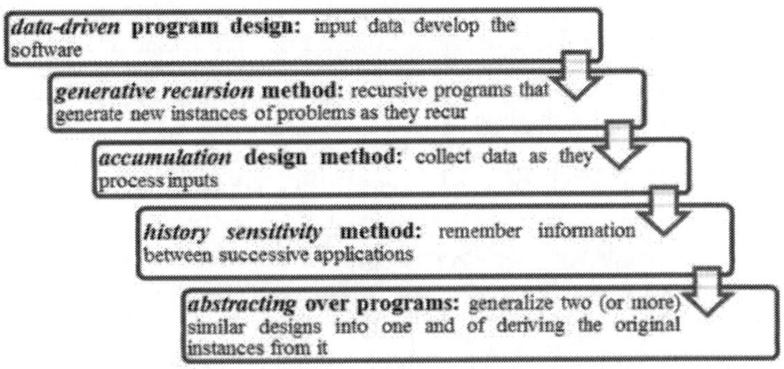

Fig. 2. Software design methods.

Developing CADOProSys Software the authors combine methods, based on program's level. Thus, at the beginning is used the data-driven method and, during the stages are used accumulation or history method.

2 CADOProSys Software general structure

CADOProSys Software was developed as educational software based on the need of both teachers and undergraduate to obtain production systems designed as accurate as possible. As any educational software, CADOProSys was specifically designed to lead a learner to develop an activity, not only to execute one. It was designed to satisfy both pedagogical and teaching setting.

There are two perspective based on which the software was developed: teachers and students. Teachers' requirements for this software were:

- support to supervise students (T_R1);
- challenging students (T_R2);
- designed for formal and informal learning (T_R3);
- determine students to argument and make decision (T_R4);
- teaching principles not formula (T_R5);
- develop the ability to design a complex production system (T_R6);
- explain the correlation between parameters (T_R7).

From students' point of view, the software should be:

- user friendly (SR1);
- have tutorials (SR2);
- comprehensible (SR3);
- steps imposed by software in designing production systems to be correlated to curriculum (SR4);
- correlation to real manufacturing systems (SR5).

Thus, the QFD chart [10] (quality function development) is shown in Table 2.

Table 2. QFD chart for CADOProSys Software.

		Teachers' requirements							
		T_R1	T_R2	T_R3	T_R4	T_R5	T_R6	T_R7	Benchmark* (Tecnomatix)
Students' requirements	S_R1	X		X				X	S
	S_R2							X	M
	S_R3	X	X	X	X	X	X	X	M
	S_R4	X	X	X		X	X		S
	S_R5		X		X		X		M
			3 levels of difficulty	>3 self-& final assessments	> 3 different options		3 levels of complexity	hint for each formula	
		Teachers' targets							

* S = strong competitive position; M = medium competitive position; W = weak competitive position

CADOProSys Software was developed based on the production systems' design principles and correlated with QFD chart identified and presented in § 2.1. of the present paper. It was designed to be able to determine: typology of manufacturing system; technical and economical parameters of manufacturing system; human resource design; layout; logistics (supply-chain and material handling); costs (Figure 3).Software allows the students to both design new production systems and develop an existing system through implementing new product, or increasing the production.

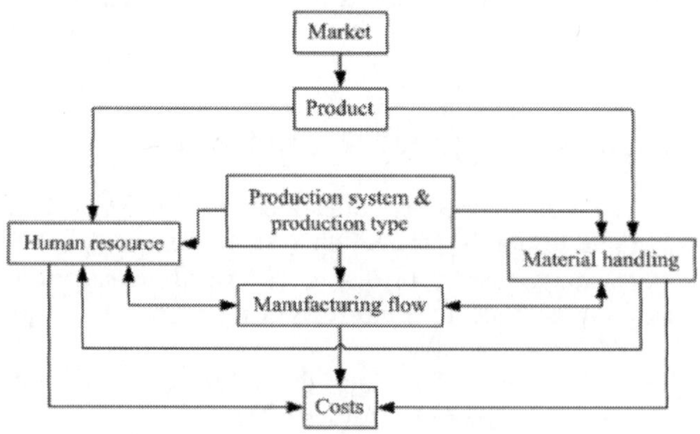

Fig. 3. Modules of CADOProSys Software.

Each module is a combination of requirements and results that interconnect to other module or direct with the conclusion page. Because CADOProSys Software is an educational one, after each selection done by students are synthetized the generated restrictions. As example, in figure 4 is presented the window that allow selection of production system type and explained what will determine in other modules (some options will be inactive). Thus, will be automatically activated the modules oriented on production type (Figure 5), in material handling module will be disabled the option "no material handling requirement". In this way, students will not be able to uncorrelate the production system type and production type or material handling equipment, increasing the quality of design and decreasing the design time.

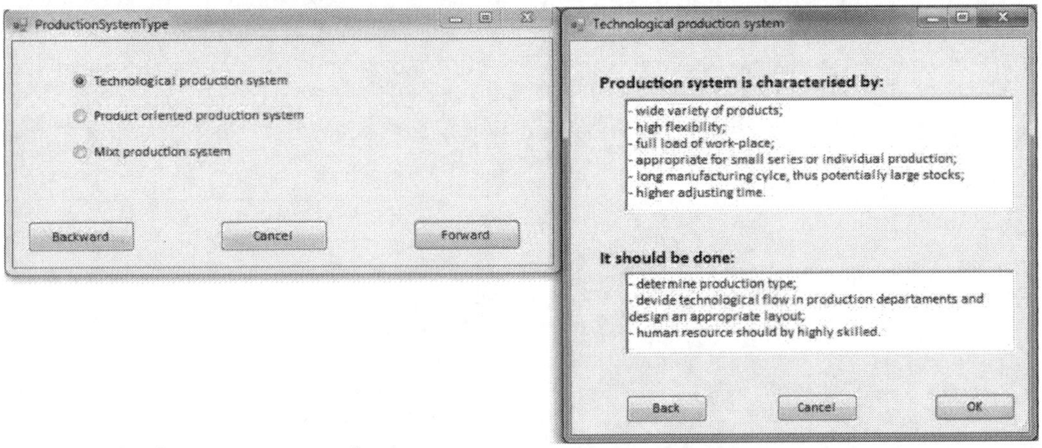

Fig. 4. Production systems type selection.

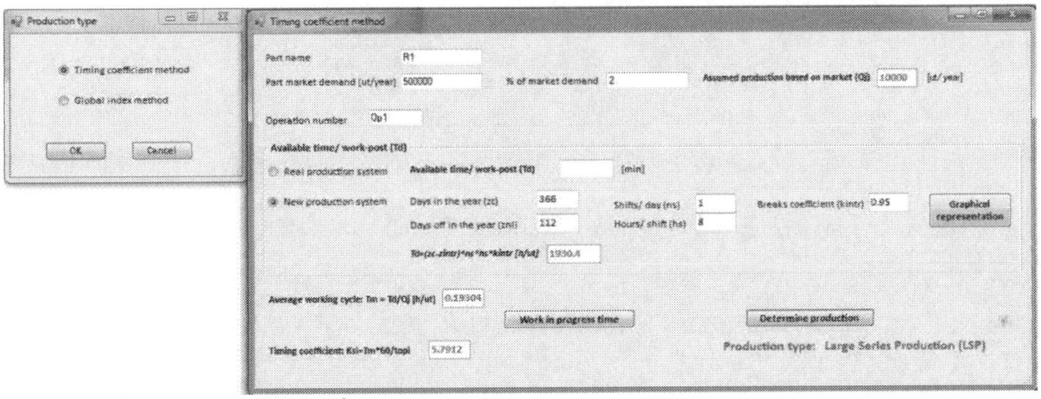

Fig. 5. Production type computation.

Modules of the CADOProSys are complex and allow the user to obtain data as final results shown on the active window and/ or Excel files that may be used for other purposes. Such an example is the production type computation module that generates an Excel file that contains data for each operation for each product (Figure 6, based on data shown in Figure 5). This module has also a submodule that allows either computation of work-in progress time or input this time based on firm's normative.

	A	B	C	D	E	F	G	H	I	J	K
1	Part name	Part market demand	% market demand	Qj							
2	R1	500000	2	10000							
3											
4	Operation number	zc	znl	ns	hs	kintr	Td	Tm	Topi	Ksi	Production type
5	Op1	366	112	1	8	0.95	1930.4	0.19304	2	5.7912	Large Series Production (LSP)
6	Op2	366	112	1	8	0.95	1930.4	0.19304	1	11.5824	Medium Series Production (MSP)
7											

Fig. 6. Production type computation: Excel file.

3 Conclusions

CADOProSys Software is in continuous development and testing, adapting to students curriculum. The development of the software was started only last year, but some of its module was already tested with the undergraduates and graduates. The tests were oriented on three directions: ability to transmit knowledge (Figure 7); capability of the students to assimilate these knowledge and time consumption. The most relevant results were obtained after tested the graduates (they already have knowledge and are more critical to the software because they are able to see better the details). The tests were done due to the main objective of the software: to be educational and adapted to the curriculum and students' requirements.

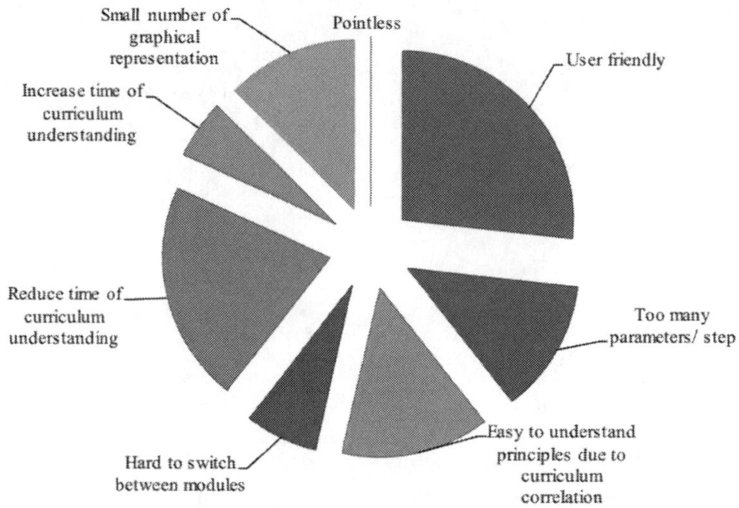

Fig. 7. Graduates and undergraduates opinions relative to CADOProSys ability to transmit knowledge.

As it can be seen from the answers given by graduates and undergraduates, CADOProSys should be improved in the following directions: reduce number of parameters/ step (by creating more windows/step); introduce more graphical representation. In conclusion, from respondents' point of view the software was useful in learning process and reduces time of computation.

References

1. W. Stevenson, *Operations Management* (McGraw-Hill PH., New York, 2012)
2. Software Advice. http://www.softwareadvice.com/manufacturing
3. Digital Manufacturing Solutions. DELMIA. http://4dmsco.com/
4. Siemens Tecnomatix https://www.plm.automation.siemens.com/en_us/products/tecnomatix/
5. E. Klimczak, *Design for Software: A Playbook for Developers* (Wiley P.H., Chichester, 2013)
6. N. Slack, A. B. Jones, R. Johnston, *Operations management* (Pearson P.H., London, 2013)
7. D. Unger, S. Eppinger, CJEN **22**, 689 (2011)
8. D. Cochran, A. Chu, M.U. Jafri, Z. Bi, JIII **2**, 65 (2016)
9. A. Thecle,Y. Benama, N. Perry, Adv. PMS **440**, (2014)
10. P. Tchounikine, *Computer science and educational software design* (Springer, Berlin, 2011)

Innovative analysis of HS 6-5-2 drills edges quality

Raluca Daicu[1,*], *Valentin* Dițu[1]

[1]Transilvania University of Brașov, Department of Manufacturing Engineering, Mihai Viteazu No.5, Brașov, Romania

Abstract. The paper is analyzing the quality of Ø8 drills edges from HS 6-5-2 steel, using an innovative technique, the measurement of electrical current at cutting. The method is useful to take the right decision about drills acquisition from different suppliers based on the quality/price ratio. Also, it can be made a fast quality comparison of drills that leads to a selection of the acquisition places. The analysis of a drill batch is done using the microscope and the measurement of the electrical current at cutting, showing the better accuracy of the last method.

1 Introduction

The estimation of the cutting tool edge wear and tool life is high due to large scale using of the numerical command machines and the manufacturing flexible lines, and these need to estimate the moment when the cutting tool has to be changed.

Another important direction about the estimation of cutting tool edge life is the quality control of the cutting edge in the series production, or when acquisitioning a batch of tools by a beneficiary, or at the comparison of different cutting tools from different producers to study the quality/price ratio and decide from where is profitable to buy the cutting tools that an industrial enterprise needs.

The methods used until now to estimate the cutting edge wear and tool life need many experiments, great material and time consumption and are expensive. Thus, it is necessary to find a method that does not have the disadvantages of the present methods.

It is known the fact that when cutting a metal with an edge of a good electrical conducting material (high speed steel, metallic carbide, etc.) it appears an electrical current as consequence of the next physical effects and phenomenon: the Seebeck effect, the Thomson effect, the phenomenon of emission in metals [1-10]. Because these physical effects and phenomenon are generally based on the heat generation in the cutting process, and the wear of the cutting tool clearly depends on it, results that the measurement of the electrical current at cutting leads to a good appreciation of the cutting tool wear state.

In paper [1] is shown the fact that at turning S275JR steel with two SPMR 150612-P30 cutting inserts, in the same conditions, there were obtained different values for the initial voltage of the electrical current at cutting, so, for the first insert it was obtained U = 12.3 mV and for the second insert U = 13 mV. If different values for the voltage of electrical

* Corresponding author: raluca.daicu@unitbv.ro

current at cutting were obtained results that in the cutting zone there are different temperatures. This can be explained by the fact that the value of the electrical current at cutting is higher due to the imperfections on the cutting edge. Normally, this type of edge should be wear out faster, so it should have a shorter life. To verify this hypothesis there were made tries at wear, on S275JR steel, for the next cutting parameters: s = 91.1 m/min, f = 0.302 mm/rev, a =2 mm. The S275JR steel has the next chemical composition: C = 0.22%, Mn = 1.15%, P = 0.055%, S = 0.055%, and the tensile strength is 440 Mpa (44 daN/mm^2). The obtained results are graphically shown in figure 1.

If the admissible wear is $VB_{B\,adm.}$ = 0.4 mm the next tool life values were obtained: T_I = 12.06 min; T_{II} = 8.55 min.

Taking as base the tool life of the cutting insert with U = 12.3 mV it can be observed a decreasing of the tool life for the cutting insert with U = 13 mV of 29%.

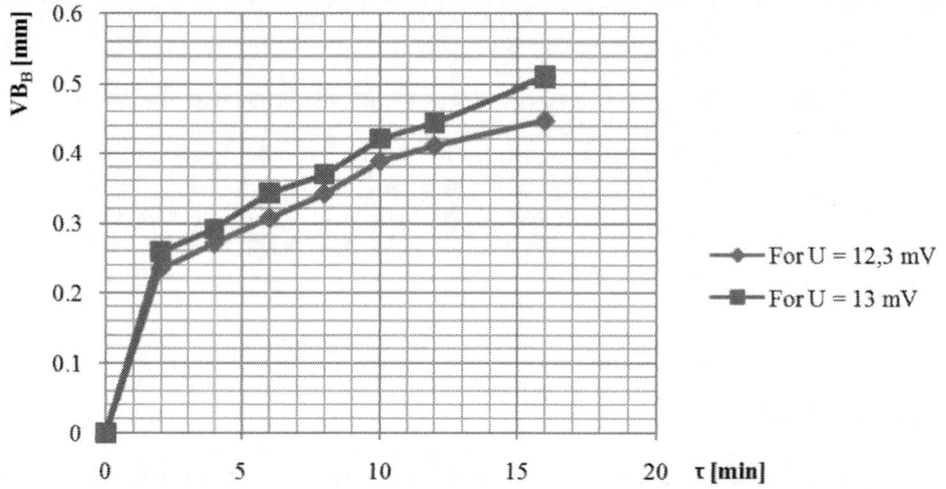

Fig. 1. Graphics for tries at wear (after [1]).

Also in paper [1] is shown the connection between electrical current at cutting and wear of cutting edge at turning 1C45 steel with a SPMR 150612 - P30 metallic carbide insert, with geometry $\alpha = 5^0$; $\gamma = 6^0$; $\chi = 45^0$ resulting the relation 1.

$$VB_B = 112.41061 * s^{0.946447} * (U - 3.567 * s^{0.306} * f^{0.117} * a^{0.097}) \quad [mm] \qquad (1)$$

where,
VB_B [mm] - wear on the side surface;
s [m/min] – cutting speed;
f [mm/rev] – cutting feed;
a [mm] – cutting depth;
U [mV] – voltage of the cutting electrical current.

The 1C45 steel has the next chemical composition: C = 0.45%, Si = 0.4%, Mn = 0.5%, P = 0.045%, S = 0.045%, Cr = 0.4%, Mo = 0.1%, Ni = 0.4%, and the tensile strength is 660 Mpa (66 daN/mm2).

In the same paper it is shown the connection between the tool life of the cutting edge and the cutting electrical current at turning the AlCu 4MgTi aluminum alloy heat treated, with an edge from HS 18-1-1-5 steel and another edge from metallic carbide K10. The obtained relations are (2) and (3).

$$T = \frac{101}{s^{0.745} \cdot f^{0.96} \cdot a^{0.2} \cdot U} \quad \text{[min]} \tag{2}$$

$$T = \frac{511.3}{s^{0.85} \cdot f^{0.39} \cdot a^{0.12} \cdot U} \quad \text{[min]} \tag{3}$$

where, T [min] - tool life.

The presented information show the fact that with the measurement of the electrical current at cutting it can be diagnosed the wear state of a cutting edge and also the fact that the specialty literature was focused more on the turning process and less on other processes like drilling.

2 The analysis of the connection between the value of the electrical current at drilling and the quality of the cutting edge

The initial quality of a cutting edge, as shown in section 1, can be appreciated by the value of the voltage of initial electrical current at cutting. The higher the voltage is, the higher the temperature is, due to the imperfections of the cutting edge. In this section is analyzed the connection between the initial imperfections of the drill's edge and the initial value of the cutting electrical current voltage.

There were purchased 30 drills with Φ8 mm diameter, the material of the cutting edge being HS 6-5-2 steel. For shortening the necessary time for experiments, the drilling was done on the 41MoC11 steel (tensile strength – 950 MPa), with the drilling machine 6GCODA1. The electrical current at cutting was measured with the installation from figure 2.

Fig. 2. Experimental setup for measuring the electrical current at drilling.

In table 1 are centralized the results obtained for initial electrical current at cutting, for all the 30 drills.

Table 1. Obtained values for initial electrical current voltage at drilling 41MoC11 steel.

No. of drill	Drill diameter [mm]	Spindle speed [rev/min]	Feed [mm/rev]	Voltage "U_0" of electrical current at cutting [mV]
1	8	560	0.25	0.5
2				0.5
3				0.4
4				0.3
5				0.3
6				0.6
7				0.2
8				0.3
9				0.4
10				0.2
11				0.4
12				0.4
13				0.5
14				0.6
15				0.6
16				0.5
17				0.3
18				0.2
19				0.2
20				0.6
21				0.4
22				0.4
23				0.6
24				0.8
25				0.4
26				0.7
27				0.6
28				0.6
29				0.5
30				0.6

Using the microscope there were examined the edges of each drill in order to find manufacturing defects and to explain the value of the cutting electrical current voltage.

For a drill there were visualized the cutting lips and the chisel edge identifying if they have defects or not.

The sequence of the experiments steps was the next: microscope visualization – take decision about drill quality – measuring the electrical current at drilling – take decision about drill quality based on the voltage value of the electrical current at cutting.

The drills were grouped in 3 categories :

- Drills quality I : $U_0 = 0.1 \div 0.3$ mV – 7 pieces
- Drills quality II : $U_0 = 0.4 \div 0.5$ mV – 13 pieces
- Drills quality III : $U_0 = 0.6 \div 0.8$ mV – 10 pieces

In figure 3 are images of the two cutting lips and the chisel edge for drill no. 10 that had the voltage of the electrical current at cutting of $U_{0,10} = 0.2$ mV (drill quality I).

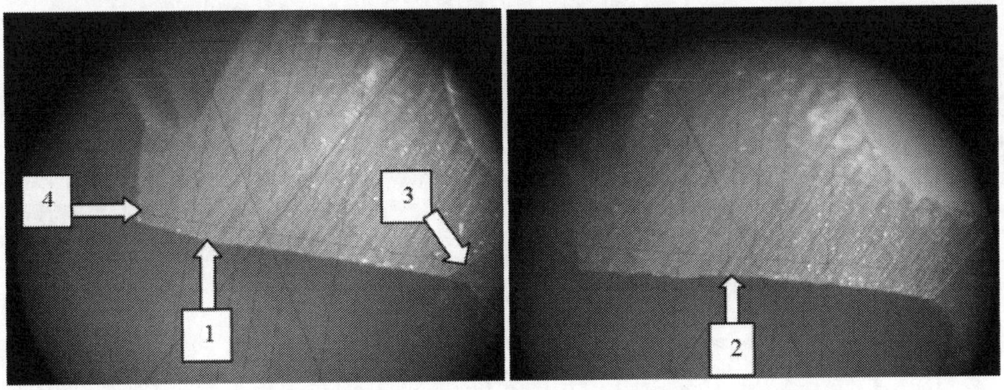

Fig. 3. Cutting lips (1, 2), chisel edge (3) and corner (4) for drill no.10 from table 1.

From figure 3 it can be seen the fact that the cutting lips and the chisel edge are adequate from the shape and roughness point of view. Also it can be seen that the corners of the drill are very good. All drills should look like this one.

In figure 4 are images of the two cutting lips and the chisel edge for drill no. 29 that had the cutting electrical current voltage of $U_{0,29} = 0.5$ mV (drill quality II).

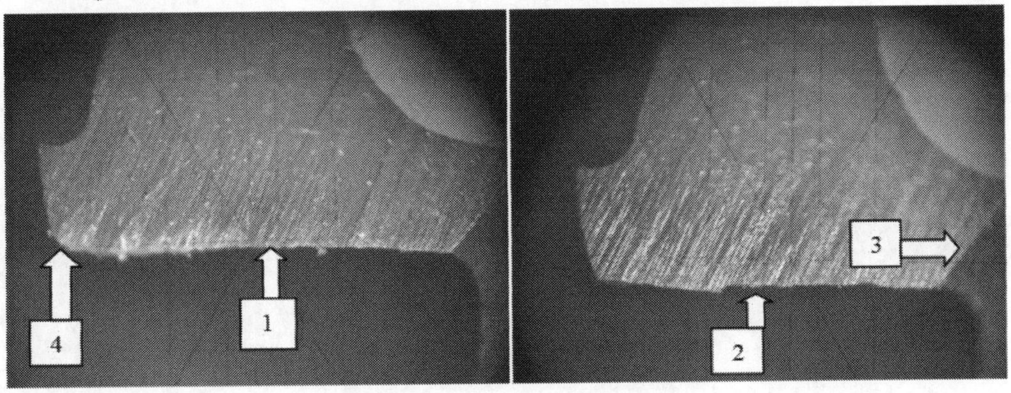

Fig. 4. Cutting lips (1, 2), chisel edge (3) and corners (4) for drill no.29 from table 1.

Analyzing the images from figure 4 it can be seen that the chisel edge (2) is very good sharpened, the drill corners (4) are one good and one rounded, fact that leads to an increase of the temperature in the cutting zone, that also leads to the increase of the electrical current at cutting, and the sharpening of the cutting lips (1, 2) is not perfectly done. All these observations explain why the voltage of electrical current at cutting has a higher value than the drill no. 10 presented in figure 3.

In figure 5 are presented images of cutting edges from drill no. 14 that had an initial electrical current of $U_{0,14} = 0.6$ mV.

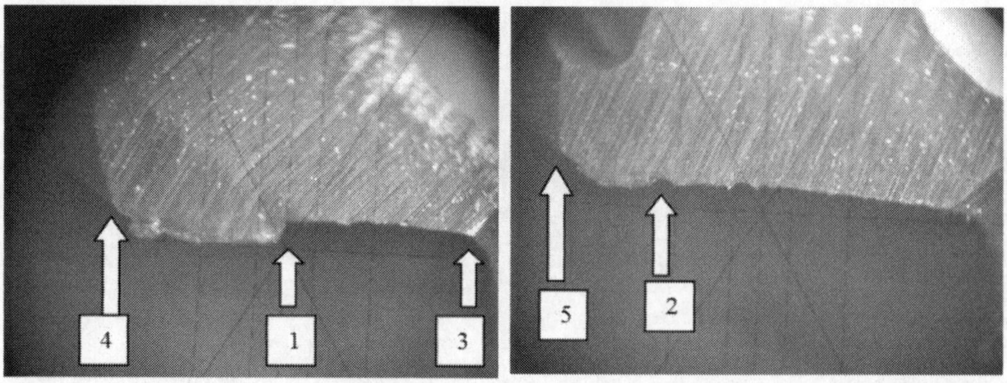

Fig. 5. Cutting lips (1, 2), chisel edge (3) and corners (4, 5) for drill no.14 from table 1.

Analyzing the images from figure 5 it can be seen that the chisel edge (3) is partially good sharpened, the cutting lips (1, 2) and corners (4,5) are improper, fact that explains the increasing temperature in the cutting zone and also of the cutting electrical current voltage.

In figure no. 6 are shown microscope images of drill no. 11 that had a voltage of the electrical current at cutting of $U_{0,11} = 0.4$ mV.

It can be seen an interesting thing, the fact that at sharpening it was used an abrasive wheel with coarse-grained and the roughness of the land surfaces clearly increased.

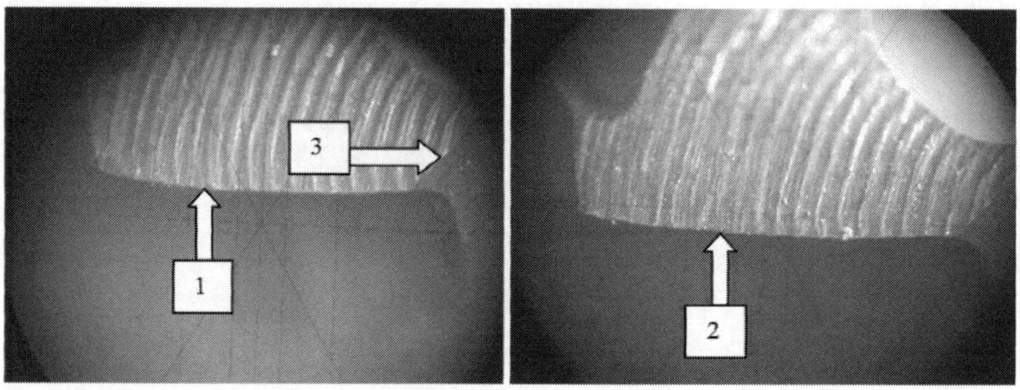

Fig. 6. Cutting lips (1, 2) and chisel edge (3) for drill no.11 from table 1.

Even if it can be seen that the sharpening was done properly, the voltage of the electrical current increased so the drill was not in the quality I group, but in the quality II group.

3 Conclusions

Due to the experiments, the next conclusions can be drawn:
- The appearance of the electrical current at cutting is known since 1950 [2], but it's utilization in practical applications was recently developed;
- The electrical current at cutting is an innovative method for the quality appreciation of the drills edges and it can be easily used;
- From the analyzed drills batch, 23.33% are quality I, 43.33% quality II and 33.34% quality III; in this situation the drills batch has a satisfying quality;

- If the drills are analyzed only at microscope it can be made an error in the appreciation of the drill batch quality of 14%;
- The electrical current at cutting is sensible to any imperfection of the drill edge;
- The electrical current at cutting has shown an improper roughness obtained at sharpening due to coarse-grain of the used abrasive wheel;
- If the life of the drills from each quality group is determined, than it can be made an analysis of the quality/price ratio.

References

1. V. Dițu, AJME **1**, 63 (2008)
2. R. Daicu, G. Oancea, AMM **808**, 40 (2015)
3. H. Bagchi, S.K. Basu, Wear **26**, 39 (1973)
4. V. Dițu, WSEAS **2**, 322 (2009)
5. A.L. Plotnikov, E.G. Krylov, E.M. Frolov STIN **1**, 2 (2009)
6. M. Murata, S. Kurokawa, O. Ohnishi, CSME **79**, 2546 (2013)
7. M. Murata, S. Kurokawa, O. Ohnishi, M. Uneda, T. Doi, CSME **6**, 958 (2012)
8. H. Hirota, M. Murata, A. Kojima, K. Yanagihara, JSPE **64**, 1191 (1998)
9. V. Dițu, AJME **8**, 18 (2010)
10. R. Tanaka, A. Hosokawa, K. Yamada, T. Ueda, Proceedings of the Seventh International Conference on Progress of Machining Technology, **786** (2004)

Researches on evaluation of smooth entrance in cutting using electrical current. A case study

Valentin Dițu[1], *Gheorghe* Oancea[1], *Raluca* Daicu[1,*]

[1]Transilvania University of Brașov, Department of Manufacturing Engineering, Mihai Viteazu No.5, Brașov, Romania

Abstract. At metal drilling, at the beginning of the process, the cutting tool in the chisel edge zone, frictions and plastically deforms the material, and then enters in cutting. If the drill is entered smoothly in cutting with a special device, its life is increased. The paper presents how the electrical current at cutting is used for the evaluation of smooth entrance in cutting in comparison with classical cutting. These are demonstrated for 41MoC11 steel drilled with a set of 30 tools having 8 mm diameter.

1 Introduction

The drill is the cutting tool that works in the most severe conditions comparatively with the other cutting tools because:
- the cutting speed varies from zero in the center of drill to the maximum value at the exterior diameter;
- due to the functional clearance angle, that has the zero value from a diameter of approximate 0.3 mm, the drill cannot cut up to its axis;
- the resulted heat at drilling is concentrated in great proportion (minimum 40%) in the cutting tool.

These reasons motivate taking any measures that can increase the tool life. One of these measures is the smooth entrance in cutting.

Smooth entrance in cutting is considered, at drilling, when at a certain diameter of the drill, the cutting force is equal with the cutting resistance, the chip is in balance and the vibrations are damped. At a certain diameter of the hole, the minimum axial force needed for cutting depends, when the other parameters are constant, on the ratio between the cutting lip and the chisel edge. If the material is not homogenous, then the axial feed should depend on the material hardness, but the cutting feed is constant at ordinary processes, fact that leads to variable cutting forces and vibrations, due to the material hardness variations. If the vibrations are damped, then the tool life of the drill increases [1].

*Corresponding author: raluca.daicu@unitbv.ro

 DOI: 10.1051/matecconf/20179402005

2 Efficiency appreciation of drill smooth entrance in cutting

It is known that the state of a cutting edge can be appreciated by the value of cutting electrical current voltage [2-8]. The higher this value is, the higher the initial temperature is [9, 10], due to the imperfections of the cutting edge.

In this section it is analyzed the value of the electrical current for classical drilling and for drilling with smooth entrance in cutting.

There were purchased 30 drills with Φ8 mm diameter, the cutting edge material being HS 6-5-2 steel. To decrease the necessary time for the experiments, the drilling was done on the 41MoC11 steel (tensile strength – 950 MPa), with the 6GCODA1 drilling machine.

The electrical current at cutting was measured with the experimental stand from figure 1 for classical drilling, where 1- mandrel for fixing the drill, 2 – collector brushes, 3 – measuring instrument for electrical current voltage at cutting, and with the experimental stand from figure 2 for smooth entrance in cutting.

Fig. 1. Experimental stand used in measuring process of the electrical current at classical drilling.

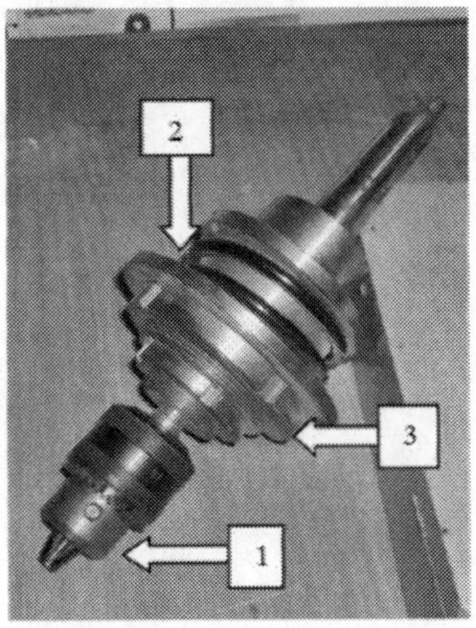

a) Experimental stand for smooth entrance in cutting b) Device for smooth entrance in cutting
Fig. 2. Experimental stand used in measuring process of the electrical current at drilling with smooth entrance in cutting.

The device for smooth entrance in cutting (fig. 2, b) has the next main elements: 1 – mandrel for fixing the drill, 2 – compression arc, 3 – element for adjusting the compression arc.

There were examined at microscope the edges of each drill in order to find manufacturing defects and to explain the value of the cutting electrical current voltage.

For a drill there were visualized the cutting lips (two) and the chisel edge identifying if they have defects or not.

In researches, the sequence of the experimental steps was the following: microscope visualization, – taking the decision about drill quality, – measuring the electrical current at normal drilling, – taking the decision about drill quality based on the voltage value of the electrical current at cutting, – measuring the electrical current at drilling with smooth entrance in cutting and establishing the final conclusion.

To measure the values of the electrical current the following cutting parameters were used: s = 14.07 m/min; f = 0.25 mm/rev; a = D/2 = 4 mm. Also the axial cutting force was estimated (F_x = 2560 N), according to [11], in order to adjust the device for smooth entrance in cutting.

In Table 1 there were centralized the obtained results for the electrical current at cutting for normal drilling and for drilling with smooth entrance in cutting, for the considered drills.

Table 1. Obtained values for cutting electrical current voltage at drilling 41MoC11 steel, classically and with smooth entrance in cutting.

No. of drill	Drill diameter [mm]	Spindle speed [rev/min]	Cutting speed [m/min]	Feed [mm/rev]	Voltage"U_0" of electrical current at normal drilling [mV]	Voltage "U_I" of electrical current for smooth entrance in cutting [mV]
1	8	560	14.07	0.25	0.5	0.3
2					0.5	0.4
3					0.4	0.3
4					0.3	0.2
5					0.3	0.3
6					0.6	0.4
7					0.2	0.1
8					0.3	0.2
9					0.4	0.3
10					0.2	0.1
11					0.4	0.2
12					0.4	0.2
13					0.5	0.3
14					0.6	0.3
15					0.6	0.6
16					0.5	0.4
17					0.3	0.2
18					0.2	0.1
19					0.2	0.1
20					0.6	0.4
21					0.4	0.3
22					0.4	0.3
23					0.6	0.4
24					0.8	0.5
25					0.4	0.2
26					0.7	0.5
27					0.6	0.4
28					0.6	0.5
29					0.5	0.4
30					0.6	0.6

DOI: 10.1051/matecconf/20179402005

Based on the obtained results the drills were grouped in 3 categories:
- drills quality I : $U_0 = 0.1 \div 0.3$ mV – 7 pieces
- drills quality II : $U_0 = 0.4 \div 0.5$ mV – 13 pieces
- drills quality III : $U_0 = 0.6 \div 0.8$ mV – 10 pieces

The data from Table 1 is graphically represented in Figures 3 and 4.

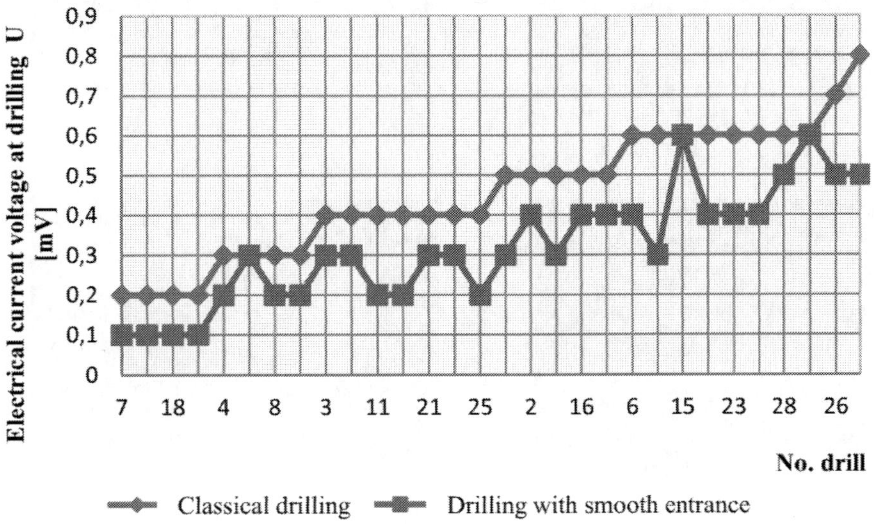

Fig. 3. Electrical current variation in ascending order of voltage.

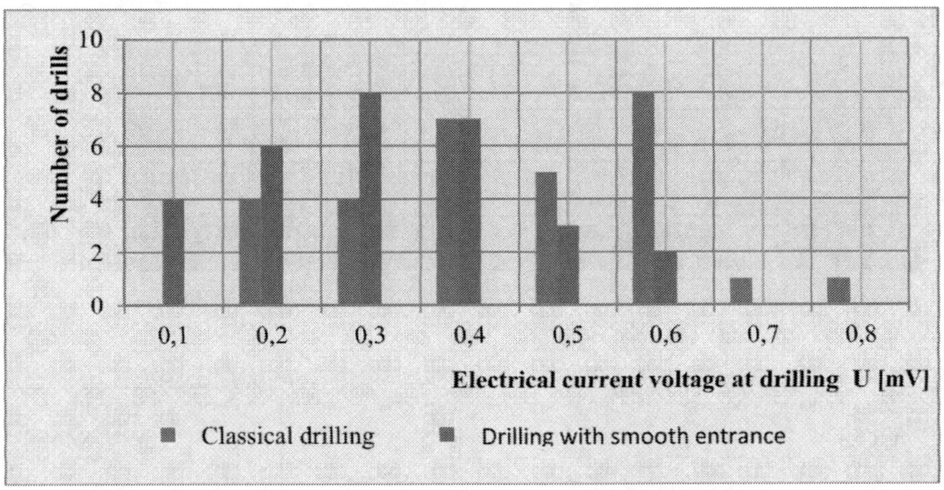

Fig. 4. Number of drills at the same value of electrical current at cutting.

The graphics presented in Figures 3 and 4 clearly show the fact that at smooth entrance of drill in cutting the voltage of the electrical current at cutting decreases, the temperature in the cutting zone also decreases, resulting the increasing of tool life.

There were observed 3 exceptions for drills no. 5, 15 and 30 for which the smooth entrance did not modify the value of the cutting electrical current voltage. To identify the reason for these exceptions, in Figures 5, 6, 7 and 8 are presented the cutting lip and chisel edges of drills no. 5 and no. 30 before normal cutting and after processing 10 mm hole length when the smooth cutting began.

 DOI: 10.1051/matecconf/20179402005

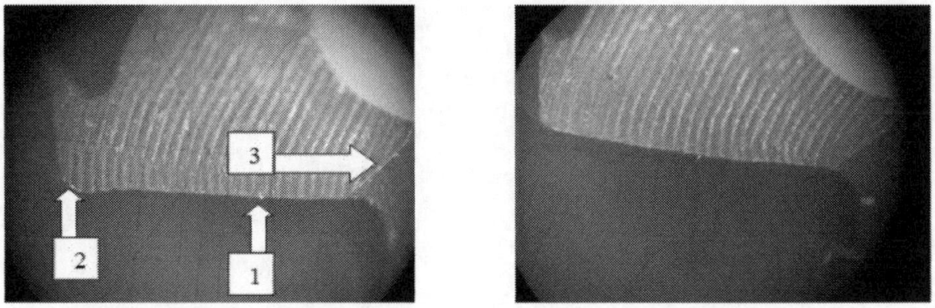

a) First edge b) Second edge

Fig. 5. Drill no. 5 at beginning of normal cutting.

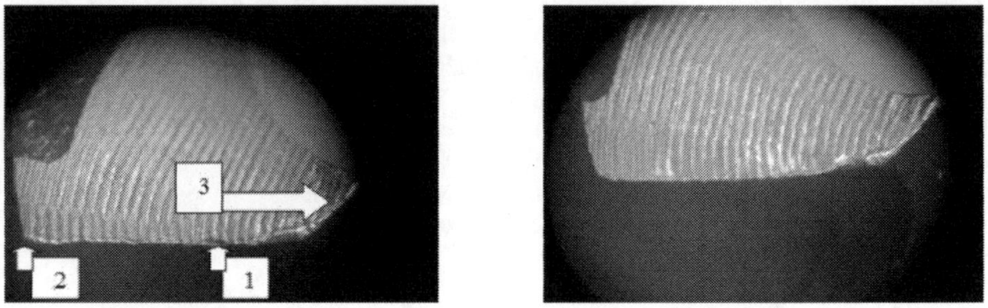

a) First edge b) Second edge

Fig. 6. Drill no. 5 at beginning of smooth entrance in cutting.

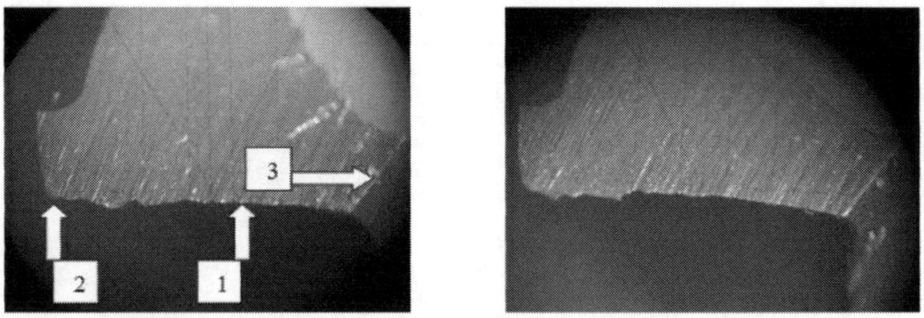

a) First edge b) Second edge

Fig. 7. Drill no. 30 at beginning of normal cutting.

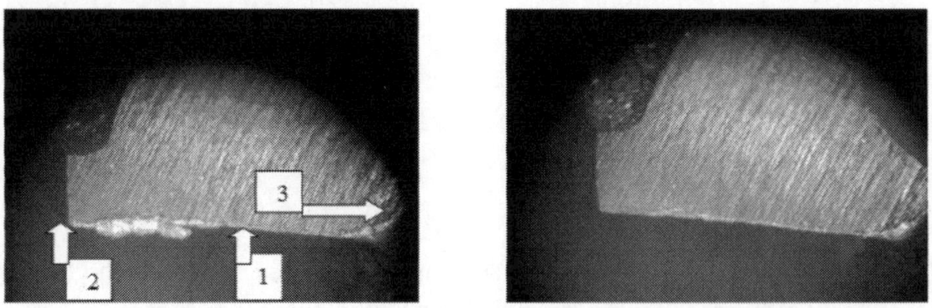

a) First edge b) Second edge

Fig. 8. Drill no. 30 at beginning of smooth entrance in cutting.

In case of drill no. 5 it can be observed that it was correctly sharpened but due to the improper roughness obtained at sharpening with a coarse-grained abrasive wheel, after 10 mm hole length it was intense worn out especially in the chisel edge zone (3).

 DOI: 10.1051/matecconf/20179402005

Drills 15 and 30 present similar characteristics. The cutting lips (1) are not properly sharpened and the chisel edge (3), at the intersection with the cutting lip (1) it is not well sharpened, reasons for which the drills were intense worn out, especially at the cutting lip zone.

For the three analyzed tools, their smooth entrance in cutting led to decrease of the temperature in the cutting zone, but the wear in the transversal edge zone has contributed to increase of the temperature, fact that had as consequence the obtaining of same values for the electrical current voltage.

3 Conclusion

Based on the researches conducted by the authors it can be concluded that the electrical current at cutting process can be used to evaluate the smooth entrance in cutting. These are demonstrated for the drilling process using a set of 30 drills with 8 mm diameter and 41MoC11 steel like for parts material.

Due to the experiments, it is easy to observe that the smooth entrance in cutting of drills contributes to increase of their life by decreasing the temperature in the cutting zone; in this case the cutting develops at the equality between the cutting resistance of the material and the cutting force; also it takes place a vibration damping. Measuring accurately the electrical current at cutting demonstrates the advantages of smooth entrance in cutting of the drill.

References

1. V. Dițu, TEHNOMUS, **264** (1987)
2. V. Dițu, AJME **1**, 63 (2008)
3. H. Bagchi, S.K. Basu, Wear **26**, 39 (1973)
4. V. Dițu, WSEAS **2**, 322 (2009)
5. A.L. Plotnikov, E.G. Krylov, E.M. Frolov STIN **1**, 2 (2009)
6. M. Murata, S. Kurokawa, O. Ohnishi, CSME **79**, 2546 (2013)
7. M. Murata, S. Kurokawa, O. Ohnishi, M. Uneda, T. Doi, CSME **6**, 958 (2012)
8. H. Hirota, M. Murata, A. Kojima, K. Yanagihara, JSPE **64**, 1191 (1998)
9. V. Dițu, AJME **8**, 18 (2010)
10. R. Tanaka, A. Hosokawa, K. Yamada, T. Ueda, Proceedings of the Seventh International Conference on Progress of Machining Technology, **786** (2004)
11. A. Vlase, *Cutting parameters, cutting allowances and technical time standards* (Technical Publishing House, Bucuresti,1983)

Influence of friction stir process parameters on surface quality of aluminum alloy A2017

Cecile Langlade[1], *Adriana* Roman[1], *Daniel* Schlegel[1,2], *Eric* Gete[1], *Patrice* Noel[1], and *Milena* Folea[3,*]

[1]Univ. Bourgogne Franche Comté, IRTES-M3M EA7274, 90010 Belfort Cedex, France
[2]ESTA, 90000 Belfort Cedex, France
[3]Transilvania University of Brasov, Department of Manufacturing Engineering, Mihai Viteazu No.5, Braşov, Romania

Abstract. Friction stir processing is an environment friendly surface engineering process that enhances mechanical properties and refines microstructure of processed zones. The aim of this paper is to study the influence of process parameters during friction stir processing of aluminium alloy A2017 plates on surface quality of the processed zones.

1 Introduction

Friction stir, originally designed as a welding method, recently became used as a surface engineering method that allows obtaining in-situ ultrafine grains in surface layers. Friction stir processing (FSP) is a thermo-mechanical technique with a very simple basic principle [1]: a rotating tool is moved along the workpiece surface while applying an axial load to increase generated heat (Fig. 1).

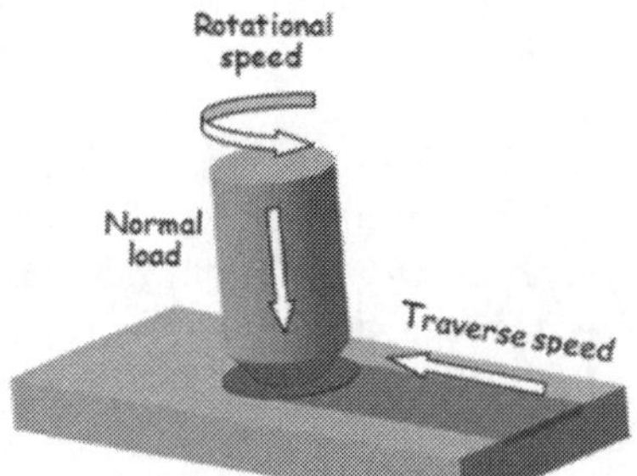

Fig. 1. Basic principle of friction stir processing [2].

Friction stir welding was patented in 1991 by British scientists of The Welding Institute (TWI) as an assembly method, but lately it has also been developed as a surface modification method due to its numerous technical, economic and environmental advantages [1, 3].

* Corresponding author: m.folea@unitbv.ro

The similarity of friction stir process kinematics with milling process kinematics makes possible the use of conventional milling machines equipped with force control devices. This is one of major advantages of friction stir processing, because it allows finishing the surface and, in the meantime, it applies in-situ localized heat treatment leading to microstructural refinement, densification, homogeneity, increased microhardness, superior wear and corrosion resistance of the surface layer. Friction stir was initially applied to aluminium alloys [1] and extended lately to other materials such as copper alloys [4, 5], magnesium alloys [6, 7], steels [8, 9] or titanium alloys [10, 11].

Until now, aluminum alloys are by far the most widely used materials in FSP industrial applications; therefore numerous studies have been dedicated to the effect of FSP conditions on the processed zone. Karthikeyan et al. [12-14] studied the friction stir behavior of the aluminum alloys A2285, A413 and A319 in order to examine effect of feed rate and rotational speed on microstructure and properties. They reported increased tensile, yield strengths and ductility along with grain size reduction due to FSP. Pasebani et al. [15] studied the effect of tool rotation rate on microstructure of friction stir processed 2024Al alloy and found significant change in size and morphology of the constituent particles after FSP. After studying the effect of friction stir processing conditions on A356-T6 cast aluminum alloy, Tajiri et al. concluded that lower rotational speed led to better results enhancing fatigue resistance and eliminating casting defects [16].

Since all the available studies on FSP are focused on microscopic changes, mechanical and tribological properties enhancement induced by FSP, the purpose of this paper is to extend research of FSP and study the effect of process parameters on surface roughness.

2 Experimental details

2.1 Materials and FSP equipment

Workpiece material used in present study was aluminium alloy A2017, an alloy with high strength and good machinability after heat treatment, generally used for structural components, airspace industry and other industrial applications. This alloy is also known as AlCu4MgSi and has the chemical composition presented in Table 1.

Table 1. Chemical composition of A2017 aluminium alloy (%wt).

Al	Si	Fe	Cu	Mn	Mg	Cr	Zn	Ti	Other each	Other total
Bal.	0.20-0.8	0.7 max	3.5-4.5	0.40-1.0	0.40-0.8	0.10 max	0.25 max	0.15 max	0.05 max	0.15 max

Friction stir tests were performed on 100x30x8 (mm) rectangular plates with 8 mm thickness on a CNC Gambin milling center (Fig. 2a). One of the major advantages of friction stir processing is that in case of horizontal plane surfaces it can be done on vertical milling machines. Due to temperature rise during friction stir processing, normal force varies unless special devices are added to the milling machine. For this study the milling center was equipped with a pneumatic tool holder for maintaining constant the normal force (Fig. 2b).

In order to minimize tool wear, an alumina tool was chosen for the tests. Traditionally, friction stir tools have a pin (probe) and a shoulder [1]. Considering the brittleness of alumina and the fact that for the present study it was intended to modify only a thin surface layer of the samples, a pinless cylindrical tool with 10 mm diameter and 0.5 mm chamfer was selected.

 DOI: 10.1051/matecconf/20179402006

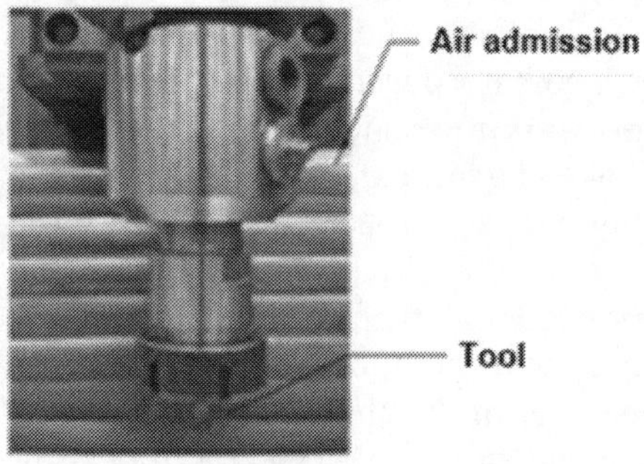

Fig. 2. Experimental stand (a) with pneumatic toolholder (b).

Samples were visually inspected after FSP in order to evaluate the surface after processing, especially the marks left by the tool. Also, surface roughness was measured using a 3D laser profilometer.

2.2 Process parameters

Three parameters were taken into account as input variables: rotation speed (rpm), traverse speed (mm/min) and normal force (N). Two variation levels were chosen for the rotation speed and the traverse speed, while for the normal force a third variation level was added considering the effect of this parameter on the deformation of the surface layer. The experiment design resulted had 12 runs (2 x 2 x 3) and the parameter variation values were set as shown in table 2.

Table 2. Experiment design.

Parameter	Test values
Rotation speed (rpm)	300
	900
Traverse speed (m/min)	50
	100
Normal force (N)	500
	1000
	1500

3 Results and discussion

The visual inspection of the samples after the tests revealed that for aluminium alloy A2017 the surface quality was highly dependent on the process parameters. In Fig. 3 are shown the extreme cases of the best two samples and most deteriorated two samples obtained after test runs with various parameters.

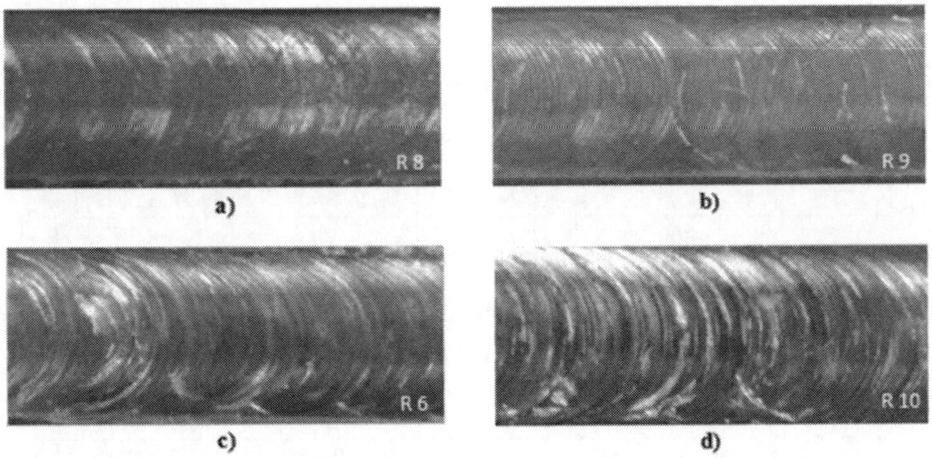

Fig. 3. Visual comparison of friction stired zones with good surface appearance (a and b) and with deteriorated surface (c and d)

For a more accurate evaluation of the surface quality each sample was studied using laser profilometer as illustrated in Fig. 4.

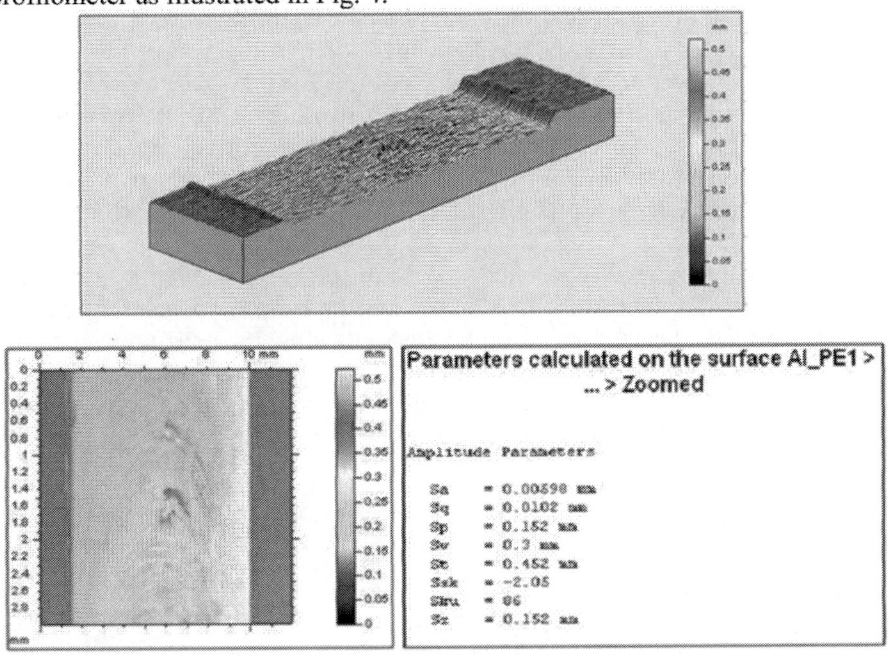

Fig. 4. Profilometric analysis for sample no.1 (50 mm/min, 500 N, 300 rpm).

To determine the influence of process parameters on the surface roughness after FSP, the surface roughness parameter S_a was used. S_a is average surface roughness and is evaluated over a 3D surface with the following formula:

$$S_a = \int\int_a |Z(x,y)| dxdy \tag{1}$$

where x and y are 2D coordinates of points of the surface a and Z is height of the surface relative to the best fitting plane.

Table 3. Surface roughness after FSP.

Test run no.	Traverse speed (mm/min)	Force (N)	Rotation speed (rpm)	Sa (µm)
1	50	500	300	5.98
2	50	500	900	12.5
3	100	500	300	5.38
4	100	500	900	8.73
5	50	1000	300	8.36
6	50	1000	900	17.4
7	100	1000	300	8.73
8	100	1000	900	2.82
9	50	1500	300	3.62
10	50	1500	900	24.1
11	100	1500	300	7.92
12	100	1500	900	14

Surface roughness measurements are presented in table 3 and illustrated in Fig. 5. In both of them it can be observed that the best surface roughness with S_a of 2.82 µm was obtained for test run no. 8 performed at high level of traverse speed, medium level of normal force and high level of rotation speed. Tool traces in this case were quite uniform as shown in Fig. 3a. A good surface quality with a close value of surface roughness parameter S_a (3.62 µm) was obtained for run no. 9 too, at lower traverse speed but at highest level of force and low level of rotation speed. In spite of the small rotation speed that reduced friction, the higher pressure of the tool and the lower traverse speed that maintained the heat longer in one spot decreased the surface quality, fact also revealed in Fig. 3b. Very poor surface quality resulted in case of test runs no. 6 (Fig. 3 c) and no. 10 (Fig. 3d) with measured values for S_a of 17.4 µm and 24.1 µm, respectively. High rotation speed, medium and low normal force, corroborated with slow traverse movement made the tool to sink in the material and leave deep traces behind.

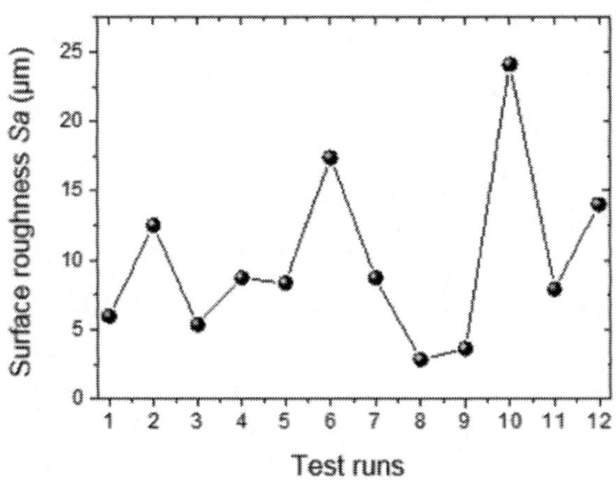

Fig. 5. Comparison of roughness (S_a)

4 Conclusions

This work aimed to determine the influence of process parameters during friction stir processing of aluminum alloy A2017 on surface quality of the processed zones.

The results obtained have shown that the optimal conditions in terms of roughness for FSP process applied to this material are represented by test conditions for runs 8 (F=1000N, f=100 mm/min, n=900 rpm) and 9 (F=1500N, f=50 mm/min, n=300 rpm) that provided the right combination between the amount of heat produced by normal force and rotation speed and the cooling rate determined by the traverse speed. Too much heat and low cooling rate seem to affect negatively the quality of the surface.

Further investigations have to be done on the influence of machining conditions, especially in what concerns the influence of FSP parameters on the stirred zone thickness and microstructure.

References

1. R.S. Mishra, .Z.Y. Ma, Mat. Sci. Eng. R **50**, 1 (2005)
2. M. Folea, C. Langlade, A. Roman, D. Schlegel, E. Gete, D. Chamoret, *Producing Nanograin Surface Layers by Friction Stir Processing - chapter in Comprehensive Guide for Nanocoatings Technology* (Nova Science Publishers Inc., New York, 2015)
3. M. K. Kulekci, U. Esme, B. Buldum, Int. J. Adv. Manuf. Technol., **85(5)**, 1687 (2016)
4. S.K. Menon, F.A. Pierce, B.P. Rosemark, K. Oh-Ishi, S. Swaminathan, T.R. McNelley, Metall. Mater. Trans. A, **43**, 3687 (2012)
5. K. Nakata, Weld. Int., **19(12)**, 929 (2005)
6. B.M. Darras, M.K. Khraisheh, F.K. Abu-Farha, M.A. Omar, J. Mater. Process. Technol., **191(1-3)**, 77 (2007)
7. D. Zhang, M. Suzuki, K. Murayama, Scripta Mater., **52(9)**, 899 (2005)
8. A. Chabok, K. Dehghani, Mat. Sci. Eng. A-Struct., **528(1)**, 309 (2010)
9. L. Wei, T. W. Nelson, Mat. Sci. Eng. A-Struct., **556**, 51 (2012)
10. Y. Zhang, Y. S. Sato, H. Kokawa, S. H. C. Park, S. Hirano, Mater. Sci. Eng. A-Struct., **485(1–2)**, 448 (2008)
11. A. Lauro, Weld. Int., **26(1)**, 8 (2012)
12. L. Karthikeyan, , V.S. Senthilkumar, V. Balasubramanian, S. Natarajan, Mater. Design, **30(6)**, 2237 (2009)
13. L. Karthikeyan, , V.S. Senthilkumar, V. Balasubramanian, S. Arul, Mater. Lett., **64(3)**, 301 (2010)
14. L. Karthikeyan, , V.S. Senthilkumar, K.A. Padmanabhan, Mater. Design, **31(2)**, 761 (2010)
15. S. Pasebani, I. Charit, R.S. Mishra, Rajiv (2015). Mater. Lett. , **160**, 64 (2015)
16. A. Tajiri, Y. Uematsu, T. Kakiuchi, Y. Tozaki, Y. Suzuki, A. Arfinaldi, Int. J. Fatigue, **80,** 192 (2015)

Characteristics of geometric structure of the surface after grinding

Stanislaw Legutko[1,*], *Krzysztof* Zak[2] , and *Jan* Kudlacek[3]

[1]Poznan University of Technology, Pl. Marii Skłodowskiej-Curie 5, 60-965 Poznan, Poland
[2]Opole University of Technology, 5 Mikolajczyka street, 45-271 Opole, Poland
[3]Czech Technical University in Prague, Faculty of Mechanical Engineering, 4 Technicka street, 166 07 Prague, Czech Republic

Abstract. This paper presents the technological and functional capabilities of surface textures produced by cutting and grinding operations on hardened steel of about 54 HRC hardness. Special focus was placed on surface textures generated by hard turning and grinding operations which are characterized by the *Sz* roughness parameter of about 3 - 4 μm and distinctly similar values of the *Sa* parameter. A set of 3D surface roughness parameters was used to characterize the superficial effects of precision grinding operations and their influence on the parts functionality.

1 Introduction

Turning, milling and grinding are the most commonly used traditional methods of materials processing, and the most frequently used operation for finishing hardened steel parts are grinding. On the other hand, the development of processing turning and milling in terms of the application of modern tool materials, such as ceramics or mixed CBN, resulted in greater use of these technologies in the development of machine components and for turning and grinding of the high hardness best results are obtained by integrating these processes in a single technological chain. By combining roughing and grinding it is possible to obtain [1]:

- high accuracy and shortening of the process by treatment in one attachment;
- carrying out of the process according to the quality requirements;
- flexibility and high process reliability;
- the possibility of dry grinding and MQL (minimum quantity lubrication).

However, it was possible to assess the quality and functionality of the surface treatment, which can affect the disorders caused by the shaping process at the time of machining or grinding hard necessary measurements of surface roughness 2D/3D. It is also known that the surface roughness analysis 3D obtain compared to 2D complete information on the state of the surface asperities is required. More details can be seen in a direction parallel to the traces of processing, and not, as in the case of 2D analysis, which refers to a profile in a direction perpendicular to the traces of processing [2, 3].

* Corresponding author: stanislaw.legutko@put.poznan.pl

In the article are presented results of research on turning and grinding steel 41Cr4 hardened to approximately 54 HRC.

2 Experimental procedure

2.1 Machining conditions and characterization of surface roughness and texture

This experimental study includes grinding operations on samples made of a 41Cr4 steel with Rockwell's hardness of 54±1 HRC. Turning was performed using CBN tools (grade CB7015 by Sandvik Coromant) with cutting parameters permitting the *Sa* roughness of about 0.42 μm. TNGA 160408 S01030 chamfered inserts with brazed-CBN tips were used. The machine tool was a CNC turning centre, Okuma Genos L200E-M. Grinding operations were performed on conventional cylindrical grinding machines using electro-corundum Al_2O_3 and CBN wheels and a water soluble emulsion as a coolant. Similarly, surfaces with the Sa near 0.2 μm were produced. Machining conditions for turning and two abrasive operations are specified in Table 1.

Table 1. Parameters of experiments

Symbol	Machining operation	Machining conditions
INIT-HT	Hard turning using CBN TNGA 160408 S01030	v_c=150 m/min, f=0.1 mm/rev, a_p=0.15 mm
GR-Ceramic	Cylindrical grinding using Al_2O_3 ceramic, 350×25×127 32A grinding wheel	v_c=11.9 m/s, a_e=0.025 mm, f_a=3.5 mm/rev
GR-CBN	Cylindrical grinding using INTER DIAMENT B107 K100 SV grinding wheel	v_c=36 m/s, a_e=0.025 mm, f_a=1.6 mm/rev

Surface profiles/topographies were recorded using a TOPO-01P contact profilometer with a diamond stylus radius of 2±0.5 μm. After their digitalization, 2D and 3D roughness parameters were estimated on the selected scanned areas of 2.4 mm×2.4 mm. Roughness parameters were determined and surface topographies were visualized using a Digital Surf, Mountains Map package.

3 Experimental results and discussion

Representative surface topographies obtained in hard turning and grinding operations performed are presented in Fig. 1. From the practical point of view, the comparison of the *Sa* and *Sz* (Fig. 2) roughness parameters which are frequently used by constructors and technologists is of fundamental importance. In terms of the surface quality criterion both operations can be classified as precision machining.

The measured values of *Sa* and *Sz* parameters are equal to 0.21 μm and 3.86 μm for grinding (GR-Ceramic) and 0.22 μm and 2.87 μm for grinding (GR-CBN). This comparison clearly depicts that ground surfaces contain a number of high sharp peaks which increase distinctly the total height *Sz*, in this case study up to 4 μm, although the *Sa* parameter is equal to about 0.2 μm.

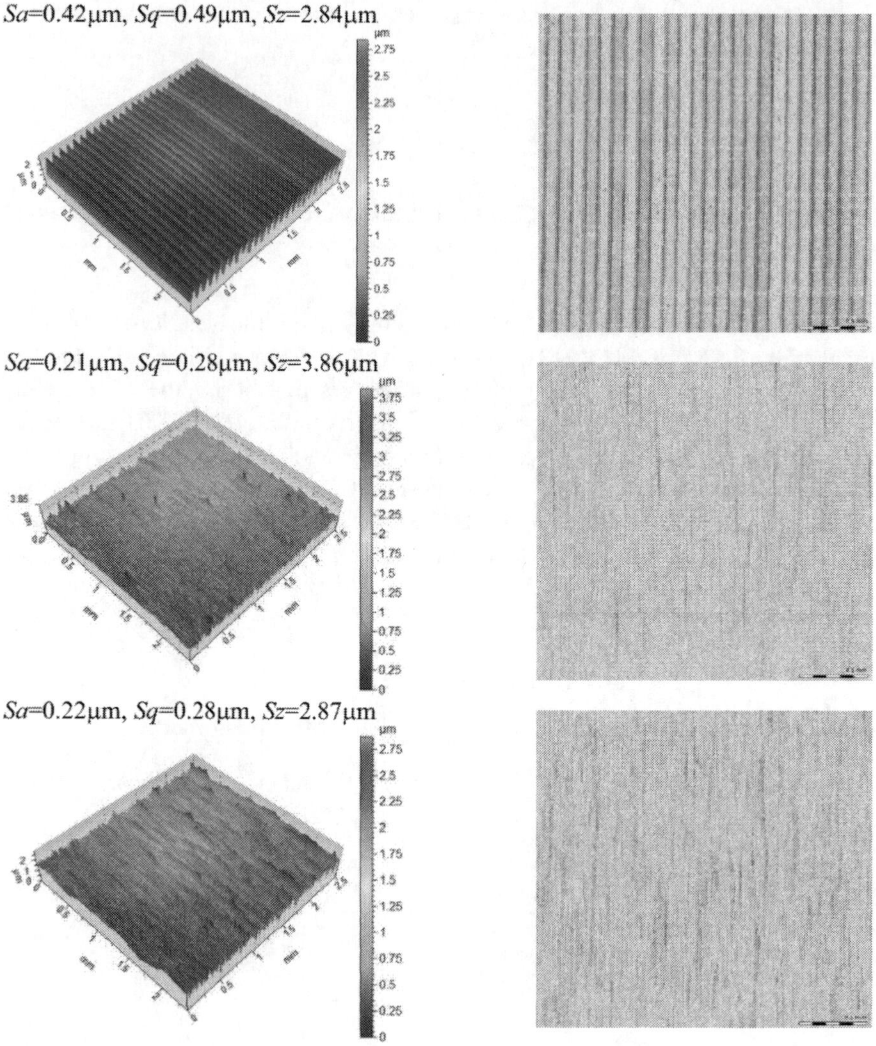

Fig. 1. Surface textures produced by turning (a), grinding with ceramic wheel (b), and grinding with CBN wheel (c)

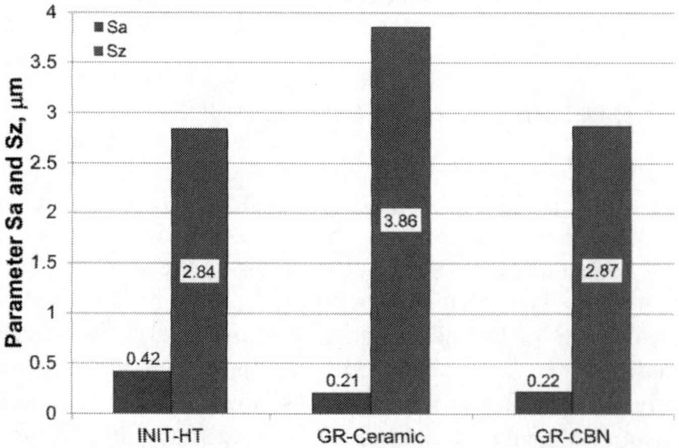

Fig. 2. Comparison of vertical parameters for 3D surface after turning and grinding

DOI: 10.1051/matecconf/20179402007

It can be observed in Fig. 1 that the ground surfaces are random anisotropic suggest the presence of periodic components.

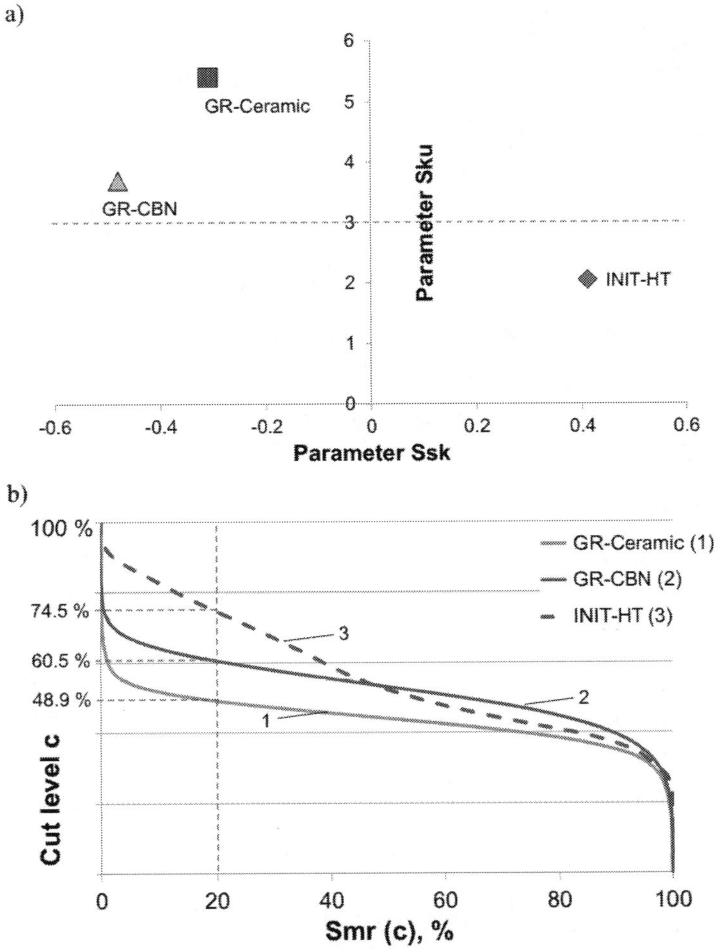

Fig. 3. Comparison of the values of the parameters associated with the distribution of ordinates 3D after turning and grinding (a) and 3D BAC shapes (b)

Analysing the roughness associated with the distribution of surface ordinates (Fig. 3a) and 3D bearing area curves (3D BAC) (Fig. 3b) compared surfaces are possible to be realized. *Ssk* and *Sku* parameters are strongly influenced by single elevations and depressions on the surface [4, 5]. Finish grinding generates surfaces with S-shape BAC (GR-Ceramic (1) and GR-CBN (2)) and negative skewness *Ssk*. It is worth noticing that the values of *Ssk* for both ground surfaces differ visibly (-0.31 for (1)) versus (-0.48 for (2)). Also using Al_2O_3 and CBN wheels results in various bearing properties of ground surfaces.

The functional topographical features of the machined surfaces can be assessed by means of the volume parameters (Fig. 4) including the peak material volume (*Vmp*), the core material volume (*Vmc*), the core void volume (*Vvc*) and the valley void volume (*Vvv*) parameters. [6, 7]. Their values measured for turned and ground topographies are equal (in order INIT-HT/GR-Ceramic/GR-CBN): *Vmp*=0.136/0.0150/0.0112 $\mu m^3/\mu m^2$; *Vmc*=0.46/0.225/0.247 $\mu m^3/\mu m^2$; *Vvc*=0.732/0.292/0.310 $\mu m^3/\mu m^2$; *Vvv*=0.0291/0.0383/0.0403 $\mu m^3/\mu m^2$. In particular, a better fluid retention ability of ground surfaces is associated with relatively higher values of *Vvv*=0.0385 and 0.0403 $\mu m^3/\mu m^2$.

DOI: 10.1051/matecconf/20179402007

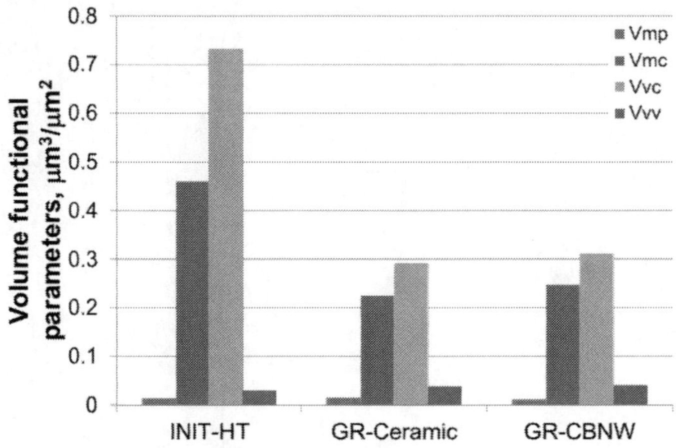

Fig. 4. Distribution of volume functional parameters for turning and grinding surfaces

4 Conclusions

The textures of ground surfaces are mixed periodic–random anisotropic respectively. This difference can be determined based on 3D surface topographies. However, the disturbances of regular surface structures by grinding can result from both the kinematical and tribological effects. 3D BAC curves and appropriate functional parameters depict that ground hard surfaces have better fluid retention abilities. This is due to higher negative *Ssk* value and higher *Vvv* volumes for ground textures. For grinding wheel Al_2O_3 and CBN ground textures have similar tribological properties, which suggest practically the same material allowance to be removed during running-in period.

References

1. W. Grzesik, Mechanik, **8-9**, 650 (2011) (in Polish)
2. W. Grzesik, K. Żak, , Archiv. Mat. Scien. Eng., **52**, 1, 46 (2011)
3. W. Grzesik, S. Brol, K. Żak, PAK, **1**, 44 (2010)
4. G.M. Krolczyk, P. Nieslony, J.B. Krolczyk, I. Samardzic, S. Legutko, S. Hloch, et al., Measurement, **70**, 203 (2015)
5. G.M. Krolczyk, J.B. Krolczyk, R.W. Maruda, S. Legutko, M. Tomaszewski, Measurement, **88**, 176 (2016)
6. B. Griffiths, *Manufacturing surface technology. Surface integrity and functional performance* (Penton Press, London, 2001)
7. R. Leach, *Characterization of areal surface texture.* (Springer-Verlag, Berlin, 2013)

MATEC Web of Conferences **94**, 02008 (2017) DOI: 10.1051/matecconf/20179402008
CoSME'16

Tool life of ceramic wedges during precise turning of tungsten

Stanislaw Legutko[1,*], *Piotr* Winiarski[1], *Tadeusz* Chwalczuk[1], *Ludmila* Marcincinova-Novakova[2] , and *Krzysztof* Zak[3]

[1]Poznan University of Technology, Pl. M. Sklodowskiej-Curie 5, 60-965 Poznan, Poland
[2]Technical University of Kosice, Faculty of Manufacturing Technologies with a seat in Presov, Bayerova 1, 08001 Presov, Slovakia
[3] Opole University of Technology, Proszkowska 76, 45-758 Opole, Poland

Abstract. Properties, application and machinability of tungsten and its alloys have been demonstrated. The comparison of the tool life and wear of the wedges made of SiAlON and whisker ceramics during the precise turning at different cutting parameters have been presented. The CNC lathe DMG CTX 310 Ecoline and tungsten of 99.7 % purity were used during the experiments. Only the wedge of whisker ceramics has proved to be sufficiently suitable and only for relatively low cutting speeds.

1 Introduction

Pure tungsten is a corrosion resistant metal which reacts with most acids only to a very small extent. Its pure form is very soft and very ductile which makes tungsten suitable for such processing as forging or pull broaching, however, even small additions of carbon or other elements make it stiff and very hard-to-process [1]. Tungsten alloys, due to their properties, are applied in many branches. In the branch of machine construction, tungsten alloys are used in manufacturing very hard cutting tools for titanium carbide is the major component of sintered carbides. Nowadays, they are composite materials consisting mainly of tungsten carbide and a bonding metal, most often cobalt; they are used in very demanding processing, with high cutting speeds where high abrasion resistance and hardness are required [2].

From the technical point of view, polycrystalline tungsten is the most important form possessing the highest content of pure metal. The properties of that material are closely related to the two factors:

- microstructure,
- presence, concentration and kind of impurities.

Tungsten microstructure which determines its properties to a high extent, depends on the way of its preparation (powder metallurgy, melting by an electron beam, refining, chemical application from the gas phase) and on subsequent operations (deformation, annealing, recrystallization). The properties of tungsten can be generalized in a number of items:

* Corresponding author: stanislaw.legutko@put.poznan.pl

- plastically processed structure consisting of elongated crystals is stronger than that of annealed and recrystallized one; this is particularly visible in the case of strength in the direction of deformation,
- the lower is the working temperature and the more energy is stored during deformation, the higher is strength,
- the smaller is the grain size, the higher is strength,
- various methods of manufacturing (drawing, rolling, pull broaching) render various effects and even after completion of the process, the differences of the properties of materials obtained by the method of powder metallurgy and those obtained by another method are clearly visible.

Plasticity and strength are the properties due which tungsten is widely used as constructional material. As late as not long ago, tungsten was considered to be a brittle and hard-to-process material. At first, brittleness was attributed to the low level of the material purity obtained by the methods of powder metallurgy. However, single crystals of the highest purity obtained by means of electron beam refining, maintained their ductility in deformation even at the temperature of - 269°C.

At the beginning of the twentieth century, the application of tungsten was limited to filaments and as alloy additive to steel. Today, its application is much more developed and includes such branches as lighting, electronics, technology of very high temperatures, medicine, aircraft industry, military applications sports and many more. About 4-5% of the world total tungsten production is used in the processes of the lighting industry. Nowadays, over 70% of artificial lighting is generated by glow-discharge tubes and the value is continuously growing. Tungsten is used for forming conductors and coils in filament lamps and as the material for electrodes in low and high pressure discharge lamps. Due to its properties, tungsten is used as the material for electric contacts, vacuum tight conducting connections. It is an ideal material for switches with high operation frequency, such as contact switches, voltage controls used in the automotive industry. Its alloys (W-Cu, W-Ag, WC-Ag) are used in manufacturing electric contacts operating at very high, medium and low voltages. In electronics, tungsten and its alloys are used as materials for electron emitters, components of vacuum lamps, sub-assemblies of laser printers and other printers, radiators of electronic devices, thin coatings in microelectronic devices and optoelectronic ones. In welding, it is used as the material for welding electrodes, heating coils for vacuum metalizing. Medicine is another area of tungsten application: anodes and cathodes in X-ray lamps used, for instance, in tomography, anti-radiation shelters or infrared reflector lamps are just a number of examples. In machine construction industry, the application of tungsten and its alloys includes, first of all: construction parts, combustion chambers in turbo engines, gyroscope rotors, servomotors in self-winding clocks, flow meter components, anti-vibration tool holders, counterweight loads and many others. Sports equipment, too, very often contains tungsten. Such components are, for example, professional hauling lines, diving loads, golf equipment, loads for runners, tennis racket loads, missiles for hunting rifles or counterweight in Formula 1 cars. In the chemical industry, tungsten plays an important role in manufacturing components for production of silica glasses, electrodes for the process of kaolinite melting, spinnerets for production of ceramic wool, crucibles for processing used nuclear fuels and melted metals or equipment for die casting. Aerospace industry is another area of tungsten application where tungsten and its alloys are used for some components of rocket engines, emitters in ion engines, parts of space nuclear reactors. Tungsten and military objects also have much in common. Its alloys are used in the production of tank ammunition reaching the speed of up to 1700 m/s, armour, grenades, counterweight in tanks. Due to its high hardness and density, tungsten is used in the production of cores of armour piercing sabot cores and composite full size missiles. In aircraft industry, the main application area is the production of aircraft wheel elements, counterweights and loads in airplanes and helicopters [3].

Sugita et al. [4] presents ultra-precision machining of tungsten alloy maximizing burnishing between the cutting tool and workpiece during cutting by combining strengths of the crystal grain and binder phase. Turning experiments conducted using workpieces of tungsten-based alloys 50 mm in diameter and 10 mm thick produced high-quality surface roughness Rz of 0.13 µm. Zebala and Kowalczyk [5] presents research of cutting process of high hardness composite materials such as sintered carbides based on the cobalt. The paper presents the influence of cutting parameters and cobalt content in a workpiece on the turning process of WC-Co with polycrystalline diamond (PCD) tool. This study was focused on the main component of cutting force F_c and surface roughness Ra. Choi et al. [6] presents analysis of the electrochemical behaviours of WC–Co alloy for micro ECM, microstructures of WC–Co with a sharp edge and good surface quality that were obtained by electrochemical milling and electrochemical drilling. The process performance can be improved by selecting the optimal combination of process parameters [7-10].

2 Methodology of experimental investigations

The purpose of the tests was to determine the suitability of selected ceramic wedges for cutting tungsten.

Within the investigation, precise turning of a shaft made of tungsten with purity of 99.7% has been performed with the use of the following wedges:

- no. 1 of whisker ceramics – NTK – SNGN120408 T01020 WA1;
- no. 2 of SiAlON ceramics – Kennametal – SNGN120408 T01020;
- no. 3 of SiAlON coated ceramics- Kennametal - SNGN120408 T01020 coated with TiC.

Turning has been performed on a tungsten shaft with the diameter of Ø20 mm, using a CNC lathe, DMG CTX 310 Ecoline. The lathe tool used in the machining process was PAFANA CSRN2 2020-12 fixed in the holder VDI B5 – 30x16/20 x40. Cutting wedge geometry was: $\lambda_s = -6°$, $\gamma_0 = -6°$, $\kappa_0 = 75°$.

Test conditions – constant parameters: feed $f = 0.08$ mm/rev, cutting depth $a_p = 0.1$ mm; variable parameters: cutting speed v_c [m/min], cutting path l_s [mm]. For wedge no. 1, cutting tests have been performed at the speeds of 40 m/min, 80 m/min and 150 mm/min; for wedge no.2 the speeds of 30 m/min, 80 m/min and 150 m/min were used; for wedge no. 3, on the other hand, the cutting speed of 150 m/min was used.

The wear of the ceramic wedges has been analysed and their life has been determined. Wear curves of the wedges have been drawn, photographs of them have been made and a mathematical model of the tested wedges life has been elaborated. VB_c - the width of the corner wear band = 0.3 mm has been adopted as a wear criterion. Prior to each exchange of the cutting edge of the wedge it was necessary to measure the tool at each fixing of the wedge in the tool holder; the measurement was realized after it has been taken out from the machine. After the lathe tool has been relocated in the tool holder and the adequate value obtained from the measurement has been entered and the required path, depth and speed of cutting have been programmed with the use of the SINUMMERIK , machining has been started. After its completion, the wedge has been inspected under microscope and the indices of its wear: VB_c, KE - cutting wedge shortening, KB - the width of the abrasion groove on the rake face have been assessed, also by means of the AxioVision program. The operation has been performed for all the edges of the wedges under investigation.

3 Results and discussion

3.1 Analysis of the wedges wear

Based on the investigation performed, it can be stated that only one of the tested wedges is suitable for turning tungsten. That is wedge no. 1 made by NTK company of whisker ceramics. As can be seen from the data presented in Table 1, wedge no. 2 made of SiAlON ceramic has significantly exceeded the blunting criterion at the first pass for the speed of v_c = 150 m/min. Only at the speed of v_c = 30 m/min the wedge durability made it possible to perform two passes. (measurement W2211 in which the wedge has not reached the criterion of blunting), which proves durability definitely too weak to machine tungsten. The wear of the wedge after the two passes is shown in Figure 1. The values of the wear indices (VB_c, *KE* and *KB*) for wedge 2 at the predetermined cutting parameters are shown in Table 1. The durability of that wedge was not sufficient for the determination of the wear curves. The marking of each measurement used in Tables 1, 2 and 3 will be explained on the example of W1234 where the individual items denote: W – kind of material to be machined, in this case tungsten, 1 - wedge no., 2 - number of the wedge corner, 3 – number of the used v_c parameter, 4 – number of the pass.

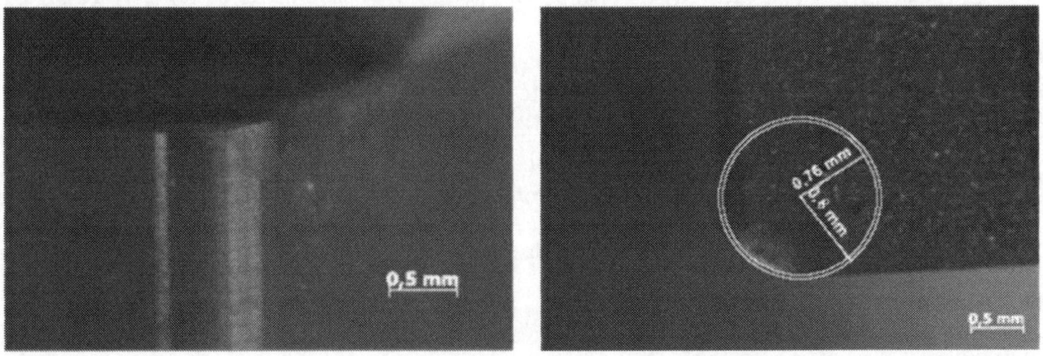

Fig. 1. The width of the wear band of wedge no. 2 after the second pass at the speed of v_c = 30 m/min, VB_c = 0.42 mmm – on the left and cutting wedge shortening KE = 0.04 mm – on the right.

Table 1. Machining parameters and wear indices of wedge no. 2.

Measurement	VB_c [mm]	*KE* [mm]	*KB* [mm]	T_s [mm]	v_c [m/min]	l_s [mm]
W2111	0.85	0.04	0.01	0.08	150	12
W2211	0.07	0.01	0.01	0.06	150	9.4
W2212	1.1	0.12	0.01	0.15	150	23
W2321	1.06	0.15	0.01	0.29	80	22.8
W2421	0.45	0.05	0.01	0.28	80	22.6
W2521	0.68	0.1	0.01	0.28	80	22.4
W2621	0.73	0.1	0.01	0.28	80	22.2
W2731	0.25	0.01	0.01	0.73	30	22
W2732	0.42	0.04	0.01	0.73	30	21.8
W2831	0.25	0.01	0.01	0.72	30	21.6
W2832	0.49	0.05	0.01	0.71	30	21.4

DOI: 10.1051/matecconf/20179402008

Wedge no.3 made of coated ceramic SiAlON, was not suitable for machining either – it showed much wear as early as after the first pass at the speed of v_c = 150 m/min. Due to the wear, was performed only one pass with that wedge (Fig. 2). The values of the wear indices (VB_c, KE, KB) for wedge 3 at that pass are presented in Table 2. The durability of that wedge, too, was not sufficient for the determination of the wear curves.

Table 2. Machining parameters and wear indices of wedge no. 3.

Measurement	VB_c [mm]	KE [mm]	KB [mm]	T_s [mm]	v_c [m/min]	l_s [mm]
W3111	0.56	0.10	0.01	0.14	150	21.2

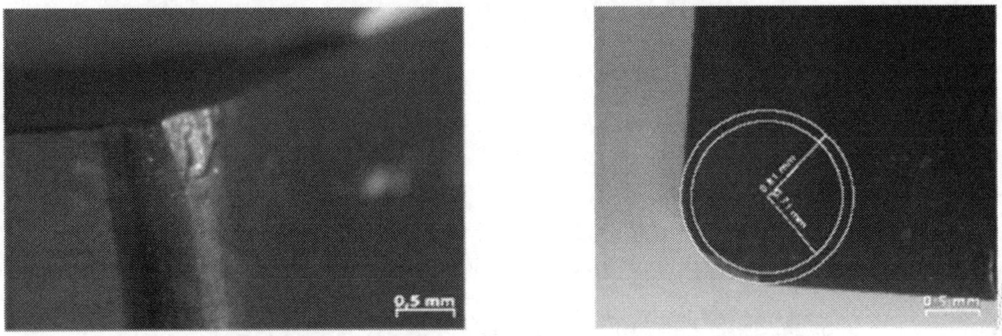

Fig. 2. The width of the wear band of wedge no.3 after the first pass at the speed of v_c = 150 m/min, VB_c – 0.56 mm – on the left and wedge shortening, KE = 0.1 mm – on the right.

Wedge no. 1 made of whisker ceramic has shown sufficient durability for performing over a dozen passes with one corner at the cutting speed of v_c = 40 m/min. The measurement results are presented in Table 3.

Table 3. Machining parameters and wear indices of wedge no. 1.

Measurement	VB_c [mm]	KE [mm]	KB [mm]	T_s [mm]	v_c [m/min]	l_s [mm]
W1111	0.01	0.01	0.01	0.165	80	13.2
W1112	0.13	0.01	0.01	0.16	80	13
W1114	0.24	0.01	0.01	0.16	80	12.7
W1116	0.35	0.05	0.01	0.16	80	12.5
W1221	0.2	0.01	0.01	0.08	150	12.4
W1222	0.43	0.01	0.01	0.08	150	12.2
W1331	0.01	0.01	0.01	0.30	40	12
W1332	0.01	0.01	0.01	0.30	40	11.8
W1333	0.01	0.01	0.01	0.29	40	11.6
W1334	0.05	0.01	0.01	0.29	40	11.4
W1335	0.06	0.01	0.24	0.28	40	11.2
W1336	0.07	0.01	0,24	0,28	40	11
W1337	0.08	0.01	0.24	0.27	40	10.8
W1338	0.11	0.01	0.24	0.27	40	10.6
W1339	0.11	0.01	0.24	0.58	40	23
W13310	0.12	0.01	0.24	0.57	40	22.8
W13311	0.13	0.01	0.24	0.57	40	22.6
W13312	0.17	0.01	0,24	0.56	40	22.4
W13313	0.19	0.01	0.24	0.56	40	22.2
W13314	0.23	0.01	0.24	0.55	40	22
W13315	0,28	0.01	0.24	0.55	40	21.8
W13316	0.33	0.03	0.24	0.54	40	21.6

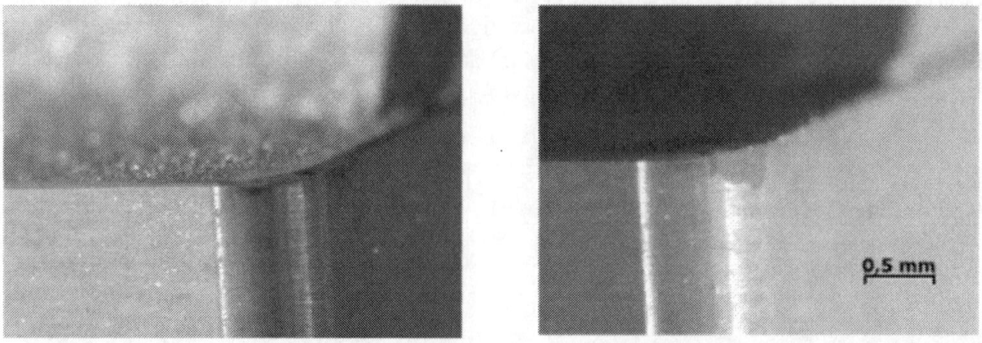

Fig. 3. The wear of the flank face of wedge no. 1 after the third pass at the cutting speed of 40 m/min – no symptoms of wear (on the left) and the width of the wear band of wedge no. 1 after the 16th pass with the same corner at the cutting speed of 40 m/min - edge blunting criterion reached $VB_c = 0.33$ mm (on the right).

As the subsequent passes were performed, one could observe a groove appearing on the rake face as result of chip friction against the wedge. Its size just before reaching the blunting criterion $VB_c = 0.30$ mm are clearly visible in Fig. 4.

Fig. 4. Groove on the rake face of wedge no.1 just before reaching the blunting criterion $VB_c = 0.30$ mm.

3.2 Determination of the wear curves

Based on the measurements of the wear index VB_c of wedge no.1 wear curve has been prepared (Fig. 5, 6) which has made it possible to determine its life period and to perform further calculations in order to make a mathematical model for predicting the wedge durability. For the speed of $v_c = 80$ m/min, the wedge corner endured 6 passes, which was sufficient for the determination of the wedge wear curve (Fig.5). Only two passes before reaching the blunting criterion of $VB_c = 0.3$ mm have been performed for the speed of $v_c = 150$ m/min, which allowed only the determination of the diagram with a straight line (Fig. 6).

DOI: 10.1051/matecconf/20179402008

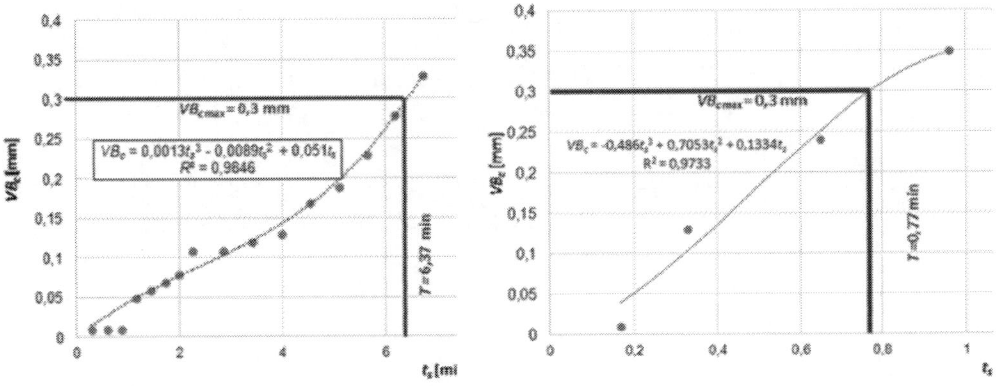

Fig. 5. Wear curve of wedge no. 1 for: v_c=40 m/min (on the left) and for v_c= 80 m/min (on the right); f= 0.08 mm/rev, a_p = 0.1 mm.

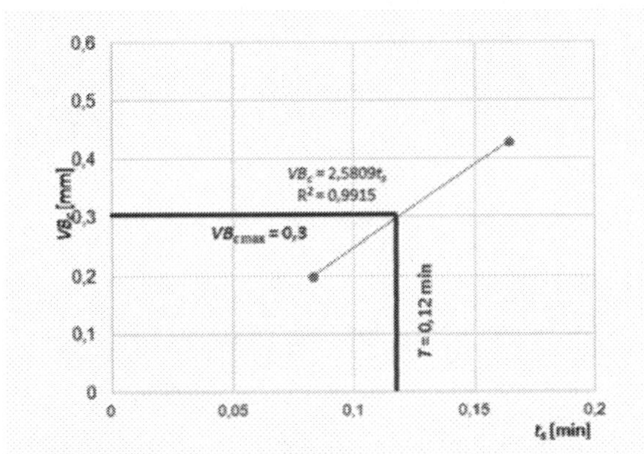

Fig. 6. Wear curve of wedge no. 1 for: v_c=150 m/min f= 0.08 mm/rev a_p = 0.1 mm.

During the analysis of the wedge shortening index, it could be noticed that the shortening occurred after the wedge blunting criterion VB_c had been reached, which is clearly visible when comparing the two diagrams presented in the Figure 7. As can be seen in Fig. 7, the *KE* wear index rises sharply when wear index VB_C reaches criterial value of 0.3 mm. In the upper diagram, the curve of wear intensity (the red one) has also been drawn.

Based on the analysis of the wear curves, the edge life periods have been determined for the individual cutting speeds: v_c= 40 m/min – T= 6.37 min; v_c = 80 m/min –T= 0.77 min; v_c = 150 m/min - T= 0.12 min.

On the basis of the results obtained, a mathematical model of the wedge life has been elaborated for wedge no. 1 using classical commonly known procedure [11]:

$$T = \frac{271034.12}{v_C^{2.91}} \quad 1)$$

Based on the mathematical model, a diagram of the wedge life period as a function of the cutting speed has been realized (Fig. 8).

DOI: 10.1051/matecconf/20179402008

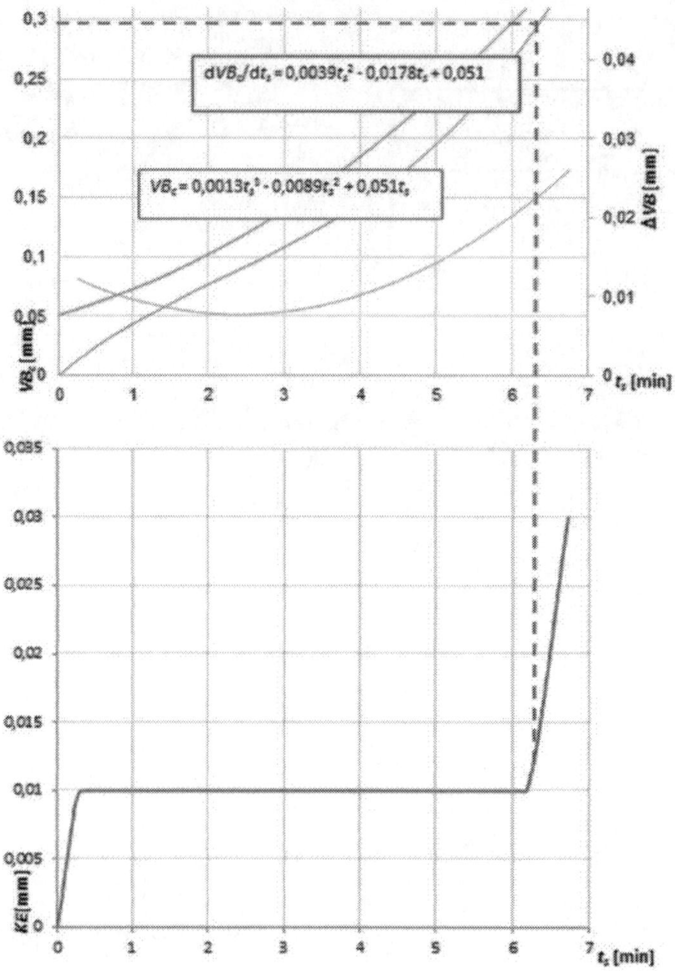

Fig. 7. A comparison of diagrams illustrating the moment of appearance of the wear index *KE*, wedge no 1, for: v_c = 40 m/min, f = 0.08 mm/rev, a_p = 0.1 mm.

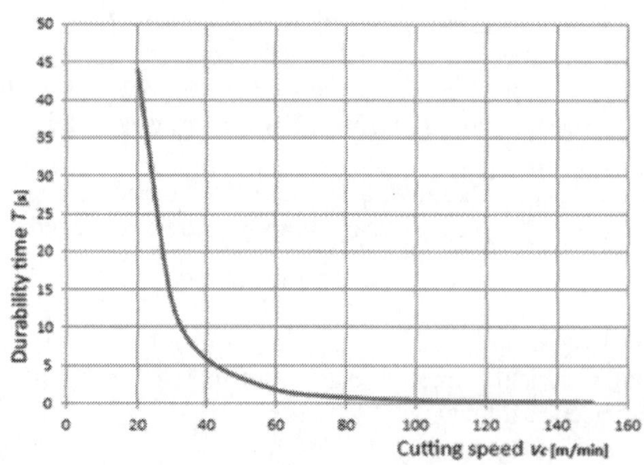

Fig. 8. Wedge life period as a function of the cutting speed determined on the basis of the mathematical wedge life model for wedge no. 1

4 Conclusions

Tungsten has proved to be a hard-to-machine material. The tool material must meet very high requirements. Only the wedge made of whisker ceramic has proved to be suitable and only for relatively low cutting speeds. On the basis of the performed tests and inspections, the following conclusions can be formulated:

- In turning of tungsten using ceramics wedges abrasive wear was prevailing. This was proved by the wear band appearing on the wedge corner and the groove on the rake face which had appeared during cutting by a wedge of whisker ceramic. Small influence of chemical wear on the total wear of the wedge was due to the high resistance of ceramic materials to this kind of wear. Most probably, adhesion, thermal and diffusion wear also occurred, but its small influence on the total wedge wear has prevented detection of its presence.
- In the process of turning no built-up-edge formation has been found due to the brittleness of the material under machining and to the formation of splinter chip. The absence of built-up-edge has also eliminated formation of dents and has influenced the small portion of adhesion wear which arises mainly in the zone of built-up-edge formation.
- Wedge no. 1 made of whisker ceramic has proved to be the most durable one. This was due to abrasion resistance better than that of the SiAlON wedges. The abrasion is the kind of wear which has dominated in turning of the tungsten shaft.
- Fine chip present in machining proves high brittleness of tungsten due to impurity. Even small quantities of impurities – in this case 0.3% - has changed the ductile form of tungsten into brittle one.
- The obtained test results allowed the elaboration of dependence of the wedge life on the cutting speed.
- Wedges made of whisker ceramic can be suitable for turning of tungsten. However, high cutting speeds result in very quick wear of those wedges.

References

1. S. Stolarz, W. Rutkowski, *Tungsten and molybdenum* (PWN, Warszawa, 1961)
2. http://www.baildonit.com.pl
3. E. Lassner, W. Schubert, *Tungsten: Properties, Chemistry, Technology of the Element, Alloys, and Chemical Compounds* (Kulwer Academic / Phenum Publ., New York, 1999)
4. N. Sugita, K. Nishioka, M. Mitsuishi, Int. J. of Autom. Techn., **5**, 3, 320 (2011)
5. W. Zebala, R. Kowalczyk, Int. J. of Adv. Man. Techn., **77**, 9-12, 2241 (2015)
6. S.H. Choi, B.H. Kim, H.S. Shin, D.K. Chung, C.N. Chu, J. of Mat. Proc. Techn., **213**, 4, 621 (2013)
7. G.M. Krolczyk, P. Nieslony, J.B. Krolczyk, I. Samardzic, S. Legutko, S. Hloch, S. Barrans, R.W. Maruda, Measurement, **70**, 203 (2015)
8. G.M. Krolczyk, J.B. Krolczyk, R.W. Maruda, S. Legutko, M. Tomaszewski, Measurement, **88**, 176 (2016)
9. R.W. Maruda, G.M. Krolczyk, E. Feldshtein, M. Szydlowski, S. Legutko, F. Pusavec, A. Sobczak-Kupiec, Int. J. of Mach. Tools and Man., **100**, 81 (2016)
10. S. Wojciechowski, T. Chwalczuk, P. Twardowski, G.M. Krolczyk, Arch. of Civ. and Mech. Eng., **15**, 4, 798 (2015)
11. G.M. Krolczyk, P. Niesłony, S. Legutko, Arch. of Civ. and Mech. Eng., **15**, 2, 347 (2015)

Experimental research about positioning correction of CNC machine with X-axis up to 12000 millimeters using laser measurement equipment

Ştefan Adrian Moldovan[1,*], *Marius* Cosma[1] , and *Vasile* Năsui[1]

[1]Technical University of Cluj-Napoca - North University Center of Baia Mare, Romania

Abstract. In this paper we present a research regarding adjustment machining accuracy on CNC machines with long axis for increasing positioning accuracy considering as a factor: thermal expansion effect. The research performed was made using laser equipment on industrial CNC machines used for production of aerospace components made of aluminium alloy extruded profile with length up to 12000 mm. Those parts have very tight tolerances and on milling process appear several factors that influence the repeatability of machining processes, the main one being the thermal expansion effect. The research results prove the improvement of positioning precision using the coefficients obtained from experimental measurements.

1 Introduction

The application methodology for this research is to check, to correct and to control the machining process, in a factory which makes parts for aerospace industry. In aerospace industry the major importance is to have a good process control for quality of the assemblies parts, to create a process repeatability, because this involves the humans, passengers safety [1].

Without a stable process control for serial production, the machining production will have a bottleneck. To have a quality assurance it will be necessary for each machined part to inspect everything, 100%, to confirm the part quality. This will involve more time and extra costs.

Using laser measurement control equipment, the positioning accuracy caused by backlash error and thermal dilatation error of CNC machines, could be checked and correct easily.

On this research the main contributions are: measurements and collecting CNC machine positioning deviation values from X-axis based on backlash and temperature influence, create a solution to solve this problem, adjust the CNC machine parameters for X-axis with the backlash correction coefficient and thermal dilatation coefficient; achievement the experimental tests [2].

* Corresponding author: moldovan.stefanadrian@gmail.com

2 The main elements that are part of the machining process

2.1 CNC machine definition

The machining process is performed in a normal processing unit, without climate control and the CNC machine used is Handtmann PBZ NT (Figure 1).

The PBZ NT is a profile machining center features impressive maximum flexibility and great cost electiveness in machining aluminum and steel. A flexible clamping system for machining different profile cross sections and up to 90 tool slots allows for a wide range of applications up to 25 meters in length. An integrated sawing unit for spatial cuts completes the highly efficient PBZ NT machine concept [3].

A few technical data for PZB NT [4]:

- 5-axis concept,
- Fixed portal,
- Part lengths up to 25 meters,
- Workpiece cross section max. 750 x 300 mm,
- Max. spindle power 37 kW, max. spindle speed 28.000 rpm,
- Control: SIEMENS Sinumerik 840D,
- CNC-controlled clamping system, variable number of clamping jaws,
- Coolant: mist.

Fig. 1. Handtmann PBZ NT [2].

2.2 Extrusion and part complexity

The research was conducted on extruded aluminium alloys work pieces used in aerospace industry (Figure 2), with a wide range of alloys and treatments. With thin walls and the section represents a high resistance to shock and breakage, develop from advanced alloys tailored for specific applications to help find the right balance of strength, damage tolerance, and corrosion resistance.

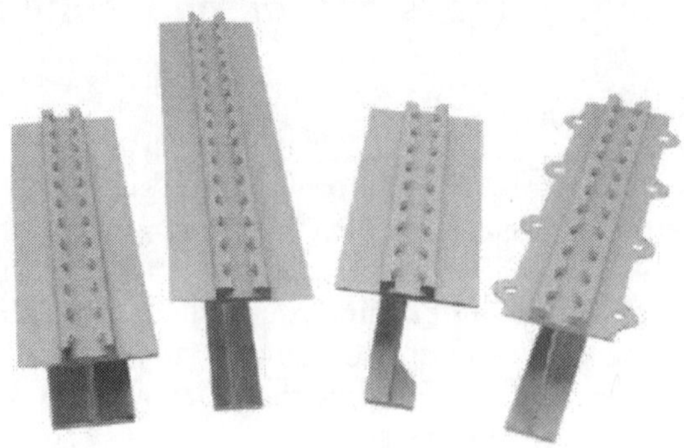

Fig. 2. Examples of aerospace parts used in researches [2].

With thin walls and the section represents a high resistance to shock and breakage [5], develop from advanced alloys tailored for specific applications to help find the right balance of strength, damage tolerance, and corrosion resistance.

Manufacturing process of those parts are define through: drilling holes with different diameters used for subassembly montage, milling different geometries, contour and additional pockets to reduce the final part weight. Most of the geometry and holes have tolerance ± 0.2 mm on 12000 mm length (Figure 3).

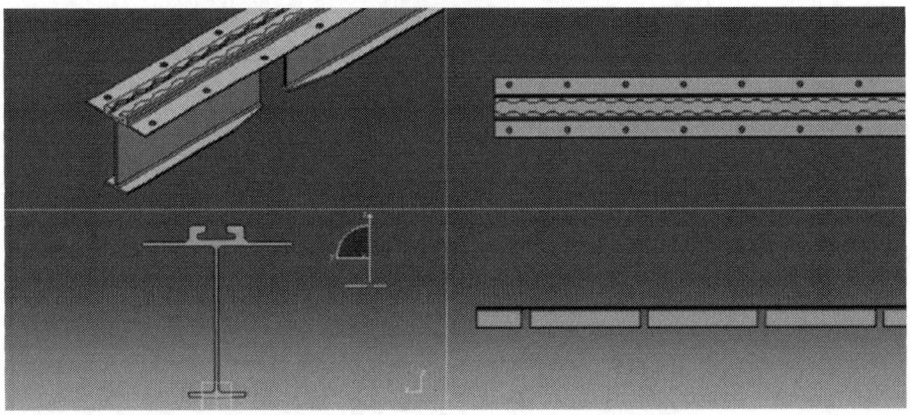

Fig. 3. Multi view of the part in 3D CAD – CATIA R21 [2].

2.3 Problem detection

All machined parts specific for aerospace industry needs to be validated by CMM (coordinate measuring machine), placed in a room with temperature control environment.
CMM is a device for measuring the physical geometrical characteristics of an object. The typical 3 "bridge" CMM (Figure 4) is composed of three axes, an X, Y and Z [6].

Fig. 4. Metris CMM, 3 "bridge" used for parts control [2].

These axes are orthogonal to each other in a typical three-dimensional coordinate system. Each axis has a scale system that indicates the location of that axis. The machine will read the input from the touch probe, as directed by the operator or programmer. The machine then uses the X, Y, Z coordinates of each of these points to determine size and position with micrometer precision typically.

Metris is an high quality and innovative metrology solutions to a wide range of industries.

Metris CMMs are supplied with CAMIO Studio and the MCC-200 controller as standard and benefit from the same broad range of hardware options including [5]:

- point to point scanning - for a Renishaw TP200 touch trigger probe,
- continuous contact scanning - Renishaw SP25 analogue probe,
- non-contact laser scanning – Metris LC and XC range of laser scanners,
- fully automated stylus and probe changers,
- motorized indexing or rotary table,
- pneumatic anti-vibration mounts,
- temperature compensation.

2.4 Methods for solving the problem

2.4.1 Laser measurement system

The Renishaw laser measurement system is an ideal solution for the complete calibration of machines (Figure 5), enabling the measurement of a wide range of geometric and dynamic characteristics.

Measurement options [7]:

- linear positioning accuracy and repeatability of an axis,
- angular pitch and jaw of an axis,
- straightness of an axis,
- squareness between axes,
- flatness of a surface,
- rotary axis/table angular positioning,

• dynamic characteristics of a machine.

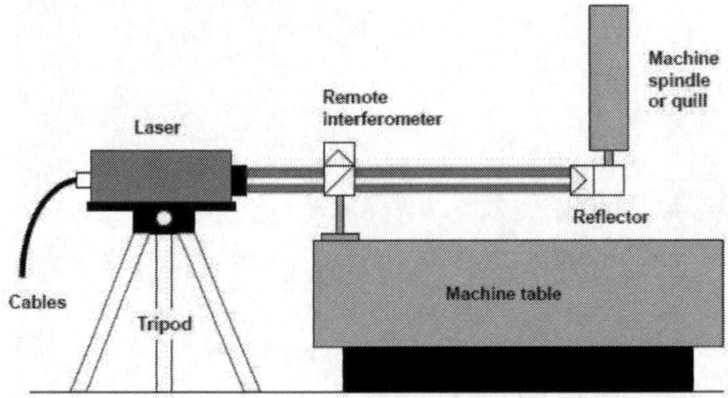

Fig. 5. Laser measurement system [7].

2.4.2 Linear measurement

The system measures linear positioning accuracy and repeatability by comparing the position displayed on a machine's readout with the true position measured by the laser. There values can then be viewed, printed and statistically analysed by the system's Renishaw plc XCal-View 2.0.0.17005 software to national and international standards. This software generate a coefficient for deviations compensation of positioning repeatability named backlash and corrects positioning for intervals from 100 to 100 mm with deviations measured from laser system (Figure 6).

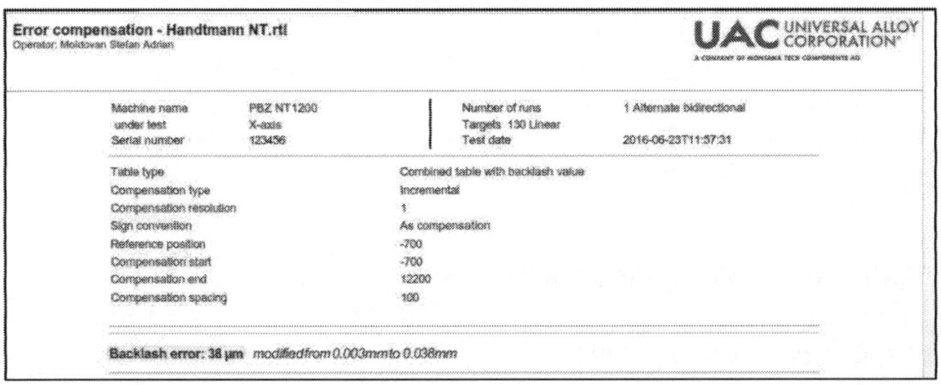

Error compensation - Handtmann NT.rtl
Operator: Moldovan Stefan Adrian

UAC UNIVERSAL ALLOY CORPORATION
A COMPANY OF MONTANA TECH COMPONENTS AG

Machine name under test	PBZ NT1200 X-axis	Number of runs	1 Alternate bidirectional
		Targets 130 Linear	
Serial number	123456	Test date	2016-06-23T11:57:31

Table type	Combined table with backlash value
Compensation type	Incremental
Compensation resolution	1
Sign convention	As compensation
Reference position	-700
Compensation start	-700
Compensation end	12200
Compensation spacing	100

Backlash error: 38 µm *modifiedfrom 0.003mmto 0.038mm*

Fig. 6. Renishaw plc XCal-View 2.0.0.17005 software [2].

For easy optical alignment the components used in this measurement comprise are linear beam-splitter and retro-reflectors (Figure 7).

DOI: 10.1051/matecconf/20179402009

Fig. 7. Linear beam-splitter and retro-reflectors [7].

In linear measurement, one retro-reflector is secured to the beam-splitter, to form the fixed length reference arm of the interferometer. The other retro-reflector moves relative to the beam-splitter and forms the variable length measurement arm. The laser system tracks any change in the separation between the measurement arm retro-reflector and beam-splitter (Figure 8).

Fig. 8. Positioning the retro-reflectors on the frame of Handtmann PBZ NT [2].

3 Results

The correction of position accuracy is developed from three steps:

- Laser measurement of X-axis positions on initial state of Handtmann CNC machine from 100 to 100 mm, the trend represented in Figure 9 with yellow colour.
- Laser measurement of X- axis positions without any coefficients, some results recorded in Table1, the trend is shown in Figure 9 with black colour. The new backlash is generated, in our case the backlash value is 0,038 mm. Measured values of deviations will be used on positioning adjustments. After this correction the measurement trend is almost straight and the machine precision is increased.
- Laser measurement of X-axis positions using the new backlash and the new correction of positioning values from 100 to 100 mm the trend represented in Figure 9 with blue colour.

Table 1. Some laser measured values of X-axis position deviations.

Index	Positions (mm)	Deviation (μm)	Index	Positions (mm)	Deviation (μm)
1	-700	0	18	1000	22
2	-600	21	28	2000	-4
3	-500	-10	38	3000	4
4	-400	9	48	4000	-3
5	-300	14	58	5000	3
6	-200	-21	68	6000	-4
7	-100	3	78	7000	-20
8	0	7	88	8000	-3
9	100	-11	98	9000	2
10	200	19	108	10000	-5
11	300	5	118	11000	-8
12	400	-23	127	11900	-25
13	500	8	128	12000	13
14	600	2	129	12100	22
15	700	11	130	12200	-2

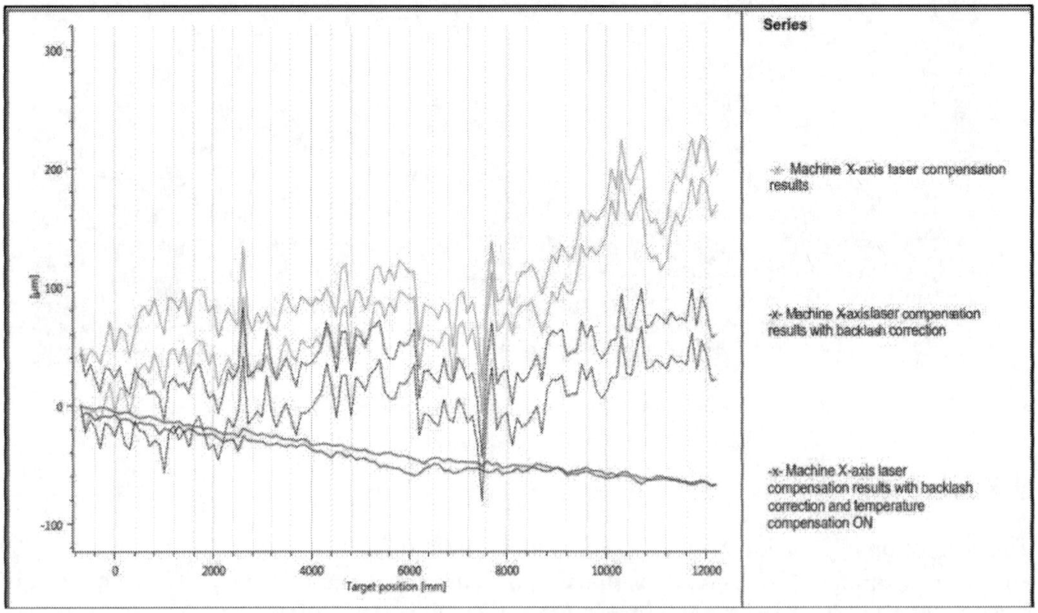

Fig. 9. The chart of measured values from Renishaw software [2].

Besides generate coefficients from Renishaw software, to can increase the positioning accuracy is necessary to take into consideration the difference between coefficients from thermal dilatation of machine frame and raw material [8]. For machining aluminium alloys the difference is $12\cdot10^{-6}$ /°C (Figure 10) coming from difference between frame dilatation coefficient $11\cdot10^{-6}$ /°C and aluminium alloy dilatation coefficient $23\cdot10^{-6}$ /°C).

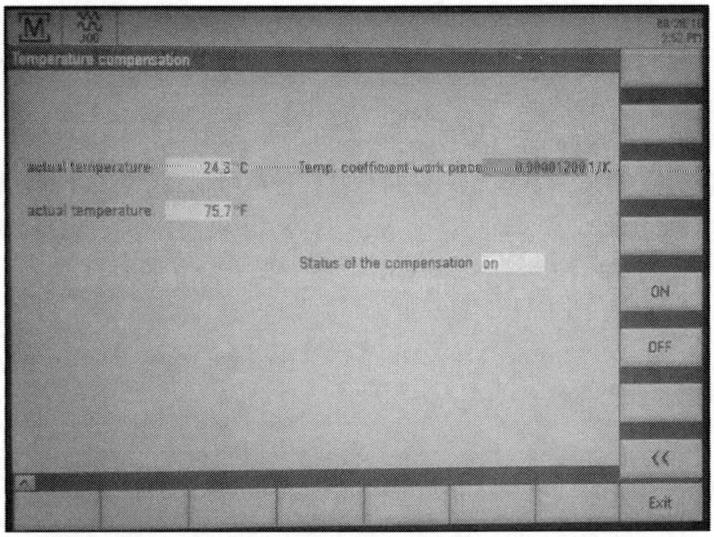

Fig. 10. Handtmann NT thermal dilatation coefficient [2].

4 Conclusions

Experimental research bring a version of thermal expansion which influence the stability of the cutting process on processing centres, which is a major concern for processing repeatability of specific parts from aerospace industry.

This research methodology improve the machining process control with are influenced by temperature. The original contribution for this research is: monitoring the X-axis positioning fluctuation and implementing through machine variables a backlash correction.

The experimental research conducted with the support of the machining department of SC Universal Alloy Corporation Europe SRL, Dumbrăviţa.

References

1. V. Năsui, *Bazele cercetării experimentale* (Editura Universităţii de Nord, Baia Mare, 2000)
2. Cercetări privind creşterea eficienţei la operaţiile de prelucrare pe maşini-unelte cu comandă numerică în cadrul S.C. Universal Alloy Corporation Europe S.R.L. (Nr.3657/09.02.2016)
3. Training Guide PBZ NT Ver. (2010)
4. https://www.handtmann.de/en/machining-centres/products/pbz-profile-machining-centres/pbz-nt/ (2016)
5. http://www.universalalloy.com/Manufacturing-Capabilities.html (2016)
6. Metris Bridge CMM User Guide (2016)
7. http://www.arc-bg.com/prod_file/81.pdf (2016)
8. Function Manual, SINUMERIK 840D sl / 828D Extended Functions (2016)

Modelling of the aluminium alloy machining process

Piotr Paszta[1,*]

[1]Faculty of Mechanical Engineering and Computer Science, Czestochowa University of Technology, Al. Armii Krajowej 21, 42-200 Czestochowa, Poland

Abstract. The article presents the possibility of modelling the process of turning aluminium alloy 2017. It reports obtained results for the distribution of stress and temperature in the machining zone. Tests for assumed machining parameters were performed on a CNC machine tool. Microscopic photographs of the turned surface are shown. Top surface measurements were taken on a computerize measuring stand including a New Form Talysurf 2D/3D 120 Taylor Hobson profilometer.

1 Introduction

Machining is one of the most widespread material working types, being part of the process of manufacturing elements of many technical devices. During machining, the workpiece obtains the required shapes and dimensions and the desired surface quality. The process progresses in a very dynamic manner, violent stresses and elastic and plastic strains form in the machined material and temperature changes occur in the machining zone, there occurs also friction, chip upsetting, material hardening, and sometimes accretion forms.

The main drawback of subtractive manufacturing is the formation of waste material in the form of chips. Very often, this is the only material working method available, impossible to be substituted or eliminated. It provides the required accuracy and top surface finish at low expenditure. A rapid development of machining tools, instrumentation, CNC machines has taken place in recent years, along with their wide dissemination, even in minor plants. A greater proportion of machining with small allowances and finishing machining is visible.

A group of aluminium alloys has found wide application in engineering. By adding alloying elements to aluminium, we can markedly improve its mechanical properties and machinability. The main aluminium alloy additions include copper, silicon, magnesium, zinc and lithium. Aluminium alloys have a low mass density and an advantageous strength to weight ratio. Aluminium alloys are categorized into wrought alloys and casting alloys. In respect of processing technology, these groups may overlap. Alloy 2017 finds wide application in subtractive manufacturing owing to its slight contents of insoluble elements, such as lead and bismuth, forming dispersed particles within the microstructure, which facilitate the chip forming process.

* Corresponding author: paszta@itm.pcz.pl

The increased concentration of copper and magnesium enhances the tensile strength and hardness and heat resistance, impairs ductility and slightly reduces the corrosion resistance. Alloys from this group are willingly used in the aircraft and automotive industries. They find application in the manufacture of car and aircraft wheel rings, aircraft bearing elements and fuselages and responsible structural parts, car and aircraft engine parts, as well as engine equipment parts, such as turbine and compressor impellers, and engine cylinder heads operating at elevated temperatures.

Aluminium alloys melt at a temperature of 550-650^{0}C. In alloys with a large content of silicon, partial melting of the machined surface may occur. Owing to the lower thermal and mechanical loads during machining of duralumin, increased machining speeds can be used. Overheating during machining may result in a drop in the strength of the part being machined. To prevent this, cooling should be used during machining [1, 2].

Aluminium alloys are well machinable. To evaluate the machinability of a given alloy, it is necessary to subject it to machining, assess the tool wear and chip shape, and then choose the appropriate machining parameters. When relaying on catalogue machining parameter selection data, they should be constantly optimized experimentally.

2 Numerical modelling results

During machining of aluminium and its alloys, a temperature of up to approx. 550^{0}C, thus being close to the melting point, can be generated. An example is modelling of the process of machining aluminium 7075 (Figure 1). A DCGT T304 solid carbide cutting plate was employed. Adopted machining parameters conforming to the catalogue data for the machined material and the cutting plate were as follows: machining speed, Vc=300 m/min, machining depth, ap=2.5 mm, synchronous feed, ft=0.25 mm/rev. The material used for machining was an aluminium alloy having high mechanical strength, comparable to that of constructional steels, and very high fatigue strength. It has very good machinability and a hardness of up to 190HB. It has moderate corrosion resistance. The alloy is used for making blow moulds, plastic foaming moulds, stamping and blanking die elements, as well as in the aircraft industry for making heavily loaded structural elements and for production of sports equipment.

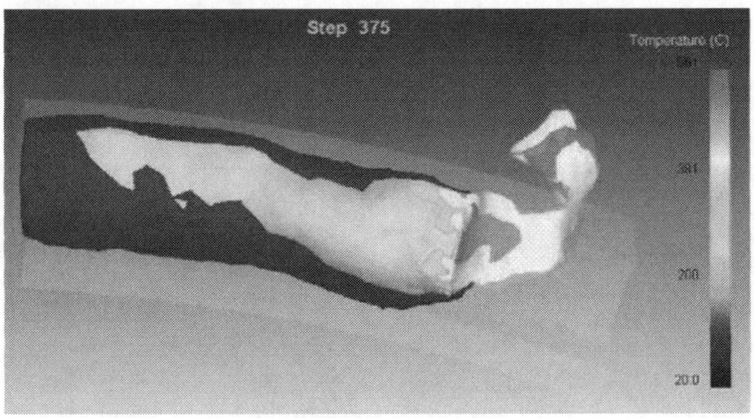

Fig. 1. The modelling outcome – aluminium alloy 7075 temperature distribution.

Nowadays, every machining process requires modelling, analysis and often verification by numerical methods. It is possible to select the appropriate tool geometry, feed parameters, rotational speed and machining depth. For machining aluminium, polycrystalline diamond and coated sintered carbide plates are used.

There many programs that utilize the Finite Element Method (FEM) aiding CAX engineering work. One of them is DEFORM 3D. It offers a number of systems and modules that enable a comprehensive analysis of manufacturing processes. This is a flexible software application that enables one to analyze a wide range of deformations, thermal processes and microstructure [3, 4].

The system can be used for modelling of machining processes. The program procedures enable modelling of engineering processes, the selection of boundary conditions, and allow the analysis of the process for any change in process conditions. With its use, for example, the following can also be computed: machining force, temperature in the tool and material contact zone, chip shapes, or material wear. The effect of machining process parameters, such as machining speed, feeds and machining depth, can be examined.

In the simulation of the duralumin machining process, the Johnson-Cook constitutive model was employed, which enables the yield stress to be calculated from the following formula [5, 6]:

$$\bar{\sigma}=\bar{\sigma}\,(\bar{\varepsilon},\dot{\bar{\varepsilon}},T) \tag{1}$$

$$\bar{\sigma}=(A+B\bar{\varepsilon}^{n})\left(1+C\ln\left(\frac{\dot{\bar{\varepsilon}}}{\dot{\bar{\varepsilon}}_{0}}\right)\right)\left(\frac{\dot{\bar{\varepsilon}}}{\dot{\bar{\varepsilon}}_{0}}\right)^{\alpha}(1-T^{*m})\;\; T^{*}=\left(\frac{T-T_{room}}{T_{melt}-T_{room}}\right) \tag{2}$$

Where: A, B, C, m, n, α are dimensionless material constants, T_{melt} – material melting point, T_{room} – reference temperature, T– workpiece temperature, $\bar{\varepsilon}$ – effective plastic strain, $\dot{\bar{\varepsilon}}$ – effective plastic strain rate, $\dot{\bar{\varepsilon}}_0$ – reference effective plastic strain rate.
The functions used in the Johnson-Cook model represent the effect of different factors on the magnitude of $\bar{\sigma}$ yield stress in the material of the workpiece shaped during machining. These are: strain hardening, plastic strain rate, material susceptibility to strain rate and the effect of temperature [7].

Material used for simulation was aluminium alloy 2017. The Usui tool wear model was adopted in the simulation. The coefficients used in the model rely on experimental relationships and are dependent on the process conditions and the materials used [6, 8, 9]. Then, a database was generated and simulation of the turning process was performed for the following parameters: machining depth, ap=1 mm, turning diameter 48 mm, rotational speed 5400 rpm, synchronous feeds ft1= 0.1 mm/rev and ft2= 0.01 mm/rev. Upon completion of the simulation, the following results were obtained. Figure 2 shows the distribution of normal stress for a feed of 0.1 mm/rev and 0.01 mm/rev, respectively. Figure 3 illustrates the distribution of machined material temperatures, while Figure 4 shows the tool cutting edge temperature.

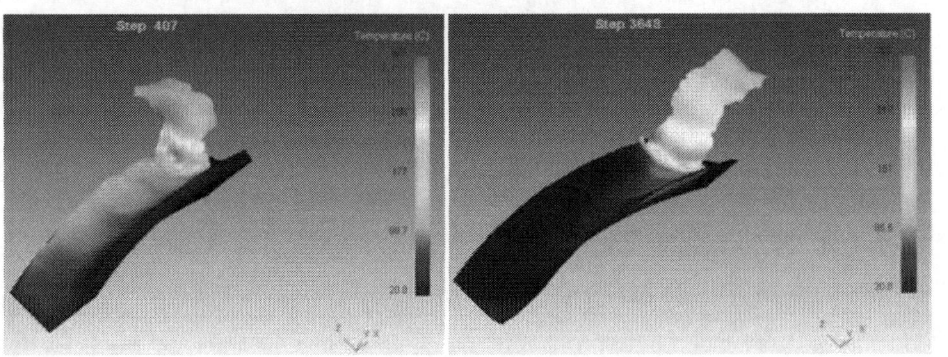

Fig. 2. The modelling outcome – machined material temperature distribution.

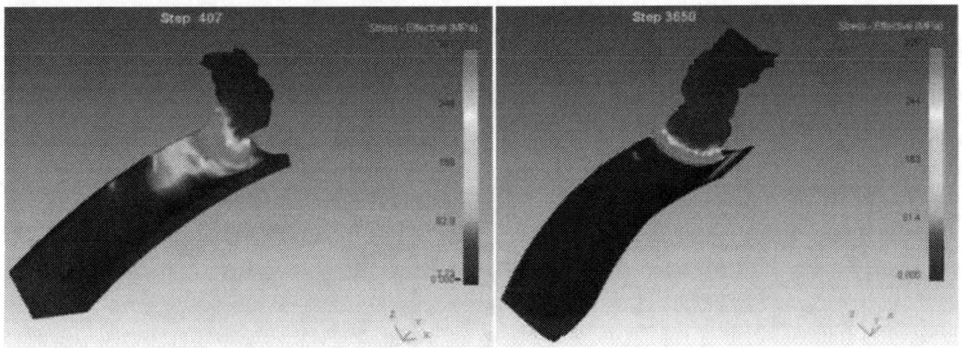

Fig. 3. The modelling outcome – normal stress distribution.

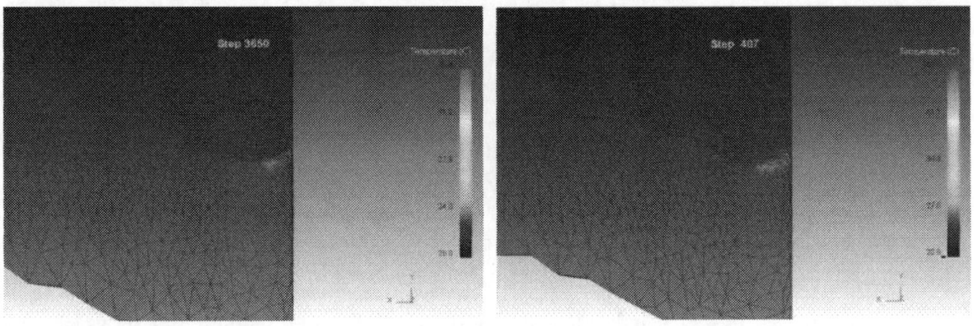

Fig. 4. The modelling outcome – tool cutting edge temperature.

3 Stereometry measurement on the profilometer

The top layer condition depends largely on the aluminium alloy type, machining parameters, tooling quality and the machine tool technical condition. With the increase in hardness and strength and the contents of hard inclusions, such as silicon, aluminium alloys are more readily machinable and form short breakable chips. At high machining speeds, a small feed and small machined material layers and with sharp cutting plate edges, additional finishing can be eliminated. Incorrectly selected machining parameters, on the other hand, may cause the formation of accretions on the cutting edge and, as a consequence, a change in its geometry, which will result in a poor surface quality. An essential role is played also by the shape of chips and their evacuation from the machining zone. Unless chips are removed effectively, scratching of the surface and an impairment of its condition may usually be caused. Should a large volume of continuous chips form, an imbalance of the workpiece, damage to the machining tool, or the complete disruption of the machining process may result.

Measurements were taken on a computerized measuring stand consisting of a New Form Talysurf 2D/3D 120 Taylor Hobson profilometer with the 'Ultra Surface 5.21' and 'TalyMap Platinium 5.1.1' software applications, which enables the comprehensive measurement of top layer surface roughness and stereometry parameters in 2D and 3D systems with a minimum measuring head resolution of 3.2 nm. The measuring stand is furnished with a table enabling measured surfaces to be precisely moved in the direction perpendicular to the gauging point displacement. It makes it possible to take measurements necessary for the three-dimensional reproduction of the stereometry of the zone in which an irregularity forms as the tool contacts the machined material (aluminium alloy 2017). A gauging tip in the form of a 2 μm-radius diamond needle was used in the measurements [10].

 DOI: 10.1051/matecconf/20179402010

Results from two turning tests performed on a CNC turning lathe are reported. Machining parameters identical to those in the numerical analysis were taken in the tests.

The three-dimensional image of the recorded deformation zone was digitally pre-processed to make it level (Figure 5).

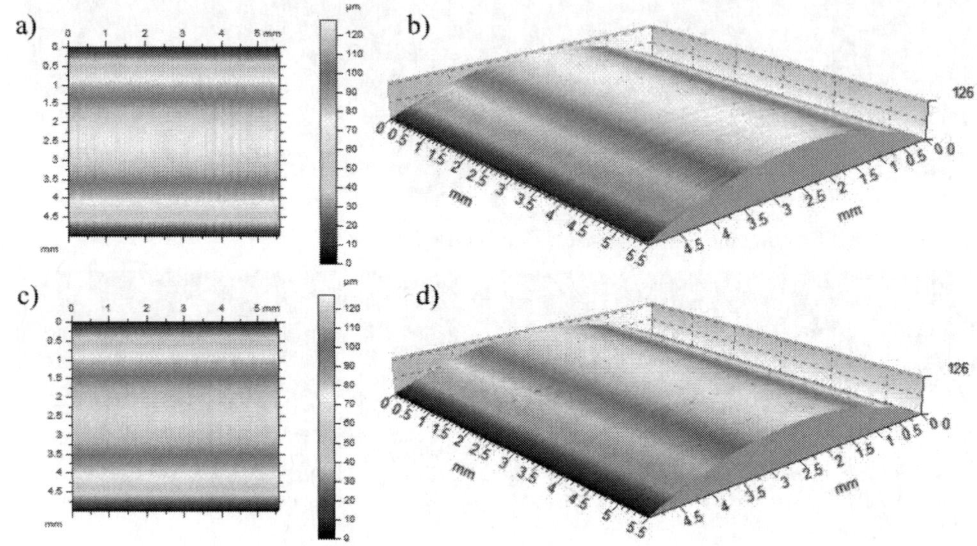

Fig. 5. The stereometry of a selected cylindrical surface area (a, c) a contour line map, (b, d) a 3D view.

The further digital processing of the area under analysis removed the shape component, the cylinder, from the image (Figure 6).

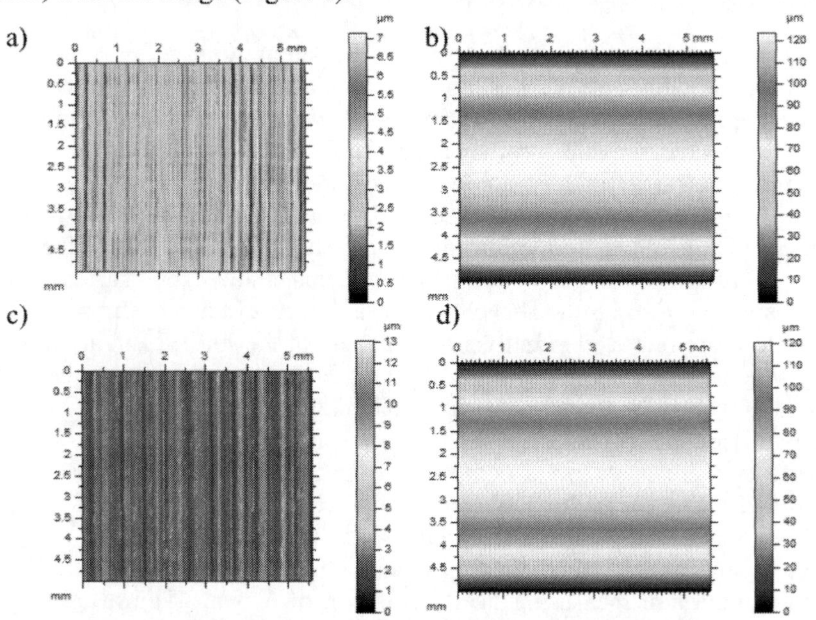

Fig. 6. An image of the stereometry of the selected cylindrical surface area: (a, c) a contour line map of the planar area, (b, d) the cylindrical shape component.

A planar image of the analyzed stereometry was formed. It is also possible to determine the three-dimensional view of the analyzed stereometry (Figure 7).

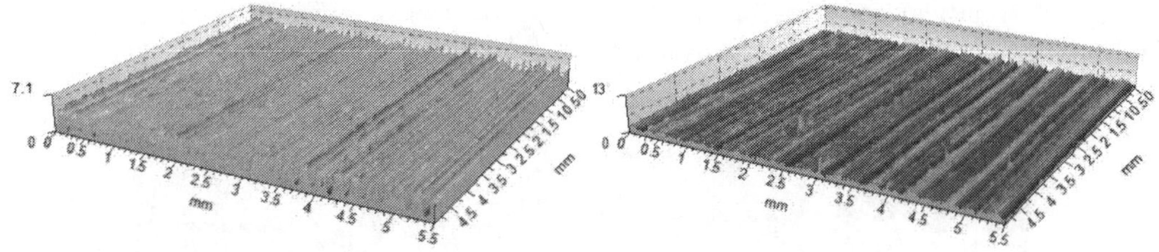

Fig. 7. A three-dimensional image of the selected area stereometry.

The selected parameters of the stereometry of the analyzed area (the height parameters of the first test: Sa=0.86 μm, Sq=1.03 μm, Sz=7.16 μm, Sp=4.41 μm, Sv=2.75 μm, and the height parameter for the second test: Sa=0.40 μm, Sq=0.54 μm, Sz=13.10 μm, Sp=10.70 μm, Sv=2.40 μm), the distribution of the area ordinates and the area load capacity curve were determined.

In addition, for the selected profile, after filtration, the roughness parameters of the analyzed profile (the amplitude parameters for the first test: Ra=0.81 μm, Rq=0.97 μm, Rz=4.03 μm, Rp=2.35 μm, Rv=1.68 μm, and the amplitude parameters for the second test: Ra=0.36 μm, Rq=0.46 μm, Rz=2.51 μm, Rp=1.47 μm, Rv=1.05 μm), as well as the distribution of the profile ordinates and the material ratio curve, can be determined (Figure 8).

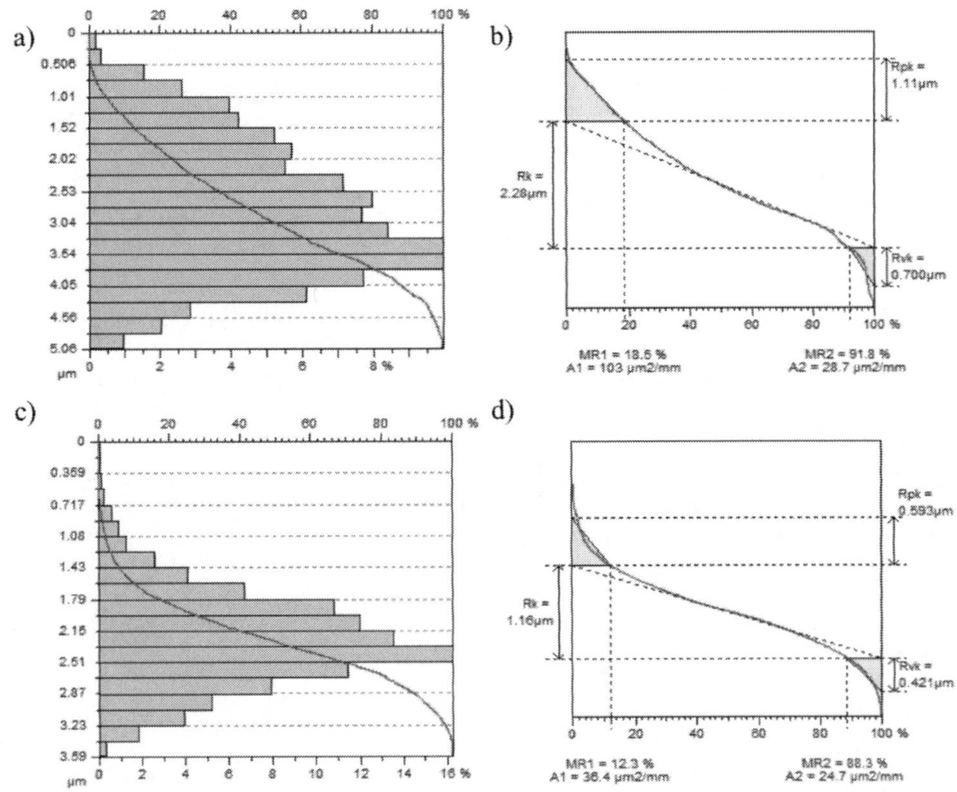

Fig. 8. The distribution of the ordinates (a, c) and the load capacity curve (b, d) for the analyzed profile.

As a complement to the analysis of the examined surface stereometries, microscopic photographs of the formed surface irregularity contours were taken. The photographs were taken utilizing the capabilities of a VMHT MOT laboratory digital microscope.

The photographs cover the fragment of the top layer area subjected to turning (Figure 9). They confirmed the results of analyses made based on surface stereometry examination.

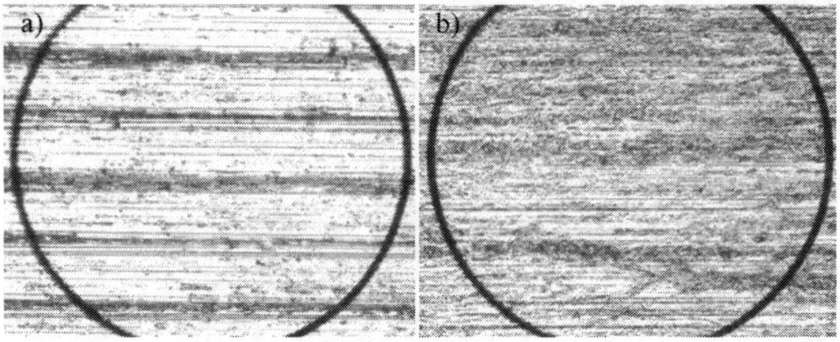

Fig. 9. A microscopic photograph of the turned surface a) ft=0.1 mm/rev, b) ft=0.01 mm/rev.

The actual roughness value of the surface machined by turning is, however, dependent on many factors, chiefly on the plastic properties of the machined material, the vibrations of the machine tool–tool system, tool wear, etc. Their impact on the surface roughness is normally determined experimentally. The basic purpose of machining is to shape a new part surface so that it meets the quantity and quality requirements specified by the designer.

4 Conclusions

The methods of modelling and simulation of technological processes are becoming essential in the design of technological processes. They allow the saving of time and indicate the direction of practical activities. It is becoming possible to test and verify the correctness of the operation conditions of new machining tools according to the user's own design, with an untypical microgeometry, without having to create them physically.

Aluminium alloys are readily machinable. The analysis showed that selecting high machining speeds, as compared to e.g. steel, did not cause any major increase in turned material temperature (approx. 200^{0}C), its fusion or structural changes. The increase in tool surface temperature is also small. The majority of heat is removed with chips.

By selecting a high machining speed, a small feed and a small material layer, as well as a sharp cutting plate, the additional finishing can be eliminated and a better surface quality can be achieved. A surface with a surface roughness of Ra=0.36 μm was obtained in the test.

The aim of the present analysis is to enable the control of the duralumin turning process and to optimally select the parameters of technological processes to be carried out, as well as to predict the operation parameters.

References

1. W. F. Hosford, *Physical Metallurgy* (CRC Press, Boca Raton, 2010)
2. L.R. Kilmetova, P.V. Sletnev, J. Eng. Sci. Technol. Rev.,**7**, 5 (2014)
3. M. K. Swamy, B. P. Raju, B. R. Teja, IOSR J. Mech. Civ. Eng., **3**, 6 (2012)
4. C. Constantin, S. M. Croitoru, G. Constantin, C. F. Bisu, *Advances in Visualization, Imaging and Simulation, Proc. 3rd WSEAS Int. Conference*, **10** (University of Algarve, Faro 2010)
5. A. A. Kawi, Al-Qadisiya J. Eng. Sci. **4**, 3 (2011)
6. *DEFORM 3D User's Manual,* (Scientific Forming Technologies Corporation, Columbus, Ohio 2007)

7. G. R. Johnson, W. H. Cook, 7th Int. Symp. Ballistics (1983)
8. J. Lorentzon, N. Jarvstrat, Arch. Comput. Mater. Sci. Surf. Eng., **1**, 1 (2009)
9. Y. R. Bhoyar, P. D. Kamble, Int. J. Mod Eng., **3**, 1 (2013)
10. A. Zaborski, Mechanik, **11** (2010)

Analyses of process parameters influence on the drilling process by using carbide drills for steel St52-3

Ioan Alexandru Popan[1,*], *Alina Ioana* Popan[1], *Sorin Cosmin* Cosma[1], and *Alexandru* Carean[1]

[1]Technical University of Cluj-Napoca, B-dul. Muncii, No. 103-105, Cluj-Napoca, Romania

Abstract. The influence of cutting parameters on the hole quality made using a new type of carbide drill, RT100 U, has been analysed experimentally and analytically. The drilling process is a common application in industrial environments and it has to be established at high accuracy and short time. The influence of process parameters (cutting speed and feed rate) on dimensional accuracy and surface roughness is analysed. The experiment was made by using external cooling and canned cycles (G81). On this experimental study was analysed the effect of feed rate onto the chip shape.

1 Introduction

In this paper a new type of carbide drill, made by Guhring Company is analysed. The RT100 U drill has 2 flutes and a FIREX coated. It has a special designed to obtain a high penetration rate, a high holes accuracy and good surface quality.

In drilling process a significant amount of heat is generated because of the friction between tool and workpiece. For increasing the productivity, intensification of cutting parameters, techniques for cooling the process is needed [1].

Flood machining is a common technique in which a large quantity of cutting fluid is applied to the cutting tool and workpiece. Dry machining, cryogenic machining and Minimum Quantity Lubrication (MQL) are some alternative methods, more environments friendly [2 - 4].

On this paper the flood machining is chose, because is one of the most often cooling method used on industry. This type of cooling comes with improving of surface quality and can increase the tool life [4, 5]. Also the manufacturer recommends flood machining for this type of drill.

A large number of drilling cycles are used on CNC drilling to improve the process. The most often met canned cycles are: (G81) used for simple drilling/spot drilling operations, (G83) peck drilling, this technique makes deep-hole drilling easy and economical and a chip-breaking canned cycle (G73). The canned cycles (G81) is used for many drilling operation because is the most efficient [6 - 8].

*Corresponding author: ioan.popan@tcm.utcluj.ro

DOI: 10.1051/matecconf/20179402011

The RT100 U drill is for general use and the manufacturer recommends some general process parameters. The manufacturer catalogues recommend using for drilling of a carbon steel St52-3 a cutting speed of 70-110 m/min and a feed rate of 0.05-0.4 mm/rev [1].

During cutting process, when the process parameters are wrong selected can appear chip control problems, low surface quality and low accuracy. By using a high cutting speed and a low feed rate can determine a bad chip evacuation and a tool broken [2, 6, 9, 10].

This paper presents the effects of process parameters (cutting speed and feed rate) on quality characteristics of holes made using this drill.

2 Cutting tool and workpiece material

A carbide drill RT100U/DIN653 (presented in fig. 1) made by Guhring Company was chosen for experimental research. The RT100 U style drill has 2 flute and FIREX coated. It is designed for high penetration rate to increase the productivity and increase the holes quality. The drill has a self-cantering 140° SU point, which eliminates the need for preliminary spot drill operations [1]. A new drill shape, RH Helix was developed to create short chips and a good chips evacuation. The wide-open flute was design in order to have short chip formation ant to eliminate the peck drilling cycle [1].

Diameter (d1) Dec. inch	Fract. inch	Wire / letter	mm	d2 mm	l1 mm	l2 mm	EDP #
0.3858		W	9.800	10.000	89.00	47.00	9055140098000

Fig. 1. The carbide drill RT100U.

The structural steel DIN ST52-3 (W.Nr 1.0580) was selected for this experiment. The chemical composition (reference value %) are: C: 0 – 0.22; Si: 0 – 0.55 Mn: 0 – 1.6 P: 0 – 0.035; and S: 0 – 0.035. This type of steel is often used for common applications and structural assemblies.

3 Experimental procedures

The Design of Experiment (DOE) methodology was used for the experiments. In this analyse a Central Composite Design (CCD) with two factors was chosen. The variation range of the input parameters is: cutting speed=70-110 [m/min]; feed rate=0.05-0.4 [mm/rev]. The values of input parameters (cutting speed and feed rate) are presented in Table 1.

Table 1. The combination of the process parameters.

No.	Cutting speed [m/min]	Feed rate [mm/rev]
1.	90	0.05
2.	90	0.23
3.	70	0.23
4.	90	0.23
5.	75.86	0.35
6.	90	0.4
7.	90	0.23
8.	90	0.23
9.	104.14	0.1
10.	90	0.23
11.	104.14	0.35
12.	75.86	0.1
13.	110	0.23

A number of 13 blind holes with Ø9.8 x15 mm were made on a 50x50x300 mm material. The experimental tests were made using a CNC vertical processing machine HAAS, with a motor power of 24 kW, operating at max. 21,000 rev/min. For the tests an external cooling system was used. The type of cutting fluid is CASTROL SYNTILO 9954.

A mechanical vise (fig. 2) was used for clamping the workpiece and a Weldon clamping device with a high rigidity, (fig. 3) to clamp the drill. The experimental stand is presented in figure 4.

Fig. 2. The clamping system used for material fixing.

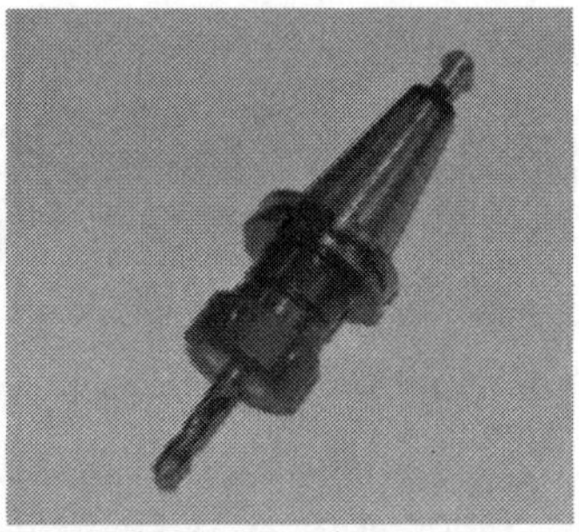

Fig. 3. The Weldon clamping system used for fixing the drill.

Fig. 4. Experimental stand (vise, drill, work piece and milling machine).

Using 3D measurement equipment, type 3D Aberlink (fig. 5), the dimensional deviation and cylindricity deviation were measured. A Mitutoyo Surftest SJ-210 was used for measuring the surface roughness (Ra), (fig. 6). The surface roughness measurements were repeated for three times for each hole. The measurements were made in axial direction for a length of 5 mm in the middle of the hole.

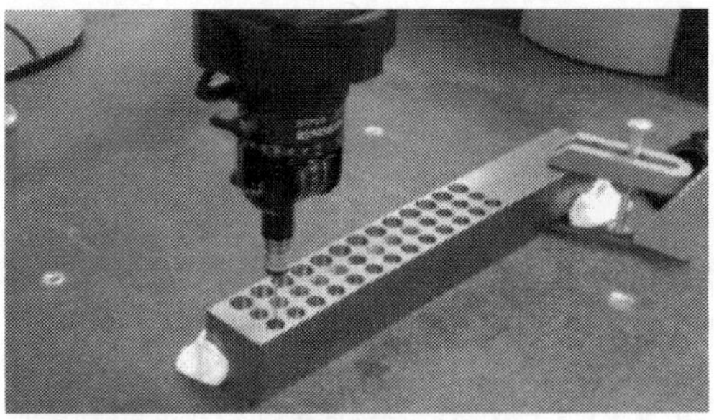

Fig. 5. Dimensional measurements.

Fig. 6. Surface roughness measurements.

The experimental results obtained in different test conditions are presented in table 2. Using Design Expert software the analysis of the experimental values and calculation of its coefficients, were made.

Table 2. Experimental results at different test conditions.

No.	Cutting speed [m/min]	Feed rate [mm/rev]:	Roughness Ra [µm]:	Dimensional deviation [mm]:	Cylindricity deviation [mm]:
1	90	0.05	0.49	0.008	0.022
2	90	0.23	1.1	0.025	0.009
3	70	0.23	1.17	0.03	0.01
4	90	0.23	1.18	0.026	0.013
5	75.86	0.35	0.26	0.054	0.032
6	90	0.4	1.17	0.04	0.006
7	90	0.23	1.08	0.022	0.009
8	90	0.23	1.09	0.023	0.01
9	104.14	0.1	0.53	0.022	0.009
10	90	0.23	1.15	0.025	0.01
11	104.14	0.35	0.31	0.083	0.05
12	75.86	0.1	0.297	0.024	0.041
13	110	0.23	0.45	0.028	0.014

 DOI: 10.1051/matecconf/20179402011

4 Results and discussions

Studying the influence of process parameters on the quality characteristics of the holes was based on mathematical models developed using RSM (Response Surface Methodology).

4.1 Effect of machining parameters on surface roughness

Starting from Pareto ANOVA table was obtained for surface roughness modulation a Quadratic Model is significant. Based on this empirical mathematical model the variation curves of surface roughness according to process parameters were outlined.

In figure 7 is presented the dependence of the cutting speed and surface roughness. It is noted that increasing the cutting speed from 75 to 104 m/min, the surface roughness increases from 0.9 to 1.1 µm (90 m/min) and decreases to 0.7 µm. The analyses were made using a 0.23 mm/rev feed rate.

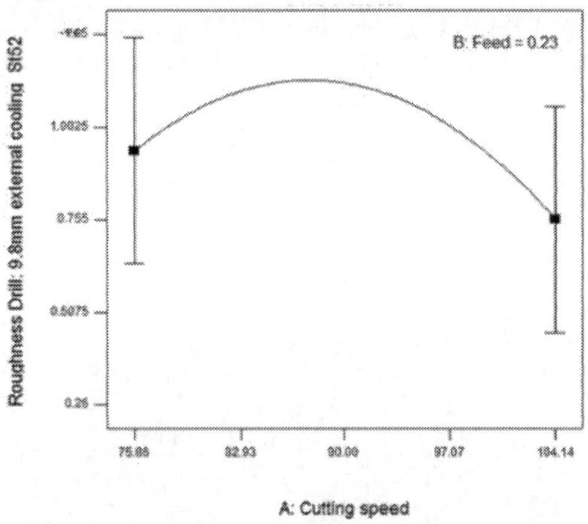

Fig.7. Effect of cutting speed (m/min) on surface roughness (Ra/µm).

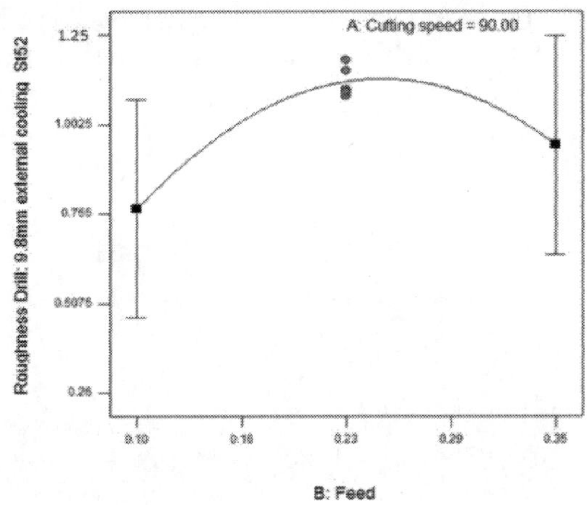

Fig.8. Effect of feed rate (mm/rev) on surface roughness (Ra/µm).

In figure 8 the effect of feed rate on surface roughness is presented for a constant cutting speed 90 m/min. Increasing the feed rate from 0.1 to 0.35 mm/rev, the surface roughness increases from 0.7 to 1.2 µm and decreases up to 0.8 µm.

4.2 Effect of machining parameters on the hole dimensional deviation

For dimensional and cylindricity deviations a linear model form ANOVA analysis was obtained. Based on this linear mathematical model the variation curves of dimensional deviation according to process parameters were drawn.

In figure 9 is presented the effect of cutting speed onto the hole dimensional deviation, the feed rate is 0.23 mm/rot. Increasing the cutting speed from 75 to 104 mm/min can be observed that dimensional deviation increases from 0.028 to 0.035 mm. In conclusion, the cutting speed does not have a strong effect on dimensional deviation.

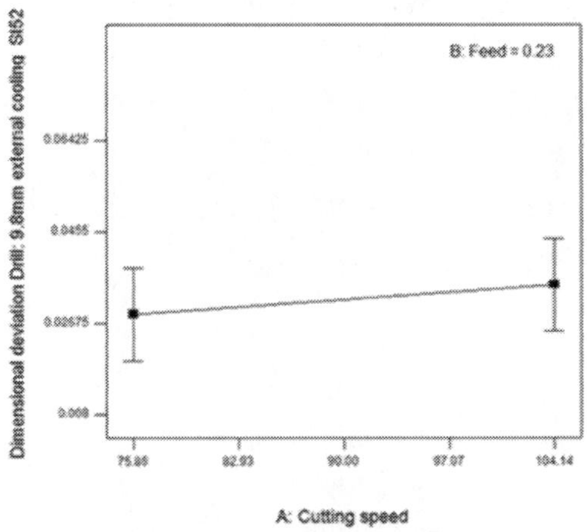

Fig.9. Effect of cutting speed (m/min) on dimensional deviation (mm).

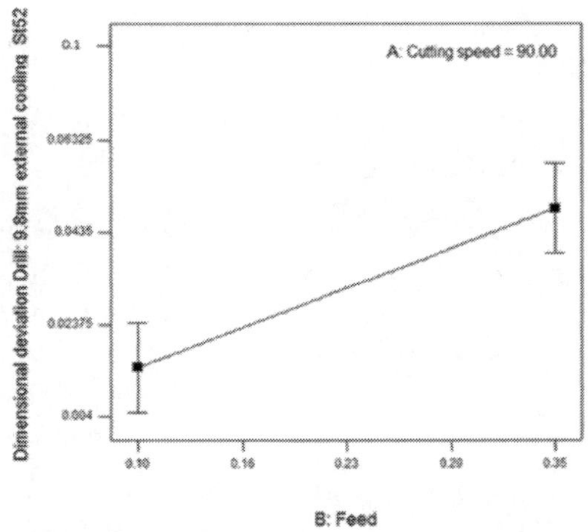

Fig.10. Effect of feed rate (mm/rev) on dimensional deviation (mm).

The effect of feed rate on dimensional deviation for a 90 m/min constant cutting speed is shown in figure 10. Increasing the feed rate from 0.1 to 0.35 mm/rev the dimensional deviation increases from 0.004 to 0.04 mm.

4.3 Effect of machining parameters on the hole cylindricity deviation

Based on a linear mathematical model the variation curves of cylindricity deviation according to process parameters were outlined.

In figure 11 is presented the effect of cutting speed on the hole cylindricity deviation, the feed rare is 0.23 mm/rot. By increasing the cutting speed from 75 to 104 mm/min can be observed the cylindricity deviation decreases from 0.02 to 0.017 mm.

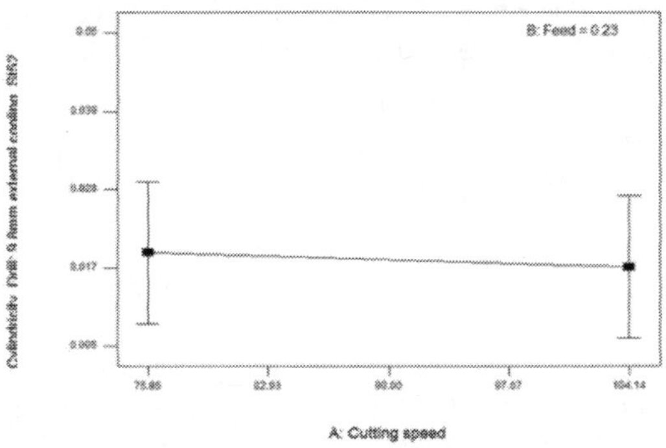

Fig.11. Effect of cutting speed (m/min) on cylindricity deviation (mm).

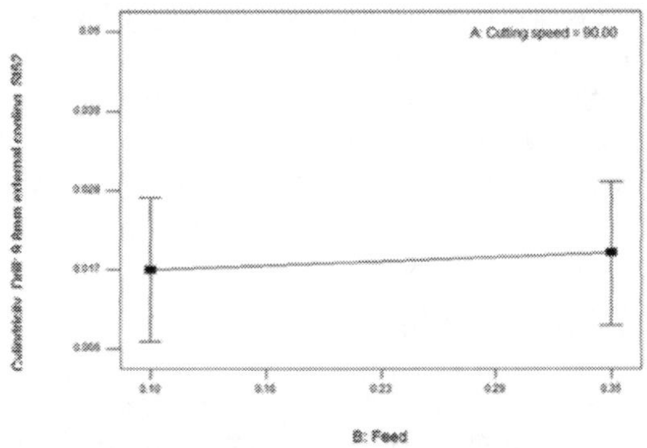

Fig.12. Effect of feed rate (mm/rev) on cylindricity deviation (mm).

In figure 12 is presented the effect of feed rate onto the hole cylindricity deviation, for 90 m/min constant cutting speed. Increasing the feed rate from 0.1 to 0.35 mm/rev can be observed the cylindricity deviation increases from 0.017 to 0.019 mm.

5 Chip shape analysis

In drilling process the chip control is important, because a wrong chip evacuation can generate a low surface quality, low dimensional accuracy or even tool brake [13].

On this experimental study was analysed the effect of feed rate onto the chip shape. The chip shape was analysed for three values of feed rate: 0.05; 0.23 and 0.4 mm/rev.

a. Feed rate: 0.05 mm/rev b. Feed rate:0.23 mm/rev c. Feed rate:0.4 mm/rev

Fig.13. The chip shape obtained for different values of the feed rate.

In the first case, using a feed rate of 0.05 mm/rev and a cutting speed of 90 m/min, a spiral long chip (longer than 100 mm), with about 0.1 mm thickness was obtained (fig. 13. a). This chip shape is considered to be difficult for evacuation in drilling process. It can stick to drill (in clamping area), resulting a damaged workpiece surface.

In the second, using a feed rate of 0.23 mm/rev and a cutting speed of 90 m/min, a short chip (about 10 mm), with about 0.3 mm thickness was obtained (fig. 13.b). For a feed rate of 0.4 mm/rev and a cutting speed of 90 m/min, a short chip (about10 mm) with about 0.5 mm thickness was obtained (fig. 13.c). In this case the chip shape was the same with the one obtained using a 0.23 mm/rev feed rate. This chip shape provides a very good chip evacuation in drilling. The result obtained from this experimental study it is concluded that to obtain a good chip evacuation it is necessary to use a feed rate over 0.2 mm/rev.

6 Conclusions

Based on the experimental results we can conclude the following:
- For improving the surface quality is recommended to increase the cutting speed and to decrease the feed rate. To obtain a surface quality around 0.8μm is recommended to use a cutting speed of 90 mm/min and a feed rate 0.23 mm/rev;
- For decreasing the dimensional deviation is recommended to decrease the cutting speed and to decrease the feed rate. The dimensional deviation around 0.04 mm is resulted from the following parameters: cutting speed 75 mm/min and feed rate 0.1 mm/rev;
- To decrease the cylindricity deviation is recommended to increase the cutting speed and decrease the feed rate. To obtain a cylindricity deviation around 0.017 mm is recommended cutting speed 104 mm/min and feed rate 0.1 mm/rev.

Using the optimal process parameters, could be processed holes at a good surface quality (up to Ra=0.8μm) and a high accuracy (0.004 mm dimensional deviation and 0.01mm cylindricity deviation). To obtain a good chip evacuation during the drilling process with this type of carbide drill, must be used a feed rate over 0.2 mm/rev.

This research was funded by the HORIZON 2020 Project – AMaTUC, No.691787/01.01.2016

References

1. www.guhring.com
2. Y. Kaynak, H.E. Karaca, R.D. Noebe, I.S. Jawahir, *Tool-wear analysis in cryogenic machining of NiTi shape memory alloys: a comparison of tool-wear performance with dry and MQL machining*. Wear **306**(1–2),51(2013)
3. S.Y. Hong, I. Markus, W.C. Jeong, *New cooling approach and tool life improvement in cryogenic machining of titanium alloy Ti-6Al-4V*. International Journal of Machine Tools & Manufacture **41**, 2245 (2001)
4. N.R. Dhar, S. Paul, A. B. Chattopadhyay, *The influence of cryogenic cooling on tool wear, dimensional accuracy and surface finish in turning AISI 1040 and E4340C steels*. Wear **249**(10), 932(2001)
5. V. Krishnaraj, A. Prabukarthi, *Optimization of machining parameter at high speed drilling of carbon fiber reinforced plastic (CFRP) laminates* , Composites: Part B, **43**: 1791(2012)
6. W.C. Chen, C.C. Tsao, *Cutting performance of different coated twist drills*. Journal of Materials Processing Technology, **88**(1), 203(1999)
7. M.N. Islam, B. Boswell, *Effects of canned cycles and cutting parameters on hole quality in cryogenic drilling of aluminium 6061-6T*, International Journal of Chemical, Molecular, Nuclear, Materials and Metallurgical Engineering, **10**(7), 857 (2016)
8. A. Prabukarthi, V. Krisharaj, *Multi-objective optimization on drilling of titanium alloy (Ti6Al4V)*, Materials Science Forum, **763**,29 (2013)
9. K.H. Park, A. Bealb, *Tool wear in drilling of Composite/titanium stacks using carbide and polycrystalline diamond tools*, Wear, **11-12**, 2826(2011)
10. T. Xia, Y. Kaynak, C. Arvin, I.S. Jawahir, *Cryogenic cooling-induced process performance and surface integrity in drilling CFRP composite material*, The International Journal of Advanced Manufacturing Technology, **82** (1), 605,(2016).
11. L. Morovič, J. Milde, *CAD model created from polygon mesh*. In Applied Mechanics and Materials : Vol. 808 = Modern Technologies in Manufacturing, Selected, Peer reviewed papers from the 12th International Conference on Modern Technologies in Manufacturing (MTeM),**808**, 233,(2015)
12. J. Peterka, L. Morovič, *Optical 3D Scanning of Cutting Tools*. In Applied Mechanics and Materials: 4th International Conference on Information Technology for Manufacturing Systems (ITMS), 663(2013)
13. L. Crişan, M. Pop Grigore, C. Neamţu, *The influence of process temperature on geometrical deviations at the drilling of carbon fiber reinforced composites*. Quality-Access to Success, (2012)
14. A. Luca, N. Balc, A. Popan, V. Ceclan, N. Panc. *Improving the quality of the parts made by rapid metal casting process*. Academic Journal of Manufacturing Engineering, **12** (1), 82(2014)

Influence of explosive welding parameters on properties of bimetal Ti-carbon steel

Mariusz Prazmowski*[,1], *Michał* Najwer[1], *Henryk* Paul[2], and *Dariusz* Andrzejewski[1]

[1]Opole University of Technology, Faculty of Mechanics, 5 Mikołajczyka St., Opole, Poland
[2]Polish Academy of Science, Institute of Metallurgy and Materials Science, 25 Reymonta St., Cracow, Poland

Abstract. Explosion welding of metals is a process of great technological significance in terms of modern metal composites manufacturing possibilities Nevertheless, selecting welding parameters is not an easy task. This paper assesses the effect of various values of distance of sheets on the quality of the bond zone in titanium (Ti Gr.1) - carbon steel (P355GH) structure. The research was carried out for initial state bonds i.e. immediately following explosion welding. The results of mechanical and structural investigations were presented. In order to determine changes in the value of strengthening, microhardness tests of both the weld and the joined plates were performed. Performed metallographic analysis shows that the standoff distance affects the quality of the bond zone boundary. Smaller distance promotes the formation of waves with lower parameters (of length and height), whereas greater distances allow forming the bond of a more pronounced, repetitive wavy character, however, increasing the quantity of the fusion zone at the same time. Also, the initial distance between the materials to be joined makes for the strengthening in the areas adjacent to bond boundary. The results received allowed to conclude that for the assumed parameters it is possible to obtain Ti -carbon steel bi-metal with properties meeting the standard's requirements.

1 Introduction

High strength, low density, and excellent corrosion resistance are the main properties that make titanium attractive for a variety of applications. Examples include aircraft (high strength in combination with low density), aero-engines (high strength, low density, and good creep resistance up to about 550°C), biomedical devices (corrosion resistance and high strength), and components in chemical processing equipment (corrosion resistance) [1]. The relatively high cost of titanium has hindered wider use, for example in automotive applications. To minimize the inherent cost problem, successful applications must take advantage of the special features and characteristics of titanium that differentiate it from competing engineering materials. The solution to this problem could be the use of layered materials [2, 3]. It is usually sufficient to apply a thin layer of a material applied on a relatively inexpensive base (construction) material to meet the operation requirements, e.g.

* Corresponding author: m.prazmowski@po.opole.pl

to obtain element characterized both by high corrosion resistance and strength [4]. However, joining materials with special properties (e.g. in the form of metal sheets) still is a big technological problem. That is why, one increasing interest in the technology of explosive treatment of materials, including the technology of explosive cladding, is observed.

The classic definition of the explosive welding defines the letter as the process of metal joining in the solid state, under the effect of a collision with high velocities caused by a controlled detonation of an explosive charge. Basically, this process can be divided into three general stages: detonation of the explosive charge, deformation and acceleration of the flyer plate and collision of the plates [5, 6]. The action of the explosive charge deforms and relocates the flyer plate (placed ad a pre-determined distance) in the direction of the base plate (Fig. 1).

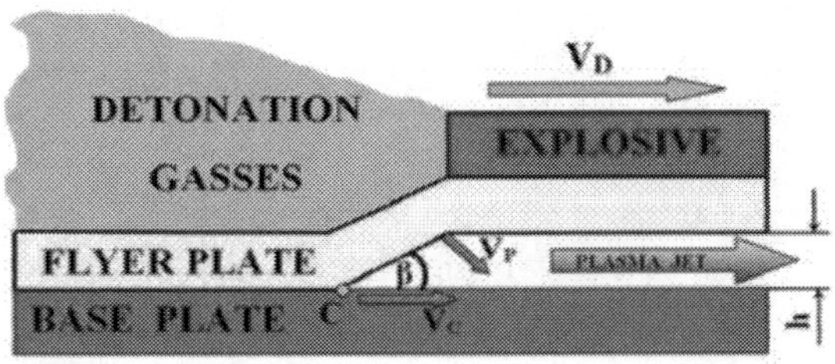

Fig. 1. Scheme of the parallel system and the shape of the ejector die in the welding process. The basic parameters are: V_D - detonation velocity, V_P - velocity of plate collision, V_c – velocity at the collision point, β - collision angle, h - distance between the plates, C - collision point.

Optimal bonding parameters, affecting equally the quality of the bond and mechanical properties of the resulting clad, must be correctly selected at the stage of designing the blast arrangement. Explosion welding parameters include physical and mechanical properties of the welded materials, the quality and quantity of the explosive, the method of detonation initiation, and the geometry of the welding structure. The event that determines the joining of materials during explosion welding is the collision of the flyer plate with the base plate. Key parameters which are decisive for obtaining a 'proper' bond during explosion welding are the collision point speed v_C and the impact angle of sheets β. They are highly dependent on the detonation velocity of the explosive v_D and the stand-off distance between the sheets (so-called bond window) h [5, 9]. The values of these settings are individually selected for particular combination of metals and alloys on the basis of equations widely available in literature. The stand-off distance is an independent variable which allows obtaining the required conditions of impact and directly influences the value of the impact angle. Literature review [5, 8] shows that the main factor deciding about the particular stand-off distance between the plates to be used is the thickness of the flyer plate (t). Usually, this parameter ranges between $0{,}5t$-$4t$. The appropriate selection of the stand-off distance will ensure obtaining the collision velocity specific for a given system of materials, e.g. [5, 8-9].

The present study focuses on the analysis of the influence of the stand-off distance between the plates on the morphological changes in the bond zone and mechanical properties of the Ti Gr.1/P355GH clad.

2 Research techniques

2.1 Research material

Research presented in this work was carried out on bimetal plates produced by explosion welding by EXPLOMET High-Energy Techniques Works in Opole. Explosion bonding was performed in a parallel arrangement where the base material was a 25mm sheet of carbon steel designed for operation in elevated temperatures whereas the deposited material (flyer plate) was a 12 mm Titanium sheet (Ti Gr.1 alloy). The chemical composition of both materials was shown in Table 1 and 2.

Table 1. The chemical composition of steel sheets, as per the supplier's certificate[10].

Basic Material	Chemical composition (%)						
P355GH	C	Mn	Si	P	S	Cr	Cu
	0.16	1.46	0.31	0.014	0.004	0.05	0.15
	Ni	Mo	Al	V	Nb	Ti	Fe
	0.06	0.02	0.033	0.004	0.009	0.004	rest

Table 2. The chemical composition of titanium sheets, as per the supplier's certificate [11].

Basic Material	Chemical composition [%]					
	C	Fe	N	O	H	Ti
Ti Gr.1	< 0.08	< 0.02	<0.03	< 0.18	< 0.015	Rest

Bonding was carried out in parallel arrangement as per the diagram shown in Fig. 1, for which a constant detonation velocity (v_D=2000 m•s^{-1}) and a variable stand-off distance between the plates h were assumed. The designation of the produced plates with process parameters were shown in Tab. 3.

Table 3. Designation of the produced plates and the relations between process parameters.

Plate	Detonation velocity v_D [m•s^{-1}]	Stand off distance h [mm]
H10	2000	10
H12		12
H15		15

2.2 Mechanical properties

Mechanical properties, based on technological tests of shearing, lateral bending and ram trials were completed according to EN13445-2 standard.

The first trial of the joint interface was a lateral bending test. Flat specimens of 10 x g mm cross-section (g is the sum of the thickness of the base and deposited material) and the length of 200mm sampled perpendicular to the detonation wave propagation were selected for the tests. The specimens were bent by 180° on a d=40mm diameter rod (Fig. 2a).

Since the bending test may be faulty in many instances due to the wavy nature of the bond, ram test Ro was additionally carried out to evaluate resistance to splitting. The samples of the ram tests were presented in Fig. 2b. The assessment of the joint strength was performed based on shearing test. The specimens were sampled from the clad in such a way that the shearing plane was parallel to the direction of the detonation propagation wave. The shape and dimensions of the specimens were given in Fig. 2c. The shearing and ram trials were performed until the total separation of the material occurred, recording the maximum force needed for the separation to happen. For ram and shearing, the spot of the sample destruction was macroscopically observed. For strong bimetals, the characteristic place of destruction is the area in the deposited or base material (typically the weaker one). The destruction of the specimen at the joint interface indicates poor strength properties of the analysed clad.

Fig. 2. Shape of the sample for: a) lateral bending test, b) ram test, c) shearing test.

Changes in the strength close to the weld zone, in the longitudinal section of the joined sheets, were analysed by microhardness measurement by Vickers method using LECO MHT Series 200 microhardness tester under the load of 100g. The measurements were performed along the line perpendicular to the interface (3 series). Impresses were made from the joint interface, outside melt regions, locating the first measurement at 0.04 mm from the bottom of the wave for each of the materials in the bimetal. The results presented in the paper are a mean value of three parallel measurements. The obtained average results for the tested bimetals were compared to the hardness of the input material as shipped (prior welding).

2.3 Optical microscopy observation

The material for microscopic examinations was sampled from the fabricated clads as well as from as-delivered sheets. The samples of output material were taken parallel and perpendicular to rolling direction. In the case of clads the polished specimens were produced as sections perpendicular to sheet surface and parallel to the detonation front motion. The analysis of the microstructural changes was conducted using an optical microscope Olympus IX 70 and an image analyzer LECO IA 32. In particular, the measurements of waves parameters were done on carefully polished surfaces using LECO IA 32 image analyzer, including: the length of the bond line *L*, height (*H*) and wave length (*n*) as well as the area of the melt surface *P* (Fig. 3). Based on the obtained results, a characteristic was done of the joined plates interface by the calculation (from equation 1) of the value of 'RGP' factor, so-called melt depth equivalent factor [6-8].

$$\mathrm{RGP} = \frac{\sum_{i=1}^{n} P_i}{L} [\mu m] \qquad (1)$$

where P_i - surface of fusion regions in μm^2.
L-the length of the bond line in μm

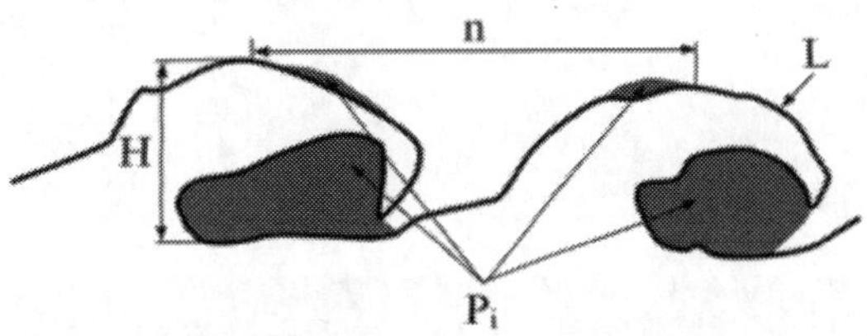

Fig. 3. Basic bond parameters: H -height of the wave, L - length of the bond line, n -wave length, P -'fusion' surface area.

3 Results and discussion

3.1 Structural observations

The preliminary observation of microstructure was performed by means of light microscopy. Fig 4a shows the structure of titanium sheet in as-is state i.e. before bonding in section perpendicular to lateral direction. It is evident that the material is characterized by the structure of compact packs of 60 to 180 μm size consisting of α phase grains. The image (Fig.4b) presents the P355GH carbon steel microstructure in as-is state characterized by an equiaxed structure of middle-sized grains in the range of 4 to 11μm for pearlite and 10-20 μm for ferrite. For pearlite, we can see a band structure of fine grains, typical for materials that underwent hot forming.

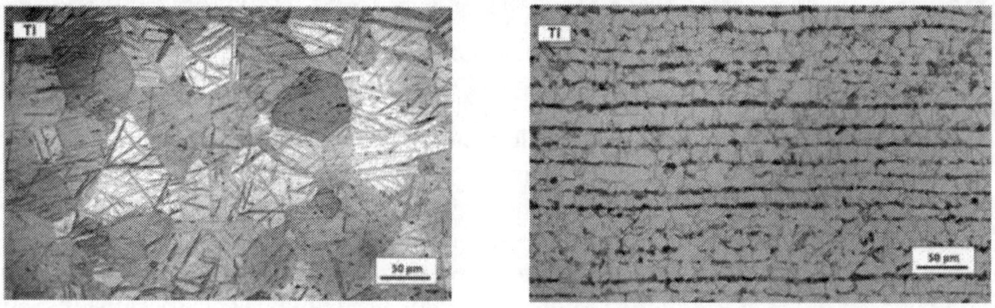

a) b)

Fig. 4. The initial microstructure of a) Ti Gr.1, b) steel P355GH.

The analysis of the bond zone was carried out by means of the optical microscopy on longitudinal section (in the direction of detonation front movement). Based on the performed measurements of the wave settings (Table 4) and microstructural analysis of the bond zone (Fig. 5) it can be concluded that a wavy bond was obtained in all 2 cases.

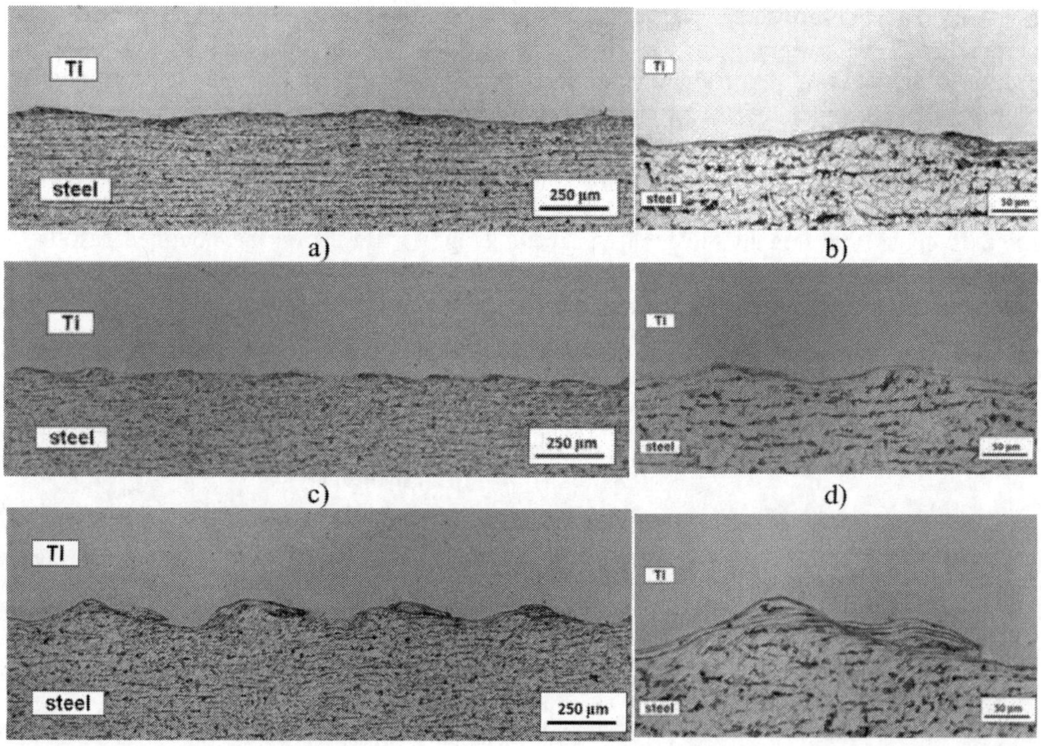

Fig. 5. The influence of stand-off distance on changes near the interface: (a-b) H10, (c-d) H12, (e-f) H15. Magnification 50x (a,c,e) and 200x (b,d,f).

For all cases (H10, H12, H15) stand-off the lengths of wave *n* was nearing (about 540 μm). For 10 mm and 12 mm stand-off (samples H10 and H12) the height of the wave was

33μm and 39μm respectively, whereas for stand-off 15mm (samples H15) the height of the wave was equal and reached 83μm. Therefore it follows that H12 sample exhibited 60% increase of the height of the wave respectively as compared to H10 and H12 (Fig. 6). In the same figure we can analyse the effect of the stand-off distance between the bonded plates on the 'intensity' of the melted zone occurrence. It was noticed that the increase in the stand-off distance caused the growth of the melted area fraction in the bond, which can be established by the analysis of 'RGP' factor. Its greatest value – 1,45 μm was obtained for the largest distance between the plates (sample H15). For the other two cases, it can be noticed that along with the distance decrease the 'RGP' factor goes down; for H12 and H10 – 0,71 μm and 0 μm respectively.

Table 4. Parameters describing wave shape and the quantity of the melted zone.

Plate	Length of the bond line L [μm]	Wave height H [μm]	Wave length n [μ m]	Melt surface area P [μm²]	Melt depth equivalent RGP [μm]
H10	4892	33	537	0	0
H12	4999	39	544	3 558	0,71
H15	5206	83	548	5 567	1,47

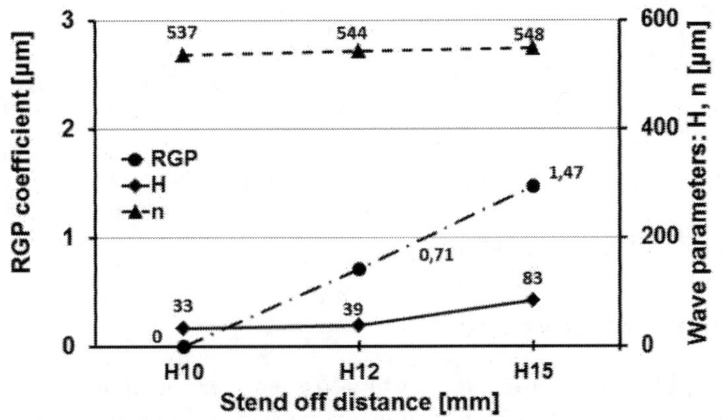

Fig. 6. Influence of the stand-off distance near the interface.

On the basis of the performed observations and measurements it can be concluded that the distance between the plates strongly influences bond-defining parameters. It is noticeable particularly when analysing the fraction of melted layers at the bond interface. Namely, with the same explosion energy, the decrease of the distance leads to a considerable reduction of the melted area.

3.2 Mechanical tests

In case of bending tests, best result was obtained for biggest stand-off distance (specimen H15), which achieved full bend (180 degrees) without any delaminating. In case of other specimens, were delaminated for bend to 40° for specimen H10 and to 90° for specimen H12.

The qualitative assessment of the obtained bond was performed based on mechanical properties determined by means of shearing tests, lateral bending and ram tests. The average values resulting from individual strength tests were presented in Fig. 7. For shearing tests, in all analysed cases, the results exceeded more than twice the minimum standard threshold of R_s = 140MPa. The highest R_s values of about 520MPa were recorded for narrow distances between the plates (H10). Larger distances added to decreasing the shear strength, which was 472MPa and 417MPa for H12 and H15 samples, respectively.

During the ram test the value of R_o was noticed to increase along with the decrease of the stand-off distance.

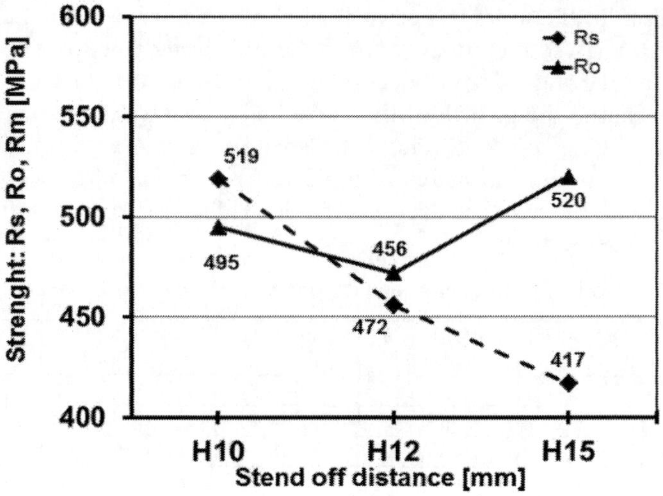

Fig. 7. The mechanical properties of Ti/P355GH bimetals.

The highest R_o values of 520 MPa was obtained for specimen cut out from plates welded with the greatest distance between the sheets (H15), Fig 7. For A8 (2*h*) specimens a 5% decrease of ram strength was recorded. In the case of bimetals fabricated with a small distance between the plates (H10 and H12) a 5-15% drop of R_o compared to the maximum result was observed. The lower quality of the weld is also confirmed by the fact that every specimens were destroyed at the bond.

3.3 Microhardness measurements

The analysis of the microhardness distribution on the section's surface revealed a considerable increase of the hardening both in the base material as well as the cladding metal compared to the initial state material hardness. Close to the bond zone in the steel and Ti plates, a 25-35% improvement of hardness was noticed, which was approximately equal to all the analysed arrangements (Fig. 8a).

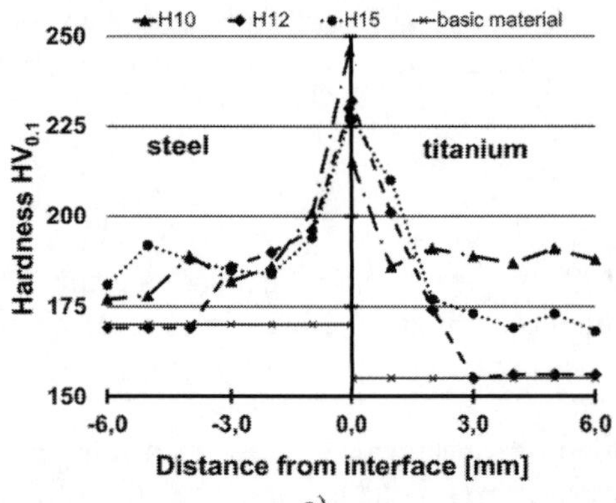

a)

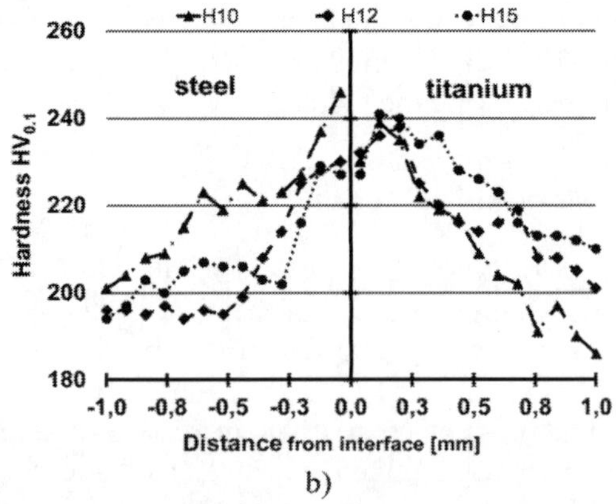

b)

Fig. 8. Microhardness measurements along 3 'scan' lines across the interface, for all analysed cases (up to 0,5mm from the interface). Load 50G.

A more detailed distribution of microhardness was analysed at the distance of 1 mm from the bond interface (Fig. 8b). The analysis of the obtained results confirms that the greatest hardening is displayed by the plate fabricated with the smaller stand-off distance between the plates, i.e. H10 and H12 samples, whereas the smallest was displayed by H15 sample. The graph shows an undeniable influence of the stand-off distance on the hardening improvement of steel. In particular, it was observed that changing the stand-off distance contributes to hardening improvement on the side of steel, but this tendency was not noticed for titanium.

4 Conclusions

The present study analyses the effect of the stand-off distance between the joined plates during explosion welding on the properties of the Ti/P355GH clad. The performed research allowed the formation of the following conclusions:

1. The mechanical properties of the fabricated bimetals, shear strength R_S strongly depend on the stand-off distance between the plates; their values rise with extending the distance.
2. Characteristic bond zone parameters, in particular the amount of melt in the bond are determined by the initial stand-off distance of the joined plates; extending the distance causes decreasing of the 'RGP' factor, i.e. the reduction of the melt volume in the bond zone occurs.
3. As a result of explosion welding, hardening of the joined materials occurs near the interface. Nevertheless, it is more pronounced in steel and less in Ti. The volume of the hardened area heavily depends on the stand-off distance. The greatest hardening, both in the steel and Ti plates, occurs in the areas adjacent to the boundary.

References

1. G. Jutjering, J.C. Williams, *Titanium. Engineering Materials and Processes* (Springer, Heidelberg, 2007)
2. P. Nieslony, P. Cichosz, G.M. Krolczyk, S. Legutko, D. Smyczek, M. Kolodziej, Measurement **78**, 129 (2016)
3. P. Nieslony, G.M. Krolczyk, K. Zak, R.W. Maruda, S. Legutko, Precis. Eng. (to be published)
4. T.Z. Blazynski (ed.), *Explosive welding, forming and compaction*, (Applied Science Publishers, London and New York 1983)
5. F. Findik, Mater. Design **32**, 1081(2011)
6. M. Prazmowski, H. Paul, J. Morgiel, Metal Forming **XXV**, 257 (2014)
7. H. Paul, M. Faryna, M. Prażmowski, R. Bański, Arch. Metall. Mater. **56**, 463 (2011)
8. H. Paul, L. Lityńska-Dobrzyńska, M. Prażmowski, Arch. Metall. Mater. **57**, 1151 (2012)
9. A. Nobili, Nobelclad Technical Bulletin Nt, Nobelclad, Rivesaltes, France (2002)
10. ISD Trade Sp. z o.o., *Inspection certificate, PN-EN 10204/06-3.1*
11. BAOIJ Titanium Industry Co. LTD, *Material certification, EN 10204 3.2.*

Dynamic study of milling low depth channels

Dorin Mircea Rosca[1], *Mircea Anton* Vasiloni, and *Milena* Folea[1,*]

[1]Transilvania University of Brasov, Department of Manufacturing Engineering, Mihai Viteazu No.5, Braşov, Romania

Abstract. This paper presents a study of dynamic aspects of the milling cutters used in particular case of low depth channels. A new calculation method was developed, taking into account the high variations of cutting forces during milling small depth channels with peripheral cutting tools. A new formula was established for the minimal value of channel depth that allows cutting process to be performed in conditions of dynamic stability.

1 Introduction

It is known that the mechanical and thermal stresses of a technological system for cutting processes are influenced by many factors. Thus, in case of milling a very important factor is the way chips are removed: peripheral milling or face milling.

Disk-cutters studied in this paper being provided with a single row of cutting edges on the cylindrical surface will fit in the first category. Chips detached through peripheral milling have variable thickness and their cross sections are defined by two cycloid curves described by points in the edge of the cutter. Evidently, the variable thickness of the chip causes the variation in time of components of total cutting force and milling torque [1]. Because from a practical point of view the most eloquent dynamic indicator in milling is torque, M_t, this variable will be studied through this paper.

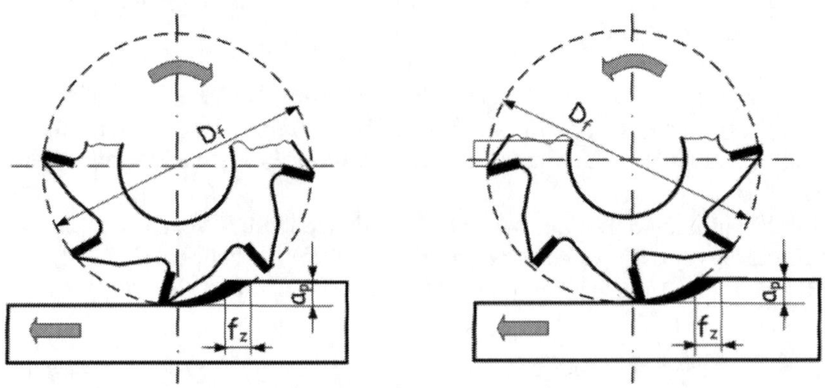

Fig. 1. Down milling (left) and up milling (right).

*Corresponding author: m.folea@unitbv.ro

This study will also focus on small width disk-type milling cutters known as slitting cutters. If this type of tool is used for the execution channels/grooves, milling method recommended in the literature [1-3] it is "down milling" and if the cutter is designed for parting operation (slitting), the optimum method is "up milling" (Fig. 1).

Whatever the method of milling, more teeth of the cutter are simultaneously in contact with the workpiece, their number, z_0, depending on the depth channel milling, on tool diameter (D_f) and the total number of teeth (z). By periodically entering and leaving the material by cutting edges results an oscillatory behaviour for cutting force and milling torque.

Since in cutting tools catalogues [1] are only indicated maximal depths of cut for slitting and considering that in case of peripheral milling narrow and small depth channels it would be appropriate to indicate also the minimal depth of cut in order to improve dynamic conditions, the aim of this paper is to determine those minimal values that avoid shock of milling cutter teeth when entering in workpiece material.

2 Calculation and analysis of milling torque

Considering one moment in time of milling process with a slitting cutter with straight teeth ($\omega = 0°$), axial cross sections (0-1, 0-2 etc.) of the chips removed by the three teeth (Fig. 2) have rectangular shapes, their thickness being calculated as functions of feed per tooth (f_z) and contact angle (φ) by the following relations:

$$a_1 = f_z \cdot \sin\varphi \tag{1}$$

$$a_2 = f_z \cdot \sin(\varphi - \delta) \tag{2}$$

$$a_3 = f_z \cdot \sin(\varphi - 2\delta) \tag{3}$$

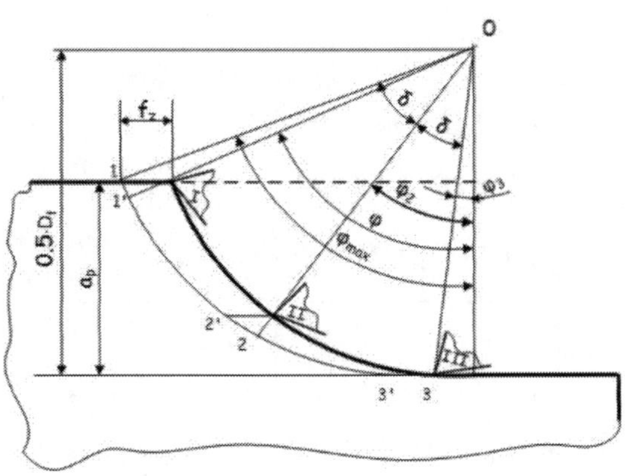

Fig. 2. Up milling with three teeth engaged in cut.

Milling process being type "up milling", the maximum value for milling torque is recorded shortly before the exit from cutting of tooth number 1and can be determined as follows:

$$M_{t\max} = \frac{D_f}{2}(B \cdot a_1 \cdot k_{s_1} + B \cdot a_2 \cdot k_{s_2} + B \cdot a_3 \cdot k_{s_3}), \tag{4}$$

where

B is processed width (channel width);

$k_{s_1}, k_{s_2}, k_{s_3}$ specific force corresponding to chip thickness a_1, a_2 and a_3.

If k_{s_1}, k_{s_2} and k_{s_3} are replaced in equation (4), defining them relative to specific force per unit $k_{s_{1.1}}$ is obtained:

$$M_{t\max} = \frac{B \cdot D_f \cdot k_{s_{1.1}}}{2}(a_1 \cdot a_1^{-z} \cdot k_{s_{1.1}} + a_2 \cdot a_2^{-z} \cdot k_{s_{1.1}} + a_2 \cdot a_2^{-z} \cdot k_{s_{1.1}}) \quad (5)$$

respectively,

$$M_{t\max} = \frac{B \cdot D_f \cdot k_{s_{1.1}}}{2}(a_1^{(1-z)} + a_2^{(1-z)} + a_3^{(1-z)}) \quad (6)$$

After in equation (6) are introduced relations (1), (2) and (3) final mathematical form of the torque results as:

$$M_{t\max} = \frac{B \cdot D_f \cdot k_{s_{1.1}} \cdot f_z^{(1-z)}}{2}[\sin^{(1-z)} \varphi + \sin^{(1-z)}(\varphi - \delta) + \sin^{(1-z)}(\varphi - 2\delta)] \quad (7)$$

for the particular case when the maximum number of teeth of a disk-type milling cutter with $\omega = 0°$ simultaneously engaged in cutting is $z_0=3$.

Generalizing the equation (7) for a number "i" of teeth simultaneously engaged in cutting is obtained:

$$M_{t\max} = \frac{B \cdot D_f \cdot k_{s_{1.1}} \cdot f_z^{(1-z)}}{2}\{\sin^{(1-z)} \varphi + \sin^{(1-z)}(\varphi - \delta) + \ldots + \sin^{(1-z)}[\varphi - (z_i - 1)\delta]\} \quad (8)$$

According to equation (8), the maximum torque is obtained by adding "i" sinusoidal functions at power (1-z). In table 1 are indicated approximate values for the exponent (1-z) for various materials processed by milling [4]:

Table 1. Coefficient z for various materials

Workpiece material	**C45**	**C70**	**15CN15**	**33MoC11**	**50VC11**	**Cast iron**	**Brass with σ_r=500 MPa**	**Aluminium alloys**
(1-z)	0.88	0.79	0.74	0.84	0.80	0.81	0.66	0.66

Cutting tool suppliers present in their catalogues various disk-type milling cutters with or without inserts that are used for slitting or peripheral slotting. Present paper refers only to disk-cutters with maximum width of 5mm and with maximum length of tooth cutting edge also of 5 mm.

In order to obtain a more uniform load during processing with disk-type milling cutters with straight teeth, a minimal condition must be satisfied [4, 5]: on the contact arc "φ" there must be at least two teeth. In the case of small depth channels there may be just one tooth engaged in cutting at a time. In this case the torque will have a pulsating and cyclic variation as in Fig.3 for down milling or as in Fig. 4 for up milling.

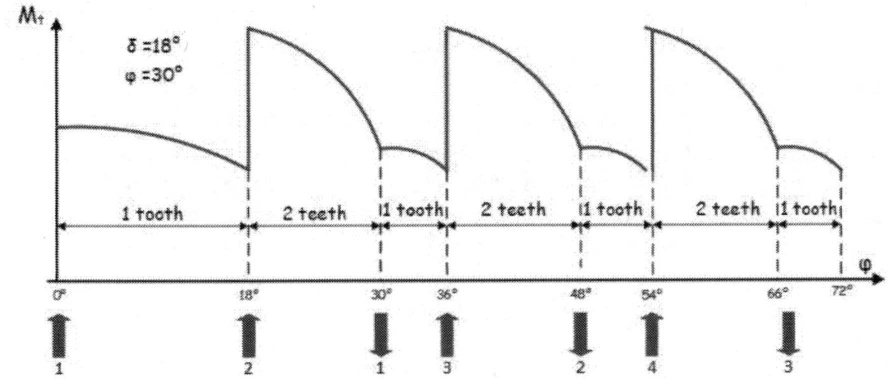

Fig. 3. Torque variation during down milling with one tooth and two teeth engaged in cutting.

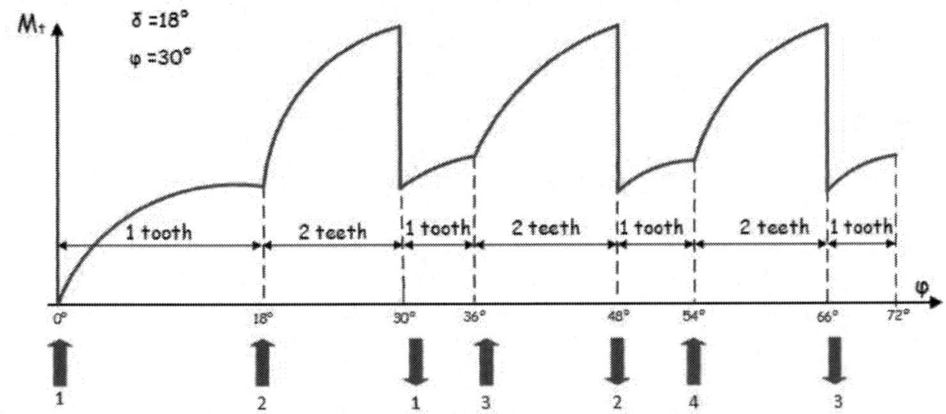

Fig. 4. Torque variation during up milling with one tooth and two teeth engaged in cutting.

It is important to reveal that by indicating in tool catalogues only the maximum depth of channel/groove processed with a slitting tool, stability issues are neglected in case of small depth channels. Until now, there is only mentioned the maximum depth of the cut [6-9].Therefore, some representative types of narrow disk-type milling cutters were selected and minimal values for the depth of the channel h_{min} were calculated for situations when at least one tooth of the milling cutter is permanently in contact with the workpiece. As can be observed in Table 2, h_{min} depends both on outer diameter D_f and on the number teeth z of the tool.

Table 2. Calculated values of h_{min}

Nr. crt.	D_f [mm]	z	h_{min} [mm]	Nr. crt.	D_f [mm]	z	h_{min} [mm]
1	80	7	15.06	6	125	11	9.922
2	80	8	11.715	7	160	13	9.163
3	100	9	11.618	8	160	14	7.922
4	100	10	9.549	9	200	17	6.753
5	125	10	11.936	10	200	18	6.030

3 Conclusions

The present study was justified by the fact that disk-type milling cutters of small width, with carbide inserts mechanically fixed, are provided with a number of teeth much lower than those made of high speed steel for a given outer diameter of the tool and stability problems might occur at low depth of processed channels/grooves.

In order to avoid shock over load of technological system during milling, an important measure to be taken is to diminish the variation of the cutting force and torque. In this respect, this paper provides for the specialists a method for calculating the maximum torque when milling using disc-type cutters. According to the new calculation formula, the maximum value of milling torque is obtained by summarizing a series of sinusoidal functions, one for the torque of each tooth engaged in cutting.

For a better protection from dynamic shocks of the technological equipment during milling narrow and small depth channels, it is strongly recommended to be indicated in the documentation of cutting tools suppliers, the minimum cutting depth which ensures at least one tooth permanently in contact with the workpiece.

References

1. L. Sauer, C. Ionescu, *Cutting tools for milling [Scule pentru frezare]* (Ed. Tehnica, Bucuresti, 1977)
2. S. Batt, *Ein Beitrag zur Entwicklung von Fräswerkzeugen mit verbesserten dynamischen Schnitteigenschaften/ Universität Magdeburg*, Dissertation Thesis, (2006)
3. N.B. Lupulescu, G. Oancea, A. Nedelcu, *Computer aided design of cutting tools for metal milling [Proiectarea informatizata a sculelor pentru frezarea metalelor]* (Ed. Lux Libris, Brasov, 2007)
4. D.M. Rosca, New Built-in-Conceptual aspects Regarding Cutting Tools Used in the Internal Surfaces Processing, AJME, **4(3)**, 53 (2006)
5. S. Enache, V. Belousov, *Cutting tools designing [Proiectarea sculelor aschietoare]* Ed. Didactica si pedagogica, Bucuresti, (1983)
6. ***PATENTSCHRIFT: ROTIERENDES Fräswerkzeug/ DE 101 24 234 B4, 2007
7. ***Cutting tools from Sandvick Coromant- Main catalogue (2012)
8. ***ISCAR tools e-catalogue (2016), accessed: May 2016, available at: http://www.iscar.com/eCatalog/Products.aspx?mapp=ML&app=57
9. ***Walter General Catalogue (2012)

MATEC Web of Conferences **94**, 02014 (2017) DOI: 10.1051/matecconf/20179402014
CoSME'16

Influence of the amount of deformation on the change in the porosity of sinters subjected to free cold upsetting

Mariusz Rosiak[1,*], and *Jerzy* Napiórkowski[2]

[1]Opole University of Technology, Faculty of Mechanical Engineering, Department of Manufacturing Engineering and Production Automation, ul. S. Mikołajczyka 5, 45-271 Opole, Poland
[2]University of Warmia and Mazury in Olsztyn, Faculty of Technical Sciences, Exploitation and Diagnostics Vehicles and Machines. St. M. Oczapowskiego 11, 10-079 Olsztyn, Poland

Abstract. The article contains a description of the technology production of cylindrical samples sintered of powder by using input material conditions of a trial deformation. The results of changes in porosity before and after the deformation are analyzed. Analysis of changes in the internal structure of the sinters were based on maps prepared porosity distribution for the states of deformation, carried out in conditions of freedom compression cold. Currently we are looking for effective methods of processing sintered towards obtaining the highest possible density of the homogeneous character the distribution volume. In the present study we analyzed the impact of the classical way of deformation sintered to changes in the internal structure due to the evolution of porosity. The study used an original approach to the assessment of porosity in the local areas, was conducted quantitative metallographic share porosity, based on the results prepared model the spatial distribution maps porosity.

1 Introduction

Unlike solid materials which are similar in terms of chemical composition, sinters are characterised by the presence of voids and discontinuities in the internal structure that are called porosity. Porosity is an important element of the internal structure of sinters, because its amount and nature of distribution in the material volume has a significant influence on final properties of the product. The amount of porosity and its diversification are influenced by a number of factors, such as e.g. properties of input powder, type of slip additives, mixing conditions of the prepared input material, pressing conditions and sintering conditions. Sinters prepared in this way under conditions typical to the powder metallurgy technology reach a minimum value of porosity within the range of 10-25%. Sintered products operating under heavy loading conditions require high strength properties that are possible to be obtained at a low evenly distributed porosity. In case of solid materials, studies on the effective deformation using plastic characteristics of materials to a greater extend, have been conducted for many years. In their papers [1-3], Bochniak and Korbel

*Corresponding author: m.rosiak@po.opole.pl

 DOI: 10.1051/matecconf/20179402014

indicate that the influence of the strain path on the efficiency of material forming is significant. Other authors, e.g. Grosman and Pawlicki also presented in their papers new possibilities of effective deformation of solid materials under complex loading states [4-9]. Rosiak [10] used a deformation method under complex loading conditions used by other researchers for the deformation of solid materials under volumetric strains of sinters. Deformation conditions that he used were the same as in the test of free compression with simultaneous reverse torsion. The aim of the paper was to find a new method for intensive and even compaction of sinters. He came to a conclusion that deformation of sinters under conditions of free compression with simultaneous reverse torsion favours an intensive and even compaction of the material volume and obtaining low and poorly diversified porosities.

This paper presents the influence of simple loading conditions created as in the test of free compression due to the possibility of compacting and obtaining low and evenly distributed porosity at the level allowing for possible heat or thermochemical treatment carried out under conditions as for solid materials.

2 Methodology and materials used for testing

2.1 Characteristics of powder for preparing samples

The input material for testing was collected on the basis of the following criteria: material susceptibility to cold forming, susceptibility to heat and thermochemical treatment, possibilities of using materials in industrial technological processes, availability in the market and economic relations.

Table 1. Chemical composition of the powder used to fabricate the sinters [11].

Chemical element	Type of powder		
	NC 100.24		
	Content of chemical element %		
	min.	max.	real
H_2 loss	-	-	0.21
C	-	-	<0.01
Ni	-	-	-
Cu	-	-	-
Mo	-	-	-
Fe	-	-	The rest

Sinters are made of NC 100.24 powder manufactured in the technology of the reduction of iron oxides contained in the ore with solid substances [11]. Chemical compositions of the powder used for the preparation of sinters are presented in table 1, the granularity of the powder is shown in table 2.

NC 100.24 powder is one of the most widely used powders in the powder metallurgy. It is characterised by a spongy structure that allows for a favourable stress distribution and high resistance of products, which makes it possible to manufacture moulded parts with low and high degrees of compaction (fig.1). Good compactibility (compressibility) of NC 100.24 powder ensures that it can be widely used in the manufacture of parts in the powder metallurgy. NC 100.24 powder was prepared by adding 0.7% of the slip additive – zinc stereate and mixing in the Y-type 2-channel mixer by Apex for 1 hour.

Table 2. Grain size of the powder used to fabricate the sinters [11].

Grain size w µm	Type of powder
	NC 100.24
	Fraction %
+212	0.0
150 – 212	1.0
106 – 150	
75 – 106	81.0
45 – 75	
- 45	18.,0

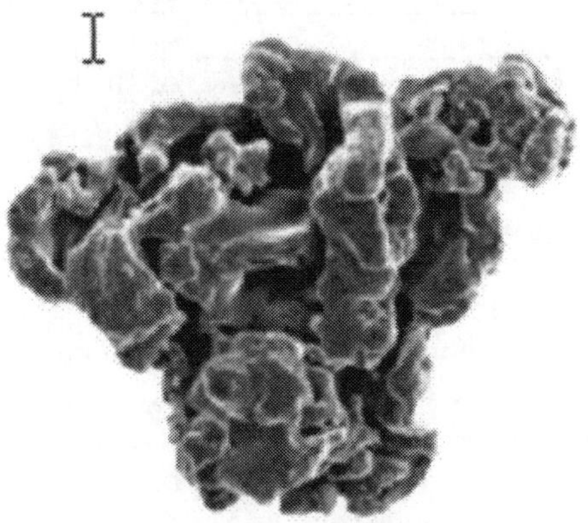

Fig.1. Image grains of iron powder NC 100.24 from Höganäs [11].

2.2 Conditions of preparing sintered samples

In order to assess the influence of the effects of plastic deformation of sinters on the porosity distribution in the finished product, the author decided to prepare sintered samples with the density $\rho_w = 0{,}80$ and having an axial-symmetric shape in the form of rolls (height h = 20 mm and diameter d = 20 mm–the shape of samples that was chosen is often used in tests conducted for this type of materials and described in literature [12-14]). Theoretical and relative densities of prepared sinters are presented in table 3 and 4.

Sintered samples were prepared, using standard conditions for conducting processes of powder metallurgy in the following order:

- preparation of weighted amounts of powder;
- compression of weighted amounts of powder in a locked die with the diameter $d_w = 20$ mm, using distance to obtain height equal to h = 20 mm, (carried out with the use of the WPM-type strength machine with a maximum pressure force of 300 kN);
- sintering of fittings at the temperature of 1100 ^{0}C under an atmosphere of endogas for 0.5 hour (endogas atmosphere was obtained as a result of the reaction of the air-propane mixture heated to 900^{0}C with hydrocarbon in the presence of a nickel catalyst; air/propane ratio was 10:1; composition of the atmosphere: 31% H_2, 23% $CO+CO_2$, 46% N_2), sintering was carried out using the SAFED belt furnace;
- annealing was carried out under an atmosphere of hydrogen at the temperature of 900 ^{0}C, sintering was carried out using the SAFED belt furnace;

• calibration process was carried out using a low die. Obtained diameters of samples were within the range from 20.05 mm to 20.07 mm.

Table 3. Theoretical density sintered (without pores) [11].

1	2				3
Type of powder	Content of chemical element [%]				Theoretical density sintered (without pores) [g/cm^3]
	Cu	Ni	Mo	Fe	
NC 100.24	-	-	-	The rest	7.800

Table 4. Density produced sintered (own work).

Type of powder	The relative density of sintered ρ_w
	0.80
	Density of sinters ρ[g/cm^3]
NC 100.24	6.24

2.3 Conditions of plastic deformation of sinters

Samples obtained after compaction and sintering of metallic powders consist of a metallic matrix and discontinuities that are geometrically differentiated and randomly distributed in the volume of the sinter. The essential feature that distinguishes the sintered metallic material (porous material) from solid metallic materials is the fact that during the deformation process, there are changes in the volume in addition to changes in the form. It is possible to form a porous material under any stress state even at the hydrostatic pressure. Porous materials are characterised by a slightly lower formability limit in comparison with solid materials. Unlike solid materials, sintered performs have the following features (that are often considered to be disadvantages): internal porosity, oxide layers in the surface layer, impurities and segregation of additives added to the basic mixture. These disadvantages are possible places of cracks that precede the damage to the material and reduce at the same time its susceptibility to forming. During plastic forming of sinters, it is required to generate such a stress state that will allow for a plastic flow of the matrix and will not cause local tensile stresses contributing to the increase in the local porosity or even fracture in the microregions of forging. Sinters under large plastic deformations are characterised by the minimum porosity that is evenly distributed in the volume and high strength properties.

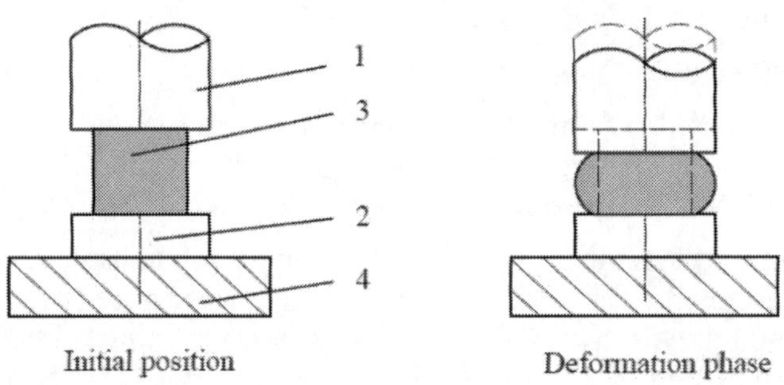

Fig. 2. Deformations phases of compressed samples in dies; 1 – upper punch, 2 – lower punch, 3 – sample, 4 – pad.

In order to assess the evolution of the porosity in the volume of forging under free upsetting conditions (Fig. 2), the author decided to use the variants of forming of sinters by means of deformations allowing for obtaining the required diameter in the plane of the maximum spinning of the barrel up to the assumed values: I – d_{1b}=22 mm, II – d_{2b}=23 mm, III – d_{3b}=24 mm, IV – d_{4b}=25 mm (Fig. 2). Preforms were formed using the hydraulic press of type LB-500. At the time of carrying out tests upsetting recorded: force F [N], the displacement Δh [mm], the time τ [s]. The maximum compressive forces which was reached during the implementation of the trial amounted to about 0.2 [MN]. On the basis of the recorded data during the compression tests were developed to strengthen the function of the material (1).

$$\sigma_{0.80} = 555.34 * (0.1015 + \varphi)^{0.5953} \quad (1)$$

3 Results and discussion

3.1 Results of porosity after the implemented deformation variations

Table 5 shows the results of the porosity and density of sinters during the deformation phase (based on the geometrical criterion); they were deformed in a way ensuring that the required diameter in the plane of the maximum diameter of the barrel corresponds to the assumed values: I – d_{1b}=22 mm, II – d_{2b}=23 mm, III – d_{3b}=24 mm, IV – d_{4b}=25 mm. The value of the ρ and ρw density was also determined using the hydrostatic weighing method [6] in relation to the whole density of sinters (Fig. 3), where the change in the density depended on the deformation phase, i.e. the size of the deformation φ. The change in the value of the average porosity in the volume of the sinter in relation to the value of the sinter θ=f(φ) is shown in Fig.4.

The porosity value of the sinter before plastic deformation, the value of the deformation in process, and the friction force in the contact plane between the tool and the deforming material etc., which all contribute to the fact that the deformation process is uneven in local areas, have a considerable impact on the value and porosity distribution of the sinter deformed in the free compression test. In order to obtain small porosities, the sinter has to be intensively compacted.

Table 5. Change in the average porosity of the volume sinters after the free compression deformation (change in the density ρ, and the relative density $\rho_w$in relation to the entire volume of the sinter).

Type in the sinter powder		NC 100.24		The average volume porosity
		ρ[g/cm³]	ρ_w	θ [%]
Before deformation		6.24	0.80	20
Density and porosity after the phase of deformation				
I	d_{1b}=22 mm	6.6621	0.85	15
II	d_{2b}=23 mm	6.8757	0.88	12
III	d_{3b}=24 mm	7.0417	0.90	10
IV	d_{4b}=25 mm	7.1797	0.92	8

$$\rho_w = \frac{\rho_s}{\rho} \tag{2}$$

ρ_w - relative density,

ρ_s - density sintered [g/cm³],

ρ - sintered density without pores [g/cm³],

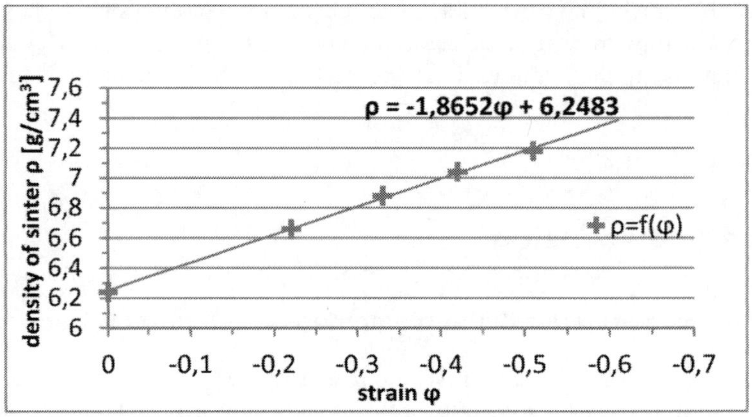

Fig.3. Variation of the average density ρ = f (φ) deformed in a free compression test.

Big deformations in the free cold compression test of sinters favour the increase in the density of the deformed material and the reduction of its porosity. If the 3-4% porosity of sinters is obtained, the heat and thermochemical treatment can be applied under the same conditions as for solid materials, which greatly facilitates the manufacture of sintered products. It is possible to deform sinters with big deformations in the range of values that do not cause the limit value φ_{gr} to be exceeded in local areas. The loss of the structural integrity and, consequently, the damage to the material may be caused if the limit value is exceeded. The increase in the average value of density is proportional to the value of the deformation under the conditions of free compression of sinters (Fig. 3).

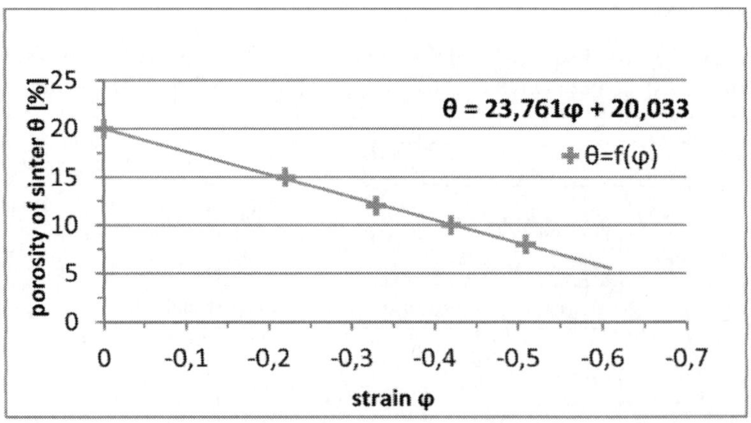

Fig.4. Variation of the average porosity θ = f (φ) deformed in a free compression test.

The average porosity value decreases proportionally to the achieved deformation as the deformation in process proceeds (Fig.4). The uneven deformation process in local areas affects the size and porosity distribution in the volume of the sinter, which is reflected in different properties of the material in particular cross-sections.

 DOI: 10.1051/matecconf/20179402014

3.2 Results of porosity distribution after the completed deformation variants

On the longitudinal section, in the axis of the cylinders, metallographic microsections were made and porosity was assessed in local areas.

Table 6. Distribution of porosity Θ in the sintered before deformation.

	Model			Contour map 2D	
	$\Theta_{śr.}$=8,07	$\Theta_{max.}$=14,4	$\Theta_{min.}$=2,39	S_{Θ}=2,318	φ_h= 0
he phase before deformation	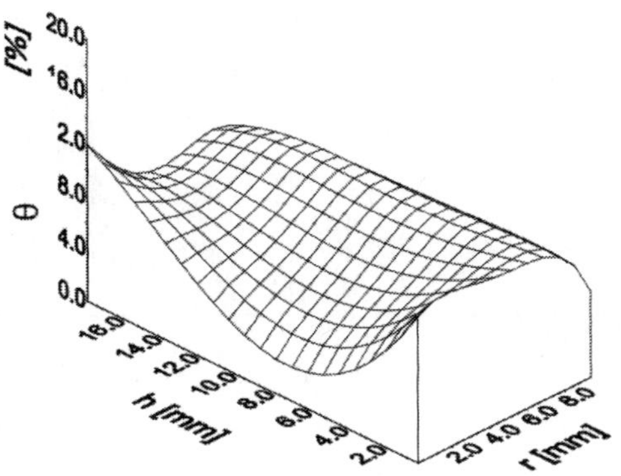			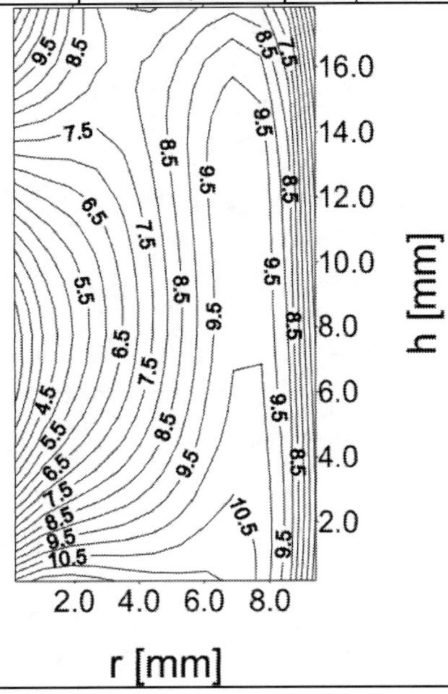	

The height of the sample was divided into 20 rows. In each of them 12 equidistant elementary areas were specified for the analysis. Using the capabilities of quantitative image analysis, the features of the objects in the form of pore sections were analysed. Maps of porosity were prepared on the basis of the obtained measurement results. Maps showing porosity distributions with isolines and plastic maps were made on the basis of a third-degree polynomial function determined by the interpolation method. Examples of porosity distribution before and after selected deformation stages are presented in tables 6-8.

The analyses carried out after the forming tests of sinters made it possible to determine the progress of compaction of these materials, expressed by the porosity distribution function. The porosity features expressed by parameters, such as the density, shape and size of pores, were also subject to significant changes during the deformation. The analysis of the porosity distribution in the zones of the emerging barrel indicates that the compaction progresses considerably less intensively and even after the fourth deformation stage porosity is high and will significantly affect the diversification of the distribution of material properties in local areas, also determining the nature of the course of phenomena taking place in the material during heat treatment or thermochemical treatment.

Porosity in sinters before deformation (table 6) carried out in the conditions of classical operations of powder metallurgy is highly varied in the entire volume of the sample. The lowest porosity is present in the axis, in the central part of the cylinder. In the zones at the base and the surface of the cylinder porosity has a considerable higher share in the internal structure of the sinter. Sinters prepared in such a way, deformed by means of free upsetting with low deformations (table 7) maintain a similarly uneven distribution of porosity, with a slightly lowers average value of the share of pores.

Table 7. Distribution of porosity Θ in the sintered after the first phase of deformation (d_{1b}=22 mm).

	Model			Contour map 2D	
	$\Theta_{śr.}$=5,74	$\Theta_{max.}$=14	$\Theta_{min.}$=1,98	S_Θ=2,503	φ_h= -0,221
The first phase deformation (d_{1b}=22 mm)	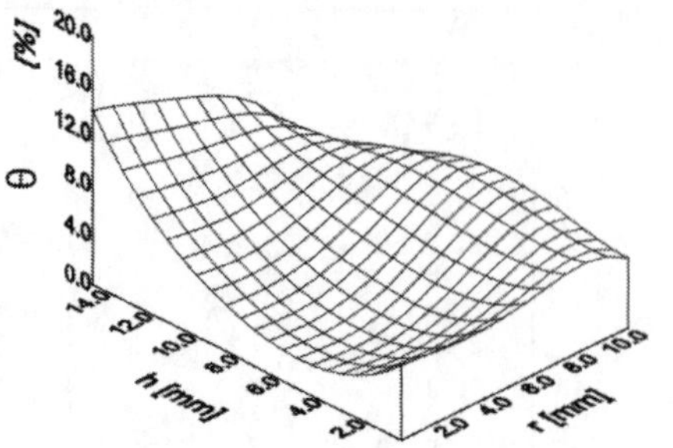			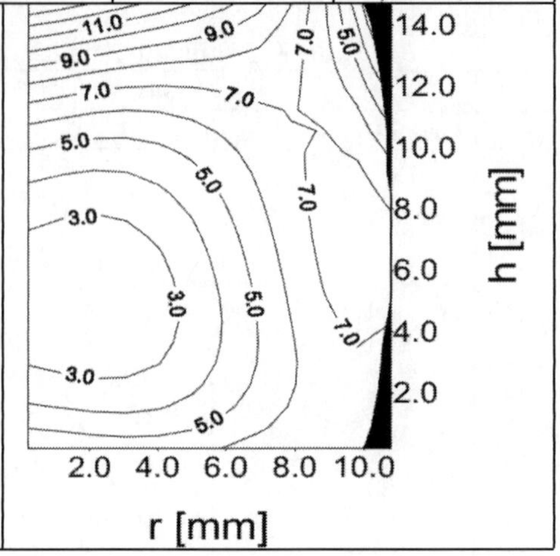	

Table 8. Distribution of porosity Θ in the sintered after the fourth phase of deformation (d_{4b}=25 mm).

	Model			Contour map 2D	
	$\Theta_{śr.}$=3,43	$\Theta_{max.}$=8,62	$\Theta_{min.}$=0,97	S_Θ=1,44	φ_h= -0,508
IV phase deformation (d_{4b}=25 mm)	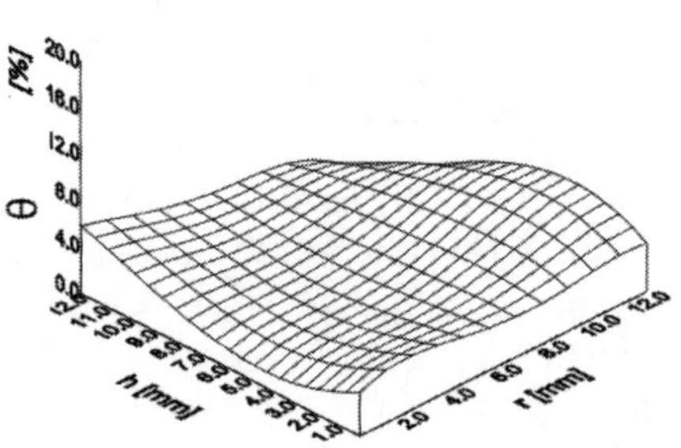			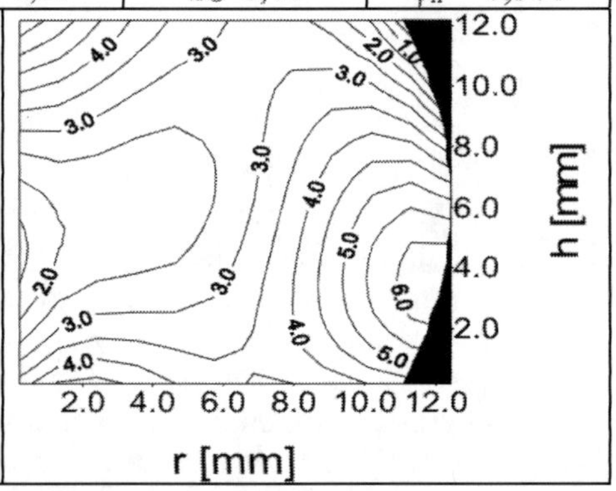	

The application of high deformations $\phi \approx -0,5$ (table 8), not exceeding φ_{gr}, makes it possible to significantly reduce the values of the average porosity and obtain low diversification of its distribution. Such values and distribution of porosity allow heat treatment and thermochemical treatment under the same conditions as for solid materials which are similar in terms of chemical composition.

4 Conclusions

The special character of porous materials requires a careful analysis of the impact of the factors that cause changes in the internal structure in the scope of the metallic matrix and changes in porosity features. During the sintering of sintered porous materials, as in the case of solid materials, the progress of deformation in local areas is not homogenous. The heterogeneity of the progress of deformation in local areas affects more or less intensive compaction of the material of the sinter, causing the reduction or even disappearance of pores. The reduction of porosity after plastic deformation is the reason for the reduction of the volume of the deformed sinter. Therefore, in contrast to solid materials, the principle of maintaining the same volume does not apply during forming of sinters.

The changes in the internal structure of the sintered material for the discussed variants of the deformation progress depend on the condition of the material before the deformation and the value of the upset as well as the nature of the deformation progress in local areas. The value and distribution of porosity in sinters before deformation determines the nature of the progress of porosity changes in the subsequent stages of deformation. The material in the central zone (in local areas of the volume) is subject to high compaction and reduction of porosity already after the deformation. After the fourth stage, the compaction becomes almost homogeneous and the material reaches almost uniform porosity. For such a distribution of porosity, the relevant literature indicates the possibility of accepting a porous material (with low porosity) for heat treatment under the conditions recommended for solid materials.

The studies permit analysis of the evolution of the distribution of porosity depending on the amount of deformation under compression freely. Most literature porosity evaluation is determined by an average value of the initial state before the deformation and the final state after deformation. Statement of the porosity distribution to the various states of deformation under compression increases substantially knowledge of changes in the internal structure of the sintered plastically deformed cold.

The examination of the assessment of the deformation progress of sinters performed under the conditions of a free upsetting test made it possible to formulate the following conclusions:

• It is possible to deform porous metallic sinters under the conditions of free compression, which results in a possibly high compaction of the material.

• However, low average value of porosity obtained after free upsetting is highly varied in local areas, which results in diverse final properties in the sinter volume. This is not a good method when you want to get big and uniform density. Low and uniform porosity is obtained by other methods, e. g. deforming sinters closed die or deforming sinters in complex load conditions - compression freely while oscillating torsion.

• Free upsetting of sintered materials does not make it possible to obtain a uniform compaction in the volume and therefore a uniform distribution of porosity.

• Forming of porous materials includes the diversification of the deformation in local areas and its impact on structural changes, as is the case of solid materials where the density is constant, and the change in the volume resulting from the change in the density.

• The size and distribution of pores depend on the evolution of the local stress and strain states.

• The research improves the knowledge of the possibility of analyzing changes in porosity in the sintered compression freely for the various stages of deformation. Most literature porosity of the sinter is analyzed for the state before and after complete deformation.

References

1. W. Bochniak, A. Korbel, Journal of Materials Processing Technology, **134**, 120-134 (2003)
2. A. Korbel, W. Bochniak, Philosophical Magazine, **93**, 15, 1883-1913 (2013)
3. W. Bochniak, A. Korbel, P. Ostachowski, A. Paliborek, International Journal of Materials Research, **104**, 10, 974–979 (2013)
4. F. Grosman, J. Pawlicki, Hutnik–Wiadomości Hutnicze R., **72**/5, 280-284 (2005)
5. F. Grosman, J. Pawlicki, Advanced Technology of Plasticity, **1**, 1219-1224 (2002)

6. J. Pawlicki, Hutnik – Wiadomości Hutnicze **8**, 438-440, (2010)
7. J. Pawlicki, Rudy I Metale Nieżelazne **11**, 550-558 (2015)
8. F. Grosman, J. Pawlicki, Hutnik – Wiadomości Hutnicze, **5**, 231-235 (2008)
9. J. Pawlicki, K. Hyc, Rudy I Metale Nieżelazne **11**, 679-683 (2013)
10. M. Rosiak,. Archives of Metallurgy and Materials **4**, 58, 1197-1205 (2013)
11. Hoganas AB, Metal Powders, www.hoganas.com
12. M. Rosiak, Grosman F. Zeszyty Naukowe Politechniki Opolskiej, **290**, seria Mechanika z. **76**, 305-309 (2003)
13. M. Rosiak, F. Grosman, Zeszyty Naukowe Politechniki Opolskiej **288**, seria Mechanika z. **75**, s. 211-218 (2002)
14. J. Łukowski, Zeszyty Naukowe Politechniki Śląskiej, seria Hutnictwo, **44** (1992)

 DOI: 10.1051/matecconf/20179402015

State space analysis of flexible manufacturing cell using timed colored Petri nets

Sanjib Kumar Saren[1,*], and *Florin* Blaga[2]

[1]Mechatronics Department, University Of Oradea, Romania
[2]Industrial Engineering Department, University Of Oradea, Romania

Abstract. In this paper we focus on analysis and evaluation of flexible manufacturing cell using time colored Petri nets. We basically model the system with respect to time to observe relation between each place and transition through arcs. The model is designed used CPN (Colored Petri nets) tools software. The FMC (Flexible Manufacturing Cell) model is studied from the Faculty of Engineering Management and Technology, University of Oradea. In this experiment we evaluate system analysis with respect to time using timed colored Petri net and describe the state space property of the flexible manufacturing cell using two cases first model with the two different place of machine and second model with compact machine place. The simulation results for the both system shows that the steps required for complete the simulation is same for both cases even though the number of place and the number of transition are different in both models. Also we discussed state space properties for both models to identify their properties during simulation.

1 Introduction

Today's world industries focus on produce quality product with less time and less money in manufacturing system. It's the best way to focus on simulation of the system to predict appropriate result before apply on real system. We focus on FMC (Flexible Manufacturing System) system to simulate the model to predict results. They described utilization of colored petri nets as a high level language system for design the system and analysis different state of the system, place invariants, occurrence graph and also use of computer tools for CPN (Colored Petri nets) [1]. Provide idea and motivation of development concurrent system using colored petri nets and CPN tools to development of model for the system [2]. Described colored petri net is discrete-event modeling language combination with functional programming language Standard ML provides graphical notation, modeling concurrency, communication, and synchronization. Also mention use of time in colored petri nets according to modeling aspect [3]. Addressed state space analysis in Colored Petri Nets to define each possible behavioral property for node and arcs in the concurrent system [4]. Described use of colored petri nets tools for modeling, editing, simulating and analyzing untimed and timed nets with state space analysis [5]. Author described FMC

* Corresponding author: sanjibksaren@gmail.com

 DOI: 10.1051/matecconf/20179402015

system development using timed colored Petri nets [6]. Derived method for state space properties in colored Petri nets [7]. Time–transition Petri net (TTPN) used to model Flexible manufacturing system based on scheduling problems. TTPN involves less transition and places for this model and to achieved the minimum Makespan in scheduling problem, they developed A* algorithm which helps to search deeper successor in few cases. [8]. In Manufacturing Systems Petri Net (PN) is useful graphical mathematical tool to analysis behavior and verifies properties in details. They mentioned state space to easily identify bounded ness, liveness and fairness in manufacturing system [9]. In this paper, they proposed systematized hybrid methodology in timed hierarchical colored Petri nets to model the flexible manufacturing system to find the better productivity setup for the system. They also explored state space to find the properties from the system likes liveness, boundedness, and reversibility [10].

2 Descriptions of the flexible manufacturing cell

The flexible manufacturing cell layout consisting of Power station, Evacuation station, and Milling machine CNC CONCEPT MILL 55, Control equipment (FANUC) of milling machine, Lathe CNC CONCEPT TURN 55, Control equipment (FANUC) the lathe, Mitsubishi RV-2AJ robot and Handle the moving robot. In Figure 1, the flexible manufacturing cell from the Faculty of Engineering Management and Technology, University of Oradea present below.

Fig 1. Flexible Manufacturing Cell (Faculty of Engineering Management and Technology, University of Oradea).

3 Descriptions of the function of the system with time

3.1 Define the time of sequences for two different FMC model

The processing time for the first model define with initial time $T_0 = 420$ s. The parts will arrive to the system according to the short proceeding time. Its mean the revolute parts (P1) arrive first in the system, then another revolute part (P2) will arrive in the CNC CONCEPT TURNS 55 (M1). After complete of revolute parts robot (R1) loading prismatic part (P3) in CNC CONCEPT MILL 55 (M2). Then last part (P4) is processing on the CNC CONCEPT MILL 55 (M2) to complete the operation. So, the sequence is **P1-P2-P3-P4** in the first model. The second model is more compact than first model. Here the initial time consider $T_0 = 0$ s. The first part (P1) will arrive in the system at time $T_0 = 0$ s. Second part (P3) will arrive at 400 s in the system, third part (P4) will arrive at 1450 s and last part (P2) will arrive at 2000 s in the system. So, the time sequence for the second model will be **P1-P3-P4-P2**. In the second model transition time is used to define the processing time. The part will be processing on transition T2. The instruction of time in transition T2 is @ + (if (p, m) = (P1, M1) then 420 else if (p, m) = (P2, M1) then 540 else if (p, m) = (P3, M2) then 1080 else 1260, all the values in seconds.

3.2 Processing time of parts for both models

Color with time will be developed on the basis of 1) Color timing (marks), 2) Transition timing, 3) Arcs timing. The time considered with the job processing time in the simulation. In the first model time for arrivals of the parts associated with the only processing time. We define the simulation time (P1, M1) = 420; (P2, M1) = 540; (P3, M2) = 1080; (P4, M2) = 1260. The processing parts are shown below in the Figure 2.

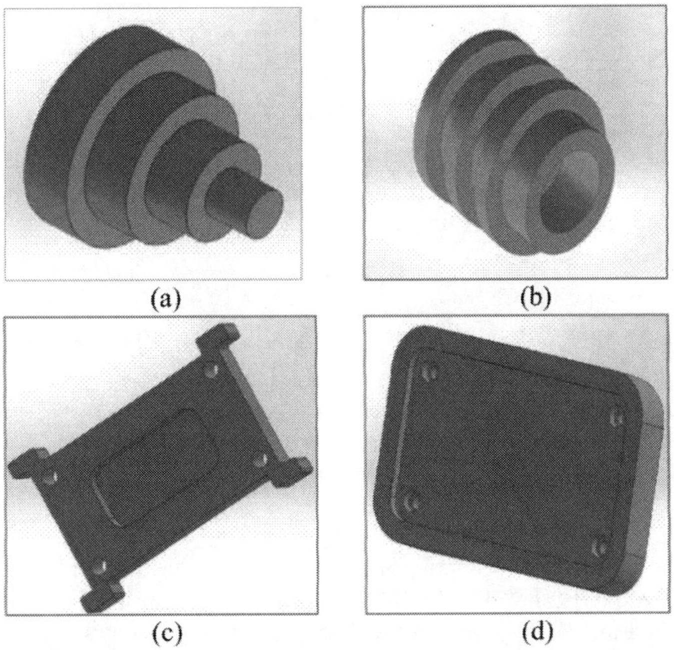

Fig. 2. (a) & (b) revolute and (c) & (d) prismatic parts are processing on CNC turning and milling.

4 Modeling of the system using colored Petri nets with respect to time

4.1 Color and variable definition for the first system

Colors define for each machine: Lathe CONCEPT TURN55 CNC (closet lathe = with M1): Lathe = {M1} Milling CNC CONCEPT MILL55 (colset Milling = with M2): Milling = {M2}.*The color set for different parts with time:* (colset pieceL = withP1|P2 timed): pieceL = {P1@420; P2@540}; (colset pieceM = withP3|P4 timed): pieceM = {P3@1080, P4@1260}. *Colors define for robot:* (closet robot = with R1 timed): Robot = {R1@0}. The color set is define for two different machine and different parts, that's cause the system is complex, for this reason color must define for each machine with piece. For lathe the color is "colourL" and for milling "colourM". (Colset colourL = product piece*lathe timed) ;(Colset colourM = product piece*Milling timed). During modeling the system, we define different machine and different parts. Each part is processed in particular machine it defines in the system, the definition for the complex color for lathe is "programL" and for milling is "programM". (Colset programL = product piece*lathe timed); (Colset programM = product piece*Milling timed). The product and machine have this complex color pairs to run the system. It can be defined by: {P1, M1}, {P2, M1} and {P3, M2}, {P4, M2}. *Color defines for Variables:* In color petri nets network connection between place and transition define by arcs. In this model arc is assigned with the variables to create connection between place and transition. The variable for turning part is defined by "pL" it has values (P1, P2) and for milling part "pM" its values are (P3, P4).

$$pL \in \{p1, p2\} \tag{1}$$

$$pM \in \{p3, p4\} \tag{2}$$

The variables for machine M1 and M2 are defined by respectively m2 and m1.

$$m1 \in \{m2\} \tag{3}$$

$$m2 \in \{m1\} \tag{4}$$

The variable for robot can be defined by "r" which has value:

$$r \in \{R\} \tag{5}$$

The model must be creating connection between the parts and machine in colored Petri nets through the variables which mention in arcs to establish the system.

$$pL\, X\, m2 \in \{P1,\, M1\}\, \{P2,\, M1\} \tag{6}$$

$$pM\, X\, m1 \in \{P3,\, M2\}\, \{P4,\, M2\} \tag{7}$$

The place position, the token position and transitions positions are mentioned below in the Table 1, Table 2 and Table 3 for the first flexible manufacturing cell with two different places for machine respectively lathe and milling. In Table 1 contains the descriptions for each place with complexity of color and interpretations. Each complex color defines name respectively "programL", "programM", "pieceL", "pieceM", "lathe", "milling" and "robot". In Table 2 describe about the initial and final token position with respect to time and Table 3 describe the transitions and arc time in the system. Arc timing consider from

 DOI: 10.1051/matecconf/20179402015

place P13 to connect with the required transitions respectively T1, T3, T4, T5, T7 and T8. During modeling the system each parameters is important to build the system.

Table 1. Place Position.

Places	Complexity of color	Interpretation
P1	complex color "programL"	Raw materials in stock for turning.
P2	complex color "programL"	Turning Parts in robot gripper (R1)
P3	complex color "programL"	Turning parts in machine for processing (M1)
P4	complex color "pieceL"	Part is unloading from lathe by robots (R1)
P5	complex color "pieceL"	Finish turning part store in storage.
P6	complex color "programM"	Raw materials in stock for milling.
P7	complex color "programM"	Milling Parts in robot gripper(R1)
P8	complex color "programM"	Milling parts in machine for processing (M2)
P9	complex color "pieceM"	Part is unloading from milling by robots (R1)
P10	complex color "pieceM"	Finish mill part store in storage.
P11	complex color "lathe"	Lathe (M1)
P12	complex color "milling"	Milling (M2)
P13	complex color "robot"	Robot (R1)

Table 2. Token Position.

Places	Initial Token position in system	Final Token position in system
P1	(P1, M1)@420;(P2,M1)@540	-
P2	-	-
P3	-	-
P4	-	-
P5	-	P1@428;P2@548
P6	(P3, M2)@1080;(P4,M2)@1260	-
P7	-	-
P8	-	-
P9	-	-
P10	-	P3@1088;P4@1268
P11	M1	M1
P12	M2	M2
P13	R1@0	R1@8

 DOI: 10.1051/matecconf/20179402015

Table 3. Transitions Position.

Transitions	Interpretation	Arc timing
T1	Turning part is load by robot (R1)	P13- T1= r@+8
T2	Part is processing in lathe (M1)	-
T3	Part is unload by robot (R1)	P13- T3= r@+8
T4	Part is ready to storage	P13- T4= r@+8
T5	milling part is load by robot (R1)	P13- T5= r@+8
T6	Part is processing in milling (M2)	-
T7	Mill Part is unload by robot(R1)	P13- T7= r@+8
T8	Mill Part is ready to storage	P13- T8= r@+8

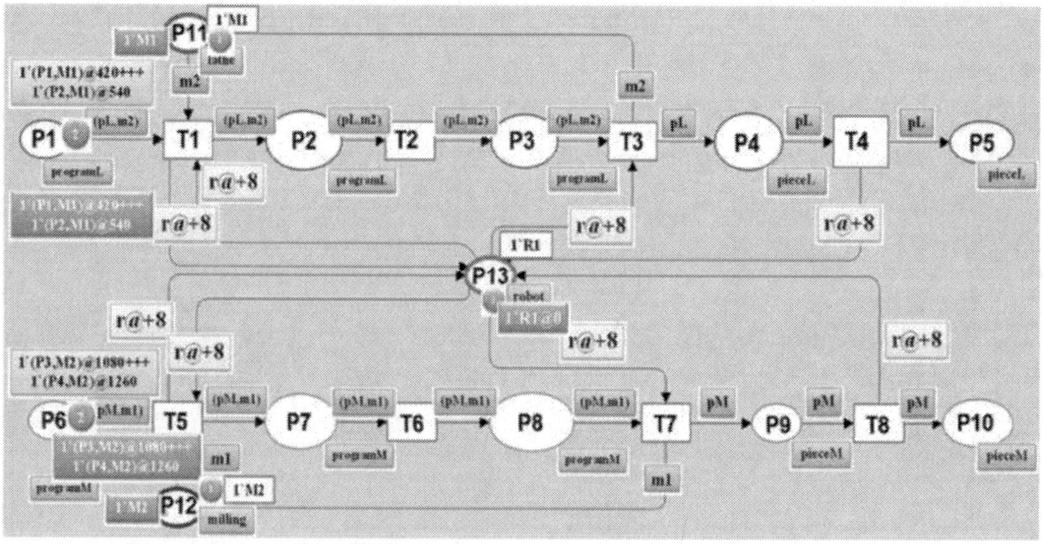

Fig. 3. Two different machine place model using timed colored petri nets condition with initial stage of simulation.

In Figure 3, present the initial stage of the simulation. It appears that during modelling place P1, P6, P11, P12 and P13 contains the processing time. in this system place P1 contain processing time for lathe parts respectively 420 s and 540 s and place P2 contain processing time for milling parts respectively 1080 s and 1260 s and initial robot time show in the system 1`R1@0. In P1, P6, P11, P12, P13 shows token position respectively 2, 2, 1, 1, 1. P11 and P12 shows the lathe machine 1`M1 and milling machine 1`M2. It is observed that when the model is ready the required program appears in green colour means the all the places, arc and transitions arc connected in proper way to simulate the manufacturing system model. In Figure 4, shows final stage of simulation. The token take move from initial place to final place to fulfil the simulation process and the time and token is appeared in the place P5, P10 and P13.

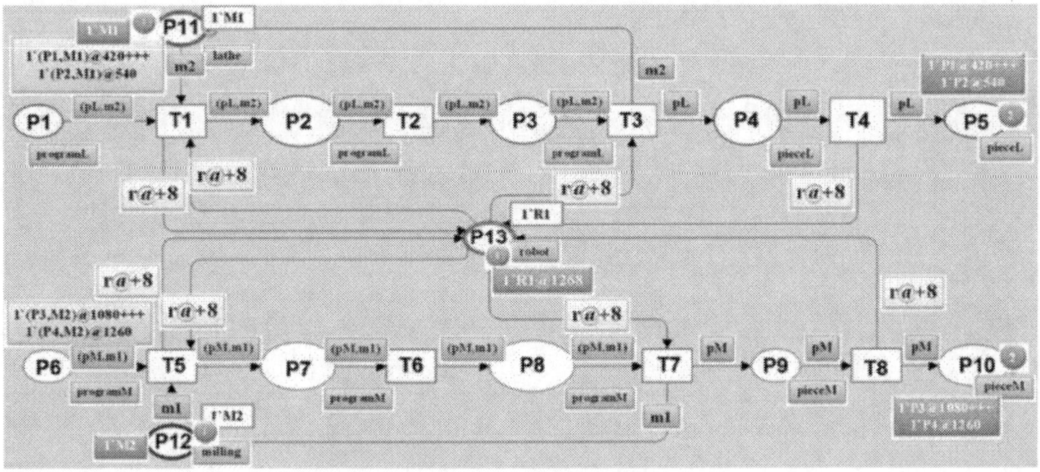

Fig. 4. Two different machine place model using timed colored petri nets condition with final stage of simulation.

4.2 Color and variable definition for the second system

Colors define for each machine: The color set for lathe CONCEPT TURN55 CNC and for milling CNC CONCEPT MILL55. So the value for machine is (colset Milling = with M1|M2), it defines: Machine = {M1, M2}. The color set for different parts are mentioned with the color. The value for the parts is: P1, P2, P3, and P4. The definitions of this value are: (colset piece = withP1|P2| P3|P4), Piece = {P1@420, P2@540, P3@1080, and P4@1260}. The robot is associated with color "robot". It has value R1. (colset robot = with R1) is the definition for robot in the model. Robot = {R1@0}. Color defines for Variables: In color petri nets network connection between place and transition define by arcs. In this model arc is assigned with the variables to create connection between place and transition. The variable for parts "P" value is:

$$p \in \{p1, p2, p3, p4\} \tag{8}$$

The variables for machine M1 and M2 are defined by respectively m2 and m1.

$$m \in \{M1, M2\} \tag{9}$$

The variable for robot can be defined by "r" which has value:

$$r \in \{R\} \tag{10}$$

The model must be creating connection between the parts and machine in colored petri nets through the variables which mention in arcs to establish the system.

$$p \times m \in \{(P1, M1), (P2, M1), (P3, M2), (P4, M2)\} \tag{11}$$

The place position, the token position and transitions positions are mentioned below in the Table 4, Table 5 and Table 6 for the first flexible manufacturing cell with compact machine place respectively lathe and milling. In Table 4, contains the descriptions for each place with complexity of color and interpretations. Each complex color defines name respectively "program", "piece", "machine" and "robot". In Table 5, describe about the initial and final token position with respect to time and Table 6 describe the transitions and arc time in the system. Arc timing consider from place P7 to connect with the required transitions respectively T1, T2, T3 and T4.

Table 4. Place Position.

Places	Complexity of color	Interpretation
P1	complex color "program"	Raw materials in stock.
P2	complex color "program"	Parts in robot gripper (R1)
P3	complex color "program"	Part in machine for processing.(M1 & M2)
P4	complex color "piece"	Part is unloading by robot (R1)
P5	complex color "piece"	Finish part store in storage.
P6	complex color "machine"	Working machine (M1 & M2)
P7	complex color "robot"	robot (R1)

Table 5. Token Position.

Places	Initial Token position in system	Final Token position in system
P1	(P1,M1);(P2,M1);(P3,M2);(P4,M2)	-
P2	-	-
P3	-	-
P4	-	-
P5	-	(P1, P2, P3, P4)
P6	(M1, M2)	(M1, M2)
P7	R1	R1

Table 6. Transitions Position.

Transitions	Interpretation	Arc timing
T1	Part is loading by robot (R1).	P7-T1=r@+8
T2	Part is processing (M1 & M2).	P7-T2=r@+8
T3	Part is unloading by robot (R1).	P7-T3=r@+8
T4	Part is ready to storage	P7-T4=r@+8

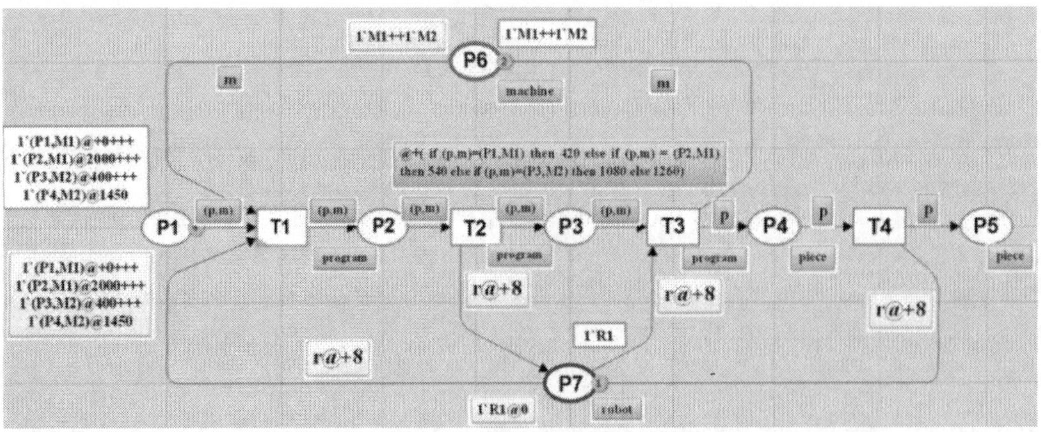

Fig. 5. Compact machine place model using CPN toots with respect to time with initial condition of simulation.

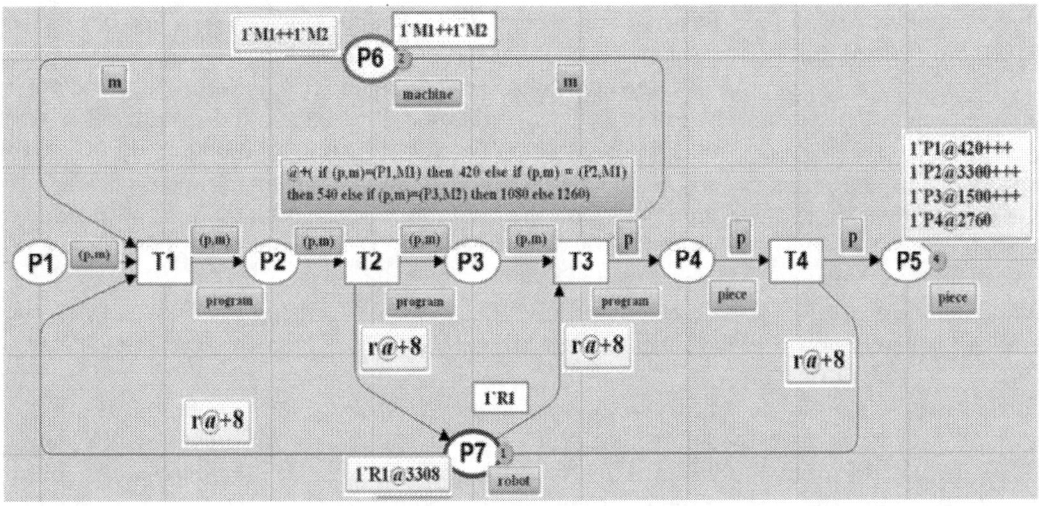

Fig. 6. Compact machine place model using CPN toots with respect to time with final condition of simulation.

In Figure 5, present the initial stage of the simulation for the compact machine place. It appears that during modelling place P1 contains the job arrival time for parts and P7 contains robot loading and unloading time. In this system processing time for lathe parts respectively 420 s, 540 s and processing time for milling parts respectively 1080 s and 1260 s and here also initial robot time show in the system 1`R1@0. In P1, P6 and P7shows token position respectively 4, 2, 1. Place P6 show the machine 1`M1+1`M2. In Figure 6, shows final simulation for the model. The token take move from P1 to P5 and it appears 4 tokens in place P5, the token position for P7 is 1 and for place the token position is 2 respectively.

5 First and second models simulation results

FMC system consisting of two machines Lathe CNC CONCEPT TURNS 55 (M1) and Milling machine CNC CONCEPT MILL 55 (M2), each machine is served by Mitsubishi RV-2AJ robot (R1) for loading and unloading for the parts and places the parts for storage area. In FMC system parts are moving through conveyor from one station to other stations. Robot (R1) loads the parts from pallet to both machines (M1 and M2) according to operation. In this simulation four parts are processing in two different machines. We consider robot loading and unloading time 8 s for each operation.

In first model Figure 4, Job is according to the less processing time in the system. According to this sequence two turning part is processing in the lathe with time 420 s and 540 s respectively. Then two different milling parts is processing according to the time with 1080 s and 1260 s. If robot loading and unloading time is consider with the job processing time then job completion time sequence will be 428 s, 548 s, 1088 s and 1268 s. The arc timing is considered r@+8 s in the system. Robot will take 1268 s complete the total simulation.

In Figure 4, we see that Place P1, both parts and machine (P1, M1) and (P2, M1) and in Place P2, (P3, M2) and (P4, M2) appears. During the simulation it is observed that only the finishing turning parts are reached in place P6 as a P1, P2 and the milling parts reached in place P10 as a P3, P4. Also we observed that the movement of the token from one place to another place. During simulation the information is transfer through the one place to another place by adding transition through arcs. This arcs contain required time and connecting program between the transition and place. When place send the information through arcs is fire and received the information in transition and send it in another place. The whole system follows the same procedure to run the system.

In second model Figure 6, we consider two machines in one place. Here we consider job arrival time in the system. The job P1 will arrived at T_0=0 seconds in the system, P3 at 400 s, P4 at 1450 s and P2 at 2000 s in the system. The robot loading and unloading time will take 8 s. The arc timing is considered to connect between the place and transition. The arc timing is r@+8. From simulation we found the total job processing time in the system. The time for P1=420 s, P3=1500 s, P4=2760 s and P2=3300 s. The robot will take 3300 s to complete the simulation. Here in second model, all the token is place in P1 and all the parts and machine are values are appeared as a (P1, M1), (P2, M1), (P3, M2), (P4, M2). At the end of the simulation only the finished parts are reached in place P5 as a parts (P1, P2, P3, P4).

In both cases, it is found that total 16 steps required finishing the simulation. If we see carefully for both cases the number of place is different. In two different machine place there are total 13 place, 8 transitions and for compact machine place there are total 7 place and 4 transitions. But the required steps are same for both models.

6 First and second models state space analysis

We analyze State space properties for both models. In state space we generate SCC graph and reachability properties. Also derives few properties during simulation which are Boundedness property, Home Properties, Liveness Properties and Fairness Properties. Basically in state space we found the behavior of the system for each place and transition. During generate state space graph it shows nodes in paths mode which shows node list in the system. In state space graph shows each nodes and arcs information of system behavior. Just click on the any nodes and arcs it shows the accurate condition of the system behavior. In Figure 4, we see in arc 1 shows information (1:1=>@420 model_1`T1 1{r=R1, pL =P1, m2=M1} so in each arc carry information about status of machine, robot and part. Each node also carries information of system .Nodes 17 shows 17@1260, place position 5 for part P1@420, P2@540 and place 10 for P3@1080, P4@1268 & place11 and 12 for machines M1 and M2.

In Figure 7, shows the system behavior of the first model. Statistics of first model observed state space and SCC graph it shows total 17 nodes and 16 arcs are required to complete the processing of system. For first model Initial Marking is not a home marking. In liveness properties there are no dead and live transitions Instances in the system. For first system Dead Markings is 17. In fairness properties no infinite occurrence sequences.

In Figure 8, shows the state space for second model. Statistics of second model verified observed state space and SCC graph here also required 74 nodes and 78 arcs to complete the system. Here also initial marking is not a home marking but here we found dead transition in the system. For second system dead markings are 74, 73, 72, 65, 64, and 63. Here we found from the both system, different nodes and arcs are required to complete the operation because the place and transition are different in both systems. The different also observed in SCC graph and dead marking in for both models. In state space analysis we found few properties in both systems. The analysis describes the information about each nodes and arcs in the system. How the both system works we got information through the state space.

 DOI: 10.1051/matecconf/20179402015

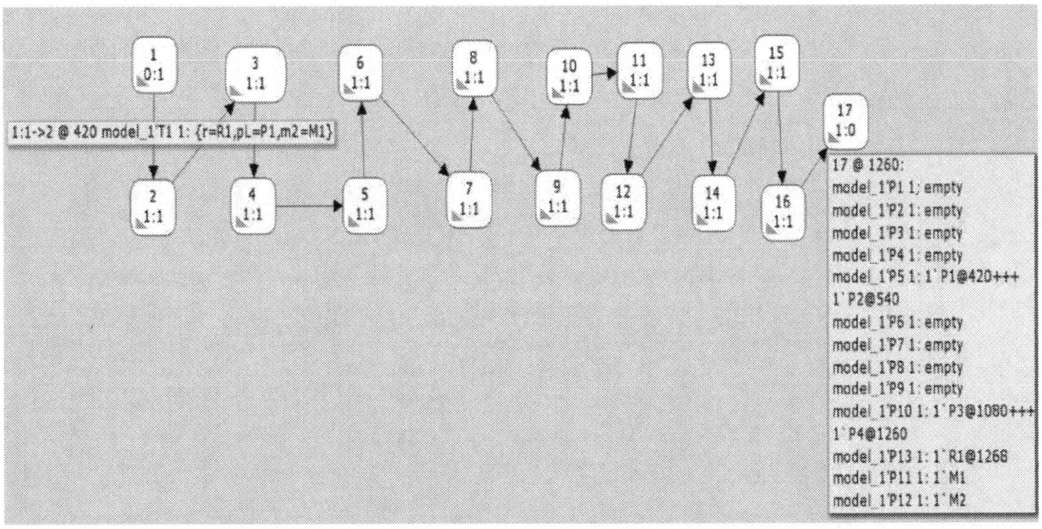

Fig. 7. State space diagram view for two different machine place model.

In Table 7, we present the boundedness properties for the two different machine place and mentioned best possible multi set bounds for the system. In place P1 appears 1\`(P1,M1)++1\`(P2,M1) with upper bound 2 means it has 2 token when its moves to the token its reached in place P5 with upper bound 2 in place P5 which appears 1\`P1++1\`P2 in the system. If we see the upper bound and lower bound in the Table 7 its means it's all token position during the simulation which observed. And it's also clear that only finishing parts are appears end of the simulation as a 1\`P1++1\`P2 and 1\`P3++1\`P4 respectively in place P5 and P9.

Table 7. Best Integer Bounds for first Model.

Best Integer Bounds	upper	lower	Best Upper Multi-set Bounds	Best Lower Multi-set Bounds
model_1'P10 1	2	0	1\`P3++1\`P4	Empty
model_1'P1 1	2	0	1\`(P1,M1)++1\`(P2,M1)	Empty
model_1'P11 1	1	0	1\`M1	Empty
model_1'P12 1	1	0	1\`M2	Empty
model_1'P13 1	1	0	1\`R1	Empty
model_1'P2 1	1	0	1\`(P1,M1)++1\`(P2,M1)	Empty
model_1'P3 1	1	0	1\`(P1,M1)++1\`(P2,M1)	Empty
model_1'P4 1	1	0	1\`P1++1\`P2	Empty
model_1'P5 1	2	0	1\`P1++1\`P2	Empty
model_1'P6 1	2	0	1\`(P3,M2)++1\`(P4,M2)	Empty
model_1'P7 1	1	0	1\`(P3,M2)++1\`(P4,M2)	Empty
model_1'P8 1	1	0	1\`(P3,M2)++1\`(P4,M2)	Empty
model_1'P9 1	1	0	1\`P3++1\`P4	Empty

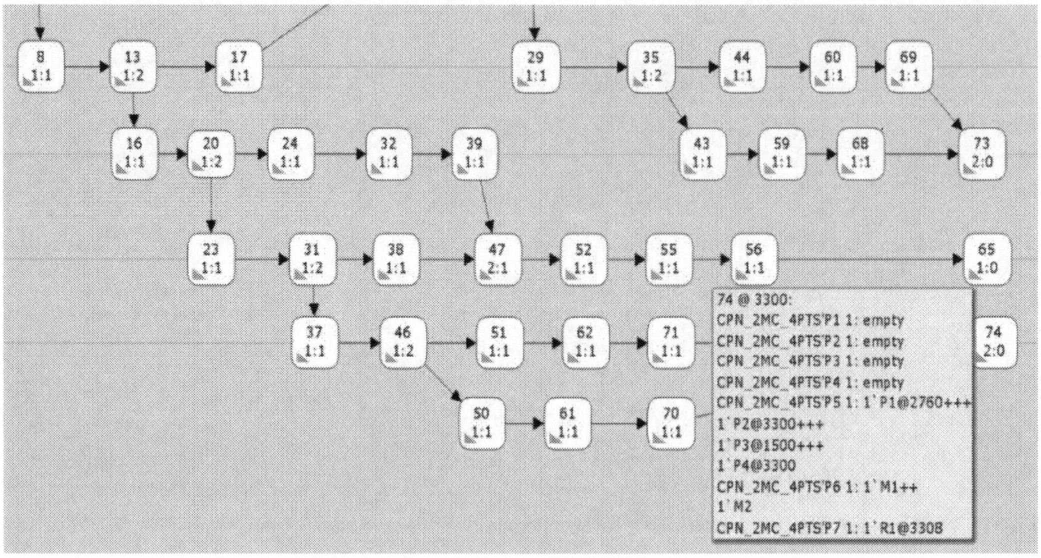

Fig. 8. State space diagram view for compact machine place model.

In Table 8, we present the boundedness properties for the compact machine place and mentioned best possible multi set bounds for the system. Upper bound present the token in each place during the simulation. If we see from the Table 8, CPN_2MC_4PTS'P1 1 its mean place P1, the token is 4 and the best multi set is 1`(P1,M1)++1`(P2,M1) ++1`(P3,M2)++1`(P4,M2). So, the upper set basically the token in the system.

Table 8. Best Integer Bounds for second Model.

Best Integer Bounds	Upper	Lower	Best Upper Multi-set Bounds	Best Lower Multi-set Bounds
CPN_2MC_4PTS'P1 1	4	0	1`(P1,M1)++1`(P2,M1) ++1`(P3,M2)++1`(P4,M2)	Empty
CPN_2MC_4PTS'P2 1	1	0	1`(P1,M1)++1`(P2,M1) ++1`(P3,M2)++1`(P4,M2)	Empty
CPN_2MC_4PTS'P3 1	2	0	1`(P1,M1)++1`(P2,M1) ++1`(P3,M2)++1`(P4,M2)	Empty
CPN_2MC_4PTS'P4 1	1	0	1`P1++1`P2++1`P3++1`P4	Empty
CPN_2MC_4PTS'P5 1	4	0	1`P1++1`P2++1`P3++1`P4	Empty
CPN_2MC_4PTS'P6.1	2	0	1`M1++1`M2	Empty
CPN_2MC_4PTS'P7.1	1	0	1`R1	Empty

7 Comparison between the models

In Table 9, we show the observation from the two different machine place and compact machine place. It's clear that both model results are around similar even the place and transitions are different for both models. So, it's possible to design same model in different way to predict the good results.

Table 9. Comparison between the models.

Parameters	Two different machine place	Compact machine place
No of machine	2	2
No of robot	1	1
No of part processing	4	4
No of place	13	7
No of transition	8	4
No of arcs (in model)	26	13
No of arcs (in state space)	16	78
No of nodes	17	74
Total processing time	1268 s	3308 s
No of steps to complete simulation	16	16
Dead marking	1	6
Home marking	Initial Marking is not a home marking.	Initial Marking is not a home marking.
Fairness Properties	No infinite occurrence sequences.	No infinite occurrence sequences.
Dead Transition Instances	None	None
Live Transition Instances	None	None

8 Conclusions

The simulation results from the first system implement with the part processing time with two different positions of machines it shows that 16 steps to take complete the process. The first model we consider only the part processing time. The completion time for the system is 1268 s. For the second model with compact machine place it also took 16 steps to complete the process but the system consider with job arrival time and job processing time. The completion time for the second system is 3308 s. The only difference is the job arrival time for the both models. But the simulation results shows the steps required to complete the simulation is same for both cases. The observation of state space analysis for two different positions of machines shows 17 nodes and 16 arcs and for compact machine place required 74 nodes and 78 arcs to complete the entire process to produce four different parts. Both cases the parts sequence is different. So it is found the system is modeling in different strategic way but the results for required steps are same for both systems. For future works it's possible to construct model using hierarchical timed colored petri net with state space analysis. In manufacturing system the number of processing parts, the number of robots and number of machines increase the quantity of token in the system to make it more complex.

The paper published has been sponsored under the Erasmus mundus partnership program agreement vide number 2014-0855/001-001 coordinated by and between University of Oradea and City University of London Under Action Plan 2 for the year 2015-18.

References

1. K. Jensen, LCNS, **483**, 75 (1990)
2. K. Jensen, L. M. Kristensen, *Coloured Petri nets: modeling and validation of concurrent systems*, (Publisher-Springer-Verlag New York Inc. 2009)
3. K. Jensen, L. M. Kristensen, L. Wells, STTT, **9**, 42 (2007)
4. K. Jensen, Form. Meth. Syst. Des., **9**, 34 (1996)
5. A. V. Ratzer, L. Wells, H. M. Lassen, M. Laursen, J. F. Qvortrup, M. S. Stissing, M. Westergaard, S. Christensen, K. Jensen, Applic. LNCS, **2679**, 13 (2003)
6. F. Blaga, I. Stanasel, A. Pop, V. Hule, T. Buidos, Annals of the Oradea University, Fascicle of Management and Technological Engineering, **1**, 4 (2014)
7. C. Søren, K. Jensen, T. Mailund, Proc. of 2nd Int. Coll. on Petri Net Technologies for Modelling Communication Based Systems, **12** (2001)
8. C. Li, W. M. Wu, G. Rong, ICNSC, **6** (2014)
9. A. Bevilaqua, E. R. R. Kato, E. C. Pedrino, R. H. Tsunaki, SMC, **6** (2013)
10. M. Aguiar, R. Barreto, R. Caldas, J. Edgar, C. Filho, ICIT, **6** (2008)

Experimental and numerical investigations of the steel sheets formability with hydroforming

Radu Vasile[1,*], *Sever-Gabriel* Racz[1], and *Octavian* Bologa[1]

[1]Department of Industrial Machinery and Equipment "Lucian Blaga" University, Sibiu, Romania

Abstract. The present paper focuses on analyzing the forming capacity of steel blanks with hydroforming process. For this research steel sheets have been in focus for numerical and experimental analysis. The main advantages for this materials are good surface finish, excellent forming capacity and close tolerances, appealing advantages for manufacturers. A finite element model has been developed from data obtained through tensile tests and forming limit curves. A newly developed hydroforming press has been used to carry out the forming experiments. Side-by-side analysis between numerical and experimental results concludes the experiment.

1 Introduction

Automotive, aviation, aerospace and medical industries focus research towards lighter materials such as aluminum alloys, magnesium alloys, high-strength steels and titanium alloys. The advantages of these materials are used to reduce the weight of the parts and to increase the structure rigidity of the future product. All these efforts will result in lower energy consumption, fewer carbon dioxide emissions and lower costs for the manufacturer and the end user. But producing high quality products forces researchers to create or improve production technologies to increase the material deformability limits [1-2]. Therefore the present research is presenting a different approach towards hydroforming press concept in forming two different steel materials.

2 Materials used in this research

The numerical and experimental researches were made in order to evaluate the forming behavior of steels DC01 and DX51D. The first material, DC01, is an uncoated mild steel used in typical deep-drawing applications, also known as 1.0330 by DIN EN 10346. The second material researched is DX51D, a mild galvanized steel grade for cold forming, also known as 1.0226. It is non-ageing steel designed for deep and extra deep drawing that is covered with a layer of zinc for good resistance to corrosion [3-4].

The chemical composition of both materials, in percent by mass, is presented in Table 1 [3-5].

* Corresponding author: r.vasile@ulbsibiu.ro

The physical and mechanical properties of both materials are presented in Table 2 [3-7], where ρ – density, E – Young's modulus, $R_{p0.2}$ – yield stress, R_m – tensile stress, r – anisotropy, ν – Poisson's ratio, n – strain hardening exponent, K – strength coefficient.

Table 1. Material chemical composition maximum values [wt%].

Element \ Material	DC01	DX51D
Fe	98	99.30
Ti	-	0.30
Si	-	0.50
Mn	0.6	0.60
P	0.045	0.10
S	0.045	0.045
C	0.12	0.12

Table 2. Material physical and mechanical properties.

Property 20[°C] \ Material	DC01	DX51D
ρ [kg/dm^3]	7.83	
E[GPa]	210	
$R_{p0.2}$ [MPa]	140-280	≥140
R_m [MPa]	270-410	270-500
A_{80} [%]	≥28	≥22
ν [-]	0.35	0.30
r [-]	1.53	0.84
n [-]	0.21	0.2
K [MPa]	619	593

2.1 Uniaxial tensile test

The uniaxial tensile tests were carried out on an Instron 5587 testing machine. The test samples of 120 x 20 x 1 [mm^3] for both materials, in accordance with ISO 6892-1:2009 [8]. The tensile machine was controlled by Blue-hill proprietary software and GOM Aramis systems was used instead of an extensometer. With this experimental configuration the main mechanical characteristics have been determined. The determined values are presented in Table 3, where σ_{max} - maximum stress and ε_{max} - maximum strain.

Table 3. Determined material characteristics.

Material	DC01			DX51D		
Property \ RD	0°	45°	90°	0°	45°	90°
E[GPa]	64.98	67.35	71.7	70.2	73.84	64.69
$R_{p0.2}$ [MPa]	181.28	190.96	188.95	349.93	349.85	341.6
R_m [MPa]	328.86	343.88	332.62	396.03	405.95	394.81
A [%]	37.43	34.65	35.39	23.74	23.41	23.33
σ_{max} [MPa]	431.47	443.97	429.32	469.35	483.78	473.18
ε_{max} [-]	0.32	0.3	0.3	0.21	0.21	0.21
n [-]	0.24	0.24	0.23	0.14	0.15	0.16
K [MPa]	590.16	614.81	591.38	605.53	626.03	621.94
r[-]	1.770	1.104	2.003	1.383	0.837	1.092
$\bar{r}$ [-]	1.495			1.037		
Δr [-]	0.783			0.401		

The true strain-stress curves for three directions from the sheet rolling direction RD have been determined, but only the samples for 0° to RD, compared side-by-side are presented in Fig. 1 and Fig. 2. As can be seen DX51D offers 20% more stress, but 34% less strain behavior compared to DC01. Also there can be noted that DX51D present an apparent flow, at almost 93% more stress compared to DC01.

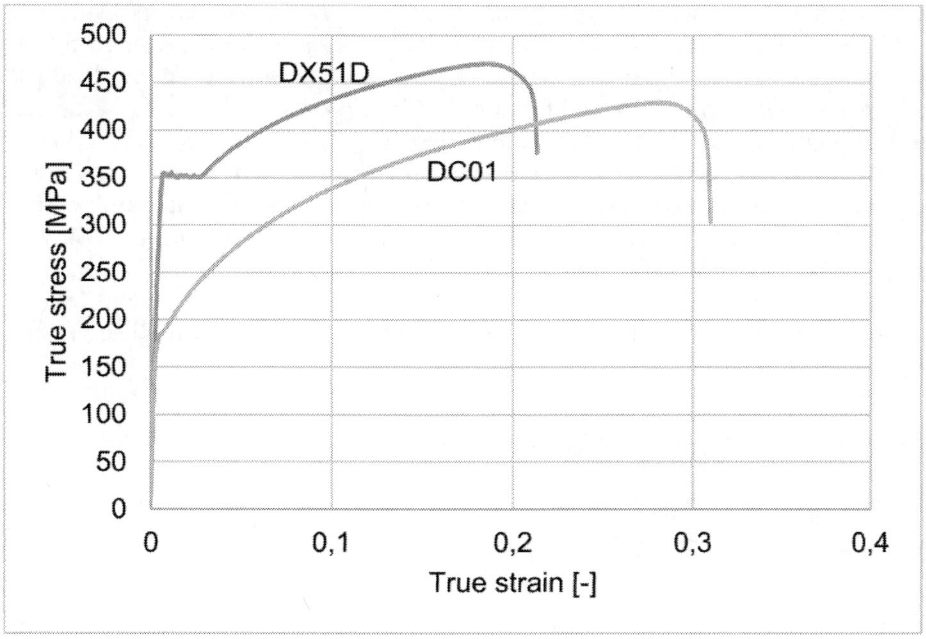

Fig. 1. True strain-stress curves for 0° to rolling direction.

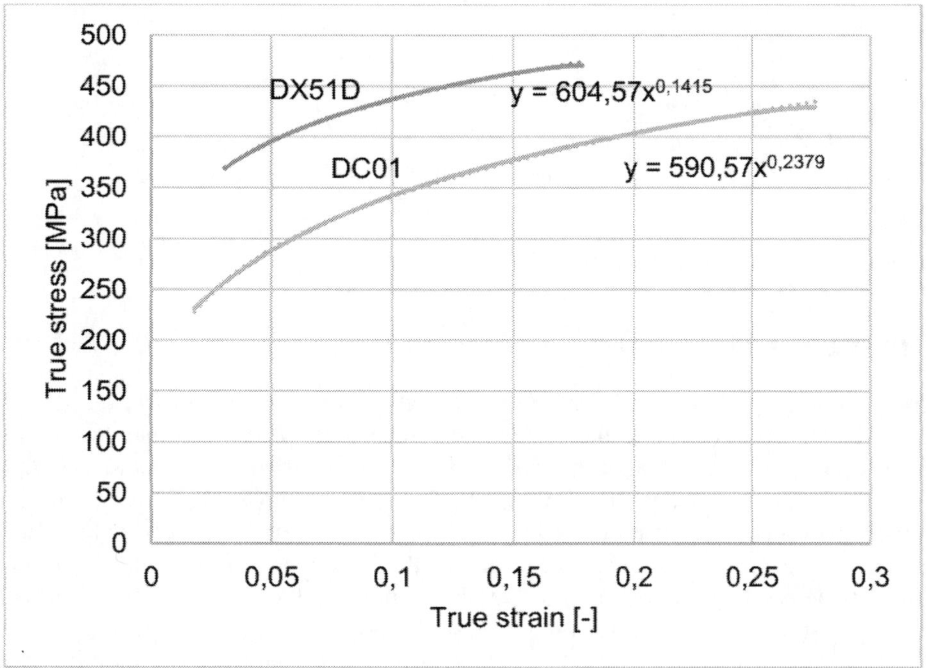

Fig. 2. Determined strain hardening curves.

2.2 Forming limit curve

The tests were carried out using the following experimental setup - a modular deep drawing unit, with exchangeable active elements, connected to a hydraulic high pressure unit and two CCD cameras connected to GOM Aramis - the optical strains measurement system. Forming limit curves have been determined experimentally by using several load paths of the test samples, between the uniaxial tension and the equibiaxial stretching. This has been obtained by realizing different test specimen geometry, suggested by Hašek [9-12]. The samples were laser cut and following the same procedure as before, the remaining burrs were removed and the samples surface covered by a layer of white matt paint to eliminate reflections and a cloud of black matt paint to obtain the diffuse network of points. Every test sample was deep drawn with a hemispherical punch until fracture and at the same time monitored by Aramis system. The test has been in compliance with standard ISO 12004-2:2008 and the results are represented in Fig. 3 [13]. It can be clearly observed a better forming capacity for DC01.

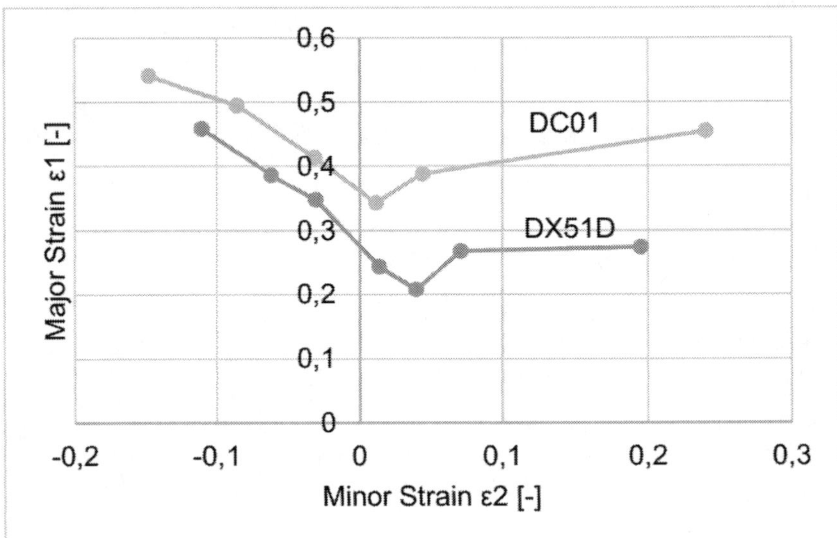

Fig. 3. Determined forming limit curves.

3 Numerical simulation

The hydroforming process has been analyzed with the help of Abaqus/Explicit software with the objective to determine the major and minor principal strains and the researched materials forming limits. A parametrized geometric model with finite elements, presented in Fig. 4, was created using the built-in tridimensional module of the program. The researched materials characteristics and parameters were experimentally determined and introduced in the program to accurately define the model in the dynamic analysis. The process parameters and values are presented in Table 4, these are R – die radius, p_s – retaining pressure, p_h – hydroforming pressure and t_0 – blank initial thickness. The Ramberg–Osgood equation was used to calculate the plastic strain and Hill yield criterion was been used in order to take into account the materials anisotropy. For evaluating the materials instability and tearing during the proposed hydroforming process, the plastic instability criterion was used. The blank was meshed with shell elements, with four nodes and reduced integration on the middle surface, with five integration points in width. The

 DOI: 10.1051/matecconf/20179402016

retaining ring and the die were meshed using tridimensional four nodes rigid elements. The retaining and the hydroforming pressure were applied five times quicker than reality.

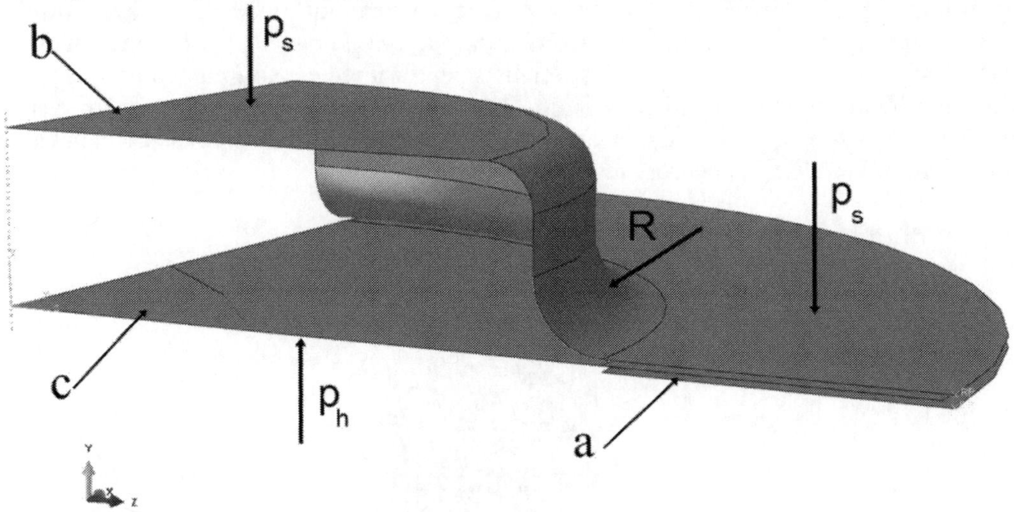

Fig. 4. Numerical simulation modelling and process parameters.

a – retaining ring, b – die, c – blank.

Table 4. Model and process parameters.

Parameter / Material	R [mm]			p_s [bar]	p_h [bar]			t_0 [mm]
DC01	4	6	8	250	150	250	350	1
DX51D								

The results obtained from numerical simulation are presented in Fig. 5 and Fig. 6. It can be noted that the biggest differences observed on the images are not the local strains compared, but the anisotropy distribution by influencing local strain values.

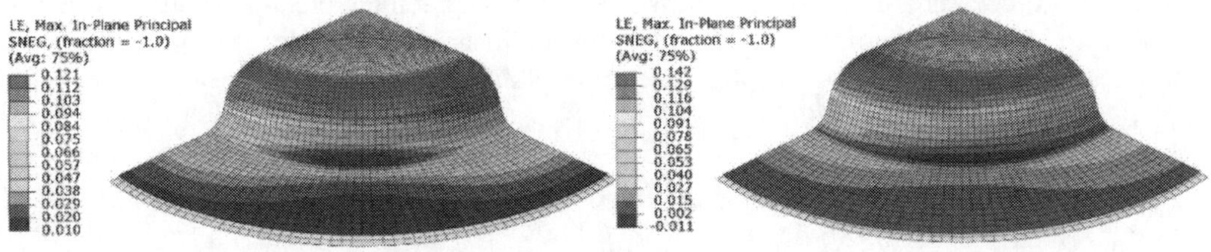

Fig. 5. Major strain results R = 6 [mm] p_h = 350 [bar].

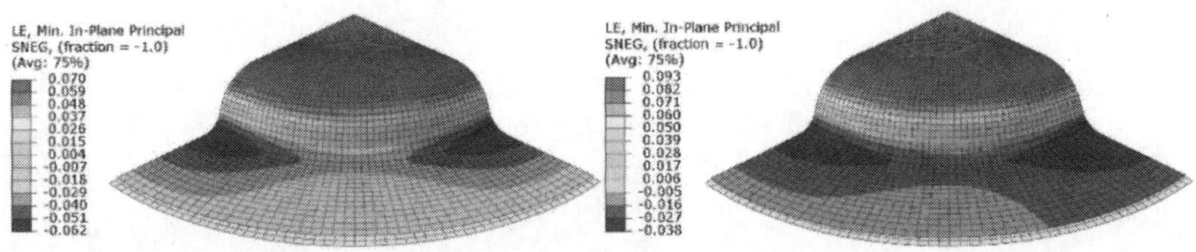

Fig. 6. Minor strain results R = 6 [mm] p_h = 350 [bar].

DOI: 10.1051/matecconf/20179402016

4 Experimental research

Hydroforming experimental investigations were carried out on a new hydroforming solution developed by the authors. Combining design knowledge with the study of hydroforming die concepts by field specialists and with the research requirements the following hydroforming solution was built. The unique hydroforming concept as seen in Fig. 7 is completely developed by the authors. The solution was built from scratch, having a new shape and modulation concept unseen in other researches.

Fig. 7. Hydroforming solution, unique concept developed by the authors.

The main process parameters - p_s and p_f - are monitored in real-time with a pressure transducer, as represented in Fig. 8. Also the blanks have been covered electrolitically with a calibrated network of points in order to optically measure the major and minor strains with GOM Argus system, image capture and computation are presented in Fig. 9.

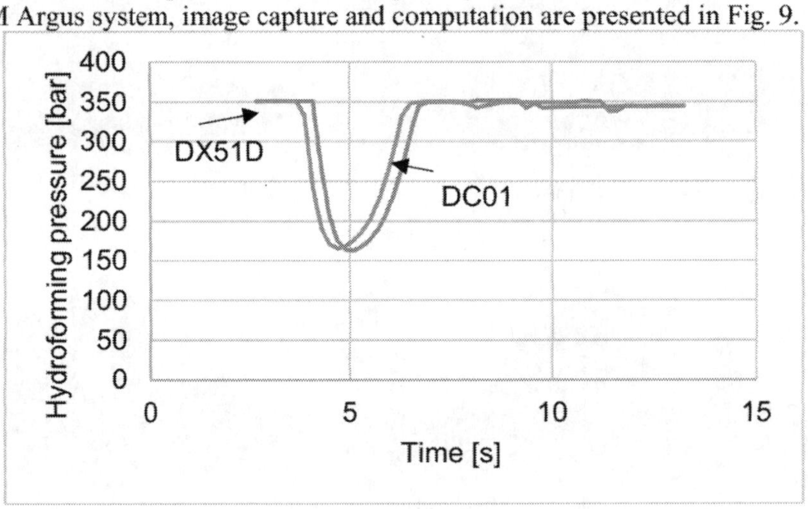

Fig. 8. Hydroforming pressure for R = 6 [mm] p_h = 350 [bar].

 DOI: 10.1051/matecconf/20179402016

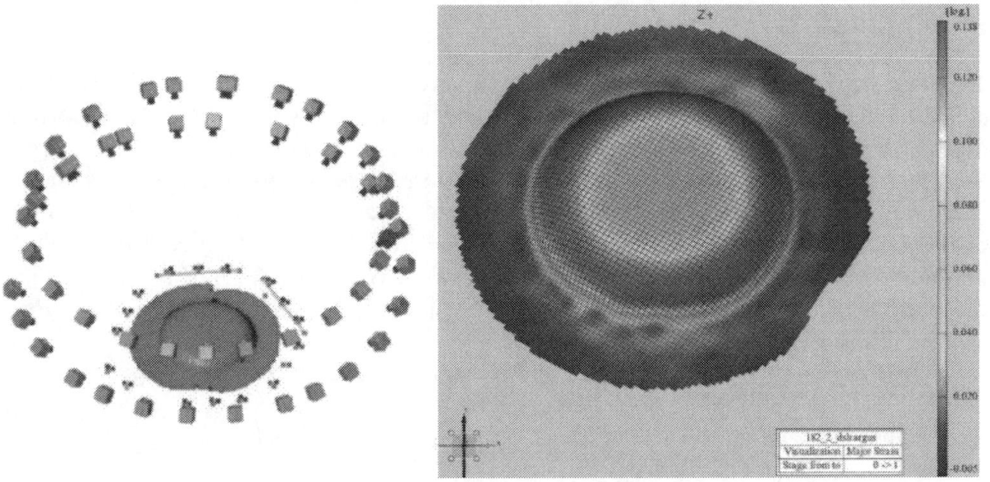

Fig. 9. Measuring local strain on the drawn part with GOM Argus System.

The main objective of the present research was to determine and study the influence of die radius and hydroforming pressure variation for two representative steel materials. The process parameters are introduced in both numerical and experimental analysis to compare the material forming limits by comparing the reached local strain.

The numerical simulation results were side-by-side compared with the experimental measured results. The maximum relative error between numerical and experimental results is below five percent. The values are represented in Table 5.

Table 5. Numerical and experimental results for hydroforming.

No.	**Material**	**Pressure p_h [bar]**	**Radius [mm]**	**Numerical simulation**		**Experimental results**		**Error [%]**
				ε_1 [-]	ε_2 [-]	ε_1 [-]	ε_2 [-]	
1	DC01	150	8	0.075	0.058	0.076	0.059	1.53
2	DX51D			0.088	0.073	0.091	0.076	3.76
3	DC01	250		0.096	0.062	0.092	0.063	2.89
4	DX51D			0.113	0.080	0.116	0.081	1.95
5	DC01	350		0.116	0.064	0.119	0.065	2.00
6	DX51D			0.131	0.083	0.127	0.084	2.31
No.	**Material**	**Pressure p_h [bar]**	**Radius [mm]**	**Numerical simulation**		**Experimental results**		**Error [%]**
				ε_1 [-]	ε_2 [-]	ε_1 [-]	ε_2 [-]	
1	DC01	150	6	0.078	0.062	0.080	0.065	3.70
2	DX51D			0.093	0.078	0.095	0.081	3.00
3	DC01	250		0.101	0.067	0.106	0.069	3.97
4	DX51D			0.123	0.090	0.126	0.092	2.33
5	DC01	350		0.121	0.070	0.126	0.073	4.38
6	DX51D			0.142	0.093	0.141	0.095	1.43
No.	**Material**	**Pressure p_h [bar]**	**Radius [mm]**	**Numerical simulation**		**Experimental results**		**Error [%]**
				ε_1 [-]	ε_2 [-]	ε_1 [-]	ε_2 [-]	
1	DC01	150	4	0.083	0.067	0.086	0.065	3.00
2	DX51D			0.098	0.084	0.095	0.074	4.56
3	DC01	250		0.108	0.073	0.112	0.076	3.91
4	DX51D			0.134	0.101	0.137	0.106	3.60
5	DC01	350		0.130	0.077	0.134	0.078	2.19
6	DX51D			0.158	0.107	0.163	0.108	2.05

5 Conclusion

The research was based on determining many material characteristics to define as accurately as possible the numerical model. This model was compared to the experimental drawn parts with an own built custom solution. All experimental results are compared and analyzed extensively with those obtained by numerical simulation and the resulted error percentage is less than 5%. In this way the accuracy of the experimental results was evaluated and the finite element model was validated. For all the studied materials, both major and minor strains and the thickness reduction have the same tendency of variation for the die radius and the hydroforming pressure, that is, the results increase with decreasing radius and increasing hydroforming pressure.

Further research can be carried out by researching variable process parameter like blank holder pressure. This can be done be designing an experiment method with three influence factors with three variable values each.

This work was carried out with the support by Flupec Hydraulics & Pneumatics Sibiu – private company where the hydroforming press was manufactured, assembled and forming tests were carried out.

References

1. I. G., D. L. R. and M. F., Sustainability issues in sheet metal forming processes, Jou. Cleaner Prod., **19**, 4, 337-347, (2011)
2. K. M., *Hydroforming for Advanced Manufacturing*, Woodhead Publishing, (2008)
3. EN 10130, *Cold rolled low carbon steel flat products for cold forming. Technical delivery conditions,* (2006)
4. E. 10346, *Continuously hot-dip coated steel flat products. Technical delivery conditions,* (2009)
5. M. Hill, *Marks' Standard Handbook for Mechanical Engineers* 8th Ed., McGraw Hill Book Company, (1978)
6. K. Siegert and S. Wagner, "TALAT Lecture 3701," (1994). [Online]. Available: http://core.materials.ac.uk/repository/eaa/talat/3701.pdf. (Accessed 04 2016)
7. D. Banabic, G. Dragos and I. Bichis, "Influence of Variability of Mechanical Data on Forming Limits Curves," (2010). [Online]. Available: http://comod.utcluj.ro/documents/articles/Banabic_Steel%20Research%20_Int.%202010.pdf. (Accessed 06 2016)
8. *Metallic materials -- Tensile testing -- Part 1: Method of test at room temperature, ISO 6892-1,* (2009)
9. J. W. Signorelli and M. d. l. A. Bertinetti, Self-Consistent Homogenization Methods for Predicting Forming Limits of Sheet Metal, in *Metal Forming - Process, Tools, Design*, InTech, 175-210, (2012)
10. D. Banabic, L. Paraianu, G. Dragos, I. Bichis and D. S. Comsa, An improved version of the modified maximum force criterion (MMFC) used for predicting the localized necking in sheet metals, *Proceeding of the Romanian Academy, Series A,* **10**, 3, (2009)
11. V. V. Hašek, Berichte aus dem Institut für Umformtechnik, Universität Stuttgart (On the stain and stress states in drawing of large un-regular sheet metal components), Essen: Giradet, (1973)
12. C. Garjob, S. G. Racz and O. Bologa, The Determination of the Forming Limit Curve

Using a Modular Device, Ac. Jou. Manuf. Eng.., **8**, 2, 39-44, (2010)
13. GOM GmbH, Gom Optical Measuring Techniques - ARAMIS FLC Computation, Braunschweig, (2007)

Changes in the mechanical properties and microstructure of anisotropic austenitic stainless sheet steel after uniaxial tensile test

Emil Yankov[1,*], and *Maria* Nikolova[1]

[1]University of Ruse, Department Material Science and Technology, Ruse, 8 Studentska Str., Bulgaria

Abstract. The aim of the investigation is to study the changes in the characteristics of an austenitic sheet material X5CrNi18-10 (1.4301, AISI 304) after a plastic deformation. Samples are cut out from the sheet material at three different directions - 0°, 45° and 90° angle to the rolling direction. The changes in the mechanical properties and microstructure of the anisotropic austenitic steel are investigated by mechanical tests (uniaxial tension tests and hardness measurements) and structural analyses (optical and scanning electron microscopy, X-ray diffraction). It is established that the strain induced phase transformation of the metastable austenite to martensite during the tension tests changes the magnetic properties of the steel. It is found out that the sheet anisotropy effect on the uniform deformation, the thickness reduction and structure of the austenite sheet material is more essential for the plastic deformation behaviour than the strain-induced $\gamma \rightarrow \alpha'$ phase transformation.

1 Introduction

Many research teams [1, 2] are dealing with the problem of the unequal deformation of sheet metal materials during the technological process of manufacturing of different parts. One of the aspects in turns of cold plastic deformation is the unequal thickness reduction during the complex loading schemes that leads to earlier destruction of the material. Another aspect is the instrument friction with the sheet material that impede the equal plastic deformation. Most of the authors are trying to explain this improper depletion in thickness and technological difficulties during plastic deformation with different surveys introduced by one direction cut (usually identical with the rolling direction (RD)) samples [3-6]. This approach does not fully explain the differences in the strain values in the other directions relative to the RD.

The texture of the austenite could not only evolve during the deformation, but also be inherited by the initial rolling. As the microstructure of the stainless steel is composed of metastable austenite (γ), the latter during cold plastic deformation transforms from paramagnetic phase into ferromagnetic martensite (α′) that has bodycentered cubic (bcc) lattice and is thought to nucleate at the intersections of the shear bands [7]. The strain induced texture of the austenite and the resultant martensite that is formed during the plastic

*Corresponding author: eyankov@uni-ruse.bg

DOI: 10.1051/matecconf/20179402017

deformation is very much influenced by the initial texture (anisotropy) of the material. Then, the enhanced work hardening, that influences the forming process and the final product received, changes the plastic technological behaviour of the sheet material.

In order to determine correctly the mechanical properties and to support the technological processes, a detailed examination of samples cut at three different directions in respect to the RD is necessary to be performed. In the present study the chosen directions in respect to the RD are 0°, 45° and 90°. Additionally, for clarification of the differences in the received results, a detailed microstructural analysis that explains the unequal changes during deformation is needed. Therefore, the aim of this study is to define in details the mechanical, structural and phase changes in three different directions with respect to the RD of the X5CrNi18-10 austenite sheet material.

2 Experimental procedures

The material was received in the form of 0.8-mm-thick sheets in cold-rolled condition. Sheets with the following dimensions: 1000×2000 mm and thickness of t_0 = 0.8 mm were cut from metal alloy X5CrNi18-10 (AISI 304, DIN 1.4301, SUS 304, GOST 07Ch18N10) with chemical composition shown in Table 1, measured by quantometer SpectroMaxx.

Table 1. Chemical composition of X5CrNi18-10 steel.

Element	C	Si	Mn	Ni	Cr	Mo	Ti	S	P	Fe
Wt.%	0.055	0.54	1.13	7.77	18.46	0.282	0.0042	0.0078	0.037	Rest

Samples are cut out from the sheet material at three different directions - 0°, 45° and 90° to the RD as shown in Fig. 1a. Specimens used for the mechanical properties measurements were determined on the base of standard ISO 6892-1:2009 (Fig. 1 a and b) and cut from the sheet with digital laser guillotine. The material was tested in as-received (called undeformed) conditions.

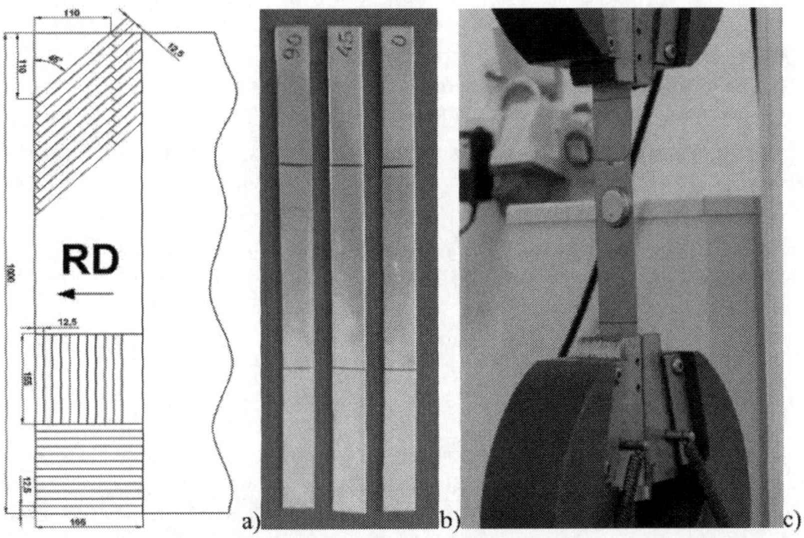

Fig. 1. Sample preparation and testing: a) a scheme of cutting of the samples; b) overall appearance of the samples; c) a fixed and tensile tested sample with the machine.

The static uniaxial tensile test was performed on a universal testing machine Instron-3384, supplied with a software Bluehill-3 at ambient temperature. For the purpose of the testing method a constant strain rate of 5 mm/min until the load fell down to 100 N (or the maximum load fell below 90%) was used. For each test a control measurement of the working length of testing, height and width was held and the values were set in the used test method via the software program. The data were saved in an appropriate digital form for subsequent analysis.

For the microstructure analysis and hardness measurements three mutually perpendicular planes from each specimens cut near the break down zone were prepared. This zone was chosen in order to monitor and analyse the maximal change in the structure and properties after the deformation. The microstructure and hardness of tensile tested samples were compared to the undeformed ones. Each sample (undeformed and tensile tested (TT)) was evaluated in this three different directions: 1) alongside the long side of the sample (plane I); 2) cross-section (plane II) and 3) top view (plane III). The surface preparation for microstructure analysis and hardness measurements of all sections of the samples before and after the tensile test involved grinding with paper of grit P400, 500, 800, 1000, 1500, 2500 and mechanical polishing with non-abrasive paste.

A Vickers Hardness tester 432 SVD by Wilson-Wilpert was used for the hardness measurements of the steel according to the standard ISO 6507-1. In order to compare the hardness changes the researches were made on the plane I, II and III of each sample (undeformed and TT) with a load of 5 kg and dwell time of 10s. Ten sequential measurements by increments of 50 μm between the imprints were made for each plane. Consequently, the data received were gathered in tables for comparative analysis. The summary results of the hardness measurements were presented in graphical form.

Optical microscopy (Epytip, Zeiss, Jena) was performed on the cross sections and top views after etching the samples with aqua regia. To ensure the comparability of the survey results for each plane of all samples one and the same magnification was utilized. 8-megapixel digital camera was adapted to the microscope and used for the image acquisition. The micrographs were process by Mountains Map premium v.7 software.

A JEOL JXCA-733 Microprobe Scanning Electron microscope (SEM) was used to study the austenite deformed grains and martensite morphology. The focused electron beam for image acquisition was operated at 19.9 kV at magnification between 600× and 1000×. The X-ray phase identification was performed with URD-6 diffractometer, applying Bragg-Brentano geometry and Fe-Kα radiation and working at 30 kV and 20 mA. The diffraction angle range scanned was 40-129° 2θ with a step size of 0.1° 2θ and counting time of 5 s/step. The lattice strain response was measured in the longitudinal and transverse directions of plane III of all samples. The qualitative phase analysis was held using Match!3 software.

The texture analysis was performed by calculating the texture coefficient [8] measuring the relative degree, of preferential orientation among the crystal planes:

$$TC_{(hkl)} = \frac{\frac{I_{(hkl)}}{I_{0(hkl)}}}{\frac{1}{N}\sum\frac{I_{(hkl)}}{I_{0(hkl)}}} \quad (1)$$

where I (hkl) and Io (hkl) are the measured intensity and standard integrated intensity (from PDF) for (hkl) reflection, respectively, and N is the number of reflections observed. The data are taken from the results process by the Match software. A TC > 1 indicates preferential orientation/texture.

3 Experimental results

The representative indicator diagrams (Fig. 2a) are result of five samples testing from one the same angle of cutting sample (0°, 45° and 90°, respectively). The experimentally determined data indicate the load at maximal charge Fm for tensile extension at maximum charge ΔLm, elongation at break ΔLbr, elongation at tensile strength at allongement A and tensile strength Rm. The diagram considered as representative is the one with the closest mechanical properties to the average results from each indicator diagram for a particular angle of cutting [9]. The highest Fm values occurs within the 45° angle of cutting samples - Fm = 6.153 kN that is 15% higher than the 90° and 18% higher than the 0° tested samples. The same trend is observed with elongation at break ΔLbr of the 45°cut sample - ΔLm = 24.934 mm that is 51% and 73% higher than the 90° and 0° tested samples, respectively.

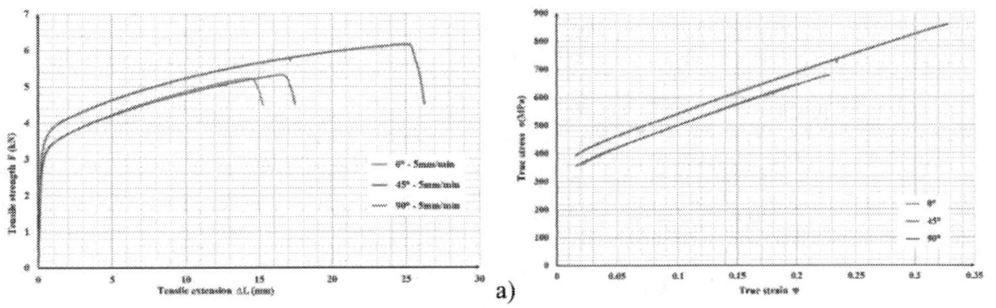

Fig. 2. Representative indicator diagrams of X5CrNi18-10; t0=0.8 mm: a) tensile strength - tensile extension, b) true stress – true strain.

As the load F and the elongation ΔL depend on the shape and dimensions of the tested pieces, quantitative comparison of the mechanical properties of the material according to the tensile strength - tensile extension diagram is not possible. Therefore, the $F = f_1(\Delta L)$ diagram (Fig. 2 a) is converted into true stress – strain diagram $\sigma = f_2(\varphi)$, Fig. 2b. The resultant curves are parallel and reconcile one another with the strain increase. The three curves vary in their maximal strain and stress before the break. The true stress maximal values occur in 45° cut samples – σ = 856.7 MPa that is 27% and 33% higher than the 90° and 0° tested samples, respectively. The true strain of the 45° cut samples (φ=0.328) exceeds that of the samples cut at 90° and 0° directions with 44.5% и 51.6%, respectively.
It is worth noting that the established variances in the mechanical properties data within the 0°, 45° and 90° cut samples could be due to the manufacturing conditions and processing during the metallurgical process. Additionally, the non-homogenous composition and defects in the structure play essential role in the initiation of breaking and its dissemination during loading of the sheet material. The various types of defects including their nature, form, size and distribution could be the reason for the material breaking at different stress values.

To clarify and confirm the differences established from the applied stress, Vickers hardness measurements with a develop method [10] is used. The results of the average hardness values of the planes I, II and III of the undeformed and TT samples cut at 0°, 45° and 90° angle to the RD are listed in Excel tables and graphically shown in Fig. 3. The ten measurements are averaged for each plane (I, II and III) of a tested specimen. As the number of the samples examined is six, the total number of the completed tables is equal to eighteen.

The comparative analysis of all planes of the samples reveals that the hardness increases irregularly for the different planes after the tensile test the samples. The hardness change of the 0° cut sample is slight in contrast to the other (45° and 90°) tensile tested samples. In plane I of the TT 0° sample the hardness increase is 47.42% higher than the undeformed plane I and for plane II and III these percentages are 52.51% and 68.44%, respectively. The reason for the higher hardness increase in plane II and III could be found in the favourable loading direction and the initial texture of the rolled material that could lead to grains fragmentation and lattice reorientation of the crystal lattice in these two planes. However, the 0° cut sample is incapable of being highly deformed and strengthened because of the exhausted ductility in this direction.

The hardness increase in plane III of the 45° cut sample is the highest – 107.44% and for plane I and II is 66.89% and 71.40% higher than the undeformed planes, respectively. These facts could be attributed to the favourable lattice reorientation and rotation of the grains in longitudinal direction and trans-crystal displacements in planes II and III. The maximal strain values set out in this TT sample is confirmed by the hardness test.

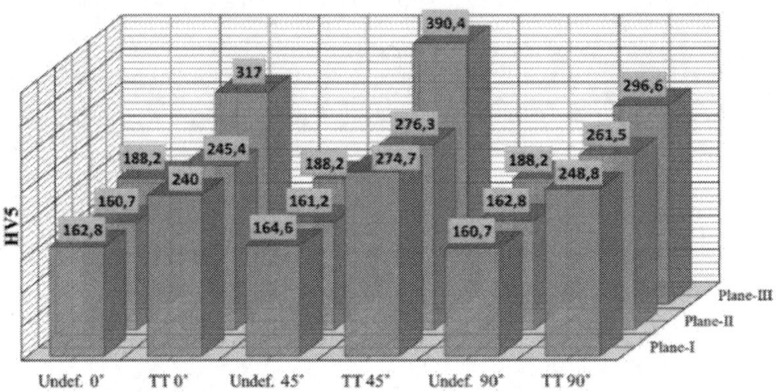

Fig. 3. Average hardness values before and after the tensile test of 0°, 45° and 90° cut samples measured at planes I, II and III.

The increase in hardness of the 90° cut sample is over 50% applied to the undeformed planes – 54.82% for plane I, 60.63% for plane II and 57.60% for plane III. The relatively balanced hardness change is due to the hindered and therefore, slight lattice reorientation and rotation of the initially textured structure, located normal to the direction of loading. Compared to the TT 0° cut sample, the hardness and maximum strain values of the cross-section to the RD cut samples are higher.

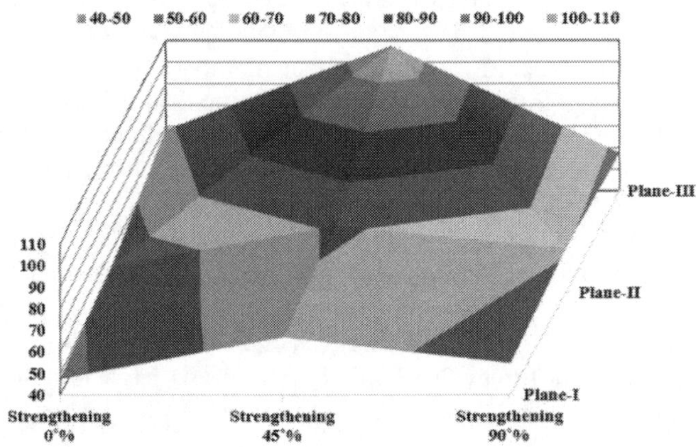

Fig. 4. Strengthening in percentage of the tensile tested samples (0°, 45° and 90°) applied to the undeformed samples for planes I, II and III.

DOI: 10.1051/matecconf/20179402017

From the former hardness values, a summary diagram of the strengthening in percentage in the three planes (I, II and III) of the three (0°, 45° and 90°) cutting directions samples is shown on Fig.4. It could be noted that the major change in the strengthening is observed in plane III of 45° cut sample – around 107%. The slightest change in the strengthening percentage is found in plane I of 0° cut samples – roughly 47%. The summary diagram received enables a faster analysis of the favourable directions for deformation and consequent strengthening of the sheet materials during more complex loading schemes. Therefore, using Vickers hardness measurements, the strain values of a sheet metal could be determined and measures could be taken to avoid undesirable effect (such as local thickness reduction, delayed cracking, folding, etc.) during tensile of the materials.

The real microstructure images of each of the three planes of the six examined samples (three undeformed and three TT samples) assembled in cubes for a better visualization is presented in Fig. 5. In general the morphology of both planes (I and II) of the undeformed samples develop deformation texture of the previous rolling. In case of austenitic sheet steel the texture development is rather complex and depends on the direction of cutting of the samples. The dominating direction of the austenite texture corresponds to the rolling direction. With increasing the deformation degree the austenite undergoes stretching and the result of this depends on the cutting direction of the samples.

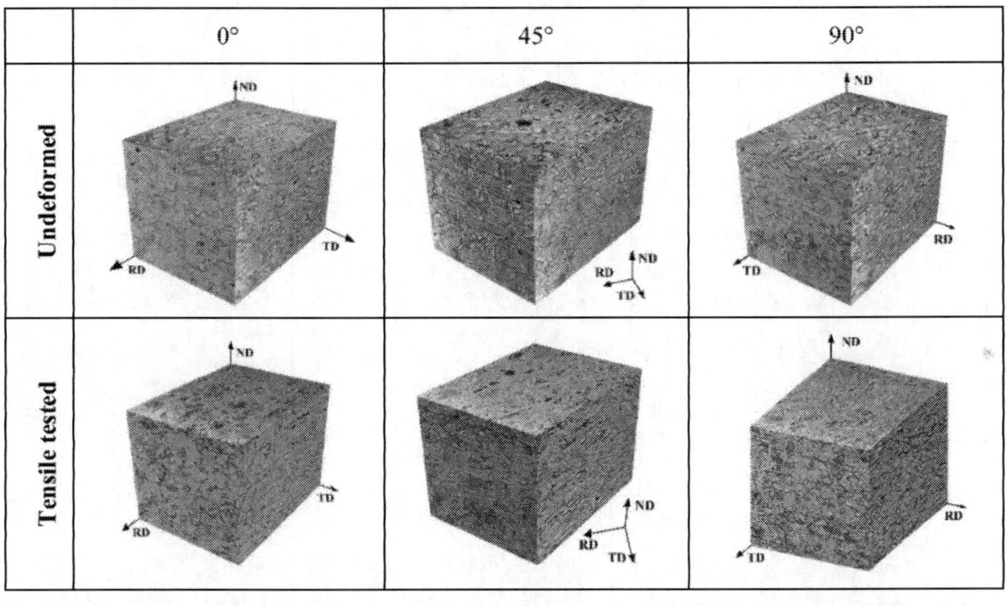

Fig. 5. Assembled 3D images of the microstructure in planes I, II and III of undeformed and tensile tested samples.

For more precise quantitative assessment of the results from the microstructure analysis, software processing of the images is used. The results obtained are shown in Fig. 6 and summarized in Table 2.

The main outcomes from the comparative microstructural analysis could be stated as follow: In the 0° cut sample the grains are slightly fragmented in the planes I and II at the expense of reorientation. In plane III the grains are elongated and fragmented because of inter-crystal creep. The grains form is insignificantly changed but there is a strong reorientation tendency in plane III and especially in plane I. The tendencies observed are consistent with the hardness values measured at planes I and II.

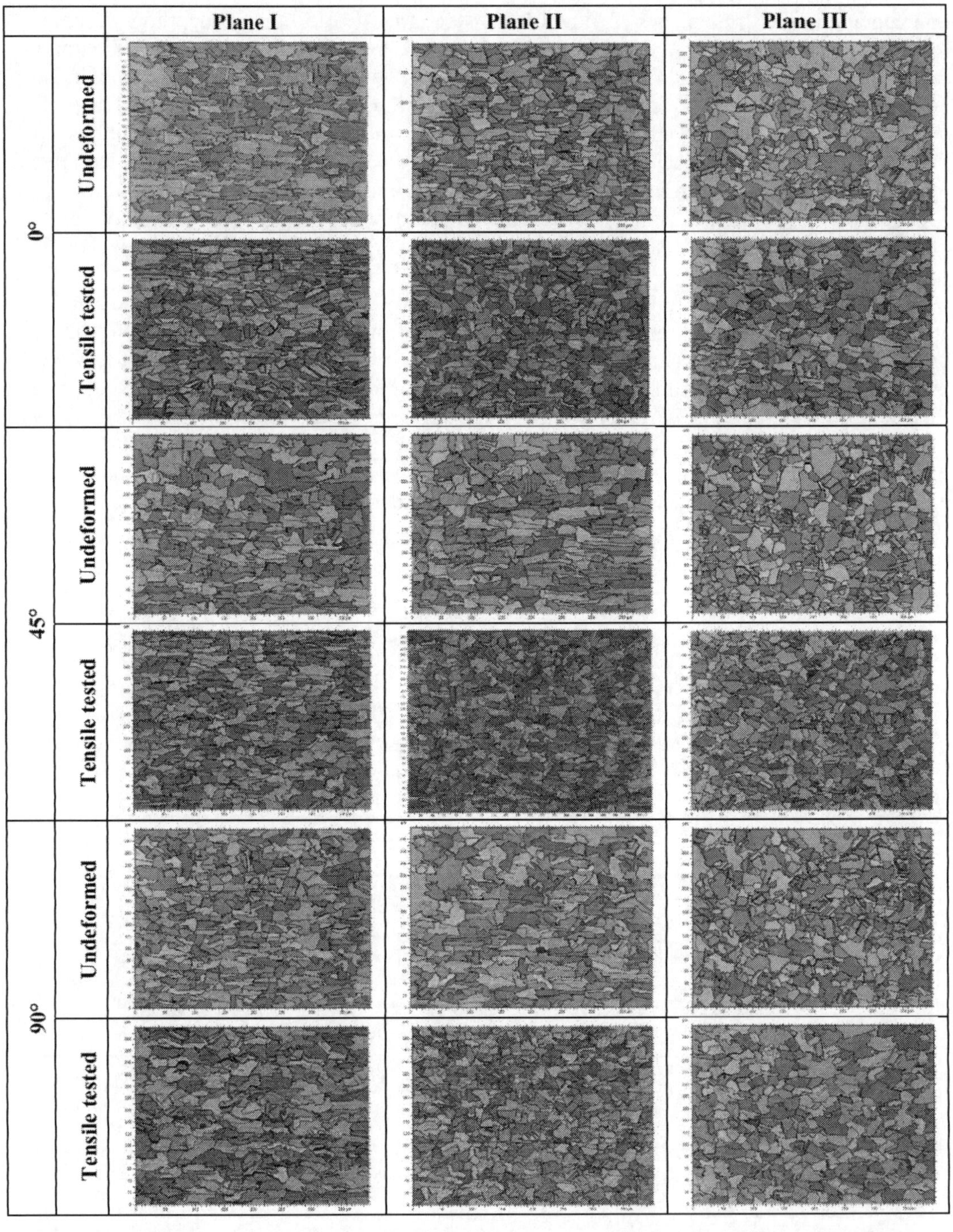

Fig. 6. A comparison of the software processed microstructures of the undeformed against tensile tested samples in planes I, II and III: a) 0° cut sample; b) 45° cut sample; c) 90° cut sample.

In the 45° cut sample there is a clearly visible grain fragmentation in all of the analysed planes confirmed by the increased number of grains, neighbours and the decrease in the mean diameter. The highest fragmentation is observed in plane II while the highest lattice reorientation is seen in plane I. This has a powerful effect on the surface roughness and hardness where the latter has the highest values (Fig. 3). The grains in the planes I and II have approximately equal form and the established reorientation in all three planes has lower values than that of the 0° cut sample. The lattice reorientation in the 45° cut sample after the tensile test is due to the pronounced grains fragmentation and trans-crystal displacement.

Table 2. Mean values of some microstructure determined parameters.

Sample	Parameter	№ of motifs	Area, [μm²]	№ of neighbors	Equivalent diameter, [μm²]	Mean diameter, [μm²]	Form factor	Round-ness	Orientation [°]
Undef. 0°	Plane I	310	386	5.90	20.6	17.4	0.276	0.423	108
	Plane II	382	313	5.66	19.1	15.9	0.269	0.431	99
	Plane III	505	237	5.64	15.5	14.1	0.340	0.441	90.7
TT 0°	Plane I	410	291	5.59	18.5	15.1	0.258	0.410	96
	Plane II	448	267	5.53	17.7	15.1	0.280	0.498	93.3
	Plane III	351	340	5.57	19.7	17.6	0.294	0.469	101
Undef. 45°	Plane I	231	517	5.57	24.6	20.5	0.232	0.400	89.1
	Plane II	211	566	5.55	26.0	21.0	0.225	0.399	99.3
	Plane III	583	205	5.71	14.8	13.7	0.348	0.464	93.6
TT 45°	Plane I	372	321	5.61	19.4	16.1	0.245	0.519	86.9
	Plane II	500	239	5.62	16.8	13.8	0.248	0.472	93.2
	Plane III	468	255	5.65	17.2	15.2	0.289	0.482	89.1
Undef. 90°	Plane I	389	307	5.61	18.9	16.1	0.272	0.422	90.6
	Plane II	246	482	5.74	23.7	20.1	0.270	0.424	91.6
	Plane III	570	210	5.81	14.8	13.2	0.333	0.483	97.1
TT 90°	Plane I	186	643	5.44	27.6	22.5	0.219	0.383	91.5
	Plane II	404	296	5.60	18.7	15.4	0.251	0.710	88.3
	Plane III	299	400	5.58	21.4	19.0	0.286	0.499	93.9

In the 90° cut sample the grains in plane I are enlarged and change their elongated form almost without reorientation. In plane II the grains are highly fragmented and change their roundness becoming more spherical. The grain enlargement could be detected also in plane III. The measured reorientation of the grains in planes II and III is low, although the former is higher than the same in plane I. The results from the microstructure of the sample are confirmed by the higher hardness values of planes II and III applied to plane I.

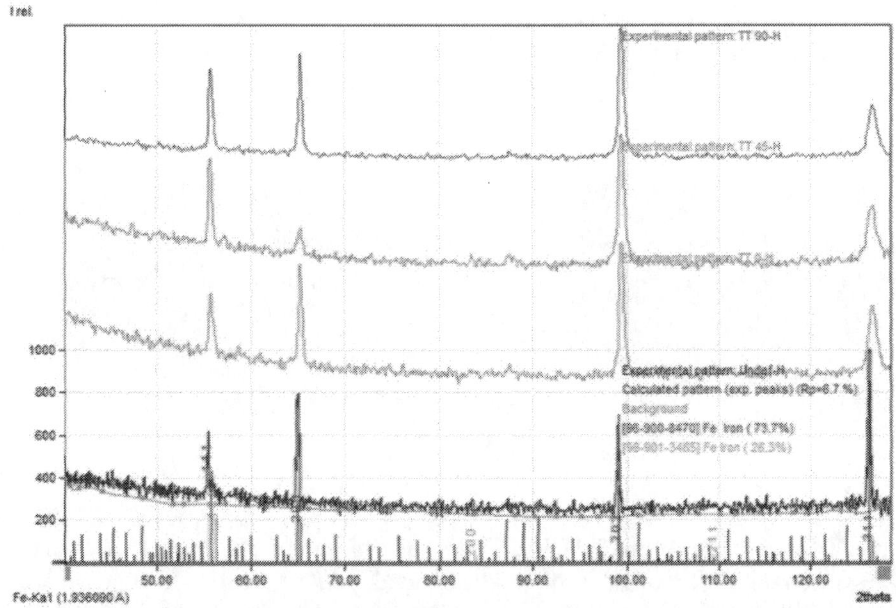

Fig. 7. Diffraction patterns of X5CrNi18-8 steel in initial state and after tensile test of the samples in horizontal direction of plane III; Bragg-Brentano geometry (BB).

The statistically representative volume elements of the microstructure analysis are less than determining average macroscopic properties (strain localized pattern) but these short-range characteristics shows a good consistency in the hardness measurements.

On the diffraction patterns of the steels in the initial and tensile tested state performed by Bragg-Brentano method (Fig. 7 and 8), reveal four strong diffraction lines from the planes (111), (200), (202) and (311) of austenite with low peak-to-peak noises. Additionally, week diffraction peaks from diffraction line (110) of α′-martensite are observed in the TT and initial state. The small quantity of the martensite phase detected in the undeformed state probably is a result of the cutting and subsequent preparation of the material for the investigations.

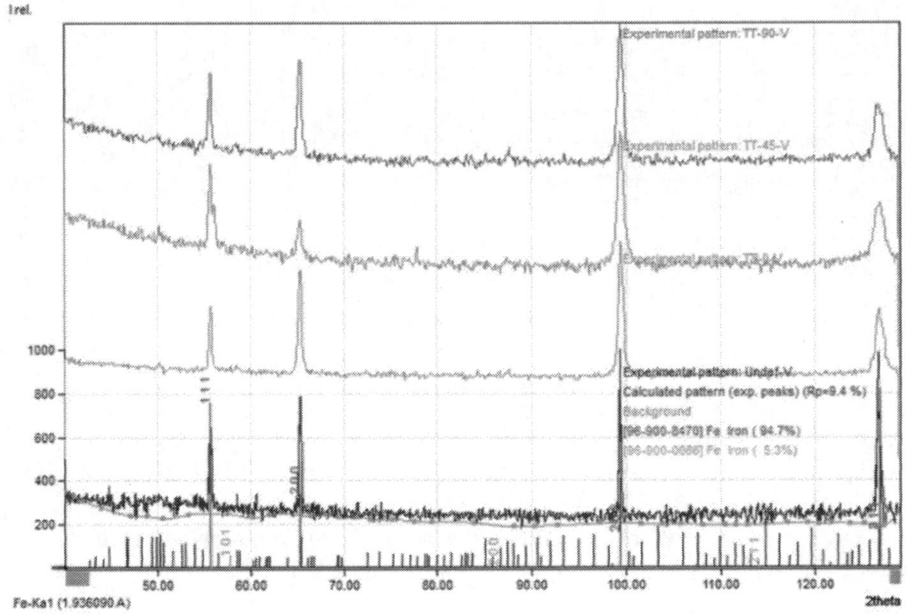

Fig. 8. Diffraction patterns of X5CrNi18-8 steel in initial state and after tensile test of the samples in vertical direction of plane III; Bragg-Brentano geometry (BB).

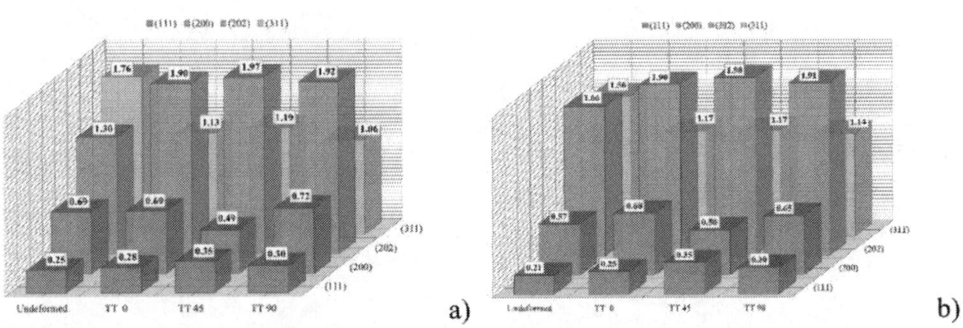

Fig. 9. Calculated texture coefficients of the undeformed and TT samples measured at planes III in: a) horizontal direction; b) vertical direction.

The results of the calculated TC (Fig. 9) presents that the undeformed sheet material is strongly textured in (311) plane both in horizontal and vertical direction of scanning. Differences are detected in the (202) plane where the TC varies, which means that this plane doesn't have strong texture character. After the tensile tests the calculated changes in the TC are very much in line with that of the intensity changes measured at planes III and shown in Fig. 10.

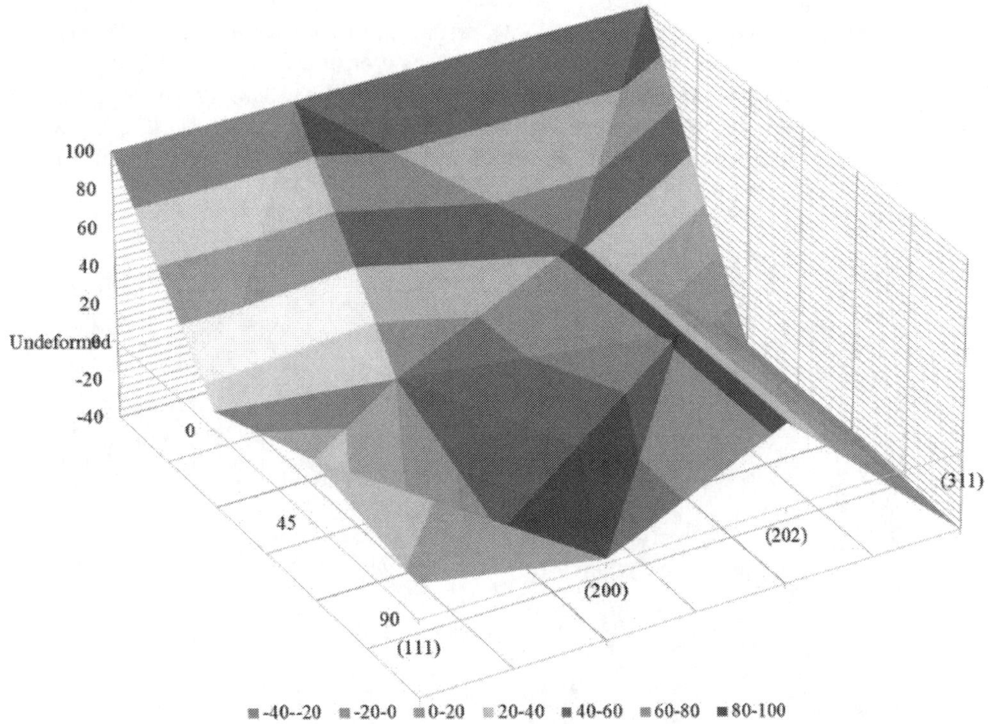

Fig. 10. Intensity changes (in percentage) of TT 0°, 45° and 90° applied to the underformed sample measured at planes III in horizontal direction.

In the horizontal direction the largest is the increase in the intensity of (202) and the highest is the decrease in (311) lattice planes. Each reflection peak corresponds to a specific family of grains and its intensity is proportional to the contribution of the number of grains of the family in the irradiated volume. Therefore, the tensile test affects the variations among the individual planes and the lattice strain and the specific grain population orientations. In this connection, after the steel deformation the following main changes in the diffraction patterns intensities could be summarized as follow:

For the 0° cut and TT sample the movements of the crystallographic planes show reorientation in austenite $(111)_\gamma$ and especially in $(202)_\gamma$ planes (44.6% increase) at the expense of decrease of $(311)_\gamma$ plane intensity (36.3 % decrease). In this case the creep of the material is not in the most favorable closed packed $(111)_\gamma$ plane, but mainly in $(202)_\gamma$ one. The results observed explains the earlier rupture of the sample because of the less ductility of the $(202)_\gamma$ plane. As a percentage applied to the undeformed sample the intensity increase in $(202)_\gamma$ plane is 44.6% and only 10.35% for the $(111)_\gamma$ plane.

All of the examined planes of 45° cut TT sample go through substantial changes in the intensities. The latter increases in $(111)_\gamma$ and $(202)_\gamma$ planes with 32.63% and 43.28%, respectively, while $(311)_\gamma$ and $(200)_\gamma$ lines become weaker (36 and 33.22% decrease, respectively). It follows that the crystals in the 45° cut sample has had a favorable initial position applied to the load direction thus contributing to the balanced reorientation in the crystallographic planes. This allows for 51.6% higher degree of deformation of this sample as compared with 0° cut TT sample.

In the 90° cut TT sample the intensity of the $(111)_\gamma$ plane increases with 18.36%, but simultaneously the creep occurs in $(202)_\gamma$ plane (43.58% increase). The reorientation in both planes is for the account of $(311)_\gamma$ which intensity decreases with 41.7%. In contrast to 45° cut sample, the $(200)_\gamma$ increases slightly with 2.19%.

Weak martensite (101) α′- lines are found in most of the tensile tested samples. The diffraction lines of ε-martensite are not disclosed that proves the direct $\gamma \rightarrow \alpha'$ transformation in this steel in such test conditions. As determined from the XRD results, the tensile test induces some transformation of the metastable austenite in magnetic α′-martensite. For that reason SEM analysis is performed and some of the results are shown in Fig. 11. The SEM observations lead to the assessment that more martensite is observed in cross-sections (plane II) of all samples but even in that plane, its quantity is small, as noted also by the XRD analysis.

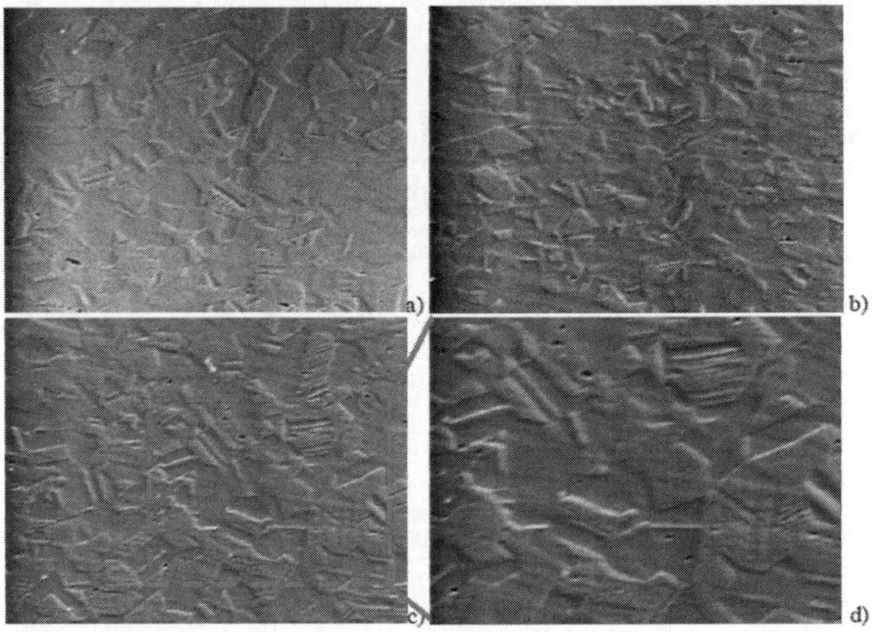

Fig. 11. Secondary image micrographs of plane II structure: a) 0° cut TT sample; Magn.600×; b) 45° cut TT sample Magn. 600×; c) 90° cut TT sample: Magn.600×; d) 90° cut TT sample: Magn. 1000×.

During the deformation, weak magnetic properties are detected as soon as 14% of deformation is reached and the former increase with the increase of the degree of deformation. This suggests that structure changes occur in the crystal lattices of the samples cut at different directions. The presence of α′-martensite, as well as the increased dislocation density will lead not only to the observed change in the magnetic permeability, but also to enhance work hardening because of the dislocation accumulation at the austenite/martensite interfaces [11]. In contrast to the assumption made by J. Talonen at al. [12] that the martensite form a "continuous network" trough the material, as seen in Fig. 11, in none of the samples the α′-phase does not form such a network that could effectively impede the flow of the soft austenite.

4 Conclusions

The uniaxial tensile test deformations at macro- and microscopic scale for an austenitic steel sheet has been examined using hardness tests, microstructure and XRD analysis. The proposed methodology for examination of an atypical behaviour of sheet metal materials clarifies the reasons why anirregular degree of deformation is observed in the three different angles of cutting. The Vickers hardness measurement is one step leading to the determination of the changes in plasticity with the material strengthening. The use of XRD analysis provides the opportunity to identify the causes of the higher plasticity in a certain directions of the austenitic sheet material. The analysis and equipment of this kind is far cheaper than other modern methods of testing (such as automated crystallographic orientation indexation in a TEM (ACOM-TEM), electron backscattered diffraction (EBSD), etc.). The identification of the changes of the crystallographic planes on the basis of the deformation scheme is a sufficiently accurate, reliable and proved over time method. The surveys carried out in this paper, show that the XRD analysis is also suitable for the determination of the anisotropy of the material. The differences in the relative intensities of the reflection lines compared to the theoretical intensity of each austenite peak indicate that the tensile test applied strongly influences the crystallographic plane $(202)_\gamma$ (and slightly the most favourable $(111)_\gamma$) at the expanse of the reduce in the intensity of the textured $(311)_\gamma$ one. This study also indicates that the grains fragmentation and austenite deformation take precedence over the martensite transformation during the uniaxial tensile test of the sheet material. The deployment of modern software products for quantitative microstructure analysis allows for rapid identification of the changes in grains shape, dimensions, orientation, fragmentation and other parameters.

The proposed examination approach with several alternative methods for analysis of the non-uniform deformation of an anisotropic sheet material will be verified for other relevant materials and other loading schemes of deformation. In that way the assurance and reliability of the proposed examination methods for analysis will be confirmed or infirmed.

The paper reflects the outcomes of the project № 16 – FMME – 02, financed by the Fund „Scientific Research" of University of Ruse "A. Kanchev".

References

1. J. Slota, M. Jurčišin, E. Spišák, M. Tomáš, M. Šiser, Acta Metall Slovaca, **21** 4, 269-277 (2015)
2. Das A., P. C. Chakraborti, S. Tarafder, H. K. D. H. Bhadeshia, J Mater Sci Tech, **27** 1 366-369 (2011)
3. D.-F. Li, Noel P. O'Dowd, C. M. Davies, S.-Y. Zhang, European J of Mech - A/Solids **30** (5) 748 (2011)
4. N. Habibi, A. Zarei-Hanzaki, H.-R. Abedi, J Mater Process Technol **224** 102–116 (2015)
5. N. Afrin, D.L. Chen, X. Cao, M. Jahazi, Mater Sci Eng A **472** 179–186 (2008)
6. M. Baig, E. El-Danaf, M. Soliman, A. Almajid, W. Alshalfan, and Khaled Alhajeri, Mater Testing **57**,10, 897-903 (2015)
7. J. Talonen, P. Nenonen, G. Pape, H. Hänninen, Metall Mater Trans A, Vol. **36**A 421 (2005)
8. Class 26_handout MTSE 3010 2013 Supplemental Texture & Grain Size Calculation.pdf, Polycrystallography, Institute of North Texas, http://engineering.unt.edu/materials/sites/default/files/sites/default/files/lamma/oldsite/documents/Class26_handout%20MTSE%203010%202013%20SUPPLEMENTAL%20TEXTURE%20%26%20GRAIN%20SIZE%20CALCULATION.pdf, (2013)
9. Yankov E., *Sheet materials testing using biaxial hydraulic bulge test* (Dissertation, Ruse, 2014)
10. YankovE., D. Gospodinov, V. Gagov, Scientific conference RU & SU'14, **53** (2014)
11. T. Narutani, Mater. Trans. JIM, **30**, 33-45 (1989)
12. J.Talonen, P. Nenonen, G. Pape, H. Hänninen, Metall Mater Trans A **36A**, 421 (2005)

MATEC Web of Conferences **94**, 03001 (2017) DOI: 10.1051/matecconf/20179403001
CoSME'16

Study of kerf geometry of multilayers materials in abrasive water jet cutting process

Bogdan Barabas[1,*] and *Andrea* Deaconescu[1]

[1]Transilvania University of Brasov, Faculty of Technological Engineering and Industrial Management, Department of Engineering and Industrial Management, Romania

Abstract. The use of abrasive jet cutting in case of multilayers materials, shows particularities and difficulties, due to different hardness and thickness of layers. This paper studies changes, occurred in the kerf geometry at passing of abrasive particles, through layers with different hardness and aims to establish a link between hardness and thickness of the layers and kerf geometry because processing accuracy is determined by the geometry of the cut. Thus, is proposed a novel algorithm based on minimization of angle of kerf, for optimization a cutting process in abrasive water jet technology, which can be utilized for materials with different hardness in layers. Direct effect of optimization of cutting consist in increasing efficiency, dimensional precision, traverse velocity, and decreasing the amount of used abrasive garnet.

1 Introduction

The cutting of welded, thermochemical treated or multilayer material shows problems in establishing the correct machining, with direct consequences on the cutting process. Form of resulting kerf to the hydro abrasive jet crossing, hence the geometry of the finished part, is influenced by different hardness of layers of cutting material, is influenced by jet characteristics (pressure, mass flow, traverse speed, abrasive type) and by the technological characteristics of the abrasive water jet cutting process (nozzle diameter, focusing tube length, distance to the work piece).

Modeling abrasive water jet cutting process of multilayer materials it is influenced by position of layers in material, appearing micro cracks to crossing limit in rough layer, breaks or micro compactions in layer with lower hardness. The elimination of these phenomena does not have an analytical solution this time, optimization of cutting jet cutting of multilayer materials are made experimental. The study of the jet behavior and the trajectory of the abrasive particles while abrasive jet cutting, has proved a complex process in three stages, before impact, during impact and during abrasive cutting. The presence of layers with different hardness increases the complexity of phenomenon modeling, by adding points to change behavior during each passage. In the context of penetration of new branches in the sphere of unconventional technologies such as milling and turning with

* Corresponding author: barabas.bogdan.florin@unitbv.ro

abrasive jet, control of geometry processing becomes a normal requirement, and optimizing dimensional precision becomes a necessity.

Choosing the incorrect processing parameters results in incomplete cutting or the reverse, a too strong cutting with high consumption of abrasive and energy, leading to a too long machining time and to an unprofitable cost price. For the user of this sort of machines, the losses become significant, the optimization of process become absolutely necessary. In this purpose the present paper proposes a new algorithm for optimization the kerf geometry of multilayer material with different hardness.

2 Theoretical consideration

Abrasive water jet cutting process is influenced by two major factors: material properties and machining parameters. The plan of abrasive water jet process is presented in Fig. 1, [1].

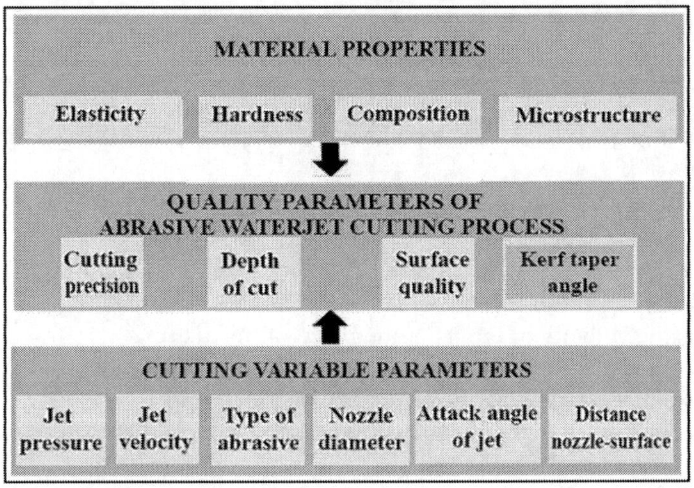

Fig. 1. Abrasive water jet cutting process.

The high speed jet is obtained by passing water at high pressure through a sapphire nozzle, applying Bernoulli's law [2-3]:

$$p_{at} + \frac{\rho_a}{2} V_j^2 + \rho_a g h_1 = p + \frac{\rho_a}{2} V_h^2 + \rho_a g h_2 \quad (1)$$

where: p_{at} is the athmospheric pressure [Pa];
p is the fluid pressure at the entrance to the nozzle [Pa];
ρ_a is water density [kg/m^3];
V_j is the velocity of the jet at the exit of the nozzle orifice [m/s];
V_h is the flow velocity of high pressure water [m/s];
$h_1 = h_2$ is the distance to the reference plane [mm].

Considering: $h_1 = h_2$, $V_j >> V_h$, $p >> p_{at}$, relation (1) become:

$$\frac{\rho_a}{2} V_j^2 = p \quad (2)$$

$$V_j = \sqrt{\frac{2p}{\rho_a}} \quad (3)$$

In reality, losses due to the friction, due to the occurrence of turbulent flow and due to compressibility of water, changes the relationship (3):

$$V_j = \mu \sqrt{\frac{2p}{\rho_a}} \quad (4)$$

DOI: 10.1051/matecconf/20179403001

where μ is a factor specific to each installation dependent on friction losses, turbulence, compressibility, the water pressure and the nozzle shape [4-6].

After embedding abrasive and air, in mixing chamber and after directing the formed jet by the focusing tube, speed can be appreciated by applying the law of conservation of moments:

$$\sum \frac{dm}{dt} \cdot V_{(initial)} = \sum \frac{dm}{dt} \cdot V_{(final)} \tag{5}$$

$$\frac{dm_w}{dt} \cdot V_j = \frac{dm_{air}}{dt} \cdot V_f + \frac{dm_w}{dt} \cdot V_f + \frac{dm_{abr}}{dt} \cdot V_f \tag{6}$$

Considering insignificant air mass and the speed of the abrasive particles and the molecules of water equal to the output speed of the focusing tube V_f, result:

$$\frac{dm_w}{dt} \cdot V_j = \left(\frac{dm_w}{dt} + \frac{dm_{abr}}{dt}\right) \cdot V_f \tag{7}$$

$$V_f = \frac{\frac{dm_w}{dt}}{\left(\frac{dm_w}{dt} + \frac{dm_{abr}}{dt}\right)} \cdot V_j \tag{8}$$

The kerf taper angle φ (Fig. 2) is a parameter that can be controlled with optimization of other parameters [7].

$$\varphi = f\left(V_f, \frac{d_j}{V_t}, h, \frac{dm_{abr}}{dt}, d_j, E, H_R\right) \tag{9}$$

where: φ is kerf taper angle [^{0}];
V_f is the velocity of the jet at the place of impact [m/s];
V_t is the traverse speed of abrasive water jet [m/s];
d_j is diameter of jet at output of focusing tube [mm] ;
h is distance of output of focusing tube to the workpiece [mm] ;
E is elastic modulus [MPa].
H_R is Rockwell hardness of piece [MPa]

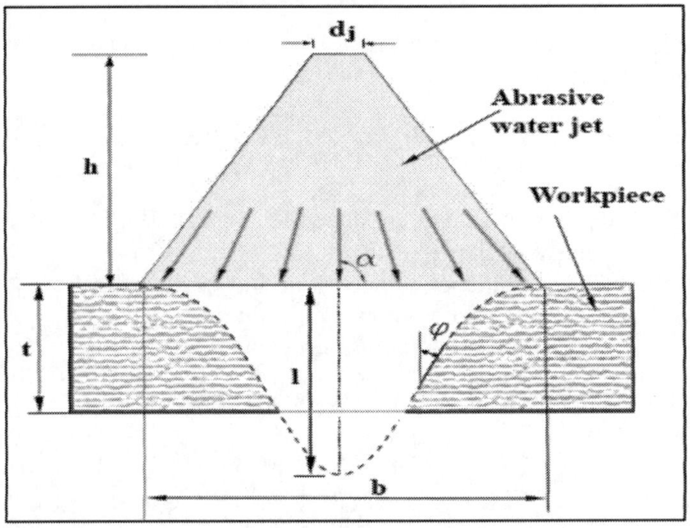

Fig. 2. The role of parameters of abrasive water jet in kerf geometry.

In the case of multilayers materials with different hardness the kerf taper angle changes value depending on the hardness of the material, switching between layers is marked by the change in angle value.

DOI: 10.1051/matecconf/20179403001

Applying relation (9) is obtained:

$$\varphi_1 = f\left(V_f, \frac{d_j}{V_t}, h, \frac{\mathrm{dm_{abr}}}{\mathrm{dt}}, d_j, E_1, H_{R1}\right) \tag{10}$$

$$\varphi_2 = f\left(V_f, \frac{d_j}{V_t}, h, \frac{\mathrm{dm_{abr}}}{\mathrm{dt}}, d_j, E_2, H_{R2}\right) \tag{11}$$

The two material parameters, E and H_R contributes directly to the cutting taper angle, determining geometric shape modification of cutting parts. Increase of precision of processing multilayers materials, becomes an important issue and process optimization is needed.

3 Modeling of kerf geometry variation in case of cutting of multilayers materials with different hardness

Increased hardness leads to increased taper angle (Fig.3) Considering $H_{R1}>H_{R2}$ and $\varphi_1>\varphi_2$ results that minimizing φ_1, optimal selection of parameters that influence the process, can lead to a significant increase quality of geometry of resulting parts. Thus decreasing of the distance h lead to a minimum of kerf taper angle.

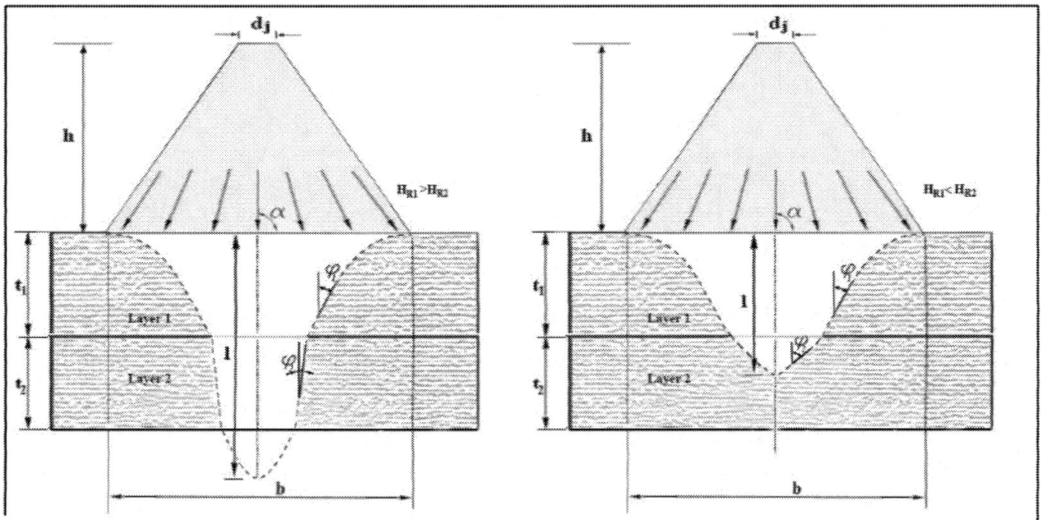

Fig. 3. The kerf taper angle in case of multilayer materials with different hardness ($H_{R1}>H_{R2}$ in left and $H_{R2}>H_{R1}$ in right of image).

Kerf geometry optimization is done considering the top layer hardness, whose taper angle is greater than the taper angle of softened layer. Initial condition in calculation algorithm is cutting condition (punching shear) of the thickness of the material $l > t$ where $t = t_1+t_2$.

Generally it is recommended that the layer with higher hardness to be cut first, but are cases (Fig. 3 right) when this could not be applied and taper angle grows stronger.

For process modelling, were considered the demands for type processing H_R, E, h, V_f si $\mathrm{dm_{abr}}/\mathrm{dt}$. Both, the speed of impact, and abrasive flow mass have less influence [8] and hardness and elasticity are material factors.

Distance h, of the head of focusing tube to the work piece, choose the minimum (1-2 mm) depending on the cutting surface [9].

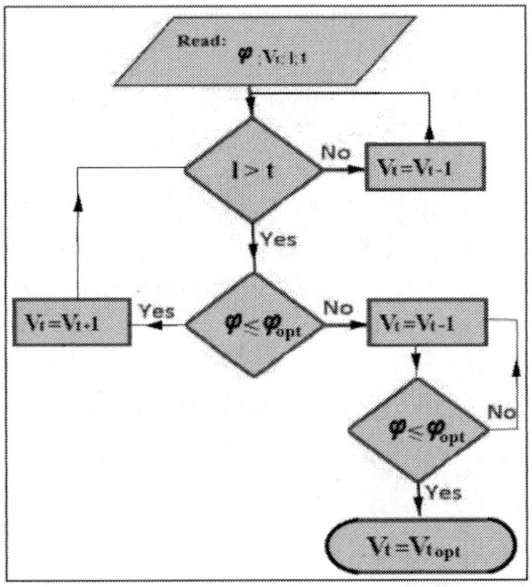

Fig. 4. Determining optimum traverse speed in order to obtain an optimum taper angle.

The algorithm presented in Fig.4, allows choosing a traverse speed that contribute to achieving a taper angle in accordance with the requirements of precision of the piece, namely φ_{opt}. The initial condition is complete cutting of material and optimum taper angle is chosen so that it ensure accuracy of workpiece. The traverse speed V_t is increased or decreased so that l>t and $\varphi \leq \varphi_{opt}$

For experiments were chosen a material composed of two plates rectified of 15NiCr13 alloy steel carburized at 63 HRC. Dimensions of cutting are: thickness $t = t_1 + t_2 = 10$ mm, where $t_1 = t_2 = 5$ mm and length of cutting, L=60 mm.

Table 1. Values of taper angle for variable process parameters.

Test no.	Pressure [MPa]	Orifice diameter [mm]	Type of garnet [Mesh]	Measured abrasive flow rate [kg/min]	Used abrasive [Kg]	Operation time [min]	Calculated taper angle [degrees]
1.	345	0.28	80	0.34	0.43	1.358	0.772
						1.525	0.572
2.			100	0.405	0.49	1.316	0.651
						1.475	0.467
3.		0.08	80	0.05	0.38	9.58	0.496
						11.06	0.378
4.			100	0.081	0.59	7.38	0.412
						9.10	0.355
5.	180	0.28	80	0.28	1.19	4.33	0.811
						6.12	0.593
6.			100	0.335	1.29	3.96	0.795
						5.82	0.588
7.		0.08	80	0.022	0.73	33.2	0.612
						42.6	0.524
8.			100	0.058	1.36	22.5	0.573
						30.4	0.402

Setting the AWJ machine, Maxiem 1530 from Omax Corporation, has been executed by changing water pressure provided by the pump for the maximum value p = 345 MPa to p = 180 MPa. Samples were executed with two pressures for each type of abrasive. Abrasive flow rate was measured in Test Mode for t=5min and the obtained value was introduced in the configuration window. Operation time is setting for two values of thickness. The taper angle was calculated as the average of three values measured on the same sample in harder layer.

4 Conclusions

The experiments shown in table 1 leads to the conclusion that increase of operation time, has positive influence to kerf geometry in terms of minimization of kerf taper angle. Increase of operation time, so decrease of traverse speed V_t causes an increase in processing cost, results that an optimal choice of traverse speed is absolutely necessary.

Develop a correct algorithm, allowed to achieve in few steps an optimal value for traverse speed. Direct effect of optimization of cutting consist in increasing of efficiency, increasing of the traverse velocity, decreasing the amount of abrasive garnet used. Optimization results in increased dimensional precision and increased surface quality.

The wide use of cutting jet do occur in the cutting process, a growing number of materials such as multilayer materials or welded materials. This research opens the door to extensive research in this area, in order to optimize cutting abrasive water jet for different types of multilayer materials with variable hardness.

References

1. B. Barabas, A. Fota, S.A. Barabas, *INTECH 2015,* 142-146 (2015).
2. D. J. Thomas, J. Man. Proc. **15,** 355–363 (2013).
3. A. Lebar, M. Junkar, J. Mod. Sim. Mat. Sci. Eng., **12**, 1159–1170 (2004).
4. J. Wang, Int. J. Mec. Sci. **49,** 306–316 (2007).
5. V. Gupta, P. M. Pandey, M. Garg, R. Khanna, N. K. Batra, Proc. Mat. Sci. **6,** 140-149 (2014).
6. M. Gent, M. Menéndez, S. Torno, J. Torano, A. Schenk, Wear J., **284,** 43– 51 (2012).
7. D. K. Shanmugam, S. H. Masood, J. Mater. Process Technol. **209,** 3887-3893 (2009).
8. S. A. Barabas, A. Florescu, G. Calefariu, F. Sarbu, Adv. Mater. Res., **1128,** 293-303 (2015).
9. D. Trainer, Int., J. Multi. Flow, **81,** 77-87 (2016).

Study on SLM manufacturing of teeth used for dental tools testing

George Razvan Buican[1,*], *Gheorghe* Oancea[1], and *Rui F.* Martins[2]

[1]*Transilvania* University of Brasov, Faculty of Technological Engineering and Industrial Management, Department of Manufacturing Engineering, Mihai Viteazu, No. 5, Brasov, Romania
[2]Universidade Nova de Lisboa, Faculty of Sciences and Technology, UNIDEMI, Campus de Caparica, 2829-516 Caparica, Portugal

Abstract. In this article are studied the possibilities of manufacturing human teeth using the Selective Laser Melting (SLM) process. The human teeth are a biological construction, and because of this, they have a very complex structure with external and internal complex surfaces, unique for each person. To manufacture such structures, even using additive technologies, it is important to find the right manufacturing strategy. In this study are presented three strategies to manufacture the first molar using a SLM process and ANSI 316 L stainless steel as the building material. The best strategy to manufacture the teeth is chosen, and once build they can be used to test dental tools.

1 Introduction

It is well-known that nowadays the additive manufacturing process is used in different domains such as: aerospace industry, automotive industry, clothing industry, medicine and even entertainment industry [1- 5].

The paper studies the manufacturing possibilities of two human teeth, from stainless steel powder, using the additive manufacturing technologies available on the SLM 250 HL machine.

The 3D models of the two teeth were realized in the CAD software SolidWorks®. In order to better design the teeth including the mesiobuccal root canal, with a 0.9 mm diameter, two X-Ray images, from a maxillary molar morphology study [6], were used, namely the buccolingual and mesiodistal views.

According to the Fig.1., in order to manufacture these teeth the following main steps need to be performed:

- generating the STL file from a CAD model;
- processing the STL file and generating the machine FAB file with the building supports and construction parameters;
- using the machine FAB file to manufacture the 3D object.

In Fig.2. and Fig.3. the 3D representation of the mandibular molar modelled in SolidWorks® is shown.

*Corresponding author: buican.george@unitbv.ro

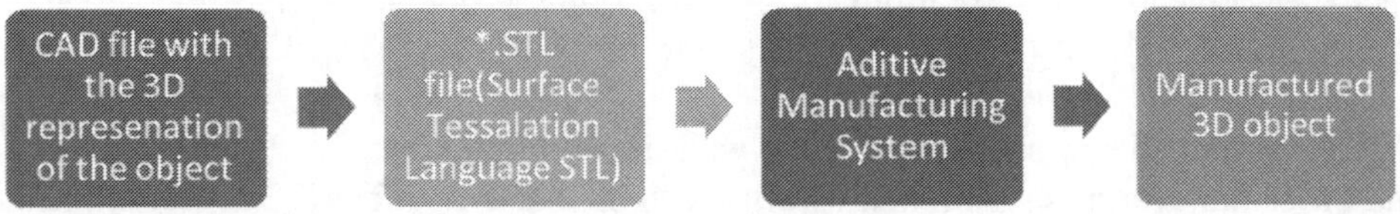

Fig. 1. Steps in creating a 3D object

In the first and second step, the STL file generated by the CAD software is loaded into Magics®. Here, the 3D construct is checked for surface tessellation errors like Manifold – leaks, intersecting faces, degenerate faces and empty spaces. Once the file is free of errors the most important step is performed. In this stage, the positioning of the object in relation to the manufacturing table and the building supports is realized [7].

The positioning of the manufacture part and the building supports have such a high importance, because, depending on the choices that are taken, the part will be completely manufactured or will become a failure. Furthermore, the positioning solution influences the mechanical properties and the residual stresses of the finished part, and many other parameters.

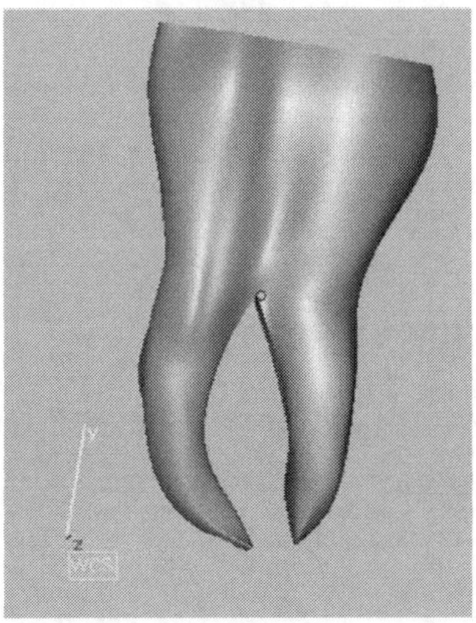

Fig. 2. Mandible molar - 3D view.

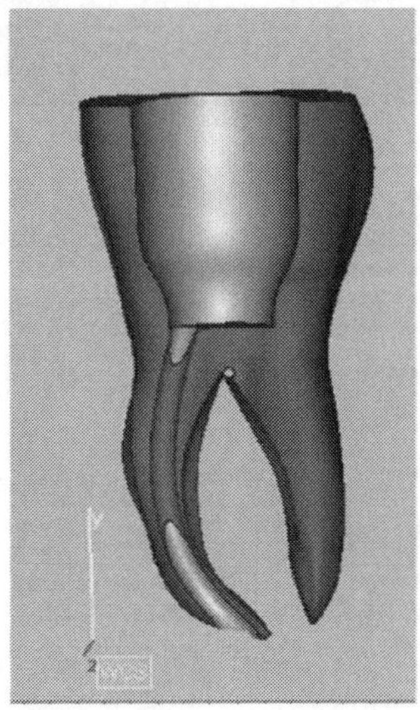

Fig. 3. Mandible molar – section.

Effects which are influenced by the part positioning solution are [8-11]:

- outer surface of the finished part (including the roughness);
- internal stresses;
- mechanical properties;
- material consumption;
- detachment of the part from the building table;
- possibility to build or not internal features;
- number of build parts;
- overall aspect;
- and even the maximum height of the manufactured object.

A lot of parameters are influenced by the positioning solution and, because of this, three solutions for the positioning problem are taken into account (there are many other solutions, but we stopped the algorithm of selection when the outcome was the most satisfactory). The considered solutions for the positioning problem are:

- mandible molar is positioned parallel with the building table (Fig.4., Fig.5., Fig.6. and Fig.7.);
- mandible molar root is positioned parallel with the building table (Fig.8. and Fig.9.);
- mandible molar crown is positioned parallel with the building table(Fig.10. and Fig.11.).

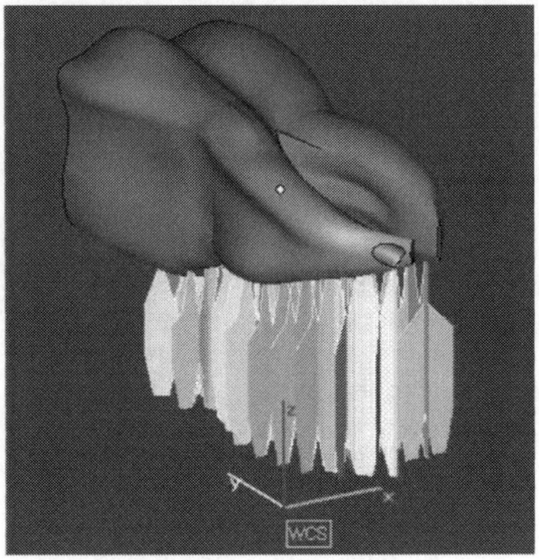

Fig. 4. First case - Mandible molar with construction supports, 3D isometric view.

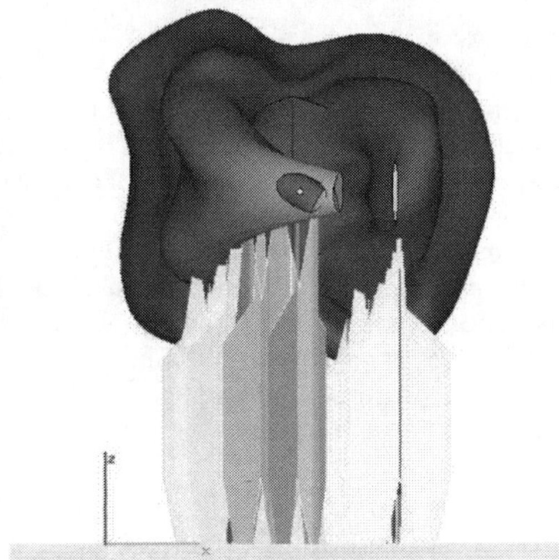

Fig. 5. First case - Mandible molar with construction supports, back view.

2 Teeth manufacturing – positioning solutions

2.1 First case - the mandible molar is positioned parallel with the building table

As the first case of positioning, for the mandible molar, it is considered the alignment of the teeth in a parallel direction with the building table (Fig.4., Fig.5., Fig.6. and Fig.7.). In this positioning solution, the construction supports must be attached perpendicular to the exterior face of the teeth that is parallel to the building table.

In trying using this positioning solution to manufacture the part were identified a series of disadvantages:

- After manufacturing the tooth, on the surface with the attached building supports it has to be applied a second machining operation to obtain a smooth surface. The surface has irregularities generated by leftover material from the building supports. But, because this surface is a reproduction of an organic growth, the shape is an irregular one and thus very hard to retouch.
- Because the final 3D object is hollow inside (it has cavities) the Magics® software attaches supports even on the interior of the part (Fig.7.). To remove the interior supports is an even more difficult operation, an operation that needs to employ special tools, thus raising the complexity of the manufacturing process and the overall costs.
- Also, the attached supports, from the exterior surface, are not enough to maintain the part fixed until the end of building process, and thus a misconstruction could occur.

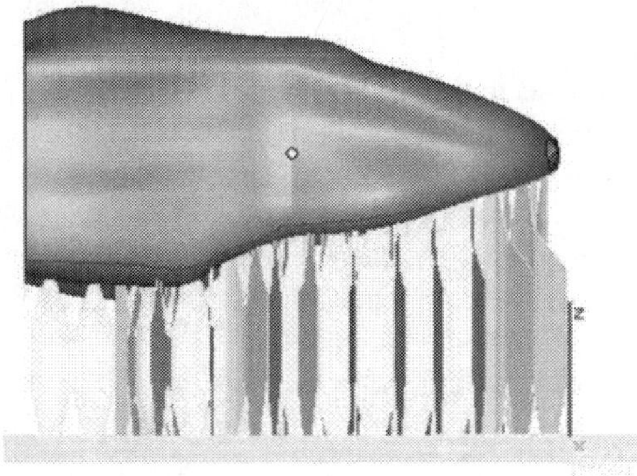

Fig. 6. First case - Mandible molar with construction supports, 3D side view.

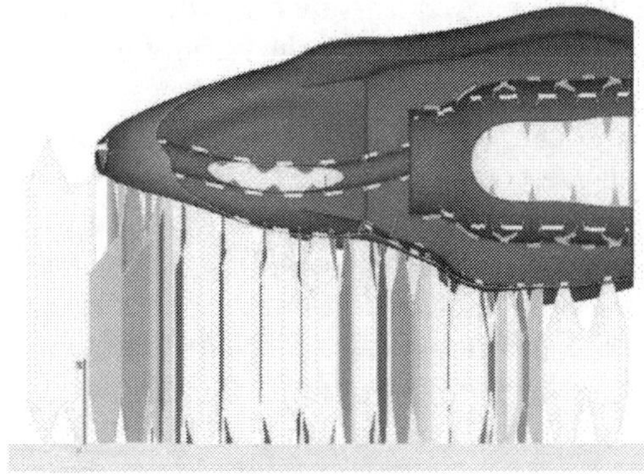

Fig. 7. First case - Mandible molar with construction supports, section.

Taking into account all drawbacks presented above, this positioning solution is considered only a theoretical one, and thus it is not used to create a real object.

2.2 Second case – the tip of the mandible molar root is positioned parallel with building table

The second positioning solution has the construction supports attached on the tip of the teeth root (Fig.8. and Fig.9.). To a first glance this could be considered as the best positioning solution because of the belief that there is no need for post processing operations.

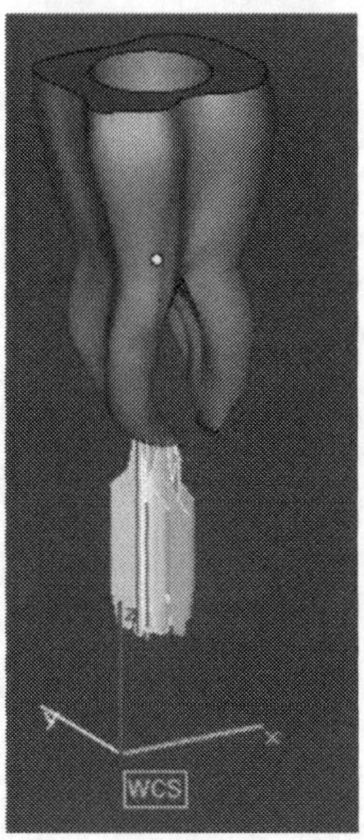

Fig. 8. Second case - Mandible molar with construction supports, 3D Isometric view.

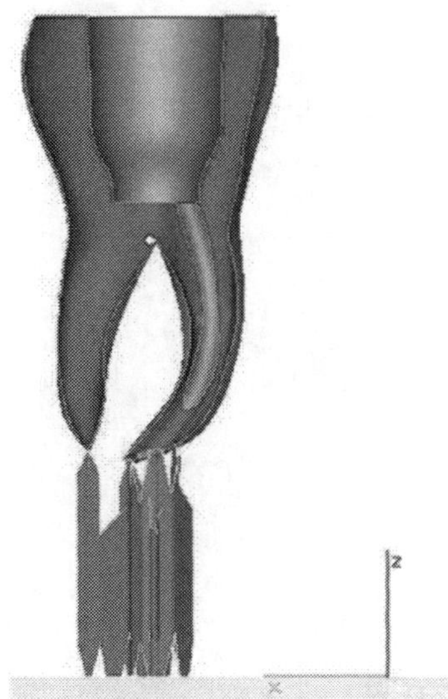

Fig. 9. Second case - Mandible molar with construction supports, section.

The area of contact between the manufacturing supports and the finished part is the smallest of all the cases, but it is well known that the process of selective laser melting is a manufacturing process that inserts in the finished part a lot of residual stresses and tensions

 DOI: 10.1051/matecconf/20179403002

due to the subsequent melting and cooling cycles [4-5,8-11]. These melting and cooling cycles are due to the way the technology works.

The selective laser melting process is an additive manufacturing method, and a layer based technology. Thus it consists in building the part in layers of fixed thickness, which are stacked one upon another, and then melted by using an energy source.

In the considered case, the metal powder (ANSI 316-L stainless steel) is laid in a thick layer of 30 μm on the building plate, and then the laser melts a predefine contour, that represents a 2D section from the part. On top of this layer, the machine lays another layer of material and the process is repeated until the part is finished.

Keeping in mind the way this technology works, and the stresses and tensions that are inserted in the part because of the heating and cooling cycles, it is concluded that the building supports in this positioning solution are not sufficient to take this job to the end. This case is a theoretical one.

2.3 Third case – the mandible molar crown is positioned parallel to the building table

The third positioning solution lay the teeth with the crown parallel to the building plate. In this position the supports are attached to the teeth crown, that, in the considered case, is a flat surface, and thus provide a sufficient surface for the part to be fixed during the entire manufacturing process (Fig.10. and Fig.11.).

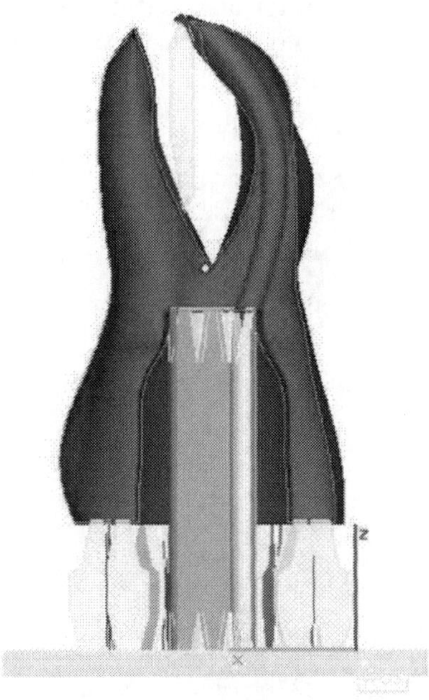

Fig. 10. Third case - Mandible molar with all construction supports, section.

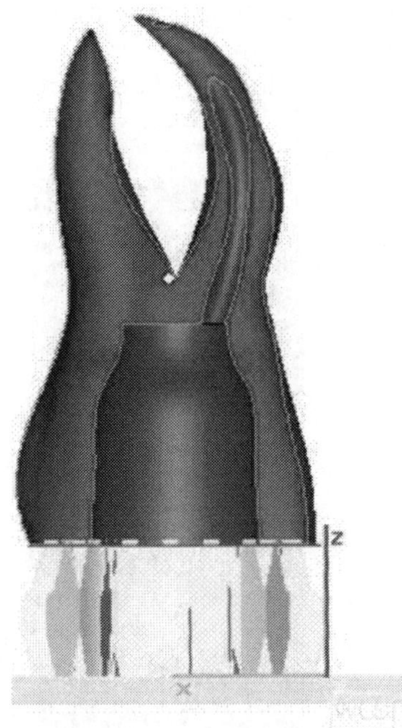

Fig. 11. Third case - Mandible molar with crown construction supports, section.

Also, on this position solution, building supports can be attached to the interior of the teeth, to further sustain the part. This is possible because here there is a flat surface.

Additionally, there is a need to attach a point support to assist the root of the teeth during the manufacturing process, and this is needed because the root has a very big curvature.

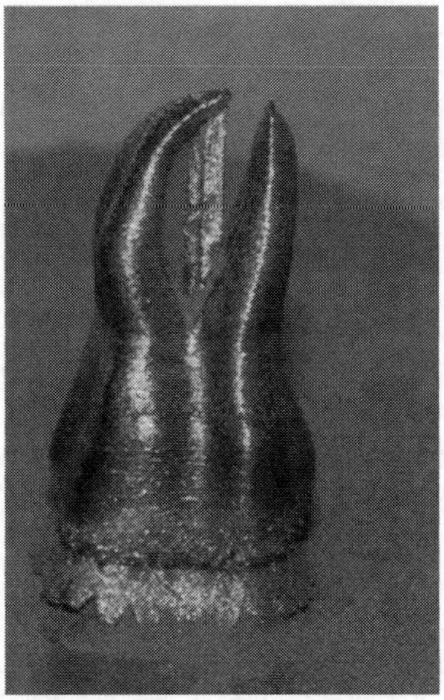

Fig. 12. Mandible molar attached to building plate, front view.

Fig. 13. Mandible molar attached to building plate, side view.

Because the building supports are attached to a plan surface, the post processing is much easier to be made. In the same time the part detaching from the building table is much improved.

The contact area between the building supports and teeth crown is situated in the same plane, but has a circular cross-section, thus providing more strength and resilience against the stresses that appear during the manufacturing process.

This case is considered the best solution for the positioning and it will be used in practice, to manufacture the teeth.

3 Teeth Manufacturing on SLM 250 HL

The mandible molar was manufactured on the SLM 250-HL equipment using a layers size of 30 μm. In order to manufacture the teeth, the positioning solutions, together with the building supports are loaded in AutoFab® and the final file with the manufacturing parameters is created.

Teeth are manufactured using a stainless steel ANSI 316-L powder. This material is laid as a thin layer of 30 μm on top the building plate, then the layers generated with AutoFab®, in the *.fab file, are melted in the metal layer using different laser powers (Fig.12, Fig.13, Fig.14., Fig.15., Fig.16. and Fig.17.).

In order to build this part it was used a strategy with different working parameters. This strategy was established according to the areas of the parts: contour, volume, overhang, skin and building supports. The meanings of the areas and the laser powers used in the manufacturing process on each area are:

- contour - it is the contour of the 2D section; the laser power is 100 W;
- volume is interior of the 2D section and it is more than just a contour is an internal surface. Here the laser power is 175 W, but in addition to the rest of the building regimes here it was defined even a path for the laser(in this case chess board, rotated to 45 degrees on each layer);

- overhang is the area where the building supports and the actual part meet. Here the used laser power was 175 W;
- skin is the outer layer of the finished part, and thus has a different set of parameters, parameters that give a smoothness surface. In this area the laser power is also 175 W, but the laser speed is changed.
- building supports are manufactured, using a laser power of 100 W, and a different pattern for the path of the laser. In this area the laser spot is at its smallest value, because the building supports can be compared to a thin wall part.

Once the parts are finished, together with the building plate they are placed in an oven for tensions eliminating process. After that they are removed from the plate and post processed to remove the building supports.

A second post processing operation can be done using a sand blasting machine to generate a fine exterior surface. This last method is also good for removing scorching marks generated by an overheating during the manufacturing process.

The mandible molar after the post processing operation is presented in the figures 16 and 17.

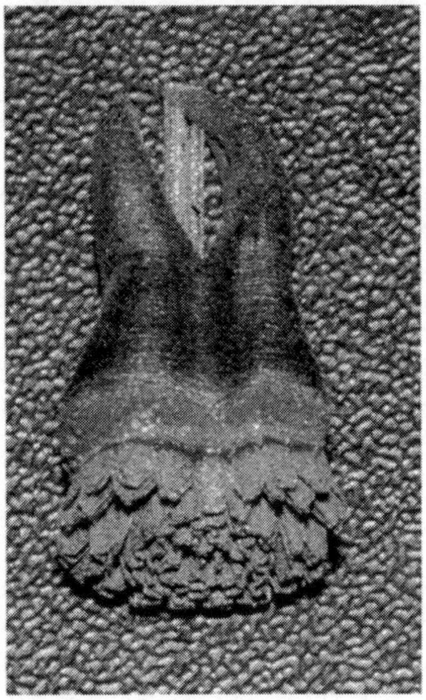

Fig. 14. Mandible molar detached from building plate, front view.

Fig. 15. Mandible molar - detached from the building table, top view.

From these studies, it is revealed that the possibility to manufacture the teeth, all the way to the last layer, is inverse proportional with the complexity of the building supports and positioning solution. This means that the intersection between the surface of the object and the surface generated by the tips of the building supports needs to be as close as possible to a plane.

DOI: 10.1051/matecconf/20179403002

Fig. 16. Mandible molar - detached from the building table with the supports removed, front view.

Fig. 17. Mandible molar – finished part, front view.

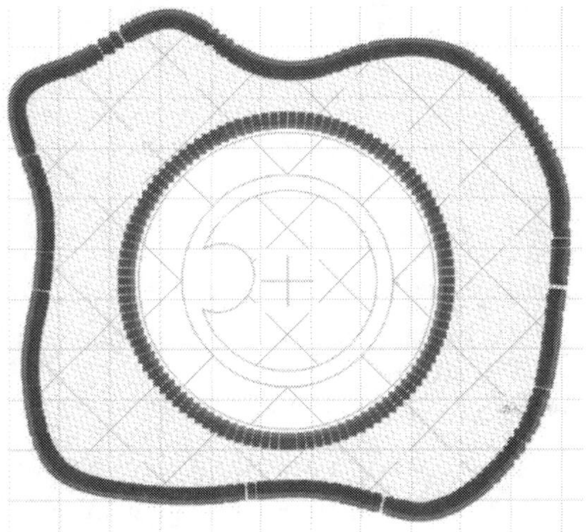

Fig. 18. Generated 2D shape from the intersections.

The intersection between the planes of the building supports tips and the surface on which they are attached, needs to generate a 2D shape that is, either parallel to the building table (for small enough objects), or inclined with a certain degree, (for bigger objects).

By studying three positioning solutions, it was identified three cases of generated shapes:

- in the first case the generated surface, from the intersection, was a 3D shape;
- the second case had generated a planar surface, but it was not a shape but rather a collection of points;
- in the third, and last case, the generated shape was a 2D shape (Fig.18.).

4 Conclusions

By studying the possibility to manufacture human teeth it was identified the possibility to reproduce organic shapes, with high structural complexity, by the means of additive manufacturing, in this case the selective laser melting technology was used. As it can be seen the organic shapes are very complex, each one unique. They have a rough overall shape, but each individual piece has its one distinctive features.

In order to build complex organic shapes with additive manufacturing it is imperative to be able to convert these shapes in 3D data, and only then the technology is able to manufacture these objects. But this is not enough, in order to secure the manufacturing of the teeth, it is necessary to further study the positioning of the objects in rapport with the building table and the building supports.

Based on the studies developed by the authors and presented in this paper, it can be concluded that complex organic shapes can be manufacture with metal additive manufacture technology. For some shapes, this being the only possibility, but in order to be able to finish the part there is a need for rigorous post processing. These post processing operations and simulation help in reducing time and money by increasing the likelihood of completely finishing the parts, or finishing the part without major defects. Using a proper strategy for a selective laser melting technology two human teeth were manufactured on SLM HL 250 machine. The manufactured teeth have been used in the testing process of dental tools.

This paper is supported by the Sectoral Operational Programme Human Resources Development (SOP HRD), ID134378 financed from the European Social Fund and by the Romanian Government.

References

1. D. Leordean, C. Dudescu, T. Marcu, P. Berce, N. Balc, Customized implants with specific properties, made by selective laser melting, Rapid Prototyping Journal, **21**, 98-104, (2016)
2. S.C. Cosma, N. Balc, D. Leordean, M. Moldovan, M. Dudescu, C. Borzan, Customized Medical Applications of Selective Laser Melting Manufacturing, Academic Journal of Manufacturing Engineering, **13**, 24-32, (2015)
3. S.A. Haba, G. Oancea, Digital manufacturing of air-cooled single-cylinder engine block, The International Journal of Advanced Manufacturing Technology, **80 (5)**, 747-759, (2015).
4. G.R. Buican, G. Oancea, A. Manolescu, Remanufacturing of Damaged Parts Using Selective Laser Melting Technology, Applied Mechanics and Materials, **693**, 285-290, (2014)
5. R. Pacurar, N. Balc, F. Prem, Research on how to improve the accuracy of the SLM metallic parts, The 14th International ESAFORM Conference on Material Forming: ESAFORM, **1353 (1)**, 1385-1390, (2011).
6. C.E. Jerome, R.J. Hanlon, Identifying Multiplanar Root Canal Curvatures Using Stainless-Steel Instruments, Journal of Endodontics, **29**, 356-358, (2003)
7. H. W. Wai, *Introduction to STL format*, (1999);
8. T. Itziar, G. Fermín, Z. Fidel, Z. Fidel, E. Aritz, Study of mechanical properties of AISI 316 stainless steel processed by "selective laser melting", following different manufacturing strategies, The International Journal of Advanced Manufacturing Technology, **51**, 639–647, (2010).

DOI: 10.1051/matecconf/20179403002

9. G.R. Buican, G. Oancea, C. Lancea, M.A. Considerations Regarding Micro Hardness of Parts Manufactured from 316-L Steel Using SLM Technology, Applied Mechanics and Materials, *Some* **760**, 515-520, (2015)
10. G.R. Buican, G. Oancea, C. Lancea, M.A. Pop, Influence of Layer Thickness on Internal Structure of Parts Manufactured from 316-L Steel Using SLM Technology, Applied Mechanics and Materials, **809-810**, 369-374, (2015)
11. S.C. Muresan, PhD Thesis, *Researches regarding the mechanical strength of complex metal parts manufactured by selective laser melting*, Technical University of Cluj-Napoca, (2012)

Experimental research on the machinability of Hardox steel by abrasive waterjet cutting

Alexandru Catalin Filip[1,*], *Mircea Anton* Vasiloni[1], and *Laurentiu Aurel* Mihail[1]

[1]Transilvania University of Brasov, Department of Manufacturing Engineering, Mihai Viteazul No.5, Romania

Abstract. One of the main present industry challenges is finding the most efficient manufacturing process for a certain part. When parts are made of strong steels like Hardox, their fabrication method is usually difficult. Abrasive waterjet cutting (AWJ) is one of the cutting processes which can be used in this case. This paper presents an experimental research on the machinability of Hardox steel by AWJ. The experiments were conducted using a factorial design model considering two of the main influence parameters like the traverse speed and the distance between the nozzle and the surface of the material. Based on the measurement of the dimensions and the roughness of the parts, the influence of the parameters was revealed and analyzed. The manufacturing time was also compared, as it directly influences the production cost. Further research is considered to develop a mathematical model which can be used for a proper choice of the process parameters depending on the initial requirements.

1 Introduction

The equipment used in heavy industries like mining or the construction of roads must meet special requirements such as a strong wear resistance and a high tenacity, leading to a good strength-to-weight ratio. Hardox steel alloys [1, 2] are some of the recent used in the field. Such steels are usually difficult to process by traditional mechanical cutting processes because of their poor machinability by cutting. In this case, a method like abrasive waterjet cutting (AWJ) can be applied.

Abrasive waterjet machining is a flexible manufacturing process. It is based [3, 4], Figure 1a, on using a high pressure waterjet (1) guided through a small calibrated orifice (2) and mixed with fine abrasive particles in a mixing chamber (3). The abrasive jet is then sent through a nozzle (10), towards the surface of the material. The high speed of the abrasive jet creates enough force able to cut the material of the part. The main advantages of the method are the capacity to cut almost any type of material, no thermal influence, easy and fast setup and a high flexibility. The process is also environmentally friendly as the "tools" – water and abrasive particles – are fully recyclable.

However, the method has some limitations. The main one is the kerf geometry which depends on the machine dynamics. If the machine does not have kerf compensation options

*Corresponding author: filipal@unitbv.ro

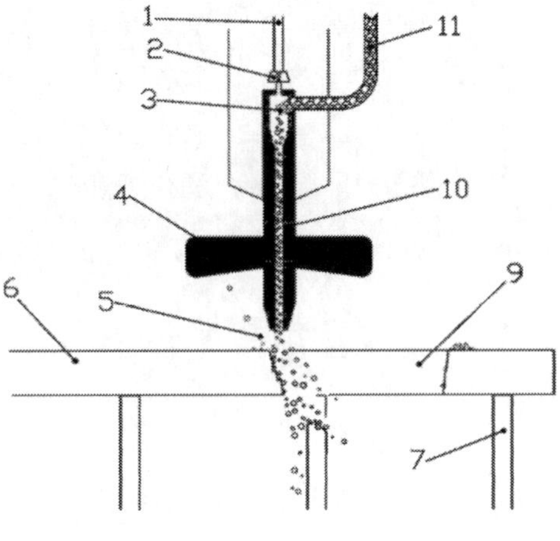

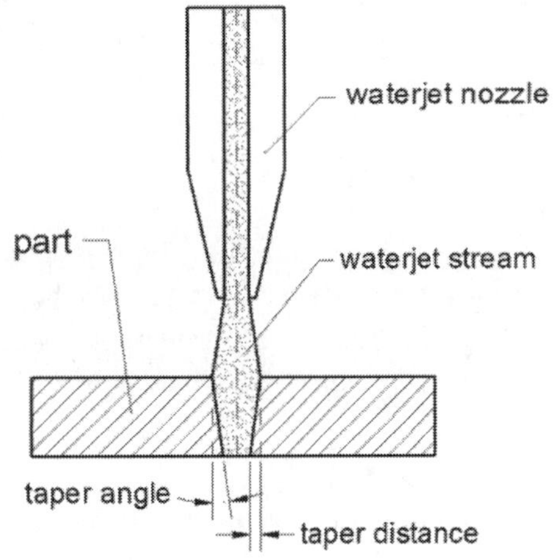

Fig. 1. a. Principle of abrasive waterjet cutting; b. Taper kerf formation.

then the kerf will be tapered, Figure 1b, and also the surface of the part will have geometry deviations due to this phenomenon.

One other issue is the model to be used on a new material. Some mathematical models have been studied and developed in the last decade [5-7]. Each of them has a specific range of validity or was developed for a certain type of material. Usually, each abrasive waterjet machine uses its own cutting model and it has a materials database from which the operator can select the one needed. If the material is not in that database, there will have to be selected a similar one, but the cutting model will not be accurate and the results may not be as expected.

Since for the Hardox steels no technological data was found for the abrasive machining method, this paper presents an experimental study on the AWJ of this material aiming to reveal the proper set of values for the process parameters depending on the quality of the cut.

2 Run of experiments and measurement of the parts

The experiments were realized in the laboratory of Advanced Manufacturing Technologies, on a waterjet machine type Maxiem, having a 20 HP pump and a maximum water pressure of 3500 bar. The dynamics of the machine does not have the possibility of automatic tilting, to compensate the effect of taper surface. The abrasive feeding system is a regular one, with constant feed rate, which was carefully measured before the experiments.

The material of the parts is steel Hardox 400 [1, 2] which has the mechanical properties according to Table 1.

Table 1. Mechanical properties of the steel Hardox 400.

Hardness HBW Brinell	Yield Strength [MPa]	Tensile Strength [MPa]	Elongation [%]
370…430	1000	1250	10%

The cutting was performed on a plate made of Hardox 400 having the thickness of 20 mm. The design of the parts is presented in Figure 2a. Some manufactured parts can be seen in Figure 2b.

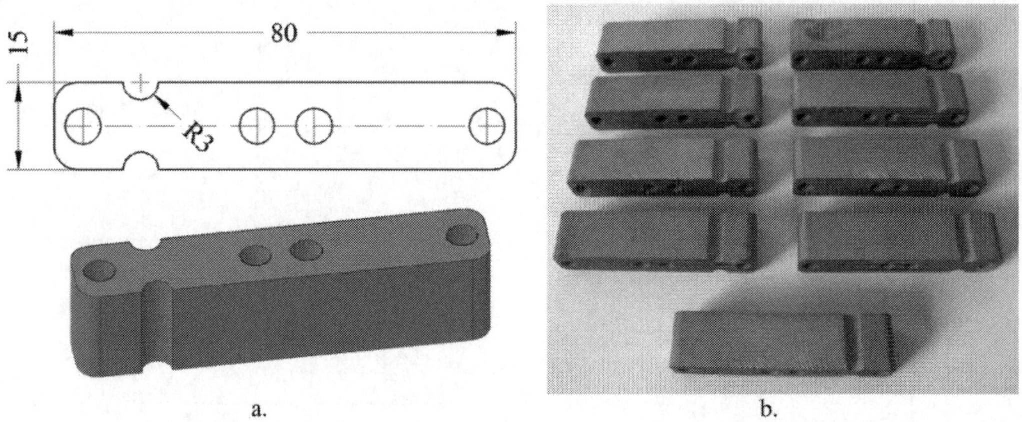

Fig. 2. a. Design of the manufactured part; b. Sample parts cut with AWJ during the tests.

The main parameters of influence which are considered by the comprehensive researches until present day [3-9] are the water pressure, the traverse speed, the abrasive flow rate and the standoff distance between the nozzle and the surface of the material to be cut. As the water pressure directly influences the process productivity and costs, almost all researches agree that it is better to use the maximum values given by the equipment. For similar reasons, there was used the default abrasive flow rate given by the equipment. The values of these last two and other constant parameters are given in Table 2.

Table 2. Values of the constant parameters used during the tests.

Water pressure [bar]	Abrasive flow rate [g/min]	Abrasive type	Jewel orifice diameter [mm]	Nozzle orifice diameter [mm]
3500	370	garnet 80 mesh	0.279	0.838

The two variable process parameters considered were the traverse speed (TRS) and the standoff distance (STD). Their range of values was established either by studying experiments done by others [9-11], either by some preliminary tests done by the authors.

Since one aim of this research is to find the proper machinability number to be used for steel Hardox 400 on Maxiem/Omax type AWJ machines, there were selected two limits for this number, 60 and 95. The machine software requires also a certain number for the quality of the cut, between five values, numbered from 1 to 5. Depending on the values of the machinability and the quality of the cut result the values of the traverse speed.

The experiments were conducted by a design plan using a factorial design based on three parameters – the machinability number, the quality of the cut and the standoff distance – and two limits of values for them. The middle range values combination was also included. Therefore, there were necessary 9 experiments into a set. Their values are shown in Table 3.

The resulting traverse speed from Table 3 is the linear speed used during the experiments, thus it is the cutting speed along the straight lines of the part.

In Table 3 there is also mentioned the cutting time, as it was calculated by the machine software.

After the trials each part was analyzed in terms of dimensional accuracy and the surface quality. Measurements were made along the two linear faces of each part.

The dimensional accuracy and the taper effect were analyzed by measuring the width of the parts in the middle of the length, Figure 3, on the two sides, named the upper and the bottom. The upper side is the entrance side of the waterjet and the bottom side is the exit side of the waterjet. The surface quality was analyzed by measuring the roughness with a portable tester Mitutoyo SJ210 Surftest, on the two linear faces of each part, approximately at 5 mm from the edge, both on the upper side and the bottom one, as shown in fig.5. The mean values were calculated for each measurement.

Table 3. Values of the variable parameters used during the tests.

Trial Nr.	Machinability Number	Quality of Cut	Standoff Distance [mm]	Resulting Traverse Speed [mm/min]	Cutting Time [min]
1	60	2	1.5	63.18	13.73
2	60	4	1.5	28.48	22.58
3	95	2	1.5	100.03	8.33
4	95	4	1.5	45.07	13.63
5	60	2	2.5	63.18	13.73
6	60	4	2.5	28.48	22.58
7	95	2	2.5	100.03	8.33
8	95	4	2.5	45.07	13.63
9	77.5	3	2	51.2	12.21

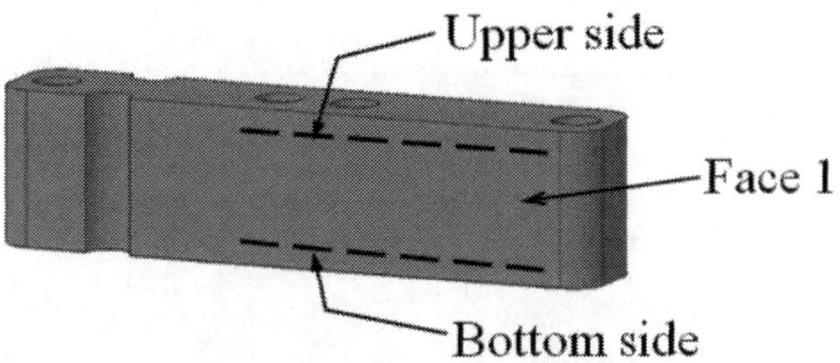

Fig. 3. Positions of the measurements on the sample part.

3 Analysis of the results

3.1 Influence of the traverse speed on the surface roughness

Figure 4 shows the variation of the surface roughness Ra depending on the traverse speed, for the upper side (Up) and the bottom one (Dw) and two values of the standoff distance (STD). The surface roughness increases with the increase of the traverse speed. This variation is in accordance with all other previous researches done by others [5-10]. But the intensity of variation is different. On the upper side of the part (Up), where the abrasive jet enters into the material, the surface roughness slightly changes with the increase of the traverse speed. There can be stated that at small thickness values, under 5 mm, the traverse speed practically does not influence the surface quality of the parts. In Figure 5 there can be seen some parts which visually confirm the statements made here. The upper left part was nr.2 of the trial (see Table 3) and the down left one was nr.3.

On the bottom side of the part (Dw), where the abrasive jet exits the material, the traverse speed has a significant influence on the surface roughness. This can be explained by the bending effect of the jet which intensifies with the increase of the traverse speed [6, 7]. At lower traverse speeds, under 45 mm/min the increase of the roughness is small, but at higher speeds the increase is significant and the quality of the part becomes unacceptable due to visible ridges which can be seen on the surface (Figure 5).

Based on the results the adequate value of the traverse speed to be used for 20 mm thick Hardox should be below 45 mm/min. This can be obtained on an Omax type machine by using a combination of the machinability number 60 and the quality number over 2.

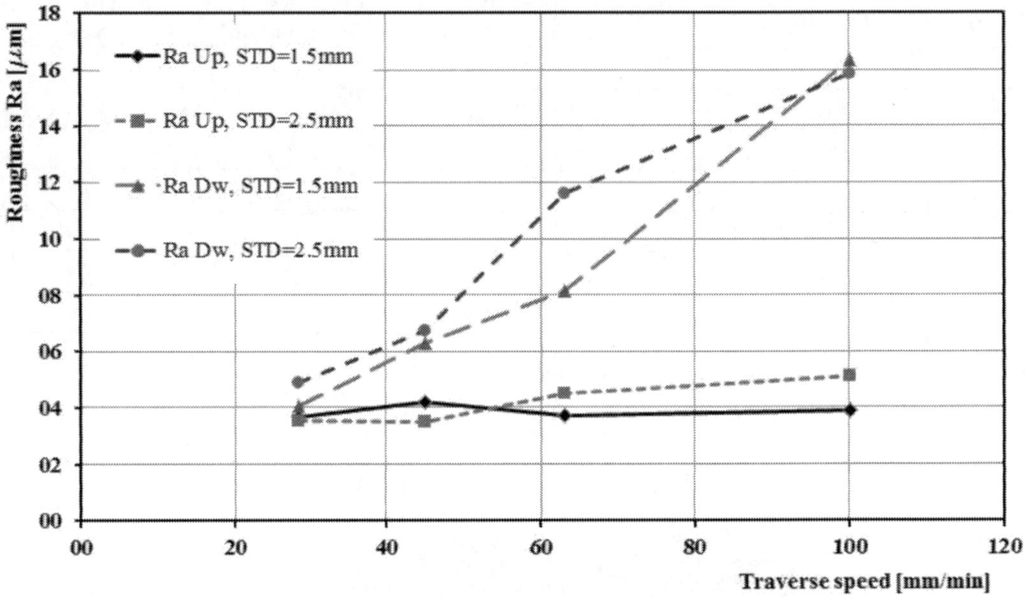

Fig. 4. The variation of surface roughness Ra with the traverse speed (TRS).

Fig. 5. Samples of parts manufactured during the trials.
Up Left – TRS=28.48 mm/min, Down Left – TRS=100 mm/min.
Up Right – TRS=63.18 mm/min, Down Right – TRS=45.07 mm/min.

3.2 Influence of the standoff distance on the surface roughness

The variation of the surface roughness depending on the standoff distance (STD) is shown in Figure 6. As seen in the diagrams, the standoff distance has a different influence on the roughness, depending on the position of the surface along the thickness of the part. Generally, the increase of STD leads to a slight increase of the surface roughness. If the standard deviation is calculated for the two sides of the part which were analyzed, the values are 0.6 for the upper side of the part (Up) and 4.8 for the bottom one (Dw). Therefore, it can be stated that the standoff distance has a more significant influence on the bottom side of the part which is the exit side of the abrasive jet.

Based on the results the recommended value of the standoff distance can be suggested in the interval 1.5…2.0 mm.

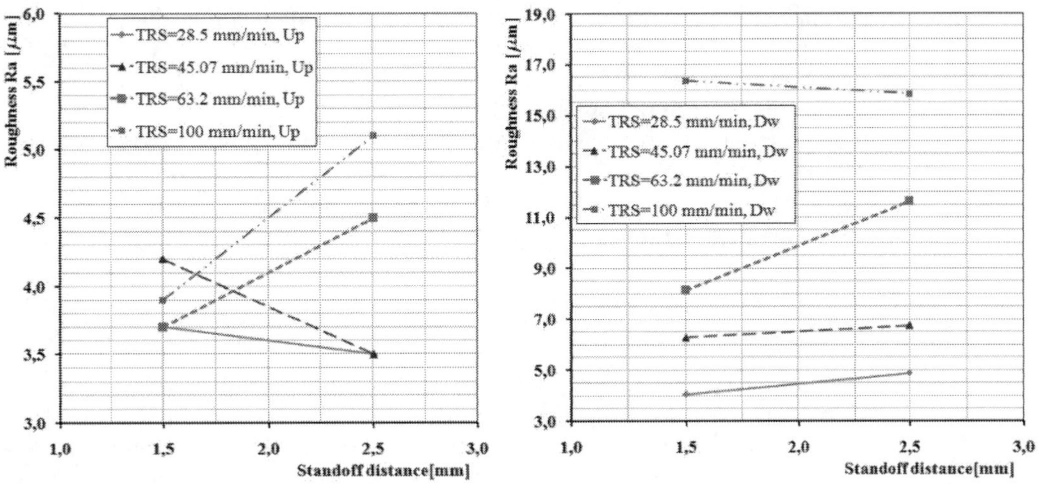

Fig. 6. The variation of surface roughness Ra with the standoff distance (STD).

3.3 Influence of the process parameters on the dimensional accuracy

The dimensional accuracy was analyzed by measuring the width of the parts, in the middle of its length, on the two faces both on the upper and the bottom sides of the part, Figure 3. The nominal value of the width was 15 mm (Figure 2a). The diagrams in Figure 7 show the variation of the mean values for the part's width on the variable process parameters considered in the experiments, the traverse speed TRS and the standoff distance STD.

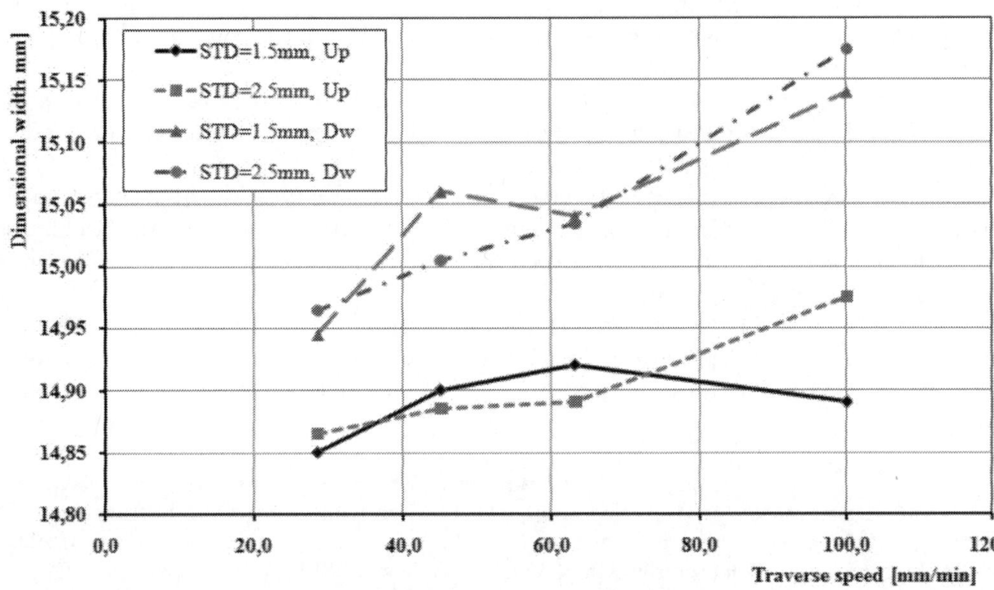

Fig. 7. The variation of the width of the parts on the process parameters.

According to the diagrams shown in Figure 7 the variation of the dimensional accuracy on the process parameters is similar to that of the surface roughness (see Figure 4). On the upper side of the part, at the entrance of the abrasive jet the increase of the dimension with the increase of the traverse speed is slightly smaller than the one found on the bottom side of the part.

Because the equipment used for experiments is not provided with automatic taper effect compensation, the dimension on the upper side of the part is smaller than the one of the bottom side. The difference is between 0.10 and 0.25 mm. For the adequate value of the traverse speed below 45 mm/min, stated in chp.3.1, the difference is below 0.14 mm, so it can be stated that the precision of the process is around this value.

The standoff distance seems to have a small influence on the dimensional accuracy of parts. The influence is slightly higher when the traverse speed has bigger values. The differences are below 0.1 mm, so it can be stated that the standoff distance has a very small influence on the dimensional precision of parts.

4 Conclusions

Abrasive waterjet cutting can be successfully used for manufacturing steels with a hard strength like Hardox [2]. The research of this paper proves this fact without any doubt.

Like all previous researches have demonstrated [4-5, 8-10], the main process parameters which affect the quality of the parts are the jet pressure, the traverse speed and the abrasive flow rate. Because the jet pressure is recommended to be used at the maximum value provided by the equipment for productivity reasons this parameters was kept constant. Because the equipment used in experiments does not have the possibility to consistently vary the abrasive flow rate, this parameter was also kept constant during the tests. The research of this paper revealed the influence of the traverse speed. It was also analysed the influence of the standoff distance even if it is not a major influence parameter.

The quality of the parts was assessed by measuring their dimensions and surface roughness.

The main conclusions of the research are:

- The traverse speed strongly influences the quality of the parts, as shown in Figs.4 and 7. If the speed is over a certain limit, the surface become extensively ridged and will be unacceptable. For Hardox 400, the type of steel used, this limit is considered around 45 mm/min. for parts having the thickness around 20 mm.
- The standoff distance has a small influence on the quality of the parts, as shown in Figure 6. Based on the results of tests, it can be stated that the adequate value of the standoff distance should be in the range 1.5…2.0 mm.
- The analysis of the cutting time (Table 3) show that the increase of traverse speed proportionally leads to a shorter time, but the quality of parts becomes worse. Usually, at AWJ a compromise has to be made between these two parameters to ensure the required quality in the smallest time possible.

Further research will be developed for different other values of the material thickness to establish a mathematical model for choosing the correct value of the machinability number in the case of Hardox 400 alloy steel when it is manufactured by abrasive waterjet cutting (AWJ).

The experiments realized in the present research were made with the support of the project PN-III-P2-2.1-BG-2016-0206, ProWaterjetTech, financed by UEFISCDI, using also the infrastructure of the "Metrology Lab Mitutoyo" from the "Institute High Tech Products for Sustainable Development: PRO-DD", lab created with the contribution of Mitutoyo Romania S.R.L.

 DOI: 10.1051/matecconf/20179403003

References

1. H. Bugłacki, M. Smajdor, Mechanical properties of abrasion-resistant Hardox 400 steel and their welded joints, Adv. In Mat. Science **4**, 2 (2003)
2. Hardox steels. http://www.ssab.com/products/brands/hardox. Accessed March 2016.
3. A.W.Momber, R.Kovacevic, *Principles of Abrasive Waterjet Machining*m (Springer London Limited ,1998)
4. A.C.Filip, H.Bulea, Surface precision on the circular abrasive waterjet machining of ceramic aluminium oxides, MTeM'2013, 8 (2013)
5. M. Hashish, A model for abrasive – waterjet (AWJ) machining, J. Eng. Mater. Technol. Transactions of the Asme **111** (1989)
6. L.Hlavac, Theoretical model of abrasive liquid jet, Transactions of the VSB **47**, 73 (2001)
7. W. Li, J. Wang, H. Zhu, H. Li, C. Huang, On ultrahigh velocity micro-particle impact on steels–a single impact study, Wear **305**, 216 (2013)
8. J. Folkes, Waterjet- An innovative tool for manufacturing, J.Mater.Process.Tech. **209** 6181 (2009)
9. H. Li, J. Wang, An experimental study of abrasive waterjet machining of Ti-6Al-4V, Int. J. Adv. Manuf. Technol. **81**, 361 (2015)
10. D. Begic-Hajdarevic, A. Cekic, M. Mehmedovic, A. Djelmic, Experimental Study on Surface Roughness in Abrasive Water Jet Cutting, Procedia Engineering, **100** (2015)

MATEC Web of Conferences **94**, 03004 (2017) DOI: 10.1051/matecconf/20179403004
CoSME'16

Additive manufacturing – a sustainable manufacturing route

Domnița Frățilă[1,*], and *Horațiu* Rotaru[2]

[1]Technical University of Cluj-Napoca, Department of Manufacturing Engineering, Bd. Muncii 103-105, 400641 Cluj-Napoca, Romania
[2]Iuliu Hatieganu University of Medicine and Pharmacy, Department of Craniomaxillofacial Surgery, Cluj-Napoca, Romania

Abstract. Additive Manufacturing (AM) technologies allow developing and manufacturing very complex shaped parts and functional products with a high level of customization, being a great alternative to Traditional Manufacturing (TM) methods like injection molding, die-casting or machining. Due to the importance of cleaner production in the field of manufacturing processes, sustainability of AM processes needs to be assessed in order to make easier its acceptance and implementation in the industry. Furthermore, the manufacturers can improve their competitiveness and profitability by considering the ecological aspects during the manufacturing step of a product. This paper gives a survey on sustainability issues related to AM.

1 Introduction

The relevance of the sustainable development and particularly the environment-friendly technologies is already well-known. Nowadays the manufacturing companies are forced to manufacture and deliver cheap and quality products while increasing environmental performance.

According to these, it is strongly necessary to correlate research subjects from Romania with topical international tendencies. Basically, it is needed to set the aspects to be considered for implementation in the practice of new techniques and methods. In the machine building area these deal with influence factors on efficiency of sustainable technologies and their efficient implementation.

A comprehensive analysis of sustainability must consider every step of the product life cycle, from raw materials to disposal at the end of product life, including manufacturing stage, when quantities of materials and energy are consumed [1-3]. Thus, green manufacturing is one step towards sustainable development (Figure 1) [4].

Several investigations have been carried out in the area of machining processes considering environmental impacts. Some studies have been conducted in order to evaluate the manufacturing processes taking into account energy consumption, lubricants use, or lubricants removal. Other works have been made for reducing the lubricants consumption

* Corresponding author: domnita.fratila@tcm.utcluj.ro

by minimal quantity lubrication method, high pressure jet-assisted machining or cryogenic machining. Some other studies developed methodologies for ecological impact assessment of TM processes [5-8].

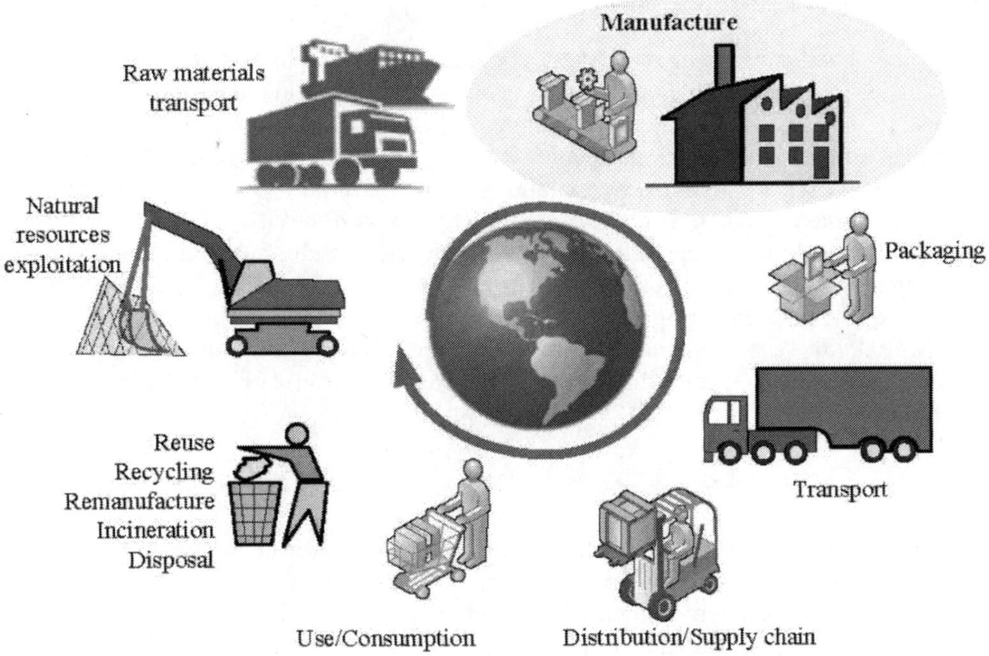

Fig. 1. Stages of product's life cycle [4].

Additive Manufacturing (AM) is a relative young technology that has the potential to fabricate geometrically complex customized products (functional parts and tooling) using less material and producing less waste. AM has the ability to build parts directly from digital models and represents an alternative to TM (e.g. machining, die-casting or injection molding) [2, 5, 9]. AM technologies are particularly suitable in the fields of medicine, automotive industry and aerospace industry where customization, parts' light weighting, and short supply chains are economically valuable [10].

Studies have shown that AM can be a viable option for the industry, but is not widely accepted yet, because of the lack of standards in AM domain. Thus, the use of AM for parts' production is impeded since in the industry the quality assurance depends on standards in selection of manufacturing processes and materials [11, 12].

Related to AM sustainability, only few researches have been conducted. Most of them are focused on electrical consumption of machines and equipment. The studies emphasize the correlation between sustainability and design quality, because the energy consumption depends on part geometry and orientation. Main AM design aspects to consider include: shape and mechanical properties of the part, surface finish, and costs [13-18].

Current practices based on human experience must be transformed into science-based practices through sustainability evaluation. A sustainable manufacturing assessment method to quantify a broad set of metrics needs to be developed using tools such as Design for Manufacturing and Assembly (DFMA) and Life Cycle Assessment (LCA).

LCA for parametric processes model has to be developed allowing estimating the environmental impact of manufacturing stage of parts/products made by AM technologies too. Such tool could improve the design of parts, processes and AM equipment with regard to reduce the environmental impacts generated through material flows, waste and energy consumption.

2 Ecological issues of AM

Sustainability characterization of AM as a part of industrial production chain is often difficult to do. The quantifiable dimension of such study should include ecological issues of AM related to materials and energy consumption, health and safety, transport and waste management and emphasizes the correlation between sustainability and design quality. The main AM design aspects to consider include: part strength, part flexibility, surface finish, enclosed voids, material cost, machine cost, and process productivity.

AM processes must demonstrate their environmental-friendly potential, by considering the sustainability principles: efficient use of material and energy, industrial waste management, low manufacturing costs, avoidance toxic emissions and materials, health and safety issues, low environmental impacts, improvement of personnel health, safety, economical efficiency, reparability, reusability, recyclability, and disposability of the products made by AM.

2.1 Energy

Despite its potential to promote cleaner manufacturing, AM cannot be regarded as an ecological-friendly manufacturing method yet, due to the high energy consumption by using heat processes or lasers to melt plastic and metal or to cure resins. AM equipment is generally not designed to be efficient. Energy loss is considerable and the heat management is poor. At mass-manufacturing scale, AM processes have higher impacts per part than TM. But this is not relevant, because they are replacing small batches of customized parts [19].

If the parts are manufactured by traditional manufacturing processes or 3D printing, the most important factor for environmental impact is the way how these methods are used. Any of these methods manufacturing only a part per week, but left on the rest of the time, could have higher impact than the same machine at maximal utilization [3, 5].

For TM, material use and waste is the largest impact. For AM electricity use dominates environmental impacts, because the energy usage per item is still very high in the manufacturing stage. The best way to reduce impacts of AM energy use is to reduce the run-time by considering some simple strategies for that: orient parts for the fastest printing, print tubular parts rather than solid; and (if possible) fill the printer bed with multiple parts [3, 5, 19].

2.2 Materials

Reducing the amount of material printed is beneficial for AM sustainability. AM uses several raw materials to create prototypes, parts or functional products based on 3D digital models by printing layers of materials, but a substantial amount of unused raw materials left behind of 3D printers.

The variety of materials used in AM includes: metals, polymers, ceramic or composite materials in forms of powders, wires and liquids. AM works with several sorts of materials including powdered or molten polymers (plastics) which are not ideal for environment (even they can be recycled) regardless of what kind of manufacturing techniques is involved. Rarely plastic by-products can be reused, but often the material properties are corrupted, making these materials no longer suitable for parts manufacturing. Some plastics are less pollutant than others. Therefore, standardized scales of flammability, toxicity, and reactivity must be consulted for choosing appropriate materials [3, 19].

The use of biodegradable printing materials or newly sand could be a solution. Using polylactic acid or polylactide (PLA, Poly) has more benefits than ABS. PLA is a biodegradable thermoplastic aliphatic polyester derived from renewable resources (corn starch, tapioca roots or sugarcane), being a bio-based polymer (less toxic) and needing lower temperatures in printing, hence affecting energy consumption. Due to all these properties, PLA is a promising bio-plastic, becoming a standard 3D printing material.

Even the toxicity may not be obvious, the health and safety issues in AM should additionally be considered. Toxic gaseous by-products are given off when plastic is heated to high temperatures and melted. The air quality inside the AM job shops needs to be analyzed in order to evaluate the impact of gaseous emissions and ultra-fine particles (UFP) emissions in industrial-scale environment. 3D printers emit low level of UFP, which could cause negative effects on human health including: lung function changes, airway inflammation, enhanced allergic problems, accelerated atherosclerosis, altered heart rate [19].

Materials for Fused Deposition Modeling (FDM) systems seem to be non-toxic developed from a wide range of commercially available thermoplastics. The melting temperatures of these materials should not be exceeded for avoiding the fumes produced during processing. The health problems caused by fumes or by post-processing operations can be: skin, eyes, and respiratory tract irritations.

2.3 Life cycle

The environmental impact of products fabrication involves several stages through product life cycle, starting with natural resources exploitation to product disposal, beyond manufacturing process [5, 19].

The transport and end of life of the machines (both 3D printers and machine tools) represent a small portion of impacts, amortized by intense utilization, but, if only few parts are made every week, those embodied impacts can be significant.

AM can change the product life cycle by shortening the supply chains and by reducing the fuel amount consumed to ship products. Traditional production target the areas of low labour costs, often far away from the markets where the products are consumed. With AM, the production can be close to the product consumer. This shortening of the supply chain reduces the transport costs associated with it and with the pollution and roads congestion [9].

2.4 Waste management

The environment state and the growing of the global consumer economy should be well balanced. Nowadays AM technologies become more widely used in many industrial sectors. Their environmental impact will depend on how these manufacturing methods are used [19].

Compared to conventional manufacturing approaches, AM may have environmental benefits because it does not require tooling. Thus, innovative designs can be created without tooling putting limits on the shapes.

Unfortunately, the opportunity to print quickly a series of variations of a product design can encourage a new kind of pollution by rapid waste generation. A critical AM issue is reuse and remanufacturing of the parts/products.

There is almost no information about waste flows associated with polymeric and metallic AM processes. Some of these flows add actually no value to the part such as: SLS powder refresh, FDM support structure materials, post-process heat treatment for reducing residual stresses or energy loss from inefficient laser and optical systems.

FDM machine can have negligible waste but only if the model does not need any support material while printing. The inkjet 3D printer wastes 40% of its ink without counting supports material [5, 19]. Depending on geometry and orientation, the support could be more mass than the final part, and this waste is difficult to be recycled.

Summarizing, 3D printers are not really less wasteful. Their waste is not necessarily recyclable and not important compared to their electricity use.

3 Potential sustainability of AM

When compared with TM methods, there are some potential sustainability benefits to AM implementation for part production in the industry [1, 3]:

- AM improves the efficiency of raw materials use (in powder, liquid or wire form) through its feature referred as net shape manufacturing;
- In addition to cost-efficiency and high freedom in design, AM could become an energy efficient and environmentally friendly manufacturing route due to high rate of material utilization;
- AM can be beneficial from an ecological perspective leading to elimination of fixed tooling, reduced material requirements and reduction of raw materials diversity needed, reduced waste from manufacturing processes;
- AM helps developing parts or products for optimal performance including reduced weight, good mechanical properties and optimal designed parts by incorporating gas flow paths, heating or cooling channels;
- Products made by AM can be 50% lighter than the same products made through TM. Performance of the final products will be also improved by increasing of energy and fuel efficiency during their life cycle (e.g. by less fuel consumption in aerospace industry) and less pollutant emissions in distribution stage;
- AM allows the fabrication of customized parts in small batches at the right moment and in the place of need (next to the customer), reducing the transportation costs within the supply chains and transportation pollution. Therefore, AM influences for logistics industry are important, with losses for transport sector but positive consequences for environment;
- Shorter product development cycles meaning: less employee-day per part, which indirectly means lower energy consumption.

4 Proposed research methodology

The main study goal consists of the predictive tool development that will allow the sustainability evaluation and comparison of the manufacturing processes. The special emphasis should be on evaluation and modeling of the potential environmental impacts in AM. Until now researches have been conducted mainly on electricity consumption of machine tools, considering "in process consumption" and "standby consumption". But a realistic study must take into account all consumed flows of materials, energy, fluids for an accurate impact assessment.

The study proposes the development of a predictive assessment method to evaluate the environmental impacts considering the materials, fluids and energy consumption during the manufacturing step of a part, unused material and parts recycling. The decision-makers need a simplification into single LCA scores by balancing energy and material use, emissions, toxicity, air pollution, water pollution, waste, etc. The single-score methodology normalizes and weights the different impacts (kg of CO_2 and NO_x, ppm of particulate matter, etc.) into generic units called "points". More points mean more environmental impacts.

The correlation between technological aspects of manufacturing processes and associated ecological issues has to be solved. The study should be conducted by a comprehensive quantitative comparison of TM processes and AM technologies to determine which method is most sustainable.

Six major elements that affect the sustainability of manufacturing processes may be taken into consideration: *manufacturing cost, energy consumption, and waste management* are easily measured, while the last three: *environmental impact, personnel health, and operator safety* are not easily to quantify.

Considering the above presented aspects, sustainability assessment method will be developed within many stages and steps, as shown in Figure 2.

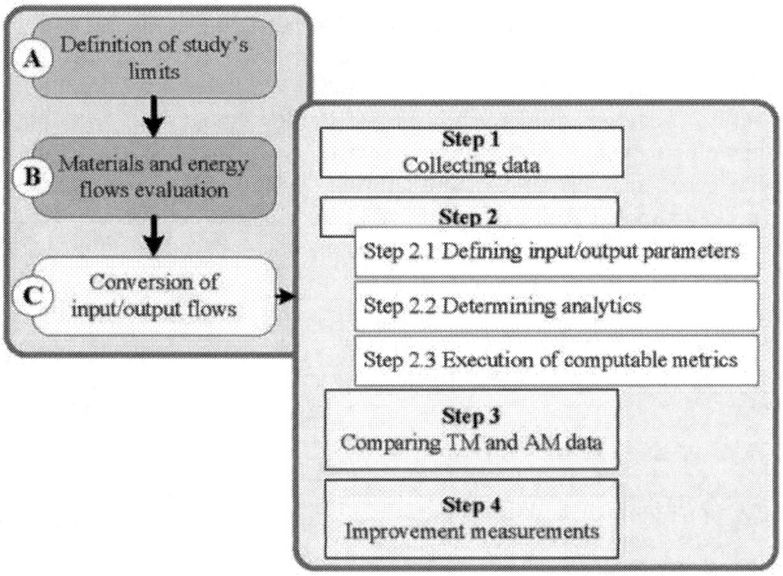

Fig. 2. Study stages.

A. Definition of study's limits. The limits of the analysis have to be clearly defined. Some elements could be neglected, otherwise all the products used in part fabrication should be considered.
B. Evaluation of consumption for all flows. All aspects of manufacturing process from raw material to final part must be examined.
C. Conversion of input/output flows into environmental impact.

The model may provide a framework for assessing and comparing sustainability for different manufacturing process including AM within 4 main steps.

Step 1 involves understanding the process mechanism and collecting relevant data.

Step 2 consists of: 2.1 Defining input (use of material, energy and other resources) and output indicators (product/part, solid waste, liquid waste, emissions) of performance and their computable metrics, 2.2 Determining analytics used for sustainability calculation, 2.3 Applying specific data of manufacturing process and enabling execution of computable metrics. Life cycle inventory data specific to the process will be considered.

Step 3 Comparing AM data generated against other manufacturing processes from industry (e.g. machining, injection molding, etc).

Step 4 Improvement measurements, based on results from previous step.

A framework for developing sustainable manufacturing metrics and potential interactions among metrics should be established. Metrics cover environmental, economic and social aspects and measure the inputs and outputs of a manufacturing process at a workstation. Workstation level measurements focus on a single machine performing one or more operations, or a piece of auxiliary equipment providing a specific function or doing a specific job with certain tools and materials under particular operations.

Preliminary LCA of AM including energy use, materials consumption in the final parts, material wasted, emissions, transportation and end-of-life disposal of the machines and parts will be performed and compared with conventional/TM processes.

5 Case study

The purpose of the case study is to analyze the environmental impacts of two additive manufacturing machines (SLS Sinter Station 2000 and FDM 1650) and a traditional computer numerical control milling machine (HAAS-VF2) from Manufacturing Engineering Department (TU Cluj-Napoca), in order to evaluate their sustainability.

The case study, considering only material and energy consumption, is the first step of methodology development for producing accurate LCA to compare various manufacturing methods. Some elements of the manufacturing systems will be neglected, but a comprehensive evaluation of their environmental performance is done by dividing the each whole process into individual elements such as: material preparation/utilization, energy consumption, material toxicity, and waste disposal. The factors taken into account in terms of environmental performance of manufacturing processes are summarized in Table 1.

Table 1. Analyse factors.

Technology	SLS	FDM	Machining
Machine	*Sinter Station DTM-2000*	*FDM-1650*	*HAAS-VF 2*
Material	powder of polyamide, ABS, titan, elastomers, polycarbonate	ABS, PLA, PVA, PC, HDPE, nylon, wax	metals, plastics (ABS), wood
Energy	high power laser beam	heat	mechanical energy
Operation	no tool, building step, infiltration	no tool, one step	cutting tools needed, multi-steps
Solid waste	material chips, unused material	removed supports, material chips	chips, tool scrap
Liquid waste	no	no	lubricants, coolants
Emissions	no	fume	tool particulate, fume, fluid vapors
Disposal	incineration, landfill, recycling	incineration, landfill, recycling	landfill, recycling

CNC machines energy use is difficult to measure and to attribute to different parts manufactured under different production condition. For CNC machines, about 80-85% of the energy used by machining equipment is constant, regardless if a part is produced or not and, thus, the energy demand per part is inversely proportional to the material removal rate. CNC machines often utilize auxiliary equipment whose energy use can exceed the cutting energy [20, 21].

The percentage of machine utilization is an important factor in its environmental impact to achieve low impacts, it is not enough that the machine processes a part quickly, but also that it spends as much of its time processing parts. The total processing time consists of basic time (stand-by mode), idle time (partial mode), and machining mode time (full mode).

Therefore the machine tool used for machining consists of various electricity-consuming elements such as: consumption in the idle state (52%), cutting energy (7-10% according to the cutting conditions), spindle rotation (20%), stage movement (1%), coolant pump (20%), numerical control, pressured air, or light [20].

For both AM equipments, Sinter Station 2000 and FDM 1650, the energy consumption is directly dependent on part geometry, finish quality and the manufacturing time that must be minimized for each case. System energy usages in terms of the energy consumption rates (ECR) have been determined. For SLS system the laser, heating, and cooling systems are the greatest contributors to energy used. The process energy is divided into non-processing energy (roller movement - 15%, piston movement by stepper motors - 25% and initial heating - 37%) and laser processing energy (16-20%). Specific energy consumption is difficult to prescribe because of the variance in build density and height [22]. In the present study case 100% build density has been considered. For the calculations of energy consumption rates (ECR), only the main processes are considered for the three manufacturing methods. The processing of the material is considered to be done continuously.

Sustainability assessment indicator is calculated by equations (Eq.1- Eq. 6). Electrical energy consumption (kWh), material consumption (kg) and fluid consumption (l) are converted in Pt using characterization factors (f_{ch}) [23] according to EcoInvent and EcoIndicator 99 databases [24, 25].

$$SI = C_E + C_M + C_F + C_W \tag{1}$$

where SI – sustainability indicator

C_E – contribution of electrical energy consumption

C_M – contribution of material consumption

C_F – contribution of fluid use

C_w – waste contribution

$$C_E = f_{ch-E} \cdot ECR \tag{2}$$

$$ECR = \frac{P}{PP} = \frac{P}{q_{mat} + \rho_{mat}} \tag{3}$$

where $f_{ch\text{-}E}$ – electricity characterization factor

ECR – energy consumption rate ($kWh \cdot kg^{-1}$ or $kWh \cdot mm^{-3}$)

P – electric power consumed during manufacturing (kW)

PP – process productivity [kg material (added or subtracted)$\cdot h^{-1}$)

q_{mat} – quantity of material (deposed or subtracted) per hour ($cm^3 \cdot h^{-1}$)

ρ_{mat} – material density ($kg \cdot cm^{-3}$)

ECR concept is used to facilitate comparison among manufacturing processes, being defined as the energy consumed in the production of material unit. For subtractive processes, ERC is defined as joule (or kWh) per unit volume of material ($kWh \cdot cm^{-3}$). For the case of SLS and FDM, ECR is usually expressed as joule (or kWh) per unit mass of material ($kWh \cdot kg^{-1}$), because the material density is influenced by the process parameters.

$$C_M = [e_n + k \cdot (1 - e_n)] \cdot d_p \cdot t_{man} \cdot f_{ch-M} \tag{4}$$

$$C_M = (V_{wp} - V_p) \cdot f_{ch-M} \tag{5}$$

where $f_{ch\text{-}M}$ – material characterization factor

e_n – nozzle efficiency

k – weighting factor to weight the impact of lost powder compared with fused powder

d_p – powder flow rate (kg material (added or subtracted)$\cdot h^{-1}$)

t_{man} – manufacturing time ($cm^3 \cdot h^{-1}$)

 DOI: 10.1051/matecconf/20179403004

ρ_{mat} – material density ($kg \cdot cm^{-3}$)
V_{wp} – workpiece material volume
V_p – part material volume

$$C_F = (d_c + d_f) \cdot t_{man} \cdot f_{ch-G} + d_{co} \cdot t_{man} \cdot f_{ch-CO} \tag{6}$$

where d_c – carrying gas flow
d_f – protecting gas flow
d_{co} – coolant flow
$f_{ch\text{-}G}$ – gas characterization factor
$f_{ch\text{-}CO}$ –coolant characterization factor

LCA analyze is conducted using EcoIndicator 99 method and SimaPro 7 software. This method considers the following impact inventory categories: carcinogens, respiratory organics, respiratory inorganics, climate changes, radiation, ozone layer, ecotoxicity, acidification/eutrophication, land use, and fossil fuels.

Environmental impacts may cause damage to human health and ecosystems, and deplete natural resources including: climate change, ozone depletion, toxicity, etc. To make appropriate decisions related to manufacturing processes, these aspects are evaluated even if they seem less concrete than data such as "kg of waste" or "kWh of energy use" and being achieved by LCA, using types of impacts normalized and weighted to calculate single score metrics [24].

The scores are expressed in Eco-Indicator points (Pt), which allows comparing the environmental impacts due materials, fluids and electrical energy consumption that have not the same unit. The size of the Pt unit represents one thousandth of the yearly environmental load of an average citizen in Europe [24]. Several variables are first normalized to disability adjusted loss of life years for humans, probable loss of species per year for environmental impacts, and loss of resource availability as measured in financial cost of resources, before the final normalization to points [25].

6 Results and discussions

Due to the limitation of materials usable by AM, it is not possible to consider the same materials for the case study, but for making fair comparison, similar materials are considered: ABS P400 for FDM process and machining and polymer PA12 for SLS. Table 2 indicates the ECRs, calculate for every manufacturing method. Processing parameters considered are: V – scanning (drawing) speed (mm/sec); W – road width size (mm); T – layer thickness (mm); ρ - material density ($kg \cdot mm^{-3}$); P - power rate (kW); k – process overhead coefficient (0.6-0.9), PP – process productivity = $V \cdot W \cdot T \cdot \rho \cdot 3600 \cdot k$ ($kg \cdot h^{-1}$). Some information from data bases and literature have been used.

The environmental performances are calculated for 1 kg material processed, making one part per day and lefting machines idling the rest of the time. The following assumptions have also been considered: ABS milling is done without cooling/lubrication (dry machining), an average consumption of support material up to 25% from part material for FDM, FDM is printing parts at 100 percent fill, SLS machine prints with 100 per cent model material, postprocessing, disposal of the parts and waste have been neglected.

 DOI: 10.1051/matecconf/20179403004

Table 2. ECRs of equipments.

	Sinterstation 2000	FDM 1650	HAAS-VF2
Material	Polymer PA12 (100% solid part)	ABS P400 (100% solid part)	ABS P400
Support material		Terpolymer (mix acid, styrene, buthyl acrylate)	
Material density, ρ (g·cm^{-3})	polymer powder	1.04	1.04
Scanning speed,V (mm·sec^{-1})	3000	2	
Road width size, W (mm)	0.4	1.4	
layer thickness , T(mm)	0.15	0.4	
Specific gravity	1.08	1.05	
K	0.6	0.9	0.9
P (kW)	16.8	1.32	22.4
PP (kg·h^{-1})	0.42	0.004	influenced by process parameters
ECR	40 (kWh·kg^{-1})	346.5 (kWh·kg^{-1})	13.9·10^{-6} (kWh·mm^{-3})
EI for energy (kWh^{-1})	0.57	0.57	0.57

Environmental performances of processes, under conditions and assumptions mentioned in the previous section and Table 2, are presented in Figures 3-5.

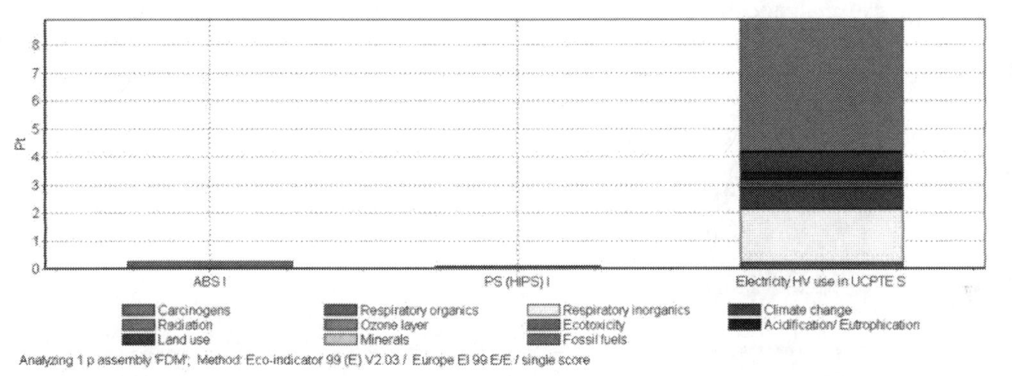

Fig. 3. FDM sustainability indicator.

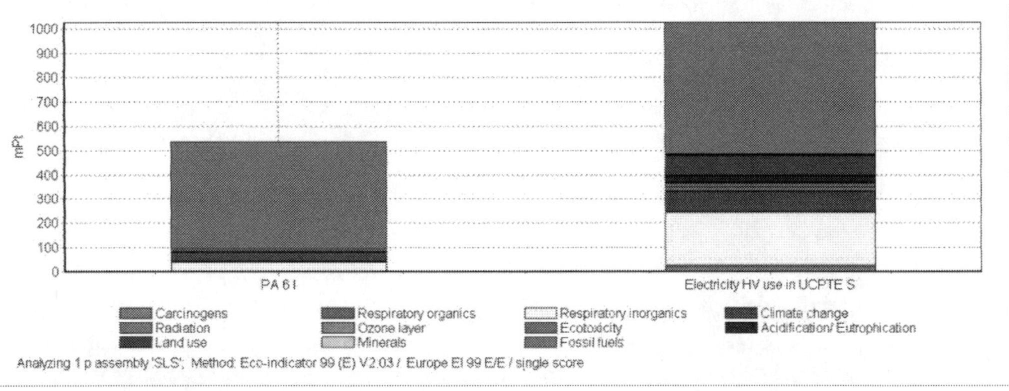

Fig. 4. SLS sustainability indicator.

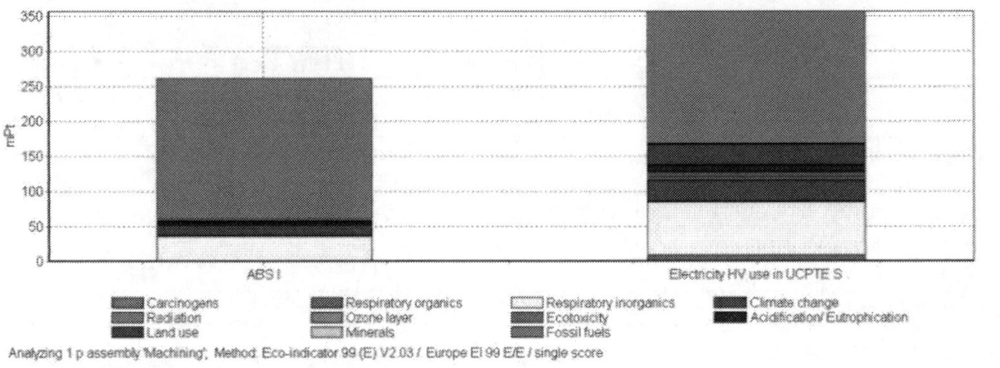

Fig. 5. Milling sustainability indicator.

Diagrams in figures presented above show the individual impacts for FDM, SLS and milling processes. It should be noted that of 11 types of impacts considered by EI 99 method, 4 are dominant for all cases: fossil fuels, respiratory inorganics, land use, and climate changes. At maximum utilization without considering the waste (chips and metalworking fluids), machining scored well for both material usage (260 mPt) and energy consumption (350 mPt) compared with SLS (550 mPt/1020 mPt) and FDM 0.2 Pt/9 Pt).

The comparison of sustainability indicators of the manufacturing methods considered is presented in Figure 6. The meaning that AM is generally more sustainable than substractive methods because it does not waste as much materials is not confirmed. The AM energy used can exceed the saving in material impacts. Some 3D printers still have high material impact because of producing about 40 per cent waste and using material with high toxicity.

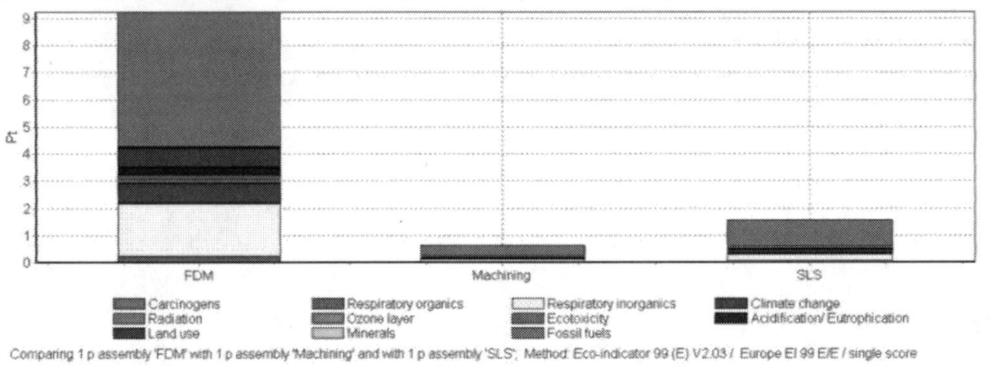

Fig. 6. Comparison of sustainability indicators per inventory subcategories.

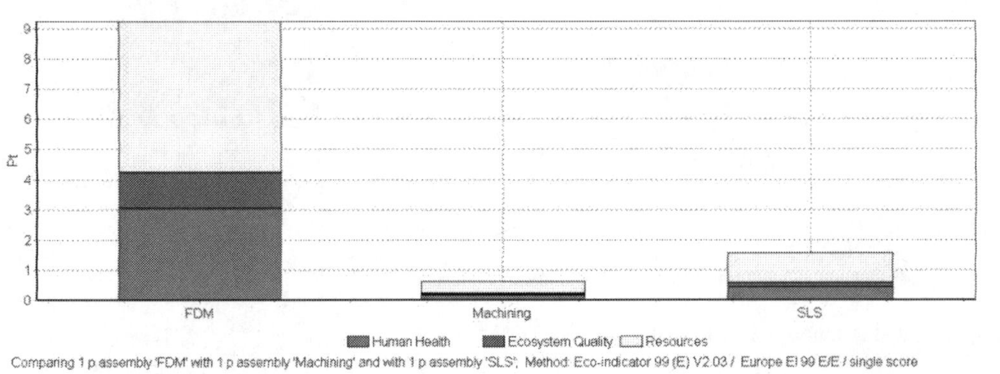

Fig. 7. Comparison of sustainability indicators per inventory main categories.

This case study indicates the relative sustainability of AM and TM manufacturing methods. As Figure 7 shows, FDM impact seems to be 10 times higher than milling's impact and 5 times than SLS. These results can be validate just across the specific scenario considered, being influenced by local conditions proposed and symplifying assumptions described above.

It can not categorically state that AM is more sustainable than machining or vice-versa. For each process the research has to be continued, because the results could vary in a wide range according to: the manufacturing strategy, usage scenario of the machines, build time related to part volume and part complexity, part orientation, etc.

7 Conclusions

The key knowledge gained from this investigation is given below.

AM has the potential to lower costs and to be more energy efficient than conventional processes. One main issue to analyze the energy efficiency of AM is the fact that 3D machines vary in energy consumption and that energy use is dependent on a several variables, including materials, load and patterns used.

It is strongly necessary an individual comprehensive evaluation of each manufacturing on order to achieve a realistic comparison of the processes.

The sustainability of AM vs TM depends primarily on the utilization rate of the machines. A high utilization reduces the idle energy use and amortizes the impact of each machine.

For both AM and CNC machines the electricity use has the dominant impact. Both cause a certain amount of material waste (from original material being removed, from support material, from unused powder at SLS), but for CNC (at maximum utilization) material waste and metalworking fluids become dominant. The electricity use is a largely time-dependent.

Within the next research step the manufacturing processes need to be analyzed considering the part geometry, part orientation, finish quality through their effects on the processing time.

The impacts and perspectives of study can be summarized as following.

Implementation of AM processes can be impeded by inability to compare AM performance against TM methods. Planning decisions for an appropriate selection of manufacturing process and material for specific application require a better knowing and assessment of these elements. Due to growing environmental concerns for manufacturing processes, it needs to analyse the potential environmental impacts of AM by understanding, comparing and characterizing the AM processes for sustainability.

The study result is an efficient tool, contributing to the requests in the field of environmental pollution related to both traditional and additive manufacturing processes, with a significant ecologic and economic impact by decreasing material and energy inputs, improving process results, minimizing waste.

Traditional manufacturers and AM job shops owners may make an informed decision about which technology to use or purchase. Furthermore, the manufacturers of 3D printers can improve their competitiveness and their profitability by considering the ecological aspects during the manufacturing step of a product.

Comprehensive assessment method, based on sustainability metrics, should address the needs of all stakeholders, facilitate growth and innovation.

This paper was supported by the AMaTUC Project, receiving funding from the European Union's Horizon 2020 Research and Innovation Programme under grant agreement No 691787.

References

1. K. Kellens, R. Renaldi, W. Dewulf, J. Kruth, J.R. Duflou, Rapid Prototyping J **20**, 459–470 (2014)
2. F. Le Bourhis, O. Kerbrat, L. Dembinski, J. Hascoet, P. Mognol, Procedia CIRP **15**, 26–31 (2014)
3. F. Le Bourhis, O. Kerbrat, J. Hascoet, P. Mognol, Int J Adv Manuf Tech **69**, 1927–1939 (2013)
4. O. Roş, D. Frăţilă, *Ecoproiectare / Ecodesign* (Casa Cărţii de Ştiinţă, Cluj-Napoca, 2007)
5. J. Faludi, C. Bayley, S. Bhogal, M. Iribarne, Rapid Prototyping J **21**, 14–33 (2015)
6. F. Pusavec, J. Kopac, J Ad Prod Eng Manag **3**, 151-160 (2009)
7. F. Pusavec, P. Krajnik, J. Kopac, J Clean Prod **18**, 174-184 (2010)
8. F. Pusavec, D. Kramar, P. Krajnik, J. Kopac, J Clean Prod **18**, 1211-1221 (2010)
9. M. Kreiger, J.M. Pearce, ACS Sust Chem Eng **1**, 1511–1519 (2013)
10. P. Berce et al, *Aplicaţiile medicale ale tehnologiilor de fabricaţie prin adăugare de material / Medical applications of Additive Manufacturing technologies* (Editura Academiei Române, Bucureşti, 2015)
11. M. Baumers, C. Tuck, R. Wildman, I. Ashcroft, E. Rosamond, R. Hague, Transparency Built-In. J Ind Ecol **17**, 418–431 (2013)
12. M. A. Krieger, M.L. Mulder, A.G Glover, J.M. Pearce, J Clean Prod **70**, 90–96 (2014)
13. D. Baechler, J. DeVuono, J.M. Pearce, Rapid Prototyping J **19**, 118-125 (2013)
14. S. S. Muthu, M.M. Savalani, *Handbook of sustainability in additive manufacturing*, (Publisher Springer Singapore, 2013)
15. P. Ratnadeep, S. Anand, J Manuf Syst **31**, 429-437 (2012)
16. R. Sreenivasan, A. Goel, A., D.A. Bourell, Physics Procedia **5**, 81–90 (2010)
17. M. D. Tabone, J.J. Cregg, E.J. Beckman, A.E. Landis, Envir Sci Tech **44**, 8264−8269 (2010)
18. B. T. Wittbrodt, A.G. Glover, J. Laureto, J., G.C. Anzalone, D. Oppliger, J.L. Irwin, J.M. Pearce, Mechatronics **23**, 713−726 (2013)
19. S.S. Muthu, M.M. Savalani, *Handbook of sustainability in additive manufacturing* (Publisher Springer Singapore, 2016)
20. J.B. Dahmus, T.G. Gutowski, ASME IMECE, **5** (2004)
21. A. A. Munoz, P Sheng, J Mater Process Tehnol **53**, 736-758 (1995)
22. R. Sreenivasan, A. Goel, D.L. Bourell, Phys Procedia **5**, 81-90 (2010)
23. Y. Luo, Z. Ji, M.C. Leu, R. Caudill, Inst Electr Electron Eng,1-6 (1999)
24. http://www.ecoinvent.org/database/database.html
25. https://www.pre-sustainability.com/download/EI99_Manual.pdf

Identification of in-line defects and failures during Additive Manufacturing Powder Bed Fusion processes

Simina Fulga[1,2,*], *Arjana* Davidescu[2], and *Ira* Effenberger[1]

[1]Fraunhofer Institute for Manufacturing Engineering and Automation IPA, Department Machine Vision and Signal Processing, 70569 Stuttgart, Germany
[2]Politehnica University Timisoara,Mechatronics Department, 300222 Timisoara, Romania

Abstract. Additive Manufacturing (AM) processes are enablers of new approaches in the field of production and design engineering, product design and business modelling. Beginning to view additive manufacturing in an industrial environment, reliable statements about the product quality are indispensable. Statements regarding compliance with geometric tolerances and exact quantifiable physical parameters, in terms of product certification are therefore imperative. The quality of the components must not only be sustainably secured but also reproducible at any time. Quality control and quality assurance are the prerequisite for highly customized unique parts, or even batch size 1 product, that can be produced by additive manufacturing as efficiently as conventional mass-produced parts. This paper will discuss an approach for the identification of in-line defects and failures during Additive Manufacturing Powder Bed Fusion processes using the example of the Selective Laser Sintering process.

1 Introduction

Additive Manufacturing (AM) processes are enablers of new approaches in the field of production and design engineering, product design and business modelling. The need for additive production is constantly increasing (see Fig. 1.), as the AM processes are the only ones which allow digital fabrication of products that simultaneously require: *functional integration* - means implementing as many technical functions as possible into as few parts as possible [1]; *complex geometries* – from free-form surfaces to weight-based topologies [2]; and *individualisation* - a strong personalization and adaptation to the customer`s needs being possible at any time.

Nowadays not only plastic and metal components are additively manufactured, but also function-integrating solutions for fibre-reinforced composites [3]. In the near future multiple materials will be used in one build process and even application-specific cavity structures and functionally gradient materials will be possible.

*Correspondingauthor: Simina.Fulga@ipa.fraunhofer.de

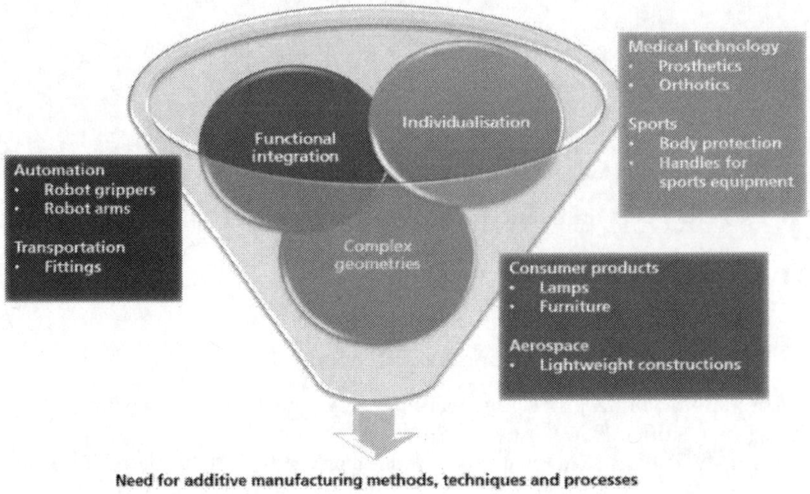

Fig. 1. Individualised products with complex geometries and integrated functionalities require an additive production.

Beginning to view additive manufacturing in an industrial environment, reliable statements about product quality are indispensable. Statements regarding compliance with geometric tolerances and exact quantifiable physical parameters, in terms of product certification are therefore imperative. The sooner these parameters can be obtained or even in-process influenced, the higher the efficiency of AM processes is. The quality of the components must not only be sustainably secured but also reproducible at any time.

Quality Control (QC) and Quality Assurance (QA), meaning the immediate product respectively process monitoring, are the prerequisite for highly customized unique parts, or even batch size 1 products, that can be additively produced as efficiently as conventional mass-produced parts.

Selective Laser Sintering (SLS) and Selective Laser Melting (SLM) are for industrial contenders, ranging from the medical field to automotive and aerospace applications, the two most important AM Powder Bed Fusion (PBF) technologies. PBF is one of the seven categories of AM processes, as defined in ISO/ASTM 52900-15 [4].

The primary focus of research in measurements for real-time control of AM PBF has been associated with determining the geometry and the temperature profile of the heat affected zone, which is the area near and including the melt-pool that is directly affected by high local temperatures [5]. For the in-line quality monitoring the question, still without a comprehensive scientific answer, is: which are the overall effects of the influencing parameters (see Fig. 2.) on AM processes having an appearance during the manufacturing process itself? Only when these effects are clearly determined, the defects and failures during AM PBF processes can be accurately defined and used for the development of an overall in-line QC system followed by an in-situ optimization of PBF processes.

It is widely known that the relationships between the parameters in the PBF process are complex [5], a fact which hinders the development of in-line QC and in-process optimization systems. In this context the scope of this paper is to identify and determine the in-line defects and failures during AM PBF processes, using the example of the SLS process. The focus on especially this AM PBF process is driven by the fact that SLS is one of the main technology that can be used to produce functional parts, even for low series production.

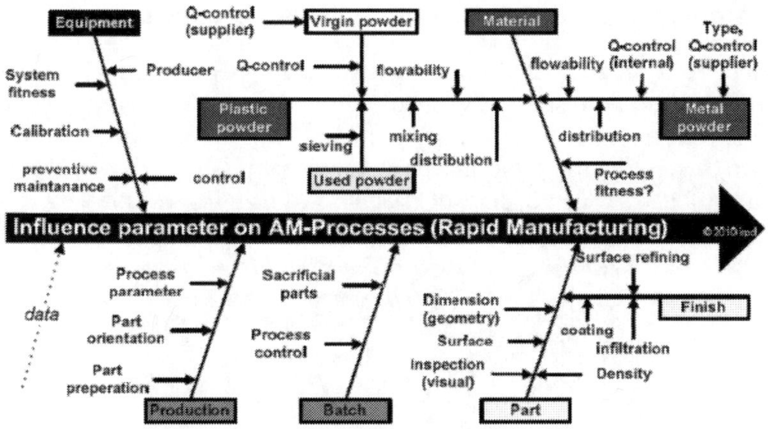

Fig. 2. AM PBF production chain - influencing parameters [6].

2 Identification and determination of in-line defects during the SLS manufacturing process

2.1 Quality management for additive manufacturing

An obstacle for a future industrial acceptance of AM is the fact that neither Quality Standards nor a generally accepted Quality Management (QM) Standard or System for the AM processes are nowadays available.

The quality of the manufactured parts and their designed functionality (e.g. grippers, see Fig. 3.) can be very easily impaired, taking into consideration the variety of the parameters that are influencing the AM process.

Fig. 3. 2-Finger angular gripper from Fraunhofer IPA.

Up to date preventing and/or detecting all defects and faults in PBF AM manufactured products is nearly impossible because the processes are influenced by numerous parameters, that can be summarized in four QM categories: *Equipment*, *Material*, *Production* including Batch and *Part* including Finish (see Fig. 2.) [6].

Starting from these QM categories, a comprehensive approach for the determination of the in-line defects and failures detectable during the manufacturing process was developed and implemented in order to set the premises for an overall in-line QC respectively for an in-process monitoring system for the SLS processes.

2.2 In-line defects during the SLS manufacturing process

Due to a deductive approach based on a qualitative top-down Fault Tree Analysis, going through each QM category, respectively subcategory, the corresponding quality influencing factors and respectively their manufacturing effects have been identified. The in-process appearances of these effects are the in-line defects (see Fig. 4.) during the SLS process, defining the overall tasks for an in-line QC system and being elementary for a future in-situ process optimisation.

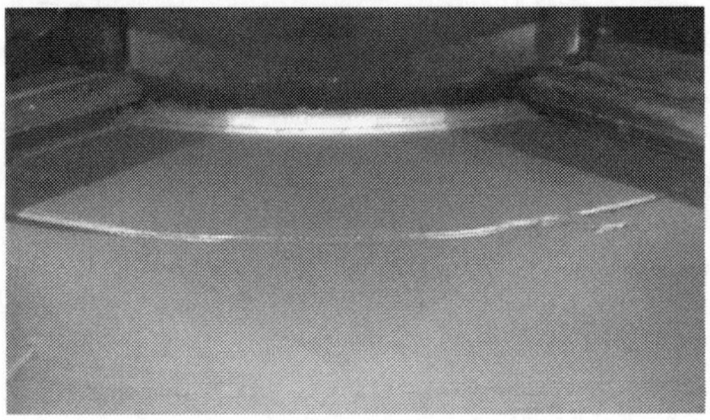

Fig. 4. Example of grave defects in the powder layer during the SLS manufacturing process: view of the powder layer after application [source: Fraunhofer IPA EOS FORMIGA P 100].

2.2.1 Equipment

Three QM subcategories were identified: uncleanness of the system, system fitness influenced by errors of system components and the system performance, influenced by inadequate functionality of the system (see Table 1). For example, taking into consideration the uncleanness of the system: one of the quality influencing factors in this case are the depositions on the laser window, which can directly influence the laser power respectively the elongation at break and the strength of the built part.

The in-process appearances of these are not only the deposits on the laser windows themselves but also geometrical distortions and even inappropriate layers' adhesion. These are the in-line defects respectively the tasks of the future in-line QC system.

Table 1. Equipment – in-line defects and failures as identified future tasks for an in-line QC system.

Quality Management aspects for AM		Quality influencing factors	Effects of the factors	In-line defects and failures
Equipment	Uncleanness of the system	deposits on laser window	low laser power, small elongation at break, low strength of the part	*inappropriate layers' adhesion, geometrical layer and part distortions, deposits on laser window*
		impurities	strength and even geometry of part impaired	*impurities in layer*
	System fitness influenced by errors of system components	laser and optical system	to low or to high laser power, small elongation at break, low strength of the part, thermal damage in case of too high laser power	*inappropriate layers' adhesion, geometrical layer and part distortions, melting of the edge zone depending on the geometry, porosity*
		temperature	geometry and	*inconstant*

		control	strength of part vitiated	*temperature*
		inert gas supply	burning of the layer	*melting of the part, black sintered*
		wear parts (e.g. seals, material supplier)	powder supplier damaged	*streaky, scaly, powder layer*
			tilted build platform	*tilted layer geometry*
			build platform got stuck or lost the height information because of defects of the stepper motor	*layers' overlapping, layers' sintering failed*
Equipment	System performance influenced by inadequate functionality of system	powder application performance	homogeneity of the generated powder layer affected	*inhomogeneity of the layer thickness*
			density variation of the powder layer	*density variation of the powder layer*
		Scaling	scaling factor variations	*geometrical deviations and distortions of the sintered layer*
		tolerances	inadequate tolerances	*geometrical deviations of the sintered layer*
		beam offset	inadequate layer sintering	*geometrical deviations and distortions of the sintered layer*
		low scan speed	high porosity and low density	*porosity, inhomogeneity*
		surface roughness	layer and part surface roughness variation	*layer and part surface roughness*

2.2.2 Feedstock

The feedstock is the bulk raw material supplied to the AM building process [4]. Two QM subcategories were identified: the virgin and the used powder. Exemplary, taking into consideration the virgin powder: the grain shape and its size distribution, as one of the quality influencing factors, are influencing the homogeneity of the applied powder layer and of the final part as well as the strength of the SLS manufactured object.

The in-process appearance of the effects of the quality influencing factors is the grain size. The quality influencing factors, their effects and the identified in-line defects and failures are presented in Table 2.

Table 2. Feedstock – in-line defects and failures as identified future tasks for an in-line QC system.

Quality Management aspects for AM		**Quality influencing factors**	**Effects of the factors**	**In-line defects and failures**
Feedstock	Virgin powder	grain shape and grain size distribution	inhomogeneity of the generated powder layer and final part, strength of the part impaired	*different grain sizes, grain shapes levels*

		thermal properties like melting point and recrystallization	inadequate layer sintering	*inconstant temperature on sintering point, inappropriate layers' adhesion*
		bulk density	affected processability	-
	Used powder	type and mesh size of sieve (preferred mesh 140 µm)	grain size variation, strength of the part vitiated	*different grain size, inhomogeneity of the powder layer*
		Melt Flow Index (MFI)	mechanical strength	-

2.2.3 Production

In this case also two QM subcategories were identified, namely: the part preparation and orientation as well as the process parameters during manufacturing (see Table 3). Also in this case, considering the atmosphere in the process chamber, as one of the quality influencing factors, the burning of the layer was deducted as effect of this factor.

The in-process appearance of these effects is the melting of the part respectively a "black sintered" layer.

Table 3. Production – in-line defects and failures as identified future tasks for an in-line QC system.

Quality Management aspects for AM		**Quality influencing factors**	**Effects of the factors**	**In-line defects and failures**
Production	Part preparation and orientation	resolution of the STL file	part accuracy affected	*high surface roughness*
		part orientation	inadequate sintering	*part and layer orientation failure*
	Process parameter during manufacturing	laser power	inadequate sintering	*inappropriate layers' adhesion, geometrical layer and part distortions, melting of the edge zone depending on the geometry, layer porosity*
		scan, sintering speed	inadequate sintering	*inconstant sintering speed*
		scan, sintering line	inadequate sintering	*scan line deviation*
		temperature profile	inadequate layer sintering	*inconstant temperature on sintering point, inappropriate layers' adhesion*
		layer thickness	part homogeneity affected	*variation of layer thickness, inhomogeneity*
		laser exposure style	-	-
Production	Process parameter during manufacturing	hatch distance	overlap area between two hatch lines	*inappropriate hatch distance, inappropriate layers' adhesion*
		atmosphere	burning of the layer	*melting of the part, black sintered*

		skywriting	melting of the edge zone is not correct	*skywriting length variations, melting of the edge zone, geometrical layer and part distortions*

2.2.4 Part

The QM subcategories, with respect to the Total Quality Management (TQM) of the part, are the part protocol and the quality check. For example, considering the recording of the entire product manufacturing process for each part produced, as one of the quality influencing factors, we deduct as an effect the fact that a part quality report is unaccounted for the end-user of the respective SLS part. If the SLS produced part has complex functionalities integrated, and not only then, a quality protocol of each layer respectively of the object as a whole is imperative. In this case the in-line task for a QC system is the recording of each and any quality part information in a protocol.

The quality influencing factors, their effects respectively the in-line tasks for a QC system are presented in Table 4.

Table 4. Part –identified future tasks for an in-line QC system.

Quality Management aspects for AM		**Quality influencing factors**	**Effects of the factors**	**In-line QC system tasks**
Part	Part protocol	recording the entire product manufacturing process for each part	no reproducibility and reliability is given	*collect any quality part information in a protocol*
	Perform quality check	incomprehensive quality control	part distortions, vitiated mechanical strength, etc.	*identification of all in-line defects and failures (see Table 1 to Table 3)*

2.2.5 Results

In Table 5 all in-line defects and failures, as future tasks for in-line QC and in-situ optimisation of SLS processes, are summed up in connection with the quality influencing factors.

These results will be the basis for correction and/or mitigation of the failures on the SLS processes with respect to the quality aspects. Therefore a systematic, rigorous Design of Experiments (DoE) approach will be implemented in order to ensure the generation of valid conclusions for the ranking of the severity and probability of the quality influencing factors, respectively of the deducted in-line defects and failures.

Table 5. In-line defects and failures during AM PBF processes using the example of SLS.

Quality influencing factors	**In-line defects and failures during the SLS process**
deposits on laser window	*inappropriate layers' adhesion, geometrical layer and part distortions, deposits on laser window*
impurities	*impurities in layer*
laser and optical system	*inappropriate layers' adhesion, geometrical layer and part distortions, melting of the edge zone depending on the geometry, porosity*

temperature control	*inconstant temperature*
inert gas supply	*melting of the part, black sintered*
wear parts (e.g. seals, material supplier)	*streaky, scaly powder layer*
	tilted layer geometry
	layers overlapping, layers sintering failed
powder application performance	*inhomogeneity of the layer thickness*
	density variation of the powder layer
scaling	*geometrical deviations and distortions of the sintered layer*
tolerances	*geometrical deviations of the sintered layer*
beam offset	*geometrical deviations and distortions of the sintered layer*
low scan speed	*porosity, inhomogeneity*
surface roughness	*layer and part surface roughness*
grain shape and grain size distribution	*different grain sizes, grain shapes levels*
thermal properties like melting point and recrystallization	*inconstant temperature on sintering point, inappropriate layers' adhesion*
type and mesh size of sieve	*different grain size, inhomogeneity of the powder layer*
resolution of STL file	*high surface roughness*
part orientation	*part and layer orientation failure*
laser power	*inappropriate layers' adhesion, geometrical layer and part distortions, melting of the edge zone depending on the geometry, layer porosity*
scan, sintering speed	*inconstant sintering speed*
scan, sintering line	*scan line deviation*
temperature profile	*inconstant temperature on sintering point, inappropriate layers' adhesion*
layer thickness	*variation of layer thickness , inhomogeneity*
hatch distance	*inappropriate hatch distance, inappropriate layers' adhesion*
atmosphere	*melting of the part, black sintered*
skywriting	*skywriting length variations, melting of the edge zone, geometrical layer and part distortions*

3 Conclusions

This paper presents an approach to identify the in-line defects and failures during Additive Manufacturing Powder Bed Fusion processes using the example of SLS.

The obtained results, having as objective to locate risks and thus to assure the quality of the final SLS manufactured product, will merge in the development of a production accompanying QC system applied in order to monitor and control in real-time SLS processes. This in-line QC system will not only cover, as up to date, the in-process measurements of surface temperature, residual stress and geometry [5], but also all other in-process appearances of quality influencing factors for SLS processes. At the same time the results are the background for an in-situ optimisation system.

Future work will be a DoE for a SLS production facility in order to rank the severity and probability of the quality influencing factors, respectively of the in-line QC and in-situ optimisation system tasks. An action plan to solve the ranked tasks, in form of a hardware and software platform will be completed.

References

1. J. Breuninger, R. Becker, A. Wolf, S. Rommel, A. Verl, *Generative Fertigung mit Kunststoffen: Konzeption und Konstruktion für Selektives Lasersintern [Additive Manufacturing withplastics: design andconstructionforSelective Laser Sintering]* (Springer Vieweg, 2012)
2. R. Becker, A. Grzesiak, A. Henning, Assem. Autom. **25**, 262-266 (2005)
3. R. Geiger, S. Rommel, J. Burkhardt, T. Bauernhansl,wt-online **3**, 169-174 (2016)
4. ISO/ASTM 52900-15, *Standard Terminology for Additive Manufacturing – General Principles – Terminology*, ASTM International (West Conshohocken, PA, 2015)
5. M. Mani, B. Lane, A. Donmez, et al., *Measurement Science Needs for Real-time Control of Additive Manufacturing Powder Bed Fusion Processes* (NIST Interagency/Internal Report-NISTIR 8036, 2015)
6. M. Schmid, G. Levy, DDMC (2012)

Conceptual approach for an in-line quality control system in Additive Manufacturing Powder Bed Fusion processes

Simina Fulga[1,2,*], *Arjana* Davidescu[2], and *Ira* Effenberger[1]

[1]Fraunhofer Institute for Manufacturing Engineering and Automation IPA, Department Machine Vision and Signal Processing, 70569 Stuttgart, Germany
[2]Politehnica University Timisoara, Mechatronics Department, 300222 Timisoara, Romania

Abstract. Additive Manufacturing is one of the genuine hopes for the forth industrial revolution since digital data is controlling the whole layered production process. At the same time the geometric freedom and tool-free production assures a high degree of individualisation. But to be the driving force behind a new industrial revolution, a qualification of additive manufacturing processes is necessary so that the resulting products meet the required quality and safety standards in the different fields of application such as in handling technology or medical technology. This paper will discuss a conceptual approach for the development of an in-line quality control system in Additive Manufacturing Powder Bed Fusion processes using the example of the Selective Laser Sintering process.

1 Introduction

Additive Manufacturing (AM) Powder Bed Fusion (PBF) processes, one of the seven categories of AM processes, as defined in ISO/ASTM 52900-15[1], open up new possibilities in terms of geometrically and functionally optimised parts and assemblies. At the same time the geometric freedom and tool-free production assures a high degree of individualisation. However, can AM technologies really find their way into industrial environments? Can highly customized unique parts be additively produced as efficiently as conventional mass-produced parts? Can we additively manufacture batch size 1 products? Nevertheless the unpredictable quality and reliability of additive manufactured parts, due to e.g. geometrical deviations, part distortions, delamination as well as the lack of reproducibility, are underlining the fact that Quality Control (QC) and Quality Assurance (QA) must be further developed in order to manufacture a product which is "fit for purpose" and "right first time“.

As far as industrial applications of additively manufactured parts are concerned, ranging from the medical field (e.g. patient specific parts, implants, instruments and even organs) to the automotive industry (e.g. ventilation inlet, see Fig.1.), the two most important AM PBF technologies are the Selective Laser Sintering (SLS) and Selective Laser Melting (SLM).

* Corresponding author: Simina.Fulga@ipa.fraunhofer.de

Fig. 1. Ventilation inlet for the automotive industry [source: RTC University Duisburg-Essen].

The scientific needs for the QA chain of the SLS and SLM product manufacturing process can be summarised with respect to:

• *Data-management*: at this time there is no uniform data management. There are no standards that define the digital process chain for the different additive manufacturing facilities. Thus, for example, geometrical and process data are widely mixed in machine-specific formats. There is no possibility to perform a recording of the entire product manufacturing process for each part produced. Therefore, the reproducibility is not provided in the AM production.

• *Real-production integration*: today additive production facilities are usually independently working laboratory machines that are not integrated in a process chain. The process variables are partly recorded during the manufacturing process but not long-term logged.

• *3D-input-data* and their effect on the quality of the AM parts: the Surface Tessellation Language (STL) format is the de facto industry standard, despite recent data formats such as "Additive Manufacturing File Format" (AMF) or the "3D Manufacturing Format" (3MF). Up to date, there are no standards or guidelines that describe and/or specify the required quality characteristics for the STL models. 3D input data affect massively the quality of the final product and must therefore be suitable prepared with the requirements of a production process [2]. This is often carried out by manual, time and costs spending, post processing steps.

• *In-line QC* and *in-process optimization*: by the parts, components or objects which are additively manufactured, quality problems appear during the production process caused by different factors. These are leading to vulnerabilities, fractures, or product failure [3]. There is no AM in-line approach which at the same time:

– can detect and classify comprehensively such errors,
– monitors and optimizes in real-time the AM processes,
– offers a complete quality report of the produced part, and
– can abort the production process in extreme cases.

In this context the scope of this paper is to develop a conceptual approach for an in-line QC system for SLS processes; SLS technology-based additive systems being nowadays the ideal solution for the production of parts with integrated functionality and even for low series production.

2 In-line QC of SLS processes

2.1 QC in AM PBF processes

Up to now the major international machine builders for AM systems are supporting completely insufficiently the efforts of the users in terms of Quality Management (QM) and QA for AM production applications. Only the aerospace industry has driven, with extremely great effort, individual solutions for the manufacturing of metallic components without finally debating this topic.

For a future industrial acceptance of the AM processes, as standard production processes, appropriate activities towards worldwide Quality Standards were started by ISO and ASTM.

The manufacturer itself is blindly dependent, at least on the equipment and on the powder quality. As service provider, always responsible towards the client for the expected quality, he has no direct control of the quality and the reproducibility of the produced parts. Therefore an overall in-line QC system during the manufacturing process is more than essential, not only for the manufacturer itself but also for all end users of the AM parts.

Typical errors and quality problems that may occur in the additive production and which especially affect the additive manufactured end-products are:

• lack of geometric accuracy of the parts which is dependent on e.g. the raw material conditioning, the temperature control, the laser offset, the cooling process, the layer thickness or the slicing procedure;
• component distortions caused e.g. by the cooling process and the temperature control in the machine;
• fluctuations in the quality depending on the placement of the parts in the space of the machine;
• reduced mechanical strength: deviating density through e.g. too low laser power or too high laser speed;
• surface defects: aged material, unsuitable material mixture, contamination by extraneous substances;
• closure of narrow or deep channels and holes in the component by partial melting of the marginal zone, dependent on the geometry.

2.2 Tasks for the in-line QC system

Starting from the defects and failures having an appearance during the manufacturing process [4], the future overall tasks for an in-line QC system for the SLS processes are:

• to identify all in-line defects and failures presented in Table 1, and
• to collect, during the additive process, all quality information in a part protocol.

Table 1. In-line defects and failures during AM PBF processes using the example of SLS [4].

Quality influencing factors	In-line defects and failures during the SLS process
deposits on laser window	*inappropriate layers' adhesion, geometrical layer and part distortions, deposits on laser window*
impurities	*impurities in layer*
laser and optical system	*inappropriate layers' adhesion, geometrical layer and part distortions, melting of the edge zone depending on the geometry, porosity*
temperature control	*inconstant temperature*
inert gas supply	*melting of the part, black sintered*
wear parts (e.g. seals, material supplier)	*streaky, scaly powder layer*

wear parts (e.g. seals, material supplier)	*tilted layer geometry*
	layers' overlapping, layers' sintering failed
powder application performance	*inhomogeneity of the layer thickness*
	density variation of the powder layer
scaling	*geometrical deviations and distortions of the sintered layer*
tolerances	*geometrical deviations of the sintered layer*
beam offset	*geometrical deviations and distortions of the sintered layer*
low scan speed	*porosity, inhomogeneity*
surface roughness	*layer and part surface roughness*
grain shape and grain size distribution	*different grain sizes, grain shapes levels*
thermal properties like melting point and recrystallization	*inconstant temperature on sintering point, inappropriate layers' adhesion*
type and mesh size of sieve	*different grain size, inhomogeneity of the powder layer*
resolution of STL file	*high surface roughness*
part orientation	*part and layer orientation failure*
laser power	*inappropriate layers' adhesion, geometrical layer and part distortions, melting of the edge zone depending on the geometry, layer porosity*
scan, sintering speed	*inconstant sintering speed*
scan, sintering line	*scan line deviation*
temperature profile	*inconstant temperature on sintering point, inappropriate layers' adhesion*
layer thickness	*variation of layer thickness , inhomogeneity*
hatch distance	*inappropriate hatch distance, inappropriate layers' adhesion*
atmosphere	*melting of the part, black sintered*
skywriting	*skywriting length variations, melting of the edge zone, geometrical layer and part distortions*

In order to develop a conceptual approach for an in-line QC system it is necessary to rank the quality influencing factors; their associated in-line defects and failures being the basis for correction of the failures in SLS processes with respect to the quality aspects.

For the ranking of the probability and severity of the quality influencing factors and of their corresponding effects [4], respectively of the in-line defects and failures, at least two factors are necessary and should be taken into account: their occurrence frequency and the sunk costs directly implicated. These two factors, especially the occurrence frequency, are strongly depending on the AM machine and on the material, the feedstock used. Therefore these two factors cannot be exactly determined for all AM PBF processes, but can be approximated for particular AM machines based on an e.g. Design of Experiments (DoE) approach.

At Fraunhofer IPA a DoE has been implemented in order to generate clear-cut conclusions for the ranking of the frequency of the in-line defects and failures. The parts' production process on a SLS machine has been monitored over a determined period of time. The overall defects and failures occurred over the screening period have been logged as well as their associated quality influencing factors and effects [4]. Thus the occurrence frequency of the quality influencing factors and their associated in-line defects and failures has been established. For each defect and failure logged over the screening period, the sunk costs in percent of one build cycle have been determined. For the sunk costs' factor the following data have been used: the year turnover of a fully loaded SLS production machine (newest generation) is about 1 Million €, calculated over an average of: 100 builds per year, with 40 parts per build and a part price of 250€; one build having a value of about 10.000€.

The ranking of the top in-line defects, to be identified by the in-line QC system, is presented in Table 2. The criteria for the ranking position were established taking into consideration two factors: the normalised value of the frequency of defects in percent of build cycles(F_1), and the normalised value of the implicated sunk costs, in percent of one build cycle turnover(F_2).The ranking scores R_s have been determined using the following function:

$$R_S = (w_1 * F_1/F_{1max}) + (w_2 * F_2/F_{2max}) \quad (1)$$

where$w_1 = 2$ and $w_2=1$ are the assigned weights, and F_{xmax} are the maximal values of the factors.

The interval of ranking scores has been split up in three. The ranking position values from Table 2 represent the interval to which the associated ranking score belongs to.

Table 2. Ranking of the top in-line defects to be identified by the in-line QC system.

Ranking position	**Frequency of defectsin percent of build cycles** (F_1)	**Sunk costs in percent of one build cycle turnover** *(F2)*	**Quality influencing factors, faults related, logged over the screening time**	**In-line defects to be identified by the in-line QC system** ⇔ tasks of the in-line QC system
1	<10%	≤100%	deposits on laser window	*inappropriate layers' adhesion, geometrical layer and part distortions, deposits on laser window*
2	<15%	<25%	impurities	*impurities in layer*
1	<10%	≤100%	laser and optical system	*inappropriate layers' adhesion, geometrical layer and part distortions, melting of the edge zone depending on the geometry, porosity*
2	< 2%	<90%	temperature control	*inconstant temperature*
2	<3%	<50%	inert gas supply	*melting of the part, black sintered*
1	<5%	100%	wear parts: material supplier	*streaky, scaly powder layer*
1	<5%	100%	wear parts: tilted build platform	*tilted layer geometry*
1	<5 %	100%	wear parts: build platform get stuck or lose the high information because defects of the stepper motor	*layers' overlapping, layers' sintering failed*
1	<7%	≤90%	powder application performance: inhomogeneity of the generated powder layer	*inhomogeneity of the layer thickness*
2	<5%	<60%	powder application performance: density variation of the powder layer	*density variation of the powder layer*

DOI: 10.1051/matecconf/20179403006

3	<1%	<50%	scaling: scaling factor variations	*geometrical deviations and distortions of the sintered layer*
3	<1%	<50%	tolerances: inadequate tolerances	*geometrical deviations of the sintered layer*
1	<20%	≤100%	improper beam offset	*geometrical deviations and distortions of the sintered layer*
2	<8%	≤80%	low scan speed	*porosity, inhomogeneity*
1	<20%	≤75%	surface roughness	*layer and part surface roughness*
1	<10%	≤90%	grain shape and grain size distribution	*different grain sizes, grain shapes levels*
1	<20%	≤100%	thermal properties like melting point and recrystallization	*inconstant temperature on sintering point, inappropriate layers' adhesion*
2	<3%	≤75%	type and mesh size of sieve	*different grain size, inhomogeneity of the powder layer*
3	<3%	≤25%	part orientation	*part and layer orientation failure*
1	<20%	≤100%	laser power	*inappropriate layers' adhesion, geometrical layer and part distortions, melting of the edge zone depending on the geometry, layer porosity*
1	<2%	≤100%	scan, sintering speed	*inconstant sintering speed*
1	<2%	≤100%	scan, sintering line	*scan line deviation*
1	<20%	≤100%	temperature profile	*inconstant temperature on sintering point, inappropriate layers' adhesion*
2	<2%	≤90%	layer thickness	*variation of layer thickness , inhomogeneity*
3	<1%	≤25%	hatch distance	*inappropriate hatch distance, inappropriate layers' adhesion*
2	<3%	<70%	atmosphere: burning of the layer	*melting of the part, black sintered*
3	<1%	<50%	skywriting	*skywriting length variations, melting of the edge zone, geometrical layer and part distortions*

2.3 Conceptual approach for solving the tasks of an in-line QC system

In order to develop a conceptual approach, the ranked tasks for the in-line QC system have been classified considering the AM PBF production steps. The results are presented in Table 3.

Table 3. In-line defects to be identified by the QC system during the AM PBF production steps⇔ basis for a firm foundation of a conceptual approach for an in-line QC system.

AM production steps	**In-line defects to be identified by the in-line QC system** ⇔ tasks of the in-line QC system (ranking position)
During all steps	deposits on laser window (1)
	inconstant temperature (2)

After each powder layer application	streaky, scaly powder layer (1)
	layer porosity (2)
	variation of layer thickness(2)
	density variations of the powder layer (2)
	impurities in powder layer(2)
	different grain size (2)
During sintering process of the powder layers	inconstant temperature on sintering point (1)
	melting of the edge zone depending on geometry (1)
	inconstant sintering speed (1)
	scan line deviation (1)
	melting of the part, black sintered (2)
	check the hatch distance (3)
	skywriting length check (3)
After sintering of each powder layer	layers' overlapping, layers' sintering failed (1)
	inappropriate layers' adhesion (1)
	layer and part surface roughness (1)
	geometrical deviations and distortions of the sintered layer (1)
	layer porosity (2)
	layer/part orientation within the build volume, only after sintering the first layer (3)
After part is finished sintered	geometrical deviations and distortions of the part (1)
	porosity analysis of the part over all layers (1)
	part surface roughness (1)

2.3.1 Hardware platform: appropriate technologies and sensors

The hardware (HW) platform must be a modular one, in order to solve all of the in-line QC system tasks identified. The modularity of the HW platform will allow in the future:
• to easily add other sensors in order to solve new quality issues that will appear with the development of the AM PBF machines
• to easily adapt the system to other AM PBF processes
• to allow a data fusion of the sensors on the Software (SW) side of the future system

The appropriate technologies, respectively the sensors' requirements for the multi-sensor analysis and field monitoring, have been identified regarding the identified tasks during the AM PBF production steps (see Table 3.) and the best price/performance ratio.

In Table 4all technologies are summed up in principle, in order to solve the classified tasks, exemplary for a SLS machine (Fa. EOS). Three IQCSM (In-line Quality Control System Modules) will be necessary. For exact requirements of the IQCSM components the inspected field, the working distance and the identified tasks must be grasped.

Table 4. Technologies and sensors necessary to solve all the classified tasks- exemplary for a SLS machine (Fa. EOS).

IQCSM no.	Sensors for the data and signal acquisition	Tasks	IQCSM *n* components
1	Machine Vision System	• must acquire *during all AM production steps* an image of the laser window as basis for ***the laser window clean check***	industrial image processing camera, lens, appropriate illumination *Obs. must be integrated in the machine, therefore needs a cooled housing*

DOI: 10.1051/matecconf/20179403006

2	Machine Vision System	• must acquire *after each powder layer application* at least one image of the powder layer, for the inspection of: ***streaky, scaly powder layer; layer porosity; impurities in powder layer and the powder grain size*** • must assure *during the sintering process* a "live stream" of the sintering process for the inspection of: ***melting of the edge zones; melting of the whole layer "black sintered"*** • must assure *after sintering of each powder layer* at least one image of the sintered layer as basis for the inspection of: ***layers' overlapping; geometrical deviations and distortions of the sintered layer; porosity; layer/part orientation within the build volume, only after sintering the first layer*** • must assure *after the whole part is finished* all the images corresponding to all sintered layers of the part in order to reconstruct from all the 2D images the 3D Real Model as basis for the 3D inspection of: ***geometrical deviations and distortions of the part on 3D level; porosity analysis of the part over all layers; part surface roughness***	industrial image processing camera, lens, appropriate illumination *Obs. If the camera is integrated in the machine needs a cooled housing; else must work through a dedicated window.*
3	Thermography system	• must assure *after each powder layer application* the necessary images for a Lock-in Thermography for the inspection of: ***homogeneity of the layer thickness; density variations of the powder layer; impurities in powder layer*** • must assure *during the sintering process* of the powder layers a "live stream" of images of the sintering process as basis for the inspection of: ***temperature variation on the sintering point, temperature distribution; scan speed variation; scan line deviation; check the hatch distance and skywriting length check*** • must assure *after sintering of each powder layer* the necessary images for a Lock-in Thermographyof the sintered layer for the inspection of: ***inappropriate layers' adhesion***	infrared camera, appropriate excitation source *Obs. If the camera is integrated in the machine needs a cooled housing; else must work through a dedicated window.*

 DOI: 10.1051/matecconf/20179403006

2.3.2 Design of the hardware architecture: hardware concept and interfaces

The hardware architecture is a modular one. New hardware modules can be added to the in-line QC system at any time in order to:

• adapt the system to other AM PBF machines
• solve new in-line quality tasks

The hardware concept and the corresponding interfaces are presented in Fig.2. Between the sensors identified and defined in chapter 2.2.1, respectively all the other sensors necessary for the system implementation (e.g. emergency stop button, machine door open), and between the PC, the communication is based on a Gigabit Ethernet (GigE) interface. The connection for the communication to the machine is established over a Programmable Logic Controller (PLC) based on PROFINET protocol and the connection for the Sensors (e.g. Cameras) with a GenICam (Generic Interface for Cameras) protocol. Both are GigE based protocols.

GigE compatible sensors will be connected directly over the ports of the GigE-Network Boards. The boards will have Power over Ethernet (POE) support. In this way the sensors with POE support (e.g. the machine vision systems) will be supplied with electricity directly over the standard Ethernet cables. A big advantage for the inline system: the data communication and the electricity will be assured only over one cable.

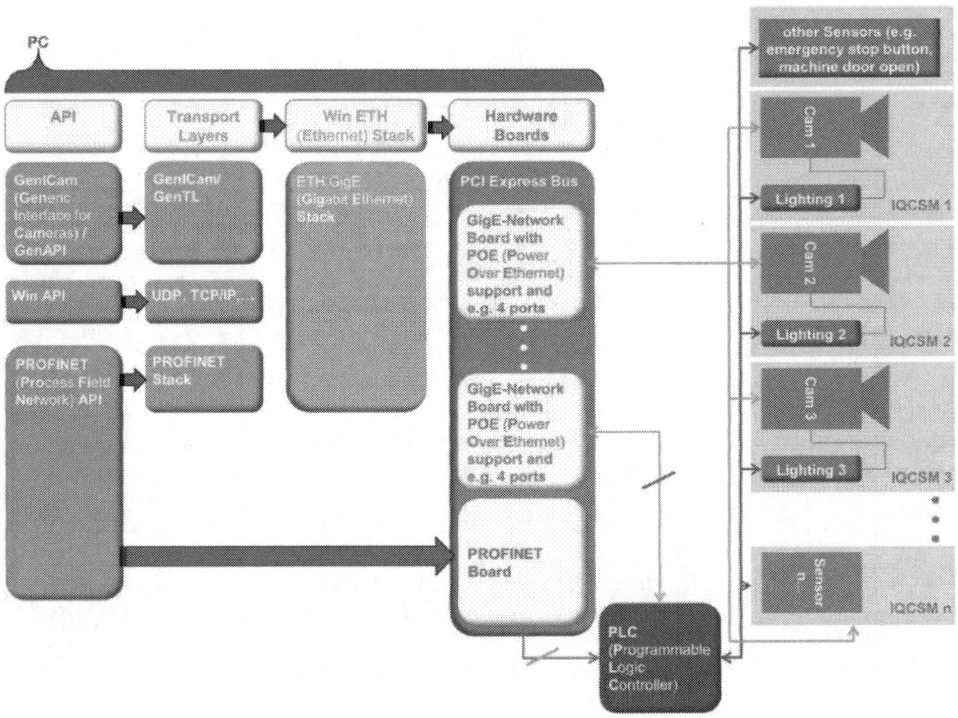

Fig. 2. Hardware concept of the modular in-line QC system for AM PBF.

The PLC, a digital computer used for the automation of the in-line QC system processes, can be connected over one port of one GigE Board or directly to a PROFINET Board. A PROFINET Board allows at least 90% of the PROFINET stack to work directly on the board. PROFINET has been described as the "all-encompassing Industrial Ethernet" since it can be used for virtually any function required in automation: discrete, process, motion, peer-to-peer integration, vertical integration, safety, and more. Because PROFINET uses standard IEEE 802.3 Ethernet, it inherently works over IEEE 802.11 wireless Ethernet.

2.3.3 Results

The premises for the implementation of an in-line QC system for AM PBF processes, using the example of the Selective Laser Sintering process, have been achieved in form of a conceptual approach. Following the developed approach, the ranked tasks of the planned system will be used as basis for the determination of the technologies and sensors necessary for the hardware platform, which will than directly flow into the concept of the hardware architecture, including the corresponding interfaces.

These results constitute the basis for the development and implementation of a software platform, including an automatic data evaluation platform, for an in-line QC system for the SLS processes.

3 Conclusions

This paper presents a conceptual approach for the development of an in-line QC system for AM PBF processes using the example of SLS.

The obtained results, having as objective to identify and determine the technologies and sensors necessary for an overall in-line detection of defects and failures during AM PBF processes, lay a firm foundation for the development and implementation of an in-line QC system on an AM PBF machine, applied to ensure the quality of the SLS manufactured parts. This QC system will assure, aside of the up to date in-process measurements of surface temperature, residual stress and geometry [5], the in-line identification of all in-process appearances of quality influencing factors.

Future work will be the development and implementation of a software platform, including an automatic data evaluation platform, for an in-line QC system for the SLS processes. All results achieved constitute the solid background of an in-line QC system, as basis for a future in-situ optimisation system, for the AM PBF processes.

References

1. ISO/ASTM 52900-15, *Standard Terminology for Additive Manufacturing – General Principles – Terminology*, ASTM International (West Conshohocken, PA, 2015)
2. J. Kroll et al., SPIE, **86500**, 86500N (2013)
3. G. Kreiseler, J. Kroll, S. Fulga, I. Effenberger, Euspen Int.Conf., **1** (2014)
4. S. Fulga, A. Davidescu, I. Effenberger, CoSME'16 (to be published)
5. M. Mani, B. Lane, A. Donmez, et al., *Measurement Science Needs for Real-time Control of Additive Manufacturing Powder Bed Fusion Processes* (NIST Interagency/Internal Report-NISTIR 8036,2015)

 DOI: 10.1051/matecconf/20179403007

Considerations on material thickness influence on the AWJ processing quality of an aluminium alloy

Eugen Herghelegiu[1,*], *Maria Crina* Radu[1], *Carol* Schnakovszky[1], and *Catalin Nicolae* Tampu[1]

[1] Vasile Alecsandri University of Bacau, Calea Marasesti 157, 600115, Bacau, Romania

Abstract. The aim of the current paper is to determine the influence of material thickness on the quality aspects of AWJ processing of the AL6061T651 alloy. The analysis is performed under different working conditions. The parameters taken into account to quantify the quality of processing are those defined in the ISO/WD/TC 44 N 1770 standard: width of the processed surface at the jet inlet (Li), width of the processed surface at the jet outlet (Lo), deviation from perpendicularity (u), inclination angle (α) and roughness (Ra). The obtained results emphasise a significant influence of material thickness on both, the surface quality and process productivity.

1 Introduction

Industrial development led to the discovery of new materials that cannot be processed by conventional methods, so that it was necessary to develop new technologies that allow the materials processing. The newest technologies are the unconventional technologies, which use the power of electricity, water, air, chemicals etc. Among them, waterjet processing is of great interest for different industrial fields since it can be used for processing of both, soft or hard metallic and non-metallic materials (the processing of hard materials requires the addition into the water jet of an abrasive material and the process is named abrasive waterjet processing - AWJ) [1,2]. Besides, this technology has many other advantages, such as: it is environmentally friendly, it does not degrade thermally the processed surface, in most of cases no additional processing is required, it allows obtaining various complex shapes (using 5-axis machines) since it is CNC controlled. However, it does has some disadvantages as the high cost of equipment, the high cost of maintenance and spare parts, the need for skilled workers, the impossibility to process pieces in package [3,4].

The parameters used to describe the quality of a cut performed by abrasive waterjet technology are defined by the ISO/WD/TC 44 N 1770 standard and are as follows (fig. 1): width of the processes surface at the je inlet (Li), width of the processed surface at the jet outlet (Lo), deviation from perpendicularity (u), inclination angle of the processed surface (α) and surface roughness (Ra) [6].

* Corresponding author: eugen.herghelegiu@ub.ro

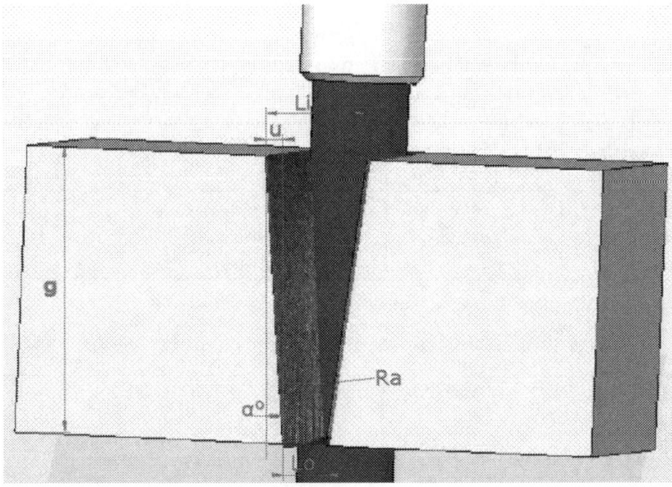

Fig. 1. Parameters used to quantify the quality of cut processed by AWJ [5, 6].

The aim of the current paper is to investigate the influence of the material thickness on the quality of cuts performed by abrasive water jet technology.

2 Experimental methodology

The experimental study was performed on the Al 6061 T651 alloy, with three thicknesses of samples: 1 mm, 6.5 mm and 10 mm. This material is widely used in the aeronautical industry for various components of the aircraft, where the parts must be of high quality, without any heat affected zone of the processed surfaces. The experimental tests consisted in performing simple linear cuts through the material thickness, which allow to measure the analysed quality parameters (presented in fig. 1). Measurements for roughness were taken in the middle of the thickness of cut surface (for the thickness of 1 mm). For 6.5 and 10mm, the measuring location was at 1 mm from the bottom surface (fig. 2).

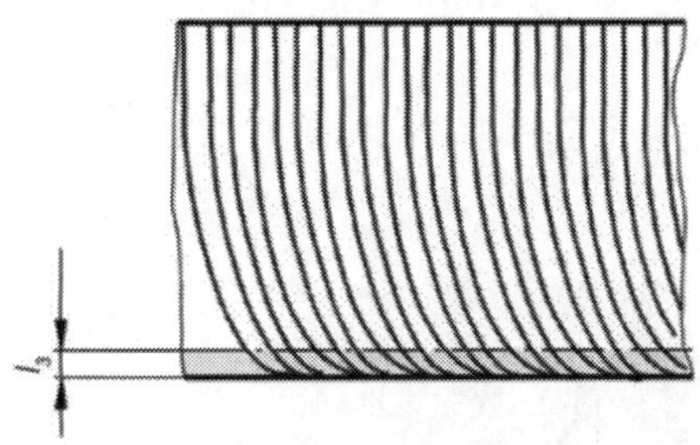

Fig. 2. Measurement of roughness (l_3 =1 mm for the thickness of cut surface equal to 6.5 and 10 mm).

The chemical composition of the material and its mechanical characteristics are presented in table 1 and table 2.

The experimental tests were performed on a Hydro-Jet Eco 0615 machine (fig. 3), equipped with a piston pump, capable of providing a maximum pressure of 150 MPa. Two process parameters were varied:

- feed rate (Vf), for which two values were used: 50mm/min and 100 mm/min;
- distance between the cutting head and sample (h), with also two values: 1mm and 3mm.

The parameters that were kept constant during the experiments are presented in table 3.

Table 1. Chemical composition of the Al 6061 T651 alloy.

Grade	Si	Fe	Cu	Mn	Mg	Cr	Zn	Ti	Other elements		Al
	%	%	%	%	%	%	%	%	each	total	%
Al 6061 T651	0.64	0.2	0.18	0.07	1	0.06	0.12	0.08	0.05	0.15	rest

Table 2. Mechanical characteristics of the Al 6061 T651 alloy.

Grade	Ultimate tensile strength, Rm		Elongation, A		Yield Strength, Rpo,2	
	[N/mm2]		[%]		[N/mm2]	
	min	max	min	max	min	max
Al 6061 T651	311.6	315.7	14	15	294.4	295.7

Table 3. Constant process parameters.

Parameter	Unit	Value
Water pressure, P	[MPa]	150
Quantity of abrasive material, Q	[g/min]	300
Orifice diameter, d_o	[mm]	0.35
Length of focusing tube, L_f	[mm]	76
Diameter of focusing tube, D_f	[mm]	1.02
Garnet Mesh	[#]	80

The quality parameters of cuts were measured by using a Leica MZ75 microscope with PC interface and a Mitutoyo SJ 201 digital roughness device.

Fig. 3. Water jet cutting machine Hydro-jet Eco 0615.

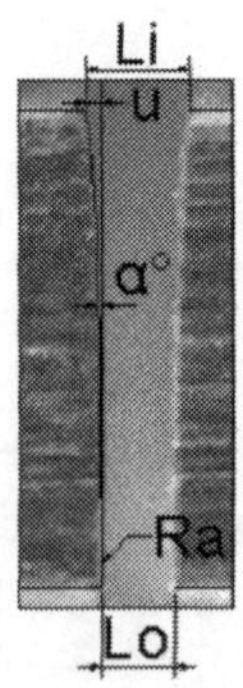

Fig. 4. Sample after cutting.

 DOI: 10.1051/matecconf/20179403007

3 Results and discussion

The results of measurements of the parameters that quantify the quality of cuts are comparatively presented in the graphs from figures 5, 6 and 7.

By analyzing the graphs from figure 5 a, b one can observe that as the material thickness increases, the width of the processed surface at the jet inlet (Li) decreases, with the same trend regardless the feed rate of the cutting head. However, the higher value of the feed rate led to narrower cuts; the explanation is related to the number of abrasive particle that get in contact with the sample material, which is becoming smaller as the feed rate increases. This effect is more visible in case of processing the sample of 1mm thickness, with the distance between sample and the cutting head of 3mm.

The graphs from figure 5 c, d show the variation of the width of the processed surface at the jet outlet (Lo) with the thickness of sample: the Lo parameter decreases as the sample is getting thicker. This effect is valid for both values of the feed rates as well as for both values of the distance between the cutting head and sample.

Distance between the cutting head and working sample, h=1mm | **Distance between the cutting head and working sample, h = 3 mm**

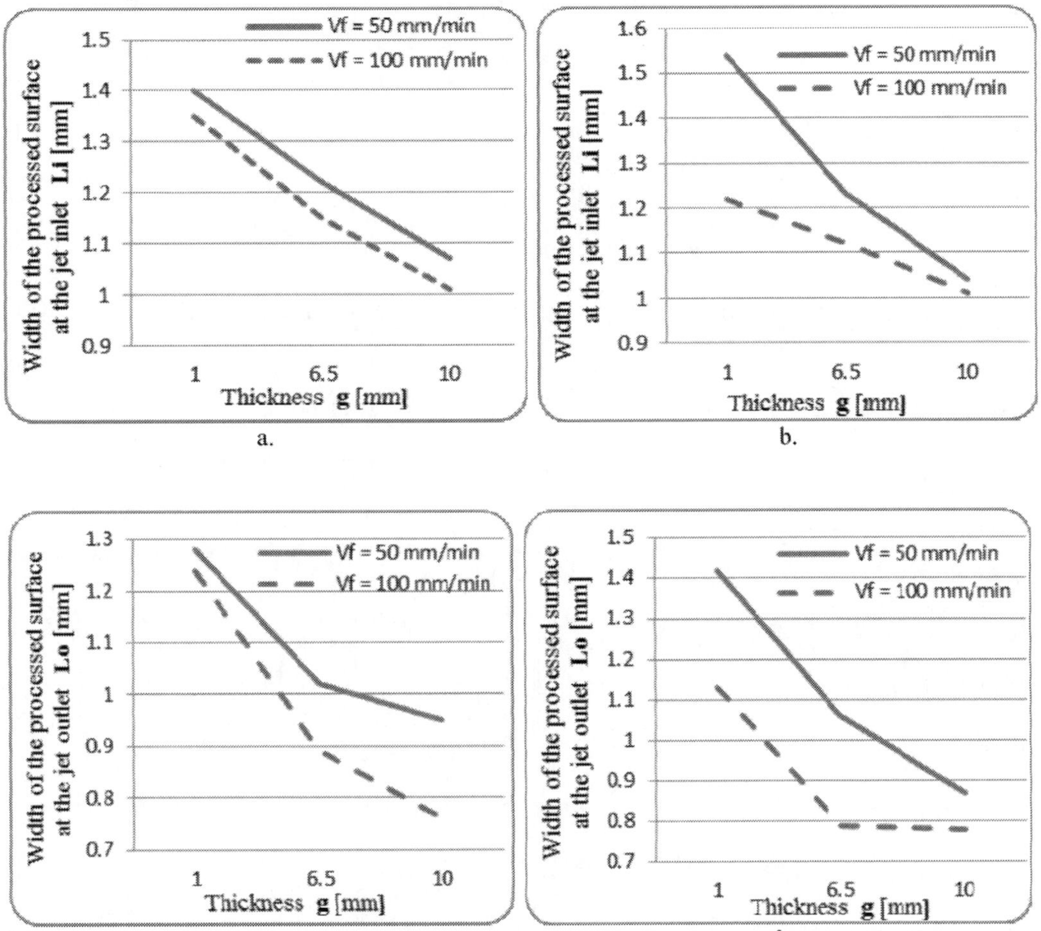

Fig. 5. Influence of the feed rate and distance between the cutting head and material on the Li and Lo.

DOI: 10.1051/matecconf/20179403007

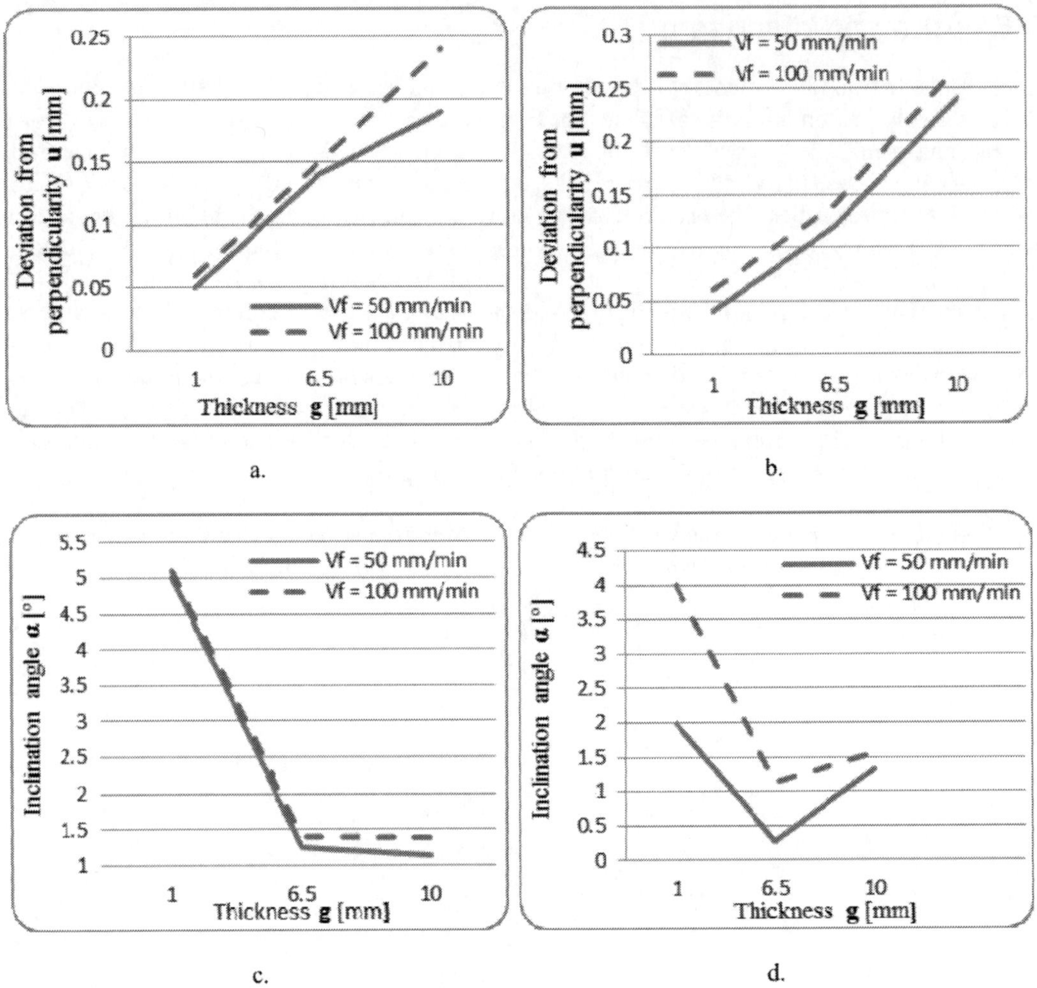

Fig. 6. Influence of the feed rate and distance between the cutting head and material on the u and α.

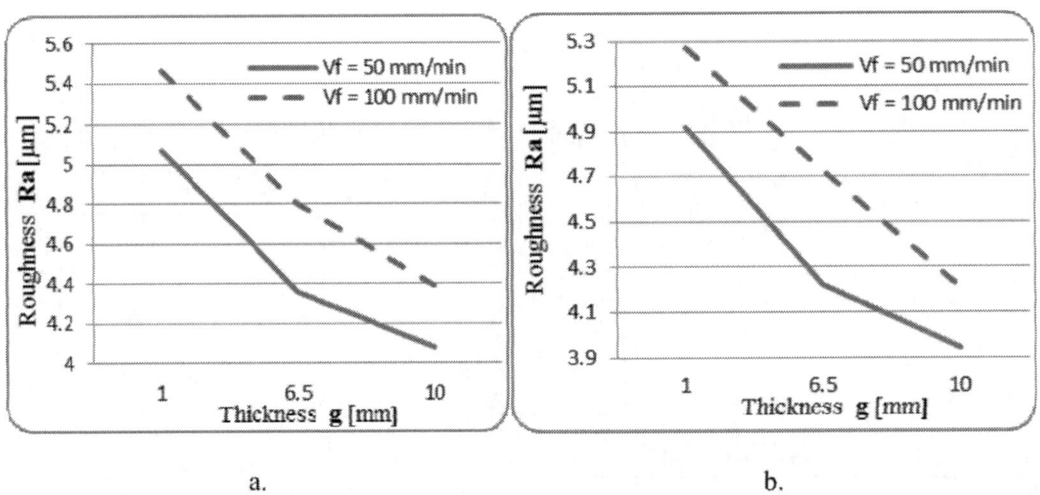

Fig. 7. Influence of the feed rate and distance between the cutting head and material on the roughness.

The deviation from perpendicularity (u) increases as the sample thickness increases (fig. 6 a, b). Smaller values of deviations were obtained in case of processing with a lower feed rate. The influence of the distance between the cutting head and sample is insignificant.

The inclination angle (α) decreases as the sample is thicker. This curious variation, (contrary to fig. 1) is due to the fact that the inclination is not the same on the cut thickness (fig. 4). The increase of the feed rate leads to an increase of the inclination angle while the effect is opposite when the distance between the cutting head and sample is getting bigger (fig. 6 c, d).

The roughness of the processed surface (Ra) is getting smaller as the thickness of sample increases. The increase of feed rate leads to an increase of roughness regardless the sample thickness, while a bigger distance between the cutting head and sample cause a slight reduction of material roughness (fig. 7 a, b).

4 Conclusions

The obtained results emphasize a significant influence of the material thickness on the quality of cuts performed by abrasive waterjet technology. All the analysed parameters are getting smaller, as the sample thickness increases, excepting the deviation from perpendicularity. This may be an important gain if we take into account the difficulty or even the impossibility of cutting hard materials with the conventional technologies. The effect is amplified (in terms of width of the processed surface at the jet inlet/outlet) when higher feed rates are used; on the contrary, the surface roughness is getting coarser as the cutting head is moving faster. The increase of feed rate means a higher productivity, leading in the meantime to good values of some parameters that quantify the quality of cuts (Li, Lo), even for thicker material. However, there are other quality parameters (u, α and Ra) whose variation with the feed rate is opposite. Therefore, the feed rate should be chosen so that to obtain an optimum between the quality of cut and process productivity. Further research will be performed in view of this. The distance between the cutting head and sample slightly affects the quality of cuts, its influence being more accentuate in case of processing with higher feed rates.

This paper was performed through the Partnerships program in PN II priority areas, developed with MEN-UEFISCDI (Romanian National Authority for Scientific Research) support, project no. 294/2014.

References

1. E. Herghelegiu, C. Radu, C. Schnakovszky, V. Zichil, Appl. Mech. Mater. **809-810**, 207 (2015)
2. P. Janković, M. Radovanović, Annals of the Oradea University. Fascicle of Management and Technological Engineering. **7**, 1528 (2008)
3. M. Chithirai Pon Selvan, N. Mohana Sundara Raju, Int. J. Eng. Sci. Technol. **7**, 254 (2011)
4. C. Schnakovszky, E. Herghelegiu, M. C. Radu, I. Cristea, Adv. Mat. Res. **837**, 196 (2014)
5. E. Herghelegiu, M. C. Radu, C. Schnakovszky, I. Cristea, Appl. Mech. Mater. **371**, 240 (2013)
6. *** ISO/ WD/ TC 44 N 1770 - 2010.

Basic research on lattice structures focused on the reliance of the cross sectional area and additional coatings

Julia Kessler[1,*], *Nicolae* Bâlc[2], *Andreas* Gebhardt[1], and *Karim* Abbas[1]

[1]Aachen University of Applied Sciences, Faculty of Mechanical Engineering and Mechatronics, 52064 Goethestr 1, Aachen, Germany
[2]Technical University of Cluj-Napoca, Department of Manufacturing Engineering, 400641 B-dul Muncii 103-105, Cluj-Napoca, Romania

Abstract. This scientific survey is about periodic lattice structures which are made by Selective Laser Melting (SLM). Selective laser melting is based on Additive Manufacturing. The increased use and increasing demand of lattice structures in different fields of applications forms the necessity of a closer look on complex structures. Lattice structures can be found in different fields of applications for example in lightweight applications, filters and heat exchangers. Because of the expanding of application areas and thus arising requirements, the quality improvement is indispensable. Additive manufacturing and especially the SLM process enable the manufacturing of highly complex shapes and structures. Further it allows the integration of lightweight structures within to be manufactured applications. These high performance structures and applications need specific boundary and process conditions [1-3]. The main aim of this survey is an extraction of important parameters concerning the shape of lattices. A first focus will be on mechanical properties and the therefore necessary tensile tests.

1 Introduction

Selective Laser Melting is a technology that is based on Additive Manufacturing. It can be divided into three process steps, the pre-process, main process and post-process. During the main process the product is printed. Therefore a metal powder is spread onto a building platform and then an outline is locally melted with a laser. On top of this layer a new layer of metal powder is applied followed by a laser melting step. This process is repeated until the whole shape is generated. The shape is defined by a CAD data. Almost no limit can be set on the complexity of shapes, what makes this kind of manufacturing technology unique compared to conventional technologies. Cavities or filigree structures like lattice structures can be generated.

There is a great potential in lattice structures like material and time saving and improving the mechanical properties of products. To exploit the maximum potential it is

*Corresponding author: kessler@fh-aachen.de

necessary to take a closer look at the process flow and shape properties of lattice structures 3-5].

Several researches already carried out a great variety of mechanical and microstructural test focusing on the properties of lattice structures. These kinds of investigations allow a fundamental understanding and the comparison of result. P. Li et al. for example analysed the deformation behaviour of lattice structures and used a reduced initial system to get the required data for FE analysis. For this purpose uniaxial tension tests were performed and the FE analysis adjusted. A comparison of a real uniaxial compression test and a prediction via FE analysis showed similar results. For more complex structures the prediction still showed disagreements [6].

A further focus was investigated by S. McKown et al. They dealt with the dynamic loading test on lattice structures. Therefore several unit cell types were tested with compression tests and blast tests. The conclusion extracted out of the survey shows that within dynamic loading, depending on the set parameters, the compression strength increases with the loading rate. Comparing the blast loading and quasi-static loading the deformation mechanism is similar [7].

The above-mentioned investigations are based on periodic lattice structures. Else to that Chunze Yan et al. took an interest in complex Schoen gyroid structures which as well are periodic with the difference that the structures are designed by circular struts and spherical cores instead of regular struts. Compression tests on this type of structures proved that there is an anisotropic behaviour depends on the orientation of the lattice structure. In context to that fact the mechanical properties are direction depended [8, 9].

These introduced investigations on different lattice structures and different types of unit cells should show the great potential and possible uses of those high performance structures.

2 Experimental set - up

The following study is aiming at the determination of a great range of impacts affecting 3D generated lattice structures as well as to link dependencies. The base for this survey is a preliminary study which is going to be expanded with different specimens and evaluation methods. In this preliminary study three different square tensile specimens (8x8x100 mm) were generated, using 316L stainless steel with an average grain size of 30-60 µm. Tests were carried out on a massive specimen, a lattice structure and a coated lattice structure. First tensile test opened up a great discrepancy within the tensile strength, which raised some new issues. Out of this survey new focuses were extracted. A particularly conspicuous aspect is the great reached tensile strength of the lattice structure. Below the experimental set up is described [10].

The Selective Laser Melting machine used to generate the specimens is the M1 Cusing which works with a 400 W fiber laser. As above described the tests are based on 316L stainless steel. In order to guarantee the same product quality and properties all specimens are built with the same process parameters. The scanning speed is set to 1450 mm/s for outlines and 1000 mm/s for surface exposure with the constant laser power of 180W. Regarding the exposure strategy the hatch distance is set to 80 µm. These adjustments have been extracted as the process optimum and will be no further discussed within this survey.

A series of different specimens and tensile test which are tested on a "Zwick" experimental machine, are the basis for this study. The range of samples covers different lattice structures, coated lattice structures, vertical struts and different rods with a similar cross sectional area. For each type four tensile specimens are printed. One specimen is intended as a reference to calibrate the tensile machine. The square rod shape is adjusted to the tensile machine and is not based on a DIN standard. Table 1 shows the four main

DOI: 10.1051/matecconf/20179403008

groups and different sections of the specimens and Table 2 shows the main characteristics of the different specimens and their unit cells.

Table 1. Sections of the specimens for the experimental set up.

Unit cells of lattice structures	Coated lattice structures	Specimens based on an almost same cross sectional area	Hollow body
A 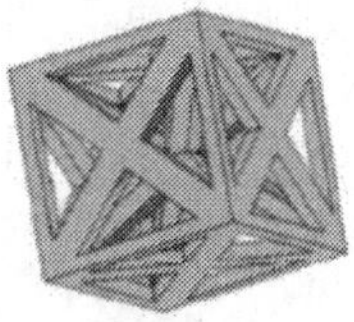	D 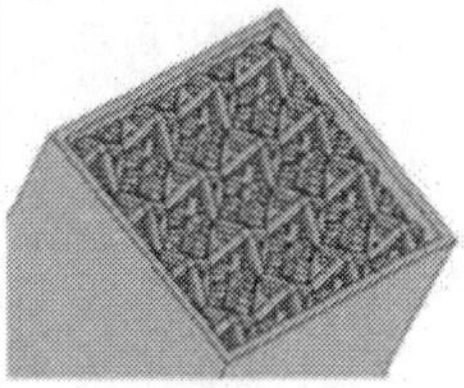	F 	J 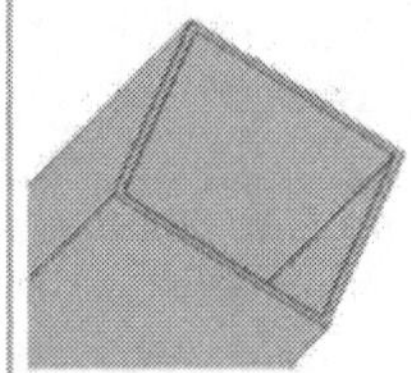
B 	E 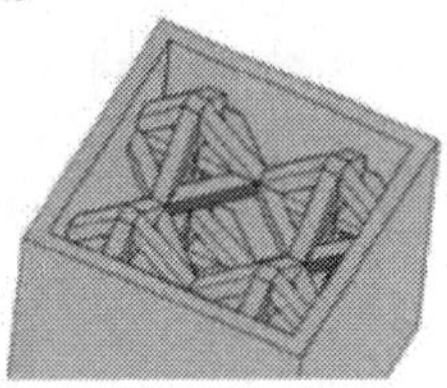	G 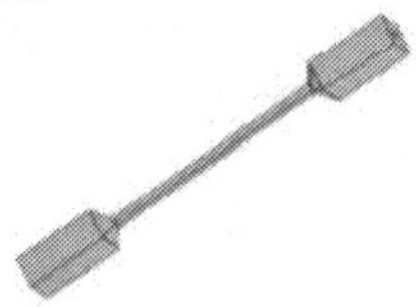	K 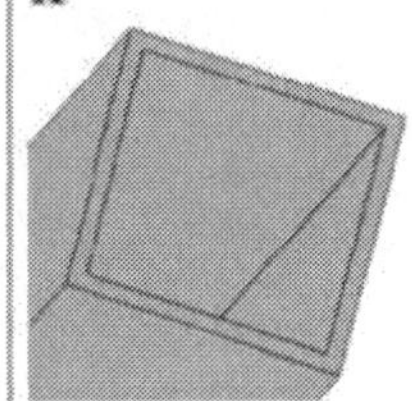
C 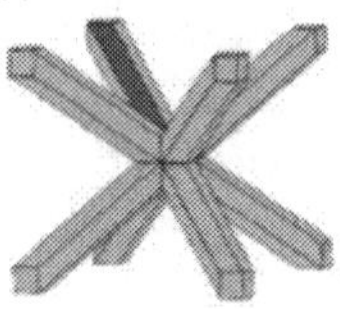		H 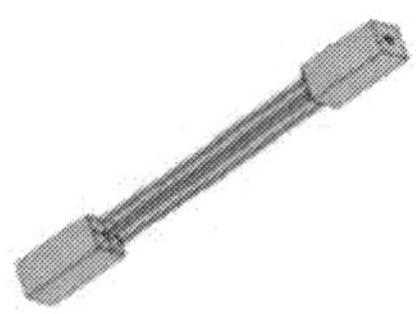	
		I 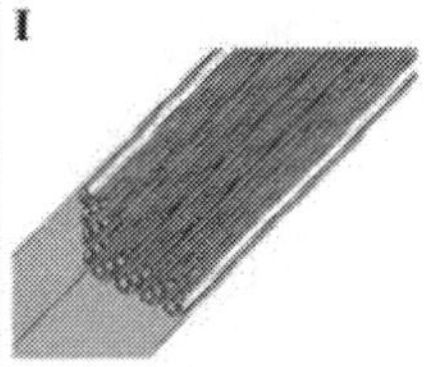	L 

Table 2. Properties of the specimens.

No.	Specimen and type of the unit cell	Coating (mm)	Strut size (mm)	Cell size (mm)	Vertical struts	Smallest cross sectional (mm^2)
A	Truss structure	-	0.3	2	yes	8.393
B	Octahedral	-	0.3	2	yes	3.69
C	Truss structure	-	0.3	2	-	2.88
D	Coated octahedral	0.3	0.3	2	yes	11.49
E	Coated octahedral	0.5	0.5	4	-	16
F	Ring face	-	-	-	-	3.675
G	Square	-	-	-	-	3.686
H	Split square 4x	-	-	-	-	3.688
I	Split square 41x	-	-	-	-	3.69
J	Hollow body	0.3	-	-	-	9.24
K	Hollow body	0.5	-	-	-	15
L	Massive	-	-	-	-	64

3 Results and discussions

The Table 3 below shows the average values of the second series of test. Within the table reached tensile strength Rm of the different specimens is listed, these become accessible with the ratio of the highest reached force F_m and the base area A. For the required calculation the base area of the specimens was used and which has a value of 64 mm^2. It is not the strained cross sectional area that is integrated to the calculation. In this case it should be noted that the calculated results represent a mixture of component test and material science. Hence the table includes the smallest strained cross sectional area that turned out to be the breaking point of the specimens and further the main major weakness. Together with those values the actual tensile strength is calculated and listed in the column $R_{m\text{-theo}}$. These calculated values finally show the significances and deviations. The results are in a descending order of the tensile strength. Like above mentioned the focus of this experimental set up are several structural properties including a range of different lattice structures. The complexity of this lattice structures is defined by their diversity of unit cell structures. Because of the fact that lattices structures are integrated within products a further, to test, aspect is the coating of lattice structures. Here the effect of a coating is evaluated. In order to interpret affects and to make conclusions out of the results, the experimental set up is further expanded with hollow body specimens. These specimens represent the coatings of the bevor tested coated lattice structures. The interaction of a coating and structure should be uncovered. A final focus is on the cross sectional area of the specimen B listed in Table 2. The great achieved tensile strength of almost 1172 MPa is a significant value that needs a special attention. Therefore the dependence and relation of the cross sectional area is tested.

Thus a variety of different specimens with almost the same cross sectional area are included in the experimental set up. Besides the cross sectional area different geometries and different distribution of the cross section are the main aspects that are looked at.

Taking a look at the Table 3, four rough divisions can be separated. Firstly the specimens that show comparable results to the reference massive probe L (E), and the once that have slight (G, I, F, H, C), great (D, K, A) and significant (B, J) deviations to the reference probe. The green marked specimens are on the one side the massive probe itself and a lattice structure E with a coating. The lattice structure has a unit cell with the size of 4x4x4mm, a strut size of 0.5x0.5 mm and a 0.5 mm coating. Both the highest reached force and theoretical cross section are a quarter of the massive specimen. Taking these values for a calculation of the theoretical tensile strength, the result is a nearly similar value as the reference. Firstly this might be an indication that the tensile strength is based on the cross section of the specimen. Continuing with the next group of probes these results are ranged in those with great deviations.

The specimen D, which is a coated lattice structure with an octahedral geometry, and the specimen A, with a truss lattice structure, have almost the same cross sectional area of the major weakness with an average deviation. Their theoretical tensile strength is comparable. Deviating from these two structures the specimen K is matched into this group because of its reached similar values of the component test. Interesting about this is the fact, that the theoretical cross section has an alike as E, but a lower theoretical calculated strength. Hence, there is further the necessity to rethink the theory of the behaviour of the cross section and the relation to the tensile strength in case of lattice structures. Further on the following specimens, marked in red, also show great deviations compared to the others. Above all the specimen B, which is a lattice structure including octahedral unit cells with further vertical struts, has an unexpected reached theoretical value of its tensile strength. This sample is the one that is the fundament for the further design of the experimental set-up. Conspicuous about the second specimen J belonging to this group is that its reached

DOI: 10.1051/matecconf/20179403008

values are significantly lower than the once described above (D + A). With a mostly alike cross section compared to D and A, just and almost half as force is achieved. Even lower strain is reached compared to the specimen B, which has an almost 2.5time smaller cross section. The shape of the last division is based on the results of the deviations of the above discussed lattice structure B. Almost every specimen has the same cross section with the difference of the chosen design and distribution.

The aim was to find out if specimens with the same cross section show the same significant deviations as the lattice structure. All in all the results show that having the same cross section does not affect the tensile strength.

Table 3. Results of the tensile test.

Specimen No.	Rm [MPa]	Fm [kN]	Atheo [mm^2]	Rm-theo [MPa]
L	621	39.75	64	621.09375
E	150	9.95	16	621.875
D	134	8.57	11.49	745.86597
K	131	8.37	15	558
A	110	7.05	8.393	839.985702
B	67.5	4.325	3.69	1172.08672
J	58	3.72	9.24	402.597403
G	33	2.1	3.686	569.723277
I	29	1.87	3.69	506.775068
F	26.75	1.71	3.675	465.306122
H	25	1.59	3.688	431.127983
C	22.5	1.45	2.88	503.472222

3.1 Curve diagram

Taking a look at the curve diagram shown in Fig.1, a few values should be described and discussed. The curves show regular behaviour with a continuous transition from the elastic to the plastic area and the presence of Hook´s law [11].Nevertheless, two specimens show extraordinary progressions. Beginning with the specimen E that has the greatest tensile test of this series, it is noticeable that there are two peaks. A high peak followed by a lower peak. It is difficult to interpret the appearance and reasoning of the second peak. A possible reason could be an early strip off of a layer and a geometry structure, which causes a temporary drop. This means that the strip off is internal, however the coat is not affected. This fact might cause a further shortly increasing tensile strength until the total break appears. On the other hand there is a specimen (C) that shows extraordinary strain behaviour. It has the greatest strain measured within this experimental set up. The cell type of this specimen is an octahedral, without any vertical struts. All in all this leads to a similar strain behaviour and great deformation of the specimen.

DOI: 10.1051/matecconf/20179403008

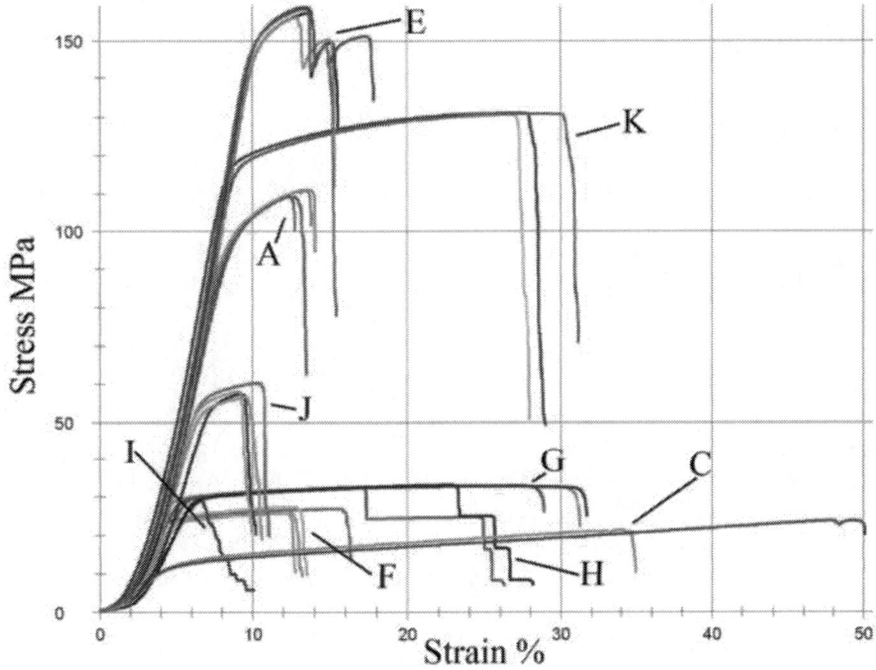

Fig. 1. Curve graphic of several tensile tests regarding this investigation.

4 Conclusions

The above mentioned results enable new starting points to verify theories regarding the properties of lattice structures. Actually the aim was to find new interactions of different factors that affect the lattice structures. In carrying out the investigation several results were extracted that need a new and adjusted test series as well as a closer look on it. A special abnormality was discovered at the specimen E. As already told there is an interesting curve progression of the tensile test. Because of the fact that reached tensile strength is similar to the tensile strength of the specimen L, it seems that neither the cell structure nor the coating has an effect on the tensile strength.

This fact is the reason for a new theory. It is necessary to have a closer look on the dimensioning of the unit cell and the coating itself. Perhaps the great cell structure opens up the possibility for a local constriction which leads to a predetermined breaking point. Moreover the comparison of the specimens B and D introduces new issues. Except for the surrounding coat of D the specimens are identical nevertheless the reached theoretical tensile strength of sample D is lower compared to B. One of the main theories was that a major weakness is at the lowest cross sectional area. Caused by the SLM process itself, generated parts show anisotropic behaviour [12, 13].

Concerning the specimen B it was necessary to reduce the structure to a specimen with just the same cross sectional area. This reduction and comparison showed that a reduction is not achieving the same results just because of having the same cross sectional area.

Regarding to this facts it is still impossible to make more concrete conclusions. There are still several parameters that might influence the gained achievements. These could be the shape, the microstructure or process parameters, which are not yet taken into consideration. Even slight deviations are an error source for calculations and misinterpretation. All these issues need to be reduces or eliminated in further studies [14].

This paper was supported by the project entitled "Boosting the scientific excellence and innovation capacity in Additive Manufacturing of the Technical University of Cluj-Napoca" (AMaTUC), founded by the European Union's Horizon 2020 research and innovation programme under grant agreement No. 691787 (www.amatuc.com).

References

1. J. Glasschröder, *Applied Research at the Laboratory for Additive Manufacturing* (iwb Application Centre Augsburg, 2013)
2. R. Poprawe, Management Circle **3**, 14-50 (2015)
3. A. Gebhardt, *Understanding Additive Manufacturing* (Carl Hanser Verlag, Munich, 2011)
4. Fraunhofer ILT, Annual Report 2015, http://www.ilt.fraunhofer.de/content/dam/ilt/de/documents/Jahresberichte/JB13/H_JB13_75.pdf, accessed 2016
5. C. Hamm, *Evolution of Lightweight Structures – Analysis and Technical Applications* (Springer Publications, 2015)
6. P. Li, Z. Wang, N. Petrinic, C.R. Siviour, Mater.Sci.& Eng. A **614**, 116 (2014)
7. S. McKown, Y. Shen, W.K. Brookes, C.J. Sutcliffe, W.J. Cantwell, G.S. Langdon, G.N. Nurick, J SANDW STRUCT MATER **13**, 479-501 (2011)
8. C. Yan, L. Hao, A. Hussein, P. Young,D. Raymont, Materials & Design **52**, 638-647 (2014)
9. A. Hussein, C.Yan, P. Young, L. Hao, Rtejournal **12**, 3 (2015)
10. J. Kessler, N. Balc, A. Gebhardt, K. Abbas, AMM **808**, 193-198 (2015)
11. H. Burkhard, *Material Testing-Determination of Material Properties* (Carl Hanser Verlag, Munich, 2003)
12. A. Schwarz, *Joining of Generatively Manufacutred Parts Made from1.4404* (Master Thesis, 2015)
13. D. Ziebura, *Development of a Systematology for Optimal Parameter Finding in the SLM-Process on the Example of IN 625* (Master Thesis, 2015)
14. J. Kessler, N. Balc, A. Gebhardt, K. Abbas, Physics Procedia **83**, 833–838 (2016)

MATEC Web of Conferences **94**, 03009 (2017)
CoSME'16 DOI: 10.1051/matecconf/20179403009

Microstructure and micro-hardness analyses of titanium alloy Ti-6Al-4V parts manufactured by selective laser melting

Camil Lancea[1,*], *Lucia-Antoneta* Chicos[1], *Sebastian Marian* Zaharia[1] and *Mihai Alin* Pop[2]

[1]Transilvania University of Brasov, Department of Manufacturing Engineering, Mihai Viteazu No.5, Braşov, Romania

[2]Transilvania University of Brasov, Faculty of Materials Science and Engineering, Colina Universitatii No. 1, Brasov, Romania

Abstract. Selective Laser Melting (SLM) is one of the powder based additive manufacturing technologies and it is, as well, the most rapidly growing technique in Rapid Prototyping. In this paper is presented a microstructure analysis using Scanning Electron Microscope (LEO 1525 SEM), of Ti6Al4V parts exposed into a corrosion environment. The corrosion environment was generated using a salt chamber with 5% and 10% NaCl concentration and an ACS-Sunrise climatic chamber. The parts were also subjected to tests in order to determine their micro-hardness, followed by a statistical processing of the obtained data. The parts, having a lattice structure, were built on a Selective Laser Melting machine.

1 Introduction

Additive Manufacturing (AM) of metal finished parts is well recognized as an interesting alternative to other conventional or unconventional processes, thanks to its capability to produce complex shapes and integrated parts of a high strength-to-weight ratio [1, 2].

In the recent years, AM has evolved from simple 3D printers used for rapid prototyping in which the most works has been built using polymer materials to sophisticated rapid manufacturing systems that can be used to create functional parts in different engineering materials directly without the use of tooling. The development of AM processes such as Selective Laser Sintering, Selective Laser Melting, Electron Beam Melting and Laser Engineered Net Shaping enabled to build parts by using metallic materials, metal matrix composites and ceramic materials. Today, Additive Manufacturing is frequently used in aerospace [3], automotive, medical fields and also in sustainable energy for building the components of small size and complex geometries [4-8].

Selective Laser Melting (SLM) is one representative of additive manufacturing processes [9, 10]. SLM is a layer-by-layer technology using a high-energy laser system. In order to manufacture a component with SLM technology the 3D CAD model is sliced into layers. The laser melts powder on a substrate plate selectively along scan tracks [11].

* Corresponding author: camil@unitbv.ro

Components manufactured with SLM offer a high geometrical flexibility and accuracy without almost any loss of material. It is well known that for obtaining components by SLM technology is used a narrow range of materials: cobalt-chrome, stainless steels, tool steels, aluminium alloys, gold alloys, nickel-based superalloys such as Inconel 625 and 718, and TiAl alloys [4, 5, 12]. The most popular for medical and aerospace applications, and not only, is the titanium alloy Ti-6Al-4V.

Among first studies of processing titanium and its alloys by Selective Laser Melting were focused on Ti-6Al-4V. Therefore, the properties of this material have been extensively examined. The influence of the process parameters like the scanning strategy on the microstructure of Ti-6Al-4V, for instance, was investigated by Thijs et al. in [13]. Vandenbroucke et al. [14] determined the mechanical properties in tensile load of SLM fabricated Ti-6Al-4V. Studies of Vrancken et al. [15] pointed out that conventional post treatments, which are used to enhance mechanical properties, appear different to additively processed Ti-6Al- 4V. Another study by Vaithilingam et al. dealt with the remelting during the SLM process of Ti-6Al-4V [16]. In [17] Vickers hardness evolution of the as-received Ti-6Al-4V alloy as a function of the post heat treatment is investigated. In [13] is revealed that the micro-hardness of as-built SLM Ti–6Al–4V varies with the processing parameters and is normally in the range of 3.43–4.90 GPa and in [18] the micro-hardness of the struts of as-made Ti–6Al–4V TPMS (Triply periodic minimal surface) structures was measured to be 4.01± 0.34 GPa [19].

In this study, micro-hardness of the specimens manufactured by selective laser melted Ti-6Al-4V alloy was evaluated by using a Vickers hardness tester for establishing the influence of corrosive environments on this property. The specimens made of titanium alloy, Ti-6Al-4V, before determining the micro-hardness were subjected to corrosive environments in saline and climatic chamber.

2 Methods samples and materials

This chapter presents the equipment used for manufacturing the specimens and the equipment used in performing corrosion tests. Finally are presented the measurement instruments used to determine the micro-hardness and the SEM analysis.

2.1 Parts manufacturing through a selective laser melting process

Selective Laser Melting (SLM), sometimes referred to additive manufacturing, or rapid manufacturing, is a 2.5 axes manufacturing process used to produce complex shape parts, starting from the CAD 3D model, saved as *.stl (STereoLithography) file. During the manufacturing process, a very fine powder of metal (usually from 20 to 100 micrometres thick) is applied in successive layers and melted, layer by layer, with a laser beam, to create a 2D image of each layer.

The principle of the method consists in sectioning the piece with a beam of parallel planes and its construction layer by layer on a SLM machine as follows: the machine table is covered with the first layer of powder; the first layer of the physical model is created by melting the metal powder along its section geometry; the machine table is then lowered with the layer thickness, and the process is repeated until completion of the model.

The system uses atomized metal powders to produce dense metal parts of aluminium alloys, titanium, stainless steel, cobalt-chromium or tool steel but the number of compatible materials increases year by year.

Selective Laser Melting (SLM) technology is used to manufacture metal parts that can be used in different fields as: engineering, aerospace, electronics, orthopaedics, dental care etc. The SLM principle is presented in figure 1.

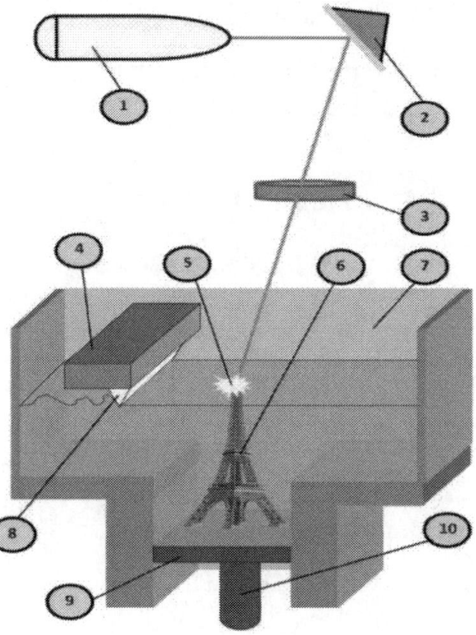

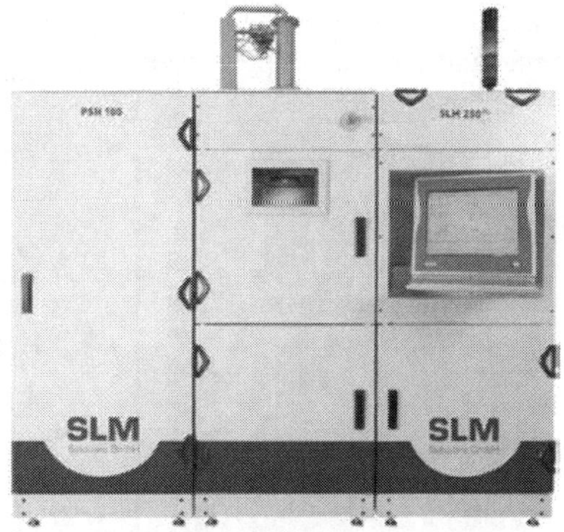

1 – Laser optics; 2 – x-y scanner, mirror;
3 – Lens; 4 – Powder supply (Recoater);
5 – Melt pool; 6 – SLM Part; 7 – Inert gas;
8 – Platform with heating system;
9 – Levelling system; 10 – Retractable platform

Fig. 1. Schematic of a typical SLM process and the SLM machine.

2.2 The geometry of samples

The samples are obtained from a Titanium alloy Ti6Al4V through selective laser melting on a SLM250HL machine. The samples have a lattice structure with the shape and dimensions presented in figure 2.

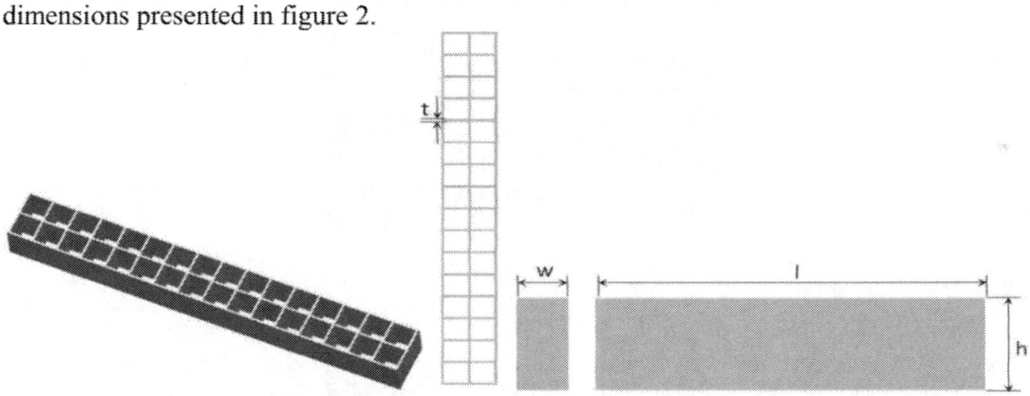

Fig. 2. The samples shape and dimensions.

2.3 Corrosive environments description

After obtaining the samples they were grouped in four sets. Each set, without the first, was placed into a corrosive environment as follows:
- the first set of specimens - without any tests;
- the second set of specimens were hold into a saline chamber with 5% NaCl concentration with an immersion time of 72h;
- the third set of specimens were hold into a saline chamber with 10% NaCl concentration with an immersion time of 144h;

- the fourth set of specimens were hold into a climatic chamber with a combined test: temperature, humidity and ultraviolet radiation (Fig 3).

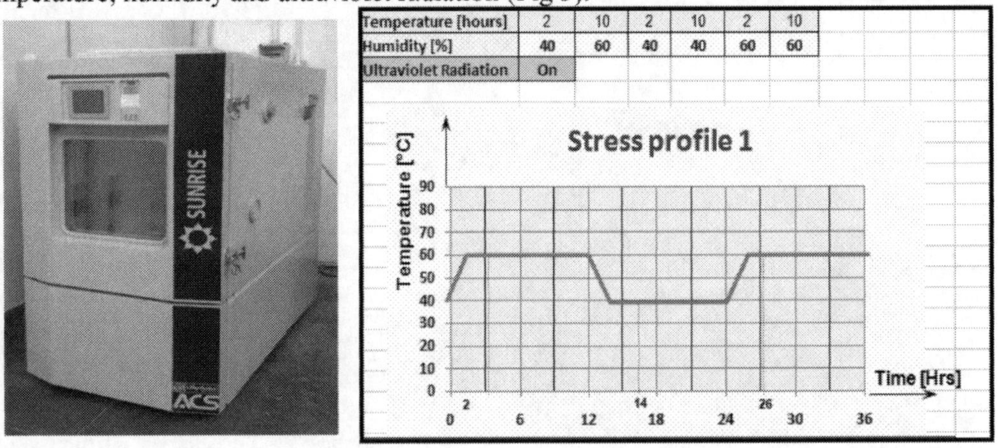

Fig. 3. The ACS-Sunrise climatic chamber and stress profile diagram (humidity, temperature, UV).

2.4 Microstructure analysis equipment

These conclusions were elaborated after analysing the micro-structural characteristics, at a scanning electron microscope (figure. 4), of all the specimens, before and after passing the testing procedures.

Fig. 4. The scanning electron microscope (LEO 1525 SEM).

A scanning electron microscope (SEM) is used for morphological, chemical and micro-structural characterizations of materials.

A high resolution, field emission LEO 1525 SEM is used. The nominal resolution of this microscope, obtained with a working distance of 2 mm, is 1.5nm at 20kV and 3.5nm at 1kV. Magnification ranges from 20x up to 500kx.

The SEM is equipped with an Energy Dispersive X-ray (EDX) microanalysis which provides both qualitative and quantitative information about the chemical composition of the sample, and with an Electron BackScattered Diffraction (EBSD) system that allows resolving local crystallographic orientations in the range of 100nm.

2.5 Micro-hardness testing equipment

The samples were cut transversally, were embedded in epoxide resin and prepared metallographic for the structural analysis used to determine the hardness values. For polishing the samples is used a Buehler Phoenix Beta Grinder apparatus (Fig. 5).

Fig. 5. BUEHLER Phoenix Beta Grinder Polisher.

When determining the micro-hardness values for the specimens of Ti6Al4V alloy is used the Vickers method which is based on the field size diagonal measurement obtained after a light load pressing of a diamond pyramid into the researched material grains. For the micro-hardness tests of titanium alloy Ti6Al4V specimens, they were transversely sliced, embedded in an epoxy resin and then was performed a grinding operation by using sandpapers with a gradual granulation change (1500, 2000 şi 2500) using Buehler-Phoenix-Beta-Grinder-polisher.

It is very important for the sliced parts to present a smooth surface without cutting signs, after the cutting operation, because these imperfections can influence the accuracy of the micro-hardness tests. The micro-hardness measurements were performed on the Micro Hardness Tester FM-700 (Fig. 6), the penetrator was applied for 15 seconds using a 100 gf load.

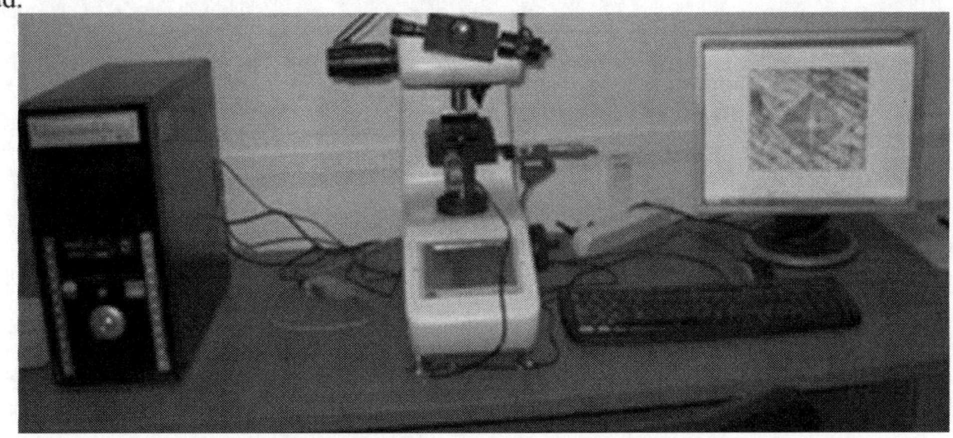

Fig. 6. Micro-hardness Tester FM 700.

3 Results and discussions

3.1 Microstructure analysis results

When finishing the corrosion tests, important information about the corrosive resistances of our specimens was obtained by using the scanning electron microscope for morphological, chemical and micro-structural characterizations of materials. The following figures (Fig. 7 Fig. 8 and Fig. 9) presents the micro-structural characteristics changes of the analysed material, before and after the immersion into the chamber with saline atmosphere, for 72 hours, using saline solution of NaCl (sodium chloride) of 5% concentration and for 144 hours, using saline solution of NaCl of 10% concentration, at different magnifying scales. The results are presented below:

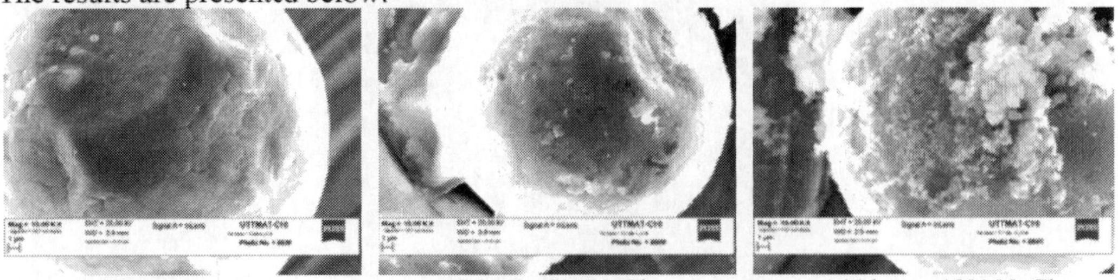

Initial structure 72 hours 5% NaCl 144 hours 10% NaCl

Fig. 7. Samples before and after the immersion into the saline chamber at 10000x magnifying scale.

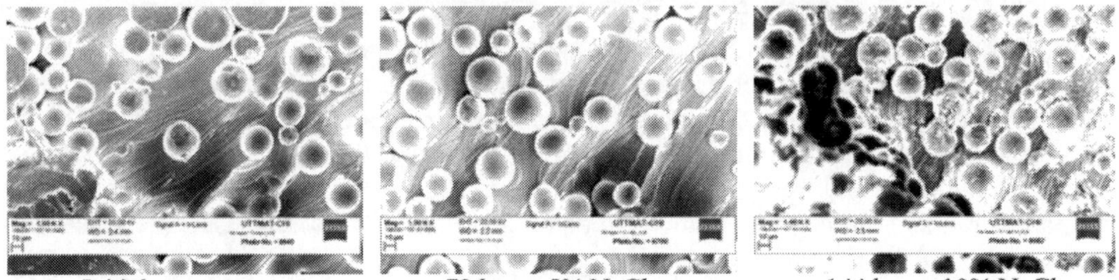

Initial structure 72 hours 5% NaCl 144 hours 10% NaCl

Fig. 8. Samples before and after the immersion into the saline chamber at 1000x magnifying scale.

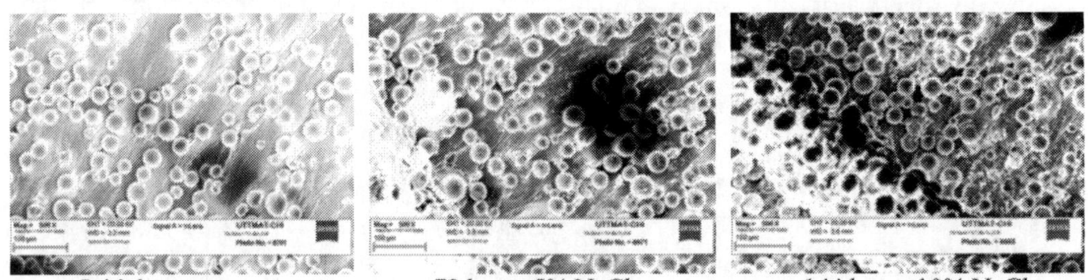

Initial structure 72 hours 5% NaCl 144 hours 10% NaCl

Fig. 9. Samples before and after the immersion into the saline chamber at 500x magnifying scale.

3.2 Micro-hardness test

Within the Vickers test, using penetrators with identical angles at the top, identical geometrical traces are obtained regardless of the loading force and, as a result, it can be said that the hardness is independent by the load size.

Due to the small depth of the penetrator ingress into the material, the method can be used also to determine the hardness of small parts, thin parts or for heat treated layers. For the micro-hardness determination of thin layers are performed several tests and only the results with closed values will be retained.

All the specimens were manufactured on a SLM machine with the objective to determine the characteristics of their micro-hardness (Table 1). The SLM machine has the following building parameters [20]:

- laser power 100 [W];

- laser scanning speed 150 [mm/s];
- hatch offset 0.175 [mm];
- layer thickness 0.05 [mm];

Table 1. The hardness values of specimens.

Micro-hardness for the first [HV]	Micro-hardness for the second specimen 5%NaCl/72h [HV]	Micro-hardness for the third specimen 10%NaCl/144h [HV]	Micro-hardness for the fourth specimen climatic chamber [HV]
380.7	383.1	445.2	400.1
390.9	388.4	350.2	430.7
420.5	386.6	384.9	472.1
411.6	413.4	386.6	420.3
414.6	407.2	389.8	416.5
393.1	362.4	347.2	407.6
449.6	357.0	389.1	457.6
426.2	436.6	315.3	443.0
440.8	397.5	434.9	462.7

3.3 The statistical analysis of the micro-hardness tests data

The statistical analysis of the micro-hardness tests data for all the specimens was accomplished using the SPSS Statistics 23 software. The obtained results were described by using the distributions of the occurrence frequencies (Fig. 10). Because the average and median values shows relatively equal for all the four tested samples, we can say that the distribution is normal, with a slight positive or negative asymmetry [21].

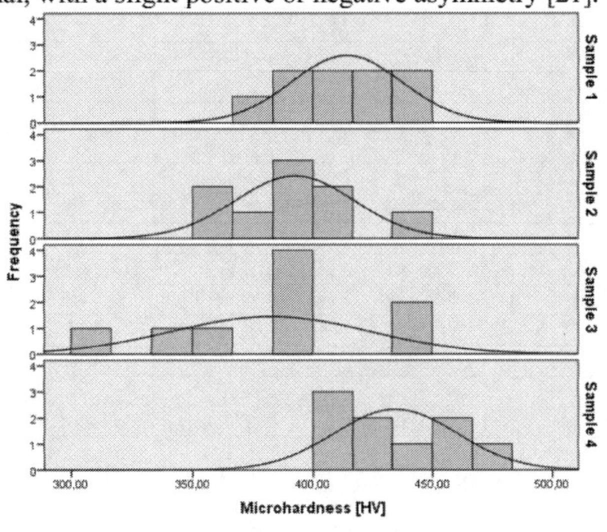

Fig. 10. Histograms of the three analysed samples at the micro-hardness tests.

After determining the micro-hardness of the titanium alloy Ti6Al4V samples, manufactured with SLM technologies, making the micro-hardness average values (Fig. 11) of the workpieces surface, was obtained an hardness average value of 405.94 HV0.1. In Fig. 10 it can be observed that a large proportion of the micro-hardness values, for the analyzed samples, are summarized in the rectangle described by the 25th percentile and the 75th percentile.

 DOI: 10.1051/matecconf/20179403009

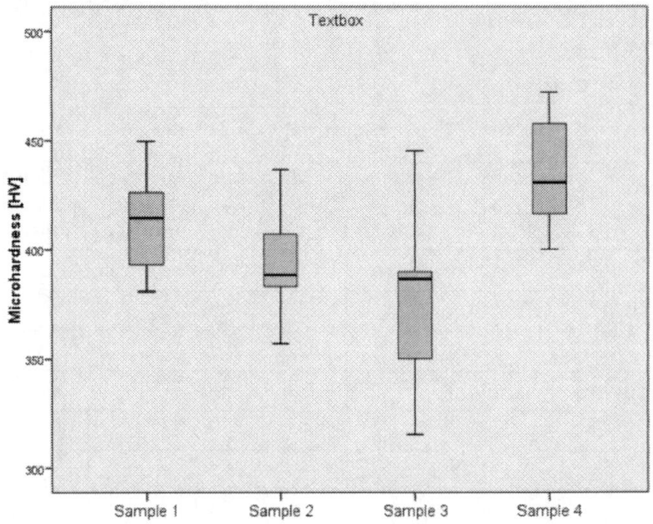

Fig. 11. Boxtype diagram representing the micro-hardness values for the tested samples.

Table 2 describes the main statistical indicators of the micro-hardness values for all the tested specimens. Each statistical indicators value transmits a certain amount of information concerning the data for which they were calculated. For the micro-hardness values of all the specimens, the statistical indicators provide information about the central tendency of the series values (mean, median and module), others give us information about scattering values (dispersion, standard deviation and coefficient of variation) and others give us an indication of symmetry values in the series (median and quartiles).

Table 2. Statistical indicators of the micro-hardness values.

Specimen		1	2	3	4
Mean		414.22	392.46	382.57	434.51
Std. Error of Mean		7.69	8.27	13.74	8.54
Median		414.60	388.40	386.60	430.70
Mode		380.70	357.00	315.30	400.10
Std. Deviation		23.08	24.82	41.23	25.64
Variance		532.709	616.11	1700.20	657.73
Range		68.90	79.60	129.90	72.00
Minimum		380.70	357.00	315.30	400.10
Maximum		449.60	436.60	445.20	472.10
Sum		3728.00	3532.20	3443.20	3910.60
Percentiles	25	392.00	372.75	348.70	412.05
	50	414.60	388.40	386.60	430.70
	75	433.50	410.30	412.35	460.15

4 Conclusions

After completing the tests and analysing the obtained results, the following conclusions can be drawn:

- the specimens tested in the Sunrise climatic chamber didn't present significant changes within their micro-structural characteristics;

- the specimens tested in the chamber with saline atmosphere presents significant changes within their micro-structural characteristics with attached salt particles on their structure, due to the corrosion phenomena;

- the micro-hardness decreases for the specimens which were placed into the saline chamber, due to the corrosion phenomena

- the micro-hardness increases for the specimens exposed to the climate chamber, where the main stress factors were temperature and humidity, due to the heat treating of the surface.

We hereby acknowledge the structural founds project PRO-DD (POS-CCE, O.2.2.1., ID 123, SMIS 2637, ctr. No 11/2009) for providing the infrastructure used in this work and the project. Financial support by the Access to Research Infrastructures activity in the 7th Framework Programme of the EU (SFERA Grant Agreement n. 228296) is gratefully acknowledged.

References

1. E. Atzeni, A. Salmi, Int. J. Adv. Manuf. Tech. **62**, 1147 (2012)
2. D. Manfredi, F. Calignano, M. Krishnan, R. Canali, E. P. Ambrosio, E. Atzeni, Materials **6**, 856 , (2013)
3. T. Wohlers, Manuf. Eng. **148**, 55 (2012)
4. D. Manfredi, F. Calignano, M. Krishnan, R. Canali, E. P. Ambrosio, S. Biamino, D. Ugues, M.Pavese, P. Fino, Light Metal Alloys Applications (InTech Press, Rijeka, 2014)
5. I. Campbell, D. Bourell, I. Gibson, Rapid Prototyping J. **18**, 255 (2012)
6. The Future of Exploration Starts with 3-D Printing. Available: http://www.nasa.gov/centers/marshall/about/star/star130327.html#.VAA_tsWSzId (accessed 2016)
7. K.G. Prashanth , S. Scudino , H.J. Klauss , K.B. Surreddi , L. Loeber , Z. Wang , A.K. Chaubey , U. Kühn , J. Eckert , Materials Science and Engineering: A **590**, 153 (2014)
8. K. Kempen, L. Thijs, J. Van Humbeeck, J. P. Kruth, Physics Procedia **39**, 439 (2012)
9. D.L. Bourell, Int. J. Powder Metall. **28**, 369 (1992)
10. J.-P. Kruth, L. Froyen, J. Van Vaerenbergh, P. Mercelis, M. Rombouts, B. Lauwers, J. Mater. Process. Technol. **149**, 616 (2004)
11. L. Löber, F.P. Schimansky, U. Kühn, F. Pyczak, J. Eckert, J. Mater. Process. Technol. **214**, 1852 (2014)
12. D. Buchbinder, H. Schleifenbaum, S. Heidrich, W. Meiners, J. Bültmann, Physics Procedia **12**, 271 (2011)
13. L.Thijs, F.Verhaeghe, T. Craeghs, J. Van Humbeeck, J.P.Kruth, Acta Biomater.**58**, 3303 (2010)
14. B. Vandenbroucke, J.P. Kruth, Rapid Prototyp. J. **13**, 196 (2007)
15. B. Vrancken, L. Thijs, J.-P. Kruth, J. Van Humbeeck, J. Alloys Compd. **541**, 177 (2012)
16. J. Vaithilingam, R.D. Goodridge, R.J.M. Hague, S.D.R. Christie, S. Edmondson, J. Mater. Process. Technol. **232**, 1 (2016)
17. S.Q. Wu, Y.J. Lu, Y.L. Gan, T.T. Huang, C.Q. Zhao, J.J. Lin, S. Guo, J.X. Lin, J. Alloys Compd. **672**, 643 (2016)
18. Yadroitsev, I., Krakhmalev, P., Yadroitsava, I., J. Alloys Compd. **583**, 404 (2014)
19. Y. Chunze, L. Hao, A. Hussein, Ph. Young, J Mech Behav Biomed **51**, 61 (2015)
20. G. R. Buican, G. Oancea, C. Lancea, M.A. Pop, Appl Mech Mater **760**, 515, (2015)
21. S. M. Zaharia, C. Lancea, L. A. Chicos, G. Caputo, Appl Mech Mater, **656**, 442 (2014)

MATEC Web of Conferences **94**, 03010 (2017) DOI: 10.1051/matecconf/20179403010
CoSME'16

Bending and compression tests for PA 2200 parts obtained using Selective Laser Sintering method

Cristina Stefana Miron-Borzan[1,*], *Mircea Cristian* Dudescu[2], and *Petru* Berce[1]

[1]Technical University of Cluj-Napoca, Manufacturing Engineering Department, 400641B-dul Muncii, no. 103-105, Cluj-Napoca, Romania
[2]Technical University of Cluj-Napoca, Mechanical Engineering Department, 400641B-dul Muncii, no. 103-105, Cluj-Napoca, Romania

Abstract. Craniofacial bone defects that produce morphological and functional disabilities can bring alteration of the quality of patient's life. For this reason, the collaboration between the field of medicine with engineering, lead to the manufacture of custom implants from biocompatible materials. In the last years it was shown that Additive Manufacturing methods are very helpful to achieve customized medical implants. All the customized implants made of biocompatible materials have to be tested for mechanical properties, in order to have the optimal characteristics for the area were will be used. This paper presents the results for bending and compression tests for a biocompatible material, PA 2200. The samples were achieved using Selective Laser Sintering method and were obtained with different laser power, starting with 4 W. Maximum load, maximum stress, specific deformation and Young's modulus were analysed. The study showed that different mechanical properties can be obtained, depending on the laser power used. At bending tests, increasing values were obtained for the investigated parameters along with laser power increasing, but at compression tests a different trend was observed.

1 Introduction

A successful result for repairing of complex craniofacial bone defects depends on the defect size, the quality of the soft tissue covering the defect, and the choice of reconstructive method. The complex craniofacial skeleton is involved with various specific functions, such as protection of the brain and optic tracts, breathing, mastication, speech, and hearing [1,2].For this reason is very important to know the forces and which are the required mechanical properties for the future implant that will take over all the forces from the area.

Recent advances in Computer-Aided Design and Computer-Aided Manufacturing have begun to revolutionize craniofacial surgery, which is frequently confronted with the reconstruction of challenging 3D anatomic structures. Additive Manufacturing technology, with widespread availability of high-resolution medical imaging, has allowed for the

*Corresponding author: borzan_cristina@ymail.com

generation of contoured 3D prostheses for craniofacial reconstruction [2].Computer technology has moved forward with the advent of Additive Manufacturing techniques which allow both the production of models of the hard tissues and custom-made prostheses from computerised scanning data [3].

Selective Laser Sintering (SLS) is a manufacturing technique which enables the final product to be made directly and rapidly, without tooling or additional machining. For biomedical applications, SLS permits the fabrication of implants and scaffolds with complex geometry accurately and economically [4]. Selective Laser Sintering technique has been identified as a promising technique capable of building complex objects with pre-defined macro and microstructures [5].

In the last decade, the bio-medical industry was focused more and more on developing and exploiting of materials that are easy accepted by the human body [6].Polyamide is a biocompatible material that has a polar molecular structure (CO– NH–) similar with the chemical structure of bone collagen and exhibits excellent mechanical properties [7].

Customized properties of parts manufactured using the Selective Laser Sintering process are achievable by variation of build parameters [8].

Static tensile test is a basic mechanical test originally designed to become the widespread and known testing method for evaluating the mechanical properties of materials [9]. But for a future cranial implant are very important the bending and compression tests, that can show the stresses in normal and accidental situations. These stresses are much closer to the real situation from the human body.

A testing sample can help to determine standard basic mechanical properties such as: tensile strength, yield strength, elastic modulus and Poisson's ratio [10].

In the literature there are few studies that reported the properties of cranial bone at bending tests. There are some researchers that presented the results at bending tests of a human bone. Hubbard in 1971 [11] and Delille in 2007 [12] presented the mechanical properties for an adult cranial bone. The infant cranial bone was analysed by other researchers: McPherson in 1980 [13], Margulies in 2000 [14], Jans in 2003 [15] and Coats in 2006 [16].

2 Method and materials

This paper is part of a larger research that is focused on the behaviour and properties of PA2200parts used for custom implants manufactured by selective laser sintering process.

The materials used: Polyamide PA 2200, on the basis of polyamide 12 (PA 12), a bio compatible material according to EN ISO 10993-1 and USP/level VI/121 °C. The material was produced by the Electro Optical Systems - EOS GmbH, Munich, Germany, especially to be used for the EOS Laser Sintering Systems.

The technical characteristics of the material are: water absorption 100°C, saturation in water: 1.93% (according with ISO 62 / DIN 53495), coefficient of linear thermal expansion: 1.09 x10-4 /K (according with ISO 11359 / DIN 53752-A), specific heat: 2.35 J/gK (DIN 51005), thermal conductivity vertical to sintered layers: 0.144 W/mK (DIN 52616).

The manufacturing process was Selective Laser Sintering, using a DTM Sinterstation 2000 System with a CO2 laser. The PA 2200 powder was brought layer by layer from the two powder suppliers on the working place using the roller and was heated to near the melting point with the help of the laser beam, which drawn the shape of the virtual model. Thus, the component particles were bonded one to others, solidifying into the shaped form drawn by beam. When a solid layer was obtained, the platform was moved down with an equal distance to the thickness of the following sintered layer. The powder was again spread and sintered in the form of a new layer. The previous layer, represented a support for the current layer of material.

The samples were achieved at temperature 170° C, with the thickness of deposited layer - 100 μm, scan speed - 1257.3 mm/s and laser power 4W, 4.5W, 6W.

Experimental researches were done in order to determine the mechanical properties of polyamide structures made of PA 2200 with different laser powers. It was manufactured sets of 5 samples for each laser power.

The samples were achieved starting with the 3D models (designed in SolidWorks 2013 software) according with SR EN ISO 178:2011 for bending tests and SR EN ISO 604:2004 for compression tests.

For bending and compression tests it was used a 10 kN testing machine (INSTRON 3366).

Figure 1 shows a picture from the bending tests of specimens.

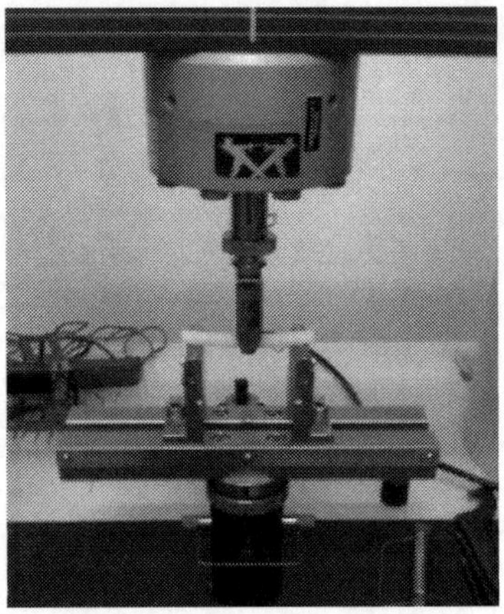

Fig. 1. Bend testing of specimens on Instron 3366 Machine.

3 Results and discussions

In tables 1, 2 and 3 are presented the mechanical properties obtained at the bending tests for PA 2200 samples obtained at 4W, 4.5W and 6W.For samples manufactured at 4W, analysing the results obtained for each characteristics, it was observed that the values were close to each other and were determined standard deviations 4.906 for maximum load, 3.098 for maximum stress and 77.464 for Young's modulus. However, the lowest values were recorded at the specimen No. 2 for all the analysed characteristics.

Table 1. The mechanical properties for PA 2200 specimens manufactured at 4 W at bending tests.

Sample	Maximum Load [N]	Maximum Stress, [MPa]	Specific Deformation, [mm/mm]	Young's modulus, [MPa]
1	106.07	66.99	0.07	1703.77
2	95.51	60.32	0.06	1535.13
3	106.65	67.36	0.07	1689.14
4	97.89	61.83	0.07	1558.41
5	101.55	64.14	0.07	1582.58
Mean Value	**101.53**	**64.13**	**0.07**	**1613.81**
Standard Deviation	**4.906**	**3.098**	**0.004**	**77.464**

The same trend it was observed also in the case of PA 2200 speciments manufactured at 4.5W, minimum values for sample No. 2 were obtanied for all the anaysed characteristics. The values were aproximatelly close, with standard deviation 6.84 for maximum load, 4.32 for maximum stress, 0.01 for specific deformation and 82.68 for Young's modulus.

Table 2. The mechanical properties for PA 2200 specimens manufactured at 4.5 W at bending tests.

Sample	Maximum Load [N]	Maximum Stress, [MPa]	Specific Deformation, [mm/mm]	Young's modulus, [MPa]
1	105.45	66.60	0.07	1699.31
2	89.53	56.55	0.06	1498.75
3	99.75	63.00	0.08	1634.58
4	104.25	65.84	0.08	1651.82
5	105.87	66.87	0.08	1701.23
Mean Value	**100.97**	**63.77**	**0.07**	**1637.14**
Standard Deviation	**6.84**	**4.32**	**0.01**	**82.68**

Also for thesamples manufactured at 6W, the minimum values for maximum load, maximum stress and specific deformation were achieved for sample No 2, but in the case of specific deformation and Young's modulus, the values were superior to the values of sample No 1.

Table 3. The mechanical properties for PA 2200 specimens manufactured at 6 W at bending tests.

Sample	Maximum Load [N]	Maximum Stress, [MPa]	Specific Deformation, [mm/mm]	Young's modulus, [MPa]
1	126.58	72.44	0.08	1591.47
2	124.64	71.98	0.08	1604.20
3	130.90	76.44	0.08	1705.20
4	126.10	74.95	0.08	1771.52
5	126.76	71.78	0.08	1655.69
Mean Value	**127.00**	**73.52**	**0.08**	**1665.62**
Standard Deviation	**2.336**	**2.068**	**0.001**	**74.416**

Analysing the mean values obtained at bending tests, for each sets of samples manufactured at 4W, 4.5W, respectively at 6W, it was observed an increasing of the values for maximum loads and stresses, once the laser power is higher (from 4.5W to 6W). The specific deformation and the Young's modulus is also increasing with the higher laser power used (from 0.06 mm/mm at 4W to 0.08 mm/mm at 6W for specific deformation and from 1613 MPa to 1665 MPa for Young's modulus).

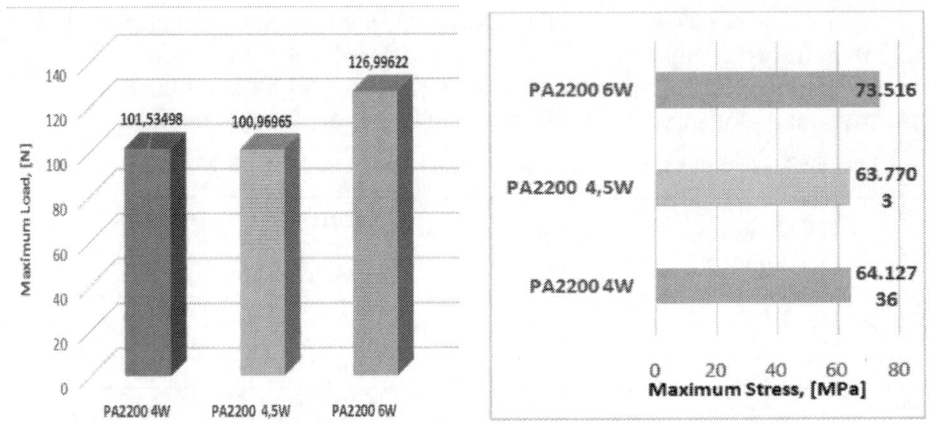

Fig. 2. Mean values of the maximum load and maximum stress for PA 2200 samples at bending tests.

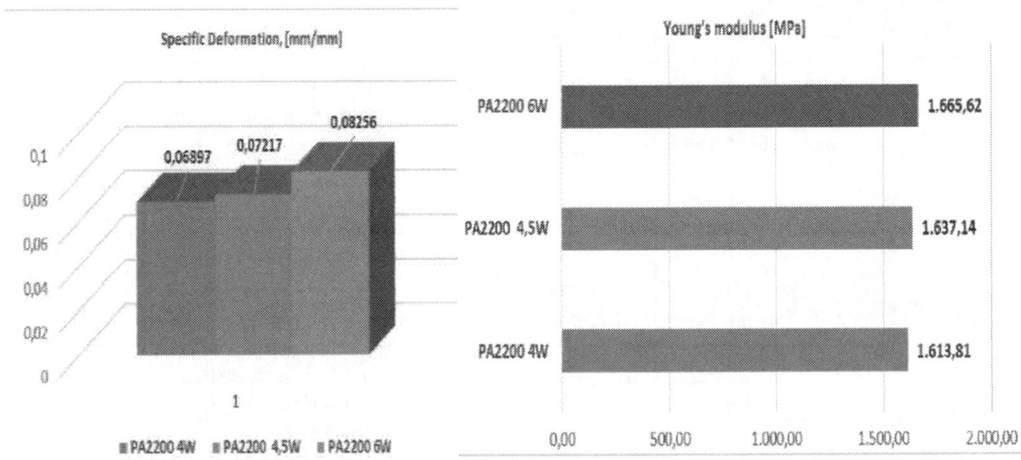

Fig. 3. Mean values of the specific deformation and Young's modulus for PA 2200 samples at bending tests.

In tables 4, 5 and 6 are presented the mechanical properties obtained at the compression tests for PA 2200 samples obtained at 4W, 4.5W, 6W.

Table 4. The mechanical properties for compression tests for PA 2200 specimens at 4 W.

Sample	Maximum Load [N]	Maximum Stress, [MPa]	Specific Deformation, [mm/mm]	Young's modulus, [MPa]
1	5634.71	62.61	0.25	1143.77
2	4968.28	55.20	0.14	1230.86
3	5413.65	60.15	0.20	1169.63
4	5214.72	57.94	0.17	1266.19
5	5404.21	60.05	0.16	1317.22
6	5288.63	58.76	0.16	1282.25
Mean Value	**5320.70**	**59.12**	**0.18**	**1234.99**
Standard Deviation	**223.91**	**2.49**	**0.04**	**67.20**

DOI: 10.1051/matecconf/20179403010

Table 5. The mechanical propertiesfor compression tests for PA 2200 specimens at 4.5 W.

Sample	Maximum Load [N]	Maximum Stress, [MPa]	Specific Deformation, [mm/mm]	Young's modulus, [MPa]
1	5452.54	60.58	0.12	1310.96
2	5522.72	61.36	0.12	1298.45
3	5151.48	57.24	0.16	1265.06
4	5560.66	61.79	0.14	1297.22
5	6080.95	67.57	0.23	1325.37
6	5361.26	59.57	0.11	1287.26
Mean Value	**5521.60**	**61.35**	**0.15**	**1297.39**
Standard Deviation	**310.44**	**3.45**	**0.05**	**20.57**

Table 6. The mechanical properties for compression tests forPA 2200 specimens at 6 W.

Sample	Maximum Load [N]	Maximum Stress, [MPa]	Specific Deformation, [mm/mm]	Young's modulus, [MPa]
1	5781.13	58.43	0.10	1395.75
2	6615.49	66.20	0.13	1385.35
3	6248.27	63.56	0.12	1406.59
4	6425.00	66.44	0.13	1433.80
5	6184.52	61.80	0.11	1394.27
Mean Value	**6250.88**	**63.29**	**0.12**	**1403.15**
Standard Deviation	**311.58**	**3.33**	**0.01**	**18.72**

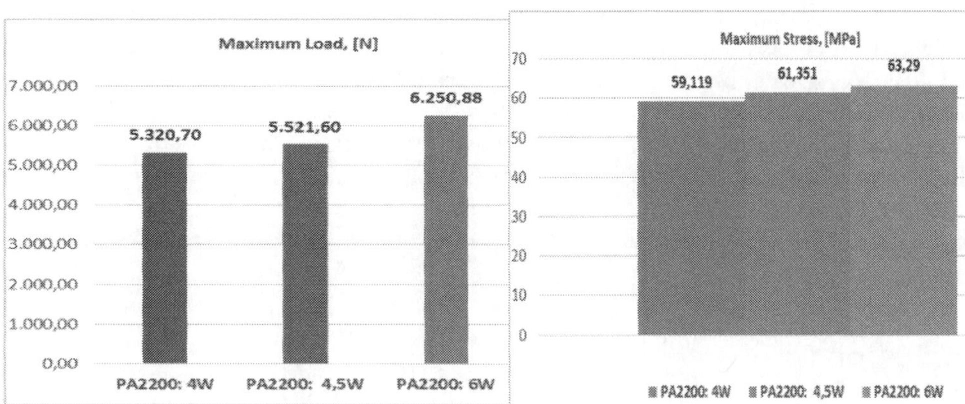

Fig. 4. Mean values of the maximum load and maximum stress for PA 2200 samples at compression tests.

Following the comparative analyse of the mean values obtained at compression test for the PA 2200 samples manufactured at 4W, 4.5W and 6W (Fig. 2, 3, 4, 5), it was observed that the forces and the stresses have an increasing trend with increasing of the laser power, but the specific deformation is decreasing. This trend can be explained by the structure of the samples achieved by selective laser sintering technology. At the lower laser powers, the parts are more porous and at higher laser powers, dense structures can be obtained. By applying forces on the porous structures, the layers are delaminated and a higher value for specific deformation is obtained than in the case of dense structures.

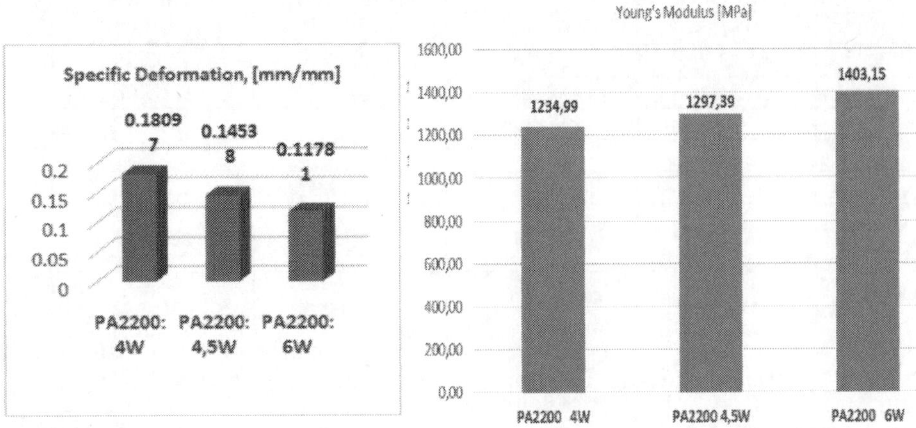

Fig. 5. Mean values of the specific deformation and Young's modulus for PA 2200 samples at compression tests.

Comparing the obtained results with the values for cranial bone from scientific literature, it is important to note that the adult and fetal cranial bones are physically different and have different mechanical properties. According with the research made by [13], the values for the elastic modulus, as a result of quasi-static bending tests, for fetal cranial bone was 3.88GPa (fibres parallel to specimen long axis) and 0.95GPa (fibres perpendicular to specimen long axis). The same study reported the elastic modulus of a single calvarium taken from a 6 year old subject as 7.38GPa (parallel fibres) and 5.86GPa (perpendicular fibres) [13, 17].

Other research made by [18] had the objective to evaluate and to compare tissue engineered cell-polymer constructs using calvarial osteoblasts and mesenchymal progenitor cells (MPC) for the reconstruction of critical-size and three-dimensionally complex cranial defects on New Zealand White rabbits. The mechanical investigations for the new grown bone, obtained using tissue engineering revealed that stiffness reached 52 ± 29 and 44 ± 16 MPa for MPC- and osteoblast-seeded scaffolds, respectively. The yield strength for the push-out tests was 180 ± 36 N for normal calvarial bone, 90 ± 1 N for unrepaired site, and 106 ± 10 N for unseeded constructs, which is about 60% of normal bone strength. MPC- and osteoblast-seeded scaffolds indicated yield strength of 149 ± 15 and 164 ± 42 N, respectively, which is about 85-90% of normal bone.

The results obtained in the present study, are in the range of the results obtained in the scientific literature.

4 Conclusions

Even the powder PA 2200 was manufactured to be used on EOS Sintering Systems, the study proved that this material can be used also for DTM Sinterstation machine if there are used optimal working parameters.

For the bending tests were observed that the results obtained for maximum load, maximum stress, specific deformation and Young's modulus are increasing together with the increasing of the laser power.

In the case of compression test it was showed that once a higher laser power is used in the manufacturing process, better mechanical properties are obtained. The values for the forces, stresses and Young's Modulus are higher with the increasing of the laser power and the specific deformation is decreasing when a higher laser power is used.

The mechanical properties obtained at bending and compression tests are in the range of the mechanical properties required for the cranial bone according with the results from the scientific literature.

References

1. Gr. Adams,*Psychological aspects of the facial form craniofacial monographno 11*(University of Michigan Press, 1981)
2. R. Tevlin, A. McArdle, D. Atashroo, G.G. Walmsley, K. Senarath-Yapa, E.R. Zielins, K.J. Paik, M.T. Longaker, D.C. Wan, J Dent Res. **93**, 12 (2014).
3. M. Mc.Gurk, A. A. Amis, P. Potamianos, N. M. Goodger, Ann R Coll Surg Engl **79**,169-174 (1997)
4. Y. Zhang, L. Hao, M. M. Savalani, R. A. Harris, K. E. Tanner, J Biomed Mater Res Part A **86A**, 3 (2008).
5. C. K. Chua, K. F. Leong, K. H. Tan, F. E. Wiria, C. M. Cheah, J. Mater. Sci. Mater. Med. **15**, 10 (2004)
6. R. Păcurar, A. Păcurar, N. Bâlc, A. Petrilak, L. Morovič, AMM, **371**, 478-482 (2013)
7. J. Li, Y. Hsu, E. Luo, A. Khadka, J. Hu, Aesth Plast Surg **35**, 636 (2011)
8. B.Caulfield, P.E. Mchugh., S. Lohfeld, J. Mater. Process. Technol. **182**, 477–488 (2007)
9. M. Ridzoň, M. Mojžiš, M. Martinkovič. P. Bella, L. Domovcová, L. Parilák, AJME **13**, 4 (2015)
10. M. Ridzoň, A. Zavacka, AMM **421**, 329-333 (2013)
11. Hubbard R. P.,J Biomech **4**, 251-263 (1971)
12. Delille R., Lesueur D., Potier P., Drazetic P., Markiewicz E., Int J Crashworthines **12**, 2 (2007)
13. McPherson G. K., Kriewall T. J., J Biomech **13**, 9-16 (1980)
14. Margulies S. S., Thibault K. L., J Biomech Eng. **122**, 364-371 (2000)
15. Jans G., *PhD thesis* (KatholiekeUniversiteit Leuven, Belgium, 2003)
16. Coats B., Margulies S., J Neurotrauma **23**, 8 (2006)
17. Motherway J. A., Verschueren P.; Van der Perre G., et al, J Biomech **42**,13 (2009)
18. J.-T. Schantz, D. W. Hutmacher, C. X. F. Lam, M. Brinkmann, K. M. Wong, T. C. Lim, N. Chou, R. E/ Guldberg, S. H. Teoh, Tissue Eng. **9**, 127-139 (2004)

Effect of vacuum oxy-nitrocarburizing on the microstructure of tool steels: an experimental and modeling study

Maria Nikolova[1,*], *Danail* Nikolov[1], *Bora* Derin[2], and *Emil* Yankov[1]

[1]University of Ruse "Angel Kanchev", Dept. of Material science and Technology, 8 Studentska Str., Ruse, Bulgaria
[2]Istanbul Technical University, Metallurgical and Materials Eng. Dept., Maslak, Istanbul, Turkey

Abstract. The thermochemical treatments of tool steels improve the performance of the components with respect to surface hardness, wear and tribological performance as well as corrosion resistance. Compared to the conventional gas ferritic nitrocarburizing process, the original vacuum oxy-nitrocarburizing is a time-, cost-effective and environmentally-friendly gas process. Because of the oxidizing nature of the gas atmosphere, there is no need to perform subsequent post-oxidation.In this study, a vacuum oxy-nitrocarburizing process was carried out onto four tool steels (AISI H10, H11, H21 and D2) at 570 °C, after hardening and single tempering. The structural analysis of the compound and diffusion layers was performed by optical and electron microscopy, X-ray diffraction and glow discharge optical emission spectrometry (GDOES) methods. A largely monophase ε-layer is formed with a carbon accumulation at the substrate adjacent area. The overlaying oxides adjacent to the ε-carbonitride phase contained Fe_3O_4 (magnetite) as a main constituent. A thermodynamic modelling approach was also performed to understand and optimize the process. The "Equilib module" of FactSage software which uses Gibbs energy minimization method, was used to estimate the possible products during vacuum oxy-nitrocarburising process.

1 Introduction

One of the cheapest and effective ways to improve the properties of the tool steels is to modify their surfaces through thermo-chemical treatments. Below the eutectoid temperature nitriding and nitrocarburizing that prevent dimensional distortions are usually applied. Because of the existence of oxidizing species (H_2O and CO_2) in the gas nitrocarburizing atmosphere, the process held at sub-critical temperature for Fe-N system in ammonia (NH_3) and carbon dioxide (CO_2) is called oxy-nitrocarburizing (ONC) [1].

The results from the thermo-chemical treatment vary considerably depending on the substrate material chosen, gas atmosphere and treatment conditions [2, 3]. During the nitrocarburizing process hard compound and supporting diffusion layers underneath are

*Corresponding author: mpnikolova@uni-ruse.bg

formed. The first one improves the corrosion resistance [4] and tribological (erosion wear and galling) properties of the die [5, 6] but when this "white" layer becomes highly porous and thick, it turns brittle. During nitrocarburizing of tool steels the compound layer with hard ε-carbonitride approaches 100% of the total volume with a certain degree of porosity [7] and oxides over it depending on the gas atmosphere used. The presence of oxygen plays important role in the process because it acts like a catalyst for ε-phase formation [8]. It is also known that the formation of some Fe_3O_4 in the surface layers improves the wear [9, 10] and corrosion resistance [11].

The underlying diffusion layer is responsible for the strengthening that enhances wear resistance and fatigue endurance [12-14] since the nitride/carbonitride phases in it are disperse and coherently bounded. In the presence of nitride stabilizing elements reactive diffusion of nitrogen with alloy carbides takes place.

Although the structure and the properties of gas nitrocarburizing steels have been subject of intense studies for a large number of years, the actual understanding of the influence of particular factors is still missing. It is not clear from the literature to what extent the external pressure and its variation during the gas nitrocarburizing could influence the structure, thickness and properties of the compound and diffusion layers. Additionally, as the gas nitrocarburizing technology of tool steels generally features long treatment, higher gas and energy consumption, thick and brittle compound layer and low environmental efficiency, the conventional gas process provides some restrictions.

In order to overcome the above mentioned weaknesses, the innovative and original gas ONC technology at low pressure is held at short cycle times that are employed to prevent surface embrittlement. The change in the gas pressure enables control on the formation of consistent surface layers with desirable thickness and properties. This is because the low pressure conditions during the first stage of the process thermodynamically destabilize the iron-nitride formation that postpones the compound layer growth [15].In contrast, a short incubation time for compound layer formation during the conventional nitrocarburizing process is needed [16]. The dense structure of this "white" layer acts like a diffusion barrier that hinders the saturation of the diffusion layer in depth and the former becomes thick and porous. During the vacuum ONC of tool steels the alloying elements nitride formation is not influenced by the change in pressure and a large microstructural and morphological modification of the surface layers could be expected.

The vacuum ONC is employed in a cycle gas flow regime at pressure values varying between 1.10^4 Pa and 8.10^4 Pa as the former is reach when pumping down the inert or exhausted gases and the second – when dissociation slowly occurs and gas reactions take place. Therefore, another advantage of the vacuum ONC process is the cycling gas flow that allows a longer stay of the reactive molecules in the chamber and their stronger tendency to decompose and interact. Moreover, the periodic pumping down the gas phase from the vacuum chamber clears the chemisorbed gas products from the surface at regular intervals. This also enables the full utilization of the reactive gases. In this way the gas consumption is significantly low and the process is cost effective, safety and environmentally-friendly. Simultaneously, the thermodynamics of the phase transformations allows certain control by the non-conventional external conditions.

The mechanical strength of both layers depends on the volume fraction and the types of the crystal lattice of the resultant products after ONC. As mentioned above the ONC process possesses attractive advantages and perspectives but its capabilities in terms of tool steel surface enhancement remains unexamined. On the other hand, the microstructure evolution prediction and control during the thermo-chemical process are thus very important. The ongoing efforts to simulate the nitriding and nitrocarburizing processes has been continuing for over 20 years because when calculations applied provide quick and reliable results, they could be very important in industrial applications. Unfortunately, there is still a lack of coincidence of the modelling the microstructure and composition evolution and the experimental results during nitrocarburizing.

The thermodynamic databases of "Equilib module" of FactSage software could be used when constructed to simulate the thermodynamic behaviour and phase transformation of the multicomponent systems depending on the composition, temperature and pressure. In order to validate the relevance of the thermodynamic calculations, the composition dependent transformations under the influence of N, C and/or O need to be experimentally determined. For this purpose, four different tool steels (one with less high temperature strength and three for hot work) are examined for the influence of the additional elements on the composition and structure of both compound and diffusion layers. These steels presents a wide variation in composition but they have some common elements such as C, Fe, Cr, V and Mo. Moreover, the four grades are commonly use nitriding/nitrocarburizing steels for extrusion dies in extruding aluminium or plastics that maintains high impact strength in quenched and tempered condition. The experimental data of the microstructure, chemical and phase composition of the ONC tool steels are presented and the applicability of the thermodynamic modelling approach is verified.

2 Experimental procedures

2.1 Materials, heat treatment, samples preparation

Four different steel grades - AISI H10 (DIN 1.2365 EN 32CrMoV12-28, 3Х3М3Ф), AISI H11 (DIN 1.2343, EN X37CrMoV5-1, 4Х5МФС), AISI H21 (DIN 1.2581, EN X30WCrV9-3, 3Х2В8Ф) and AISI D2 (DIN 1.2379, EN X153CrMoV12, Н12МF) with composition shown in Table 1 (measured by quantometer SpectroMaxx) were subject to the vacuum ONC process. All the experiments were carried out with round or quadrangular specimens with sizes of about 25 mm and a thickness of 10 mm. The samples were used in quenched in agitated mineral oil and tempered condition, which was obtained by conventional heat treatment. The heat treatment regimens were shown in Table 2. The second time tempering was physically coupled with the thermochemical treatment. The microstructure of the tool steel samples prior to the thermochemical treatment was tempered martensite with carbides.

Table 1. Chemical composition of the steel samples (wt.%) before the thermochemical treatment.

Element, wt.%	**C**	**Si**	**Mn**	**Cr**	**Mo**	**V**	**W**	**P**	**S**	**Fe**
AISI H10	0.332	0.52	0.409	3.57	3.73	0.70	0.114	0.023	0.0099	Bal
AISI H11	0.36	1.18	0.423	5.38	1.32	0.33	0.034	0.034	0.0051	Bal
AISI H21	0.292	0.436	0.373	2.57	0.053	0.212	8.38	0.022	0.013	Bal
AISI D2	1.51	0.275	0.386	13.14	0.61	0.203	0.04	0.021	0.0075	Bal

Table 2. Preliminary heat treatment of the samples.

Steel/Heat treatment	**AISI H10**	**AISI H10**	**AISI H21**	**AISI D2**
Quenching temperature,°Ctime	1000-1020 10 min	1040-1050 10 min	1130-1140 10 min	1040-1050 10 min
Tempering temperature, °C hour	540-550 1 hour	540-550 1 hour	540-550 1 hour	540-550 1 hour

2.2 Thermo-chemical treatment

A hole of 2 mm in diameter was drilled near the edge of the samples for holding them during the process. The decarburized layer that formed at the surface during the heat treatment was removed by grinding. Before the thermochemical process, the steel samples were mirror polished, lapped and degreased. The oxy-nitrocarburized layers were prepared by a vacuum process carried out in industrial equipment for 7 hours at 570°C in NH_3 and CO_2 atmosphere (ratio 90:10 vol. %, respectively) in a cycling gas flow rate. During stage I of the process the pressure in the vacuum chamber was 7-8.10^4 Pa for the first five hours and increased up to 1.10^5 Pa during the last 2 hours (stage II) [17]. The cooling off was carried out in the NH_3 atmosphere till 400°C and after that in N_2 in air.

2.3 Characterization

Cross sections of the oxy-nitrocarburized samples were prepared by conventional metallographic techniques. The light optical microscopy (LOM) was performed on cross sections of the samples using a Neophot 21 (Zeiss, Jena) microscope after etching the samples with picral reagent. An oil immersion objective for reflected light was applied. The average thickness of the compound and diffusion layers was determined in cross-section using a digital camera images.

The metallographically prepared sections were also used for scanning electron microscopy (SEM) analysis. The secondary electron micrograph of the surface of the ONC samples are taken by JEOL JSM-5510 (Japan) and accelerating beam voltage of 10 kV, vacuum pressure below 1,33.10^{-5} mbar. Prior the analysis each sample was coated with thin layer of gold to reduce electric charge of the specimens scanned and to increase the image resolution.

For the phase analysis X-ray diffraction (XRD) was performed on the top surface of the samples using URD-6 diffractometer with Fe-Kα line-focused radiation applying Bragg-Brentano geometry. The diffractograms were analysed over a 2θ range of 30° to 120 ° degree with a step size of 0.1° 2θ and continuing time of 1.5 sec/step. The Fe-Kα radiation source was used at 30 kV and 20 mA. XRD Match!3 software was used to determine the phases of the resulting data.

The main metalloid elements participating in the ONC process – N, C and O composition-depth profiles were determined by glow discharge optical emission spectroscopy (GDOES) of LECO GDS-850A.

3 Results and discussion

3.1 Modelling

Before the ONC experiments, Fe-N and Fe-N-C phase diagram calculations were performed by using advanced "phase diagram" module of FactSage 7.0 [18]. In the calculations, SGTE2014 database was selected to detect the liquid and solid solutions in the system.

Iron-nitrogen interaction is important to understand the nitriding process phenomena. Figure 1 illustrates a Fe-N binary phase diagram calculated by FactSage at 1 bar. As expected, nitrogen is less soluble in the ferritic α-Fe phase (BCC_A2) rather than austenitic γ-Fe phase (FCC_A1). Below eutectoid temperature (~587 °C), iron nitride γ′ phase (M_4N) occurs when the nitrogen level exceeds solubility limit of α-Fe. Subsequent Fe_2N_{1-x} phase, which is so called ε phase (HCP_A3), emerges after N/(Fe+N) mole ratio exceeds 0.2.

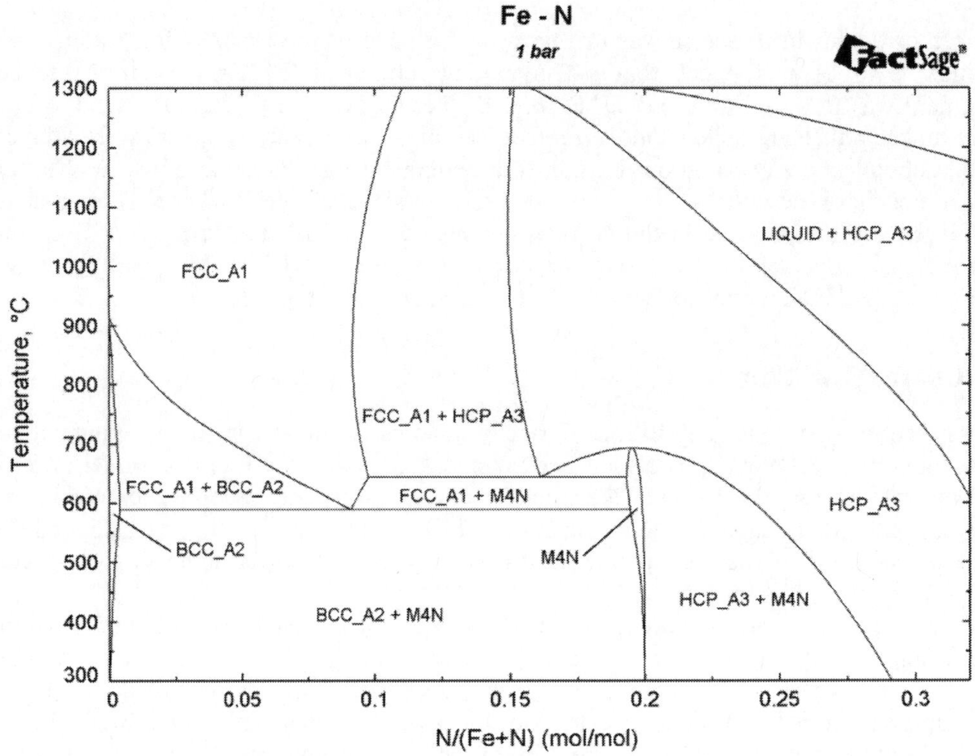

Fig. 1. Fe-N binary phase diagram calculated by FactSage at 1.10^5 Pa (1 bar).

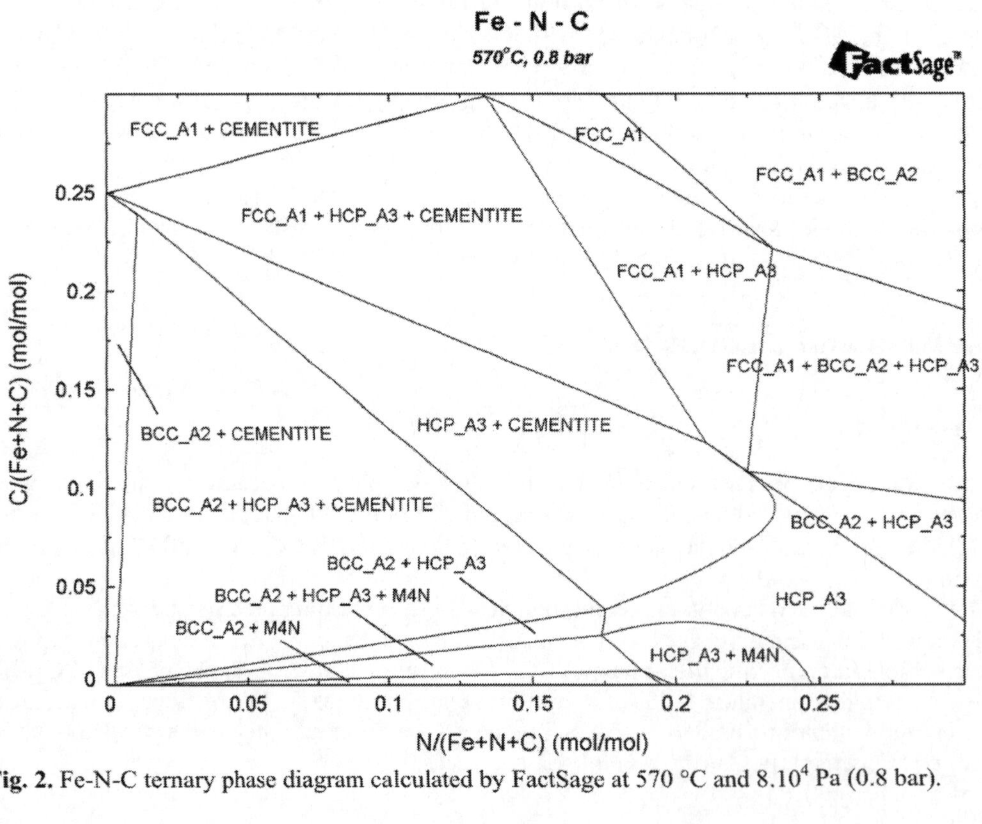

Fig. 2. Fe-N-C ternary phase diagram calculated by FactSage at 570 °C and 8.10^4 Pa (0.8 bar).

DOI: 10.1051/matecconf/20179403011

In order to simulate the ONC process in equilibrium condition, an Fe-N-C ternary phase diagram was calculated by FactSage at 570 °C and 8.10^4 Pa. Figure 2 shows that, very high concentrations of N and low concentrations of carbon in the iron, which may represent the top layer of the transformed area, stimulate the formations of mixture of $\gamma'(M_4N)$ and ε (HCP_A3) phases or only single ε (HCP_A3) phase in the system. When carbon concentration is increased at these levels, cementite or α-Fe phases (BCC_A2) are formed, respectively, together with ε (HCP_A3) phase.

3.2 Experimental verification

The microstructure of AISI H11 (Fig. 3b) steel is very similar to that of the ONC AISI H10 (Fig. 3a) steel, but the compound layer of the former appears to be thinner and the carbides in the diffusion layer looks slightly coarser. The quantity of the alloying elements in the solid solution (tempered martensite) is pre-determined by the initial heat treatment of the steel. Beneath the compound layer, the microstructure of the diffusion one that contains coarse laths of the tempered martensite outlined by the grains of the former austenite, is clearly seen. The lighter tempered martensite is represented by the darker looking C-rich regions (Fig. 3a and b). The carbon atoms released within the carbide-nitride transformation are expected to diffuse within the grain boundaries, where Fe_3C is formed (Fig. 3a and b).

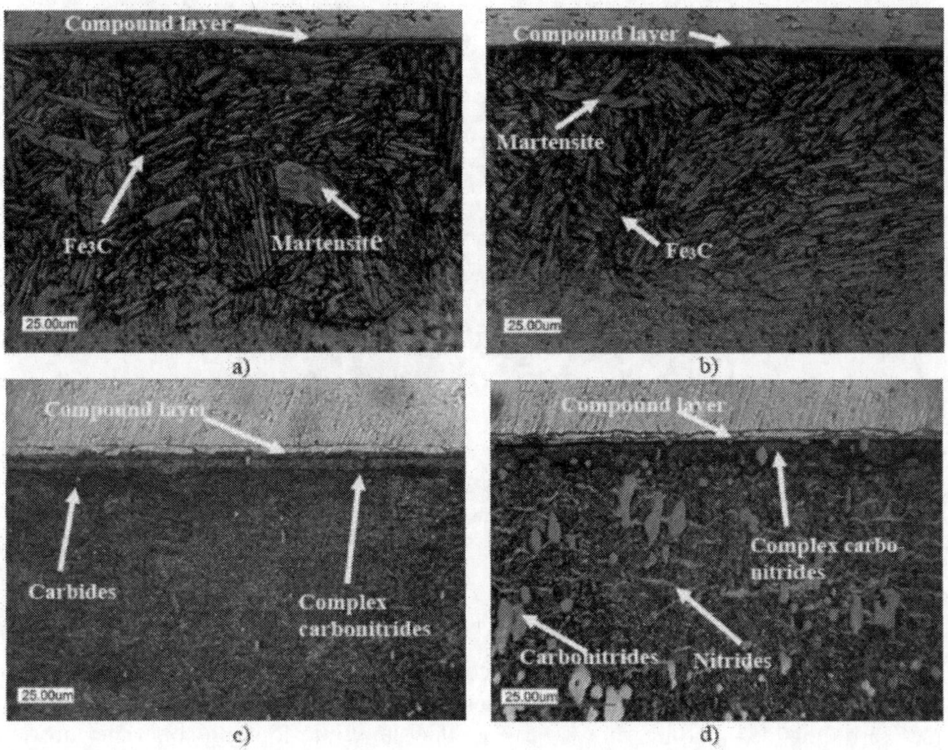

Fig.3. Optical micrograph of cross-section of: a) ONC H11 steel; b) ONC H10 steel; c) ONC H21 steel; d) ONC H10 steel. All the samples are etched in picral.

In the ONC H21 steel (Fig. 3c) the interface between the compound and diffusion layers is well defined and seems dark and thick (beyond 5 μm in depth). The tempered martensite structure of the diffusion layer seems more disperse than that of the previous samples. In this steel white-looking globular carbide phases are present in the diffusion zone. These are likely to be W-containing carbides since although the WN formation may provide a higher structural stability, the small diffusivity of tungsten hinders this process. The compound layer of the ONC AISI D2 steel (Fig. 3d) appears to be somewhat thicker than that of the other samples. The former primary carbides in the diffusion layer that have become carbonitrides (with high carbon content), look darker than the phases with the same morphology located deeper in the diffusion layer. The bright nearly parallel with respect to the surface tubular phases are the Cr-containing nitirdes. Apparently, Cr in the ONC D2 alloy could be considered as mobile because of the conversion of the primary carbides ($M_{23}C_6$) into carbonitrides/nitrides and the dissolution of carbides close to the compound layer interface. The "aging" of the nitrides in the diffusion layer is obvious because of the formation of large nitride crystallites.

Fig.4. Secondary electron micrographs of the ONC samples: a) structure of the diffusion layer of the ONC H10 steel - tempered martensite and carbonitreds; b, c) diffusion layer structure of the ONC H21 steel.

Since the carbon content in H10 steel is much lower than the equivalent to convert all the alloying elements in the V-, Cr- and Mo-containing carbides, the martensite contains a substantial amount of dissolved alloying elements that are available for the development of disperse (carbo)nitrides that are shown on Fig. 4a. The white-looking spherical nitrides in the diffusion layer of the H21 sample (Fig. 4b and c) have similar shape with the primary carbides, but there are far more disperse (Fig. 4c). It is reasonably to be considered that these are transformed $(Fe\text{-}Cr)_{23}C_6$ because the latter carbide nucleates faster during the tempering procedure phases when W is added [19]. During the prolonged ONC treatment and when the alloy element concentration is above ~ 2 wt.% [20] the discontinuous precipitates coarsen into lamellae and spheres.

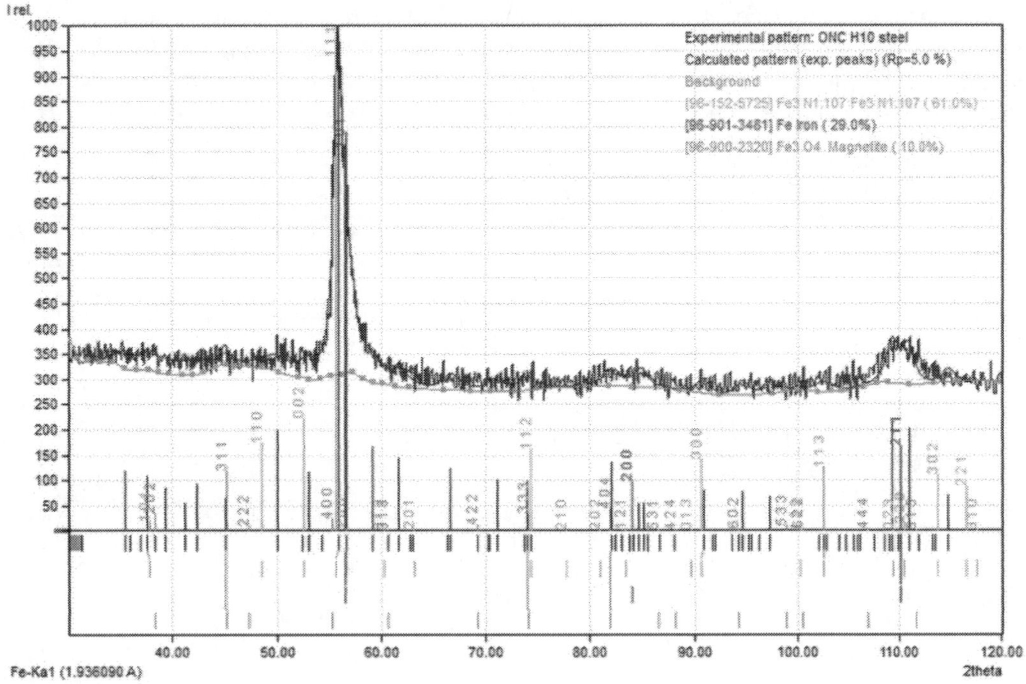

Fig.5. XRD diffractograms of the ONC AISI H10.

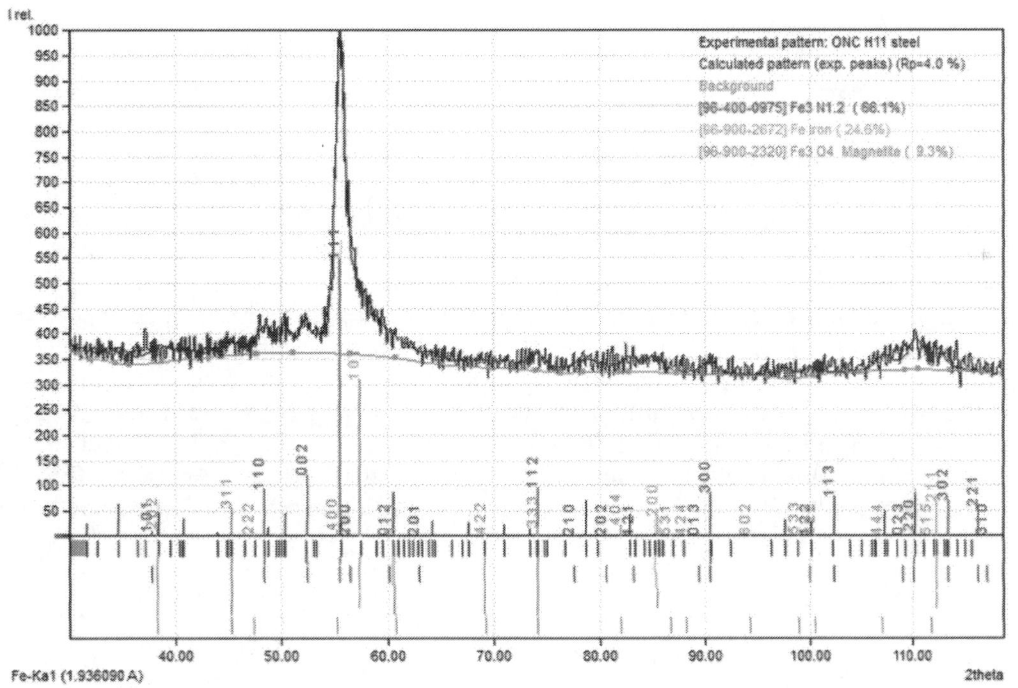

Fig.6. XRD diffractogram of the ONC AISI H11 steel.

The XRD analysis of the ONC AISI H10 steel reveals that ε-$Fe_{2\text{-}3}N$ is the dominant phase together with minor peaks of ferrite (α-Fe) and magnetite (Fe_3O_4). The diffraction lines of the ε-peak determines a phase stoichiometry close to $Fe_3N_{1.107}$ (Fig. 5) with unit cell parameters a= 4.7126 Å c= 4.3752 Å. The unite cell parameter of the ferrite (tempered martensite) is determined to be near a= 2.8920 Å. The surface ε-carbonitride is not oversaturated with nitrogen (slightly more than 12 wt.%) and the compound layer is approximately 2 μm thick (Fig. 9). At a distance of about 0.8 μm the N and C concentration become merely equal (Fig. 10). The inside porosity of the ε-Fe_{2-3}(N-C) reaches a depth of about 0.5-0.8 μm judged from the oxygen concentration peak (Fig. 11), where Fe_3O_4 is formed (see XRD results).

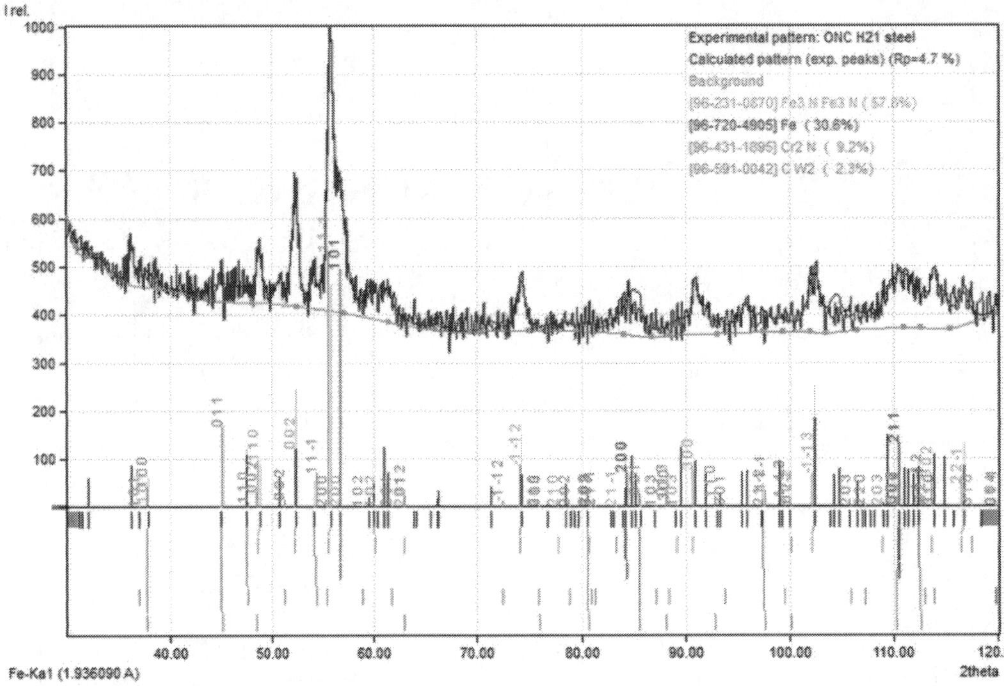

Fig. 7. XRD diffractogram of the ONC AISI H21 steel.

Taking account of the XRD results of the ONC H11 sample (Fig. 6), the composition of the compound layer constitutes monophase carbonitride with a stoichiometry close to $Fe_3N_{1.2}$ and larger cell parameters (a = 4.7241 Å c = 4.3862 Å) than the AISI H10 ε-phase. This indicates a higher interstitial atom content in the ε-phase. The tempered martensite cell parameter is smaller (approximately a= 2.8550 Å). Within the ε-compound layer of the H11 steel, a significant decrease of the nitrogen concentration at a depth of 0.4-0.5μm has occurred and beyond it rises again up to 2 wt. % (Fig. 9). This minimum in the nitrogen concentration underneath could be attributed to the diffusion of nitrogen to a larger depth evoked by nitride forming elements and the hindered nitrogen uptake and inward diffusion beyond 0.4 μm depth where the carbon and oxygen concentration prevails (Fig. 10 and 11).

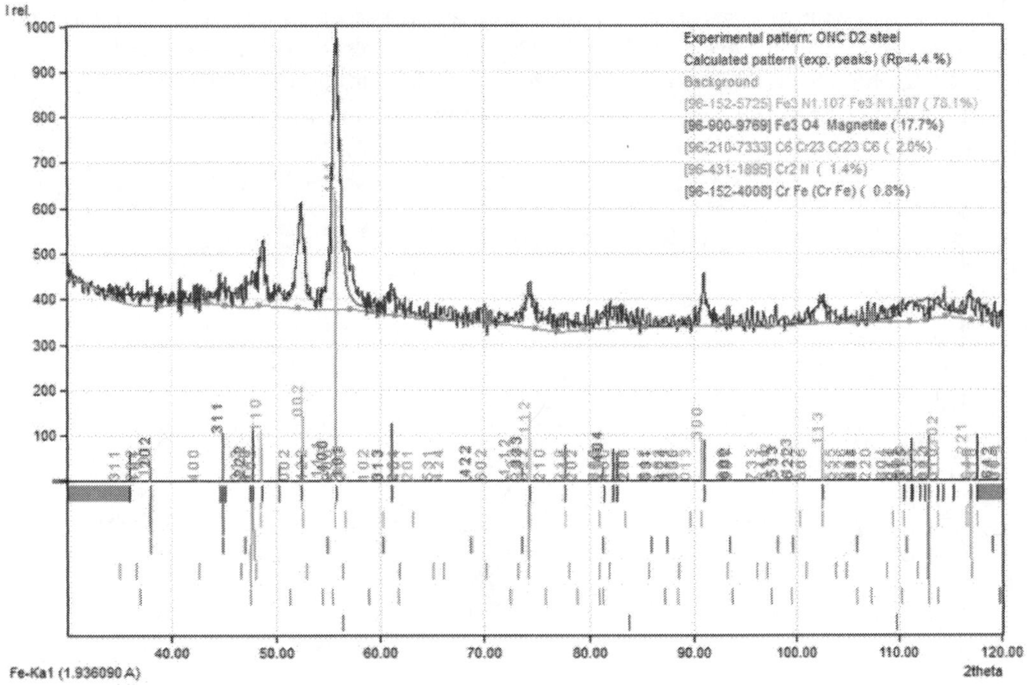

Fig.8. XRD diffractogram of the ONC AISI D2 steel.

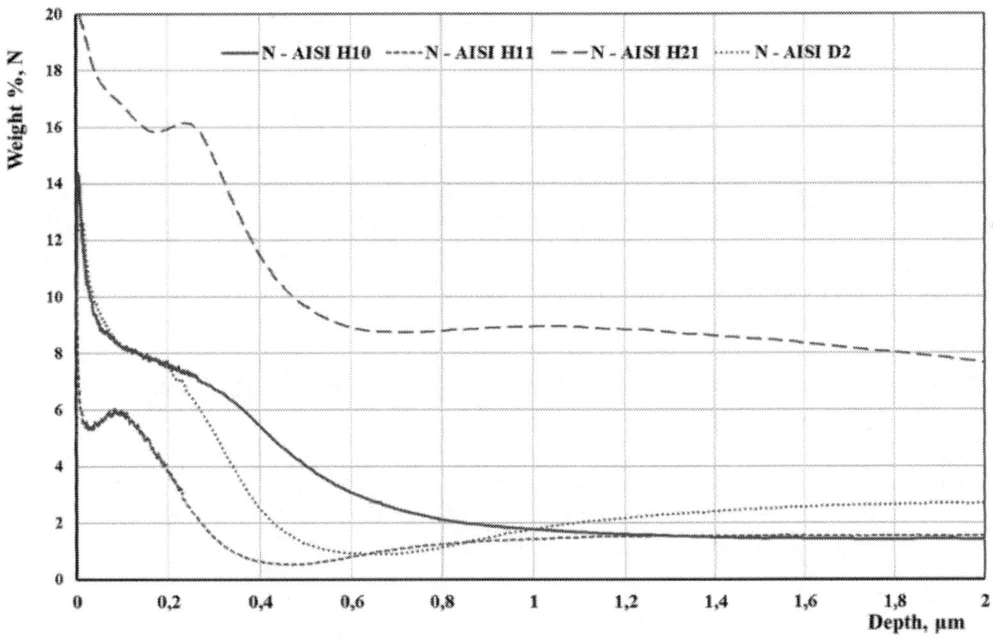

Fig. 9. GDOES concentration–depth profiles of nitrogen of the ONC samples.

The XRD spectra of the AISI H21 steel (Fig. 7) indicate that the main phases after ONC are the ε-carbonitride and magnetite. The ε-carbonitride formula corresponds to Fe_3Nwithunit cell parameters a= 4.7160 Å c= 4.3940 Å. In the compositionally complex surface region, magnetite is formed on the inside of the ε-carbinitride pores at level beneath 0.5 µm equal to the depth of the porosity determined by the GDOES results (Fig. 11).In the XRD diffractogram of the ONC AISI H21steel(Fig. 7) minor peaks of the detected phases

DOI: 10.1051/matecconf/20179403011

belong to Cr_2N with hexagonal lattice structure. Small percentage of W_2C phase is also determined using XRD analysis. In the AISI H21 steel the registered N concentration at the surface is the highest of all samples and a smoother drop in the nitrogen concentration in depth is observed (Fig. 9). The carbon concentration does not exceed that of the nitrogen but forms two separate carbon hills (Fig. 10).

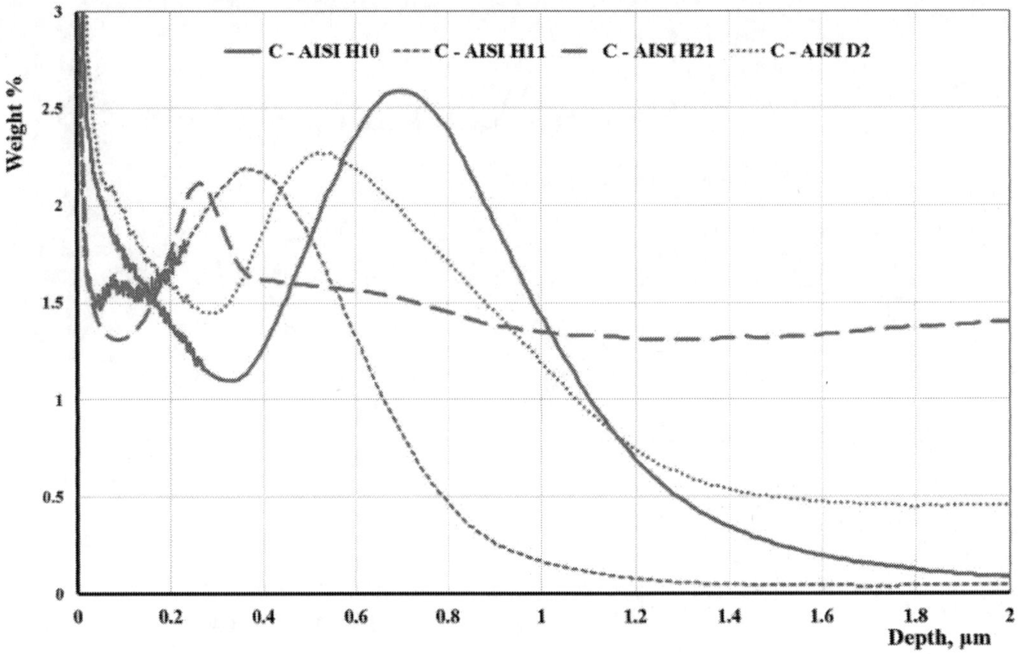

Fig. 10. GDOES concentration–depth profiles of carbon of the ONC samples.

The XRD analysis of the ONC D2 sample (Fig. 8) demonstrates intense peaks of ε-$Fe_3N_{1.107}$and Fe_3O_4 and minor peaks from $Cr_{23}C_6$, Cr_2N and Cr-Fe ferrite. The ε-compound layer shows limited amount of porosity as a consequence of the ONC treatment. The thicker compound layer is evidenced by the presence of more than six ε-$Fe_{2\text{-}3}$(N,C) distinct reflections. Their positions correspond to ε-$Fe_3N_{1.107}$ with lattice cell parameters a= 4.7126 Å c= 4.3752 Å. The detected reflections show the presence of hexagonal Cr_2N and cubic $Cr_{23}C_6$ phases. The Cr-containing tempered martensite (Cr-Fe) is found to have larger (a = 2.9000 Å) cubic unit cell relative to the others.

From the GDOES results of the ONC H21 and D2 steel (Fig. 9) is clear that the higher the content of the nitride forming elements, the larger the capacity of the diffusion layer for binding nitrogen in it. Therefore, the nitrogen concentration in the diffusion layer is relevant to the alloying elements concentration.

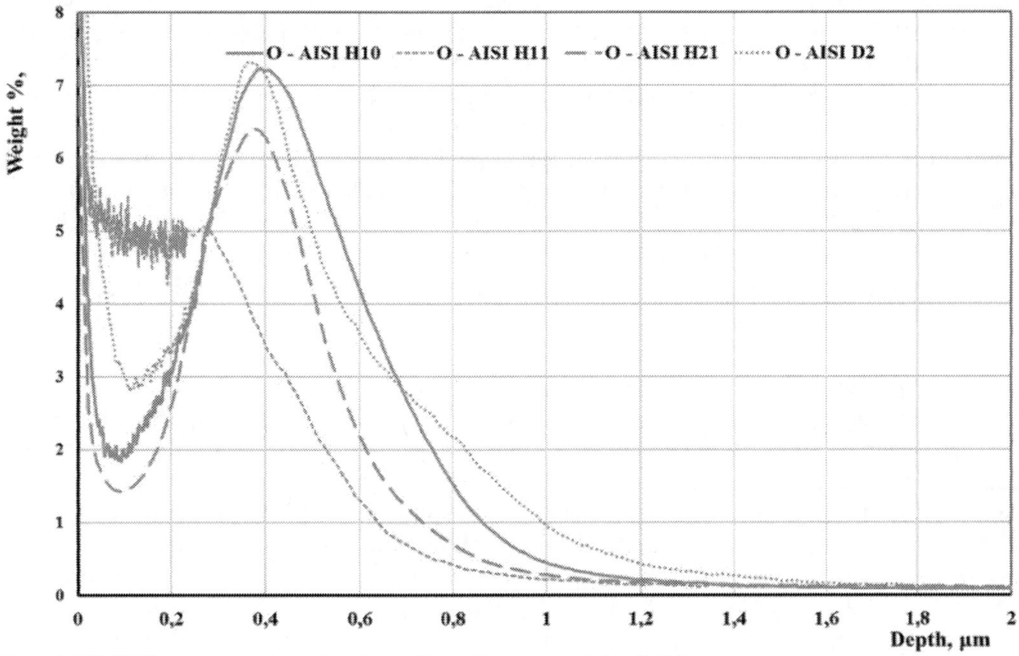

Fig. 11.GDOES concentration–depth profiles of oxygen of the ONC samples.

4 Discussions

At low pressure conditions (stage I), where unstable compound layer nuclei are formed, a large period of time is needed to establish local equilibrium of gas phase atmosphere with the solid substrate. This effect is highlighted during vacuum ONC of the tool steels because much more nitrogen is taken by the alloying elements before attaining saturation at the surface since the alloy nitride phases develops in direct contact with the gas atmosphere. The stronger nitride forming elements are supposed to combine first with nitrogen until they are exhausted [21]. However, although thermodynamically possible, a certain solid state reaction, where the number of variables is high, may not be kinetically favorable. As the nitrogen concentration in the diffusion layer grows, the driving force for precipitation of the alloying elements nitrides certainly increases. In that event, since the Cr diffuses more rapidly in ferrite than the other metallic alloying elements (such as V, Mo, W), Cr-containing nitride precipitates are encountered even in the ONC H21 – the tool steels with lower Cr content. The Cr in this area participates in the formation of Cr_2N phase where the ordered arrangement of metal atoms corresponds to hcp lattice and the nitrogen atoms are distributed in the interstitial voids such as in ε-$Fe_{2-3}(N,C)$. Simultaneously, the precipitation of the nitrides of the alloying elements has no considerable effect on the concentration distribution in depth of the diffusion layer of all samples, although the coarsening process is additionally benefited from the preliminary single tempering and, therefore, the lower amount of nitride forming elements tied as carbides.

The discontinuous coarsening of fine nitrides involves also the growth of the cubic martensite from the different nucleation sides. The quantitative information for the full phase composition drown by the XRD results could be misleading to a certain extend because the lattice structure of many nitrides is similar to that of the carbides.

The increase in pressure value (stage II) increases the thermodynamic stability of the iron nitrides together with the surface nitrogen concentration that leads to the occurrence of iron nitrides at the surface. The results obtained from the surface composition and phase analysis of the compound layer demonstrate that the reaction product after 7 hours of ONC is composed of pure ε – iron carbonitride. As opposed to alloying niritrides, only the diffraction line of the ε-Fe_{2-3}(N,C) is considered to be unaffected by the superimpositions. Concerning the lack of γ′ - nitride, this phase appears to be absent during the ONC of the tool steels because of four main reasons: 1) crystallographic resemblance of ε, Cr_2N, orthorhombic carbides; 2) the higher carbon content in the surface area (coming from C, released from the alloy carbides transformation[22] together with that from the external source) that stabilizes the ε-phase [23]; 3) the low solubility range of N and C in γ′- Fe_4N [24]; 4) the delayed compound layer formation because M. Somers at al. [25] proposed that in contrast to the treatment of iron, the γ'-formation during nitrocarburizing needs prolong process time. From the results of the present study is clear that the thickness of the compound layer within the vacuum ONC depends on the pressure, composition and phase structure of the underlying carbonitride film.

Based on the thermodynamic calculation it is seen that the ε-phase is becoming stable with increasing of nitrogen and a very small increase of carbon. From the diagram is clearly seen that no single γ′-phase regions could be formed during the vacuum ONC process compared to the pure iron. The computational results show the large phase stability of coexisting ε, cementite and ferrite phases and their stability increase with the nitrogen and carbon concentration even at very low nitrogen concentrations. The microstructure and XRD results are consistent with these FactSage predictions. The increase of carbon stabilizes the ε-phase that is in a good agreement with the isopleths plot. However the mole fractions of these phases depends on both N and C content in the steel substrate that may lead to difficulties in their experimental detection. Since the stage II involves higher pressure in the chamber, the final phase stabilities should be determined by this stage. The oxygen bearing atmosphere allows the oxygen to dissolve into the ε-iron carbonitride but at 570 °C oxides are not likely to be formed due to the presence of highly active hydrogen in the atmosphere. When the oxygen activity increases (during the cooling stage [15]) the magnetite grows in the pore structure of the surface carbonitride layer. The discrepancy between the experimental and predicted phase composition may be due to neglecting the occurrence of oxides when lowering the temperature.

The ε-carbonitride with relatively low nitrogen content (as in the ONC H11 steel) may tend to decompose into ε-phase with higher nitrogen content at the surface and carbides and/or oxides at the innermost area of the layer. Close to the compound/diffusion layer interface judging from the composition some intermediate ε-carbonitride-magnetite phases are likely to be formed. Depending on the concentration of the interstitially dissolved in ε elements, the excess nitrogen from the carbonitride-oxide transformation could either dissolves in the ε-Fe_{2-3}(N,C) without causing local nitrogen enrichment but displaying a steeper nitrogen concentration gradient (as in AISI H10, H11 and D2 steels) or participate in nitrogen recombination (as in AISI H21 steel). This results are in agreement with earlier observations showing that microvoids are observed because of the iron vacancies formed during oxidation [26], but more experimental work is needed to verify this interpretation.

5 Conclusions

The present study demonstrates a new solution for successful surface modification of tool steels and its efficacy. According to the results it could be claimed that during the first stage of the vacuum ONC local equilibrium condition is likely to prevail at phase interface in the diffusion layers that appeared to be time, pressure and substrate composition dependent. Concerning the overall alloy elements content, it could be concluded that the higher concentration of the nitride forming elements increases the driving force for the coarsening reaction. The coarsening process is additionally enhanced by the preliminary single tempering and, therefore, the lower amount of nitride forming elements tied as carbides. The increase in pressure value that increases the thermodynamic stability of the iron nitrides results in the formation of thin ε – iron carbonitride compound layer. The thickness of the compound layer during ONC depends on not only on the volume fraction of the nitride forming elements but on their type and interphase distribution. Close to the compound/diffusion layer interface an intermediate ε-carbonitride-oxide phase region is likely to be formed that turn to magnetite containing compound layer during cooling.

A customized diagram is successfully developed by computational thermodynamics for the vacuum ONC of tool steels. This diagram predicts a large area of coexisting ε, cementite and ferrite phases under the low pressure conditions of the process. It is seen that the model satisfactorily describes the experimental data to the detected nitrogen and carbon contents. The discrepancy between the experimental and predicted phase composition is due to neglecting the occurrence of iron oxides. The isotherms do not describe equilibrium gas mixture composition and substrate composition but when incorporating the thermodynamic predictions with the kinetic simulations, the ONC process of different alloys could be optimized.

The paper reflects the outcomes of the project № 16 – FMME – 02, financed by the Fund "Scientific Research"of the University of Ruse "A. Kanchev". The authors would like to thank D-r Slavcho Topalski for his help with the execution of GDOES analysis in Fraunhofer Institute (für Werkstoff- und Strahltechnik IWS).

References

1. M. Nikolova, P. Danev, I. Dermendzhiev, J Mat Sci Tech (Bulgaria), **21** 2, 84-96 (2013)
2. K. Shetty, S. Kumar, P. R. Rao, Surf Coat Tech **203**10, 1530-1536 (2009)
3. J. Wang, J. Xiong, Q. Peng, H. Fan, Y. Wang, G. Li, B. Shen,Mater Charact **60,**197 – 203 (2009)
4. M. Teimouri, M. Ahmadi, N. Pirayesh, M. Khoee, H. Khorsand, S. Mirzamohammadi, JAlloys Compd, **477** 1-2, 591-595 (2009)
5. M. HeydarzadehSohi, M. Ebrahimi, A. HonarbakhshRaouf, F. Mahboubi, Surf Coat Tech **205,** 84–89(2010)
6. M. Tercelj, A. Smolej, P. Fajfar, R. Turk, Tribol Intl **40**, 374–384(2007)
7. P. C. King, R. W. Reynoldson, A. Brownrigg, J. M. Long, Surf Eng**21** 2, 86-98 (2005)
8. T. Bell, Heat Treat. Met., **2** 2,39–49 (1975)
9. S.-H. Chang, K.-Ts. Huang, Y.-Hs. Wang, Mater Trans, **53** 4, 745 – 751 (2012)
10. A. Khadar Syed, J. Hardell, B.Prakash, Estonian J. of Eng **16** 2, 123–134 (2010)
11. K. H. Lee, K. S. Nam, P. W. Shin, D.Y. Lee, Yo-S. Song, Mater Lett.**57,** 2060 – 2065 (2003)
12. G.H. Farrahi, H. Ghadbeigi, JMater. Process. Technol.**174,** 318–324 (2006)
13. F. Ashrafizadeh, Surf Coat Tech**174–175,** 1196–1200 (2003)
14. M. Rosso, A. Bennani,Trans on Eng Sci**17,**25-134 (1997)
15. D. Nikolov, M. Nikolova, Scientific Technical J Mater Sci. Non-equilibrium phase transformations **3**, 29-33, (2015)
16. E.J. Mittemeijer, ASM Handbook, Vol. 4A, *Steel Heat Treating Fundamentals and Processes*, J. Dossett and G.E. Totten (editors), 619-646 (2013)
17. P. Danev, Proc. of Scientific Session, *RU&SU'2004*, Eng. Sci. (in Bulgarian) **41** 2, 73-77 (2004)
18. C.W. Bale, E. Bélisle, P. Chartrand, S.A. Decterov, G. Eriksson, A.E. Gheribi, K. Hack, I.-H. Jung, Y.-B. Kang, J. Melançon, A.D. Pelton, S. Petersen, C. Robelin, J. Sangster,

P. Spencer, M-A. Van Ende, FactSage thermochemical software and databases, 2010–2016 Calphad, **54,** 35–53 (2016)
19. O. Salas, J. Oseguera, N. Garci, U. Figueroe, J.Mat Eng &Perf, **10** 6, 649-655 (2001)
20. S. S. Hosmani, Metal Mater.Eng.**19** 1, 65-84 (2013)
21. S.M. Hassani-Gangaraj, M.Guagliano, Appl. Surf.Sci**271,** 56–163(2013)
22. P.C King, R. W. Reynoldson, A. Brownrigg, J. M. Long, Surf Eng **21**2, 86-96 (2005)
23. Mei Yang, Dissertation, Worcester Polytechnic Institute, 1-139 (2012)
24. M.A.J. Somers, International Heat Treatment and Surface Engineering IFHTSE Global 21, **5**1, 7-16 (2011)
25. M.A.J. Somers, E.J. Mittemeijer, Heat Treating, *Proc. 17th ASM Heat Treating Society Conference*, ASM International, 321-330 (1998)
26. R. H. Jutte, B.J. Kooi, M. A. J. Somers, E.J. Mittemeijer, Oxid. Met **48** 1/2, 87-109 (1997).

Research on the mechanical behaviour of an airplane component made by selective laser melting technology

Răzvan Păcurar[1,*], *Ancuța* Păcurar[1], and *Nicolae* Bâlc[1]

[1]Technical University of Cluj-Napoca, Faculty of Machine Building, Department of Manufacturing Engineering, B-dul Muncii no. 103-105, 400641, Cluj-Napoca, Romania

Abstract. The main objective of the presented research consists in the redesign of an airplane component to decrease its weight, without affecting the mechanical behaviour of the component, at the end. Femap NX Nastran and ANSYS FEA programs were used for the shape optimization and for the estimation of the mechanical behaviour of a fixing clamp that was used to sustain the hydraulic pipes that are passing through an airplane fuselage, taking into consideration two types of raw materials – Ti6Al4V and AlSi12 powder from which this component could be manufactured by using the selective laser melting (SLM) technology. Based on the obtained results, the airplane component was finally manufactured from titanium alloy using the SLM 250 HL equipment that is available at SLM Solutions GmbH company from Luebeck, in Germany.

1 Introduction

Airplane industry is one of the most spectacular fields of research regarding the Selective Laser Melting (SLM) technology. New types of materials were successfully developed in the last years in this field, being machinable on the SLM modern equipment [1-3]. The mechanical behaviour of the airplane components produced from different type of materials was one of the most important researches developed in this field by different researchers in the world [4-5]. Shape optimization and FEA programs are often used in order to estimate the mechanical behaviour of different type of airplane components made by SLM technology [6-9]. The research performed by the authors in this field of research was mainly focused on the possibility of redesigning of an airplane component - the fixing clamp used for sustaining the hydraulic pipes that are passing through an airplane fuselage, in such way that the weight will be significantly decreased, without affecting the mechanical behaviour of the component, at the end. Femap NX Nastran program was successfully used for optimizing the shape of the component and ANSYS FEA program was successfully used for estimating the stress and strain values of the component to be made by aluminium or titanium alloy by SLM. Based on the obtained results, the fixing clamp was manufactured from Ti6Al4V material by selective laser melting technology at SLM Solutions GmbH company from Luebeck, Germany, at the end.

[*]Corresponding author: razvan.pacurar@tcm.utcluj.ro

2 Redesign of an airplane component made by SLM technology

The research performed in the field of aerospace industry was focused on the redesigning and analysis of the airplane component presented in Fig. 1.a), with the aim of decreasing the weight of the component without affecting its mechanical strength at the end. For the redesigning process of the airplane component, a dedicated topology optimization program called Femap NX Nastran was used, knowing that in the bore area of the component (Ø50 [mm] - size diameter), a force will be applied on the entire circumference and knowing that the base area of the component will be fixed with 8 screws on the airplane fuselage. These were the main constrains that were applied in the case of the airplane component, as presented in Fig. 1.b).

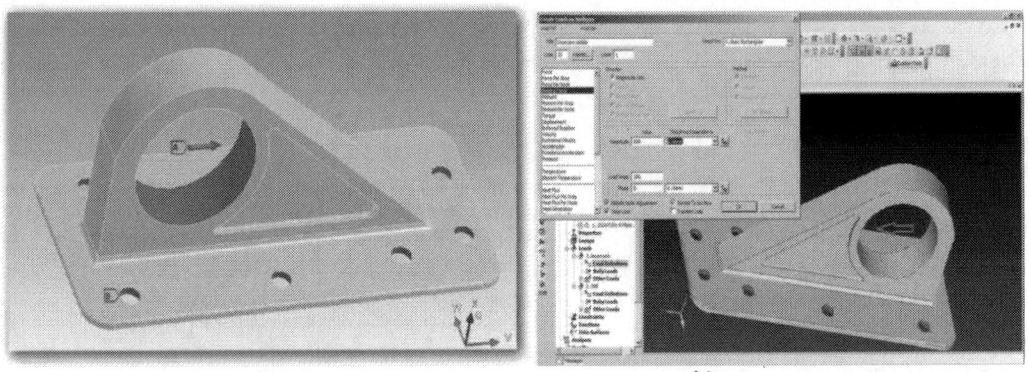

a) b)

Fig.1. Kinematical and loading constrains applied on the component surface - Femap NX Nastran.

Two types of materials machinable by Selective Laser Melting (SLM) process were considered for the made analysis – TiAl6V4 and AlSi12, the mechanical characteristics, such as Tensile strength, Yield strength or Poisson ratio, undertaken from the material file of the supplier company (SLM Solutions GmbH from Germany) being defined manually in the library of the FEMAP program, as well. After the mesh was generated, by specifying the percent volume reduction in the FEMAP program (variable percent), it was possible to observe that a significant volume of the material can be removed in the connecting area of the base plate and supporting leg of the component (see Fig. 2 a, b and c). Same observation was valid in the case when the TiAl6V4 or AlSi12 material was considered for the manufacturing of the airplane component and with the increasing of loading force from 500 [N] to 3000 [N], range that was considered for the made analysis.

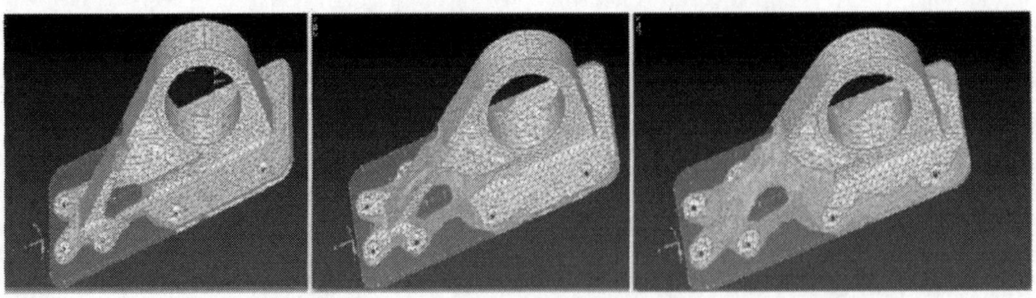

a) b) c)

Fig.2. Results of the topology optimization for the percent volume reduction of: a) 20 %; b) 30 %; c) 45 % .

 DOI: 10.1051/matecconf/20179403012

Based on the results obtained using the Femap NX Nastran program, the optimized shape of the airplane component has been designed using SolidWorks program, as presented in Fig. 3. Further analyses regarding the mechanical behaviour of the fixing clamp were required to be made using the ANSYS FEA program.

Fig.3. Shape of the fixing clamp after the topology optimization performed by Femap NX Nastran.

3 Finite element analyses to estimate the mechanical behaviour of the airplane component made by SLM

For the made analysis performed by ANSYS program, first of all it was necessary to specify the thermal and mechanical characteristics of the material, as shown in Fig. 4.a). Thermal expansion coefficient, Young's modulus, Poisson's ratio and Yield Strength are only few of the characteristics that were defined in the ANSYS material loading window, as they were specified on the material file from the supplier company website (SLM Solutions GmbH from Luebeck, Germany) [10] (see Fig. 4.b).

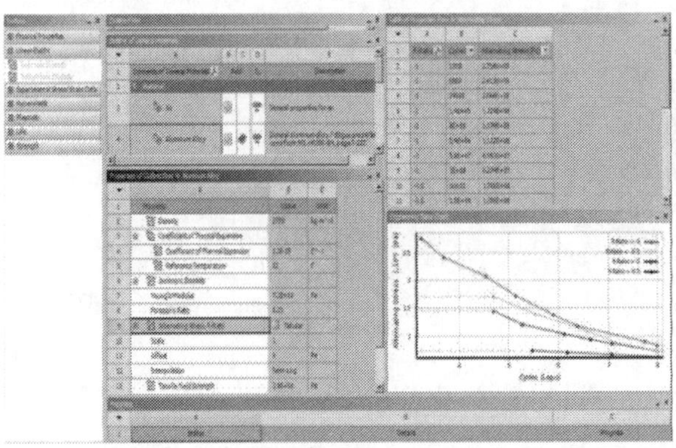

a)

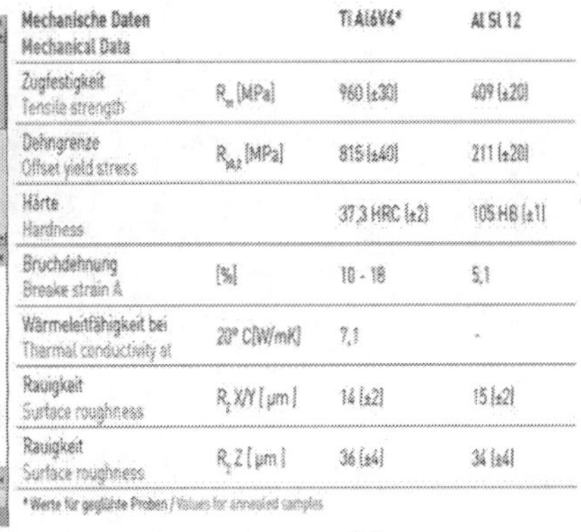

Mechanische Daten Mechanical Data		TiAl6V4*	Al Si 12
Zugfestigkeit Tensile strength	R_m [MPa]	960 [±30]	409 [±20]
Dehngrenze Offset yield stress	$R_{p0,2}$ [MPa]	815 [±40]	211 [±20]
Härte Hardness		37,3 HRC [±2]	105 HB [±1]
Bruchdehnung Breake strain A	[%]	10 - 18	5,1
Wärmeleitfähigkeit bei Thermal conductivity at	20° C[W/mK]	7,1	-
Rauigkeit Surface roughness	R_z X/Y [µm]	14 [±2]	15 [±2]
Rauigkeit Surface roughness	R_z Z [µm]	36 [±4]	34 [±4]

* Werte für geglühte Proben / Values for annealed samples

b)

Fig.4. Thermal and mechanical characteristics of the materials defined in the ANSYS FEA program.

The next step of the made analyses consisted in the loading of the 3D model in "*.igs" format and the establishment of the kinematical and technological constrains. As shown in Fig. 5. a) , the part has been considered fixed in the 8 bores from the base plate area, while a force of 500 [N] has been considered acting on the circumference of the larger bore area, indicated with arrow in this figure. Last step before running the analyses consisted in the mesh generation. A mesh having a total number of 122.037 nods and 77.349 elements was generated as presented in Fig. 5. b). After running the made analyses, both in the case when part has been considered made of TiAl6V4 and AlSi12 materials, some appreciations were made regarding the stress and strain values of the component made of these specific materials (see Fig. 6. a and b).

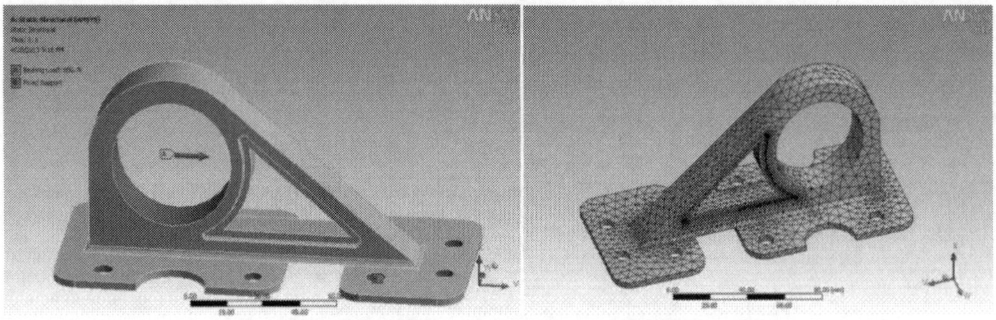

a) b)

Fig.5. Kinematical, technological constrains and mesh of the analyzed component.

As it was possible to notice by analyzing the equivalent von Mises stresses values that were obtained in the case of AlSi12 and Ti6Al4V materials, the maximum value was 13.122 [MPa] in the case of part made of titanium (see Fig. 6.a), as compared to the value of 35.982 [MPa] that has been obtained in the case of part made of aluminium (see Fig. 6.b). In both analyzed cases, the maximum stress value resulted in the base plate area of the component (closed to the stressed bore area of ϕ 50mm). Even if the loading force of 500 [N] was distributed on the entire circumference of the bore, due to the vertical positioning of the component, the action of the gravitational force on the applied force direction determined the apparition of stresses in the area of bores from the base plate that are used for fixing the component on the airplane fuselage. Modifications of the component shape were required to be done in this area in order to minimize the value of the resulted stresses in the structure of the material.

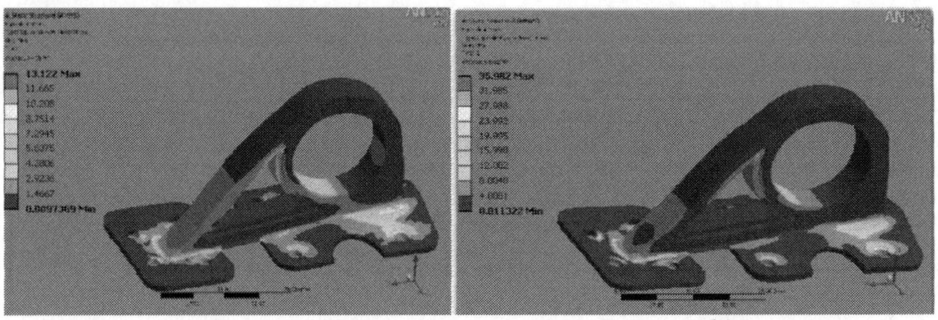

a) b)

Fig.6. Equivalent stress von Mises, for the component to be made ofTiAl6V4 and AlSi12 material by SLM.

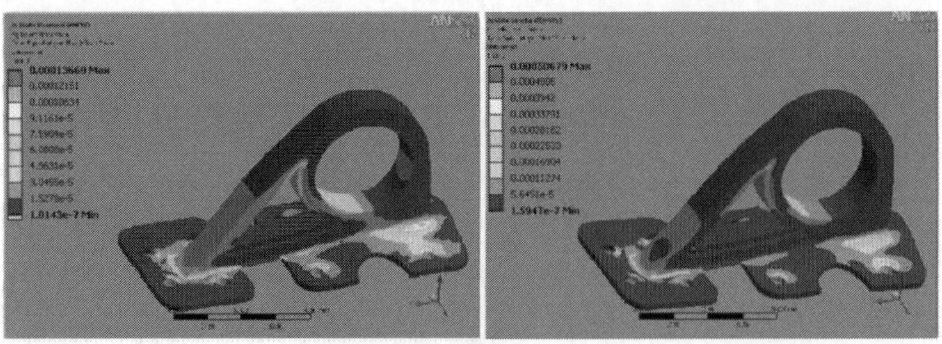

a) b)

Fig.7. Equivalent strain von Mises, for the component to be made ofTiAl6V4 and AlSi12 material by SLM.

The equivalent strains von Mises obtained in the case of the component to be made of Ti6Al4V and AlSi12 material by SLM were low, as shown in Fig. 7. a) and b), both in the case of titanium (0.00013 mm) and aluminium (0.00056 mm).
Taking into account the preliminary results that were obtained after the finite element analyses that were made by using the ANSYS FEA program and the fact that the stress and strain values were relatively low, both in the case when part has been considered to be made of Ti6Al4V and AlSi12 material, it was possible to redesign the airplane component in order to reduce its weight. The changes that were made at this stage of the analysis in term of reducing the weight of the component were not as significant as it were done using the Femap NX Nastran program, but were possible to be realized without affecting the mechanical behaviour of the component mainly in the area of the base plate in general and in the area of horizontal traverse used for connecting the two sides of the plate, in particular (the stress and strain values were negligible in these areas, in the case of both analyzed materials). As one may notice in Fig. 8.a) and b), the horizontal traverse has been eliminated in the new optimized version of the component that has been designed using SolidWorks.

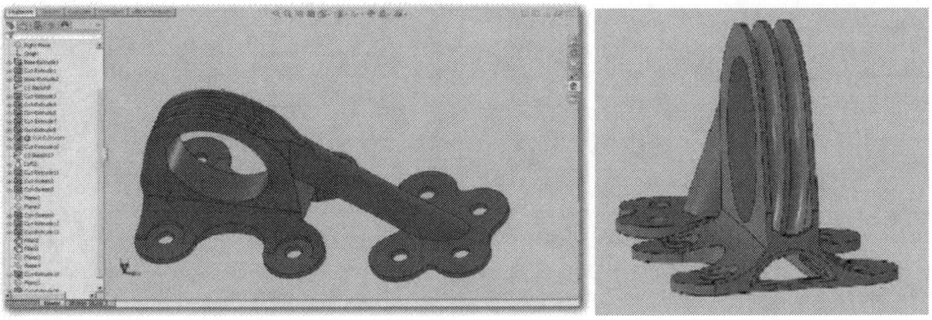

a) b)

Fig.8. Optimized version of the component designed using SolidWorks.

As shown in Fig. 8. b), there were also realized a series of modifications regarding the external circumference of the ϕ 50 [mm] bore, by realizing a cutting chamfer into the component's wall. The optimized version of the component presented in Fig. 8 represents not only an optimization of the structure of the part, but also an optimization by the technological point of view. The main advantage of the part considered to be made by SLM consists in the fact that the open areas of the part in the wall region of the bore will avoid the cleaning of closed elements from supports in this area, at the end. By realizing all these modifications it was possible to reduce the weight of the component at the end, as well. Of course, in order to validate the final shape of the component, analyses similar to the one described in this chapter was required to be done for the new shape of the component, using the ANSYS FEA program. Not only the shape of the component was considered changed for the made analyses, but also the stressing force in the area of the ϕ 50 [mm] bore, as well. The results that were obtained in the case of optimized shape stressed with different values of forces (from 500 [N] to 3000 [N]) are briefly presented in Fig. 9. a), b), c) and d).

Taking into consideration the value of the Yield strength which is 815 [MPa] in the case of Ti6Al4V material and 211 [MPa] in the case of AlSi12 (see Fig. 4 a and b) and by taking into consideration that the component will be hypothetically stressed with a force higher than 500 [N], considering a safety coefficient of C=4, it was possible to determine by dividing the value of the Yield strength to the value of safety coefficient that σ_a would be 203.75 [MPa] in the case of Ti6Al4V material and 52.75 [MPa] in the case of AlSi12. In this case, if the component is stressed with a force value of 1500 [N] the aluminium component will not resist to such a value ($\sigma_{max} \leq \sigma_a$ condition will not be respected in this

 DOI: 10.1051/matecconf/20179403012

way). This was one of the main reason why Ti6Al4V material has been taken into consideration for manufacturing of the component by selective laser melting technology, being considered as having a properly response by the mechanical behaviour (stress-strain) point of view, taking into consideration the results obtained after the performed analyses realized by using the ANSYS FEA program, as well.

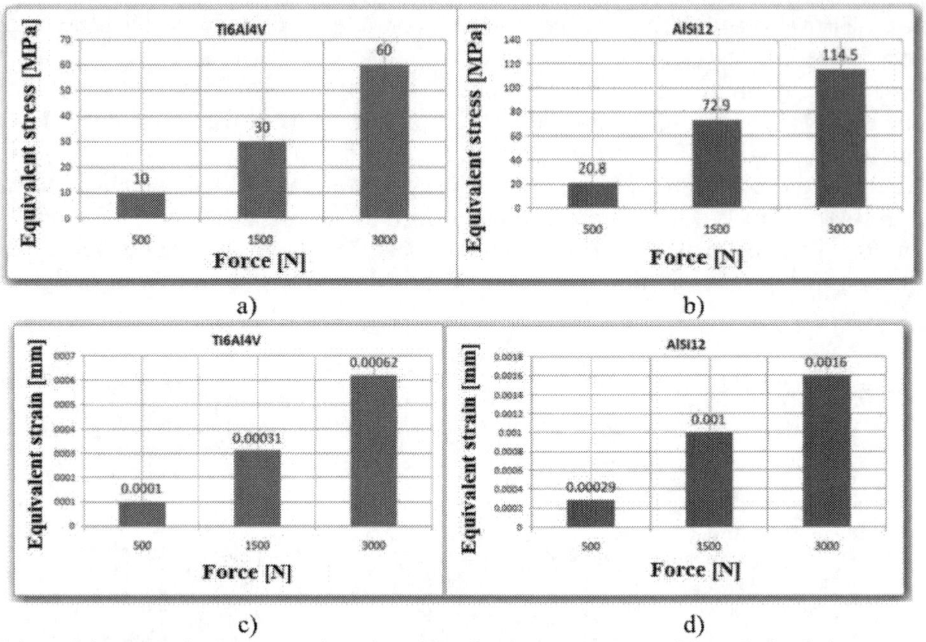

Fig.9. Value of the equivalent stress and strain values for the component with optimized shape.

4 Manufacturing of the airplane component by SLM technology

The airplane component was manufactured from Ti6Al4V powder material, using the SLM 250 HL equipment from the SLM Solutions GmbH Company in Luebeck, Germany, presented in Fig. 10. a).

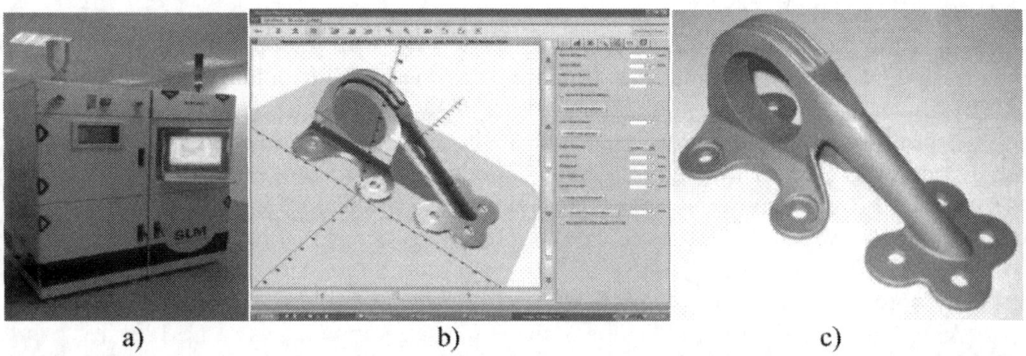

Fig.10. Airplane component manufactured at SLM Solutions GmbH Company from Luebeck, Germany.

Magics software program was used in order to prepare the "*.stl" file and for the support generation to be used for sustaining the part onto the building platform of the machine during the manufacturing process. After the preparation of the "*.stl" file in Magics, the Autofab program was further on used in order to "slice" the model (the thickness of one layer was 50 [μm], 1526 layers were obtained in total) and in order to define the technological parameters to be used during the SLM manufacturing process (see Fig. 10. b). A laser power of 175 [W], a scanning speed of 400 [mm/s] and a powder bed temperature of 180 [°C] were used for the realization of solid part, while a value of 100 [W] for the laser power, 250 [mm/s] for the scanning speed and 200 [°C] were used for the realization of the supports. The maximum level of oxygen admissible inside the working chamber of the machine was 0.2 [%], with a pressure level of the inert gas (argon) having approximately 10 bars. The weight of the component manufactured from Ti6Al4V material by using the SLM 250 HL equipment from SLM GmbH company from Luebeck, Germany had 337.17 grams (see Fig. 10. c).

5 Conclusions

Femap NX Nastran program was successfully used for the shape optimization of the fixing clamp component that was used for the sustaining of the hydraulic pipes that are passing through an airplane fuselage, the most important quantity of the material being removed in the connecting area of the base plate and supporting leg of the component. A further shape optimization was required to be done after the finite element analyses that were made using the ANSYS FEA program, in the area of the base plate in general and in the area of horizontal traverse used for connecting the two sides of the plate, in particular. By increasing the value of the stressing force in the bore area of ϕ 50 mm, from 500 [N] to 3000 [N] and by considering a safety coefficient having the value C=4, it was possible to notice that the values of the stress and strain values are too high in the case of aluminium material (AlSi12). This was the main reason why titanium material (Ti6Al4V) has been taken into consideration for the manufacturing of the component by selective laser melting (SLM) technology. The airplane component was manufactured from Ti6Al4V material on the SLM 250 HL equipment at SLM Solutions GmbH company from Luebeck, in Germany, at the end.

This paper was supported by the project entitled "Boosting the scientific excellence and innovation capacity in additive manufacturing of the Technical University of Cluj-Napoca" (AMaTUC), founded by the European Union's Horizon 2020 research and innovation programme under grant agreement No. 691787 (www.amatuc.com).

References

1. D. Tomus, Y. Tian, P. A. Rometsch, M. Heilmaier, X. Wu, Mater. Sci. & Eng.: A **667**, 42 (2016)
2. Q. Shi, D. Gua, M. Xia, S. Cao, T. Rong, Optics & Laser Technology **84**, 9(2016)
3. B. Zhang, , G. Bi, S. Nai, C. Sun, J. Wei, Optics & Laser Technology **80**, 186 (2016)
4. T. M. Mower, M. J. Long, Mater. Sci. & Eng.: A **651,** 198 (2016)
5. S. Dadbakhsh, B. Vrancken, J.-P. Kruth, J. Luyten, J. V. Humbeeck, Mater. Sci. & Eng.: A **650,** 225(2016)
6. A. Weigang, L. Weiji, Chinese Journal of Aeronautics **20**, 524 (2007)
7. D. I. Papadimitriou, C. Papadimitriou, Aero. Sci. & Tech. **55,** 24 (2016)
8. H. Huang, K. Ekici, Aero. Sci. and Tech. **39**, 481 (2014)
9. A. M. Gadade, A. Lal, B.N. Singh, Aero. Sci.& Tech. **55**, 227 (2016)
10. SLM Solutions GmbH, SLM Materials Characteristics, 2016, available at: http://slm-solutions.com/sites/default/files/attachment/page/2016/01/151023_slm_metal_powder.pdf, accessed: 10.05.2016

MATEC Web of Conferences **94**, 03013 (2017) DOI: 10.1051/matecconf/20179403013
CoSME'16

Tool electrode wear in electrical discharge of small diameter holes

Laurențiu Slătineanu[1], *Oana* Dodun[1,*], *Iulia* Carp[1], *Margareta* Coteaţă[1], and *Irina* Beşliu[2]

[1]Department of Machine Manufacturing Technology, Gheorghe Asachi Technical University of Iasi, Blvd. D. Mangeron, 39A, 700050, Iasi, Romania
[2]Department of Mechanics and Technologies, University Stefan cel Mare of Suceava, Universitatii Street 13, 720229 Suceava, Romania

Abstract. In the paper, the problem of obtaining small diameter holes in workpieces made of high speed steel by electrical discharge machining was formulated. A fulfil factorial experiment was designed and materialized, taking into consideration the tool electrode diameter, pulse on time and pulse off time as independent variables. The tool electrode wear was evaluated by means of the decrease of tool electrode mass and length. On the base of experimental results, power type empirical mathematical models were determined. One noticed the higher influence exerted by the too electrode diameter, whose increase determine the decrease of the tool electrode.

1 Introduction

Holes with distinct diameters and lengths must be sometimes achieved in metallic workpieces, in order to obtain certain pieces; these holes could be used when mobile or fix joins has to be ensured. It is known that the surfaces specific to cylindrical joins are easier obtained by machining processes if compared with joins involving other types of surfaces.

If the workpiece material has an acceptably machinability, the small diameter holes could be efficiently obtained by drilling. If the workpiece material has high mechanical properties (high ultimate strength, high hardness etc.) and when the holes diameters are lower than 1 mm, the problem of obtaining these holes by nonconventional machining methods could be formulated. Thus, small diameter holes could be achieved by electrical discharge machining, electrochemical machining, laser beam machining, electron beam machining etc. Due to the presence in mechanical enterprises of equipment for electrical discharge machining, the electrical discharge machining method could be an acceptable and accessible solution applied in order to obtain small diameter holes.

The electrical discharge machining method is based on material removal from workpiece as a consequence of electrical discharges developed between the closest asperities existing on the active surface of tool electrode and surfaces to be machined of the workpiece. The particles detached from tool electrode and workpiece during electrical discharge machining process are removed from the work gap due to the circulation of a

* Corresponding author: oanad@tcm.tuiasi.ro

dielectric fluid. Usually, there is at least a work movement achieved by the tool electrode or by the workpiece, so that the machined surface is an envelope of distinct positions of the tool electrode active surfaces during the machining process.
Jiang et al. considered [1] that an efficient small hole EDM process needs an accurate pulse monitoring strategy and a precise control. They used an adaptive control system and the wavelet transform detecting method in order to accurately identify the pulses. They appreciated that the machining process involves two stages, one in which the debris generated by the discharges can be easily removed from the discharge gap and a second stage, when the adaptive control system could reduce the arc pulse ratio, in order to improve the process efficiency and the quality of small-hole drilling process.

D'Urso et al. developed [2] an investigation aiming to highlight the influence of peak current, voltage and pulse frequency at micro-EDM drilling of stainless steel plates by using tubular tool electrodes having the outer diameter of 0.3 mm and inner diameter of 0.12 mm. They established some mathematical models able to show how the process input factors affect the material removal rate and tool wear ratio. Two technological windows corresponding to the influence exerted by hole depth on tool wear ratio and material removal rate were elaborated.

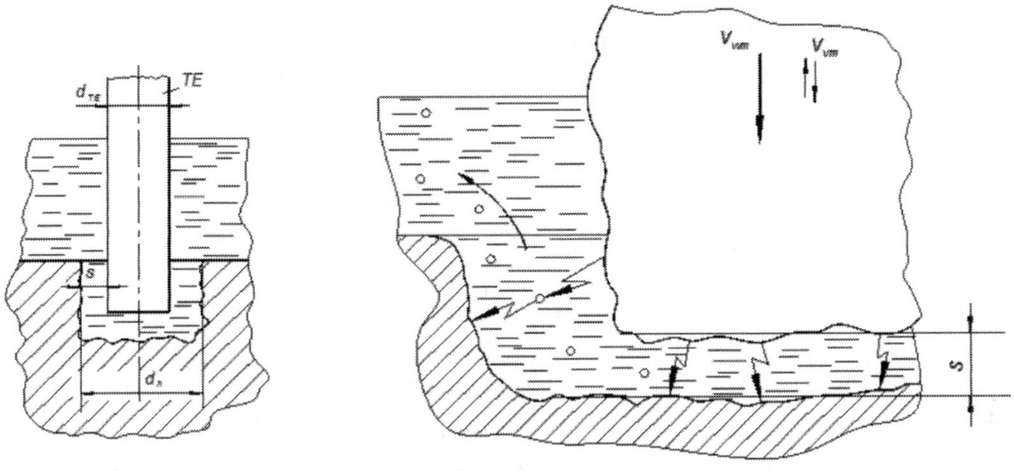

Fig. 1. Work zone in the case of obtaining small diameter holes by electrical discharge machining: a) principle scheme; b) tapering of hole entering zone as a consequence of spurious electrical discharges.

2 Theoretical premises

In order to obtain small diameter holes by electrical discharge machining, wire type tool electrode could be used. The machining schema corresponding to electrical discharge machining of small diameter holes is presented in Fig. 1. Usually the wire tool electrode is clamped in a chuck attached to the work head of the electrical discharge machine, while the workpiece is placed directly on the table of machine-tool or it can be clamped in a vice placed on the machine tool table.

The tool electrode achieves a linear movement together with the machine tool head to the workpiece; the subsystem for achieving the work movement ensures maintaining of a pre-established distance between the tool electrode active surface and the surface to be machined of the workpiece. Over the low speed work movement v_{wm} of the machine work head, a rectilinear reciprocating movements v_{vm} is overlapped , in order to contribute to the removal of particles detached from tool electrode and workpiece by phenomena of absorption - pressing.

The tool electrode achieves a linear movement together with the machine tool head to the workpiece; the subsystem for achieving the work movement ensures maintaining of a pre-established distance between the tool electrode active surface and the surface to be machined of the workpiece. Over the low speed work movement v_{wm} of the machine work head, a rectilinear reciprocating movements v_{vm} is overlapped , in order to contribute to the removal of particles detached from tool electrode and workpiece by phenomena of absorption - pressing.

As a consequence of the electrical discharges developed between tool electrode and workpiece, a material removal process develops not only from the workpiece, but also from the tool electrode. Usually, the tool electrode is achieved of materials characterized by a low machinability by electrical discharge machining, just in order to diminish the tool electrode wear. An image about the machinability by electrical discharge machining is offered [3] by the value of the so called *Palatnik's criterion:*

$$\pi = c\rho\,\lambda\,\theta_m^2 \qquad (1)$$

where c [J/(kg·K)] is the specific heat of the workpiece material, ρ [kg/m^3] – the workpiece material density, λ [W/(m·K)]– the coefficient of thermal conductivity, θ_m [K] – the absolute melting temperature.

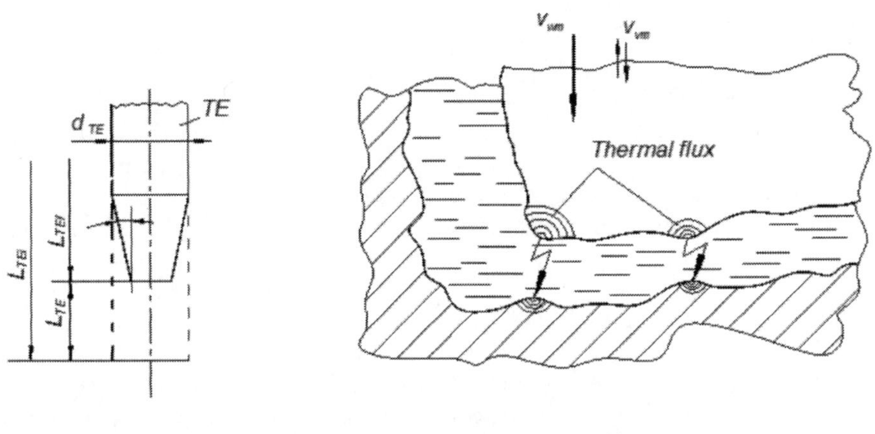

Fig. 2. More intense wear of tool electrode edges and shape change as a consequence of wear phenomenon:
a) active zone of tool electrode;
b) heat dissipation at the tool electrode edge and near the tool electrode axis.

If one considers two materials for tool electrode (steel and copper) and takes into consideration the values specific to these materials (in the case of steel, c=420 J/(kg·K), ρ=7850 kg/m3, λ=54 W/(m·K), θm=1755 K, and in the case of copper, c=385 J/(kg·K), ρ=8933 kg/m3, λ=401 W/(m·K), θm=1358 K), one can find πsteel=5.48361·1014 and πcopper=2.543·1015; because the ratio πcooper/ πsteel=4.638, this means that a tool electrode made of steel will be affected by a wear more intense than 4.5 times compared with the wear of tool electrode made of copper.

A supplementary factor able to affect the development of the material removal from tool electrode is the possibility of evacuation the high quantity of heat generated by each electrical discharge. If in the zones close to the tool electrode axis there is the possibility of heat dissipation in the tool electrode metallic mass, in the zones found near the edge existing between the tool electrode frontal place surface and the cylindrical surface, there is not a possibility of efficient dissipation of the heat (Fig. 2) and such a fact determines a more intense process of heating and of material removal. Thus, in time, a decrease of tool electrode length and a tapering of the former sharp edge could be expected (Fig. 2, *a*).

Problems could be also generated by the metallic particles movements between the side cylindrical surface of the tool electrode and the cylindrical surface gradually generated in the workpiece. Along their motions in the side gap, the metallic particles could generate spurious electrical discharges along the trajectory including the tool electrode, metallic particle and workpiece. The spurious electrical discharges could contribute to a supplementary material removal from the workpiece and this could generate a tapering of the hole entering zone (Fig. 1, *b*).

The problem of hole shape error appears especially when long holes must be obtained by electrical discharge machining.

As one can see, the diameter of the tool electrode d_{TE} could be determined as the following difference:

$$d_{TE} = d_h - 2s \quad (2)$$

where s is the size of the work gap and d_h – hole diameter.

Being an important output parameter, the tool electrode wear was studied, in order to determine which are the influence factors and how the tool electrode wear could be diminished.

Table 1. Values of the input factors and output parameters in the case of obtaining small diameter holes by electrical discharge machining (machining process duration: 40 min, test piece made of high speed steel HS14-0-1-4).

Exp. No.	Input factors						Decrease of tool electrode mass, Δm_{TE}, [g]	Decrease of tool electrode length, Δl_{TE} [mm]
	Tool diameter		Pulse on time		Pulse off time			
	Coded value	d_{TE}, [mm]	Coded value	t_p, [s]	Coded value	t_b, [s]		
1	0	0.6	0	0.11	0	0.03	0.011166	4.4286
2	0	0.6	0	0.11	1	0.06	0.01279	5.1254
3	0	0.6	1	0.22	0	0.03	0.013808	5.5242
4	0	0.6	1	0.22	1	0.06	0.015344	6.0772
5	1	2.6	0	0.11	0	0.03	0.00096	0.6588
6	1	2.6	0	0.11	1	0.06	0.001134	0.4525
7	1	2.6	1	0.22	0	0.03	0.000082	1.1619
8	1	2.6	1	0.22	1	0.06	0.000274	0.7717

3 Experimental research

The experiments were developed on a Sodick A3DL ram electrical discharge machine. As input factors, one considered the tool electrode diameter d_{TE}, pulse on time t_p, the pulse off time t_b.

The test pieces were made of high speed steel (HS14-0-1-4), containing 14.31 % tungsten, 4.64 % cobalt, 4.09 % chrome, 1.43 % vanadium, 0.69 % manganese.

The material of tool electrode was electrotechnical copper. Two diameters of tool electrode were taken into consideration, d_{TEmin}=0.6 mm and d_{TEmax}=2.6 mm. As output factors, one considered the decrease of tool diameter length Δl_{TE} and decrease of tool mass Δm_{TE}, respectively. The lengths of tool electrodes were measured just by means of the machine tool subsystem for control of distinct linear parameters, while a Partner type analytical balance was used in order to determine the tool electrode mass before and after each experiment.

The values of the input factors and of the output parameters were included in Table 1; one inscribed both the coded values and the proper values of the input factors.

The experimental results presented in Table 1 were mathematically processed by means of adequate software [5] and the following empirical mathematical relationships were determined:

- In the case of tool electrode massic wear:

$$u_m = 0.002358\, d^{\,2.391} t_p^{\,-1.257} t_b^{\,0.582}, \tag{3}$$

for which the value of the Gauss's criterion is:

$$S_G = 4.15566 \cdot 10^{-5} \tag{4}$$

- In the case of tool electrode linear wear:

$$u_l = 3.707 d^{-1.368} t_p^{\,0.563} t_b^{\,-0.220}, \tag{5}$$

for which the value of the Gauss's criterion is :

$$S_G = 0.547565 \tag{6}$$

The values of the Gauss's criterion is determined as a sum of the least squares corresponding to the differences between the measured values and the values calculated by means of the selected model, for the same experimental points [5].

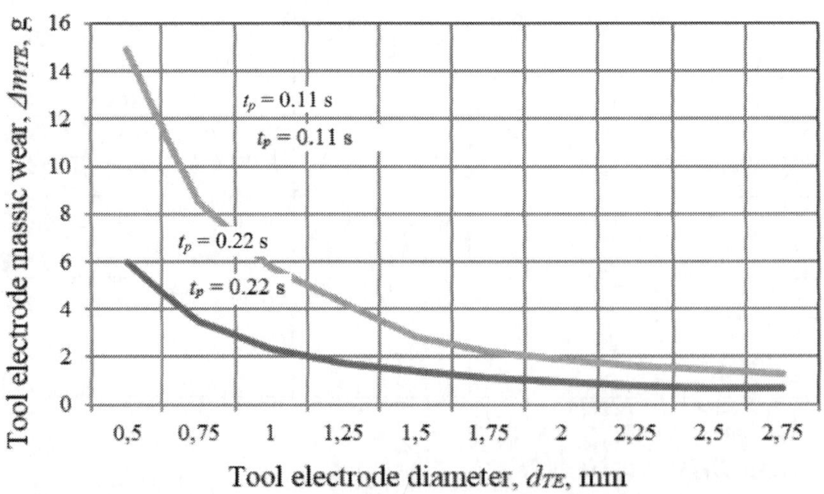

Fig. 3. Influence exerted by the tool electrode diameter on the massic tool electrode wear (t_b=0.03 s, test piece made of high speed steel HS14-0-1-4, tool electrode material: electrotechnical copper, process duration: 40 min).

On the base of the empirical mathematical models constituted by the relationships (3) and (5), the graphical representations from Fig. 3 and Fig. 4 were elaborated, in order to highlight the intensity of the influence exerted by the values of the input factors on the values of the output parameters.

As one can see by the examination of the empirical mathematical models corresponding to the tool electrode massic and linear wear, the increase of the tool electrode diameter determines a decrease of the tool electrode wear and the fact could be explained by improving the process of heat evacuation from the tool electrode active zone when the tool electrode diameter is higher. Indeed, the higher mass of the tool electrode and the higher conductivity of the copper (compared with the electric conductivity of the dielectric liquid) ensure the conditions of faster decrease of the temperature immediately after the electrical discharge is finished and, thus, a decrease of the tool electrode wear could be identified.

If in the case of the empirical mathematical models valid for the tool electrode massic and linear wear a similar influence exerted by the tool electrode diameter could be observed, in the case of the influence of pulse on time and pulse off time there are some differences and these could be generated by the necessity to use the power type empirical functions, instead of more adequate empirical functions.

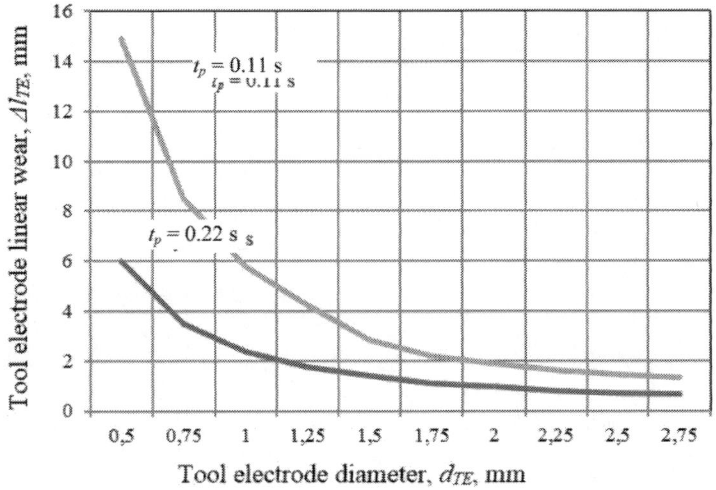

Fig. 4. Influence exerted by the tool electrode diameter d on the linear tool electrode wear (t_b=0.03 s, test piece made of high speed steel HS14-0- 1-4, tool electrode material: electrotechnical copper, process duration: 40 min).

4 Conclusions

In the case of workpieces made of materials characterized by a low machinability by classical machining methods, the ram electrical discharge machining could constitute a solution applicable in order to obtain small diameter holes. The problem of obtaining small diameter holes in workpieces made of high speed steel was formulated. A full factorial experiment with three independent variables (tool electrode diameter, pulse on time and pulse off time) was designed in order to highlight the influence exerted by the input factors on the massic and linear tool electrode wear. By mathematical processing of the experimental results, power type empirical mathematical models were determined. These models confirmed the expected decrease of the tool electrode wear when the tool electrode diameter increases, due to the improvement of the conditions for the heat evacuation from the work zone. In the future, there is the intention to extend the experimental research concerning the tool electrode wear, in order to better understand the process of wear evolution in the case of electrical discharge obtaining of small diameter holes.

References

1. Y. Jiang, W. Zhao, X. Xi, L. Gu, , J. of Mechanical Science and Technology, **26**, 6, 1885 (2012)
2. G. D'Urso, G. Maccarini, C. Ravasio, , Int J Adv Manuf Technol, **72,** 1287 (2014)
3. B.A. Artamonov, A.L., Vishnitskiy, I.S. Volkov, A.B. Glazkov, *Dimensional electrical metalworking* (Vysşhaia Şkola, Moskva, 1978)
4. I. Carp, *Theoretical and experimental study of processing small dimension surfaces by electrical erosion process,* PhD Thesis (Gheorghe Asachi Technical University of Iaşi, 2010)
5. G. Creţu, *Basics of experimental research. Laboratory guide,* (Gheorghe Asachi Technical University of Iasi, 1992)

MATEC Web of Conferences **94**, 04001 (2017) DOI: 10.1051/matecconf/20179404001
CoSME'16

Risk reduction in dimension inspection of the plastic injection-molded parts from mechatronic devices by using optical 3D measuring techniques

Ion Cristian Braga[1,*], *Anisor* Nedelcu[1], and *Razvan* Udroiu[1]

[1]Transilvania University of Braşov, Department of Manufacturing Engineering, Mihai Viteazu No.5, Braşov, Romania

Abstract. As of the definition presented by Harashima, Tomizuka, and Fukada in 1996, the mechatronics is the synergistic combination of precision mechanical engineering, electronic control and systems thinking in the design of products and manufacturing processes. The most of the mechatronic devices need the precise dimensions of the plastic parts, as long as the combination of those parts leads to a final haptic characteristic defined within specific limits or when the certain travel way is linked with an electrical contact. The increasing of the risks to produce bad mechatronic devices are directly related to the combination of the plastic injection-molded parts out of different cavities. The paper's aim is to present reducing of the risks to have bad final parts assembled with the components out of plastic injection-molded parts by using optical 3D measuring techniques at first validation of the parts out of the tool and setting parameters in the injection machines. The shrinkage and the warpage are more easily detected in that way and this will support first article inspection, but also during the entire production process. A case study presents the analysis of the data coming from the measurements of the plastic parts from each cavity and the combination of those parts, by using the ATOS inspection software. The CAD data are compared with the measured ones and the differences will be visible in the colored plotted areas, also the differences of the parts out of distinct cavities will be also displayed by overlaying of the measurements.

1 Introduction

The plastic material is more or less a modern material, even the first manufacturing of one kind of plastic, cellulose nitrate, known as Celluloid, was discovered in 1856 and refined in 1877. In the ninetieth century, Bakelite was the other plastic invented and used more in the manufacturing of the telephones or radios, the inventor being Leo Baekeland. The twentieth century, starting with 1935, the other inventions brought new materials as Styrofoam, polyvinyl chloride (PVC), acrylic, polyurethane, epoxy, and nylon, even new materials as synthetic rubber or textile fibers [1].

* Corresponding author: jbc69mc@gmail.com

The automotive industry starts to use the plastic material in the 1950th when the ABS was first used then polyamide and polycarbonate, also, alloys and blends of various polymers. Nowadays, a lot of automotive components are made from plastic materials, and can be mentioned the bumpers, interior trim, seating, dashboard, exterior trim, under-bonnet components, lighting, liquid reservoirs, fuel systems [2].

The challenges are bigger now when the future of the automobile is focused on reducing the CO_2 emissions to reach limit 95 g CO_2/km in 2020, so the reducing of the weight become a priority. And this reduction of the weight can be obtained by replacing the actual materials, like metal, with plastic materials, because almost 75% of all materials used in manufacturing of the automobile are metal, as a study published in the magazine of the Plastics Europe AISBL, based on the study of Association Française de Mécanique (figure 1).

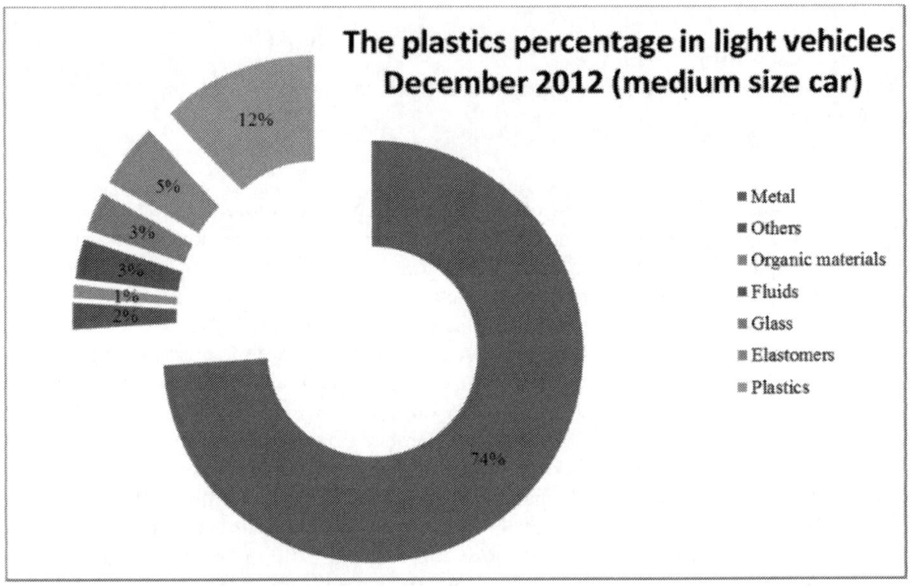

Fig. 1. Material percentage in light vehicles.

The same magazine, using the data from 2012, presents also the percentage of the using for the plastic materials in the automobile, under four categories: interior, exterior, under the hood and electronics (figure 2). As in interior a lot of mechatronic devices are used in order to realize the comfort of the driver and the passengers, as well as the safety aids using the sensors, buttons and touch screens [3], the parts manufactured by plastic injection molded processes need the precise dimension to fit with the other components and to reach the specific goals of each one.

And this precision dimensioning of the plastic parts will be obtained by early development of the product and the processes, but also in the serial production. The serial production processes of the plastic injection molded parts must be checked at the start of production and the validation of the injection molding machine and the molding tools must be validated prior to this. Of course, also the working methods and the training of the people who adjust or operate the machine are to be made prior. Then the monitoring of some dedicated characteristics will ensure that the process is stable and the outputs are according to the expectation.

This paper presents the possibility to reduce the risks in the serial production of the plastic injection molded parts by using the optical 3D measuring techniques at validation of the start of the production of a batch, and not using only the usual 2D measurement equipment as caliper or dial indicator or microscope.

2 Validation of the plastic injection molding processes

2.1 The plastic injection molding process

The plastic injection molding processes is a repetitive process using the plastic material, usual pellets or granules, which is melted and heated with the reciprocating screw into a barrel and injected into a mold cavity through a nozzle, then held under pressure until is removed in a solid state and ejected.

The mold tool could be with one or more cavities, which can be similar cavities or not. This mold is made from two parts: one is fixed where the material is injected and the other is moveable, on the moveable platen, as is presented in figure 2 [4].

Generally, there are two types of the plastic materials molded: the thermoplastics and the thermosets. The thermoplastics, more often used, can be melted and solidified in a repeating cycles, heated at around 260 Celsius degrees. On the other hand, the thermosets are heated on a temperature around 120 Celsius degrees, upon their final heating, become insoluble and infusible permanently [5]

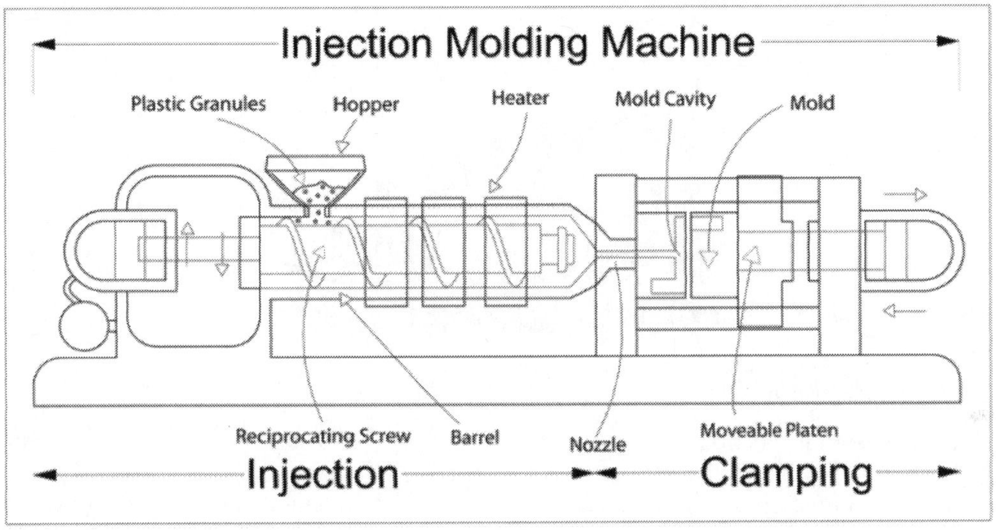

Fig. 2. An injection molding machine.

Many factors can influence the product dimension tolerances and are very important to keep these factors under control. Can be mentioned here the design of the product, the plastic material, the design of the mold, the capability of the injection molding machine and the process parameters such as cooling temperature, demolding temperature, mold temperature, melt temperature, injection time, holding time, mold close or open time, injection pressure, back pressure or holding pressure.

2.2 The validation of the plastic injection molding process

Even the design of the product and the mold will take care to prevent the shrinkage or the warpage on the parts, by choosing the location of the gate and the material selection [6], the issues can be occurred also in the serial production and these effects will generate issues in the next assembly processes of a mechatronic device. Therefore, the selection of the process parameters depend on the type of the injection machine used, and the setting operator will work with some changes of these parameters, at the end the product can have deviation from the expected dimensions.

In order to prevent this, after the validating of the product design and the validation of the machine capability, it is necessary to validate the injection molding process at each start of production and to monitor some selected characteristics along the entire batch production. The statistical process control can support the team to detect the variation in the process and the trend of the processes, in that way to prevent to produce parts put of tolerance and to generate the scrap or bad assembled mechatronic devices.

3 Study case

3.1 The validation of the start of production by using 1D/2D measurement devices

For the study case was selected a part from a mechatronic device manufactured in the processes of the plastic injection molding, a cover used afterwards in an assembly where will have some movements in the functioning, a rotation movement in a combination with a housing (figure 3).

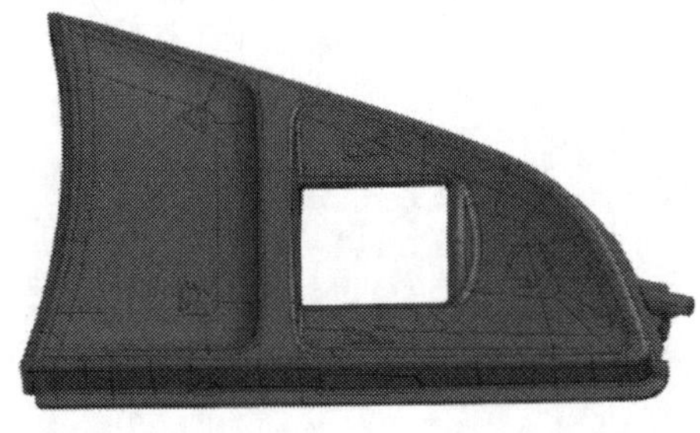

Fig. 3. A plastic injection molded cover.

The dimension inspection of the part at the start of production and in the monitoring visual failures at each four hours is made by measurement of a characteristic, dimension K1, with a dial indicator in the laboratory, by the operator, as in figure 4, but also by visual checking of the part to detect the issues as incomplete molding, sink mark, scratches, burrs or other visual failures.

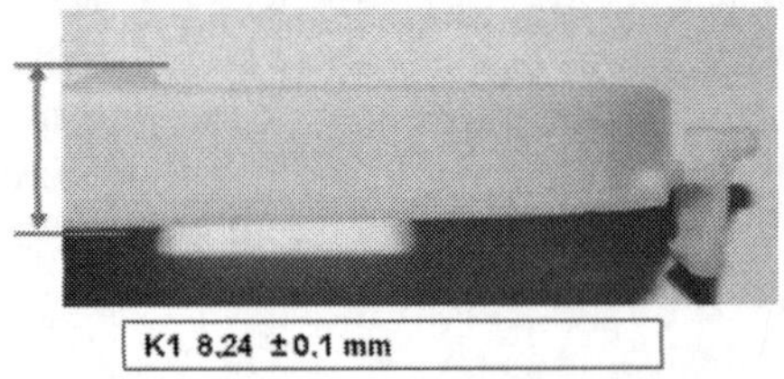

Fig. 4. Measuring method for the characteristic.

The dimension inspection of the part at the start of production and in the process monitoring at each four hours is made by measurement of a characteristic, dimension K1, with a dial indicator, in the laboratory, by the operator, as is presented in figure 4, but also is made by visual checking of the part to detect the possible issues as incomplete molding, sink mark, scratches, burrs or other visual failures.

 DOI: 10.1051/matecconf/20179404001

Based on the measurements of the K1 with the dial indicator and referring to the specification, the sample parts are within the tolerances, the same the attributive visual characteristics. In order to monitor the variability and stability of the process, the capability of the machine was evaluated using the software Qs-Stat, Machine Acceptance Cm/Cmk, from Q-Das [7], and the result shows a very good machine capability index, Cmk, and the model distribution being a normal distribution (figure 5).

Model distribution			Normal Distribution
Calculation method			$M4_1$ Percentile (0,135%-$\bar{x}$-99,865%)
Potential Capability index	C_m	$8{,}73 \le \mathbf{10{,}88} \le 13{,}03$	0 … 2
Critical capability index	C_{mk}	$5{,}14 \le \mathbf{6{,}41} \le 7{,}69$	0 … 1,67
	The requirements were met (C_m,C_{mk})		

Fig. 5. Capability index and model distribution.

The Gauss curve shows a narrower distribution, as in figure 6, it mean, the process is precise enough. At the end, the decision made on the sampling is that the serial production is according with specification and parts can be used in the next steps in the processes, and there are not detected any risks in link with the functioning of the entire mechatronic device.

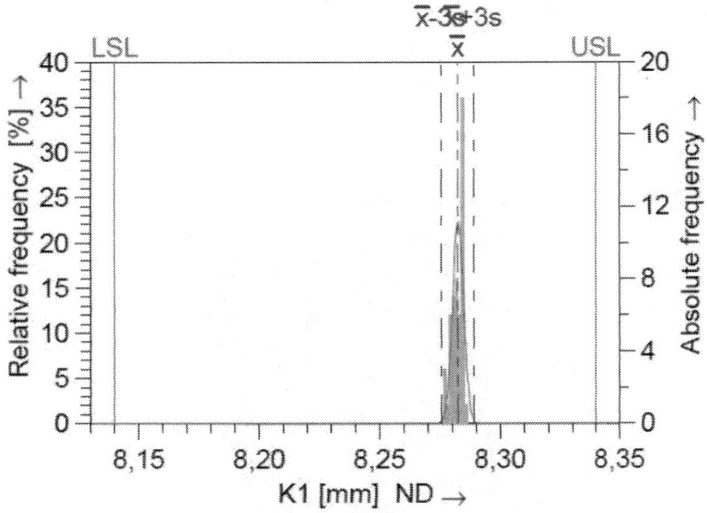

Fig. 6. Gauss curve.

As was mentioned in the chapter 2.2, from the design of the product and design of the mold was considered the warpage limitation, but also the selected process parameters can influent the final dimension of the part and the warpage or even shrinkage could occur, and there could be some supplementary studies by using Taguchi method to establish the optimal combination of the process parameters [8]. And the 1D/2D measurement devices and the measurement system established only on these measurements, even combined with the statistical process control will not secure enough all the possible risks to deliver the plastic injected molded products in the assembly and there to detect some issues from functioning of the mechatronic device.

3.2 The validation of the start of production by using optical 3D measuring technics

In the above paragraph was presented the possible risks, not detectable with the measuring system established, and the probability of generating scrap in the assembly processes can be considerable. Because of that, the study case presents the using of the optical 3D measuring technics.

The paper proposes to design a combined measuring system based on the 1D/2D device measurements to monitor the process through entire batch using statistical process control, but validating the start of production of the batch by using optical 3D measurements.

Hence, the possible shrinkage or warpage could be easily detected by the appraisal and the quick decision will lead to checking of the parameters and optimizing them so that the parts will be free of these kinds of issues.

3.2.1 Optical 3D Coordinate Measuring Machine

The 3D Coordinate Measuring Machine is an optical 3D measuring machine developed for a quality control in the production and manufacturing process. More models are available for different part sizes and applications, the case study used *ATOS ScanBox - Optical 3D Coordinate Measuring Machine, Series 4,* ATOS CORE MV 135.

The core of the systems is the 3D scanner, whose sensors provide full-field 3D coordinates for each individual measurement and up to 16 million independent measuring points are captured within some seconds. The scanning system is based on the triangulation principle: the sensors projects different fringe patterns over the parts to be measured, then two cameras will record this, as in figure 7.

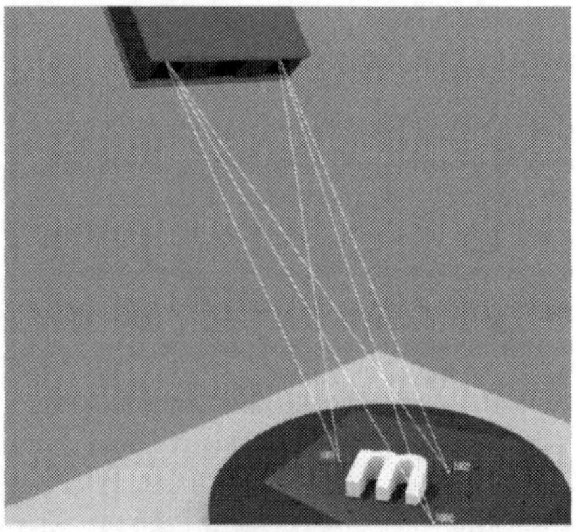

Fig. 7. Measuring principle: fringe and triangulation.

The visible points seen by both cameras in a single scan are recorded and in order to have complete measured part digitized, the different angle measurements will be done. Some reference circular points which are fixed on the frame for part fixation, as in figure 8, are used in order to transform each measurement into the global coordinate system [9].

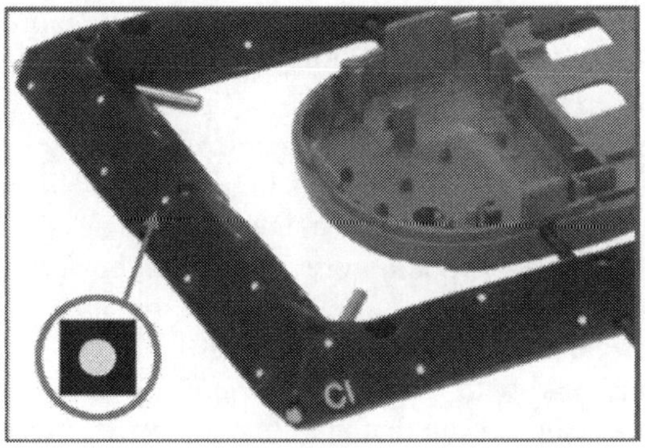

Fig. 8. Measuring principle: reference circular points.

3.2.2 Comparing the scanned part with the CAD model

The case study is made on a part coming out from a mold with 2 cavities. The first measurement will have the aim to compare the scanned result of the part with the 3D model from CAD. In figure 9 is presented the overlapping of the measurement from the scanned part with this CAD model. As can be seen, in the upper right corner and in the lower left corner are some deviation from the model and there the warpage is present on the measured part: in the left lower corner, are two points in blue colour, 0.079 and 0.063, the sign minus in front of the points shows that the part seems to have less material than the requirements. In the same way, it can be seen in the right upper corner another two points in blue colour, 0.071 and 0.048. Based on the assembly tests and the studies for the combination of the cover with the other assembled parts (housing, electronic, rubber etc) the assembly processes could be approved or not, depend on the results positive or negative of the tests.

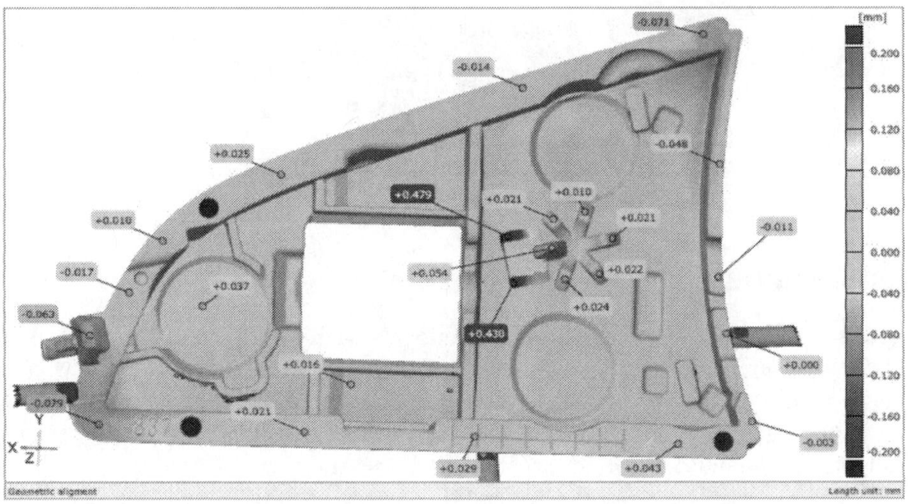

Fig. 9. Scanned part cavity 2 versus 3D model.

3.2.3 Comparing the scanned part from cavity 1 with the scanned part from cavity 2

While the measurement by comparing the scanning with the CAD model shows to the assessor the areas where the part is out of tolerance or even not out of the tolerance but warpage or shrinkage detected, by comparing the scanning of the part from cavity 1 with scanning of the part from cavity 2 will let the assessor to take more quick the decision if the assembly test for the validation of the start of the production must include all the combination between the cavities for all assembly components (figure 10). As can be seen in this picture, there are some very small deviation between the parts from cavity 1 versus cavity 2, therefore, the parts are almost identical and the decision is to be ordered only a sample using one cavity for the following assembly tests. It will save money and the risks to produce scrap in the next step are prevented, also, the cost of quality prevention will be decreased, thanks to a decision made on the comparison between the cavities.

And the reducing of the prevention cost will be as bigger as numbers of injected molded parts are coming from a mold with more cavities. For instance, based on the samples from the presented study, there can be seen that are almost any differences between both cavities, so for the assembly tests is enough to use one of the cavities and, if the detected warpage on the corners will not lead to a malfunction, the plastic injection molding process could be approved and the serial production of the batch could start and run until some possible specific deviation detected in the statistical process control monitoring.

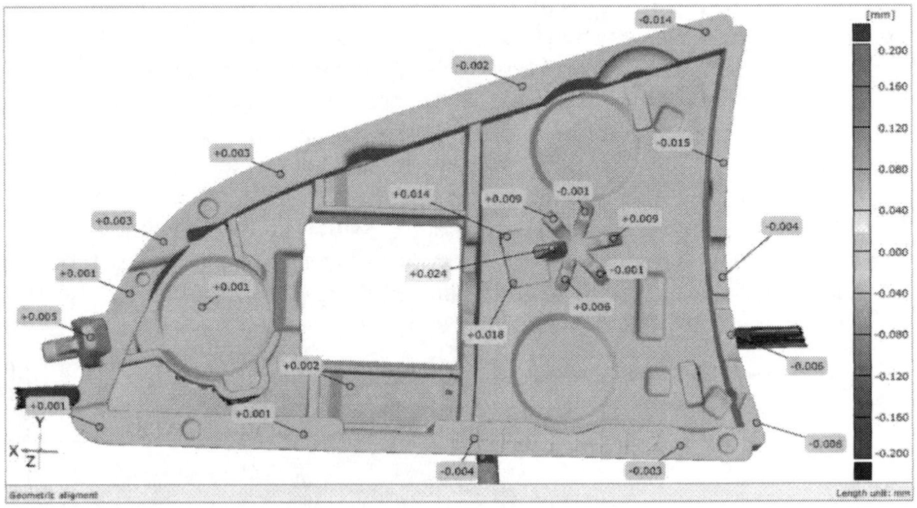

Fig. 10. Scanned part cavity 2 versus part cavity 1.

4 Conclusions

Nowadays, in almost all of the organizations which have the plastic molding injection processes, the method used to validate the start of production is by choosing some characteristics, then measuring these characteristics with 1D/2D measurement devices and follow up the capability of the process. Only at validation or revalidation after repairing and maintenance of the mold it is used the 3D optical measurements

The proposed method to use the optical 3D measuring in the dimension inspection of the plastic injected molded parts at starting of the production of each batch will prevent the risks to produce parts affected by warpage or shrinkage, even the selected characteristics to be measured shows good result and the statistical process controls shows very precise, based on the distribution and capability index.

Another advantage of this method is represented by the saving of the quality prevention cost through comparing the scanned parts coming from different cavities of the same mold or even different molds, also the saving of the rework cost from next step operation in assembly processes. As the method is more precise and accurate, could be used also in the other inspection areas and to replace the conventional measuring system or to be used complementary at validation of the processes.

References

1. R. Mustalish, *Modern Materials: Plastics*. In Heilbrunn Timeline of Art History. New York: The Metropolitan Museum of Art (2000) http://www.metmuseum.org/toah/hd/mome/hd_mome.htm (2004)
2. K. Szeteiova, *Automotive Materials: Plastics in Automotive Markets Today*, Slovak University of Technology, (Bratislava, 2010)
3. Automotive - The world moves with plastics http://www.plasticseurope.org/Document/automotive---the-world-moves-with-plastics-brochure.aspx?Page=DOCUMENT&FolID=2 (2013)
4. AV Plastics, http://www.avplastics.co.uk/what-is-injection-moulding (2016)
5. D.V. Rosato, M. G. Rosato, *Injection Molding Handbook*, (Springer Science+Business Media New York, 2000)
6. M. Huszar, F. Belblidia, H. M. Davies, C. Arnold, D. Bould, J. Sienz, Sustainable injection moulding: The impact of materials selection and gate location on part warpage and injection pressure, Sustainable Materials and Technologies **5** (2015)
7. Q-Das, Experst in statistics, http://www.q-das.com/en/applications/ (2016)
8. P. Postawa, J. Koszkul, *Influence of processing conditions on changing of shrinkage and mass POM injection molding parts*, COMMENT, (Gliwice Poland, 2005)
9. GOM, http://www.gom.com/metrology-systems/atos-scanbox.html (2016)
10. R. H. Bishop, *The Mechatronics Handbook*, (CRC Press, 2002)

Statistical analysis of the measurement errors in an installation of water meters. Study on the volume of the water loss in the installation

Mihaela Rodica Clinciu[1,*], and *Ramona* Clinciu[1]

[1]Transilvania University of Braşov, Department of Manufacturing Engineering, Mihai Viteazu No. 5, Braşov, Romania

Abstract. The paper presents the estimation of the statistical parameters for experimental data collected in an installation consisting of several water meters and also a study on the volume of the water loss in the installation, aiming to emphasize ways in which errors in metering the consumed water are produced. The water meters analysed are of three types, belonging to two different precision classes. Data analysis was performed on data collected from the water meters considered, data representing the measurement errors.

1 Introduction

Estimation theory is a branch of statistic and signal processing that deals with estimating the values of parameters based on measured/empirical data. The parameters describe an underlying physical setting in such a way that the value of the parameters affects the distribution of the measured data. An estimator attempts to approximate the unknown parameters using the measurements [1, 2].

The paper presents the estimation of the statistical parameters for experimental data collected in an installation of water meters and a study on the volume of the water loss in the installation. The water meters considered in the present research are of three types, belonging to two different precision classes, connected in series arrangement [3].

2 Statistical analysis of experimental data

General metrology uses frequently the normal distribution as a statistical model, although usually in many situations asymmetrical statistical distributions, of Weibull type, occure [4]. In these cases one must verify the hypothesis that the data analyzed really comes from a normal population. Determining the appropriate statistical model is realized by verifying that the empirical distribution is consistent with the theoretical distribution assumed properly in the case studied.

The algorithm that performs the statistical processing of the experimental data involves the following steps:

* Corresponding author: clinciumr@unitbv.ro

1. The experimental data collection.

In the experimental research, there were considered three types of water meters belonging to two different precision classes. The considered water meters were used to measure a constant volume of water (10 liters).

The experimental installation (Figure 1) contains water meters of three types, connected in series arrangement: three water meters of type I (precision class B), three water meters of type II (precision class B) and three water meters of type III (precision class C). As to the precision classes of the water meters, class C represents high precision and class B represents mediate precision. Water meters of precision class C are in general branching water meters, while water meters of precision class B are apartment water meters, with a more accesible price, but with a lower precision [3].

Water meters of type I are apartment water meters for cold water or hot water, of precision class B, with nominal diameter Dn15 and minimum starting flow 30 l/h.

Water meters of type II are apartment water meters for cold water, of precision class B, with nominal diameter Dn15 and minimum starting flow 30 l/h.

Water meters of type III are single-jet water meters, of precision class C, with nominal diameter Dn15 and minimum starting flow 15l/h.

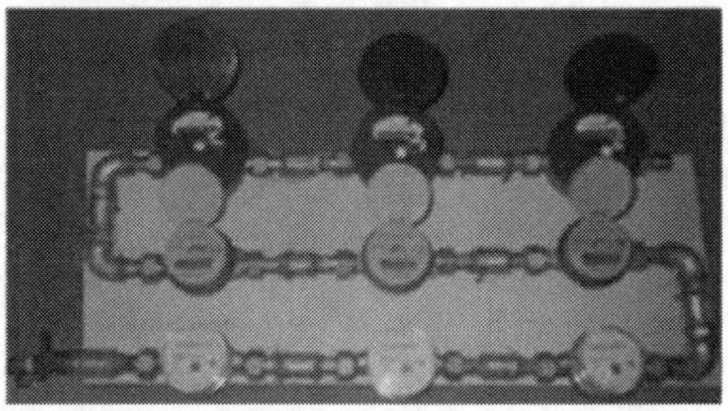

Fig. 1. Experimental installation.

The experimental data is represented by the measurement errors resulting by measuring a constant volume of water (10 liters), using the water meters considered. The operation was performed in laboratory conditions, by a single person, for n times. Experimental data is grouped into samples of a certain volume, n=18, for each water meter considered.

The maximum admisible errors of water meters are included in the Legal Metrology Norm NML 003-05 "Water Meters" [5], where it is specified that measurement errors on volumes delivered at flowrates between the transitional flowrate (included) and the overload flowrate may not exceed errors ± 2% for water having a temperature of ≤30 °C.

2. Verification of the homogeneity of the samples of experimental data. During this stage, one must perform:

- the verification of the random feature of selection, using the method of iterrations;
- the removment of aberrant values, using the Grubbs general test.

3. The graphical representation of experimental data and the assumption of a theoretical distribution. This stage includes the construction of the probability plot and then, on its basis, the assumption of a statistical model presumed to be adequate for describing the experimental data.

4. Estimation of the parameters of the statistical model selected for each sample.

During this stage, according to the distribution considered, one must apply an analytical method of estimation (least squares, maximum likelihood) in order to determine the parameters of the selected model for each sample.

5. Validation of the statistical model. This stage consists in applying a goodness of fit test to all samples representing the experimental data. The goodness of fit test used are: the

general Kolmogorov-Smirnov test and the normality goodness of fit test Lilliefors. Distribution functions validated at a risk level $\alpha = 0.05$ are considered acceptable.

3 Case studies – water meters of precision classes B and C

3.1 Water meters of type I, precision class B

The experimental data was collected by reading the indications of the water meters of type I, precision class B. It represents the measurement errors obtained by measuring a constant volume of water (10 liters) using the water meters considered.

The measurement errors of the water meters of type I, precision class B are shown in Table 1; on the basis of the probability plots presented in Figure 2, a, b and c, normal distribution is assumed for the samples considered. Table 2 presents the estimated values of the parameters of the normal distribution.

Table 1. Measurement errors of water meters of type I, precision class B.

Water meter	**Measurement errors** (m^3)
1- I B	-0.00004, -0.00013, 0.00012, 0.00013, 0.00021, 0.0001, 0.00022, -0.0002, -0.00013, -0.0007, -0.00005, 0.00039, -0.00005, -0.00015, 0.0001, -0.0005, -0.00002, -0.00001
2 -I B	0.00007, 0.00011, -0.00006, -0.0002, -0.00011, 0.00004, 0.0002, 0.00008, -0.00001, 0.00011, -0.00012, 0.00002, -0.00007, -0.00012,-0.00003, -0.00001, 0.00003, 0.00002
3 -I B	-0.00008, 0.00005, 0.00006, 0.00008, 0.00003, 0.00018, 0.00007, 0.00003, 0.00004, 0.00002, 0.00004, -0.00003, 0.00002, 0.00015, -0.00002, 0.00005, 0.00001, 0.00005

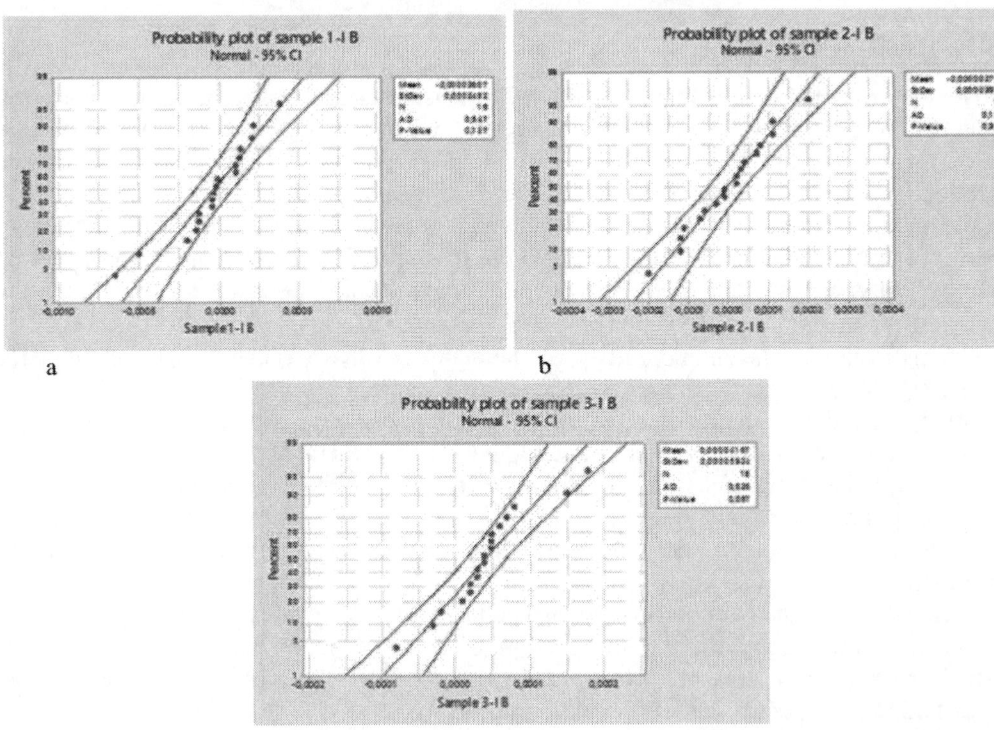

a b c

Fig. 2. Probability plots of the samples 1, 2, 3 collected by using water meters of type I, precision class B.

 DOI: 10.1051/matecconf/20179404002

Table 3 presents the values of the statistics of the goodness of fit tests applied [3]. These values are compared to the critical values of the statistics of the goodness of fit tests, $d_{n,\alpha}$ for the Kolmogorov-Smirnov goodness of fit test and $L_{n,\alpha}$ for the Lilliefors goodness of fit test.

Table 2. Parameter estimation - water meters of type I, precision class B.

Water meter	1-I B	2-I B	3-I B
Mean	$m_1 = 0.00004$	$m_2 = 0.0$	$m_3 = 0.00004$
Standard deviation	$\sigma_1 = 0.00025$	$\sigma_2 = 0.0001$	$\sigma_3 = 0.00006$

Table 3. Validation of the statistical model - water meters of type I, precision class B.

Kolmogorov-Smirnov test: for $d_j \le d_{n,\alpha} = 0.309 \Rightarrow$ the hypothesis of normal distribution is accepted		
$d_1 = 0.1661$	$d_2 = 0.0903$	$d_3 = 0.1663$
Lilliefors test: for $L_j \le L_{n,\alpha} = 0.200 \Rightarrow$ the hypothesis of normal distribution is accepted		
$L_1 = 0.1105$	$L_2 = 0.0810$	$L_3 = 0.1663$

3.2 Water meters of type II, precision class B

The experimental data represents the measurement errors shown in Table 4, resulting by measuring a constant volume of water (10 liters) using the water meters of type II, precision class B.. The appropriate probability plots are presented in Figure 3, a to c, and normal distribution is assumed for each sample. Table 5 presents the estimated values of the parameters of the normal distribution and Table 6 presents the values of the statistics of the goodness of fit tests applied, which are compared to the critical values of the statistics of the goodness of fit tests, $d_{n,\alpha}$ and $L_{n,\alpha}$ [3].

Table 4. Measurement errors of water meters of type II, precision class B.

Water meter	**Measurement errors** (m^3)
1-II B	0.0005, 0.00014, 0.00023, 0.00022, 0.00012, 0.00016, 0.00012, 0.00011, 0.00023, 0.00004, 0.00022, 0 0.00006, 0.00002, 0.00028, 0.00017, 0.00005, 0.00007
2-II B	-0.00008, 0, -0.00003, 0, 0.00005, 0.00001, -0.00002, -0.0003, 0, -0.00002, -0.00006, -0.00009, -0.00005, -0.00005, -0.00003, -0.00007, -0.00011, -0.00001
3-II B	-0.0001, -0.00008, 0, 0.00002, -0.00005, 0.00004, 0, -0.00002, -0.00002, 0.00001, -0.00002, -0.00001, 0.00001, -0.00006, -0.00003, 0.00001, -0.00002, -0.00001

Table 5. Parameter estimation - water meters of type II, precision class B.

Water meter	1-II B	2-II B	3-II B
Mean	$m_1 = 0.00014$	$m_2 = -0.00004$	$m_3 = 0.00002$
Standard deviation	$\sigma_1 = 0.00010$	$\sigma_2 = 0.00005$	$\sigma_3 = 0.00003$

Table 6. Validation of the statistical model - water meters of type II, precision class B.

Kolmogorov-Smirnov test: for $d_j \le d_{n,\alpha} = 0.309 \Rightarrow$ the hypothesis of normal distribution is accepted		
$d_1 = 0.1461$	$d_2 = 0.1743$	$d_3 = 0.2035$
Lilliefors test: for $L_j \le L_{n,\alpha} = 0.200 \Rightarrow$ the hypothesis of normal distribution is accepted		
$L_1 = 0.1461$	$L_2 = 0.1634$	$L_3 = 0.148$

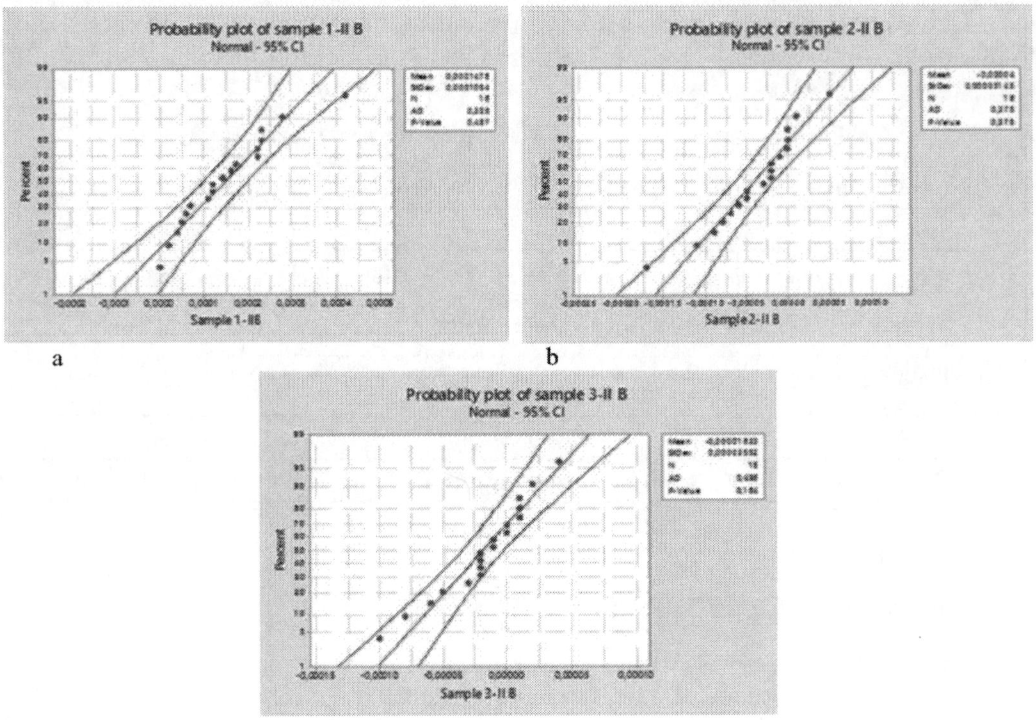

a b c

Fig. 3. Probability plots of the samples 1, 2, 3 collected by using water meters of type II, precision class B.

3.3 Water meters of type III, precision class C

The experimental data was collected by reading the indications of the water meters of type III, precision class C. It represents the measurement errors which are resulting by measuring a constant volume of water (10 liters) using the water meters considered.

The measurement errors of the water meters of type III, precision class C, are shown in Table 7. The probability plots for each sample are presented in Figure 4, a to c, and normal distribution is assumed in each case. Table 8 presents the estimated values of the parameters of the normal distribution and Table 9 presents the values of the statistics of the goodness of fit tests applied, which are compared to the critical values of the statistics of the goodness of fit tests, $d_{n,\alpha}$ and $L_{n,\alpha}$ [3].

Table 7. Measurement errors of water meters of type III, precision class C.

Water meter	Measurement errors (m^3)
1-III C	0, 0.00001, -0.00003, 0.00006, 0.00001, 0.00003, 0.00003, 0.00002, 0.00003, 0.00002, -0.00002, 0.00006, 0.00001, 0.00002, -0.00001, 0.00002, 0.00005, 0.00004
2-III C	0, 0, 0.00003, 0.00024, 0.0001, 0.00002, 0.00011, 0.00011, 0.00015, 0.00013, 0.0001, 0.00004, 0.00006, -0.00004, 0.00005, 0.00009, 0.00002, 0.00006
3-III C	0.00002, 0.00015, 0.00043, 0, 0.00008, 0.00052, 0.00052, 0.00038, 0.0003, 0.00048, 0.00063, 0.00023, 0.00037, 0.00036, 0.0004, 0.00029, 0.00045, 0.00052

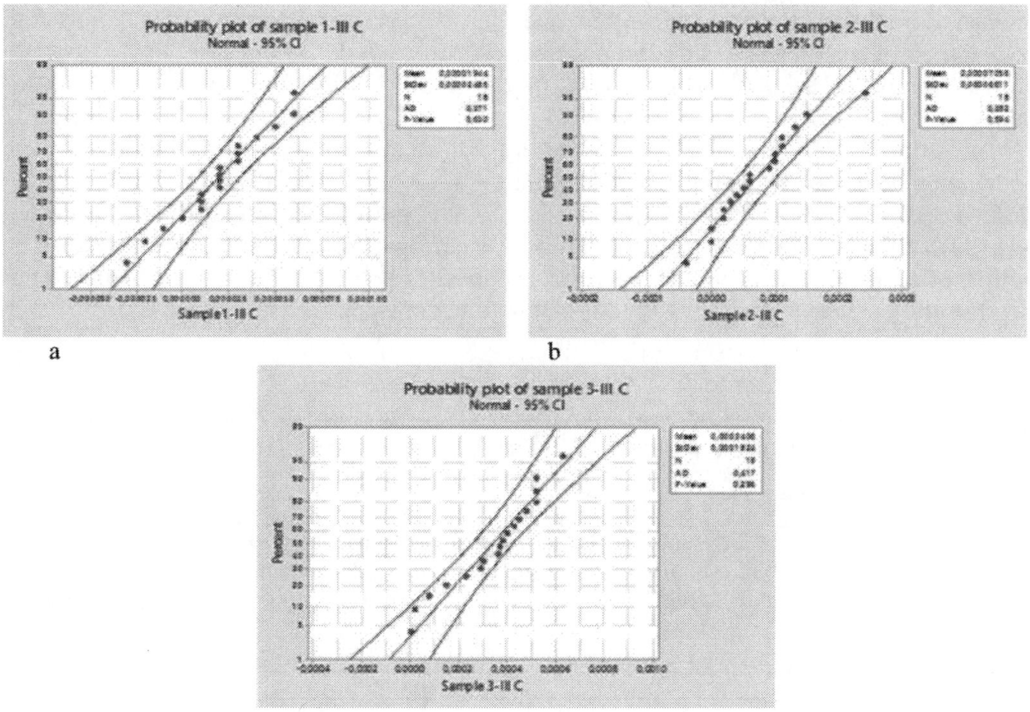

a b

c

Fig. 4. Probability plots of the samples 1, 2, 3 collected by using water meters of type III, precision class C.

Table 8. Parameter estimation - water meters of type III, precision class C.

Water meter	1-III C	2-III C	3-III C
Mean	$m_1 = 0.00002$	$m_2 = 0.00007$	$m_3 = 0.00034$
Standard deviation	$\sigma_1 = 0.0002$	$\sigma_2 = 0.00006$	$\sigma_3 = 0.00018$

Table 9. Validation of the statistical model - water meters of type III, precision class C.

Kolmogorov-Smirnov test: for $d_j \leq d_{n,\alpha} = 0.309 \Rightarrow$ the hypothesis of normal distribution is accepted		
$d_1 = 0.1009$	$d_2 = 0.1190$	$d_3 = 0.1536$
Lilliefors test: for $L_j \leq L_{n,\alpha} = 0.200 \Rightarrow$ the hypothesis of normal distribution is accepted		
$L_1 = 0.1009$	$L_2 = 0.1190$	$L_3 = 0.1070$

4 Study on the volume of the water loss in the installation

Water meters of precision class C are designed such as to start the correct registration of the water consumption from the minimum starting flow *Qmin* = 15l/h, while in case of the water meters of precision class B, the minimum starting flow is *Qmin* = 30l/h. As a consequence, since water meters of precision class B are apartment water meters, it results that any water consumption which is less than 30l/h will not be correctly registered by the apartment water meter. In case of a block of flats, although apartment water meters might not register correctly the water consumption due to a reduced flow in the apartment installation, the proper consumption will be correcly registered by the branching water meter which is the general water meter for the block of flats. Thus, the water losses are not registered by the apartment water meters and the water consumption is not paid correctly, so that the water company has financial losses.

The volume of the water loss depends on the water flow in the installation. The cases presented below prove this fact.

As to those presented above, a study was conducted using the water meters in the installation described in Figure 1, for determining the volume of the water loss in case the flow is less than *Qmin*=15l/h. It was considered the case of a leakage difficult to detect, when, from time to time, a very small drop of water appeares and falls from the tap. The period of time in which a water loss has occurred is t = 32'53", and the actual water volume collected drop by drop is V=2 l, such as the corresponding water flow Q is:

$$Q = \frac{V}{t} = 0{,}0036\,\mathrm{m}^3/\mathrm{h}\,. \tag{1}$$

Table 10 presents the volume of the water loss, which is not measured by the water meters and the volume of the water loss corresponding to one hour, for water flow Q.

Table 10. Volumes of the water loss registered by the three types of water meters for water flow Q.

Type of water meter	The volume of water registrated (l)	Mean of the volume of water registrated (l)	Actual volume of water (l), corresponding to time t	Volume of water loss (l), corresponding to time t	Volume of water loss (l), corresponding to one hour (1h)
I	0,05 0,04 0	0,03	2	1,97	3,594
II	0,03 0,01 0,02	0,02	2	1,98	3,612
III	0,17 0,15 0,13	0,15	2	1,85	3,375

A second case considers a leakage, still difficult to detect, which ocurred in time t' = 16'35", the volume of water loss being V=2 l, too. The corresponding water flow Q' is:

$$Q' = \frac{V}{t'} = 0{,}0072\,\mathrm{m}^3/\mathrm{h}\,. \tag{2}$$

Table 11 presents the volume of the water loss, which is not measured by the water meters and the volume of the water loss corresponding to one hour, for water flow Q'.

Table 11. Volumes of the water loss registered by the three types of water meters for water flow Q'.

Type of water meter	The volume of water registrated (l)	Mean of the volume of water registrated (l)	Actual volume of water (l), corresponding to time t'	Volume of water loss (l), corresponding to time t'	Volume of water loss (l), corresponding to one hour (1h)
I	0.15 0.17 0.16	0.16	2	1.84	6.65
II	0.15 0.13 0.14	0.14	2	1.86	6.72
III	0.56 0.58 0,57	0.57	2	1.43	5.17

As seen from Table 10 and Table 11, the volume of the water loss depends directly proportional with the water flow in the installation.

5 Conclusions

Water meters, as measuring devices, may measure with errors that are greater than the admissible ones; this leads to errors in the metering of the consumed water and the financial losses are carried by the water company or by the client; on the other side, conditions for the irrational utilization of the potable water resources are created.

The tolerated maximum measuring errors are ±2% for cold water (if water temperature <= 30°C) [5]. As seen in the cases considered in this research, there are measuring errors exceeding the tolerance interval for each type of the water meters analysed and they lead to errors in metering the consumed water.

The errors in metering the consumed water may appear in the period of their utilisation. As to the measuring errors, in the period in which water meters are used, it is possible for these errors to exceed the tolerance interval and if this exceeding represents a certain percentage, corresponding to the established level of significance, a metrological control of the water meters should be performed. As a consequence, it would be necessary to develop an algorithm which would provide information on the moment when the verification would be required for the water meters to remain within the appropriate performance specifications.

Other cause which could lead to errors in metering the consumed volume of water is represented by a water flow in an installation, less than the minimum starting flow of the water meters, which creates conditions for the incorrect registration of the water consumption. In order to solve this problem, one needs to use accurate water meters and leakage detecting equipment

References

1. I. Hohan, *Technology and reliability of the systems* (Ed. Didactică şi Pedagogică, Bucureşti, 1982)
2. I. Martinescu, I. Popescu, *Analysis of the reliability and of the security of the systems* (Ed. Universităţii „Transilvania" Braşov, Braşov, 2002)
3. M.R. Clinciu, *Research on the reliability of the water meters* (Ph.D. Thesis, Transilvania University of Braşov, Braşov, 2009)
4. V.G. Voda, A. Dorin, *Statistical reliability problems* (Ed. Economică, Bucureşti, 1994)
5. NML 003-05, *Water meters* (Monitorul oficial al României, Bucuresti, 2005)

Method for determining the calibration life of measuring devices. Case study – water meters

Mihaela Rodica Clinciu[1], and *Ramona* Clinciu[1,*]

[1]Transilvania University of Braşov, Department of Manufacturing Engineering, Mihai Viteazu No. 5, Braşov, Romania

Abstract. The paper presents the results of an experimental study conducted on specific measuring devices - water meters. The aims of the study are to determine the calibration life of the measuring devices as a function of the metrological reliability and to emphasize the parameter deviation as an important research tool for the quality assurance in metrology. The experimental study conducted is aimed at determining the calibration life of water meters of the same type and of precision class B. The research performed on the considered water meters can lead to the conclusion that the appropriate moment for performing the calibration for a specific set of measuring devices can be estimated by using the parameter deviation functions, these being obtained by means of regression analysis.

1 Introduction

Each measuring device is characterized by a certain tolerance interval. In case the measuring device is correctly adjusted, the measurement errors are located inside the tolerance interval. In time, by using the measuring device, the measurement errors might start to grow and, at a certain moment, they may exceed the tolerance interval. If the probability density function (pdf) for the measurement errors exceeds the tolerance interval with a certain percentage, it is considered that a metrological disturbance has appeared [1]. Metrological disturbances are difficult to discover at the moment they are produced, they can be discovered only during the periodical checking of the measuring device or by discovering the negative consequences generated by using the measuring device. In these conditions, it becomes very important to predict, with a certain precision, the moment when the measuring device measures with greater errors than the tolerated ones [1].

Parameter deviation represents an important research tool for the estimation of the calibration life of the measuring devices. It helps to predict the moment in time when the probability density function for the measurement errors exceeds the tolerance interval with a certain percentage, corresponding to the established level of significance. This moment corresponds to the stage in which the considered measuring device needs to be checked and, if necessary, calibrated [1, 2]. This experimental research may be added to former researches performed on different types of measuring devices [3-5].

* Corresponding author: r.clinciu@unitbv.ro

The present paper presents the results of an experimental study conducted on specific measuring devices - water meters of precision class B, a medium precision for the water meters. The experimental data is obtained in laboratory conditions and it is represented by the measuring errors, obtained by measuring the same volume of water by a single person, experimental data being grouped in samples produced at a certain interval of time.

The calibration life of the water meters, based on the parameter deviation functions, was estimated by using an appropriate algorithm. The estimated calibration life, obtained as a function of the metrological reliability, is compared to the requirements set in the appropriate standards. In the case of the water meters analysed, the estimated calibration life proved to be less than the one set in the standards, this fact meaning that the use of the respective water meters might lead to errors in metering.

2 Estimating the calibration life of water meters

General metrology uses especially the normal distribution, although positive and asymmetrical distributions, as Weibull distribution, are frequently met in the analysis of the metrological reliability [6].

The algorithm developed for estimation of the calibration life of the measuring devices uses the following notations: α - the level of significance; m - the mean; σ - the standard deviation; n - the volume of the sample; k - the number of samples; x_{ji} - the experimental data, which is assumed to be normally distributed, according to the normal distribution function $N(m, \sigma, t)$ ($j = 1,..., k$, $i = 1,..., n$) having both parameters m and σ variable; m_0 - the initial value of the mean; a - the variation coefficient of the mean; σ_0 - the initial value of the standard deviation; b - the variation coefficient of the standard deviation; T_ε - the tolerance interval for variable ε, as a measurement error; ti - the initial moment; dt - the period of time at which samples are collected (it is considered that dt is constant); t_r - the calibration time [2-4].

The case of a normal distribution $N(m_0, \sigma_0, t)$ is considered. In time, due to the internal parameter deviation, the parameters of the distribution become m and σ and the normal distribution becomes $N(m, \sigma, t)$. It is considered that the mean m and the standard deviation σ may have a linear or an exponential variation, according to the equations:

$$m = m_0 + a \cdot t \tag{1}$$

$$\sigma = \sigma_0 + b \cdot t \tag{2}$$

$$m = m_0 \cdot \exp(a \cdot t) \tag{3}$$

$$\sigma = \sigma_0 \cdot \exp(b \cdot t). \tag{4}$$

The algorithm for estimating the calibration life of the measuring devices as a function of metrological reliability includes the steps presented below [2]:

- Collecting experimental data (measurement errors);
- Performing data outlier tests- Grubbs test [7];
- Graphical representation (probability plot) and distribution assumption – normal distribution [7, 8];
- Parameter estimation for the normal distribution [7];
- Goodness-of-fit tests; it is considered a general goodness-of-fit test (Kolmogorov-Smirnov) and a normality goodness-of-fit test (Lilliefors). The optimum distribution is choosen by comparing the values obtained for the statistics of the tests considered with the critical values of the tests: d_j - the value of the statistics of the Kolmogorov-Smirnov goodness-of-fit test is to be compared to $d_{n,\alpha}$ - the critical value of the statistics of the Kolmogorov-Smirnov goodness-of-fit test; L_j - the value of the statistics of the Lilliefors

goodness-of-fit test is to be compared to $L_{n,\ \alpha}$ - the critical value of the statistics of the Lilliefors goodness-of-fit test [6-8];

- Determining the parameter deviation functions by using regression analysis and the correlation coefficients for the linear variation and for the exponential variation [7, 8]. The type of the parameter deviation function, linear or exponential, for parameters m and σ, is indicated by the highest value of the two calculated correlation coefficients for each parameter;

- Determining the calibration life, using the parameter deviation functions and the method described in [2]. The period for the metrological control of water meters is set to seven years, as stated in standards [9].

3 Experimental study

The experimental study is conducted on water meters of the same type, of precision class B, a medium precision for the water meters, connected in series arrangement, at a water source (Figure 1). At the time the experimental study was conducted, all water meters were in their specified working period, they have been used in installations for a specific period of time, 34 months, and were still in their working period according to standards prescriptions [9].

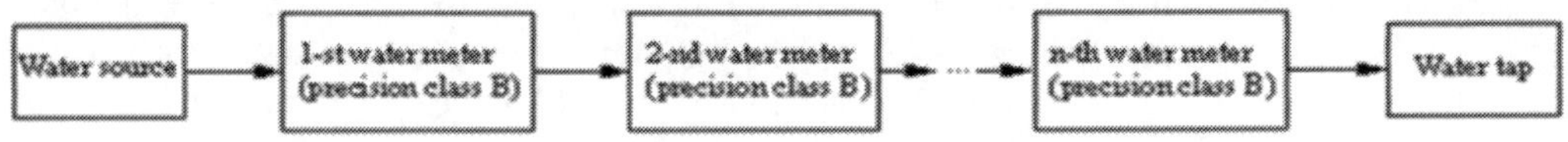

Fig. 1. Block scheme of the experimental installation.

The experimental data is represented by the measuring errors, obtained by measuring a certain volume of water, 0.01 m^3. The measuring errors were obtained as the difference between the indications of each water meter used and the true value of the volume of water (0.01 m^3). Data is grouped in eight samples produced at a certain period of time (a month), by measuring the same volume of water with the group of water meters, in laboratory conditions. Table 1 presents the measuring errors, obtained by measuring the same volume of water with the water meters considered. By analyzing the values of the errors, one can notice that they are all located within the appropriate tolerance interval.

Table 1. Measuring errors.

Sample no.	**Measuring errors** (m^3)
1	-0.00007, -0.00011, -0.00006, -0.0002, -0.00011, -0.00004, -0.00016, 0.00008, -0.00001, -0.00011, -0.00002, 0.00002, -0.00007, -0.00012, -0.00003, -0.00001
2	0.00011, 0, .00011, -0.00006, -0.0002, -0.00011, 0.0001, 0, 0.00005, -0.00008, 0, -0.00008, 0, 0.00011, 0.00007, 0, 0.00001
3	0.00017, 0.00011, -0.00006, -0.0002, -0.00011, 0.00004, 0.00019, 0.00008, -0.00001, 0.00011, -0.00002, 0.00002, -0.00007, -0.00012, -0.00003, 0.00002
4	-0.00011, 0.00015, 0.00016, 0.00008, 0.00003, 0.00018, 0.00007, 0.00025, -0.00004, 0.00004, -0.00003, 0.00002, -0.00002, 0.00005, -0.00011, -0.00015
5	-0.00011, 0.00015, 0.00016, 0.00008, -0.00013, -0.00018, -0.00007, -0.00015, -0.00014, 0.00014, 0.00013, 0.0002, 0.0002, 0.00015, 0.00018, 0.00019
6	0.00017, 0.0002, 0.00018, 0.00019, 0.0002, 0.00019, -0.00019, 0, 0.0001, 0.00012, -0.00011, -0.00005, -0.00009, 0.0002, -0.00002, -0.00005
7	0.00019, 0.0002, 0.00017, 0.0002, 0.00018, 0.00019, -0.00019, 0, 0.0001, 0.00012, -0.00011, -0.00005, -0.00001, 0.0002, -0.00009, -0.00002
8	0.0002, 0.0002, 0.00018, -0.00019, 0.0002, 0.00019, -0.00019, 0.0002, -0.00001, 0.00012, -0.00011, 0.00016, -0.00003, 0.0002, -0.00007, 0.00019

The tolerated maximum measuring errors are ±2% for cold water (if water temperature <= 30°C) [9]. The tolerance interval is $T_{\varepsilon} = 0.0004$ m^3, the level of significance is $\alpha = 0.05$, the volume of the sample is $n = 16$, the number of samples $k = 8$, initial moment $ti = 0$ and the interval between the moments at which samples are collected $dt = 1$.

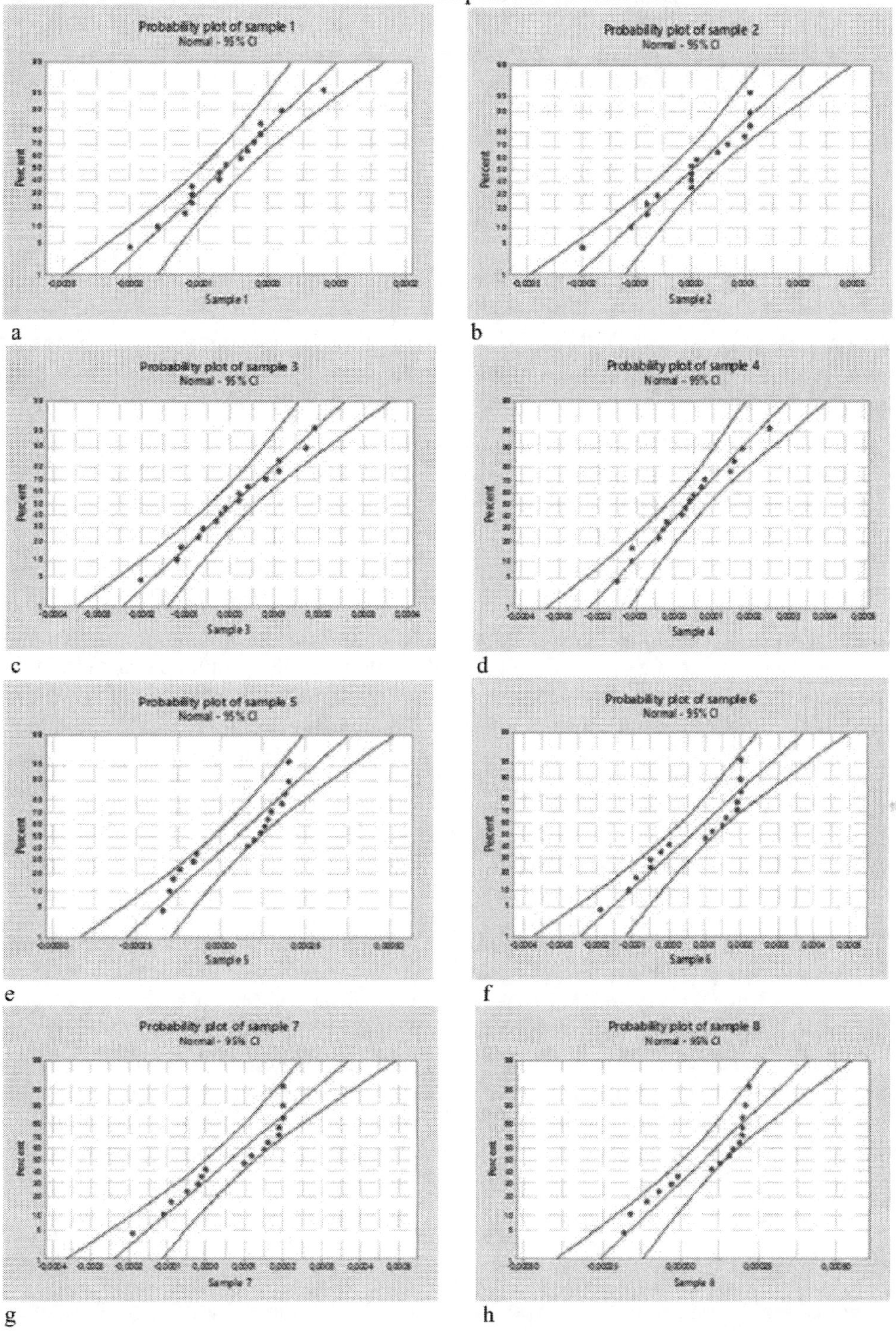

Fig. 2. Probability plots of the samples 1 to 8.

The probability plots of the samples are presented in Figure 2, a to h. By analyzing the probability plots, one can assume that the distribution is the normal distribution and this fact is to be checked by using the goodness-of-fit tests.

The critical value of the statistic of the Kolmogorov-Smirnov goodness-of-fit test for the considered sample size is $d_{n,\alpha} = 0.328$ and the critical value of the statistic of the Lilliefors goodness-of-fit test for the considered sample size is $L_{n,\alpha} = 0.213$ [6-8].

Table 2. Values of the statistics of the goodness-of-fit tests.

	Sample							
	1	2	3	4	5	6	7	8
d_j	0.1191	0.1792	0.0797	0.0963	0.2676	0.2197	0.2175	0.2710
L_j	0.1191	0.1167	0.0787	0.0963	0.2051	0.1580	0.1591	0.2085

Table 2 presents the values obtained for the statistics of both Kolmogorov-Smirnov and Lilliefors goodness-of-fit tests, for each sample. These values are according to both conditions: $d_j \leq d_{n,\alpha} = 0.328$, for Kolmogorov-Smirnov goodness-of-fit test and respectively $L_j \leq L_{n,\alpha} = 0.213$ for Lilliefors goodness-of-fit and this indicates normal distributions [7, 8].

Table 3. Values of the estimated parameters of the samples.

	Sample							
	1	2	3	4	5	6	7	8
m	-0.00006	0	0.00001	0.00004	0.00005	0.00006	0.00007	0.00008
σ	0.00007	0.00009	0.00010	0.00011	0.00014	0.00013	0.00013	0.00014

The values of the estimated parameters m and σ for each sample are given in Table 3. By using the values of the means and standard deviations of the samples, presented in Table 3, one can find the parameter deviation functions, by means of regression analysis [7, 8]. The correlation coefficients for linear variation of the parameters are $r_{m\text{-}l} = 0.9421$ and $r_{\sigma\text{-}l} = 0.9197$ and the correlation coefficients for exponential variation of the parameters are $r_{m\text{-}exp} = 0.9228$ and $r_{\sigma\text{-}exp} = 0.9133$; these values indicate a linear variation for both parameters, according to the equations:

$$m = -0.00005 + 0.00002 \cdot t. \quad (5)$$

$$\sigma = 0.00007 + 0.000001 \cdot t. \quad (6)$$

The calibration life for the considered case is calculated by means of parameter deviation functions presented in equations (5) and (6), according to the method described in [2]. The value which is obtained for the calibration time for the case considered is $t_r = 2.4029$ time units (months), value which is less than the remaining period for the use of the water meters (water meters have been used for 34 months before the experiment), according to the pre-set period for the metrological control of the water meters, which is set to 7 years in the appropriate standards [9]. This fact means that the use of the respective water meters, more than the period indicated by the value t_r, is possible to lead to errors in metering.

4 Conclusions

The method for determining the calibration life based on parameter deviation functions can be used for estimating the right moment for performing the metrological control/calibration for a specific set of measuring devices. This moment might be later than the pre-set one, and in this case financial costs with the metrological control are diminished, or earlier than

DOI: 10.1051/matecconf/20179404003

the pre-set calibration time, and in this case greater errors than the tolerated ones are prevented.

By estimating the calibration life of the measuring devices as a function of metrological reliability, based on parameter deviation functions, it might be possible to test whether, for a certain type of measuring devices, the values obtained for the calibration life t_r are according to the requirements set in the appropriate standards. As future research, it is intended to consider other types of metrological devices, to test whether the values obtained for the calibration life meet the requirements set in the appropriate standards.

References

1. U. Wiener, F. Vereş, Quality of products and metrology J. **5**, 265-271 (1972)
2. R. Clinciu, Bulletin of the Transilvania University of Braşov. **7(42)**, 65-68 (2000)
3. R. Clinciu, AMM. **371**, 695-699 (2013)
4. R. Clinciu, AMM. **657**, 694-698 (2014)
5. R. Clinciu, AJME. **12/1**, 76-81 (2014)
6. V. Gh. Vodă, Applied metrology **XXVI(1)**, 11-14 (1979)
7. I. Martinescu, I. Popescu, *Reliability* (Gryphon, Brasov, 1995)
8. D. Kececioglu, *Reliability Engineering Handbook* (Prentice Hall, Englewood Cliffs, N.J., 1991)
9. NML 003-05, *Water meters* (Monitorul Oficial al Romaniei, Bucuresti, 2005)

Identification and analysis of incompatibilities occurring in the granite slabs manufacturing

Agnieszka Czajkowska[1], and *Renata* Stasiak-Betlejewska[2,*]

[1]Department of Strength of Materials, Concrete Structures and Bridges, Faculty of Civil Engineering and Architecture, Kielce University of Technology, Poland
[2]Institute of Production Engineering, Faculty of Management, Częstochowa University of Technology, Poland

Abstract. The research object of the study presented in the paper is a granite that, as a raw material valued worldwide, has a comprehensive application in the construction industry. It is characterized by durability and scratch resistance. Authors presents research findings on nonconformities identified in the granite slabs manufacturing that are caused by natural phenomena and result from the manufacturing process specificity. The quality level of the granite slabs was assessed by authors in order to identify nonconformities caused by the production process factors and to find causes of its occurrence and its minimizing. Quality management tools were applied in the research analysis to obtain proposal of the quality level improvement that can be introduced in the analysed manufacturing process.

1 Introduction

Nowadays, having a quality certificates and its monitoring is the norm. The quality requirements obligate companies to implement the TQM tools [1, 2]. The only way to differentiate their products is the continuous improvement and the conviction of its potential customers with high quality products and reliability of the manufacturer. Keeping the stone products quality is difficult because its quality level does not only depend on the efficiency of the production process factors but to a large extent it is determined by the nature of the raw material which often suffers from defects that are difficult to detect during a visual inspection. In the case of natural stone there are two groups of incompatibilities:
- natural, formed by the action of the climate and the environment,
- resulting from the production process as a result of improper operation of machines and people.

It often happens that as a result of visual control, the raw material is not flawless. The incompatibility occurs only during the production process. Trends that govern the stone industry does not allow companies to not follow quality requirements. The variety of designs and shapes of the granite products forces companies to keep the high quality of the manufactured products in accordance to individual customer requirements.

* Corresponding author: renatastasiak@wp.pl

In caring about the quality of products there are used reliable quality management tool [2]. The quality level of the granite slabs produced in the stone company was analysed in the paper with Pareto diagram - Lorenz using [1-4] to identify incompatibilities that have the greatest influence on the granite slab final quality.

2 The research object characteristics

2.1 The granite characteristics

Granite is a solid deep magma rock composed mainly of the alkali feldspar, the quartz, and the plagioclase. The average particle size is 1-25 mm [4]. Given the chemical composition of the granites can be divided into three groups: calcium-alkali peraluminowe and peralkaliczne. Granite has a clear blow, usually in three orthogonal directions, which makes it easy to weathering, as well as operation. Taking into account the chemical composition of the granites can be divided into three groups: calcium-alkali peraluminowe and peralkaliczne. Taking into account the chemical composition of the granites it can be divided into three groups: calcium-alkali, peralumin and peralkali. Granite has a clear stroke, usually in three orthogonal directions, which makes it easy to weathering, as well as operation. Granite comes in different colours, such as grey, white and pink, green, red, black and other (Fig. 1).

Fig. 1. Selected examples of granite slabs used in the enterprise a) ruby star (India), b) bohus (Scandinavia), c) Impala (South Africa) [own study.]

Due to the ease of receiving large blocks, cutting and polishing, and rich and beautiful colours the granite is used in the building industry as a building and decorative stone.

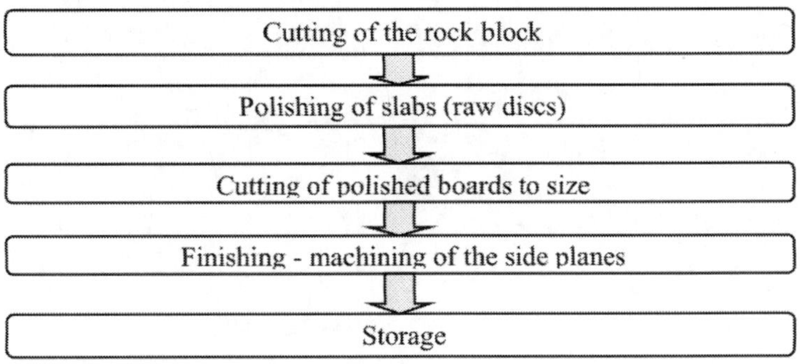

Fig. 2. The main stages of the granite slabs production process.

The extensive use of granite is provided not only by the aesthetic qualities but also by its practical features such as: high abrasion resistance, scratch resistance, and corrosive chemicals, high frost resistance, well led and accumulates heat. The richness of textures and colours gives unlimited possibilities of shaping space, both indoors and the buildings architecture. In the production process of granite slabs there are identified five major operations, where incompatibilities occur (Fig. 2).

Weather conditions play a very large role, especially during the storage of polished boards in the open air. Some weather factors (sun, wind, rain) negatively affect the polished boards quality and long-acting cause its partial destruction. Polished board after several years of exposure to air becomes rough and loses its natural aesthetic. This problem can be remedied in two ways : by storage of slabs with its covering with a layer of transparent film and by slabs covering with special chemical agent, whose properties allow for the extraction of colour depth, complementary micropores in the structure of the stone and the creation of an artificial protective coating. All finished granite products are covered by protective coating compulsorily.

2.2 The analyzed enterprise characteristics

The analyzed enterprise was founded in 1980 in the Southern part of Poland. At the beginning it employed 12 people. The company currently has 8 branches in Poland as well as partner companies around the world. The business core of the company is natural stones processing as well as wholesale and retail sale of stone products. The company offers construction elements such as window sills, thresholds, stairs, floors, housing fireplace, kitchen and bathroom countertops, elements of the facades of buildings and fences, and polished granite and marble slabs. The above-mentioned products are made from all kinds of granites, marbles and sienites available on the Polish market of Polish origin and imported from around the world, both in the form of blocks, slabs and finished pieces with a unique colour and design. The production are also used in selected types of sandstone and travertine.

3 Research findings and discussion

The analyzed enterprise produces about 350 finished granite slabs per month. The number of detected incompatibilities disqualifying the product including approximately 12%. Based on the data of the quality control department there was noted a significant increase in the incompatibilities level in from March to August in 2015. Data shown in Figure 3 present the level of incompatibilities that has increased from approximately 11% in March to 21% in July (the highest number of incompatibilities in the analyzed period).

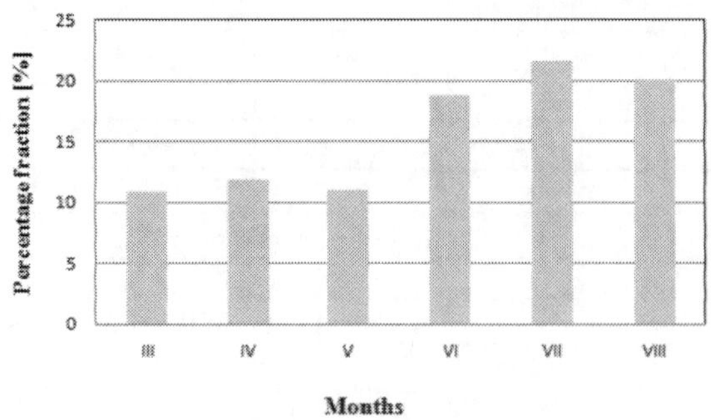

Fig. 3. The level of incompatibilities from March to August in 2015 in the analyzed enterprise. [own study].

The analysis shows the change in the incompatibilities structure in the analyzed period (Fig. 4).

Taking into account the lowest incompatibilities number in March (10.86%) and the highest incompatibilities number in July (21.6%) presented in Figure 4, it was noted a significant increase of the incompatibilities number for N6 (scratches), N5 (unpolishing) and N1 (microcraks).

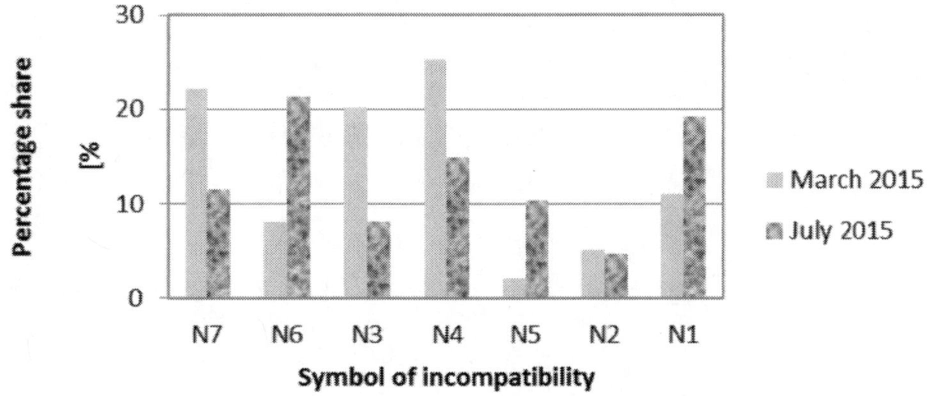

Fig. 4. Comparison of the incompatibilities structure in March and July 2015 in the analyzed enterprise.

For the incompatibilities identified in July there was prepared Pareto – Lorenz diagram which was aimed to the identified incompatibilities hierarchy.

Data from the quality control department and its percentage share are presented in Table 1.

Table 1. Incompatibilities identified in July 2015 in the analyzed enterprise in order of its frequency

Symbol	Name of the incompatibility	Percentage share	The cumulative percentage share
N6	Scratches	31,76	31,76
N1	Microcraks	16,47	48,24
N4	Incompatibility caused by the shear blade	15,29	63,53
N7	Borers	12,94	76,47
N5	Unpolishing	9,41	85,88
N3	Discoloration	8,24	94,12
N2	Stains	5,88	100,00

Pareto – Lorenz diagram (Fig. 5) was based on data from Table 1 and it shows the incompatibilities levels in July 2015.

There are two significant cases of incompatibilities (N6 and N1) that represent almost half (48.24%) identified incompatibilities.

 DOI: 10.1051/matecconf/20179404004

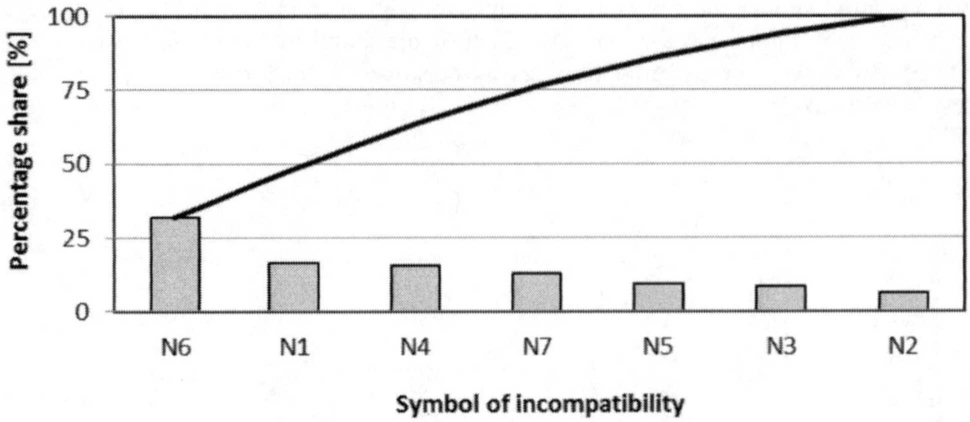

Fig. 5. Pareto – Lorenz diagram for incompatibilities identified in July 2015 in the analyzed enterprise.

There are several causes of the granite slab cracking (Fig. 6) :
- incorrect setting of the machine by an operator (the administration of improper slab thickness causes uneven pressure head polishing),
- improper sequence of the polishing wagons (if the wagon wheels of a smaller grain are mixed with such abrasive grains coarser, then it causes a distortion of the cycle and, consequently, formation of scratches on a polished plate),
- dirty polishing wagons (if the wagons, in particular the two last with the smallest grains suitable final gloss polishing are dirty, it already smooth the granite slab surface during its polishing).

Cracks largely depends on the technical condition of used machinery and equipment and attention and commitment of employees. This incompatibility can be avoided by increasing employee involvement.

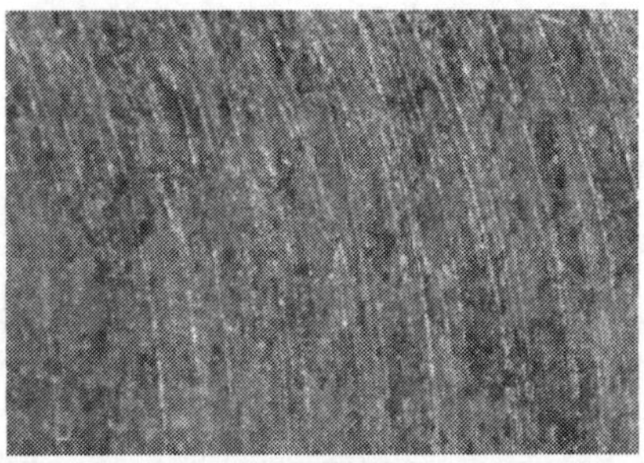

Fig. 6. Cracks produced on the surface of the granite slabs (repairable incompatibility).

Shortcommings affected by microcracks are much hurder to avoid (Fig. 7). Unfortunately, this incompatibility is often hidden at the beginning of the production process and it is revealed only during its lifetime on the already polished granite slab. Unfortunately, in most cases it is irreparable incompatibility.

Microcracks are among the most serious incompatibility of the stone slabs. The granite slab, which has the microcracks, is much more susceptible to cracks, just along these microcracks. The use of granite slab with microcracks in products such as countertops or stairs expose the user to the big danger.

DOI: 10.1051/matecconf/20179404004

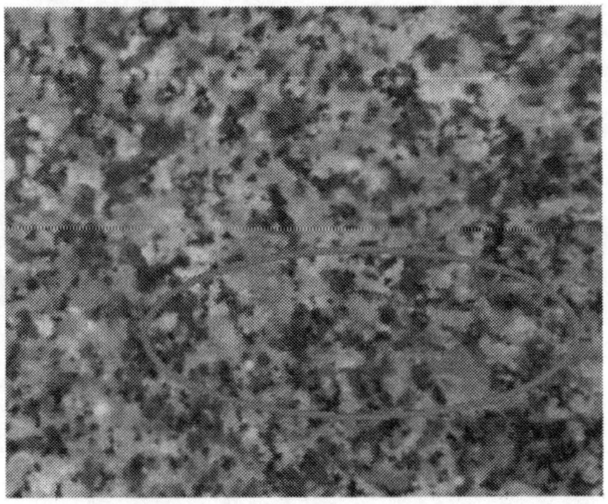

Fig. 7. The stone microcraks [own study].

Searching causes of such a large incompatibilities increase affected by microcracks, the stone supply should be checked. Microcracks are natural incompatibility group that become apparent only during the manufacturing process. Analysis of the microcracks causes results in the identification of the incompatibilities increase reason that was connected with the second and the third quality level stone supply ordering by the analzyed enterprise. The reason was the purchase of worst quality stones, which are cheaper, but suffer from defects.

4 Conclusion

In the analyzed enterprise the production is based on the standards applicable to building materials and ISO standards. Both the company executives and individual employees undergo training aimed at preventing irregularities and improving the quality of manufactured products. Despite all quality activities there are still some imperfections related to the production process as well as those which are the source of raw material defects impossible to detect in the visual control. The incompatibilities analysis in March and August 2015 shown that the main causes of the granite slabs low quality level were cracks and microcracks. The first defect is closely related to the production process, the machine and its operator. Large growth of defects associated with microcracks involves raw material and its natural disadvantages.

The main recommendation in this case is a series of training courses for officers and employees of the analysed company in the field of methods and tools of TQM and compliance with the maintenance schedule of machinery and equipment.

References

1. J.J. Dahlgaard, K. Kristensen, G.K. Kanji, *Fundamentals of Quality Management*, (PWN, Warszawa, 2000)
2. A. Hamrol, W. Mantura, *Quality Management. Theory and Practice*, (PWN, Warszawa-Poznań, 2011)
3. A. Hamrol, *Quality Management by Examples,* (PWN, Warszawa, 2015)
4. R. Duda, L. Rejl, *Encyclopaedia-Minerals*, (Elipsa 2, Warszawa, 1994)
5. L. Parv, *Data management in innovative engineering*, (Transilvania University Publishing House, 2015)

 DOI: 10.1051/matecconf/20179404005

Method of using of the Box-Cox transformation at the application of the xbar and s chart

Nicolae Eftimie[1,*]

[1]Transilvania University of Brasov, Department of Manufacturing Engineering, Mihai Viteazu No.5, Brasov, Romania

Abstract. The application of the most statistical process control techniques is based on the assumption that the distribution of the measurements is normal. However, there are many situations in practice when the process data distribution is not normal. In certain cases, the Box-Cox transformation can be used for converting the process data distribution into a normal distribution. Considering these aspects, the paper presents a method of application for the xbar and s chart that can be used in the case when the measurements distribution is not normal. The proposed method consists of the following stages: the testing of normality of the process data, the application of the Box-Cox transformation and the testing of normality of the transformed data. In the case when the distribution of the transformed data is normal, they are used at the application of the xbar and s control chart.

1 Introduction

The most of control charts that are used in practice are based on the assumption that the measurements distribution is normal. A tool often used in practice for testing this assumption is the Anderson-Darling normality test. A description of the procedure of application for this test is given in [1].

However in practice there are situations when the measurements are not normally distributed. One approach that can be applied in these cases is the transformation of the process data in order to obtain a set of normally distributed values [2].

Box and Cox [3] developed a method that can be used for converting the process data distribution into a normal distribution. According to this method, each measurement x_i will be transformed by means of the following system of equations:

$$y_i = \begin{cases} x_i^{\lambda} & \lambda \neq 0 \\ \ln(x_i) & \lambda = 0 \end{cases} \tag{1}$$

The λ parameter is determined as follows:

- The standardized values of the transformed data are calculated using the system of equations (2):

* Corresponding author : eftimien@unitbv.ro

DOI: 10.1051/matecconf/20179404005

$$z_i = \begin{cases} \dfrac{x_i^{\lambda} - 1}{\lambda \cdot G^{\lambda - 1}} & \lambda \neq 0 \\ G \cdot \ln(x_i) & \lambda = 0 \end{cases} \tag{2}$$

where G represents the geometric mean of the process data.
- It is calculated the standard deviation for the distribution of the z_i variable, by means of the equation (3):

$$\hat{\sigma}_z = \frac{\bar{s}_z}{c_4} \tag{3}$$

where the c_4 coefficient is computed using the relation (4) [2]:

$$c_4 = \sqrt{\frac{2}{n-1}} \frac{\Gamma\left(\frac{n}{2}\right)}{\Gamma\left(\frac{n-1}{2}\right)} \tag{4}$$

where n is the size of the sample.
- It is determined the value of λ as being that for which the standard deviation of the z_i variable distribution is minimal.

2 Problem formulation

In the case when the measurements distribution is not normal, the procedure of application for the *xbar* and *s* chart becomes difficult.

The objective of the paper is to develop a method of application for this type of control chart in the cases when the distribution of the process data can be converted into a normal distribution by means of the Box-Cox transformation.

The method consists of the following stages:
- The testing of normality of the process data. If the measurements distribution is normal, they can be used at the control chart application, otherwise the data must be transformed.
- The application of the Box-Cox transformation.
- The testing of normality of the transformed data.

In the case when the distribution of the transformed data is normal, they are used at the application of the *xbar* and *s* control chart, otherwise another method must be applied for the control chart application.

3 Problem solution

The proposed method of application of the *xbar* and *s* chart using the Box-Cox transformation is based on a computer program, which has been developed in Excel 2013 and Visual Basic for Applications for Excel 2013.

Each stage of the proposed method is performed on a separate worksheet.

The paper presents the operation of the program by means of an example.

The input data for the presented example were generated by computer according to a Weibull distribution, with the shape parameter β=2.2, scale parameter η=2.5 and threshold parameter γ=150.

The first worksheet of the application, which is labelled "*Data*", contains a table that is used by the program user for recording the measurements (Figure 1).

Each sample may contain up to 10 measurements. In the presented example, the size of the sample is equal to 5.

	A	B	C	D	E	F	G	H	I	J
1	x_1	x_2	x_3	x_4	x_5	x_6	x_7	x_8	x_9	x_{10}
2	151.192	152.566	152.455	151.103	151.017					
3	155.37	153.32	151.722	150.577	150.543					
4	151.697	151.784	151.936	152.59	151.481					
5	151.112	153.757	152.842	153.689	152.47					
...										
24	150.934	152.005	152.033	151.485	154.713					
25	152.688	151.573	151.478	154.037	153.359					
26	152.535	151.921	151.163	151.969	151.141					

Fig. 1. The Input data area.

The second worksheet of the program, which is labelled "*AD*", is used by the program for applying the Anderson-Darling normality test for the process data according to the theoretical notions presented in [1]. The values for the number of measurements *N*, the average *xbar* and the standard deviation *s* are calculated for the measurements recorded in the worksheet labelled "*Data*" (Figure 2).

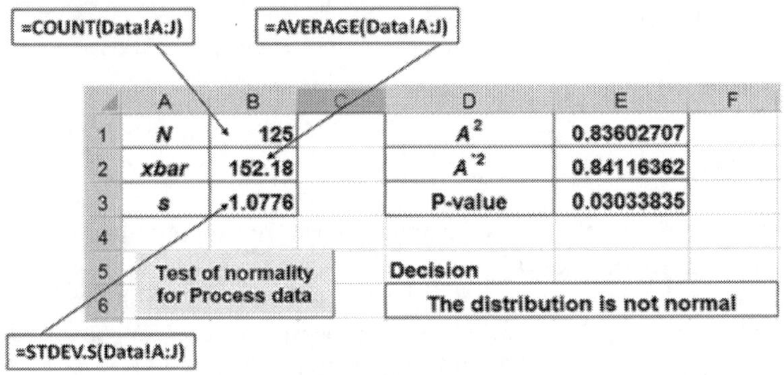

Fig. 2. The Anderson-Darling test for the process data.

The user of the program must click the control button labelled "*Test of normality for Process data*". As a result of this action, the procedure *A_D_x*() will be run. The listing for the first part of this procedure is presented by Figure 3.

```
Sub A_D_x()
Dim N As Integer, I As Integer
Dim xbar As Double, s As Double
Dim A2 As Double, A2M As Double, P As Double
Dim x() As Double, F() As Double
N = Range("B1").Value
xbar = Range("B2").Value
s = Range("B3").Value
For I = 1 To N
    ReDim Preserve x(I)
    x(I) = Application.Small(Sheets(1).Range("$A:$J"), I)
    ReDim Preserve F(I)
    F(I) = Application.Norm_Dist(x(I), xbar, s, 1)
Next I
```

Fig. 3. The procedure A_D_x() – Part 1.

After the declaration of the variables, the procedure collects from the worksheet the values for the number of measurements *N*, the average *xbar* and the standard deviation *s*. By means of a *FOR* structure, the measurements are ordered and for each of them, the cumulative distribution function is calculated. The measurements and the corresponding values of the cumulative distribution function are recorded into two arrays.

The listing for the second part of the procedure *A_D_x*() is presented in Figure 4.

```
A2 = 0
For I = 1 To N
   A2 = A2 + (2 * I - 1) * (Application.Ln(F(I)) + Application.Ln(1 - F(N + 1 - I)))
Next I
A2 = -A2 / N - N
A2M = A2 * (1 + 0.75 / N + 2.25 / N ^ 2)
If A2M < 0.2 Then
   P = 1 - Exp(-13.436 + 101.14 * A2M - 223.73 * A2M ^ 2)
ElseIf A2M >= 0.2 And A2M < 0.34 Then
   P = 1 - Exp(-8.318 + 42.796 * A2M - 59.938 * A2M ^ 2)
ElseIf A2M >= 0.34 And A2M < 0.6 Then
   P = Exp(0.9177 - 4.279 * A2M - 1.38 * A2M ^ 2)
ElseIf A2M >= 0.6 And A2M < 13 Then
   P = Exp(1.2937 - 5.709 * A2M + 0.0186 * A2M ^ 2)
End If
Range("E1").Value = A2
Range("E2").Value = A2M
Range("E3").Value = P
If P >= 0.05 Then
   Range("D6").Value = "The distribution is normal"
Else
   Range("D6").Value = "The distribution is not normal"
End If
End Sub
```

Fig. 4. The procedure A_D_x() – Part 2.

First, it is calculated the value of A^2, which is used for computing the value of A^{*2}, denoted into the procedure by *A2M*. The next stage is the calculation of the *P*-value, which depends on the value of A^{*2}. Using the *P*-value, the procedure makes the decision about the normality of the process data distribution.

The third worksheet of the program, which is labelled "*BoxCox*", is used by the program for transforming the process data. The first columns, from *A* to *J*, are used for collecting the measurements from the worksheet "*Data*". By means of the formula (2), there are calculated the standardized values of the transformed data z_i, for different values of the λ parameter.

For each sample, the value of the standard deviation, s_z is computed (Figure 5).

=IF(ISNUMBER(Data!A2),Data!A2,"")

=IF(ISNUMBER(A2),IF($X$5=0,GEOMEAN($A$2:$D$26)*LN(A2),(A2^$X$5-1)/($X$5*GEOMEAN($A$2:$E$26)^($X$5-1))),"")

	A	B	C	D	E	K	L	M	N	O	U
1	x_1	x_2	x_3	x_4	x_5	z_1	z_2	z_3	z_4	z_5	s_z
2	151.192	152.566	152.455	151.103	151.017	29.46377	30.82713	30.71515	29.37715	29.29365	0.766339899
3	155.37	153.32	151.722	150.577	150.543	33.76604	31.59645	29.98383	28.86938	28.8368	2.091634336
4	151.697	151.784	151.936	152.59	151.481	29.95913	30.04514	30.19588	30.85138	29.74645	0.419483093
...											
25	152.688	151.573	151.478	154.037	153.359	30.95058	29.83689	29.7435	32.3422	31.63666	1.128366154
26	152.535	151.921	151.163	151.969	151.141	30.79582	30.18098	29.43552	30.22869	29.41411	0.587297822

=IF(ISNUMBER(K2),STDEV.S(K2:T2),"")

Fig. 5. The worksheet BoxCox – Part 1.

The standard deviation σ_z for the distribution of the variable z_i is calculated according to formula (3), by means of the values of $\bar{s}_z$ and c_4 (Figure 6).

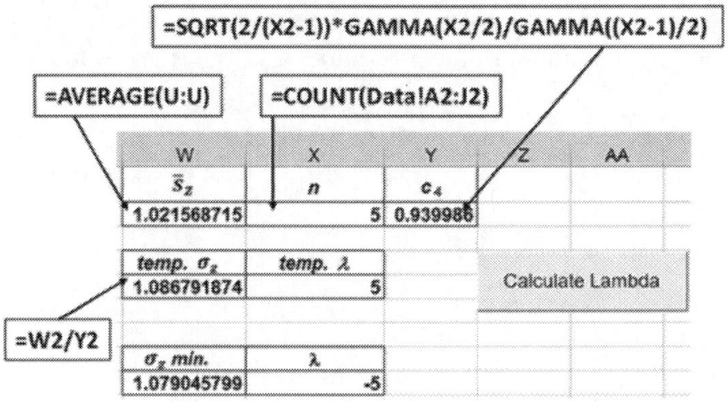

Fig. 6. The worksheet BoxCox – Part 2.

The value of the λ parameter is calculated by means of the procedure *Calculate_Lambda*(). The listing of this procedure is presented by Figure 7.

```
Sub Calculate_Lambda()
   Dim Lambda As Double, temp As Double
   Dim sigma_min As Double
   Lambda = -5
   Range("X5").Value = Lambda
   sigma_min = Range("W5").Value
   temp = -4.99
   Do While temp <= 5
      Range("X5").Value = temp
      If Range("W5").Value < sigma_min Then
         sigma_min = Range("W5").Value
         Lambda = temp
      End If
      temp = temp + 0.01
   Loop
   Range("W9").Value = sigma_min
   Range("X9").Value = Lambda
End Sub
```

Fig. 7. The procedure Calculate_Lambda().

This procedure will repeat all the calculations on the worksheet "*BoxCox*" for the λ parameter taking values into the interval [-5, 5] with a step value of 0.01. In this manner, the value of λ that correspond to the minimal value of σ_z will be found. By means of this value, the transformed data are calculated using the formula (1) (Figure 8).

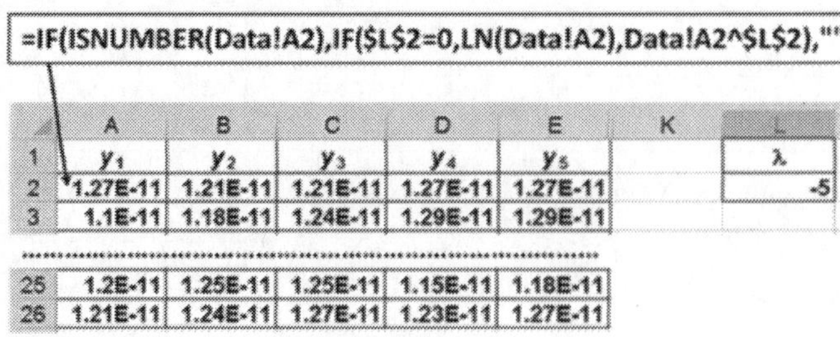

Fig. 8. The transformed data.

 DOI: 10.1051/matecconf/20179404005

The distribution of transformed data will not always result as being normal. Therefore, the next stage that must be performed, consists of application of normality test for the transformed data. This operation will be achieved in the same manner as the test of normality for the process data (Figure 9).

	A	B	C	D	E	F
1	*N*	125		A^2	0.73417296	
2	*ybar*	1.22613E-11		A^{*2}	0.73868372	
3	*s*	4.30917E-13		P-value	0.05429646	
4						
5		Test of normality for Transformed data		Decision		
6				The distribution is normal		

Fig. 9. The test of normality for the transformed data.

For the presented example, the distribution of the transformed data has resulted as being normal.

Therefore, in this case, the classic procedure of application of the *xbar* and *s* chart can be applied for the transformed data. This operation is performed on the worksheet labelled "*Chart*" (Figure 10).

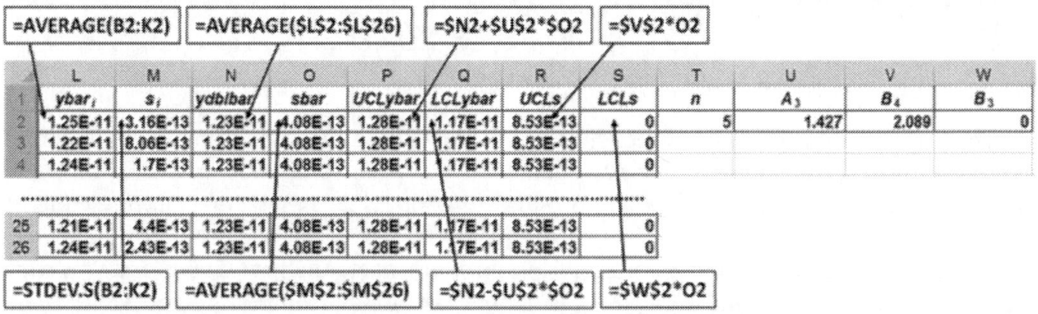

	L	M	N	O	P	Q	R	S	T	U	V	W
1	$ybar_i$	s_i	ydblbar	sbar	UCLybar	LCLybar	UCLs	LCLs	n	A_3	B_4	B_3
2	1.25E-11	3.16E-13	1.23E-11	4.08E-13	1.28E-11	1.17E-11	8.53E-13	0	5	1.427	2.089	0
3	1.22E-11	8.06E-13	1.23E-11	4.08E-13	1.28E-11	1.17E-11	8.53E-13	0				
4	1.24E-11	1.7E-13	1.23E-11	4.08E-13	1.28E-11	1.17E-11	8.53E-13	0				
...												
25	1.21E-11	4.4E-13	1.23E-11	4.08E-13	1.28E-11	1.17E-11	8.53E-13	0				
26	1.24E-11	2.43E-13	1.23E-11	4.08E-13	1.28E-11	1.17E-11	8.53E-13	0				

Fig. 10. The calculations that are needed for the representation of the control chart.

The columns from *B* to *K* contain the transformed data, which are used for calculating the points of the chart, as well as the central lines and the control limits. The control chart is presented by Figure 11.

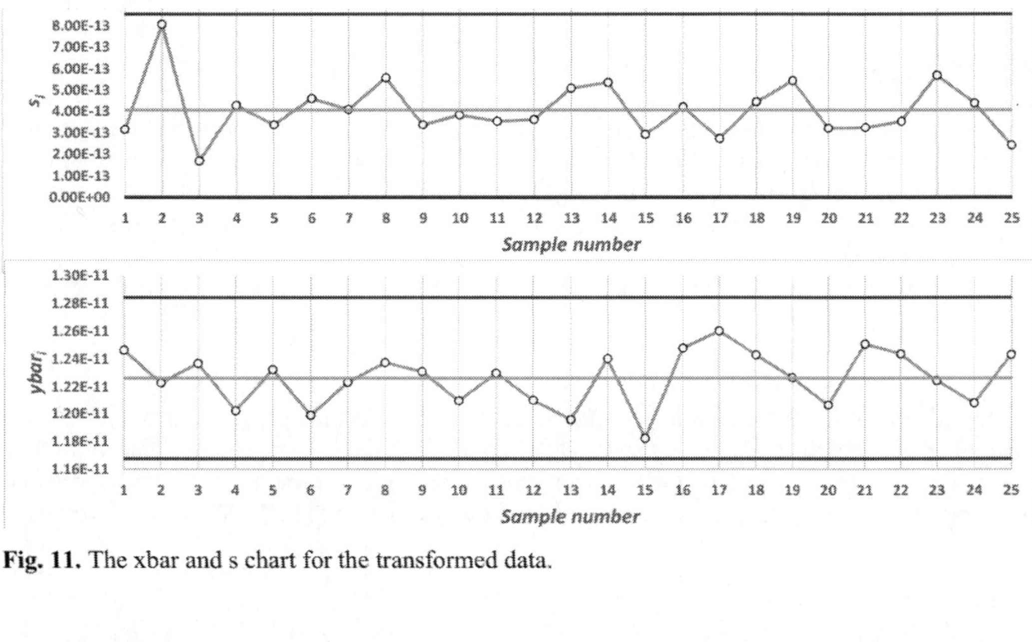

Fig. 11. The xbar and s chart for the transformed data.

It can be observed that for the presented example the process is under control.

The process capability analysis is presented by Figure 12.

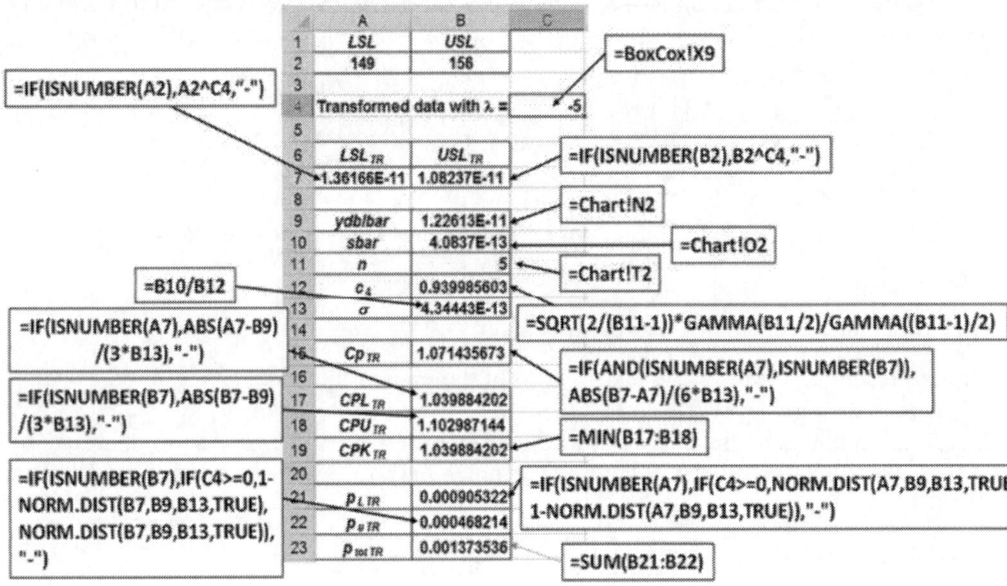

Fig. 12. The process capability analysis.

The program uses the value of the λ parameter and the specification limits for calculating the transformed specification limits: LSL_{TR} and USL_{TR}. For the presented example, the value of the λ parameter is less than 0, therefore the value of LSL_{TR} is greater than the value of USL_{TR}.

In the next stage, the program collects from the worksheet labelled "*Chart*" the values of the center lines: *ydblbar* and *sbar*, as well as the size of the sample, *n*. The value of c_4 coefficient is calculated according to the formula (4).

After the calculation of the standard deviation, σ the program computes the values of the capability indices for the transformed data: Cp_{TR}, CPL_{TR}, CPU_{TR} and CPK_{TR}.

The program also calculates the proportions of nonconforming items that correspond to the transformed data: p_{LTR}, p_{UTR} and p_{totTR}.

4 Conclusions

The presented method can be used in plant practice for controlling the processes in the case when the process data distribution is not normal, but can be converted into a normal distribution by means of the Box-Cox transformation.

The method is applied by means of a computer program which was developed so that the users do not have to possess computer programming knowledge.

Each stage of application of the proposed method is implemented on a separate worksheet.

The user of the program must record the measurements of the samples and all the calculations are performed by the program.

In order to assess the normality of process data distribution, the program uses a procedure developed on the base of the Anderson-Darling test. If the measurements distribution is normal, they can be used at the control chart application, otherwise the data will be transformed by means of a procedure developed on the base of the Box-Cox transformation.

The Anderson-Darling normality test is applied again, this time for the transformed data. In the case when the distribution of the transformed data is normal, they are used at the application of the *xbar* and *s* control chart, otherwise another method must be applied for the control chart application.

References

1. A.D. Sleeper, *Design for six sigma statistics* (The McGraw-Hill Companies, Inc., New York, 2006)
2. A. Mitra, *Fundamentals of quality control and improvement, third edition* (John Wiley & Sons, Inc, New Jersey, 2008)
3. G.E.P. Box, D.R. Cox, J ROY STAT SOC B MET, **26**, 211 (1964)

Choice and assessment of improvement measures critical for process operation (in reference to the requirements of ISO 9001:2015)

Adam Górny[1,*]

[1]Poznań University of Technology, Faculty of Management Engineering, 11 Strzelecka St., 60-965 Poznań, Poland

Abstract. Improvement measures should be seen as a *sine qua non* condition for improving operating efficiency. The measures may vary in nature and range from minor process enhancements to continuous improvements. Improvement measures were selected and evaluated by the FMEA process methodology which helped identify irregularities and their causes as well as indicate which irregularities are most likely to hamper proper process performance. In defining the relationships, reference was made to the requirements of the ISO 9001:2015 standard [1]. Process performance disruptions were assumed to reduce the efficiency and effectiveness of task performance and, most importantly, put process participants at risk of property damage and health impairment. Particular emphasis was placed on occupational safety with a view to reducing the damage caused by failures to secure adequate working conditions. It is only after the scope of analysis has been extended to include criteria related to occupational safety and the ability to ensure worker-friendly working environments that an organization gains the ability to rectify the related errors and, ultimately, attain compliance with ISO9001:2015.

1 Introduction

Whenever irregularities occur in any field of business, improvement measures become imperative. Improvement may be seen as the achievement of a greater capacity to satisfy requirements. Improvement measures need to be aligned with the specific nature of the tasks at hand with proper account taken of any inconsistencies and the viability of the improvements that are undertaken [2, 3].

The goal in assessing improvement measures by identifying the potential benefits is to provide a rationale for their application. A crucial factor is the practical viability of the improvement measures and, ultimately, the ability to attain improvements in every area that is of importance for ensuring the efficient operation of an organization. This presumption may be applied to working conditions, which are at the heart of this paper and which underpin

* Corresponding author: adam.gorny@put.poznan.pl

the operation of processes [4, 5]. The working environment may be seen as a critical area for innovation and improvements that contribute to creating better conditions for process performance [6].

Process efficiency hinges on the effort exerted to improve process functionalities. To achieve it, one needs to carefully select measures based on their evaluation in view of critical aspects of deployments that are essential for the efficient overall operation of an organization [7].

Once working environment issues have been incorporated into the scope of improvement measures, it becomes necessary to recognize human factor requirements. The human factor can be characterized indirectly in terms of an organization's ability to provide its employees with optimal conditions for operating in the working environment. Ultimately, therefore, ensuring adequate conditions for the operation of humans in the working environment will become a systemic requirement that critically affects process efficiency. FMEA enables one to identify irregularities and propose a loss reduction strategy [8].

The application of process FMEA helps identify the root causes behind the emergence of barriers that prevent the satisfaction of requirements associated with specific processes (such as manufacturing). Such factors may be linked to pivotal issues affecting the performance of processes such as the operation of production facilities, machinery and equipment [9, 10] as well as the conditions for implementing production processes. Process FMEA may also be employed to shape working environments [11, 12] and consequently, in connection with ergonomic requirements, ensure worker comfort [6]. Such comfort, in turn, is bound to boost the ability to carry out processes. Such considerations should be recognized as one the areas for improvement listed in the 2015 edition of the ISO 9000 series of standards.

2 Improvement measures

2.1 Nature of improvement measures

Improvement measures may be defined as enhancements of the tasks carried out by an organization. Such measures range from the minor tweaking of processes to continuous improvements constituting part and parcel of the systemic approach to management [3]. The ISO 9000 standard [13] defines improvement measures as actions taken to make an organization more efficient. Such measures may help reduce the number of failures and complaints, boost production efficiency, cut operating expenses, etc. Once a non-conformity occurs, corrective action is required to remedy the problem and minimize its consequences. To this aim, an evaluation is in order of the need to remedy the non-conformities and, most of all, prevent their recurrence in the same or another location. The next step is to implement such remedies and appraise their effectiveness. Any improvement measures should adequately address the effects of any non-conformities that have been discovered [1, 14].

The measures taken should focus on ensuring the lasting success of an organization, defined as the achievement and retention over the long haul of the ability to accomplish the adopted objectives [11, 15]. Such measures should extend to all factors and conditions that influence the attainment of an adopted goal. Therefore, decisions must be based on factual evidence, which in turn makes it necessary to conduct a thorough study of the irregularity in question [16].

To ensure that improvement measures are effectively carried out and that the anticipated benefits result from an organization's adopted operating strategy, it is often advisable to employ the tools and improvement methods that help select the proper improvement measures and properly evaluate improvement results [12, 14, 16, 17].

2.2 Measures improving process operation

Improvement measures are a *sine qua non* condition for improving an organization's processes [18]. Improvement measures are only effective if they are consistent with the nature of any non-conformities. If chosen properly, such measures help carry out improvement tasks according to plan and in a duplicable manner based on scientific presumptions as well as gather such data and information as is necessary to select adequate improvement methods. To solve the problem effectively, one needs knowledge of the concerned process, access to proper identification and improvement tools and the ability to apply them. As improvement outcomes should be measurable, it is critical to define indicators that will enable an organization to e.g. identify the possible benefits, specify the verification timeline and assign responsibilities for specific measures [3]. Such methods include process FMEA which can be used to influence the quality of performing specified tasks [16, 19]. The fundamental aim in applying process FMEA is to raise error detection rates at early stages of process design and consequently generate savings on the cost of future improvement measures [14].

While conducting an analysis, one should bear in mind that its successful completion depends on the nature of the existing non-conformities. Therefore, improvement measures must be consistent with the nature of such non-conformities. The choice of improvement measures can only be correct if the issues at hand are properly analyzed so as to adequately define their characteristics. The tools employed must help find all significant issues that affect non-conformities and assess their impact [2].

The FMEA analysis employed to define the nature of a problem may be carried out [16, 17]:

- Systemically, in which case processes will be addressed in their entirety as a system made up of subsystems which in turn comprise their own lower level subsystems,
- In a problem-centered approach, in which the analysis extends to all areas in which irregularities are found.

This helps identify the problem in areas of significance for improvement measures. The improvements entail rectifying discrepancies between the existing and the desired states. They may also be seen as a way to increase process effectiveness and efficiency. One result of improvement measures is the elimination of irregularities which stand in the way of process operation.

Process performance verification can be described as the assurance of [16, 17]:

- Proper process planning that helps answer the question of what is the appropriate course of process operation that guarantees the fulfillment of requirements,
- Proper process performance, especially where such performance and the achievement of the desired improvement outcomes are at risk,
- Achieving improvement outcomes regardless of the non-conformities faced.

The FMEA technique is a teamwork-based method which can be used to detect, prevent, eliminate and control potential process errors [18]. Its application helps consistently and lastingly eliminate process weaknesses. This is achieved by recognizing the real causes of irregularities and employing adequate preventive measures of proven effectiveness.

An additional option of using process FMEA relies on the principle of "continuous improvement" and specifically on successive analyses of the process which help identify irregularities (failures) and make adjustments and adopt solutions that effectively eliminate the causes of defects. Such analyses may also give rise to new ideas for measures that improve the operation of processes.

3 Choice and application of improvement measures

3.1 Process FMEA methodology as a method of identifying and selecting improvement measures

The evaluation and choice of solutions suited for the process FMEA methodology requires a three-staged sequence of actions [3, 14, 16, 17]:

• Stage 1 covers a preparation which involves defining the process relationships that arise in the course of performing a specified task. This typically entails a prior breakdown of the task into processes or elementary operations on an as-needed basis. To ensure that the measures taken are effective, it is critical to employ teamwork and adopt as many presumptions as possible that will enable the organization to properly analyze the process at hand.

Despite the breakdown, processes should be viewed as a complex amalgam of subsystems, each of which comprises its own lower-level subsystems. One should recognize that each part of the system plays its specific role. A proper breakdown reflecting to the complexity of the system makes it possible to commence the analysis at any level, which is particularly important when dealing with complex processes.

• Stage 2 represents the process FMEA analysis proper. At this stage, the organization needs to identify any potential process failures. A failure is defined as any improper operation of a process, i.e. such process performance as does not lead to the desired outcome. Failures of this kind are described with reference to their type, impact and cause. A failure cause is an improper operation of a lower level subsystem. Its impacts affect a subsystem of a higher level.

Once an organization has identified the sequence of events that trigger the occurrence of a failure, it is in a position to assess the likelihood of the failure in specific processes, specify the rate of failure detection prior to process completion and prior to the occurrence of the adverse effects that influence the next sub-process and assess the significance of the failure for the proper operation of processes and for process performance outcomes. This kind of numeric evaluation comes on a scale from 1 to 10 depending on the severity of the impact of a given criterion.

Descriptions of evaluation criteria and examples of tendencies in the evaluation factors which influence the numerical scores are summarized in Figure 1.

Once calculated, the numbers describing a failure's impact on the end result of a process can be used to arrive at the failure risk priority number ($RPN = O \times D \times C$), which ranks on a scale from 1 to 1000. The selection of critical process failures is possible by reference to the number representing the priority of the failure risk. High failure risk priorities point to the need to take preventive action aimed at modifying process performance.

• Stage 3 concerns the adoption of and supervision over preventive measures. Stage 3 focuses on completing the tasks identified at Stage 2. The findings of analyses are used as a basis for modifying processes with a view to eliminating failure causes or mitigating their impacts.

In order to carry out improvement measures, it is critical to define the necessary means, identify persons responsible for taking action and define the related timeline. Commonly, costing figures are additionally provided with an eye to establishing the benefit to cost ratio [14, 20].

The indicator (priority number) identified at Stage 2 is used to define the sequence of improvement measures in the order of criticality. The higher the indicator, the more severe the concerned failure. Where differences between criticality levels are minor, more subtle criteria should be used to rank failure causes. In such cases, it is recommended that

 DOI: 10.1051/matecconf/20179404006

component evaluations be used as ranking weights starting with failure severity, followed by failure occurrence frequency and failure detectability [17, 21].

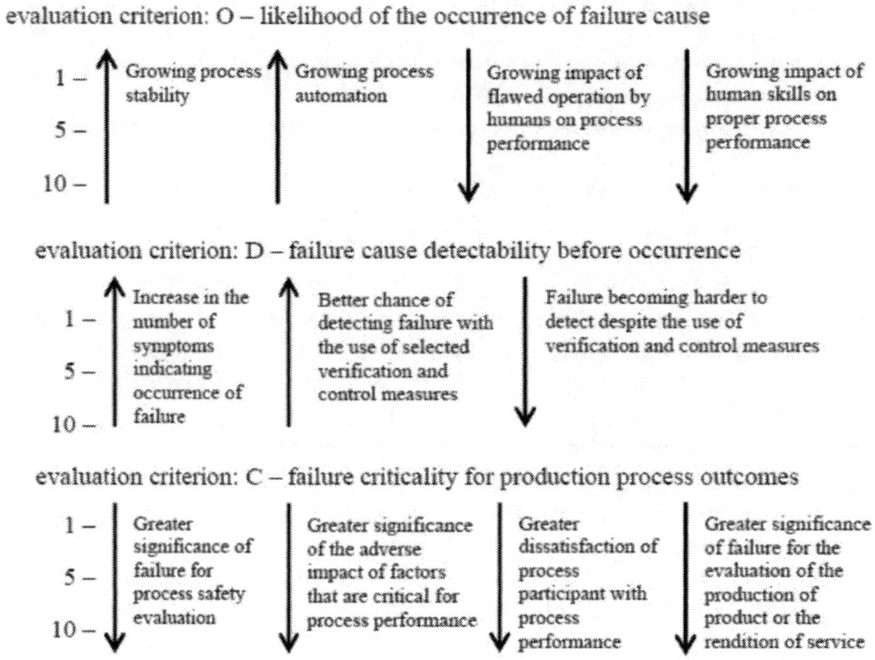

Fig. 1. Description of evaluations employed in the process FMEA method.
Source: Own research based on [14, 17, 21].

3.2 Applicable of process FMEA to select improvement measures (example)

The methodology described in Section 3.1 was applied to assess the criticality of failures and the choice of improvement measures in a company that carries out welding work. The procedure helped link process disruptions with such effects as reduced labor efficiency, lower performance accuracy, material waste and employee health impairment.

The analysis focused on the process of tank cover welding. Such welding was conducted in a manufacturer of pressure containers that comply with harmonized European legislation, including Directive 2014/68/EU [22]. The company has been awarded permits to manufacture pressure containers from Polish supervisory authorities, including the Office of Technical Inspection (UDT).

A critical precondition for producing weld joints of satisfactory quality is to put in place a proper technological regime covering the entire welding process. Due to the nature of the work, there is also a risk of workers being poisoned by the gases and fumes from electrode melting, which will result in a substandard joint.

The evaluation was conducted by a team appointed by the company's management. Its members were in-house technological engineers, constructors and occupational health and safety experts. Also employed in an auxiliary capacity was an external expert tasked with the methodological supervision over the evaluation process. The irregularities involved and the severity of their impact are shown in Table 1.

DOI: 10.1051/matecconf/20179404006

Table 1. Irregularities assessed and evaluation scores

Item assessed: the process of tank cover welding										
Evaluation aspect:							Evaluation criterion:			
Function		Effect	Failure	Cause	Prevention	Detection	O	D	C	RPN
a)	Tight sealing of tank	Employee poisoning (health impairment)	Loss of ability to work	Work without the use of protective items	Work performed in accordance with instructions manual	Failure to employ controls during work performance	8	8	9	576
b)		Poorly performed joint (leaky joint)	Inability to performance expected function	Failure to comply with technological process specifications during joint welding	Work performed in accordance with instructions manual	Failure to employ controls during work performance	8	9	7	504

Table 2. Process evaluation indicators after the application of improvement measures.

	Employed improvement measures increasing the ability to perform the process of welding tank cover		Evaluation criterion:			
	Preventing the occurrence of defect causes	Detectability of defect causes	O	D	C	RPN
a)	Ensuring effective ventilation at workstation and making the use of PPE mandatory	Adoption of control and monitoring principles and their application	4	5	9	180
b)	Ensuring proper parameters and compliance with technological joint welding regimes	Adoption of control and monitoring principles and their application	3	7	8	168

4 Possibilities of and potential benefits from applying the process FMEA method

Due to the universal nature of the process FMEA method, it is applicable in practically any industry in which a high degree of reliability is required. The use of the method is recommended particularly at the stage of designing difficult-to-control processes and during their performance for the purposes of improvement. A significant benefit of the process FMEA method is that it can be employed in the manufacturing operations of any kind, wherever inefficient processes may result in substantial losses [20]. This benefit can be attributed to the method's nature. The evaluation that forms a part of the method can only be appropriate if extensive information has been gathered about the facility in question [14, 16, 19, 21].

The research has found the need to examine links between the factors that are critical for process operation and any improvement measures. This is necessary for the FMEA to produce the expected benefits.

The ability to rectify a problem classified for priority (with an RPN) depends heavily on:
- The likelihood of the occurrence of the causes of a specific defect,
- The detectability of the causes of such a defect,
- The ability to apply measures that eliminate defect causes and are well suited for the company's operating methods,
- Access to resources required to deploy measures that eliminate defect causes,
- Defect significance for the relevant process.

Irregularity analysis should begin with indicating the nature of any irregularities. These are commonly described as the inability to perform a vital function. This step is followed by a causal analysis aimed at identifying the root cause of the issue [2, 23].

The steps taken in analyzing a problem and the measures adopted to rectify it are shown in Figure 2.

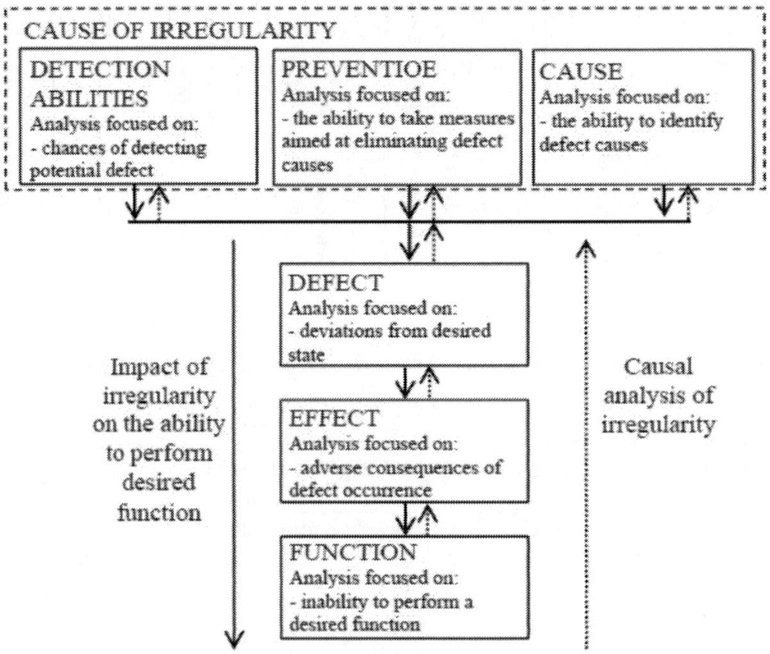

Fig. 2. Steps taken to identify and rectify irregularities.

The potential benefits depend on the approach taken [14]. In the problem-centered approach, analysis is extended to all areas in which irregularities were discovered. The drawback of this approach is its limited capacity for correcting potential risks throughout process performance. The analysis is limited to the risks which have already surfaced and attracted attention. An inexperienced evaluation team may easily overlook the risks that are rare but nevertheless generate severe consequences. The advantage of the problem-centered approach is the simplicity and conciseness of the analysis it involves [17].

The systemic approach makes the analysis general as well as more transparent. The approach is better at identifying such potential risks in the process as have not yet surfaced or that are relatively rare [17]. The systemic approach requires a well-trained evaluation team and strict compliance with the adopted procedures.

Regardless of which approach is taken, the FMEA method does not generate new solutions. All it does is point to the parts of the process that are critical and suggest why it seriously hinders efficient performance [3, 11, 19]. This requires engineering (and also technological) knowledge, the awareness of standards and legal regulations and experience working with the processes at hand. It may be viewed as a key resource to be used in carrying out improvement measures.

Some of the most significant areas of management in today's enterprise are quality, environment and occupational health and safety. By incorporating the above occupational health and safety requirements and issues into the management system, an organization will gain the ability to view the problem as an interdisciplinary area of improvement. An essential part of the improvement is to modify not only process performance but also the environment in which the process takes place. This follows from ISO 9001 which emphasizes the importance of focusing not only on process implementation but also on ensuring that the conditions for process performance are adequate. A study of the case discussed in section 3.2 shows that failures to ensure an adequate working environment, use personal protection equipment and ensure correct process specifications (proper technological specifications for welding) will all lead to the same consequences, i.e. the production of leaky tanks. By adopting proper prevention and control measures that guarantee the detection of defects, the manufacturer will accomplish the same result, which can be described as improved process effectiveness. This has been demonstrated by the process assessment results shown in Table 2. The resulting benefits can be evaluated by comparing the results of process assessments (RPN) carried out before and after improvement measures (see Table 1). In both case a) and case b), the measures taken have helped reduce the risk of tank dysfunctionality (caused by leaky joints) by a factor of ca. 3, i.e. down to an acceptable level.

The acceptable and unacceptable ranges of non-conformity as well as suggested measures to be taken to maintain or improve current status have been summarized in Table 3.

Table 3. Acceptable and unacceptable ranges of non-conformity [24, 25].

RPN value	Level of irregularity	Suggested measures
Acceptable irregularity range		
(0 – 200)	Minor (tolerable)	No process improvements are necessary
<200 – 500)	Mild (moderate)	Process improvements are recommended
<500 – 800)	Medium (significant)	Process improvements are necessary
Unacceptable irregularity range		
<800 – 1000>	Major	Processes must be suspended until improvements are made

The assessment methodology requires that specific causes of irregularities be identified by independently describing the relevant processes [26, 27]. Simultaneous evaluations carried out at multiple process points will reveal a number of issues associated with process impacts on workers.

The requirements in these fields are often mutually complementary. Therefore, to identify issues, one needs to employ methods and tools that make it possible to gain a global perspective on the issue at hand [6, 12]. One such tool is process FMEA which helps [7, 12, 27]:

- Recognize the factors that hinder the satisfaction of requirements or disrupt processes; such factors may be associated with the conditions that affect process operation and that are vital to their improvement,
- By consistently identifying and eliminating failures in e.g. production processes, applying the right preventive measures and avoiding the process failures that have been identified on a preliminary basis but that remain unknown, it is possible to achieve the state of "zero failures" [21],
- Identify problems encountered at various stages of process performance,
- Utilize the process approach structure which enables one to link inputs (such as data and information) with outputs (such as decisions).

Without a doubt, one of the drawbacks of the method is that it often leads to wrong conclusions unless the evaluation team employed is highly experienced. To relate the issue at hand to working conditions, one needs to account for the relationships between man – technical item – working environment in the process in question. An inexperienced evaluation team may find it difficult to define the problems clearly and link them correctly with specific aspects of process performance [4, 20, 26]. To obtain objective conclusions, it is not enough to demonstrate the direct impact of given risks. What needs to be done as well is to look into the risk of the existing irregularities generating further effects (losses) at a later time. As losses of this kind usually follow a definable loss emergence pattern, timely improvement measures will help effectively break the cycle of problem escalation.

5 Conclusions

Early on (in the early 20th century), the FMEA method continued to gain popularity in an ever greater number of applications in preparing complex processes in astronautics, nuclear engineering and aviation. By today, the scope of its use has grown substantially. FMEA is a preventive method which helps identify defects and their causes sufficiently in advance. Organizations relying on complex manufacturing processes employ FMEA to identify issues associated with many areas of their operations. This is particularly critical where multiple problems coincide in such varying fields as:

- Ensuring continuous production at a required efficiency level,
- Maintaining technical machinery and equipment in proper working condition,
- Managing the impacts on the environment as well as on occupational health and safety,
- Assuring the required quality of products and services,
- Ensuring the activities are sufficiently profitable.

No matter how complex an organization's processes are, it is always possible to identify issues having the greatest influence on final outputs. The approach helps detect key problems and assess their impacts. In the case discussed in section 3.2, the requirement must be met to describe:

- The effects of the company's inability to produce tanks that meet the desired functionality specifications,
- The way in which the defects render the organization incapable of achieving the required functionality,
- The immediate causes of defects as well as the underlying causes of defect occurrence,
- The possibility of employing improvement measures,
- The conditions in which a defect can be detected, the likelihood of its detection and the measures available to prevent defect occurrence.

In this process, the organization will also detect defects having the most adverse impact on process efficiency and effectiveness. Such impacts include the risk of health impairment and, in the most extreme cases, life-threatening risks.

Process FMEA helps identify potential threats to process performance. The method should be recognized as useful for recording and monitoring/observing measures aimed at averting any potential impacts. If used properly, FMEA will lower process costs while ensuring the required efficiency and quality. Where ISO 9001:2015 applies, process FMEA helps identify issues and make improvements against a range of requirements which translate into process efficiency. As a consequence, the conditions needed for adequate process performance will be created.

In addition, organizations rely on simulations to help define the potential effects of improvement measures. However, significant limitations must be taken into account in applying such techniques.

To ensure that evaluations are objective, it is critical to obtain information on the nature of irregularities as well as their causes and effects. The identification process needs to be conducted in a way that ensures that all irregularities are detected and properly described. During an FMEA analysis, account must be taken of the long-term consequences of the occurrence of specific types of errors.

The use of the FMEA method in process analysis brings into focus a number of requirements that could be overlooked without such comprehensive analysis. These include requirements associated with the management of the working environment that help ensure working conditions that are friendly to production floor workers. Where a broad approach to an issue is critical for the fulfillment of systemic requirements, process FMEA will significantly facilitate the evaluation and selection of best improvement measures.

References

1. EN ISO 9001:2015, Quality management systems. Requirements (ISO 9001:2015), European Committee for Standardization, Brussels
2. T.A. Carbone and D.D. Tippett, Engineering Management J., **16(4)**, 28-35 (2004)
3. C.J. Price, N.S. Taylor, Reliability Engineering & System Safety, **76(1)**, 1-10 (2012)
4. A. Górny, The Elements of Work Environment in the Improvement Process of Quality Management System Structure, In: W. Karwowski, G. Salvendy (Eds.), *Advances in Human Factors, Ergonomics and Safety in Manufacturing and Service Industries*, 599-606 (AHFE 2011)
5. A. Kawecka-Endler, B. Mrugalska, Human-Computer Interaction: Applications and Services, III, **8512**, 700-709 (2014)
6. A. Górny, Foundation of Control and Management Sciences, **11**, 127-138 (2008)
7. A. Sutrisnoa, I. Gunawanb, S. Tangkumanc, Procedia Manufacturing **4**, 23-29 (2015)
8. C. Estorilio, R.K. Posso, Int. J. of Quality and Reliability Management, **27(6)**, 721-733 (2010)
9. A. Górny, Evaluation of compliance with minimum safety requirements in machine operations: A case of assessing the control devices of a press, In: P.M. Arezes, at al. (Eds.), *Occupational Safety and Hygiene*, 497-501 (Taylor and Francis Group, London 2013)
10. B. Mrugalska, P.M. Arezes, Safety Requirements for Machinery in Practice, In: P.M. Arezes, at al. (Eds.), *Occupational Safety and Hygiene*, 97-101 (Taylor and Francis Group, London 2013)
11. A. Kawecka-Endler, B. Mrugalska, Analysis of Changes in Work Processes, In: W. Karwowski (Ed.), *Advances in Social and Organizational Factors*, 672-681 (AHFE 2012)
12. A. Górny, Management Systems in Production Engineering, **3**, 106-111 (2014)
13. EN ISO 9000:2015, Quality management systems. Fundamentals and vocabulary (ISO 9000:2015), European Committee for Standardization, Brussels
14. M. Dudek-Burlikowska, J. of Achievements in Materials and Manufacturing Engineering, **45(1)**, 89-102 (2011)
15. EN ISO 9004:2009, Managing for the sustained success of an organization. A quality management approach (ISO 9004:2009), European Committee for Standardization, Brussels
16. D.H. Stamatis, *Failure Mode and Effect Analysis. FMEA from Theory to Execution*, 2nd edition, (ASQ Quality Press, Milwaukee 2003)
17. A. Hamrol, *Quality management with examples* (Wydawnictwo Naukowe PWN, Warszawa 2005)

18. M. Bahrami, D. H. Bazzaz, S. M. Mojtaba, Procedia - Social and Behavioral Sciences, **41**, 418-425 (2012)
19. C. S. Carlso, *Effective FMEAs. Achieving Safe, Reliable, and Economical Products and Processes Using Failure Mode and Effects Analysis* (John Willey and Sons, New Jersey 2012)
20. S. Kmenta, K. Ishii, J. of Mechanical Design, **12(6)**, 1027 (2004)
21. B. Gajdzik, A. Wyciślik, *Quality, environment and safety in enterprise management* (Wydawnictwo Politechniki Śląskiej, Gliwice 2008)
22. Directive 2014/68/EU of the European Parliament and of the Council of 15 May 2014 on the harmonization of the laws of the Member States relating to the making available on the market of pressure equipment, OJ L 189, 27.6.2014, pp. 164-259
23. A. Shahin, Int. J. of Quality and Reliability Management, **21(7)**, 731-746 (2004)
24. A. Górny, Wykorzystanie FMEA w procesie identyfikacji zagrożeń i oceny ryzyka zawodowego, In: J. Charytonowicz, W. Pilecki (Eds.), *Zastosowania Ergonomii/ Wybrane kierunki badań ergonomicznych w 2008 roku*, 75-86 (Wyd. Polskiego Towarzystwa Ergonomicznego PTErg, Oddział we Wrocławiu, Wrocław 2008)
25. A. Górny, Wykorzystanie FMEA procesu w analizie zdarzeń wypadkowych i doskonaleniu warunków wykonywania pracy, In: J. Charytonowicz (Ed.), *Zastosowania Ergonomii. Wybrane kierunki badań ergonomicznych w 2016 roku*, 41-55 (Wyd. Polskiego Towarzystwa Ergonomicznego PTErg, Oddział we Wrocławiu, Wrocław 2016)
26. H.S. Bunney, B.G. Dale, The TQM Magazine, **9(3)**, 183-189 (1997)
27. S.G. Teng, S.M. Ho, D. Shumar, P.C. Liu, Int. J. of Quality and Reliability Management, **23(2)**, 179-196 (2006)

Investigations of friction-relevant parameters to ensure reliable bolted joints

Max Klöcker[1,*], and *Sebastian* Westphal[2]

[1]TH Köln, D-50679 Köln, / member of GFAV e. V., D-40724 Hilden, Germany
[2]formerly TH Köln, D-50679 Köln, / now Hosokawa Kolb GmbH, D-53859 Niederkassel, Germany

Abstract. During the process of torque-controlled tightening of bolted joints, the reliability of load transmission is limited by the achieved accurate preload. The preload is mainly influenced by head- and thread friction. The knowledge of reliable coefficients of friction and the dependence on assembly parameters are essential. Although a variety of friction coefficients were published but the overall ranges ask for more precision. So called windows of friction coefficients are available in known guidelines, to define permissible scattering. However, there are practical doubts about useful friction coefficients, especially if newly developed lubricants, corrosion-resistant coatings and securing devices against bolt-loosening are used. In this study, different parameters of a bolted joint connection, such as lubrication, the surface protection of the screw connection or used security systems, have been investigated on their frictional influence. To make a reliable conclusion about the coefficients of friction, only a measured examination like DIN EN ISO 16047:2013-01 can help. By means of a designed test bench those values could be detected by measurement and supplemented with the collected data of vibration test results at the TH Köln. On this way practically useful coefficients of friction and their scattering have been made available.

1 Introduction – Task of the investigations

High duty bolted joints are tightened torque-controlled in most of all cases. Both the thread flanks as well as the head bearing areas are subject to sliding friction. In the elastic deformation range and under constant friction conditions between the tightening torque T and the produced preload F_V exists a linear relationship.

The tightening torque T is composed of the head friction moment T_b. and the thread torque T_{th}. The thread torque T_{th} is required for obtaining the assembly preload F as well as for overcoming the friction forces at the thread flanks (fig.1).

As shown in fig. 2 the generated preload is substantially dependent on the friction in the bearing areas of the thread and the head. Usually about 85% of the tightening torque T is used to overcome the friction forces and only a small amount is left for generating the assembly preload F.

[*]Corresponding author: max.kloecker@th-koeln.de

 DOI: 10.1051/matecconf/20179404007

$$T = \overbrace{T_{th,P} + T_{th,F}}^{T_{th}} + T_b = F_V \cdot \left[\frac{P}{2 \cdot \pi} + \frac{d_2}{2 \cdot \cos\left(\frac{\alpha}{2}\right)} \cdot \mu_{th} + \frac{D_b}{2} \cdot \mu_b \right] \quad (1)$$

$$T_{th,P} = F_V \cdot \frac{d_2}{2} \cdot \tan(\varphi)$$

$$T_{th,F} = F_V \cdot \frac{d_2}{2} \cdot \tan(\rho_{th})$$

$$T_b = F_V \cdot \frac{D_b}{2} \cdot \mu_b$$

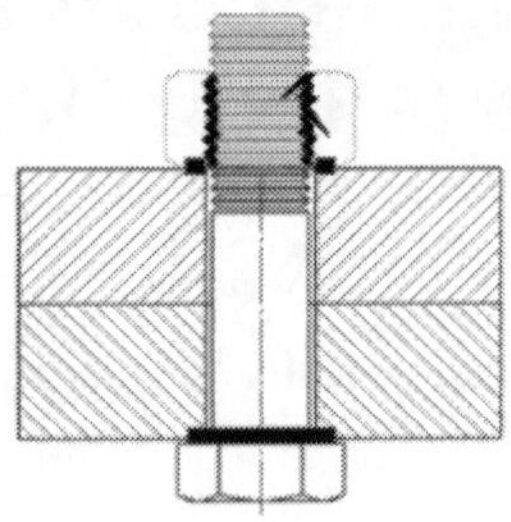

$T_{th,P}$	[Nm]	thread pitch torque
$T_{th,F}$	[Nm]	thread friction torque
T_{th}	[Nm]	thread tourque
T_b	[Nm]	head friction torque

Fig. 1. Relations between assembly preload Fv and tightening torque T.

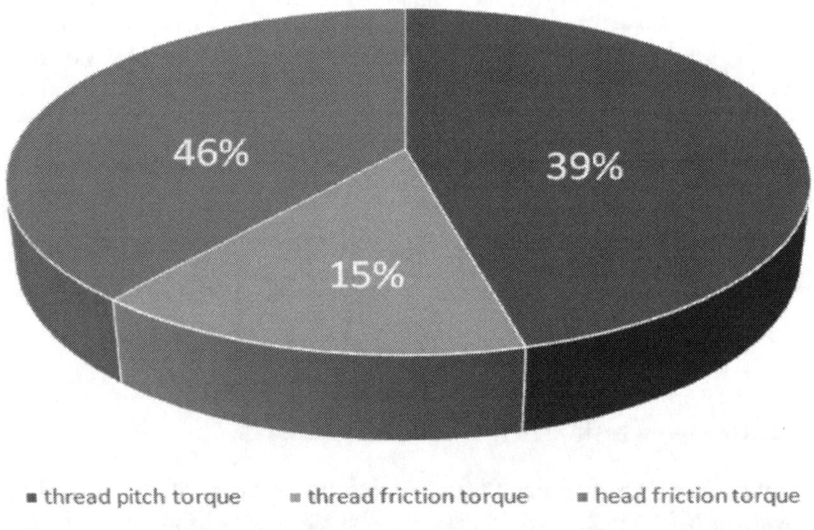

Fig. 2. Typical components of torque generating the preload, thread friction torque and head friction torque (example: hex. bolt ISO 4017, M12 – 8.8, T = 83 Nm; F_V= 43 kN μ_{tot}= μ_{th}= μ_b= 0,12).

The accurate preload is the essential factor for reliable load transmission of high duty bolted joints. In practice the preload is not measurable. Therefore, the preload is only adjustable indirectly via the torque. The friction coefficients in the thread and head contact areas are uncertain quantities. As the following chart fig. 3 shows different preloads are achieved with a determined tightening torque. With a low coefficient of friction which is lower than the value adopted for the mounting, the bolt can be broken or overstressed. Therefore minimal possible friction values for the calculation of the tightening torques according to VDI 2230 [1] should be used. If the actual coefficient of friction is too high, the required minimum assembly preload is not achieved. For this reasons, the knowledge of actual range of the friction coefficients taking into account the impact parameter is of vital importance for the durability of the bolt connection.

 DOI: 10.1051/matecconf/20179404007

This paper deals with the evaluation of an extensive data pool of TH Köln [2-3], as well as specific parameters investigated with the testing apparatus for determination of coefficients of friction between threads and head bearing surfaces [4]. The result shows a differentiated picture of the friction effect of different parameters for different bolt connections.

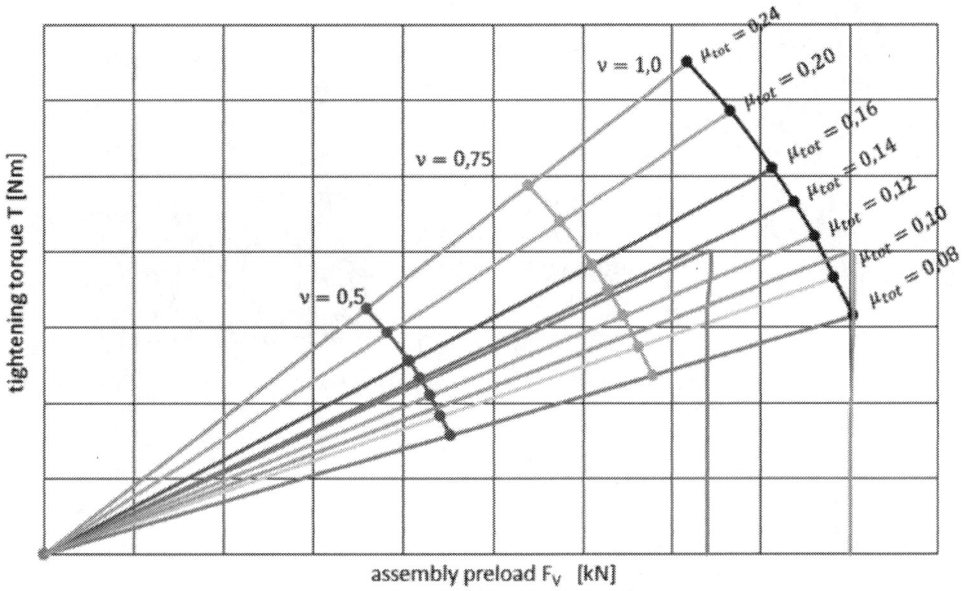

Fig. 3. Effect of different coefficients of friction on the assembly preload F_V at a determined tightening torque T.

2 Testing apparatus for determining coefficients of friction between threads and head bearing surfaces

In order to describe the general friction conditions without regarding shape and dimension of the connecting elements, it is expedient to determine the different coefficients of friction such as thread friction and head friction. The standard method for determining these coefficients of friction is described in the standard DIN EN ISO 16047 [5].

The measurements required are relatively complex because of the need of sensors for the tightening force and at least two different torques. If all data are measured in one setting, the device is very expensive.

The test rig designed at the TH Köln is far less costly [4]. The values are determined in three steps:

1. Experimental determination of the total coefficient of friction: μ_{tot}.

2. Experimental of thread friction μ_{th} (the head friction is isolated by use of an axially acting antifriction bearing).

3. Evaluation of head friction coefficients μ_b using the values of steps no. 1 and 2 and equation (2).

$$\mu_b = \frac{\frac{T}{F_V \cdot d} - 0{,}0222 - 0{,}528 \cdot \mu_{th}}{0{,}668} \quad (2)$$

The design of the device as shown in fig. 4 complies to the requirements of DIN EN ISO 16047. The mating test parts and the test standard conditions corresponds to this standard. The test bolt is inserted through a replaceable, hardened test washer and connected with the test nut. The test nut is secured by a milled groove against rotation. The installed axial cylindrical roller bearing beneath the removable test washer allows separate consideration between the thread and the total friction of the examined bolt assembly.

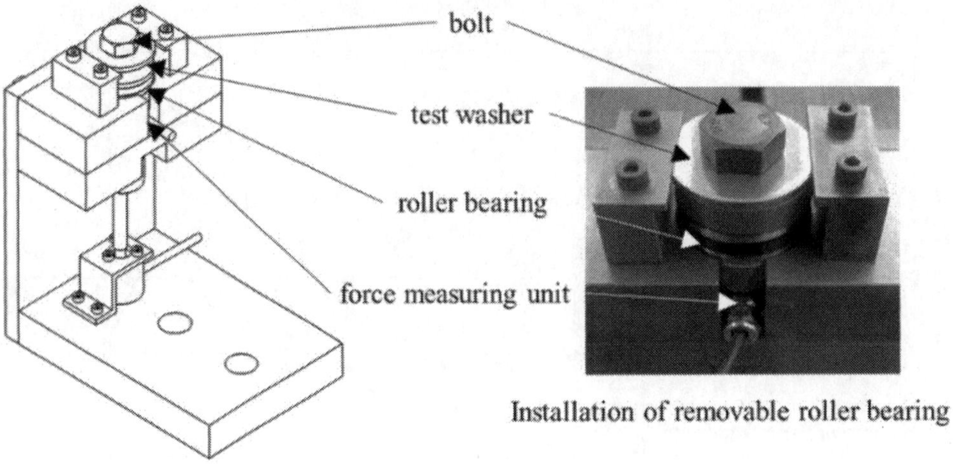

Fig. 4. Test apparatus.

3 Experimental results

The following friction-relevant parameters were isolated during preceding studies: lubricants, corrosion preventing coatings and a choice of securing elements. Table 1 shows the selected test conditions.

Table 1. Test conditions.

bolt	ISO 4017 M12 x 70 – 8.8
nut	ISO 4032 M12 – 8, bright
assembly load	100 % (F=42,3 kN)
lubricants	varies
coatings	varies
securing elements	varies

3.1 Parameter study: lubricants

The examined lubricants described are typically used for the assembly of bolted joints in the industry. The coating of all bolts was black oxide. Nuts were in bright condition.

ENGINE OIL HD 30

Conventional heavy dutyoil for passenger cars without additives, SAE viscosity grade 30.This oil is applied with a brush on the fasteners. In the test series of this lubrication state it is described by the term "lightly oiled".

VOLER AC

The non-soap grease paste is mixed with additives of aluminium particles, which are intended to compensate for the unevenness of the mounting surface. Thus the influence of friction is lower. The paste is able to prevent contact corrosion even at high surface pressures.

GLEITMO 805®

The high-performance grease paste is based on mineral oil with a synergistic effective combination of active-reaction white solid lubricants (means various inorganic compounds in the form of smooth, soft powders).

MOLYKOTE 1000®

This lubricant is on molybdenum disulfide base. It belongs to the group of fatty pastes. It consists of highly concentrated solid lubricants in oil, which can easily be applied with a brush.

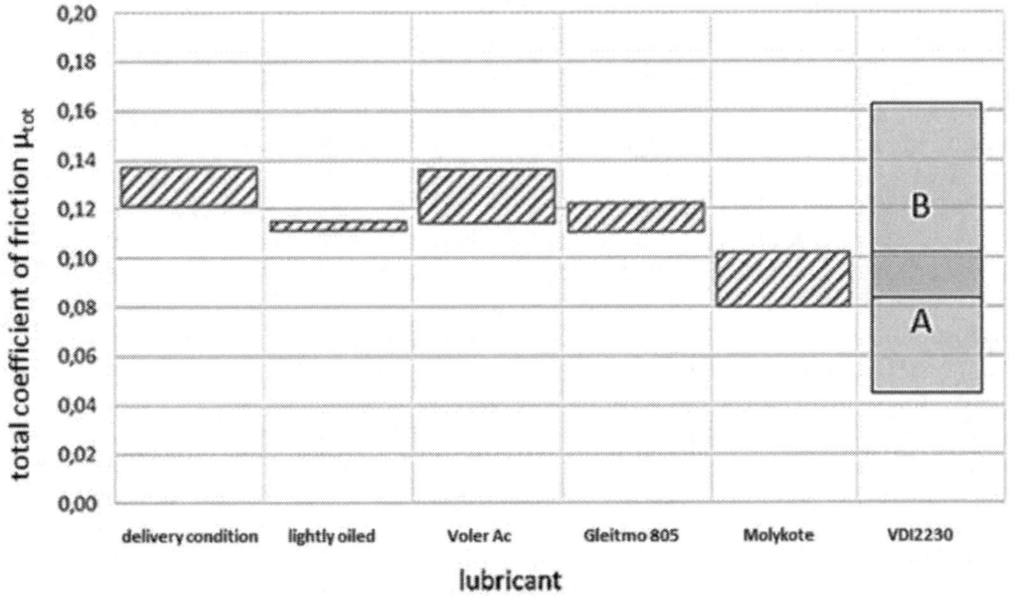

Fig. 5. Experimentally determined total friction coefficients μ_{tot} depending on the examined lubricants.

SUMMARY OF THE EXPERIMENTAL RESULTS

By applying a lubricant to the connecting elements lower friction factors are generally achieved. It can be stated that the friction coefficient classes A (0,04 to 0,10) and B (0,08 to 0,16) of VDI 2230-1: 2014-12 [1]as well as the VDA-standard 235-101 [6] (0,09 to 0,14) cover a very large range. It can be observed that some lubricants are found on the top, while others are found at the lower limit of this classes.

3.2 Parameter study: corrosion preventing coatings

Typically in the industry used corrosion preventing coatings were examined. Contribution was made to the fact that in automotive industry only CrVI-free coatings are allowed by EU-acts. During these examinations all interacting surfaces were lightly oiled with engine oil HD30.

BLACK OXIDE

Blackening or "black oxide" is one of the non-metallic surface coatings. The thin, oil-carbonaceous branded oxide leaves a dense surface film in oiled condition. The connecting members have sufficient corrosion protection mainly for transport and storage purposes.

EOMET 321 A +VL

The coating is a chromium-free thin dry-film non-electrolytic corrosion protection system. It consists of passivated zinc and aluminium flakes in a water-based binder. The topcoat +VL ensures low friction. Designation according to DIN EN ISO 10683: flZnnc - 480 h – L.

DELTA PROTEKT

It is a micro-layer corrosion layer system. The base coat is a layer having a high content of zinc and aluminum flakes embedded in a resin-free matrix. The resin-free structure of the layer ensures the electrical conductivity and thus the cathodic remote protective layer and is therefore essential for the corrosion protection. As topcoat organic layers, inorganic sealants or lubricants are used. Designation according to DIN EN ISO 10683: flZnnc - 480 h.

DARCOMET 500 B

The coating is non-electrolytic. It contains zinc and aluminium flakes in a chromium binder and integral lubrication. PTFE in the basecoat is used to ensure consistent torque-tension characteristics without a supplemental sealer. Designation according to DIN EN ISO 10683: flZnycL - 720 h.

ELECTROGALVANIZATION

Electrogalvanized bolts are often used alternatively to blackening. The steel is immersed in an aqueous bath and electricity is used to move electrons from the anode to the cathode which induces the zinc anodes to be "oxidized" and dissolve as zinc ions in the aqueous solution, to be transported as ions through the solution and to be "reduced" as metal onto the work (the cathode).

SUMMARY OF THE EXPERIMENTAL RESULTS

The use of different corrosion preventing coatings combined with light lubricated bolts has no influence on the total friction coefficients without regard to electrogalvanized bolts. It is known that electrogalvanized bolts have the tendency to fret because the pressure between the mating surfaces is high. For different coatings it can be stated that the friction coefficients are approximately in the center of friction class B (0,08 to 0,16) of VDI 2230-1: 2014-12 as well of the VDA-standard 235-101 (0,09 to 0,14)

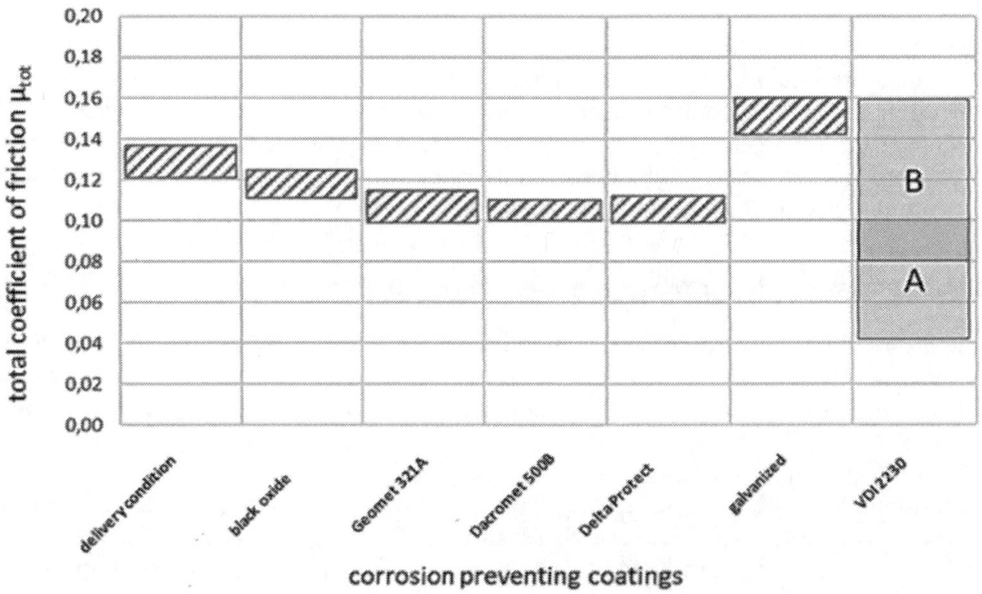

Fig. 6. Experimentally determined total friction coefficients μ_{ges} depending on the examined coatings.

 DOI: 10.1051/matecconf/20179404007

3.3 Parameter study: securing elements

A variety of common securing elements were examined. The locking effect of the investigated elements against loosening is based on the principles of frictional connection and form closure. The shape of the various security elements, for example profiled surface and wedge effect is specific to increase the frictional connection in the head or nut [7].

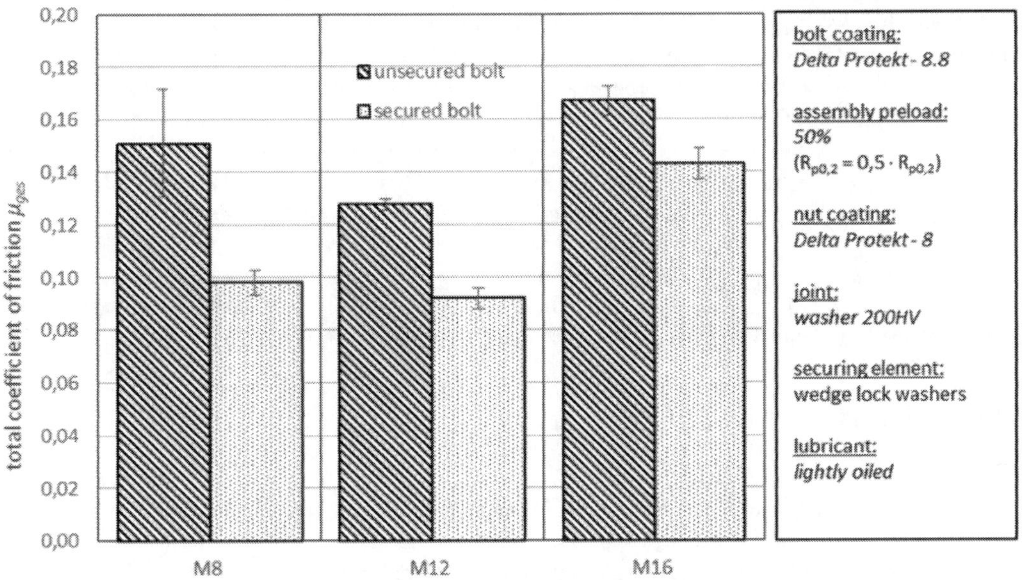

Fig. 7. Experimentally determined total friction coefficients μ_{ges} for unsecured bolts and secured bolts with wedge lock washers.

SUMMARY OF THE EXPERIMENTAL RESULTS

When elements to prevent loosening by rotation as wedge lock washers for example are used, the friction moment may increase. In this case the coefficient of total friction increases significantly, as shown in fig. 7. The results for the various sizes show that the coefficient of friction at small sizes increases more than at larger sizes. In fact attention has to be payed to the increasing total friction if any securing elements are used.

During the study experience has shown that the amount of lubricant applied by each user depends. This may also lead to deviations of the respective friction conditions. For this reason tests were performed on the reproducibility of the results.

3.4 Comments to the measurement results

The data collected with the friction apparatus compare quite well with those obtained on the vibration test rig which covers measurements over some years. For example the test range for a hexagon bolt ISO 4017 M12 x 45 -8.8 with the surface coating "Black oxide" in lubrication condition "Molykote" provides a deviation of the total friction coefficient μ_{tot} of about 5%. Using wedge lock washers with surface coating "Delta Protekt" and lubrication condition "lightly oiled" the deviation of the total friction coefficient μ_{tot} is about 7%. It should be noted that with increasing assembly load the friction decreases. This is about 10% from 25 % of full load.

4 Conclusions

For a safe design of bolted joints an orientation on the friction classes of VDA and VDI standards is not sufficient in each case. The investigations which relate to both experiments and on data pools collected over years at the TH Köln show the wide range of the parameters, in particular of types of lubricants, which cover the values from one window to the other. Even if the evaluated friction coefficients are more in class B than in class A, it must nevertheless be noted that the class B covers a range of 1:2. That results in a relationship of preload F_V of more than 1:0,9 and a range of assembly torque T of 1:1.6, if the preload is limited by the tensile strength. E. g. bolts M12-8.8: F= 45,2 kN to 40,7 kN, T = 63 Nm to 102 Nm. Conclusions are that bolts with a higher coefficient of friction shall be designed with a lower preload. Then a higher assembly torque has to been taken into account. Therefore measurements for determining correct friction coefficients are essential optimizing the design of bolted connections and the required assembly data.

The authors wish to thank Gesellschaft zur Förderung Angewandter Verbindungstechnik e. V. (GFAV), D-40724 Hilden, for their kind funding support of the investigations and HEICO Befestigungstechnik GmbH, D-54963 Ense, for the manufacture of the test apparatus.

References

1. VDI 2230-1:2014-12(2014)
2. R. Franke, E. Gärtner, M. Klöcker, ZEV-Rail, Glasers Annalen **132** (2008)
3. T. Hübing, M. Klöcker, R. Franke, VDI-report **2153** (2011)
4. S. Westphal, ZEV-Rail, Glasers Annalen, (to be published)
5. DIN EN 16047 :2013-01 (2013)
6. VDA 235-101 (1999)
7. DIN 25201-4: 2010-03E (2010)

MATEC Web of Conferences **94**, 04008 (2017) DOI: 10.1051/matecconf/20179404008
CoSME'16

Robust development of a CNC turning process

Laurentiu – Aurel Mihail[1,*]

[1]Transilvania University of Brasov, Manufacturing Engineering Department, 500466 Mihai Viteazul 5, Romania

Abstract. The article present an optimization study developed on a CNC HSM center for the external turning of spheroidal graphite iron precision mechanics parts. For isolating as much as possible the thermal error effects on the machining accuracy study, there were developed studies for statistically controlling the machining process and for mastering the measuring process, also. For assessing the process there were measured the dimensional critical quality characteristics. As analytical tools, were designed experiments applying the Taguchi Method, and the results are analyzed with ANOM method.

1 Introduction

The cutting processes are a base technology, despite of the new trends in manufacturing, as it is the additive manufacturing concept. Even if the cutting technologies are generally considered matured (especially within the industrial field), there it is enough room for optimizations taking account of the developments trends: cutting machines with increased dynamics, optimized or new cutting tools materials and geometries, new materials to be machined, etc.

In this context, the current study presents an optimization research for improving the turning process in a real industrial environment case.

The paper is organized in the following manner. An overview of the parameter design based on the Taguchi method is given first. Then, the parameter design with the characteristics is introduced. The experimental detail of using the parameter design to determine and analyze the optimal cutting parameters in turning operations is described next. Finally, the paper concludes with a summary of this study.

2 Robust engineering by Taguchi Method

2.1 The concept of Robust Engineering / Taguchi Method

The Robust Engineering approach, as an innovative concept applied through the Taguchi Method, can be used for the optimization of the industrial processes starting with the earliest product and process development lifecycle phases.

*Corresponding author: L.Mihail@unitbv.ro

Generally, within the organizational management, product development is considered a major tactic for profitability. It consists of the following two aspects [1]:
- (1) Product quality: what consumers desire (functions or appearance);
- (2) Engineering quality: what consumers do not want (functional variability, running cost, pollution).

The manufacturing process must take in considerations the aspects regarding the product development, and, for achieving these goals, it can be designed in an optimal manner using the Robust Engineering.

Quality engineering is robust design based on the following three procedures:
- (1) Orthogonal array,
- (2) SN ratio, and
- (3) Loss function.

These three procedures are not robust design per se because they are used to evaluate technical means, products or processes. To design robustness, it is required to understand the meaning of robust design before considering technical or management means [1]. Those who consider practical means are engineers and production people.

Robust design (including product and process design) generally means designing a product that can function properly under various conditions of use or a process that is non-sensitive at the noise (see the Fig. 1 and Fig. 2) [2]. The noise factors being external controllable or not controllable factors.

The objective of the parameter design is to optimize the settings of the process parameter values for improving performance characteristics and to identify the product parameter values under the optimal process parameter values.

In addition, it is expected that the optimal process parameter values obtained from the parameter design are insensitive to the variation of environmental conditions and other noise factors. Therefore, the parameter design is the key step in the Taguchi method to achieving high quality without increasing cost.

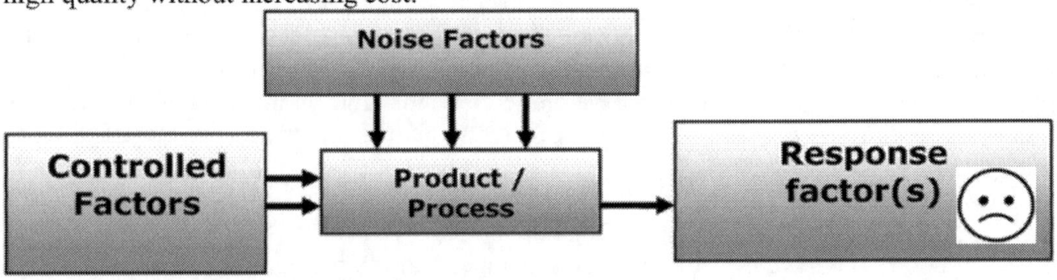

Fig. 1. The generic Parameter Diagram for the non-robust processes.

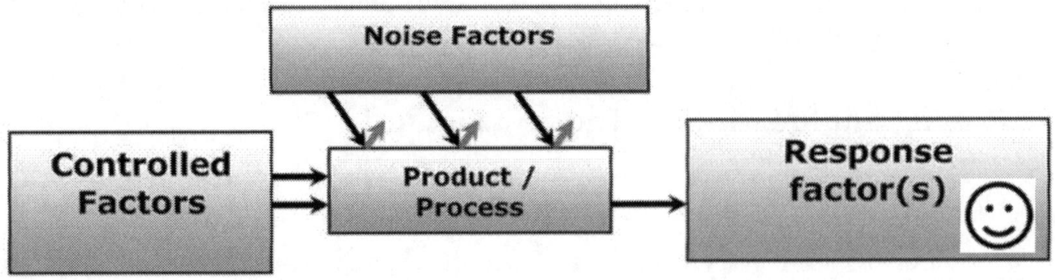

Fig. 2. The generic Parameter Diagram for the robust processes.

 DOI: 10.1051/matecconf/20179404008

2.2 State-of-the-art of using Robust Engineering on CNC turning process optimization

Dedicated state – of – the art literature present extensive approaches regarding the methods used for optimization of the cutting process. Therefore, the [3] reference present researches regarding the next techniques for the optimization of industrial processes:

- (1) Lagrange's method;
- (2) Geometric programming;
- (3) Goal programming;
- (4) Dynamic programming;
- (5) Fuzzy logic;
- (6) Genetic algorithm;
- (7) Scatter search;
- (8) Taguchi method;
- (9) Response surface methodology.

The Robust Engineering is becoming more and more used for the optimization of the manufacturing by cutting processes, too. One of the most used cutting process within the worldwide manufacturing industry is the turning process, a relatively simple operation but that continues to need applied research especially because of the cutting high performance dynamics, workpiece materials and cutting tools. Therefore, within the Table 1 it is presented a brief state of the art containing some studies on the same argument, treated within this paper.

Table 1. State of the art of several researches regarding the use of Taguchi Method on optimization of turning operation.

Reference	**Optimization method**	**Optimization criteria**	**Optimization parameters**	**Part's material**	**Factorial lattice**	**Data analysis**
[4]	Hybrid: Taguchi Method& Differential evolution algorithm	Unit cost	Cutting parameters:cutting speed, cutting feed, cutting depth	Metal	L_{16} 5^4	SNR(STB)& ANOVA
[5]	Hybrid: Taguchi Method &Grey relational analysis	Surface quality: R_a andR_t roughness and circularity	Cutting parameters: cutting speed, cutting feed, cutting depth & cutting fluid	high carbon high chromium alloy tool steel	L_9 3^4	SNR(STB)& ANOVA
[6]	Taguchi Method	Ra roughness, cutting force, cutting tool wear	Cutting parameters: cutting speed, cutting feed	nitrogen alloyed duplex stainless steel DSS ASTM A 995 grade 5A and grade 4A	L_9 2^3	SN (STB)& ANOVA

[7]	Taguchi Method	Surface roughness Ra	Cutting parameters: cutting speed, cutting feed, cutting depth & cutting tool material	High-alloy white cast iron (Ni-Hard) at two different hardness levels (50 HRC and 62 HRC)	L_{18} 4^3 (2 levels for cutting tool material)	SNR(STB)& ANOVA
[8]	Taguchi Method& RSM	Surface roughness R_a and R_z	Cutting parameters: cutting speed, cutting feed, cutting depth	AISI 304 austenitic stainless	L_{27} 3^3	SNR (STB)& ANOVA and RSM with regression alanalysis
[9]	Taguchi Method	Surface roughness R_a and R_z and flank wear (VB)	Cutting parameters: cutting speed, feed rate and depth of cut	Hardened AISI 4140 (51 HRC)	L_9 3^3	SNR & ANOVA
[10]	Taguchi dynamic approach coupled with a proposed ideal function model	Dimensional precision and accuracy and the surface roughness R_a	Coolant, cutting speed, feed, depth of cut, coating type, chip-breaker geometry, nose radius and shape of the insert	A range of steels	L_{18} $2^1 \times 3^7$	SNR & ANOVA
[11]	Taguchi Method	Surface roughness (Ra)	Insert radius, feed rate, and depth of cut	AISI 1030 steel bars	L_9 3^3	SNR & ANOVA
[12]	Taguchi Method	Surface roughness (Ra), tool wear, power consumption and chip reduction co-efficient	Cutting speed, feed rate and depth of cut	Hard metal	L_9 3^3	SNR & ANOVA, multiple regression analysis

2.3 The design of the experiment

2.3.1 The process to be optimized

The process to be optimized is a turning process of a cast iron part (a forged shaft used in a mechanical mechanism), on a high performance CNC turning machine. Because the part is a precision part, it is very important to master the dimensional accuracy at different levels. For that scope, it was studied the influence of the two main cutting parameters on the external diameter accuracy, at different temperature measurement levels. It is very well known that the cutting process it is developed with heat generation and propagation as a gradient within the part's material, also. That temperature gradient changes the dimensional accuracy within the time. For underlining and researching this phenomenon, the external diameters were measured at 2 temperatures. The first diameter measurement was developed at the moment of machining, when the part is discharged automatically from the chuck (at 33,5°C), and the second diameter measurement was measured at after several hours, after the coldness of the part, within the metrological laboratory environment (at 21,4 °C). The part's superficial temperature was measured with an infrared pyrometer.

The material machined is a continuously cast spheroidal graphite iron EN (-1563)-GJS-500-7 (GGG50). This kind of spheroidal graphite offers higher wear resistance, strength & heat-treatment response compared to other materials while still possessing good machinability and excellent surface finishes. Noise and vibration damping are good in this grade. For further detailed data regarding the material, within the Table 2 and the Table 3 there are presented another properties.

Table 2. Chemistry(Typical Ranges) (Subordinate to Mechanical Properties) [13].

Element	**Carbon**	**Silicon**	**Manganese**	**Sulphur**	**Phosphorous**	**Magnesium**	**Others / Alloying**	**Iron**
Typical values	3.25 – 3.70	2.40 – 3.00	0.10 – 0.30	0.005 – 0.020	0.015 – 0.08	0.04 – 0.07	Residual	Balance

Table 3. Mechanical Properties (Anticipated Values) [13].

EN-GJS 500-7:1997 (GGG40)	**Tensile strength UTS N/mm2**	**0.2% Proof Stress N/mm2**	**Elongation %**
25mm – 50mm	500	320	7
55mm – 100mm	450	300	7
105mm – 200mm	420	290	5
205mm – 600mm +	420	290	5
Another characteristics:			
- Brinell Hardness 170 … 230			
- Microstructure - Contains Type I & II nodular (spheroidal) graphite in accordance with ASTM A247. The rim contains approximately 200/250 nodules/ mm2, and is predominately ferritic with the core containing 90/150 nodules/mm2. The core matrix is mixed ferritic/pearlitic (10–50% pearlite). Chill carbides will be less than 5%, well dispersed.			
- Heat Treat Response - EN-GJS-500-7 can be hardened by conventional methods but EN-GJS 600-3 is recommended.			
- Density - 7.3 g/cc			

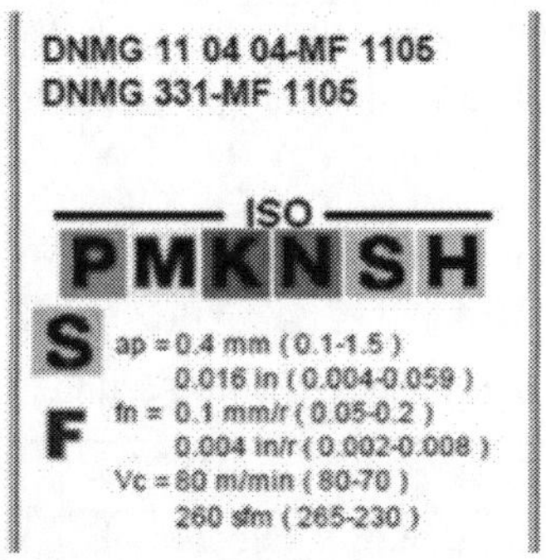

Fig. 3. The cutting insert used (DNMG 110404 MF 1105) for the turning process and the cutting data recommended ranges [14].

The turning process is developed with a DNMG 11 04 04-MF 1105 cutting insert, withcorner radius (R_E) of 0.3969 mm and made of a carbide that is grade 1105, coated by PVD with Aluminum oxide (Fig. 3). According with the same Fig. 3, the feed is recommended to be in the range 0.05 … 0.2 mm / rot and the cutting speed is recommended to be adopted in the rage 80 … 70 m /min.

The cutting tool wear was not took in consideration because the entire experiment was developed within the prescript range and the level of machining constituted by the 25 test parts does not influence on the wear.

Fig. 4. The machine tool used for the experimentation [15].

Nakamura Super NTX is a CNC machine tool with twin spindle and twin turret. The used configuration can hold 24 tools on stock, the main spindle (for milling) turret can arrive at up to 8000 rot / min and the indexing range is on 230 degree (±115 degrees). The left and right spindles (for turning) can arrive up to 4500 rot/min. The turrets can arrive at 3600 rot / min for each of the 12 individual rotation tools stations from the total of 24 tools [15].
The forged workpiece initial diameter was 12 ± 0.4 mm and the final turned nominal diameter is 10.9 ± 0.1 mm. Therefore, the radial cutting depth is 0.55 mm.

2.3.2 The process parameters and DoE

Very accurate prescriptions for the cutting data it was not available for that particular application. Therefore, starting from the cutting tool producer's prescriptions (Fig. 3) there were considered 5 levels for the cutting speed parameter and 5 levels for the cutting feed parameter. The machine tool is a Nakamura high performance turning and milling HSM CNC (Fig.4).

As experimental matrix, it was selected an $L_{25}2^5$ fractioned factorial lattice. That means that the 2 experimental factors are variated at five levels each one, and the entire fractionated factorial design has 25 trials (Table4).

Table 4. The DoE - $L_{25}2^5$ fractioned factorial lattice and the correspondent parameter setup.

Run order	Level of cutting speed	v_c	RPM	Level of cutting feed	f
		[m/min]	[rot/min]		[mm/rot]
1	1	70,0	1857,7	5	0,2000

2	2	72,5	1924,1	4	0,1625
3	5	80,0	2123,1	2	0,0875
4	4	77,5	2056,8	4	0,1625
5	1	70,0	1857,7	4	0,1625
6	5	80,0	2123,1	3	0,1250
7	5	80,0	2123,1	4	0,1625
8	2	72,5	1924,1	2	0,0875
9	1	70,0	1857,7	1	0,0500
10	4	77,5	2056,8	5	0,2000
11	2	72,5	1924,1	5	0,2000
12	5	80,0	2123,1	1	0,0500
13	3	75,0	1990,4	1	0,0500
14	4	77,5	2056,8	2	0,0875
15	5	80,0	2123,1	5	0,2000
16	3	75,0	1990,4	4	0,1625
17	1	70,0	1857,7	3	0,1250
18	3	75,0	1990,4	2	0,0875
19	1	70,0	1857,7	2	0,0875
20	2	72,5	1924,1	3	0,1250
21	3	75,0	1990,4	3	0,1250
22	3	75,0	1990,4	5	0,2000
23	2	72,5	1924,1	1	0,0500
24	4	77,5	2056,8	3	0,1250
25	4	77,5	2056,8	1	0,0500

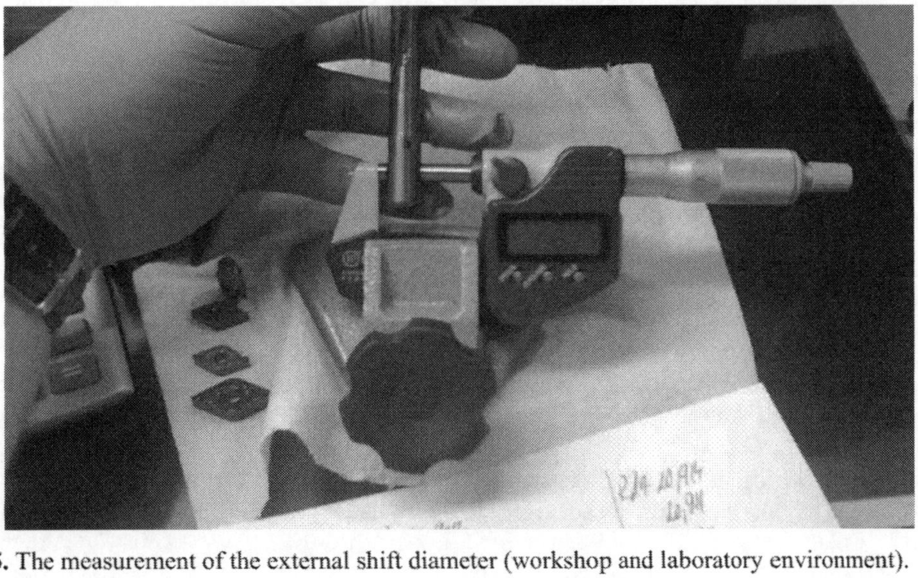

Fig. 5. The measurement of the external shift diameter (workshop and laboratory environment).

2.3.3 The metrological validation of the gage used for measuring

The measurement system used is a QuantuMike Mitutoyo micrometer (Fig. 5), IP 65 Coolant Proof wired directly at a PC station with MeasurLink version 7.4.10 full configuration [16].

An important moment for the measuring process is the measurement instrument and process metrological validation, done by applying the GRR methodology (Fig. 6). The overall GRR index has the value of 2.7% and the number of the distinct categories is 8, fact that validate the measuring process (the repeatability, reproducibility and the resolution are appropriate for that particular case).

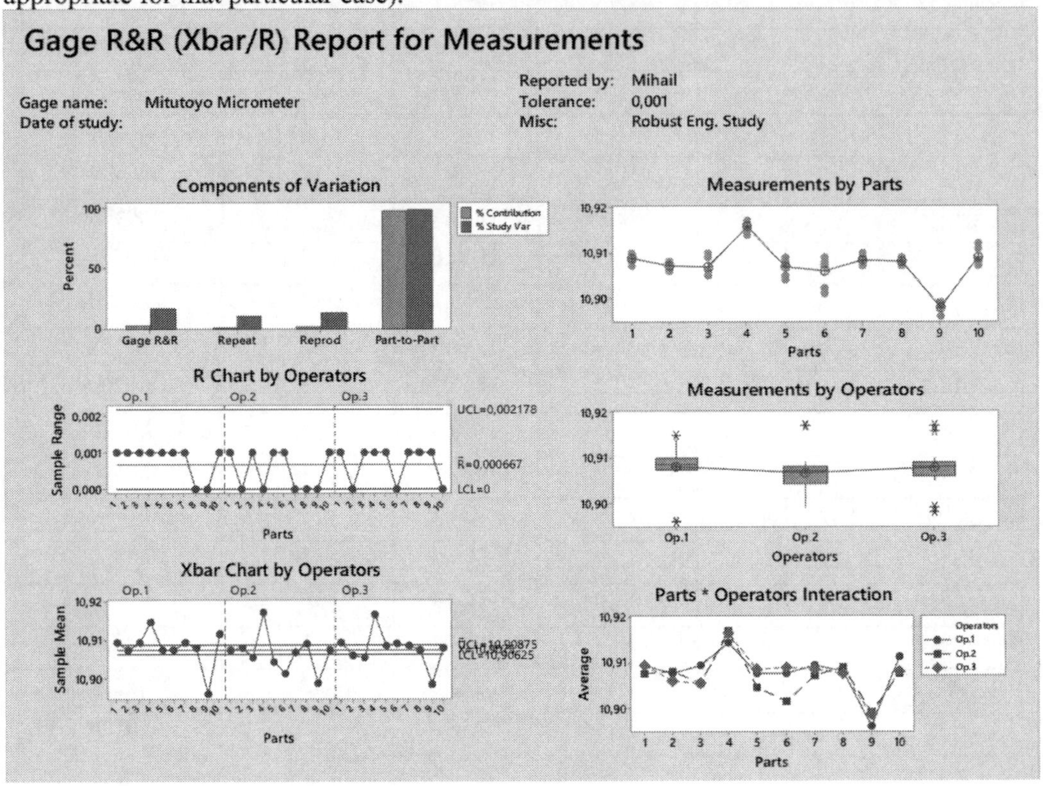

Fig. 6. Gage Repeatability and Reproducibility study results.

3 Results, data analysis and conclusions

3.1 Experimental results, the data analysis and conclusions

The 25 machined parts were measured at 2 different material temperatures and the final results are presented graphically analyzed by ANOM (see Fig. 7, Fig. 8 and Fig. 9).

The Fig.7 presents the analysis of means (ANOM) of the influence on the test piece external diameter measured immediately after being machined (at a material temperature of 33,5°C). Therefore, there are presented the influence on the specified characteristic of the

"A" factor (cutting speed) and the "B" factor (cutting feed). For each factor, by ANOM, there are computed the means of the values for each one of the 5 experimental levels of the researched parameters. With dotted line it is represented the mean of means.

The Fig. 8 represents an ANOM graphic obtained by the same approach as the one from Fig. 7, but it is for the measurements obtained by measuring the test parts in laboratory

 DOI: 10.1051/matecconf/20179404008

environment, after that the parts material arrived at 21,4 ˚C (after 15 hours of thermal stabilization in an ambient temperature of 20,0 ˚C).

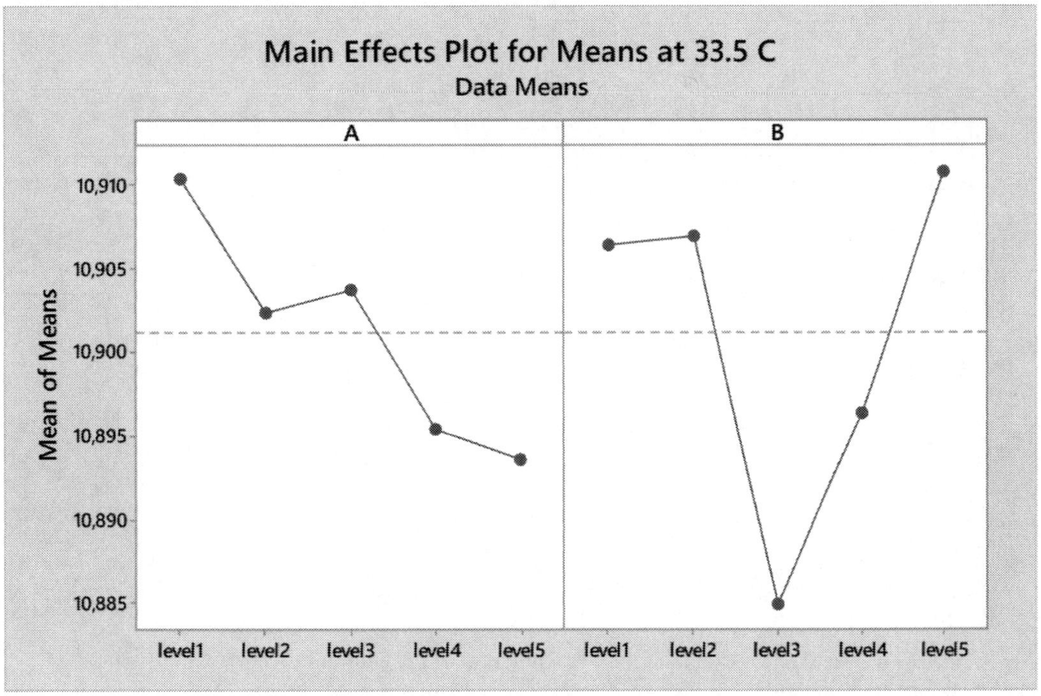

Fig. 7. ANOM for the measurements of the external shift diameter (workshop environment).

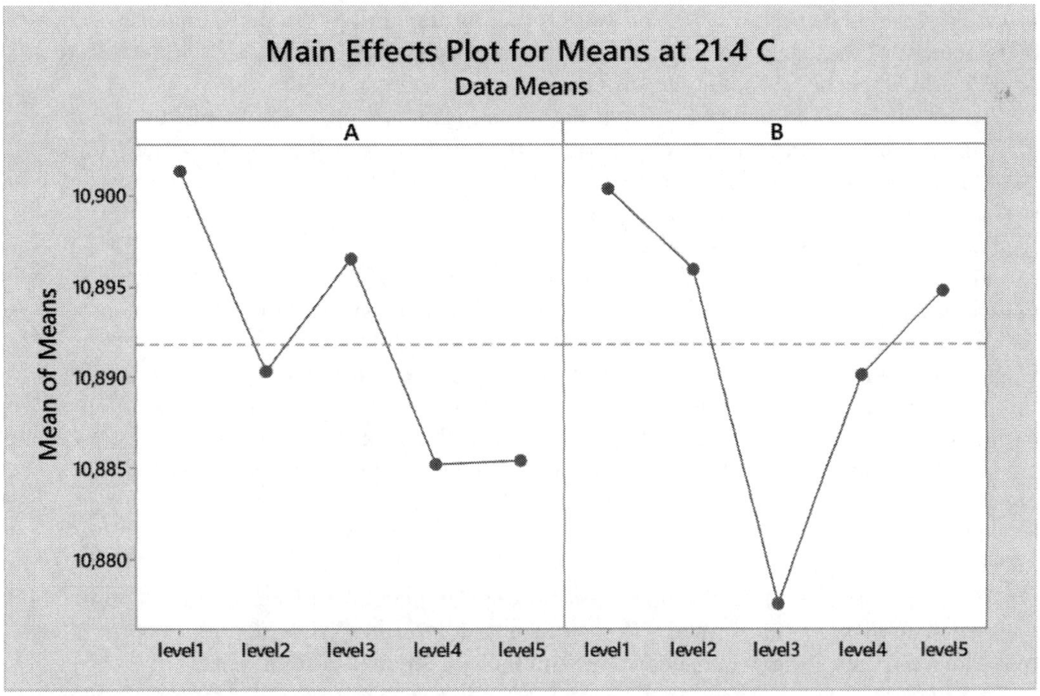

Fig. 8. ANOM for the measurements of the external shift diameter (laboratory environment).

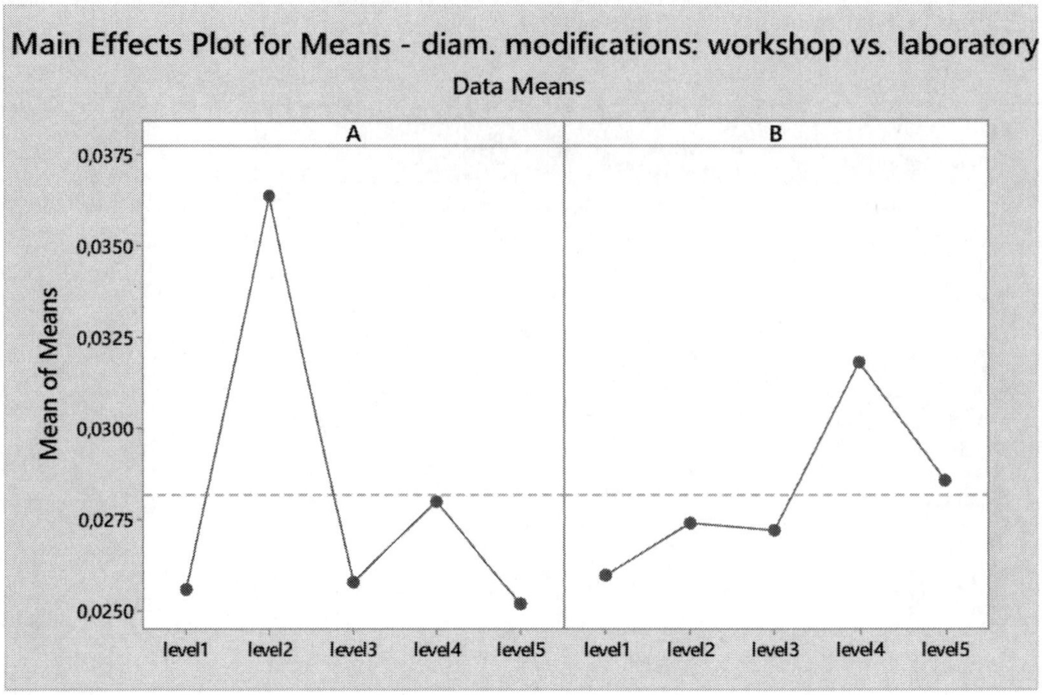

Fig. 9. ANOM for the range of the dilatation / contraction of the external shift diameter (workshop and laboratory environment).

According with Fig. 7 and Fig. 8, the general trend for the external diameter is to decrease once the cutting speed ("A" parameter) is increasing. The phenomenon is considered to be validated by theory knowing that the heat gradient is growing in intensity once the cutting speed increases. The dilatation of the cutting tool system and of the part's material generate that decrease of part's external diameter. The same trend is underlined by the measurements developed within the laboratory environment.

According with the same Fig. 7 and Fig. 8, the influence of the raise of the cutting feed ("B" parameter) present an inflexion point at the third level of the experimental parameter. Also, it can be observed by comparing the two graphics that the influence of the two researched parameters on the diameter's variation range is a little bit higher for the cutting feed case than of the cutting speed case.

Also, it can be observed that the level of influence on the external dimeter of the two experimental parameters range is practically the same (approximatively 6 μm), by comparing the diameters measured before and after contraction.

The final remarks can be extracted from the Fig. 9 that present the dilatation / contraction range. Therefore, the best cutting process setup would be at the first levels of the two parameters: v_c = 70 m / min and s = 0,05 mm / rot. Because of economic reasons, it can be considered to be better to higher the level of two parameters, reason for which it can be choose their third levels v_c = 75 m / min and f = 0,125 mm / rot. To continue the compromise by gaining more productivity and losing some accuracy, it can be choose the fifth levels v_c = 80 m / min and f = 0,2 mm / rot.

The results were took in consideration for the next batch of pieces and the optimization was, in that manner, confirmed. The surface quality (dimensional and roughness) was measured and the results were according with the customer's requirements.

3.2 Future development

As future developments, it is interesting to take in consideration the roughness quality characteristic for a simultaneous optimization of the turning process. Also, it may be opportune the change of the tool's insert grade and taking this parameter in consideration as a distinctive experimental parameter within the factorial lattice, for future optimization developments.

The measurements realized in the present research were made using the infrastructure of the"Mitutoyo Metrology Laboratory" from the " Institute High Tech Products for Sustainable Development: PRO-DD" of "Transilvania" University of Brasov, laboratory created with the contribution of Mitutoyo Romania S.R.L.

Table 5. Abbreviations.

Abbreviation	Meaning
PVD	Physical Vapor Deposition
DoE	Design of Experiments
HSM	High Speed Machining
CNC	Computer Numerically Controlled
$L_a b^c$	Lattice with a runs for b factors variated at c levels
SNR	Signal to Noise Ratio
STB	Smaller The Best (Better)
ANOVA	ANalysis Of VAriation
ANOM	ANalysis Of Means
IP	Instrument Protection
GRR	Gage Repeatability and Reproducibility
Xbar / R	Average and Range

References

1. G.Taguchi, S. Chowdhury, Y. Wu, *Taguchi's Quality Engineering Handbook* (John Willey & Sons and ASI, 2004)
2. L.–A. Mihail, Proceedings of MEQAPS '13, 60 (2013)
3. A. Aggarwal, H. Singh, Sadhana Vol. 30, Part 6, 699 (2005)
4. A. R. Yildiz, Applied Soft Computing 13, 1433 (2013)
5. C.-J.Tzenga, Y.-H.Linb, Y.-K. Yanga, M.-C.Jeng, Journal of Materials Processing Technology 209 (2009) 2753–2759
6. D. P.Selvaraj, P. Chandramohan, M. Mohanraj, Measurement **49**, 205 (2014)
7. M.Günay, E.Yücel, Measurement **46**, 913 (2013)
8. I.Asiltürk, S.Neseli , Measurement **45**, 785 (2012)
9. I.Asiltürk, H.Akkus, Measurement **44**, 1697 (2011)
10. Y.-F. Tzeng, Materials and Design **27**, 665 (2006)
11. M. Nalbant, H. Gokkaya, G. Sur, Materials and Design **28**, 1379 (2007)
12. A. Das, A. Khan, S. Mohanty, Lean Thinking 6 (2015)
13. www.gefebe.com, accessed at 30 Aug. 2016
14. http://www.sandvik.coromant.com/en-us/products/Pages/productdetails.aspx?c=dnmg %2B11%2B04%2B04-mf%2B1105&m=5732258&Country=ro accessed at 30 Aug. 2016
15. http://www.nakamura-tome.co.jp/products/en/atc/super-ntx.html, accessed at 30 Aug. 2016
16. www.measurlink.com,accessed at 30 Aug. 2016

Reliability analysis of scarifying tooth for trailed scarifier

Cristin-Olimpiu Morariu[1] , *Adela-Eliza* Dumitrascu[1,*], and *Doina Valentina* Ciobanu[2]

[1]Transilvania University of Brasov, Department of Manufacturing Engineering, Mihai Viteazul No.5, Brasov, Romania
[2]Transilvania University of Brasov, Department of Forest Engineering, Forest Management and Terrestrial Measurements, Sirul Beethoven Street, No. 1, Brasov, Romania

Abstract. In this paper are estimated the reliability indicators of scarifying tooth for trailed scarifier. Taking into account the constructive and technological factors, as well as the duration of exploitation, the determined values of Anderson-Darling statistics indicates that the Weibull distribution permits to model with maximum accuracy the lifetime of such attachment. The location parameter is estimated using the correlation coefficient method and minimum lifetimes are determined experimentally. Analyzing the obtained parameters, it can be shown that it can be used the normal distribution for reliability estimation. In this respects, it was applied the point and confidence interval estimation of the average and standard deviation lifetime of scarifying teeth of trailed scarifier.

1 Constructive characteristic of the scarifying teeth

A scarifier attachment is designed to break up compacted material, asphalt and rocky subgrade to facilitate grading [1]. The scarifier is a rear mounted tool used for tearing up asphalt, hard-pack or other obstacles to aid the grading operation [2].

The scarifiers are terracing equipment used at deep loosening of hard and very hard soils, of rocks relatively weak (which cannot be loosened directly with another equipment, type bucket or blade), for loosening of soils with medium hardness, with boulders and roots, or for removing of the old pavements [3].

From constructive point of view, scarifiers can be [3]:

- Trailed by a crawler tractor or other machine for construction;
- Mounted at the front of a tractor or as attachment on a construction machine (bulldozer, motor grader etc.); in the case of motor grader, the scarifiers can be mounted in front of or between the axles.

The active component of the scarifiers (tooth) can be of different types (figure 1), depending on the actual working conditions [3]: (a) straight teeth; (b), (c) curved teeth; (d), (e), (f) teeth of special construction (d. - with heel; e - without heel; f - with removable tip). The teeth can be mounted rigid or articulated on a beam (frame).

[*] Corresponding author: dumitrascu_a@unitbv.ro

 DOI: 10.1051/matecconf/20179404009

In figure 1 is presented the common scarifying teeth used as auxiliary attachment for bulldozer for execution and rehabilitation of forest roads. The main characteristics of the attachment according to the tensile force of the basic machine are shown in table 1.

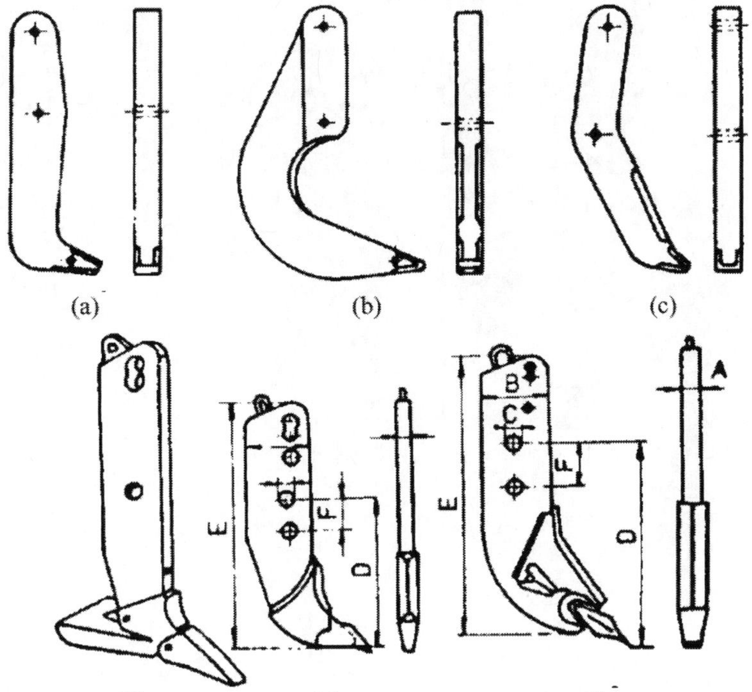

Fig. 1. Constructive shapes of scarifying teeth.

Table 1. Scarifier characteristics.

Parameters	Traction force [kN]					
	30	45	60	100	150	250
Numbers of teeth	3÷5	3÷5	3÷5	1÷5	1÷3	1÷3
Horizontal rotation angle [deg]	-	-	-	10÷15	10÷15	10÷15
The length of the tooth [cm]	45÷80	50÷80	60÷100	80÷140	80÷150	100÷190
The maximum depth of cut [cm]	20÷50	40÷60	35÷45	40÷100	50÷120	60÷150
Cutting angle [deg]	40÷60	40÷60	35÷45	40÷80	40÷80	40÷80
The weight of equipment [t]	0.07÷0.19	0.12÷0.20	0.16÷0.21	0.18÷0.21	0.17÷0.24	0.20÷0.27

Compared to usual activities, in the case of rehabilitation and maintenance of forest roads, the scarifying teeth gauge dimensions are lower. In this manner, in Figure 2 it were examined a 3D model of trailed scarifier that has constructive characteristics able to support the traction force of 100 kN (table 1).

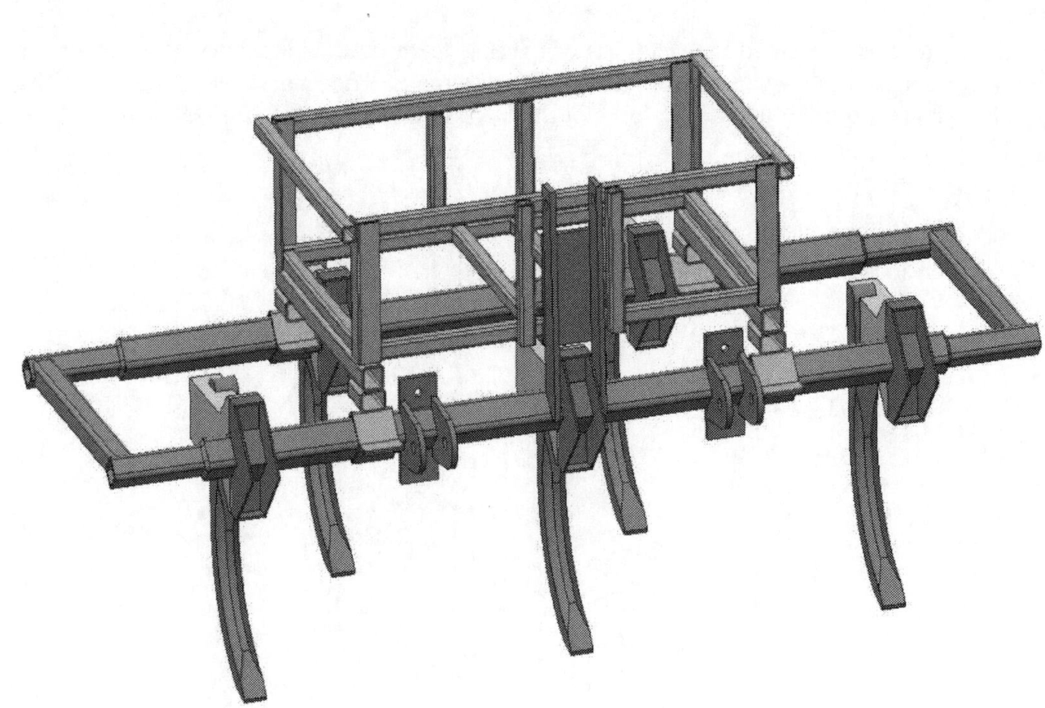

Fig. 2. 3D model of the auxiliary attachment.

2 Nomenclature

$\mathcal{W}(x,\beta,\eta,\gamma)$ = three-parameter Weibull distribution;
$\mathcal{W}(x,\beta,\eta)$ = two-parameter Weibull distribution;
β = shape parameter of Weibull distribution;
η = scale parameter of Weibull distribution;
γ = location parameter of Weibull distribution;
$\mathcal{N}(x,\mu,\sigma)$ = normal distribution (Gauss-Laplace);
μ = mean;
σ = standard deviation;
L = lower limit of confidence interval;
U = upper limit of confidence interval;
$\rho(X,Y)$ = correlation coefficient of random variables X and Y;
CL = confidence limits;
$1-\alpha$ = confidence level;
ML = maximum likelihood;
$\hat{x}$ = estimated value;
$v(r,n)$ = random variable $\frac{\hat{\beta}}{\beta}$;
$k(r,n)$ = random variable $\hat{\beta}\cdot ln\frac{\hat{\eta}}{\eta}$;
m = sample average;
s = standard deviation of the sample;
$t_{p,v}$ = the quantile p of Student's t-distribution, with ν degrees of freedom;
$\chi^2_{p,v}$ = the quantile p of Chi-squared distribution, with v degrees of freedom;

$F_{p;m;q}$ = the quantile p of Fisher-Snedecor distribution, with m and p degrees of freedom.

3 Reliability analyses of the scarifying teeth

In order to analyze the reliability of scarifying tooth, it was applied the Anderson-Darling test, considering the four statistical distributions: Weibull distribution, log-normal distribution, exponential distribution and normal distribution.

The calculation consists in applying statistical relationship [4]:

$$A_n^2 = n \cdot \int_{-\infty}^{\infty} \frac{[Q(y) - F(y)]^2}{F(y) \cdot [1 - F(y)]} \cdot dF(y) \quad (1)$$

The obtained results are a measure of the discrepancy between the empirical distribution, sampling values and the considered distribution.

Anderson-Darling test statistic results as:

$$A_n^2 = -\sum_{i=1}^{n} \frac{2 \cdot i}{n} \cdot \{lnF(y_{(i)}) + ln[1 - F(y_{(n+1-i)})]\} - n \quad (2)$$

considering the calculus relation of empirical distribution:

$$Q[y_{(i)}] = \begin{cases} 0, \text{if } Y < y_{(1)} \\ \frac{i}{n}, \text{if } y_{(i)} < Y < y_{(i+1)}, \quad i = \overline{1, n-1} \\ 1, \text{if } Y > y_{(i)} \end{cases} \quad (3)$$

The lifetime estimation of analyzed auxiliary attachment it was achieved by applying the finite element method (FEA) and 3D modeling was simulated using Ansys Workbench program. Based on the results, the determined values of Anderson-Darling statistics (table 2) indicate that the Weibull distribution permits the modeling with maximum accuracy of lifetime of such attachment.

Table 2. Anderson-Darling statistics.

Distribution	Anderson-Darling statistic values, A_n^2
Weibull	2.095
Log-normal	2.113
Exponential	3.391
Normal	2.106

The statistical model is validated applying the Anderson-Darling test that permits to compare the date set for considered distributions (figure 3).

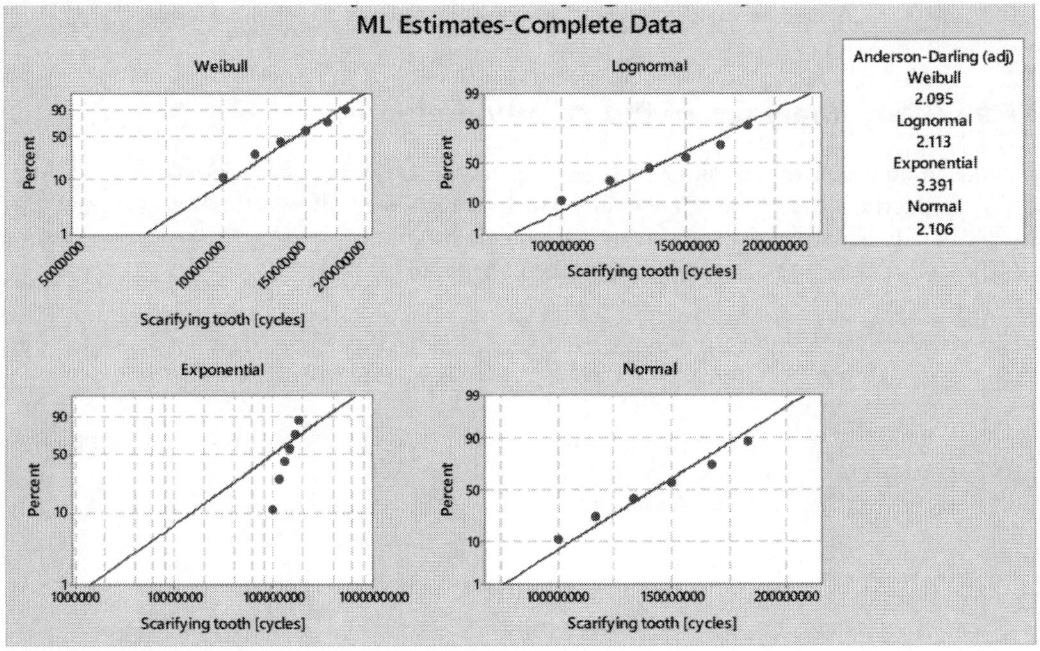

Fig. 3. Probability plot of Anderson-Darling test for scarifying tooth.

Also, it is estimated the reliability of trailed scarifier teeth using three-parameters Weibull distribution.

The location parameter (or failure free life) γ is estimated by the correlation coefficient method and minimum lifetimes are determined experimentally (table 3). The value of location parameter indicates the minimum lifetime value of each subassembly estimated under real operating conditions.

Table 3. Minimum lifetime estimation.

Parameter	Estimated value
Correlation coefficient (ρ)	0.99818
Location parameter ($\hat{\gamma}$)	$5.813 \cdot 10^7$

The estimation method consists of determining the location parameter values, γ, under maximum conditions of correlation coefficient values, [5-6]:

$$\rho(Y,C) = \frac{\sum_{i=1}^{n} ln(y_i - \gamma) \cdot c_i - \frac{\sum_{i=1}^{n} ln(y_i - \gamma) \cdot \sum_{i=1}^{n} c_i \cdot}{n}}{\left\{\left[\sum_{i=1}^{n} ln^2(y_i - \gamma) - \frac{[\sum_{i=1}^{n} ln(y_i - \gamma)]^2}{n}\right] \cdot \left[\sum_{i=1}^{n} c_i^2 - \frac{(\sum_{i=1}^{n} c_i)^2}{n}\right]\right\}^{1/2}} \quad (4)$$

where the values of deterioration probabilities are determined using the approximate calculation relationship, [4]:

$$c_i = \frac{1}{1 + \frac{n-i+1}{i} \cdot F_{0.50;2\cdot(n-i+1);2\cdot i}}, i = \overline{1,n} \quad (5)$$

The results were obtained by using an application in Mathcad 14 software. Knowing the values of the location parameter, by means of a transformation of random variable of the form:

$$X = Y - \gamma \tag{6}$$

the obtained values are two-parameter Weibull distributed (table 4).

Table 4. Lifetime estimation of scarifying tooth.

Issue	Lifetime [cycle], $x_i, i = \overline{1, n}$
1	$4.1872 \cdot 10^7$
2	$5.8872 \cdot 10^7$
3	$7.4872 \cdot 10^7$
4	$9.1872 \cdot 10^7$
5	$1.0887 \cdot 10^8$
6	$1.2487 \cdot 10^8$

The point estimate method used is maximum likelihood. In this case, the estimated values of the shape and scale parameters of the two-parameter Weibull distribution are obtained as a solution of the equations system [6-7]:

$$\begin{cases} \frac{1}{\hat{\beta}} + \frac{1}{n} \cdot \sum_{i=1}^{n} lnx_i - \frac{\sum_{i=1}^{n} x_i^{\hat{\beta}} \cdot lnx_i}{\sum_{i=1}^{n} x_i^{\hat{\beta}}} = 0 \\ \hat{\eta}^{\hat{\beta}} = \frac{1}{n} \cdot \sum_{i=1}^{n} x_i^{\hat{\beta}} \end{cases} \tag{7}$$

Because the sample size is very small, $n = 6$, the asymptotically properties of maximum likelihood estimators cannot be applied in this case, the estimations with confidence intervals were calculated on the basis of two random variables, independent of the sample size (n) and the type of test ($n = r$, for the complete test), [8-11]:

$$\begin{cases} v(r,n) = \frac{\hat{\beta}}{\beta} \\ k(r,n) = \hat{\beta} \cdot ln\left(\frac{\hat{\eta}}{\eta}\right) \end{cases} \tag{8}$$

Thus, for the shape parameter:
- corresponding to a confidence level *1-α*, it is obtained as the solution of the equation of the probability:

$$Pr\left[v_{\alpha/2}(n,n) \le \frac{\hat{\beta}}{\beta} \le v_{1-\alpha/2}(n,n)\right] = 1 - \alpha \tag{9}$$

and it has the form:

$$\frac{\hat{\beta}}{v_{1-\alpha/2}(n,n)} \le \beta \le \frac{\hat{\beta}}{v_{\alpha/2}(n,n)} \tag{10}$$

- unbiased mean estimate:

$$\hat{\hat{\beta}} = \frac{\hat{\beta}}{\bar{v}(n,n)} \tag{11}$$

- unbiased median estimate:

$$\hat{\beta}_{Me} = \frac{\hat{\beta}}{v_{0.50}(n,n)} \tag{12}$$

For scale parameter:
- corresponding to a confidence level *1-α*, it is obtained as the solution of the equation of the probability:

$$Pr\left[k_{\alpha/2}(n,n) \le \hat{\beta} \cdot ln\frac{\hat{\eta}}{\eta} \le k_{1-\alpha/2}(n,n)\right] = 1-\alpha \tag{13}$$

and it has the form:

$$\hat{\eta} \cdot \left[e^{\frac{-k_{1-\alpha/2}(n,n)}{\hat{\beta}}}\right] \le \eta \le \hat{\eta} \cdot \left[e^{\frac{-k_{\alpha/2}(n,n)}{\hat{\beta}}}\right] \tag{14}$$

- unbiased mean estimate:

$$\hat{\hat{\eta}} = \hat{\eta} \cdot e^{\frac{-\bar{k}(n,n)}{\hat{\beta}}} \tag{15}$$

- unbiased median estimate:

$$\hat{\eta}_{Me} = \hat{\eta} \cdot e^{\frac{-k_{0.50}(n,n)}{\hat{\beta}}} \tag{16}$$

The point estimate and with confidence intervals for Weibull distribution parameters it was determined based on experimental results (table 5).

The confidence level used is *1-α* = *0.95* and the corresponding values of the random variables $v(r,n)$ and $k(r,n)$ were taken from [6], (table 6).

Table 5. Parametric estimation with 95% confidence interval – Weibull distribution.

Shape parameter	$\hat{\beta}$	$\hat{\hat{\beta}}$	$\hat{\beta}_{Me}$	$\hat{\beta}_L$	$\hat{\beta}_U$
	3.3322	2.4905	2.8026	1.1322	5.2526
Scale parameter	$\hat{\eta}$	$\hat{\hat{\eta}}$	$\hat{\eta}_{Me}$	$\hat{\eta}_L$	$\hat{\eta}_U$
	$9.344 \cdot 10^7$	$9.457 \cdot 10^7$	$9.456 \cdot 10^7$	$6.502 \cdot 10^7$	$1.381 \cdot 10^8$

Table 6. Corresponding values of the random variables.

$\boldsymbol{v(r,n)}$	$\boldsymbol{k(r,n)}$
$\bar{v}(6,6) = 1.338$	$\bar{k}(6,6) = -0.0402$
$v_{0.50}(6,6) = 2.802$	$k_{0.50}(6,6) = -0.0399$
$v_{0.975}(6,6) = 5.252$	$k_{0.975}(6,6) = 1.208$
$v_{0.025}(6,6) = 1.132$	$k_{0.025}(6,6) = -1.301$

High values of shape parameters, β = 3.3322, indicates that it can be used the normal distribution for reliability modeling. The same conclusion can be reached from Anderson-Darling statistics analysis, as shown in Figure 3.

Point estimate of the normal distribution parameters is performed using the equations:

$$\begin{cases} \hat{\mu} = \dfrac{\sum_{i=1}^{n} x_i}{n} \\ \hat{\sigma} = \sqrt{\dfrac{\sum_{i=1}^{n}(x_i - \hat{\mu})^2}{n-1}} \end{cases} \tag{17}$$

where μ represents the mean of operating time and σ is standard deviation of operating time.

The calculation of confidence intervals for the parameters provides also valuable information on the two reliability indicators.

The calculation formulas used are:

- corresponding to a confidence level *1-α*, for the mean is:

$$\hat{\mu} + t_{\frac{\alpha}{2},n-1} \cdot \hat{\sigma}/\sqrt{n} \leq \mu \leq \hat{\mu} + t_{1-\frac{\alpha}{2},n-1} \cdot \hat{\sigma}/\sqrt{n} \tag{18}$$

- corresponding to a confidence level *1-α*, for the standard deviation is:

$$\sqrt{\frac{(n-1)\cdot\hat{\sigma}^2}{\chi^2_{1-\frac{\alpha}{2},n-1}}} \leq \sigma \leq \sqrt{\frac{(n-1)\cdot\hat{\sigma}^2}{\chi^2_{\frac{\alpha}{2},n-1}}} \tag{19}$$

The point estimate and with confidence intervals for normal distribution parameters it was determined based on experimental results and there are shown in table 7.

Table 7. Parametric estimation with 95% confidence interval - normal distribution.

Mean	$\hat{\mu}$	$\hat{\mu}_L$	$\hat{\mu}_U$
	$1.4167 \cdot 10^8$	$1.0902 \cdot 10^8$	$1.7432 \cdot 10^8$
Standard deviation	$\hat{\sigma}$	$\hat{\sigma}_L$	$\hat{\sigma}_U$
	$3.1111 \cdot 10^7$	$1.9419 \cdot 10^7$	$7.6302 \cdot 10^7$

4 Results and discussions

The reliability of scarifying teeth can be modeled by using the Weibull distribution model. In table 8 are presented the analytical expressions of reliability functions and the values of the main reliability indicators that are estimated based on experimental results collected from the operation of these attachments (figure 4).

Table 8. The main reliability indicators.

	$F(t)$	μ	σ
Scarifying tooth for trailed scarifier	$1 - e^{-\left(\frac{x-5.813\cdot10^7}{9.344\cdot10^7}\right)^{3.3322}}$	$1.416 \cdot 10^8$	$3.111 \cdot 10^7$

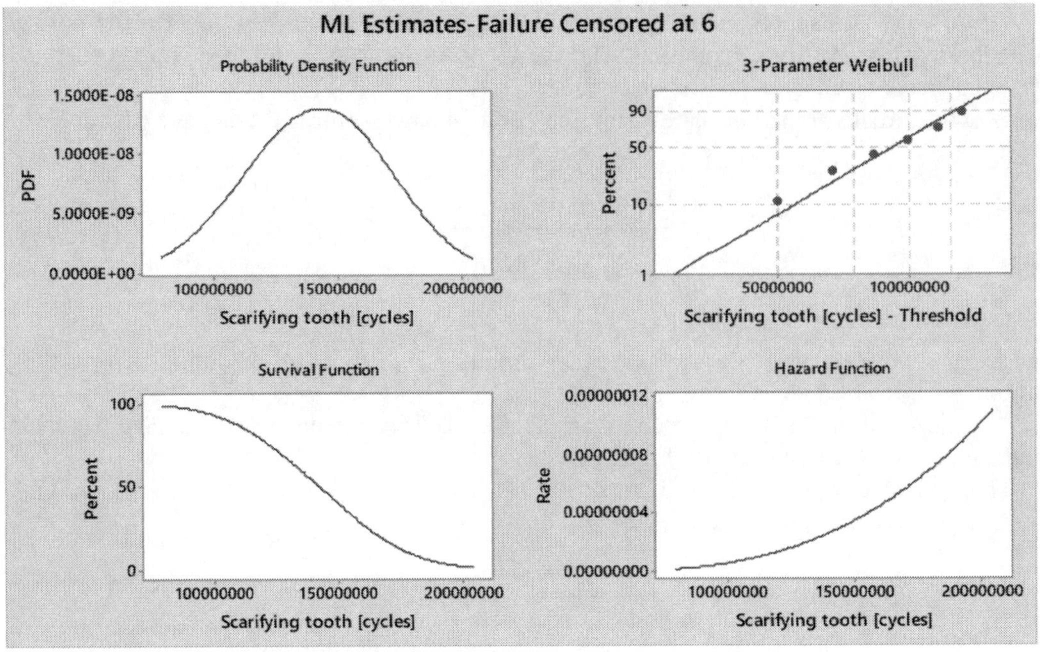

Fig. 4. Reliability distribution overview plot for scarifying tooth.

5 Conclusions

The study conducted was focused on the reliability analysis of the scarifying teeth of scarifier trailed in real rehabilitation and maintenance conditions of forest roads. This analysis allows the following conclusions:

- Modeling of reliability can be achieved with a good accuracy using three-parameters Weibull distribution;
- The value of the minimum lifetime estimated for actual operating conditions is $5.813 \cdot 10^7$ (table 3);
- High values of shape parameter, allow us to use normal distribution for reliability estimation. This has led to the possibility of punctual estimation and confidence interval of the average lifetime of scarifying teeth of trailed scarifier and to estimate the standard deviation of their lifetime.
- Knowledge of parameter values of reliability statistical model and the values of the main indicators of reliability, enabling a much better approach to preventive maintenance activities and realistic planning them.

References

1. Volvo motor grader. Front mounted scarifier attachment. Volvo Construction Equipment, North America, Inc., One Volvo Drive, Asheville, NC 28803-3447, https://www.volvoce.com/SiteCollectionDocuments/VCE/History/10__motor%20graders/07%20Volvo/All%20common/V%20G%20FrMountScarif%202234341034-0505.pdf, accessed 05.03.2016
2. Volvo motor grader. Ripper / scarifier attachment. Volvo Construction Equipment Group, Ref. No. 21 2 434 1036, Canada 1/2001 - 10,0, https://www.volvoce.com/SiteCollectionDocuments/VCE/History/10__motor%20grad

ers/07%20Volvo/All%20common/V%20G%20RipperS%202124341036-0101.pdf, accessed 25.02.2016
3. D.V. Ciobanu, *Utilaje si instalatii folosite in cariere si balastiere* (Transilvania University Publishing House, Brasov, 2006)
4. D.B. Kececioglu, *Reliability & Life Testing Handbook, vol. I* (PTR Prentice-Hall, Inc., Englewood Cliffs, New Jersey, 1993)
5. C.O. Morariu, A.-E. Simon, Studies Concerning the Estimation of the Location Parameter for Weibull Distribution, *International Conference "Challenges in Higher Education and Research in 21st Century"*, June 2-5, Sozopol, Bulgaria, 323-326 (2004)
6. C.O. Morariu, *Optimizarea incercarilor de fiabilitate a rulmentilor* (Transilvania University of Brasov, doctoral thesis, 1999)
7. D.B. Kececioglu, *Reliability Engineering Handbook, vol. I, vol. II* (PTR Prentice-Hall, Inc., Englewood Cliffs, New Jersey, 1991)
8. N.R. Mann et al., M*ethods for Statistical Analysis of Reliability & Life Test Data* (John Wiley & Sons, Inc., 1974)
9. J. John, I. McCool, *Using the Weibull Distribution. Reliability, Modeling, and Inference* (John Wiley & Sons, Inc., Hoboken, New Jersey, 2012)
10. I.N. Gibra, *Probability and Statistical Inference for Scientist and Engineers* (Prentice-Hall, Inc., Englewood Cliffs, New Jersey, 1976)
11. D.B. Kececioglu, *Reliability & Life Testing Handbook, vol. II* (PTR Prentice-Hall, Inc., Englewood Cliffs, New Jersey, 1994)

Statistical inferences for bearings life using sudden death test

Cristin-Olimpiu Morariu[1,*], and *Sebastian-Marian* ZAHARIA[1]

[1]Transilvania University of Brasov, Department of Manufacturing Engineering, 29 Eroilor Street, Braşov, Romania

Abstract. In this paper we propose a calculus method for reliability indicators estimation and a complete statistical inferences for three parameters Weibull distribution of bearings life. Using experimental values regarding the durability of bearings tested on stands by the sudden death tests involves a series of particularities of the estimation using maximum likelihood method and statistical inference accomplishment. The paper detailing these features and also provides an example calculation.

1 Introduction

The durability / reliability tests of the bearings are made with specialized tests machines in order to meet the following conditions:

– ensuring identical operating conditions for all bearings tried;
– to reproduce the operating conditions of the bearings in accordance with the assumptions for calculating the basic radial load (moderate speeds, charging with a pure radial/axial load for ensuring a proper lubrication by using recommended lubricants, that are uncontaminated or chemically and mechanically degraded, to provide cooling, the inner ring rotating and outer ring fixed) and to permit the easy mounting and dismounting of bearings;
– to be provided with equipment for measuring temperature and vibration, in order to detect the moment when the damage begins;

Also, when manufacturing the bearings, their testing is achieved by trying to control the indicators of reliability. This test is called type test and it is done at well-defined intervals of time. This type of test verifies if the nominal bearing durability is maintained at the levels required by regulations. Their role is to conduct the audit of production quality through the global assessment of the quality of materials used and technological methods applied.

The main indicator of reliability of bearings, besides the Weibull distribution parameters, is the basic rating life, L_{10}, for a bearing or for a group of identical bearings which operate under the same conditions and durability associated with a reliability of 90%, at a conventional material quality manufacturing and operating conditions:

* Corresponding author: c.morariu@unitbv.ro

DOI: 10.1051/matecconf/20179404010

$$L_{10} = \gamma + \eta \cdot \left[ln \left(\frac{1}{1 - 0.10} \right) \right]^{1/\beta} \tag{1}$$

In other words, nominal durability is the 10th quantile of the functioning period, $t_{0,10}$, depending on the test procedure used in the manufacturing process of bearings, there can be used:

- Censored tests. An attempt is considered censored at the level r, if it ends at the moment t_r, when the rth failure occurs (the number of failures r is established prior to the beginning of attempts).
- Truncated tests. In this case of reliability test, the attempts end after covering a period t_c, which was previously established.
- Sudden death attempts, [1,2,3]. The method consists in random grouping of the n products that are tested, in ℓ subsets and taking every subset as an assemble of m ($n = \ell \cdot m$) series elements. (Figure 1.).

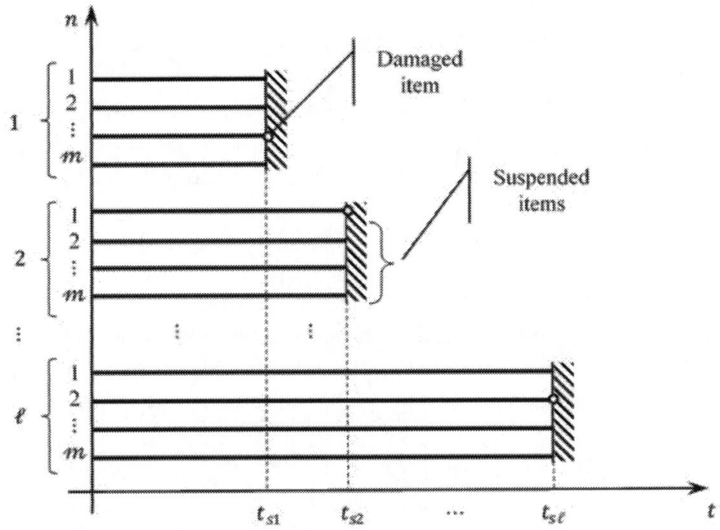

Fig. 1. Deployment scheme of sudden death tests

After making such a grouping, the tests continues until the first appearance and registration of an damaged element from each assemble, the other e $m - 1$ elements will be suspended.

The registered deterioration times $t_{s1}, t_{s2}, \ldots, t_{s\ell}$ represent the first statistic order in the case of an sample of volume m, taken from an Weibull population. If the n elements reveal a Weibull population, then the deterioration times that correspond to the sudden death tests have the distribution function [3]:

$$F_s(t) = 1 - [1 - F(t)]^m = 1 - e^{-\left(\frac{t}{m^{-\frac{1}{\beta}} \cdot \eta} \right)^{\beta}} \tag{2}$$

If the random variable T has the distribution $\mathcal{W}(t, \beta, \eta)$, then the first statistics of order T_s corresponding to an sample of volume m has as distribution $\mathcal{W}\left(t_s, \beta, m^{-\frac{1}{\beta}} \cdot \eta\right)$.

Between the population parameters and the parameters of the first order statistic distribution exist the following relationships [3]:

$$\begin{cases} \beta_s = \beta \\ \eta_s = m^{-\frac{1}{\beta}} \cdot \eta \\ t_{sp} = m^{-\frac{1}{\beta}} \cdot t_p \end{cases} \quad (3)$$

In [3] it is shown that in the case of the maximum likelihood estimators, applied for deterioration times obtained by primary fault line method, there can be established the following random variables, independent of the population parameters:

$$\begin{cases} \dfrac{\hat{\beta}_s}{\beta} = v(\ell,\ell) \\ \hat{\beta}_s \cdot ln\left(\dfrac{\hat{\eta}}{\eta}\right) = s(\ell,m) \\ \hat{\beta}_s \cdot ln\left(\dfrac{\hat{t}_{0.10}}{t_{0.10}}\right) = q(\ell,m,0.10) \end{cases} \quad (4)$$

This procedure is an alternative method used to test the reliability of the bearings [1,2,3,4] which ensures an even distribution of the lower estimated values of reliability indicators. This can be demonstrated by the ratio between the primary line faults dispersion and the dispersion of population [1]:

$$\frac{V(T_s)}{V(T)} = m^{-\frac{1}{\beta}} \quad (5)$$

According to the working regime in the case of the bearings reliability tests are used the accelerated attempts. For a shorter attempts period, the constraints are increased, taking into account the operating conditions [1,2,5]. The acceleration factor is determined by the formula:

$$L_{10h} = \frac{1000000}{60 \cdot n} \cdot \left(\frac{C}{P}\right)^q \quad (6)$$

where:

n = speed test, [rpm].

According to the volume of products, the specific attempts for bearing manufacturing are attempts done on low volumes of samples.

2 Special features of estimating the statistical model

Estimating the Weibull distribution parameters of the statistical model and making inferences for the main statistical indicators of reliability the bearings presents a number of features for tests sudden death.

For an accurate estimation is mainly used the triparametric Weibull distribution.

The location parameter, γ, is estimated using the correlation coefficient with the condition of using the maximum values possible [5]:

$$\rho(Y,C) = \frac{\sum_{i=1}^{\ell} ln(t_{si}-\gamma)\cdot c_i - \dfrac{\sum_{i=1}^{\ell} ln(t_{si}-\gamma)\cdot\sum_{i=1}^{n} c_i\cdot}{n}}{\left\{\left[\sum_{i=1}^{\ell} ln^2(t_{si}-\gamma) - \dfrac{\left[\sum_{i=1}^{\ell} ln(t_{si}-\gamma)\right]^2}{n}\right]\cdot\left[\sum_{i=1}^{\ell} c_i^2 - \dfrac{\left(\sum_{i=1}^{\ell} c_i\right)^2}{n}\right]\right\}^{1/2}} \quad (7)$$

where the deterioration probability values are calculated using the approximate computing relationship, [6,7]:

$$c_i = \frac{1}{1+\frac{n-i+1}{i}\cdot F_{0.50;2\cdot(n-i+1);2\cdot i}}, i=\overline{1,n} \tag{8}$$

Knowing the values of the location parameter, through a transformation like the following one:

$$X = T_s - \gamma \tag{9}$$

the obtained values are biparametric Weibull distributed.

The punctual estimation method used for finding out the parameters for the Weibull biparametric distribution is the maximum likelihood method. In this case, the estimated values of the shape and scale parameters of the Weibull biparametric distribution are obtained as a solution of the equation, [6,8]:

$$\begin{cases} \frac{1}{\hat{\beta}_s}+\frac{1}{n}\cdot\sum_{i=1}^{\ell} lnx_i - \frac{\sum_{i=1}^{\ell} x_i^{\hat{\beta}_s}\cdot lnx_i}{\sum_{i=1}^{n} x_i^{\hat{\beta}_s}} = 0 \\ \hat{\eta}_s{}^{\hat{\beta}_s} = \frac{1}{n}\cdot\sum_{i=1}^{\ell} x_i^{\hat{\beta}_s} \end{cases} \tag{10}$$

Based on these estimated values ($\hat{\beta}_s$ *şi* $\hat{\eta}_s$) and using the equations (3), there are determined the population parameters of the tested bearings, ($\hat{\beta}$ *şi* $\hat{\eta}$). The estimated nominal sustainability, $\hat{L}_{10}$, s computed using the equation (1).

For making statistical inferences for estimated parameters are used the distributions of statistics (4). The determination of these distributions is performed by the numerical simulation Monte Carlo. The algorithm used consists of the following steps:

A. Entry data:

- n – sample size;
- ℓ - the number of groups into which the sample is divided;
- $m = \frac{n}{\ell}$ - effective sample size;
- N_{sim} – number of simulations.

B. Generating n random uniform numbers ($n_{ai}, i = \overline{1,n}$) in the interval [0,1].

C. Calculating the damaging times (t_i) by using the inverse distribution function:

$$t_i = ln\left(\frac{1}{1-n_{ai}}\right). \tag{11}$$

D. Dividing the n random values calculated at the previous step in ℓ groups, each one of them having m damaging times.

E. Computing the ℓ deterioration times corresponding to the primary fault line from the condition regarding the minimum m durations, corresponding to each group.

F. Estimating the Weibull distribution parameters by taking into account the ℓ values previously determined.
Obtaining the maximum likelihood estimates ($\hat{\beta}_1$ *şi* $\hat{\eta}_1$) by iterative numerical solving the system of equations (9) and using the relationships (3);

G. Calculus of 0.10 quantiles, by using the equation (1).

H. Computing the specific random variables:

$$\hat{\beta}_{1,s} = v(\ell,\ell);\ \hat{\beta}_{1,s}\cdot ln\hat{\eta}_{1,s} = s(\ell,m);\ \hat{\beta}_{1,s}\cdot ln\hat{t}_{1,0.10} = q(\ell,m,0.10) \tag{12}$$

I. Repeat the steps A,…,**H** for N_{sim} times.
J. Determining the random variables quantiles by taking into consideration the N_{sim} made for each case.
K. Output data:
- ASCII file (*.prn) which contains the quantiles values.

On the basis of the original algorithm was carried out a Mathcad program.

If the random variables (4), independent of population parameters, there are known statistical inferences for the Weibull distribution parameters and nominal durability is achieved using the equations:

1. For the shape parameter:

– bilateral symmetric interval, corresponding to a confidence level of $1-\alpha$, is obtained as the solution of the probability equation:

$$Pr\left[v_{\alpha/2}(\ell,\ell) \le \frac{\hat{\beta}_s}{\beta} \le v_{1-\alpha/2}(\ell,\ell)\right] = 1-\alpha \tag{13}$$

and it has the form:

$$\frac{\hat{\beta}_s}{v_{1-\alpha/2}(\ell,\ell)} \le \beta \le \frac{\hat{\beta}_s}{v_{\alpha/2}(\ell,\ell)} \tag{14}$$

– unbiased average estimate:

$$\hat{\bar{\beta}} = \frac{\hat{\beta}_s}{\bar{v}(\ell,\ell)} \tag{15}$$

– median unbiased estimation:

$$\hat{\beta}_{Me} = \frac{\hat{\beta}_s}{v_{0.50}(\ell,\ell)} \tag{16}$$

2. For the scale parameter:

– bilateral symmetric interval, corresponding to a confidence level of $1-\alpha$, is obtained as the solution of the probability equation:

$$Pr\left[s_{\alpha/2}(\ell,m) \le \hat{\beta}_s \cdot ln\left(\frac{\hat{\eta}}{\eta}\right) \le s_{1-\alpha/2}(\ell,m)\right] = 1-\alpha \tag{17}$$

and it has the form:

$$\hat{\eta} \cdot \left[e^{\frac{-s_{1-\alpha/2}(\ell,m)}{\hat{\beta}_s}}\right] \le \eta \le \hat{\eta} \cdot \left[e^{\frac{-s_{\alpha/2}(\ell,m)}{\hat{\beta}_s}}\right] \tag{18}$$

– unbiased average estimate:

$$\hat{\bar{\eta}} = \hat{\eta} \cdot e^{\frac{-\bar{s}(\ell,m)}{\hat{\beta}_s}} \tag{19}$$

– median unbiased estimation:

$$\hat{\eta}_{Me} = \hat{\eta} \cdot e^{\frac{-s_{0.50}(\ell,m)}{\hat{\beta}_s}} \tag{20}$$

3. For basic rating life, $L_{10} = t_{0.10}$

– bilateral symmetric interval, corresponding to a confidence level of $1-\alpha$, is obtained as the solution of the probability equation:

$$Pr\left[q_{\alpha/2}(\ell, m, 0.10) \leq \hat{\beta}_s \cdot ln\left(\frac{\hat{t}_{0.10}}{t_{0.10}}\right) \leq q_{1-\alpha/2}(\ell, m, 0.10)\right] = 1 - \alpha \tag{21}$$

and it has the form:

$$\hat{t}_{0.10} \cdot \left[e^{\frac{-q_{1-\alpha/2}(\ell,m,0.10)}{\hat{\beta}_s}}\right] \leq t_{0.10} \leq \hat{t}_{0.10} \cdot \left[e^{\frac{-q_{\alpha/2}(\ell,m,0.10)}{\hat{\beta}_s}}\right] \tag{22}$$

– unbiased average estimate:

$$\hat{\hat{t}}_{0.10} = \hat{t}_{0.10} \cdot e^{\frac{-\bar{q}(\ell,m,0.10)}{\hat{\beta}_s}} \tag{23}$$

– median unbiased estimation:

$$\hat{t}_{0.10,Me} = \hat{t}_{0.10} \cdot e^{\frac{-q_{0.50}(\ell,m,0.10)}{\hat{\beta}_s}} \tag{24}$$

3 Results and discussions

After testing on the stands a lot of tapered roller bearing ($n = 20$), type LM11949/LM11910, random grouped having $\ell = 5$ and $m = 4$, requested with axial load $Pa = 450\ [daN]$ at $4000\ [rpm]$, based on the experimental results, the estimated value of the positioning parameter is:

$$\hat{\gamma} = 63.759\ [h]$$

This value was determined in accordance with a coefficient of variation equal to:

$$\rho(\hat{\gamma}) = 0.995$$

which indicates a very good correlation with the Weibull biparametric model.

To achieve statistical inference, further on the random quantiles for the random variables needed to be computed (4).

The following values were obtained:

$\bar{v}(5,5) = 1.448,$	$\bar{s}(5,4) = -0.686,$	$\bar{q}(5,4,0.10) = 0.328,$
$v_{0.50}(5,5) = 1.011,$	$s_{0.50}(5,4) = -0.347,$	$q_{0.50}(5,4,0.10) = 0.131,$
$v_{0.975}(5,5) = 3.522,$	$s_{0.975}(5,4) = 0.683,$	$q_{0.975}(5,4,0.10) = 3.038,$
$v_{0.025}(5,5) = 0.617,$	$s_{0.025}(5,4) = -4.248,$	$q_{0.025}(5,4,0.10) = -1.149,$

under the conditions of $N_{sim} = 10000$ simulations.

Point estimates and confidence intervals for Weibull distribution parameters estimated based on experimental results are presented in Table 1. The confidence level used is:

$$1 - \alpha = 0.95.$$

4 Conclusions

The paper presents a way to achieving statistical inference for the sudden death tests, more convenient in terms of the operations of assembly and disassembly of bearings on the stand bearings and the accuracy of the experimental results obtained.

Because testing bearings on the stands is done using samples of small volume, making inferences statistics (building confidence intervals and testing statistical hypothesis) is not recommended to use the asymptotic properties of estimates, obtained by maximum likelihood method, because the confidence intervals will result having very large amplitudes.

Table 1. Parametric estimation with 95% confidence interval – Weibull distribution

Weibull distribution parameters	Parametric estimation with 95% confidence interval – Weibull distribution					
Shape parameter, β	$\hat{\beta}_s$	$\hat{\beta}$	$\hat{\hat{\beta}}$	$\hat{\beta}_{Me}$	$\hat{\beta}_L$	$\hat{\beta}_U$
	1.817	1.817	1.255	1.463	0.516	2.942
Scale parameter, η	$\hat{\eta}_s$	$\hat{\eta}$	$\hat{\hat{\eta}}$	$\hat{\eta}_{Me}$	$\hat{\eta}_L$	$\hat{\eta}_U$
	157.909	338.568	494.026	409.967	232.502	3504.769
Basic rating life $t_{0.10}$	$\hat{t}_{0.10,s}$	$\hat{t}_{0.10}$	$\hat{\hat{t}}_{0.10}$	$\hat{t}_{0.10,Me}$	$\hat{t}_{0.10,L}$	$\hat{t}_{0.10,U}$
	109.543	161.922	135.165	150.655	30.437	304.683

In this case, are recommended the random variables (4). Unfortunately the literature does not contain quantiles of these random variables. Therefore, this paper presented an original algorithm for easily calculating the quantiles of these random variables.

Also, the method described is based on Monte Carlo simulation techniques. For this reason, in order to obtain more precise values we need to increase greatly the number of simulations. The minimum recommended number of simulations is $N_{sim} = 1000$ of simulations. The results presented in this paper were obtained on the basis of 10000 of simulations.

References

1. W. Changsen, *Analysis of Rolling Element Bearings, (*Mechanical Engineering Publications LTD, London, 1991)
2. T.A. Harris, *Rolling Bearing Analysis*, 3-rd. Edition (John Wiley & Sons, Inc., 1991)
3. J.I. McCool, IEEE Transactions on Reliability, **19**, 127 (1974)
4. J.I. McCool, *Using the Weibull Distribution. Reliability, Modeling, and Inference (*John Wiley & Sons, Inc., 2012)
5. C.O. Morariu, *Optimizing the reliability tests of bearings, doctoral thesis*, Transilvania University of Brasov (1999)
6. D.B. Kececioglu, *Reliability Engineering Handbook* (PTR Prentice-Hall, Inc. Englewood Cliffs, New Jersey, 1991)
7. D.B. Kececioglu, *Reliability & Life Testing Handbook* (PTR Prentice-Hall, Inc. Englewood Cliffs, New Jersey, 1993)
8. D.B. Kececioglu, *Reliability & Life Testing Handbook* (PTR Prentice-Hall, Inc. Englewood Cliffs, New Jersey, 1994)

Quantification of the quality problems in the construction machinery production

Renata Stasiak-Betlejewsk[1,*] and *Agnieszka* Czajkowska[2]

[1]Institute of Production Engineering, Faculty of Management, Częstochowa University of Technology, Poland
[2]Department of Strength of Materials, Concrete Structures and Bridges, Faculty of Civil Engineering and Architecture, Kielce University of Technology, Poland

Abstract. Paper presents results of the research findings on the quality problems analysis conducted in the chosen enterprise producing a machinery for the construction industry. The aim of the study was to identify the causes and consequences of the quality problems in the production by using selected methods and tools of the quality management, as well as to identify opportunities to minimize the existing risks in the production process. The results show that approximately 77% of production losses are caused by incompatibilities resulting from two quality problems caused by employee errors and the low quality level of supplied screw taps. The results of the study indicate the corrective measures that may be taken in the company in order to eliminate and reduce identified production problems contributing to the production of machines that do not meet building standards.

1 Introduction

The construction machines industry in Poland experienced the biggest boom in the 70's and 80's and afterwards there was a collapse in the early 90s around the world noted the significant decline in a demand for all kinds of cranes, excavators, bulldozers and more specialized equipment [1]. Polish construction industry crisis has been stopped in 1994, when the statistics show the construction machines production increase by more than 30% annually. Polish construction industry has undergone the structural transformation. Some manufacturers has started to certify their products with ISO 9000 (PN-EN-29000).

The fulfilment of European standards is important to approximate the technical-operational parameters of the construction machinery. Currently, the construction machines users pay special attention to the following factors such as: a safety, a quality and environmental operating conditions and the need to meet the requirements of ergonomic work (i.e. human factors). The first two requirements are governed by the relevant European standards EN and ISO. Environmental performance of the machines driving machines is regulated by the strict Euro II Directive since January 1996. Regulations on human factors (HF), which include among others: noise, vibration and ergonomics of the machine, are more complex. Cushioned cabin of materials absorbing a part of the vibration

*Corresponding author: renatastasiak@wp.pl

energy seats is matched to the shape of the human body is only one of several means already used on a large scale.

Polish construction industry includes almost 140 plants of various sizes, employing some 50,000 workers. The sale of the construction machinery, including the road construction, is now about 10% of total sale in Polish engineering industry. Turnover of this sector since the early 90's steadily growing and it is estimated to exceed 500 million USD. Resources domestic construction machinery market consists of about 60 000 of the machinery and equipment - including about 10 000 of the mobile cranes, 11 000 of excavators, 10 000 of the cat track and 3 000 of the wheel loaders. About half of these devices was manufactured before 1997, 40% - in 1997-2000, 30% - in 2001-2005. Machinery and equipment imports constitute about 70% of the total construction machinery number. In the foreign observers' opinion, Polish construction machines and equipment are competitive in terms of the quality and the price [2, 3].

The growing number of the road construction and the boom in the construction industry involves more the heavy construction machines and the equipment. There is also much smaller machinery and equipment involved in the completion of the road construction investment such as the plate compactors. The plate compactors are particularly suitable for the soil consolidation in a rather wide pits, but also a surfactant. It is also applied successfully for compacting asphalt and paving of cobblestones [2].

The aim of the paper is analysis of the quality problems identified in the manufacturing of the plate compactors machines. Obtained research findings have been analysed with the quality management tools and quality management method (Ishikawa diagram, Pareto – Lorenz diagram and FMEA method) applying to identify the most significant problems in the manufacturing process to improve the final quality of the plate compactors used in the road construction investments.

2 Characteristics of the research object

The research on the plate compactors manufacturing was conducted in the chosen Polish construction machine manufacturing enterprise. The plant is engaged in designing and manufacturing of small machines and equipment for the construction industry, repair and maintenance of the roads. Currently the company offers 14 different products, including 9 types of the soil plate compactors, 2 types of the surface cutters, the slurry pump with the combustion engine, a generator and a concrete cutter profiles. The study was discussed the petrol plate compactor ZGS-12/500 that is used to vibratory compaction, e.g. sand, gravel, asphalt or paving stones.

2.1 The quality issues in the analysed enterprise manufacturing profile

The analysed enterprise is the innovative and technical Polish enterprise that was established in 1987. Since the beginning of its activity it is engaged in designing and manufacturing of small and medium-sized machinery and equipment for construction, repair and maintenance of roads. The subject of the research conducted in the analysed enterprise is the petrol plate soil compactor type ZGS-12/500 that was designed by the enterprise. The machine consists of a number of assemblies, subassemblies and components, which are made of different grades of steel, epoxy resins and elastomeric components. Compactor type ZGS-12/500 is used to work on the road construction and repair of roads, the laying of pavements, car parks, squares, sports fields, compaction of narrow trenches in industrial and hydraulic engineering. Due to the large decrease in efficiency and effectiveness, it is pointless to use it, the share of clay fraction exceeds 10%, a fraction of dust is greater than 30%. Through the use of special plates elastomer, which is an accessory, there is a possibility of levelling the pavement and concrete blocks. Compactor provides immediate stabilization of the soil and it help to obtain the surface durability. The petrol plate compactor analysed in the paper is presented in Figure 1.

Fig. 1. The petrol plate compactor type ZGS-12/500.

The liability for the quality of the product and its components occurring in the manufacturing process is provided by the Quality Controller and Representative of the Enterprise Quality System. The control process of the production process is described in the Technical Conditions. There are also specific control methods at the various stages of the process. The whole quality control process and its components are strictly inscribed in the Quality Assurance System as described in the corresponding Procedures. The materials used for the production of individual parts of the plate compactor should comply with the technical documentation and material standards (important quality feature of the plate compactor is that castings made of gravy cast iron shall comply with BN74/19046-04 standard and dimensions of parts, assemblies and complete compactors for which to design drawings not specified dimensional tolerances should be done in 13 accuracy class according to PN-79/M-02139 standard). Another important quality feature of the plate compactor are screw connections that should be tightened by hand or mechanically correct key. The size of allowable tightening torques should comply with PN-81/M-82056 standard with a maximum tolerance of 10%. Bolted connections should be secured against self-loosening in accordance with the technical documentation. Welded joints should be consistent with the requirements of BN-73/1610-1603 standard. Every compactor should be tested (uptime during the test shall be 15 minutes) to check the correct operation of the machine and all cooperating assemblies and components. There are tested: correct function of the speed and the engine start, turnover inclusion centrifugal clutch, rated speed of the engine (DTR = 3600 rev/min), the rate of entry into the cold vibrator turn (tdop = 10 s), operating temperature of the vibrator (at the end of the test should not exceed 95°C), any grease leaks, any metallic rattles and creaks during operation, CO content in the exhaust gas (perm. content <4.5%). The critical quality feature of the plate compactor quality is related to the issue of the protection against corrosion. All connecting elements (screws, bolts, nuts, washers, pins and steel components) specified in the design documentation must be protected against corrosion by galvanizing. The surfaces to be coated should be proportionate to the degree of purity B2a in accordance to PN-70/H-97050 standard. Lacquered surfaces should be covered with a layer of a pre-primer. Cover painted shall meet the conditions workmanship min. 3 class according to PN-79/H-97070 standards. All painted surfaces should be covered with double layer of varnish [4, 5].

3 Research findings and discussion

In recent times, the main problem occurring in the enterprise is a production losses increase. In order to identify and systematize the causes of manufacturing problem there was created Ishikawa diagram with using the brainstorming method. There were identified five manufacturing causes areas i.e. : 'materials', 'machine', 'method', 'man', 'organization' (Fig. 2). Analysing data presented in the Ishikawa diagram (Fig. 2), it can be stated that 'man' and 'method' areas generated the majority of the manufacturing problems. Improper personnel policy was identified as the main problem cause that is linked with quality issues performance in the analysed enterprise.

Research findings presented in the Ishikawa diagram (Fig. 2) also allows linking of the manufacturing losses increase with the following reasons causing this situation:
-company seeking to reduce the production costs has decided to change supplier of taps on the supplier offering a lower price (however this change, despite the fact that led to the restrictions mentioned above costs, is the main reason for picking threads and making losses);
-a very important factor is partly used machinery;
-persons employed directly at the production process does not comply with the quality standards;
-the lack of a smooth flow of information between the different levels of the manufacturing process;
-poor application of technology, particularly in the manufacturing of new products.

In order to identify the significant factors affecting the final quality problems occurring in the construction machines manufacturing process the Pareto – Lorenz diagram was applied to evaluate the hierarchy of the manufacturing incompatibilities in the plate compactor manufacturing. Pareto – Lorenz analysis is carried out to identify and to establish the hierarchy of the incompatibilities identified in the manufacturing process. Research results are used to elaborate the corrective and preventive actions for the production process within the quality issues on the analysed product [6 - 8].

In order to arrange the causes of production losses in terms of their frequency they were carried out analysis of Pareto-Lorenz. Authors identified the following causes of the manufacturing process losses as the following incompatibilities:
N1 - poorly bent shaft;
N2 - unevenly welded worktop;
N3 - broken threads in the mounting vibrator;
N4 - damaged paint cover of the engine plates;
N5 - improper angle of the shaft.

The frequency of incompatibilities' occurrence in the analysed enterprise in 2014 and the cumulative percentage of identified incompatibilities were presented in Table 1 and Figure 3.

Analysing data presented in Figure 3, it can be stated that over 77% of incompatibilities are caused by two causes (N4 and N3). Reduction of the production losses and thereby the enterprise profits increase can be achieved by eliminating of the failures in the engine plate painting and searching supplier of the highest quality taps. Acquisition of the low quality sheets and pipes happens occasionally, which it is certainly a result of long-term relationships with reliable suppliers. Cause like poorly bent shaft due to the rarity of its occurrence has little effect on the resulting waste.

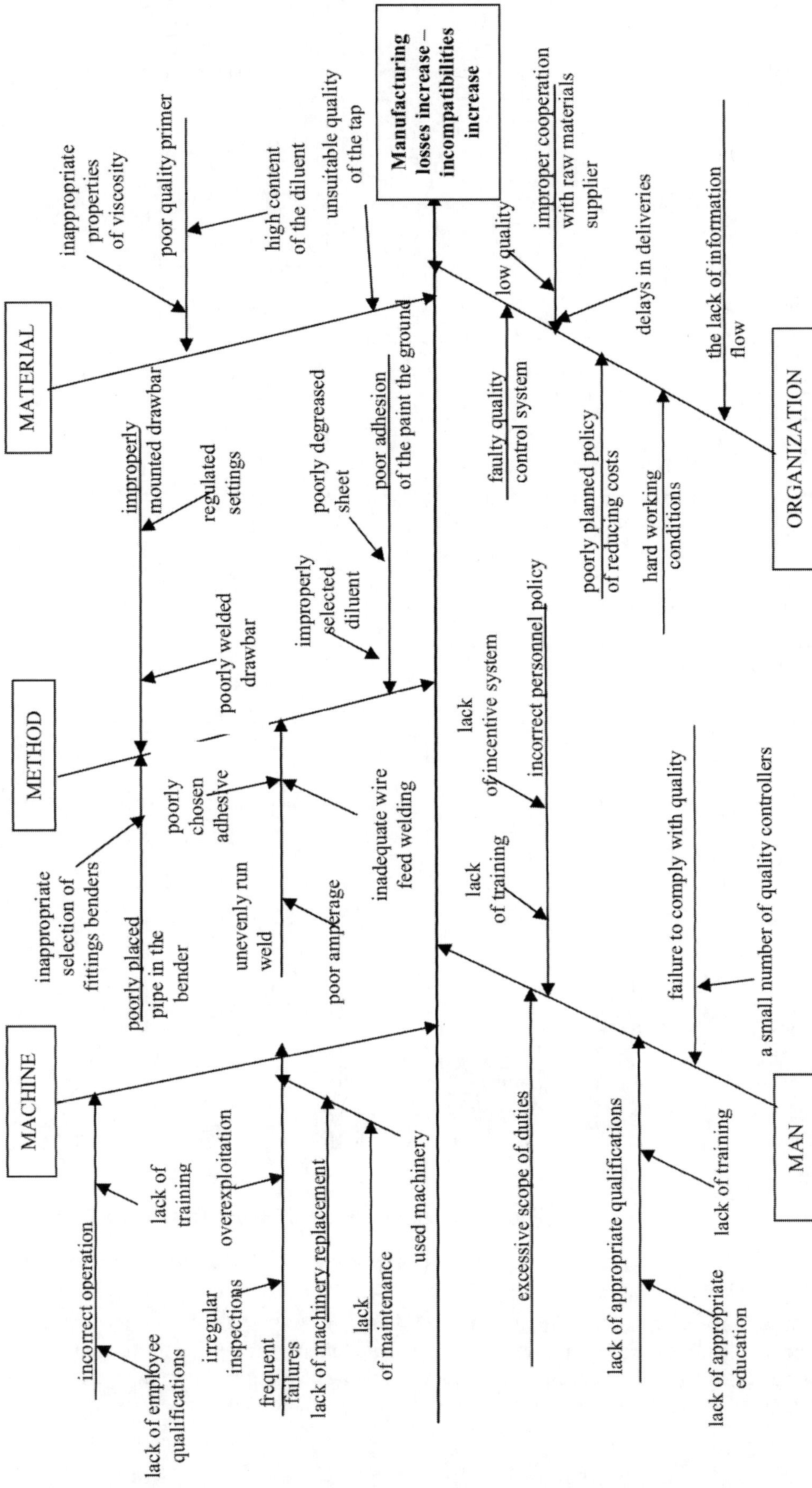

Fig. 2. The Ishikawa diagram for the factors affecting the manufacturing losses increase in the analysed enterprise.

Table 1. Incompatibilities identified in the petrol plate compactor manufacturing process in the analysed enterprise in 2014.

Denotation of the incompatibility	The name of the incompatibility	The frequency of incompatibilities occurrence in the research period	The percentage of incompatibilities [%]	The cumulative percentage of incompatibilities [%]
N4	damaged paint cover of the engine plates	39	50,65	50,65
N3	broken threads in the mounting vibrator	21	27,27	77,92
N5	improper angle of the shaft	9	11,69	89,61
N2	unevenly welded worktop	5	6,49	96,10
N1	poorly bent shaft	3	3,90	100
Total		77	100	

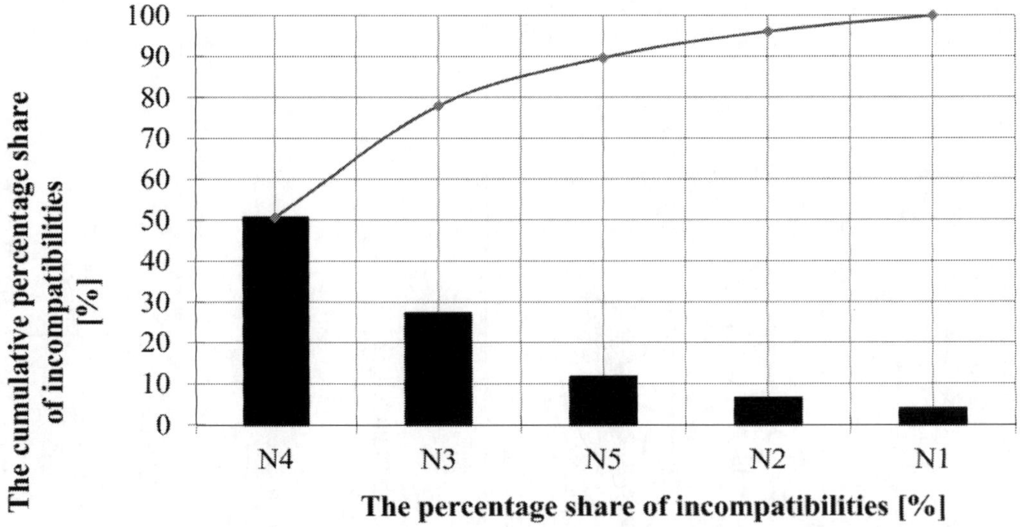

Fig. 3. Pareto-Lorenz diagram for the incompatibilities identified in the petrol plate compactor manufacturing process in the analysed enterprise in 2014.

Failure Modes and Effects Analysis (FMEA) method was applied in order to analyse possible failures, their causes and effects of identified incompatibilities in the petrol plate compactor manufacturing process. This method requires defining the relationship cause-effect-defect by scoring points on a scale 1÷10 due to three criteria: the risk of defects

 DOI: 10.1051/matecconf/20179404011

appearance (R), the probability of defects detecting (W), the importance of defect appearance (Z). The product of these three values multiplying is called the priority number P, which is as a measure of the validity of individual failures and it provides the basis for their hierarchy. Priority number of the Risk (LPR) accounts according to the formula: LPR = LPW × LPZ × LPO (1) where: LPW - Priority Number of Appearing. LPZ - Priority Number of Meaning (for customer). LPO - Priority Number of Detectability [6, 9].

Table 2 presenting the graduating system in the research findings analysis was applied in order to elaborate results for table 3.

Table 2. Scale of grading R, W, Z in FMEA method.

No	Risk of defects appearance (R)		Probability of defects detecting (W)		Effects/meaning of defects (Z)	
1	Unlikely	1	High	1	Hardly perceptible	1
2	Very unlikely	2-3	Moderate	2-5	Few perceptible	2-3
3	Unlikely	4-6	Low	6-8	Moderately perceptible	4-6
4	Moderately likely	7-8	Very low	9	Hardly perceptible	7-8
5	High probability	9-10	Impossible	10	Very strongly perceptible	9-10

Identifying quality problems in the manufacturing process (Table 3) allows using of preventive measures, implementing of corrective actions and testing their effectiveness. Quality problems, which are the source of the greatest threats, should be solved as the first step. Table 3 describes the preventive measures for the failures occurrence. The technology change is the key factor in this step since the improper technology increases production losses. There are suggested also staff training which is why the company's management should focus on elaborating of the training system for employees.

An important factor in the production process is also used machinery, which is why the company should pay attention to more frequent inspections of the equipment, as well as reflect on the partial replacement of the machine park. For this company the majority of the faults and their causes are adequately easily detectable, and therefore should be already in the design phase the production of all identify them and take any action that will ensure that the probability of failures and thus the number of risk will be minimized. This is in line with the motto FMEA method, which reads: "prevent the emergence of errors rather than detection."

As Table 3 shows the introduction of any corrective measures led to a significant reduction in the number of risk for individual failures. An analysis of incompatibilities carried out by using Pareto-Lorenz diagram shows that over 77% production loss is caused by two incompatibilities (damaged paint cover of the engine plate and broken threads in the mounting vibrator). The first one is caused by the most common workers' failure while the second - the poor quality of supplied taps.

 DOI: 10.1051/matecconf/20179404011

Table 3. FMEA method for defects identified in the petrol plate compactor manufacturing process in the analysed enterprise in 2014.

Name of defect	The location of the defect	Cause	Effect	R	W	Z	P	Corrective actions	R	W	Z	P
damaged paint cover of the engine plates	Production warehouse	poor quality primer, improperly degreased sheet	waste, complaint, return, customer dissatisfaction	7	4	10	280	information flow improvement training, change of the supplier	3	4	10	120
broken threads in the mounting vibrator	Production	worker's failure, poor quality of the tap	waste	8	3	8	192	change of the supplier, change of the technology, training	5	2	8	80
improper angle of the shaft	Production	worker's failure, breakdown of the machine, improper technology	waste, complaint, return, customer dissatisfaction	5	2	8	112	information flow improvement , training, changes in production technology, more frequent inspections of machines	3	2	8	48
unevenly welded worktop	Production	worker's failure, breakdown of the machine, improper technology	waste	5	1	9	45	staff training, change of the technology, quality control extension	3	1	9	27
poorly bent shaft	Production	worker's failure, breakdown of the machine, improper technology	waste	5	1	9	45	staff training, changing of the technology	2	1	9	18

4 Conclusion

Analysis of the research findings and corrective and preventive actions obtained as the result of the applied quality management tools and method allows to summarize suggestions for the analysed enterprise manufacturing process:

-greater focus on "improving" their employees, the introduction of a workers training system, skills upgrading, which will reduce the number of mistakes made by the crew;

-consider the possibility of changing taps' supplier;

-the production technology improvement;

-the information flow improvement both between employees and their superiors, and between different departments in the company;

 DOI: 10.1051/matecconf/20179404011

-intensified control measures.

Another important factor contributing to the improvement of the production process in the analysed enterprise is machines capacity utilization level increase, which will ensure increase in production, as well as reducing maintenance costs of the equipment.

References

1. Report Polish Roads, January (2007)
2. Report News of the building technique, June (2008)
3. Report Construction, maintenance and operation, (2007)
4. P.I.T. Project technical conditions, ZGS-12
5. S. Borkowski, J. Selejdak, Sz. Salamon, *Efektywność eksploatacji maszyn i urządzeń*, (Czestochowa University of Technology, Częstochowa,2007).
6. S. Borkowski, *Mierzenie poziomu jakości*, (Higher School of Management and Marketing, Sosnowiec, 2004)
7. J. Bank, *Zarządzanie przez jakość*, (Gebethner i S-ka, Warszawa, 1996)
8. L. Costinescu, C. Cojocariu, M. Dudița, L. Parv, I.L. Velicu, D. Munteanu. *JOAM*, **17**, 1-2, (2015)
9. L. Parv. *Data management in innovative engineering* (Transilvania University Publishing House, 2015)

MATEC Web of Conferences **94**, 04012 (2017) DOI: 10.1051/matecconf/20179404012
CoSME'16

Comparison of the methods for determination of calibration and verification intervals of measuring devices

Pavlina Toteva[1,*], *Stoyan* Slavov[1] , and *Dimka* Vasileva[1]

[1]Technical University of Varna, Department of Technology of Machine Tools and Manufacturing, 9000 Varna, 1 Studentska str., Bulgaria

Abstract. The paper presents different determination and optimisation methods for verification intervals of technical devices for monitoring and measurement based on the requirements of some widely used international standards, e.g. ISO 9001, ISO/IEC 17020, ISO/IEC 17025 etc., maintained by various organizations implementing measuring devices in practice. Comparative analysis of the reviewed methods is conducted in terms of opportunities for assessing the adequacy of interval(s) for calibration of measuring devices and their optimisation accepted by an organization – an extension or reduction depending on the obtained results. The advantages and disadvantages of the reviewed methods are discussed, and recommendations for their applicability are provided.

1 Introduction

Many organizations use technical devices for monitoring and measurement (TDMM) of processes, products and services. This includes most of the enterprises involved in processing and manufacturing industry, as well as testing laboratories and authorities for technical control. In order to prove a result of measurement reliable, it is necessary to use only a calibrated TDMM. This is a basic prerequisite in widely applied standards with requirements for management systems, such as ISO 9001, ISO/IEC17020, ISO/IEC17025, ISO14001, OHSAS 18001 (ISO 45001) in terms of ensuring adequate management and maintenance of TDMMs [1-6].

According to the Legal Metrology in the Republic of Bulgaria, to ensure the accuracy and reliability of measurements in health care service and public safety measurements, environmental protection, state and municipal receivables and trade payments metrological control should also be carried out. For this purpose the State Agency for Metrology and Technical Surveillance has been founded to regulate the frequency of verification of measuring instruments in use that are subject to metrological control.

The accuracy of the technical devices used in each accredited laboratory is regulated in clause 5.5.2 of ISO/IEC 17025: 2006, namely: *"Technical devices and related software used*

* Corresponding author: pavlina_toteva@tu-varna.bg

 DOI: 10.1051/matecconf/20179404012

for testing, calibrations and sampling must provide the accuracy required and must be in compliance with the requirements for these tests and/or calibrations."

In organizations, technical devices for monitoring and measurement used to check the required conformity of products and services must provide valid and reliable results [1].

Calibration of measuring instruments should be carried out at appropriate intervals, the calibration periods depending on different factors, such as required uncertainty, frequency of use, method of use, stability of measuring instruments, the specific guidance of the manufacturer of TDMM, etc. [8].

Before being put into operation a technical device should be calibrated or checked in order to be determined whether it satisfies the requirements set out by the laboratory and whether it is in accordance with the relevant essential requirements [4].

The periods of verification of measuring instruments are determined in the Ordinance on measuring instruments subject to metrological control [7] and by order of the Chairman of the State Agency for Metrology and Technical Surveillance. For other types of TDMMs Bulgarian legislation does not specify periods for calibration. The organizations using TDMMs are responsible for determining the time intervals themselves to ensure reliable results. Moreover, it requires the presence of documented methodology by which the organization determines the periods for calibration [1, 5] relevant for the applied TDMM.

This leads to difficulties in small and middle-size companies because of the use of a large number of TDMMs undergoing calibration. Furthermore, it should be taken into account that TDMMs are usually obtained from different manufacturers (suppliers); they are introduced in operation at different time points, so frequency and duration of use in process monitoring and control is not identical. This often requires an individual approach to the metrological maintenance of each TDMM (or a group of TDMMs) which further complicates the task of implementing appropriate methodology that is adequate for all TDMMs used in the organization.

Inadequate definition of calibration intervals carries risks for the organization in two ways: too short an interval results in higher costs for calibration of the respective TDMM for the organization. If a longer than necessary period is considered, one faces the risk of getting unreliable results using TDMMs in processes for monitoring and/or production evaluation. Consequently, the organization can also suffer significant losses limited not only to the financial aspect.

For these reasons, the issue of choosing one or another method for determining the periods to calibrate TDMMs is not unequivocal and it is relevant for all organizations. Therefore, the paper presents an overview of the most common methods as well as their advantages and disadvantages and their practical applications.

2 Existing methods for determining the intervals for the calibration of TDMMs

It is well known that in meteorology it is practically impossible to predict the exact time point at which the indications of a TDMM will go beyond the margin of error of measurement. Therefore, multiple methods for determining the calibration interval of a TDMM are developed aiming at the guarantee that the proportion of measurements that fall outside the margin of error remains as small as possible. The methods for determining calibration intervals are introduced by the international organizations OIMLH and ILAC [10, 11].

2.1 Methods for determining the intervals for the calibration of TDMMs recommended by International Organization of Legal Metrology (OIML) and International Laboratory Accreditation Cooperation (ILAC)

OIML and ILAC offer general guidelines for the criteria and methods for determination of calibration intervals [10]. In general, the methods for determining the time interval between calibrations of TDMMs can be divided into two main groups - statistical and algorithmic techniques. Statistical methods are based on the statistical modelling of the relationship between the probabilities of indications of a TDMM to be in the tolerance of the metrological characteristics for a certain period after the last calibration. It aims at predicting the duration of the interval to the next calibration based on the probability of reaching a percentage of measurements within acceptable limits at the end of this period. They usually require a significant amount of data from previous calibrations and they are relatively challenging in terms of their practical application (see Table 1).

Table 1. Comparison of methods for determining ranges of calibration of TDMMs [12].

Main indicators	1st method – Automatic adjustment or "staircase" (based on calendar time)	2nd method - Checklist (based on calendar time)	3rd method – Duration of measurement device usage	4th method - Control during operation or testing with "black box"	5th method - Statistical approaches
Reliability	medium	high	medium	high	medium
Difficulty of application	low	high	medium	high	high
Balanced load	medium	medium	low	medium	bad
Applicability in specific measurement devices	medium	low	high	high	low
Availability of measurement devices	medium	medium	medium	high	medium

The second group of methods is based on both relatively simple and complex logical algorithms for determining the calibration interval depending on the results and the conditions of TDMMs during calibration. Usually these approaches contain logically reasoned "decision" to extend or shorten the calibration interval depending on observations of current or previous calibrations. Due to their nature, these methods are called algorithmic methods [13]. Currently, algorithmic methods are relatively widely used in practice because of their uncomplicated character and the low cost of implementation. However, they have some shortcomings, some of the most important ones being the following:

-Change in the calibration interval can be induced by even a single evidence of a TDMM outside the margin of error, which can be caused by random factors, i.e. by random fluctuations in readings.

-Algorithmic methods usually do not include the impact of climate uncertainty of measurement of a TDMM in determining the calibration interval.

-Algorithmic methods cannot be easily adapted to the determination of target reliability related to the quality of measurements. The level of reliability achievable with algorithmic methods can only be determined experimentally or by simulation.

-If the prescribed interval is consistent with the desired level of reliability, the results of next calibration or next few calibrations often lead to moving away from the optimal interval.

DOI: 10.1051/matecconf/20179404012

-The time span necessary to achieve reliable determination of time intervals could reach up to 40-50 years.
-Automation of algorithmic methods is usually demanding. Therefore, they are often implemented manually by calibration specialists which increases the time for their realization, leading to increased costs of the process.

2.2 Comparative analysis of the methods for determining and optimizing the intervals for the calibration of TDMMs given in Recommendations on Interstate Standardization (RMG 74-2004), State Measurement System (GSI)

In RMG 74-2004 GSI "Methods for determining the intervals of verification and calibration of measuring devices" [11] six methods for determining and optimizing calibration intervals of TDMMs are described. Table 2 depicts the comparative analysis of the advantages and disadvantages of these methods for the optimization of calibration intervals of TDMMs carried out in different organizations [11]. The results of the analysis indicate that none of the discussed methods provides comprehensive information about the duration of time interval between calibrations of TDMMs. To obtain the most accurate time interval it is necessary to conduct empirical research within the life-cycle of a TDMM. Empirical tests should be carried out in not too long intervals of the operation period as these tests are labour-intensive and they are associated with increased expenditure of time and human resources. Regarding this, it is appropriate to conduct accelerated testing of a TDMM allowing prediction of the importance of starting parameters that determine the performance of a TDMM.

Based on the performed comparison and taking into account the data from the source [14] the authors propose a combined method for statistical check and update (if necessary) of calibration intervals of TDMMs using an algorithmic approach.

Table 2. Methods for determining verification intervals of TDMMs.

Methods for setting the verification intervals	Advantages	Disadvantages
1. Based on statistical implicit and explicit rejections.	High degree of reliability of the test results.	A large amount of experimental data for the process of change over time; the survey is labour-intensive.
2. Economic criteria.	Minimize operating costs of a TDMM; remove the consequences of possible errors caused by measurement errors.	Use of approximate models which relate mostly to one type of a TDMM.
3. Arbitrarily set the initial period of calibration with subsequent adjustments during operation of a TDMM.	Minimal financial and time costs.	Incorrect determination of the initial interval; there are no recommendations for all types of TDMMs about the initial interval; lack of reliability data of TDMMs.
4. Set the intervals for the calibration of a TDMM based on analogy.	Lack of financial and time costs.	The results obtained for analogues are not always applicable to newly developed or purchased TDMMs. There are no data on the reliability of the elements of analogue.

Methods for setting the verification intervals	Advantages	Disadvantages
5. Determination of the intervals for the calibration of a TDMM on indicators of reliability (failure rates or time limit for metrological operation).	High degree of reliability of test results.	A large amount of experimental data for the process of change over time; the survey is labour-intensive.
6. Determination of calibration intervals for analysis of the progressive component of the inaccuracy of a TDMM.	Low financial costs and high reliability of test results.	Need data availability for the longest period of inspection as well as costs for constant monitoring of the need for verification.

3 A combinatorial method for determining the calibration interval of TDMMs

3.1 The essence of the method

The methodology described [13-15], aims at minimizing most of the discussed shortcomings of the algorithmic methods in determining or verifying a calibration interval of a TDMM.

In the analysis of calibration intervals, the term "reliability" of indications of a TDMM is linked to the probability of a parameter measurement of a TDMM within the tolerance range for the measured value. The observed reliability is defined as the proportion of measurements that have been found to be within the tolerances when performing measurements, calibration or other checks.

In this method, the observed reliability is introduced as the variable R_{obs}. The verification of a calibration interval is based on the comparison between the observed values of reliability (R_{obs}) and the specified target reliability of indications of a TDMM (R_{targ}). R_{obs} is defined as the total number of tests (n) carried out during the current calibration against the number of tests (x) where the readings were within tolerance. For example, if n = 24 and x = 19, the value of the observed reliability R_{obs} will be:

$$R_{obs} = \frac{x}{n} = \frac{19}{24} = 0.792 = 79.2\% \quad (1)$$

The current interval is tested by comparing the value obtained for R_{obs} and the target reliability R_{targ} set to verify whether there is a significant difference between these two parameters. The size of difference and confidence levels determines the level of significance. If the evaluation of the observed reliability R_{obs} and the set target reliability R_{targ} reveals a significant difference, the selected current calibration interval of a TDMM is rejected and a shorter or longer interval is recommended (Fig. 1).

The algorithm used for verifying the current calibration interval calculates whether R_{obs} is within the confidence limits set by the value of the target reliability R_{targ}. If so, the current interval is considered to be properly defined and a recommendation for its preservation should be done. Confidence limits for Q are determined based on the function of the binomial distribution. With the pre-set total number of tests (*n*) and the number of tests within the tolerances of the measured value (x), the upper confidence limit p_U and the lower confidence limit p_L are determined by the dependencies below:

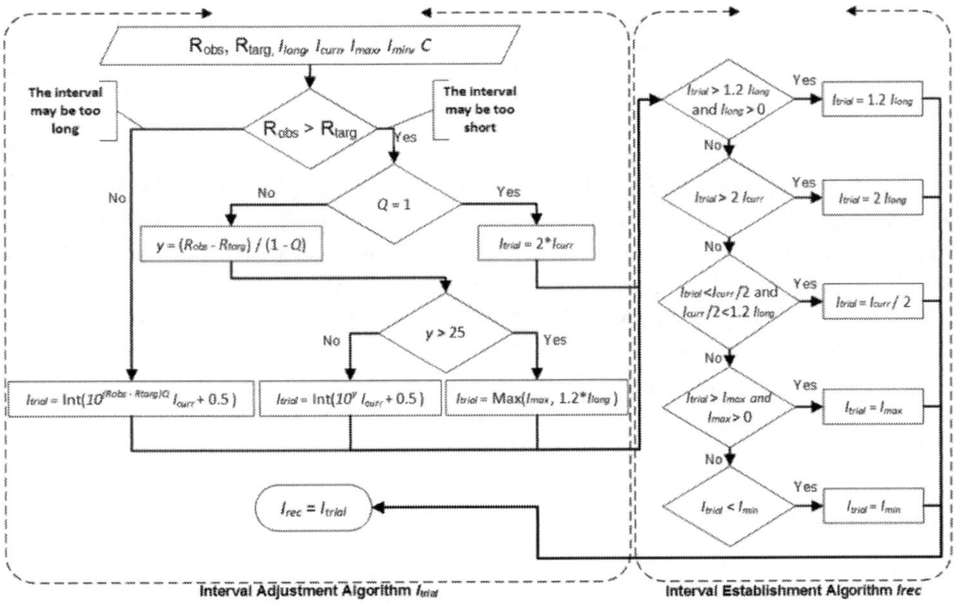

Fig. 1. Algorithm to check and determine the recommended interval I_{rec}.

$$\sum_{k=0}^{x}\binom{n}{k}p_U^k(1-p_U)^{n-k} = \begin{cases} \frac{1-C}{2}, 0<x<n \\ 1, x=n \\ 1-C, x=0 \end{cases} \tag{2}$$

$$\sum_{k=0}^{x}\binom{n}{k}p_L^k(1-p_L)^{n-k} = \begin{cases} \frac{1-C}{2}, 0<x<n \\ 1-C, x=n \\ 0, x=0 \end{cases} \tag{3}$$

where C is the given confidence level of changing the examination interval. If the requirement $p_L \leq R_{targ} \leq p_U$ is fulfilled, the selected current interval satisfies the check and may be kept unchanged. Otherwise, it does not satisfy the verification and should be adjusted (increased or decreased).

3.2 Calculating the confidence level Q for the rejection of the selected interval

The value of Q is determined by a numerical iterative method based on the pre-set confidence level C and the resulting values for p_1 (lower limit) and p_2 (upper limit) upon satisfaction of the following equations:

$$\sum_{k=0}^{x}\binom{n}{k}p_1^k(1-p_1)^{n-k} = \frac{1-C}{2} \tag{4}$$

and

$$\sum_{k=0}^{x}\binom{n}{k}p_2^k(1-p_2)^{n-k} = \frac{1-C}{2} \tag{5}$$

The process of the iterative calculation begins by setting the variable C to a value close to 1.0 (for example, 0.9999999). Iterations continue until the lower or upper limit (p_1 or p_2)

satisfy one of the following:

$$p_1 < R_{targ} \text{ and } R_{targ} - p_1 < \varepsilon, \tag{6}$$

$$p_2 > R_{targ} \text{ and } p_2 - R_{targ} < \varepsilon \tag{7}$$

where ε is the pre-set level of accuracy (ε = 0.00001). If any of the conditions (6) or (7) is satisfied in the course of iterations, the confidence level Q for the rejection of the inspected interval takes the value of C. For the calculation of parameter C, an iteration method of interval bisection is used, the so called "bisection method" [16]. Gradually, approximate values for the variables p_1 and p_2 are set. These are obtained as follows: First, the variables P and y_p are defined by the expression $P = (1 - C)/2$ and $y_P = \Phi^{-1} \cdot (1 - P)$, where Φ is the inverse function normal distribution. Next, the variable p is calculated by the equation:

$$p \cong \frac{a}{a + b \cdot e^{2 \cdot w}}, \tag{8}$$

where:

$$w = \frac{y_p (h + \lambda)^{1/2}}{h} - \left(\frac{1}{2b-1} - \frac{1}{2a-1} \right) \left(\lambda + \frac{5}{6} - \frac{2}{3h} \right)$$

$$h = 2 \cdot \left(\frac{1}{2b-1} + \frac{1}{2a-1} \right)^{-1} \text{ and } \lambda = \frac{y_P^2 - 3}{6}$$

Once the variable p is determined, the preliminary values p_1 and p_2 are defined as:

$$p_1 = p - dp \text{ and } p_2 = p + dp \tag{9}$$

where dp is set conditionally (e.g. $dp = p / 100$). During each iteration step, the values of p_1 and p_2 are recalculated by:

$$I_{P1}(x, n - x + 1) = 1 - C \text{ for } p_1 \text{, and} \tag{10}$$

$$I_{P2}(x + 1, n - x) = C \text{ for } p_2. \tag{11}$$

Upon receipt of the recalculated values for p_1 and p_2, they are tested and the iterative process for the calculation of C is terminated, (6) or (7) are fulfilled and Q takes the value of C for the current iteration step.

3.3 Study of the influence of "target reliability" (R_{targ}) and "set confidence level" (C) parameters

To examine the results of the method, it is necessary to set specific values of the confidence level C and the target reliability R_{targ}. Upon further consideration it is assumed that methodology is set at nonzero current interval $I_{curr} \neq 0$, and at zero minimum permitted interval $I_{min} = 0$, for the maximum acceptable interval I_{max} and for the longest interval I_{long} the set values being 10 000.

When verifying statistical significance, the level of confidence is generally assumed to be 95%. If the value of the confidence level C is set to be 95% by methodology, the confidence limits p_L and p_U determine 95% confidence level for the parameter Robs, and the tested

 DOI: 10.1051/matecconf/20179404012

interval will be rejected with 95% confidence if R_{targ} is out of the calculated tolerance (p_L, p_U).

It is important to take into consideration that the inspected interval will be rejected if R_{obs} significantly differs from R_{targ}, as the level of importance is assigned by the chosen level of confidence (95%). Therefore, if the maximum possible value of R_{obs} is 1.0 (i.e. 100%) and R_{targ} is high, no extension of current interval will be recommended. This will hold true until enough experimental data are collected, resulting in its rejection. Furthermore, the following should be taken into account:

Assuming that R_{targ} = 95% and confidence level is 95%, then the methodology will not recommend an extension of the interval until the following condition is not satisfied, i.e. 59 measurements with a TDMM should be within the permissible scope of measurement parameter's dispersion for 59 measurements. If R_{targ} = 90%, then the recommendation to extend the interval will occur if 29 measurements are within the permissible dispersion scope of measurement parameter from all 29 measurements. If R_{targ} = 85%, then the condition for extension will be fulfilled in 19 measurements in the permissible field of all 19 measurements.

Changing the confidence level also affects the conditions for the extension of the inspected interval. For example, if R_{targ} = 95% and the confidence level C is set to be 80%, a recommendation for an extension of the current interval will occur in 32 measurements within the permissible limits of 32 measurements. If the confidence level is 70%, a recommendation to extend the interval will occur if 24 measurements are within the permissible limits of 24 measurements carried out with a TDMM.

As can be seen above, the higher the target level of reliability R_{targ}, the more difficult it is to obtain an extension of the interval for calibration. This also holds true for shortening the interval. For example, if R_{targ} = 95% and confidence level C is also 95% recommendation to shorten the current interval will occur if there are two measurements that are outside the tolerance of the measured parameter. If the confidence level is 95% and R_{targ} = 64% recommendation to shorten the current interval will occur if three measurements are outside the tolerance of the measured parameter.

Similar to the extension of the interval, shortening of the verified calibration interval is recommended more often at a lower confidence level C. For example, assuming that R_{targ} = 90% and the confidence level C is set to be 95%, a recommendation to shorten the current interval calibration will occur at two measurements outside the tolerance parameter measurements. However, if the confidence level is set to 89%, then only one measurement outside the tolerance is enough to recommend a shorter current interval.

These results demonstrate that a reduction in the confidence level C leads to a more frequent rejection of current interval and to recommendations for its change. In addition, for higher values of R_{targ}, a decrease in current interval is more often recommended than its increase.

Since the test intervals are likely to decrease rather than increase, the confidence level C can be raised in order to achieve higher target reliability R_{targ} for the inspected TDMM, bearing in mind that the excessive increase more frequently leads to the recommendation for changing the calibration interval.

Under similar additional conditions, it could be concluded that higher values of target reliability (R_{targ}) lead to a lower level of confidence (Q) in the change of the interval.

Figure 2 shows the relationship between the parameters' alterations "confidence level" Q and "target reliability" R_{targ} - Q = f (R_{targ}) for three types of tests: Q1 - for 20 measurements; Q2 - for 30 measurements and Q3 - for 50 measurements.

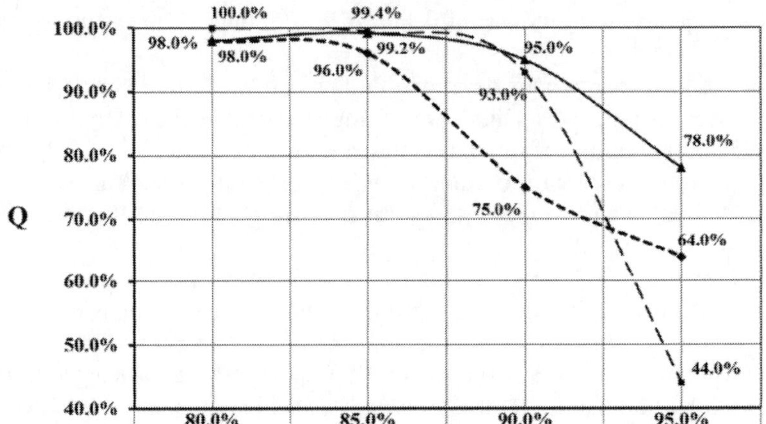

	80.0%	85.0%	90.0%	95.0%
Q1 (20 tests)	0.98	0.992	0.95	0.78
Q2 (30 tests)	0.98	0.96	0.75	0.64
Q3 (50 tests)	1	0.994	0.93	0.44

R_{targ}

Fig. 2. Interrelationship between "level of confidence" Q and "target reliability" R_{targ} parameters.

The figure illustrates clearly for all three tests that an increase in the value of R_{targ} beyond a certain threshold (R_{targ}> 80-85%) leads to abrupt decrease in the confidence level Q regarding the change of calibration interval. Therefore, using this methodology it is recommended that the target reliability should be selected within 80% $\geq R_{targ} \leq$ 90%.

4 Conclusion

The implementation of a large number of methods for the determination of calibration intervals within a company is not recommended since this complicates the supporting process of TDMMs and results in higher costs on the one hand, and requires highly qualified human resources on the other.

The methods described in [10, 11] could be problematic for application in small and medium-size enterprises.

Despite its belonging to the so called algorithmic methods, the proposed methodology for verifying and modifying the calibration interval of a TDMM overcomes some of their shortcomings.

The main advantage of the method is that there are no changes in the calibration interval unless it is rejected based on statistical data checks in the calibration of a TDMM. Furthermore, in case of rejection of the inspected interval, the methodology contains a simple algorithm for calculating the recommended calibration interval with regard to the restricted conditions pre-set by the user (I_{min}, I_{max}).

The methodology uses relatively simple logical and mathematical algorithms which make it manageable for a wider range of specialists, at the same time allowing easy automation with modern software products.

This makes it possible for the proposed methodology to be implemented as an independent tool for the control and regulation of calibration intervals in organizations that use and manage a few TDMMs and to be integrated as a module of more complex systems for metrological analysis.

In the future, we will focus on the methods for determining calibration intervals applied in sectorial management systems, e.g. in automotive and aerospace industry and medical devices manufacturing. Our research work will be based on standards such as 16949, 9100, 13485, VDA standards, OEM CSR, etc. regarding the management of the measuring and

 DOI: 10.1051/matecconf/20179404012

monitoring devices. We will also elaborate on the specific approaches for metrology confirmations like Gauge repeatability and reproducibility (GRR) and Capability of Gauges (Cg & Cgk) for new and repaired ones.

References

1. Quality Management Systems - Requirements (ISO 9001:2015), ISO, 44
2. Environmental Management Systems - Requirements with Guidance for use (ISO 14001:2015), ISO, 48
3. Conformity assessment - Requirements for the operation of various types of bodies performing inspection (ISO/IEC 17020:2012), ISO, 28
4. General requirements for the competence of testing and calibration laboratories (ISO/IEC 17025:2005/Cor.1:2006), ISO, 39
5. Measurement management systems - Requirements for measurement processes and measuring equipment (ISO 10012:2003), ISO, 28
6. ISO/DIS 45001, Occupational health and safety management systems - Requirements with guidance for use, ISO, (to be published)
7. Ordinance on measuring instruments subject to metrological control, promulgated in the State Gazette of the Republic of Bulgaria **22** (2015)
8. A. Williams, SLR Ellison, M Roesslein, *EURACHEM/CITAC Guide Quantifying Uncertainty in Analytical Measurement* (LGC, UK, 2000)
9. *EURACHEM/CITAC Guide: Use of uncertainty information in compliance Assessment* (LGC, UK, 2007)
10. ILAC-G24/OIML D 10: Guidelines for the determination of calibration intervals of measuring equipment used in testing laboratories
11. Recommendations on interstate standardization, RMG 74-2004. State system for ensuring the uniformity of measurements, State measurement system. Methods for determining the intervals of verification and calibration of measuring instruments
12. Radev Hr., Dyukendzhiev G., Zakharov I. et al., Metrology and Measuring Equipment, 1, SOFTRADE, **489**, (2008)
13. http://www.isgmax.com - The Integrated Sciences Group Website
14. Establishment and Adjustment of Calibration Intervals, Recommended Practice RP-1, National Conference of Standards Laboratories, (1996)
15. Jackson, D., Ferling, J. and Castrup, H., Concept Analysis for Serial Number Based Calibration Intervals, Proc. Meas. Sci. Conf., **23** (1986)
16. *Numerical Recipes* (Cambridge University Press, 1992)

MATEC Web of Conferences 94, 05001 (2017) DOI: 10.1051/matecconf/20179405001
CoSME'16

An approach to support entries for a design failure modes and effects analysis

Felicia Veronica Banciu[1,*]

[1]Politehnica University Timişoara, Mechanical Faculty, IMF Department, 300222 Timişoara, România

Abstract. The design of a fixture device is very laborious, requiring consideration of many issues from series production, information on the raw part material, required precision surfaces, ensuring an appropriate orientation scheme, continue with requirements related device itself as number of parts, parts type, complexity, manageability, easy operation and maintenance. That is why realising a proper analyse regarding the possible failures associated to these issues is overwhelming. This paper's aim is to propose an approach for collection of items necessary for starting a Design FMEA, using Functional Analysis, TRIZ functional modelling and axiomatic design approaches. The proposed approach – to extract items for a FMEA analyse – is applied to an example of an expanding mandrel regarding its internal components and theirs relations in order to accomplish the main requirement: the accurate locating and clamping of a part.

1 Introduction

There exists many tools, methods and approaches to manage the design process and some of them are general ones; they have general rules applicable in a wide range of products. This is also the case of FMEA which is an approach done by a cross multidisciplinary team and developed during the product and its associated processes design, it is a common approach where its applied and addressed elements are different but the logic is the same.

In first stages of product's design it is important to find ways to support and argument decisions making in order to avoid to recall some functions later in design process. It is also important "to see" what happens in case of a choice or another or to see what are the consequences of a failure to a certain component at a certain level of design, its implications in proper product functioning or its main functions achievement.

The function is the core element that "guides" the product's design and functional analysis is a common tool used for functions inventorying. A functional analysis helps to list product's internal and external functions but without emphasizing their harmful effects if they exist.

The design of a fixture device is very laborious, requiring consideration of many issues from series production, information on the raw part material, required precision surfaces,

* Corresponding author: felicia.banciu@upt.ro

ensuring an appropriate orientation scheme, continue with requirements related device itself as number of parts, parts type, complexity, manageability, easy operation and maintenance. That is why realising a proper analyse regarding the possible failures associated to these issues is requiring a lot of time, multidisciplinary teams and could be subject to numerous failures.

This paper's aim is to propose an approach for collection of items necessary for starting a FMEA. The idea of the approach is to enrich the approach from [1, 2] using Functional Analysis, TRIZ functional modelling and axiomatic design approaches, the latest for tracking the existing couplings and their impacts of propagation breakdowns.

2 Approaches used in product's design

2.1 FMEA

FMEA is a method that is applied at any detail level in design-component, function/sub-assemblies and assemblies and implies a very good understanding of systems functions and design and it implies the whole product design process [3].

To realize a FMEA following activities are done: an inventory of failure modes, what effects they have on the product, process and corrective measures that are to be applied. If analyse of failure criticity is followed, the RPN are calculated using existent data, existent reliability informations and external environment. They are very important and should be inventoried, the potential safety implications in system failures, its performance indicators, maintainability indicators extraction and not least the generation of demands for maintenance service, assembly, packaging, materials, safety, service operation. It is essentially an evaluation technique that can be applied at any level of detail following an approach as well as bottom-up or top down up. It applies equally to functions, subsystems, parts and assemblies, as a basic requirement without which cannot be achieved a good functions understanding and system operation. FMEA can be applied to recognize probable failure modes, conclude their effect on the process of the product, and categorize actions to diminish the failures. A vital step is anticipating what might go incorrect with a product. It is a popular tool for reliability and failure-mode analysis [4]. Process FMEA is described as having the goal to identify failures associated with the fabrication, assembly processes, machine tools, fixtures and Production Methods. Design FMEA deals with possible components failure; consider the integration of components and their interactions effects on product functions, subassemblies functions [3, 5].

2.2 Functional analysis

FAST - tool used for functional analysis-is a diagram that facilitates graphically and logic view of what functions are and how they work together. Thinking a product in terms of functions does not hinder / constrain thinking related to the physical / processes features and characteristics and therefore can lead to a better problem definition and an easier way to solution. FAST describes the logic relations that must exist between the various functions that must be met in order to satisfy the main product's functionality [6]. FAST diagrams and thinking is the base for different studies, approaches or technical analyses.

2.3 TRIZ functional modeling

Related to product's functions also functional TRIZ modeling [7, 8] could be used. The HU Diagrams (Harmful-Useful Diagrams) are designed around the concept that all systems

DOI: 10.1051/matecconf/20179405001

have positive aspects (useful functions) and all systems have negative aspects (harmful functions). A function is defined as capturing some aspect of a system including function, activity, state, process, condition, transformation. There are eight possible interactions/links between–useful-harmful functions ones producing or counteracting another.

2.4 Axiomatic design framework

The design is interplay between what we want to achieve and how we choose to satisfy the need. To systematize the thinking process in this interplay, it was created the domain concept, foundation of axiomatic design and it is guided by two axioms. These axioms are: the Independence Axiom - maintain the independence of the functional requirements (FRs) sections of the design should be separable so that changes in one have no (or as little as possible) effect on the other and the Information Axiom - minimize the information content of the design. The four domains are: Customer Domain, Functional Domain, Physical Domain and Process Domain and for each domain there are corresponding Customer Needs (CNs), Functional Requirements (FRs), Design Parameters (DPs) and Process Variables (PVs). The relations between the functional requirements FRs and design parameters DPs at a given level of the design hierarchy are captured in the Design Matrix [9]. The axiomatic design methodology is based on functional analysis and minimization of information content – aiming a robust design [9, 10].

3 The proposed approach

The general view of proposed approach is in figure 1.

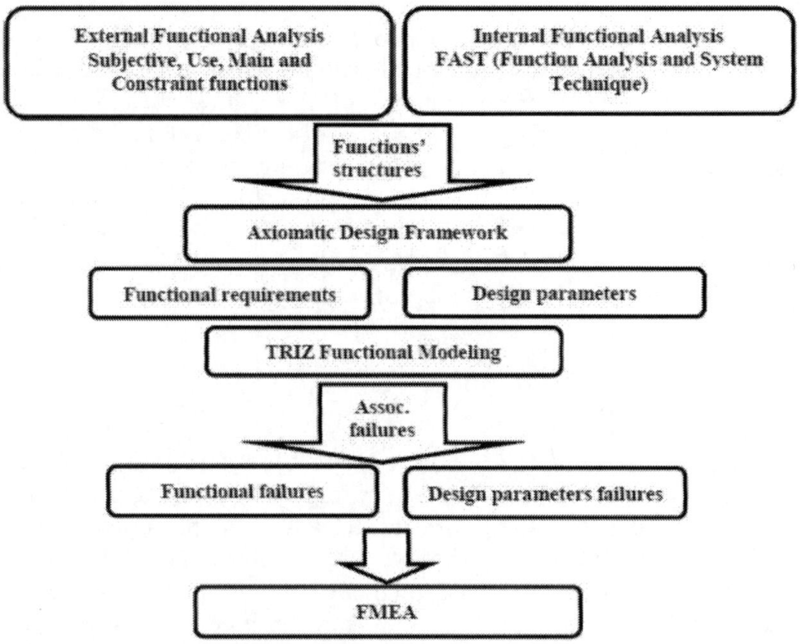

Fig. 1. The proposed approach.

As a work procedure/approach, first applying a functional analysis - the technical functions and elements that conducts to achieve the main functions, also the external environment elements will emphases the functional links and flows.

The results of a functional analysis – identification of functions – are entries in

Functional Requirements domain and for all of these functions are inventoried the physical parts/subassemblies which accomplish these functions - the design parameters (axiomatic design).

The design matrix contains the FRs-DPs elements. Also here are emphasised the eventual couplings that exist between design parameters. A framework to systematize and have an overall view on functions/components/potential failures and their consequences on different detail levels could be that given by axiomatic design.

The possible couplings will allow following the failure modes track, namely a failure at a certain level can influence occurrence of failures to lower levels of detail or on the same level. In this paper is presented the overall approach without a deep detailed view on each of its steps-design methods/tools.

A functional analysis helps to list product's internal and external functions but without emphasizing their harmful effects if they exist and to solve this, applied to functional requirements and design parameters a TRIZ functional modeling could be useful. We can obtain a complete function structure regarding the useful and harmful function effects if the TRIZ functional modeling is applied to those functions identified through a functional analysis. Thinking in terms of functional modeling, for these functions there can be identified their useful and harmful effects associated to these will be inventoried and possible useful or harmful effects of design parameters. This will represent a part of the entry data needed for a FMEA. In this paper is presented only the general approach without a complete modelling technique that deals only with useful-harmful functions and their effects on each other, this being a subject for further work. In addition, the design matrix associated to FR-DP is not represented as a matrix but the couplings between design parameters are mentioned and explained.

4 Proposed approach applied to a self-centring fixture device

Fixtures are considered as a subsystem with a functional independent activity that provides accurate locating and clamping for part during the machining operations. The acting elements that intervene in design process and generates inputs and constraints are:

a) external elements

- **part** material class, type, properties; part configuration (Number and dimension of holes, surface types, symmetry, slots, bosses etc.), shape, size, -tolerances (linear and angular) and accuracy, roughness, surface coatings,
- **machine tool** (machine tool type (its equipment and also its characteristics), interference check (machine tool has to be able to perform all movements with fixture, tool and part in their places) Forces and moments acting during machining process, chips removal,
- **cutting tool-operator (**accident proof concept, fool-proof concept, ease of use, loading and unloading efficiency**)**,
- **manufacturing process** regarding the operations (machining) on parts, - type of machining, number of machining surfaces (single surface, more identical surfaces in linear displacement, more identical surfaces in circular displacement);
- **batch size**.

b) internal elements:

- a great importance should be spent on clamping device itself, namely on its internal components and theirs relations in order to accomplish the main requirement: accurate locating and clamp of a part.

The proposed approach will be applied to an example of expanding mandrel (figure 2) - regarding the internal components and theirs relations in order to accomplish the main requirement: accurate locating and clamp of a part. The functions inventoried during an internal functional analysis correspond to each of the mandrel parts or subassemblies roles.

DOI: 10.1051/matecconf/20179405001

This expanding mandrel in figure 2 is used for centring and fixing an inner cylindrical part. It is centred and fixed using the disk jaws 2 displaced in axial movement by bushing 8. This bushing is mounted on stem 9 using a retaining 6 and locking nut 5 and thus is made the link with the actuating/operating system. The disks (plate fin) 2 are radially displaced and guided using the multiple wedge 4. Springs with ring shape located in disks slots assure the contact between disks and wedge. To avoid the rotation of bushing 8 against the wedge is used the sliding key 7. Centring and fixing the part is done actuating the stem in left direction and for loosening the part in the right direction.

Against the part are working radial forces S/n and in the same time, there appear frontal forces – a good axial part positioning and friction torque.

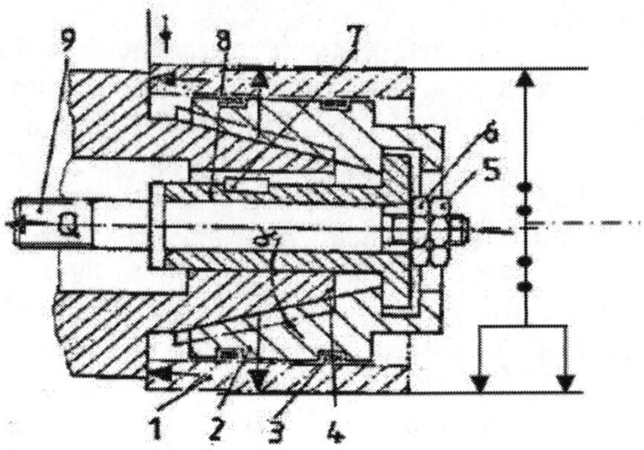

Fig. 2. Expanding mandrel [11].

Analysing the mandrel operation and its components and applying the Internal Functional Analysis FAST (Function Analysis and System Technique) (first part from approach in figure 1) were identified the functional requirements. These functions generates the functional requirements entries and their associated design parameters (way of their achievement) that are the base of the axiomatic design approach.

Next, using the TRIZ modeling technique and thinking from a useful/harmful perspective the functional requirements and their associated design parameters it were analysed the possible positive and negative aspects. All of these functional requirements and theirs associated design parameters- existent solutions and parts – have attached the useful-harmful perspective with their associated useful and harmful aspects are presented in table 1. This table offers an overview on possible issues and harmful/negative aspects that are main entries in a Design FMEA analyse regarding the possible failure modes.

Table 1. Functional requirements and design parameters – useful and harmful aspects.

Functional requirements FRs	Design parameters DPs
FR0 Centring and hold the WP on its inner cylindrical surface - useful function, useful effect	DP0 Self-centring mandrel
FR1 Centring WP on inner cylindrical surface – useful function	DP1 Materialize WP axis
FR11 Use up clearance between work piece and elements - useful function , useful effect	DP11 Elements that contact the WP
FR111 Contact between the work piece and centring/fixing elements - useful function with useful and harmful effects	DP111 Disks Harmful effects : - Disks wear - Contact friction

It implies that work piece to be in contact with disks in order to receive the force- useful function	
FR12 Transmit the movement synchronous to disks - useful function	DP12 Mechanism with actuation/operation rod/stem - one harmful effect is the fact that the stem is pulled and has to be tensile-
FR121 Disks movement - useful function	DP121 Multiple wedge with sloped surfaces in contact with disks Harmful effect - friction forces and wear
FR1211 Contact between multiple wedges and disks - useful function - the contact between multiple wedges and disks implies surface wear and friction issues	DP1211 elastic force - The disks under the spring's elastic force has to move so a harmful effect here can be introduced by friction force that has to be lower; - Another negative aspect here can be that the springs lose their properties, the improper elastic force developed by springs.
	DP 1212 ring type springs
	DP 1213 slots for ring springs in wedges body
FR1212 Guide disks in radial movement	DP1212 Guiding slots on surface of multiple wedge - it implies that two pieces to be connected in order allow the movement ; - the contact area, the type of fit between disks and wedge
FR 13 Moving disks on sloped surfaces of multiple wedge - useful function	DP13 Operating disks, using a bushing mounted on stem/rod - it implies that two pieces to be connected in order to transmit the movement. This raises two aspects that have to be considered: - the contact area bushing and disks - the fact that the stem is pulled and has to be tensile
FR131 Transmitting the operating force to bushing	DP 131 retaining nut and safety nut mounted on screw stem/rod Harmful effects: - Contact area between safety nut and retaining nut - Contact area between nuts and stem - Contact between retaining nut and bushing - Wear issues related with Contact between retaining nut and bushing
FR 1311 Contact between bushing and stem - useful function with both useful and harmful effects	DP 1311 close fit between bushing and stem - contact pressure assures the force and movement transmitting and in the same time there could be a harmful effect related to high pressure in contact area
FR 132 Contact between disks and bushing -useful function	DP 132 Bushing shape Harmful effect - contact area-friction, wear
FR 133 Multiple wedges sliding against stem/rod and bushing	DP 133 Free fit between bushing and multiple wedge
FR 134 Avoid bushing rotation against multiple wedges - useful function	DP 134 Using a sliding key Harmful aspects - Contact between sliding key and bushing - Contact pressure - Wear and friction between sliding key and its guiding slot in multiple wedge
FR15 transmit the movement to the operation stem	DP 15: operator/machine tool

- useful function –harmful effect is the fact that the stem is pulled and has to be tensile	
FR2 prevent the WP movement during working	DP 2 provide the holding forces
FR21 self - locking mechanism	DP21 wedges functional characteristics
FR22 prevent WP rotation	DP22 the mechanism with wedges and plungers, provide F*n force - friction forces that has both positive as negative aspects
FR23 hold the piece in axial direction - Useful function	DP23 WP propped in mandrel's body and its shape
FR3 loosen, un holding the WP	DP3 assure the reverse movement
FR 31 transmit the reverse movement to the operation rod/stem	DP 31 machine tool
FR 32 reverse disks movement	DP32 operating stem and bushing in the right side
FR33 Disks inward movement	DP33 elastic force, springs

In this table, in the first column are functional requirements achieved using one or more than one design parameter. In this second case when a functional requirement is achieved using more than one design parameter, the line containing these design parameters has its correspondent first column blank in functional requirements place.

The physical contact between mandrel's parts generates friction forces that has both useful and negative aspects. In addition, aspects related to surfaces characteristics, surface wear, stress could conduct to failure and could be considered as entrance elements in analysing possible failure modes. In axiomatic design (and in designing, generally), the link between functional requirements (what is needed) and design parameters (solutions) are given by mathematical equations expressing their dependencies.

The contact between the work piece and centring/fixing elements is a function that has both useful and harmful effects - useful is that friction helps part fixture but also it has to be considered the contact area between parts – the wear and contact issues are the harmful effect.

The force F depends on the value of the wedge's parameters- angle, sliding angles between wedges and disks, friction coefficients between all the parts being in relative movement. Even if the design is functional from a mathematical point of view, with imposed physical parameters, it can happen that force F cannot be achieved due to other components e.g. springs marked on Fig. 2 with 3. If is considered the FR121 with its DP 121, the corresponding sub-matrix is uncoupled, not dependent on other assemblies (it is about a solution to keep together the disks against the multiple wedge) but the fail generated by springs is major because it leads to not achieve the FR0, higher level requirement. The design parameter for example "guiding slots on surface of multiple wedge" does not appear directly in calculating the force F but its fail (the contact area, the type of fit between disks and wedge) can affect force. Taking into account that this expanding mandrel is with moving elements, its accuracy can be discussed relatively to another types of mandrels proper to deliver the main functions. This is another negative aspect.

5 Conclusions

The advantage of applying this approach is the fact that splitting the product (device) into functions and their associated parts/subassemblies and thinking to their useful and harmful effects offers a deeper and detailed view on whole device/product from different point of views - functional/parts/effects- and could be used as entries for a Design FMEA.

In future work the presented approach could be continued through a complete TRIZ modelling technique that deals only with useful-harmful functions and their effects on each other and with design matrix. All the methods/tools presented in approach need a thorough teamwork.

The approach presented in this paper is a general one and its novelty and originality are that it could be adapted to various designs being useful in systematizing the knowledge about functional role/functions/components/process/useful-harmful sides of functions/components used as entries in analysing of potential failures modes and their causes. Analysing the harmful functions effects will offer a broader perspective of issues related to fixture design and associated failure modes thus leading to improvements, to discover failure at its earliest possible point in product design process. Simultaneously, broadening the perspective on failure modes issues, it can be a support for continuous innovation.

References

1. F. Banciu, G. Drăghici, A New Functional Approach to Support FMEA Method in Product Conceptual Design, in Proceedings of the 16th International Conference Modern Technologies, Quality and Innovation, **1**, (2012)
2. F. Banciu, G. Draghici, C. Turc, A Point of View on Functional Approaches Used in Product Design in Applied Mechanics and Materials **371**, (2013)
3. Failure Mode Effects Analysis (FMEA) at http://asq.org/learn-about-quality/process-analysis-tools/overview/fmea.html
4. L. Lipol, J. Haq, Risk analysis method: FMEA/FMECA in the organizations. Int. J. of Basic & Appl. Sci. IJBAS-IJENS **11**, 74, (2014)
5. S.-H. (Gary) Teng, S.-Y Ho, Failure Mode and Effects Analysis: An integrated approach for product design and process control. Int. J. of Quality & Reliability Management **13**, 12 (1996)
6. J. Borza, FAST Diagrames: The Foundation for Creating Effective Function Models, *Trizcon 2011*, 1–10, (2011)
7. J. Terninko, A. Zusman, B. Zlotin, *Systematic Innovation: An Introduction to TRIZ (Theory of Inventive Problem Solving)*, (Saint Lucie Press, Boca Raton, Florida, USA, 1998)
8. S. D. Savransky, *Engineering of Creativity: Introduction to TRIZ Methodology of Inventive Problem Solving*, (CRC Press 2009)
9. N. P. Suh, "Axiomatic Design-Advances and Applications", (New York, Oxford University Press, 2001)
10. D. G. Lee, N. P. Suh, *Axiomatic Design and Fabrication of Composite Structures,* (Oxford University Press, 2006)
11. N. Gojinețchi, N. Gherghel, *Workholding Devices Design* (Iasi, 1983)

MATEC Web of Conferences **94**, 05002 (2017) DOI: 10.1051/matecconf/20179405002
CoSME'16

The CAD drawing as a source of data for robot programming purposes – a review

Krzysztof Foit[1,*], and *Grzegorz* Ćwikła[1]

[1]Silesian University of Technology, Faculty of Mechanical Engineering, Institute of Engineering Processes Automation and Integrated Manufacturing Systems, Konarskiego 18A St., 44-100 Gliwice, Poland

Abstract. Industrial robots are widely used in mass production, successfully replacing the workers during repetitive tasks. The lack of direct human participation in the production process involves the necessity of programming of the machines, which is mostly the time-consuming process. Most of the operations are well described in the process documentation, but some actions require the use of complicated paths, like for example welding, laser or water cutting, edge grinding, application of liquid sealants or films of paint etc. All of the mentioned activities need the precise movements of robot's effector and this is - in many cases - connected with the necessity of entering of many points that approximate the path. One of the ways to solve this problem is the possibility of generating the complex path automatically, using the information from the technical documentation. This paper presents such approach that uses the CAD drawings as a data source. The available literature presents this problem mainly in the aspects of large software applications, but this paper was focused on presenting the problems of the proper analysis of the drawing and how to process the data to the form that can be used during robot programming.

1 Introduction

The history of industrial robotics started in the mid of 20th century, together with the first robot, developed by George Devol. At the beginning robots were used for manipulation task only: the first Unimate machines were working in GM factory and were used to stack the hot pieces of metal moved from the die-casting machine. Soon the robots were adapted to the other works, like painting, welding etc.

Performing of handling tasks is relatively simple activity that is carried out by robots. Basically, it is required only to define particular points that should be reached during the movement of the gripper carrying out the part. Of course, an important issue is to ensure collision-free movement of the individual links of the manipulator, while performing the transportation task. If there are only a small number of obstacles in the robot's workspace, the basic knowledge of the rules of motion programming is often enough to do the job properly. These tasks often use the simplified method of trajectory definition called "point-

* Corresponding author: krzysztof.foit@polsl.pl

to-point." In this method, the operator defines the key points that must be reached in exact or approximate way. The movement of the particular links of manipulator's kinematic chain is calculated by inner algorithm of robot's controller and the operator has very limited influence on the shape of resulting trajectory. The "point-to-point" method could use both on-line and the off-line programming. On the other hand, tasks related to painting, welding, cutting, etc. usually require precise definition of the path followed by a moving tool. Such approach is called "continuous point" method of trajectory definition and requires a large number of points to be generated and stored in the controller memory. In order to define the path manually, the demonstration is used: the wrist of the robot is led by the operator along the path, while the controller samples the position of kinematic chain links and store them in memory. The off-line path generation involves the use of special, dedicated software. Because the use of complex path increased during time, the modern robot controllers often offer support for advanced paths, like parametric curve definition, curve fitting and interactive curve design.

The relationship between CAD software and robot programming has not always been clear. In the early years of robotization the tasks conducted by robots used simple trajectories with a small number of points. Another milestone was the use of robots for painting or welding. It soon became apparent that programming of complex trajectory is very time-consuming process, and the quality of the paint coating or weld is largely dependent on operator skills. Thus began the search for efficient methods of designing the complex paths. Market launch of the PC-class computers and the rapid development of software for engineers, made possible to develop the methods for off-line programming of the robot, which also had a beneficial effect of short-term exclusion of the robot from the production process. Over the time, the researchers began to take into account the use of CAD software as one of the tools for preparing a program for the robot.

This paper is focused on presenting the possibility of using CAD software and the documents created by these applications as a source of data used in robot programming process. The research trends and problems are presented. Finally there are presented the selected solutions that are the results of works conducted in the Institute of Engineering Processes Automation and Integrated Manufacturing Systems.

2 The background of the idea

Off-line programming of a robot brings a number of advantages. The most important is the shortest time of machine inactivity during the realization of the production process. Others, less significant from the manufacturing process point of view, are [1]:

- Increased safety – programmer is not performing in the real environment and the developed program can be tested using simulation software,
- More comfortable conditions for programming – the work is moved from the factory floor to the office room.

Of course there are also disadvantages, like for example [1]:

- the necessity to buy a proper software and costs of personnel training,
- differences between robots that result in differences in accuracy and positioning errors,
- the necessity to perform the tests on real machine in the isolated environment.

These disadvantages are the background for the other problem, which is the strict connection between robot and off-line programming software – robots manufacturers often deliver dedicated application for the particular type of robot. Of course, there are some generic off-line programming and simulation systems on the market, like RobCAD or Robotmaster, but they are rather expensive.

It is worth to notice that CAD software has became more popular over the years. Today the electronic documentation replaced the paper documents and the same is in the field of engineering. Neto and Mendes say that « it is very common to see a human being explaining something to another human being with base on a CAD drawing » [1]. The increased interest in CAD techniques has started the search for the relationship between electronic drawings and creation of programs for the machines involved in the production process – the very popular manifestation of such activity is the elaboration of the G-code, which has became a standard for programming of the CNC machines. This is the clear example of how the data from the CAD drawings could be used in the manufacturing process. However, doing the same in the field of robotic is almost impossible, because there is no standard language that could be used to program any robot – almost every manufacturer uses its own programming language. Such fact makes the use of CAD data for robot programming purposes quite hard but not impossible. The need for such research is justified, among other things, by high prevalence of CAD systems that offer ample opportunities to extend the functionality of application by using APIs and scripting programming languages. This topic is also the subject of works conducted by scientist from all around the world.

3 The examples of CAD – robot interfaces implementation

The main purpose of the CAD – Robot Interface is to use the data entered into the CAD program during creation of technical drawing as a source of coordinates used in the robot program. It should be noted that the drawing of the item is not used directly, but as a base to determine the characteristic points of the robot path (data for the trajectory planner software) or to designate the path directly in the CAD program. The most commonly used methods of obtaining coordinates from CAD drawings include:

- processing the drawing data by running scripts within the CAD application,
- passing on the data from the CAD application to the external program by using the CAD application API,
- exporting a drawing to the one of universal formats (the most frequently used is the DXF standard) and processing the resulting file in an external application.

The mentioned methods could be mixed in order to achieve optimal results, depending on the type of a CAD program and its capabilities.

An interesting example of the use of data from CAD system for programming the robot has been proposed by Kim [2], who has described the system for application of the glue during the process of connection of soles to the shoes. The author has elaborated the script for AutoCAD, written in AutoLisp language. The trajectory was generated on the basis of four polylines that describes the points, separately for the XY and YZ plane. The polylines has been extracted from the 2D drawing that consists of three views of the sole. Moreover, the author has also proposed the second solution, where the path was generated on the basis of data acquired from the 3D scan of the shoe. The data from the scanner has been imported to the CATIA software, where the trajectory line has been drawn point by point, using the mouse. After saving the path to the IGES file, the data for the robot program have been extracted using author's "Iges to Point" program. Of course, there is the need to do some additional work, like calibration and scaling, however author has claimed that conducted experiments confirmed the usefulness of the methods.

Another approach has been presented by Neto et al. in the works [1,3-4]. Authors have recounted the process of acquiring the data from the CAD system using scripts and API. As a source application they have used Autodesk Inventor. The software has been supported by external programs, written by authors of the paper. From the inside, Inventor application has been extended by scripts, written in Visual Basic for Application language. On the other hand, the definition of the trajectory was realized by drawing a line (a virtual path), like in the previously mentioned example. Authors also have pointed out the main limitation of their interface that is the collaboration only with the particular type of robot, because of lack of the universal language that could program any robot.

The third example of CAD-based Human-Robot Interface has been developed for a welding robot. Pires et al. [5] have presented the system based on AutoCAD. Again the desired trajectory was represented by line placed on the drawing, however there is a possibility to insert all the necessary parameters (like voltage, velocity, torch distance from plate etc.) directly – they were introduced as labels with the corresponding value. After exporting to the DXF format, the drawing was processed by two applications: the first one for extracting the point data from DXF, while the second one for adjusting the data for performing the simulation of welding.

The presented examples have some common properties. First of all, in all cases the trajectory should be directly mapped in drawing as a segmented polyline or a curve. Moreover, all of the mentioned examples are dedicated to a particular type of robot. Finally, it should be noted that the authors put emphasis on data adjusting process – particularly on accuracy and calibration.

4 Research activities

Research on the use of CAD systems in the process of programming the robots have been also carried out at the Institute of Engineering Processes Automation and Integrated Manufacturing Systems. In parallel, the study related to the milling with the use of robot arm [6] and related to the planning of collision-free path of robot's effector [7-8] have been conducted.

In regard to the use of CAD systems for robot programming, three concepts have been developed. All tests were conducted on a Mitsubishi Movemaster RV-M1 robot.

The Mitsubishi Movemaster RV-M1 robot has a small, five axis manipulator with about 1 KG of lifting capability. The controller has the ability to communicate with the computer using Centronics port for one-way data flow or RS-232 port for two-way data flow. It means that using the RS-232 port, the operator can send the commands to the controller, but – in addition – can also download the data. The controller's memory is divided into the two partition: one is used for storing the points of trajectory, while the second stores the program. The robot is using its own programming language, so any software must be adapted to support it. For this reason, all the software developed during the experiments work only with Movemaster RV-M1 robot so far.

4.1 The "CAD TO RV-M1 DXF" translator for AutoCAD

The first CAD – robot Interface for the Movemaster robot has been developed as a result of the simple experiment. The interface named CAD TO RV-M1 is a simple, external processor of the DXF file. The main idea of the approach was to draw the trajectory in the AutoCAD, using lines for approximating the point-to point trajectory and complement it with the point entities, which represent the state of the gripper. The processor uses a « dictionary » file for translating the DXF file entities into the simple program, which could be realized by the robot. The concept of the dictionary was also intended to allow the adaptation of the processor program for using it with other types of robots. The block diagram of the developed CAD TO RV-M1 processor is shown in Figure 1.

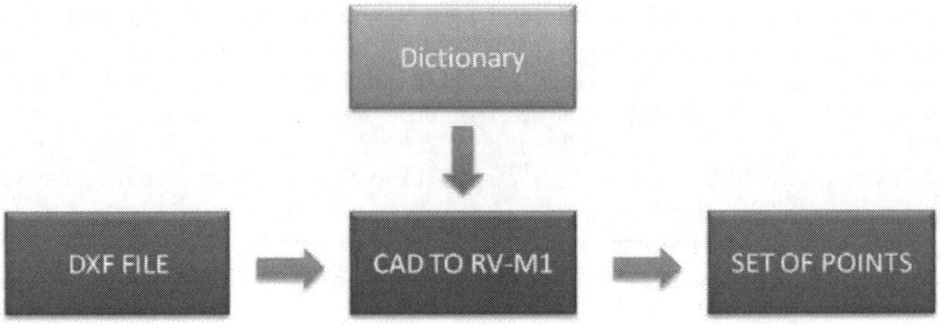

Fig. 1. Block model of CAD TO RV-M1 program data flow

Later, the CAD TO RV-M1 program has became the part of ROBBO package, which is a Mitsubishi Movemaster programming and simulation system, developed mainly for didactic purposes [9].

4.2 Retrieving the trajectory data from the SVG file

The method of retrieving the trajectory data from the SVG file has been created as the alternative to the method that use DXF file, which was described in section 4.1. The SVG acronym stands for Scalable Vector Graphics and it is the standard created mainly for World Wide Web, but it is open, free to use and it continuously evolves. The file contains a drawing coded in the text form, using the XML (Extended Markup Language) standard. In fact, many graphics programs can import and export the SVG format. The XML notation introduces the opportunity to use XML parsers to extract the trajectory data from the drawing. The SVG standard has been also used as the part of Human-Robot Interfaces [10-12].

The idea to use SVG graphics for robot programming has been presented in [13-14]. The standard is supported by many application, including the ones distributed on free licence (GNU GPL) – this definitely lowers the costs of implementation of the method. The drawing coded in the SVG standard could be easily converted to the other format – vector or raster one – using one of widely available converters. The significant limitation of the SVG manifests in its 2D character – the points are defined on the plane – but it is not a problem as long as the effector also moves on the plane. The second limitation is that the SVG cannot be easily converted to the set of points, which could be included in the robot's program. In [14] the author proposed the solution of this problem by doing prior conversion of SVG file to the raster graphics (bitmap) file or HPGL file. The first solution assumes that the plane, where the robot operates, should be divided into quadratic cells that will be mapped to corresponding pixels of raster file.

a) b)

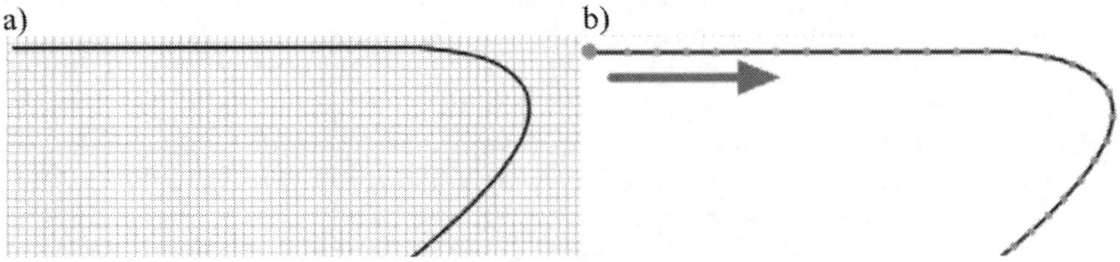

Fig. 2. Extracting path data from SVG file: rasterization (a) and sampling (b) [14]

Using the conversion to HPGL could give better results because the sampling and optimization of the path is done by the application that converts SVG to HPGL. The elaborated methods are in the experimental stage. The both ideas are shown in Figure 2.

4.3 Definition of the path using mixed reality

The method described in this section uses the photo image of the part to help the operator to define the path of robot's effector during the operation carried out on the plane, e.g. applying the glue or sealants, deburring, drilling. The photo is taken by the camera mounted above the plane on which the robot operates. Acquired image is initially processed in order to be clear and properly readable. Then it is imported to the CAD program, where is scaled and orientated according to the robot's coordinate system [15]. In order to define the path, the operator inserts the drawing entities to the CAD program, using the upper layer, above the photo of the part. The Figure 3 shows the example of the path defined with the help of mixed reality[15]. Further details connected with obtaining of the coordinates for the robot are described in [15]. The presented method has a little in common with the one presented in [2], but it is using the photo instead of 3D scan and operates in 2D space. The main advantage of the presented method is that it uses only the free and open source software.

Fig. 3. Example path of effector. In the right-bottom corner of plate the calibration cross is visible

4.4 Current research activities

Each of three methods, developed in the Institute of Engineering Processes Automation and Integrated Manufacturing Systems have the common aim, which is to help the user to design a complicated path that will be recreated by the robot's effector. The previous researches have led to the development of method that uses the mixed reality along with the CAD system in order to generate the data for the robot's controller. The use of free and open source software has significantly reduced the costs and has made the method readily available for the target user. This is the significant factor, which encourages to do the further study of this topic using such sort of the software.

The important topic of the current research is the planning of trajectory in the 3D virtual environment. This issue is only partially solved because the current state of knowledge allows to design the 3D trajectory in the CAD application and export it to the DXF file, so the coordinates could be interpreted by translator. On the other hand, this task could be simplified, when the 3D model of the object could be used in order to adjust the path along the object's edges or surfaces. So far, the 3D model of the object should be drawn in CAD

 DOI: 10.1051/matecconf/20179405002

application or scanned using 3D scanner. Taking into account that the 3D scanner is not always available, it is advisable to search for other effective and low-cost solutions. It is considered to use the mixed reality in the process of three-dimensional trajectory planning. The aim of the current research is also the evaluation of the usefulness of different methods of converting the series of object's photos to the 3D model.

As the third path of research, the problems described in the next section and connected with the robot-based machining will be taken into consideration. The measurements that will be done during machining may lead to valuable conclusions, which may be useful during the process of optimization of the generated trajectory.

5 Open problems

Because the CAD – Robot Interfaces are often used in experimental stages, there are still some open problems.

One of the problem is the stability of movement during the following of the path by the effector. This problem is connected with the number of points that are generated by the application when the arc, circle or curve should be mapped. The points are spread over the path in unequal distances – less densely on the straight sections and more densely on curves. This leads to the changes of effector's speed, which may be critical parameter during applying sealants, glues or operation of cutting. Moreover, it manifests in the changes of tool's load during milling or deburring. If the points are located too densely, the effector may vibrate if it should achieve the points accurately, because there is no enough time to accelerate and decelerate the whole kinematic chain.

The second issue is connected with the previously described one and it is the result of « zero-tolerance », CAD-generated path that must be mapped by the tool, while the different robots of the same type may have different accuracy and also the parts may differ one from another because of manufacturing tolerances. In some cases this may lead to the overloading of the tool or even to its destruction. These types of problems could be solved by using the force-sensing technique described in [16], but it requires the special system that will introduce the correction during the realization of the robot's program.

6 Conclusions

The CAD-Robot Interfaces are still the subject of interest of many scientists. As a justification for conducting such work, a large popularity of the CAD systems is indicated. On the other hand, the lack of standardized programming language for robots makes hard to elaborate the methods that will have the real application in the industry. Without developing the standards, similar to the one used in e.g. PLC programming, there will be no possibility to engage a wider range of potential recipients in the research process.

The solutions presented in the paper are still in development phase, but the scientists had shown the potential possibilities of reducing the gap between the process of designing and the production. There are still open problems that come from the differences between the accurate world of the CAD documentation and the real limitations, like construction differences between robots or tolerances in manufacturing. Beside taking into account these issues, further research should focus on the decreasing of the operator's participation in the process of aquisition of the coordinates from the CAD documentation.

References

1. P. Neto and N. Mendes, Rob. Auton. Syst. **61**, 896 (2013)

2. J. Y. Kim, J. Robot. Syst. **21**, 625 (2004).
3. P. Neto, N. Mendes, R. Araújo, J. Norberto Pires, and A. Paulo Moreira, Ind. Robot An Int. J. **39**, 294 (2012)
4. P. Neto, J. N. Pires, and A. P. Moreira, in *2010 IEEE Conf. Robot. Autom. Mechatronics* (IEEE, 2010), pp. 516
5. J. N. Pires, T. Godinho, and P. Ferreira, Ind. Robot An Int. J. **31**, 71 (2004)
6. G. Gołda and A. Kampa, Przegląd Mech. **3**, 35 (2015).
7. D. Reclik and G. G. Kost, J. Autom. Mob. Robot. Intell. Syst. **2**, 38 (2008).
8. K. Foit, G. G. Kost, and D. Reclik, J. Achiev. Mater. Manuf. Eng. **47**, 57 (2011).
9. J. Świder, K. Foit, G. Wszołek, and D. Mastrowski, J. Achiev. Mater. Manuf. Eng. **25**, 7 (2007).
10. K.-W. Jeon, J. Ki, Y.-M. Kwon, and H. Ko, in *Front. Converg. Biosci. Inf. Technol. 2007. FBIT 2007* (IEEE, 2007), pp. 622
11. S. Jo, J. Ki, K.-W. Jeon, Y.-M. Kwon, and S. C. Ahn, in *RO-MAN 2007-The 16th IEEE Int. Symp. Robot Hum. Interact. Commun.* (IEEE, 2007), pp. 493
12. C. Reinicke, M. Buss, in *The Internet Challenge: Technology and Applications*, (Springer Netherlands, 2002), pp. 55
13. K. Foit, Adv. Mater. Res. **837**, 577 (2014)
14. K. Foit, Sel. Eng. Probl. **5**, 27 (2014)
15. K. Foit, Adv. Mater. Res. **1036**, 737 (2014).
16. K. M. Murphy, R. J. Norcross, and F. M. Proctor, in *Proc. Second Int. Symp. Robot. Manuf. Res. Educ. Appl. Albuquerque, NM* (Citeseer, 1988).

 DOI: 10.1051/matecconf/20179405003

Particularities of fully-parallel manipulators in 6-DOFs robots design: a review of critical aspects

Lucian Milica[1,*], and *Gabriel* Andrei[1]

[1]"Dunărea de Jos" University, Faculty of Engineering, 800201, Galati, Romania

Abstract. A whole range of industrial applications requires the presence of parallel mechanisms with six degrees of freedom (6-DOF) which have been developed in the last fifteen years, and one of the reasons why they still are a current topic is that present-day computers are capable of performing real-time motion laws of great complexity associated with these types of parallel mechanisms. The present work underlines particularities of parallel manipulators and their importance in the design of 6-DOF robots. The paper reveals the progress made in the last twenty years in the development of 6-DOF parallel manipulators, which increasingly find a wide scope of applications in different industrial areas such as robotics, manufacture and assisted medicine. It also emphasizes the need to determine singular configurations and the effect of cinematic redundancy which can increase the working space of the manipulators by adding active joints in one or more branches of the manipulator. Throughout the work, there were outlined three types of singularities encountered in the modelling of different types of parallel manipulators, and three types of redundancy. Furthermore, an analysis was made of the dimension of the workspace for a series of parallel manipulators, highlighting a number of factors that influence its size.

1 Introduction

In industry, the need to perform manifold operations under accurate handling and indefinite repetition has led to the emergence of automated devices with numerical control, which took over the monotonous work of human operators. The increase in flexibility of industrial processes through frequent and fast modification of the operating plan carried out by these devices has required their rescheduling. Thus, the first industrial robots appeared in the very sense of the term. With the emergence and development of computing, the numerical control of these devices was achieved by easier to-use and low-cost computer means.

It is well known that parallel manipulators are much stiffer and more accurate than series manipulators [1]. Thus, in recent years, in more technical applications parallel manipulators have been introduced, such as flight simulators, radar antennas, telescopes, pointing devices [2]. In addition, parallel manipulators have also been introduced in the

* Corresponding author: milica.lucian@ugal.ro

processing centers with multiple degrees of freedom and high speed. The development of parallel manipulators with 6-DOFs took a greater momentum especially in the last fifteen years, starting as a basis for the research work carried out by a number of researchers in the development of parallel mechanisms with three degrees of freedom. Some researchers [3, 4] identified as such three types of planar parallel manipulators (Fig.1), consisting of three kinematic chains which link the mobile platform with the fixed one. These manipulators are distinguished among themselves through different types of kinematic joints which make the three kinematic chains mentioned above. Thus, it can be distinguished parallel manipulators type 3-RPR (revolute joint-prismatic joint-revolute joint), 3-RRR (three revolute joints) and 3-PRR (a prismatic joint and two revolute joints).

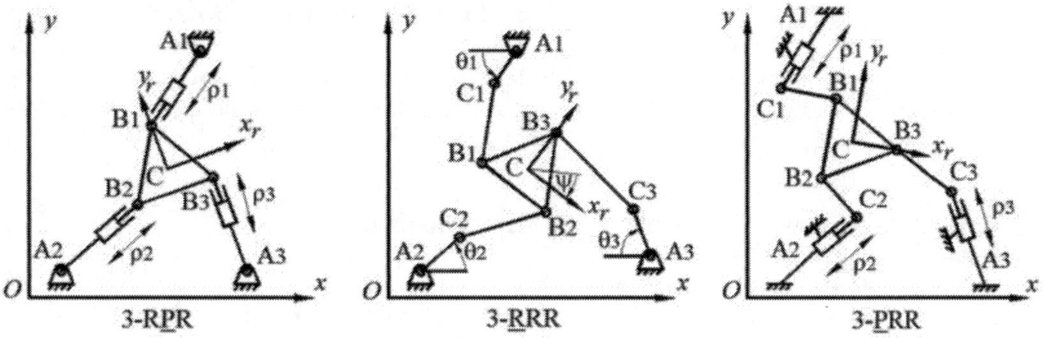

Fig. 1. Different types of planar parallel manipulators [3, 4]

Jaime Gallardo et al. [5] studied spherical parallel manipulators, making reference to the families of spherical parallel manipulators with two arms. The main characteristic of this family is a compact asymmetric topology consisting of two arms and a spherical joint. The topology of the spherical manipulators is best expressed through the definition, advanced by the International Federation for Theory of Machines and Mechanisms (IFToMM), of this type of parallel manipulator: "manipulator parallel, which controls movement of the end effector via at least two linkages, starting from the end effector to the base ". Jaime Gallardo et al. described a transformation of a Stewart platform, into a spherical parallel manipulator (SPM), with a simplified structure. The position of the mobile platform is controlled by three arms with a common spherical joint attached to the mobile platform, while its orientation is controlled by the remaining arms.

A spherical parallel manipulator is a parallel manipulator whose mobile platform is an element with fixed point. Therefore all points attached to the mobile platform are moving on concentric spheres (Di Gregorio [6]). Several SPM with different architectures have been analyzed in the literature. However, due to two factors, to simplify the manufacturing process and the belief that a symmetric architecture is absolutely necessary for the proper performance of a manipulator, most of these contributions propose parallel manipulators with identical arms. On the other hand, since the robotic systems must be both flexible and capable of operating under different conditions, the natural question arises whether that "symmetry" is indeed essential in the design process. A typical SPM is a so-called "agile eye" which has followed the pioneering work of researchers Gosselin and Angeles [7] and contributed significantly to the further development of a SPM. In their work, Zhang et al. [8], using spherical passive joints along with four connectors, were able to determine, due to redundancy, the only possible orientation of the moving platform. In his work, Wohlhart [9], analyses the movement of a Gough-Stewart platform, where the displacement of the mobile platform is defined by a translation and a rotation movement.

In order to generate a class of spherical joints, Innocenti and Parenti-Castelli [10] have removed three translation joints, preserving the spherical passive joint, thus connecting the mobile platform with the fixed platform. However, such an architecture requires a strategic positioning of the passive spherical joint. For example, if the center of the spherical passive joint is placed in the geometric center of the triangle formed by spherical joints remaining attached to the mobile platform, the mobility mechanism can be badly affected and it can be easily transformed into a structure. Parallel manipulators type RPS (revolute joint-prismatic joint-spherical joint) represent another class of parallel manipulators, which have been dealt with by a number of researchers in different aspects related to the analysis of dynamic and kinematic behavior [11-13]. A particular approach to kinematic and dynamic problems of the manipulators type RPS is that of using the screw theory which proved to be an effective method for determining kinematic characteristics, including instant movements of the mechanism [14-19]. Also, one should not miss the fact that the parallel manipulators present some disadvantages such as: a limited workspace, different kinematic characteristics for each axis of the joints, the presence of singularities inside the workspace. For these reasons, many researchers sought to explore new approaches that would lead to an increase of the workspace, maximizing mechanism isotropy and avoiding internal singularities [20-23].

2 Advantages of using parallel mechanisms

Many applications in precision engineering (lithography, electron beam microscopy, microscopic assembly, aerospace industry, medical applications) require high-precision positioning in order to manipulate an object in different environments (vacuum or moisture). In the medical field, precision surgical instruments are required to avoid tissue damage during operation [24]. All the applications aforementioned refer to different environments, with certain characteristics. That is why, for example in wet environments, it is difficult to use conventional bearings for lubrication reasons. To overcome these problems, a number of mechanisms have been developed to satisfy both the requirements of accuracy and reliability. In most of the mentioned areas, it is necessary to manipulate an object within a space with six dimensions, which implies that the manipulator should have six degrees of freedom.

In general, this type of parallel robots is made up of two significant parts connected to each other through several elements. One is arbitrarily designated as fixed and is called the base plate, while the other one is connected via multiple kinematic chains at the base and is called motion platform or mobile platform. Robot elements are connected to each other through the spherical, cylindrical, revolute or prismatic joints. Usually, a parallel mechanism is symmetrical if it meets the following conditions: the number of branches is equal to the number of degrees of freedom corresponding to the mobile platform. An actuating device controls every branch, and the location and number of active joints of the arms are the same [25].

Parallel robots are generally defined as closed kinematic chain mechanisms comprising a terminal body with n degrees of freedom (mobile platform) linked to a fixed platform through n independent kinematic chains. According to some authors, some restrictions are necessary: each kinematic chain consists of maximum two articulated segments; the operation is performed by n simple driving elements, one for each kinematic chain excluding redundant parallel manipulators where the number of actuators is higher than the spatiality degree of the mobile platform movement. For over two decades, parallel robots have attracted the attention of many researchers who consider them as being particularly valuable in terms of construction alternatives for robotic mechanisms [26, 27]. Various types of architectures of these mechanisms [28] have recently been studied and many are regularly used in the industrial world, as is the case of various types of machine tools [29] and industrial robots [30]. Parallel robots can be encountered especially in technical applications where a high speed motion of a rigid body in space is desired. Positioning and precision in executing the tasks are essential, because these robots are generally intended for handling of fragile objects; any error in the positioning of the tool could result in costly damages.

Parallel robots can be driven through actuators based on electric, hydraulic or pneumatic effect. Regardless of the type of actuation - electric, pneumatic or hydraulic - the presence of transducers is compulsory because it is absolutely essential to know accurately the relative position of the kinematic chain elements. The reason why parallel mechanisms are of great topicality is that present-day computers are able to solve complex calculations in real-time as required by the algorithms related to the parallel mechanisms.

Parallel mechanisms have a robust construction and can move bodies of considerable mass and size at high speeds, establishing a translational motion or rotation of the mobile platform. Compared to serial robots, parallel robots have some special features:

- high rigidity;
- running;
- high dynamic loading capacity;
- chain closed structure ensuring the acquisition of passive and active tasks much more efficiently;
- high positional accuracy.

Thanks to these features, parallel manipulators have been introduced within the machining centers with numerical control. It has been demonstrated by some researchers that a full analysis of the performance of parallel manipulators requires the determination of static and dynamic stiffness. In this direction, by analyzing the performance of a tripod-type parallel manipulator, Zhou and Xi [31] have carried out a finite element analysis to determine the vibration occurring in a technological process while Hong et al. [32] have investigated the stability of a machine tool whose table is positioned through a parallel manipulator.

3 Kinematic and dynamic analysis

The class of robots known as the Stewart platform has been implemented in the development of flight simulators [33-35]. Parallel Delta robot prototype [36, 37], the Tsai mechanism [38], as well as parallel robot star-type [39], are equipped with three actuators to move the mobile platform, giving it its three degrees of freedom.

Mobility is the main parameter of a structural mechanism and also one of the fundamental concepts in modeling dynamic and cinematic mechanisms [40]. IFToMM defines mobility as "the necessary number of independent coordinate system to define the configuration of a kinematic chain or mechanism" [41]. Mobility is, in fact, that which verifies the existence of a mechanism to impose the condition $M > 0$.

In calculating the mobility mechanism the following parameters are generally used:

- number of joints p;
- number of kinematic elements n, $(n = m\text{-}1)$;
- total number of elements m, including fixed base;
- mobility M;
- degrees of freedom of kinematic joints f_i.

Gogu has advanced a new formula for the calculation of mobility parallel mechanisms, with the number of arms $t \geq 2$ [42].

$$M= \sum_{i=1}^{p} f_i - \sum_{i=1}^{t} S_i + S_p \tag{1}$$

where f_i represents degrees of mobility of kinematic pairs i $(i = 1, ..., p)$, S_j represents the degree of spatiality of j arms considered isolated from the rest of the mechanism, and S_p represents the degree of spatiality of the mobile platform. The degree of spatiality of the mobile platform depends on its links with the fixed base and it is given by the maximum number of independent motion parameters between the two. Mobility indicates the number of independent parameters both from the kinematic model and dynamic model; on the basis of these parameters it may be determined the number of inputs required by the actuation mechanism.

Gosselin and Angeles [43] have developed kinematic and dynamic straight model Agile Wrist, which is characterized by three concurrent rotations.

Kinematic analysis is carried out in two ways: via direct kinematic method and by inverse kinematic method. The transformation by which one can determine the instantaneous movement of mobile platform, as a function of relative velocities of the actuated joints, represents direct kinematic method. The transformation has the expression:

$$\dot{\mathbf{q}} \rightarrow \dot{\mathbf{x}} \tag{2}$$

where vector $\dot{\mathbf{q}} = [q_1 \ldots q_m]^T$ is the vector of generalized coordinates (parameters of its components being q_j joints), and the vector $\mathbf{x} = [\alpha\ \beta\ \gamma\ x\ y\ z]^T$ describes the position of the mobile platform, where x, y, z are the coordinates of the characteristic point in relation to the fixed reference system, and α, β, γ are the angles of orientation in relation to the same fixed reference system.

Inverse kinematic method consists in determining the movements of the actuated joints, knowing the movement imposed on the mobile platform. The transformation in this case has the expression:

$$\dot{\mathbf{x}} \rightarrow \dot{\mathbf{q}} \tag{3}$$

Dynamic analysis of parallel robots is usually approached by the analytical method of classical mechanics. Dasgupta and Mruthyunjaya [44] used Newton-Euler method to develop dynamic equations of Stewart platform, taking into account both the gravity and dynamic effect and friction that occur in joints.

Tsai [45] has also presented an algorithm to solve the inverse dynamics of a Stewart platform, using the same Newton-Euler method.

It is known that this approach leads to a set of equations, but they are very difficult to use in order to obtain the laws of advanced control. At the same time, quite few results were obtained by specific dynamic modeling of the parallel mechanisms.

Kane and Levinson [46] have obtained some recursive relations with respect to the balance of the generalized forces which apply to an arm of the serial robot. Kane's dynamic equations are described by means of a Stanford robot. Sorli et al. [47] have also provided a dynamic modeling for a Turin parallel robot, which has three identical legs and six degrees of freedom.

In some simplifying assumptions on the geometry and distribution of robot inertia, Geng et al. [48] and Tsai and Stamper [49] have developed a system of Lagrange equations of motion.

Determining the mechanical behavior of the manipulator according to the mechanism configuration calls for an optimal design of the operation and control system. Lagrange equations are systems of second-degree differential equations that determine the dynamics of parallel manipulator. Starting from the basic equations:

$$\frac{d}{dt}\left(\frac{\partial L}{\partial \dot{\mathbf{x}}}\right) - \frac{\partial L}{\partial \mathbf{x}} + \frac{\partial \Phi^T}{\partial \mathbf{x}}\lambda = \mathbf{f} \tag{4}$$

Nguyen, Bouzgarrou et al. [50] determined all structural parameters of the robot (mutual kinematic constraints, size of components). Also, the results thus obtained were compared with data provided by ANSYS finite element analysis software.

Analytical calculation involving Lagrange method is still too long and the risk of errors is high [51]. It turns out that the time for numerical calculation increases with the number of elements. Therefore, this calculation method is impossible to be done in real-time, being practically unusable.

4 Singular configurations

In fact, most existing parallel robots have limited workspace and its calculation is quite complicated because of singularities (those configurations in which instantaneous complete movement of mechanism cannot be determined). Many of the recent works focuses on avoiding critical positions that can handle the mechanism where the workspace is reduced and the problem of finding an algorithm to determine a route that avoids singular configurations becomes complex [52-54].

Whenever is wanted to design a parallel manipulator, it is very important to understand the intrinsic nature of singularities and their relationships with the robot kinematics and configurations [55]. Currently, there are many different points of view on the singularity phenomena encountered in the kinematics of the parallel mechanisms. The first study on the positions of singularity encountered in parallel mechanisms date back to 1970 when Hunt [56], in his study, defined two types of singular configurations: stationary configurations and uncertain configurations. Merlet [57] proposed another approach for the determination of singular configurations through the geometry of Grassmann, one of the founders of the vector geometry and the multidimensional geometry, who assumed the existence of a space of n dimensions. Later, Gosselin and Angeles [58] defined three types of singularities in terms of the rank of matrix. These authors brought important contributions on this issue. In general, singular configurations for parallel manipulators are achieved when the determinants of different Jacobian matrixes are zero. It has been illustrated and classified different types of singular configurations. Thus, two types of outstanding singularities were highlighted: the inverse kinematical singularity and direct kinematical singularity which are used to study and analyze the parallel manipulators [59]. Another approach to the problem of singular positions was highlighted by Zlatanov, Bonev et al. [60], using the screw theory while Wolf and Shoham [61] have investigated these issues through the linear complex approximation.

Singularities are important aspects to be considered when talking about parallel robots. To determine these critical positions is to identify them and avoid them by ensuring the kinematic stability and accuracy of the system. Many robot manipulators have six degrees of freedom and are able in nonsingular positions to generate a space with six dimensions for an infinitesimal motion to end effector /mobile platform [62]. In a singular position the axes of the joints generate a movement space for five of the six axes so that even an infinitesimal motion of translation or rotation is not possible. This single position can be determined by placing the condition that the determinant of the Jacobian matrix (6×6) be equal to 0. Another way to describe singularities is through geometric line providing a qualitative approach to them [63, 64]. Singular positions correspond to the positions in which the Plücker coordinates robot of the axes of the joints form a dependent linear system, and these situations can be characterized in a pure geometric way [65]. In this context, a prismatic joints axis, for example, is defined as an "infinite line". The algorithm used for instance, by Stoica et al. [66] for the analysis of singular configurations is based on the determinants of the two Jacobian matrixes, **A** and **B**, derived from the geometric model and the inverse geometric model. Using the matrix representation, there is the following expression:

$$\frac{\partial f}{\partial \mathbf{x}}\dot{\mathbf{x}} - \frac{\partial f}{\partial \mathbf{q}}\dot{\mathbf{q}} = 0 \Rightarrow \mathbf{A} \cdot \dot{\mathbf{x}} + \mathbf{B} \cdot \dot{\mathbf{q}} = 0 \qquad (5)$$

where:

- **q** is generalized coordinates vector with its components forming the robot space;
- **x** is the position vector of a point of the mobile platform and angles ϕ, θ and ψ give its orientation;

- **f** is an *n*- dimensional vector, including geometric constraints of each kinematic chain *i*, for *i* = 1, 2..., *n*.

Thus, some authors have identified three types of singularities [67, 68]:

- singularity of type I, is the first type of singularity which occurs when the determinant of the Jacobian matrix **B** is zero. If the matrix determinant **B** is zero, the manipulator loses one or more degrees of freedom ;
- a second type of singularity (singularity of type II) occurs when the determinant of the Jacobian matrix **A** equals 0. In this case the robot wins one or more degrees of freedom.
- singularity of type III occurs when determinants require both matrixes to be zero. These types of singularities are encountered in the literature as the architectural singularities.

In these particular positions, the degree of spatiality of the effector motion decreases, and therefore some of the input parameters have no effect on its movement. The problem of critical position is the subject of special studies on the manipulator mechanisms. Thus Ben-Horin and Shoham [69] performed an analysis of singularities, for a series of parallel robots through Grassmann-Cayley algebra. Tong Z. et al. [70] highlight the appearance of singularities for a category of symmetric Gough-Stewart manipulators.

These studies determine the set of critical positions which sometimes form "surfaces" on the workspace, thereby delimiting the size of this space.

5 Redundancy of parallel mechanisms

A new development direction for 6-DOFs parallel manipulators is the micro-manipulators. Since 1989, Hara [71] proposed a 6-SPS micro-manipulator parallel which uses a piezoelectric ceramic actuator. In 1990, Taniguchi [72] developed a new parallel micro-manipulator type 6-PSS. In his research Yun and Li [73] refer to the features of a 6-DOFs parallel manipulator with piezoelectric actuators and dual redundancy (Fig. 2).

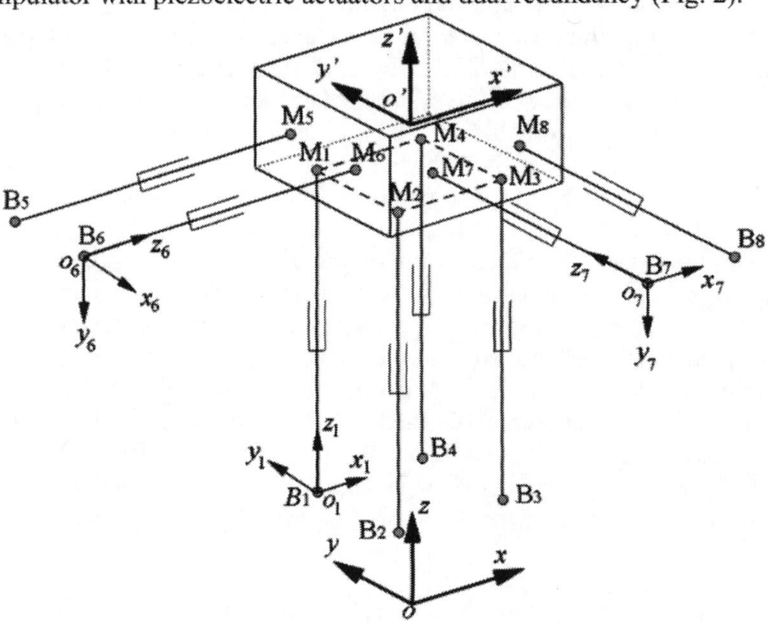

Fig. 2. Parallel manipulator with dual redundancy, type 8-PSS [73]

The work of Yun and Li offers suggestions for improving the structure of the parallel redundant mechanism and gives solutions for the optimization of a control algorithm, in order to achieve a possible workspace, more accurate motion and dynamically superior features. The existence of redundant movements enables the execution of a desired movement in several ways, which creates the possibility of optimizing the operation of the robot. The degree of redundancy is the consequence of the particular relative positions of axes couplings [74]. Moreover, according to recent research, the redundancy is a way to avoid singular configurations [75].

The redundancy of parallel manipulators can be of three types: actuators redundancy, arms redundancy and kinematic redundancy [76, 77]. Actuators redundancy consist in replacing passive joints with active joints [78, 80] in which case both the number of degrees of freedom and the manipulator's mobility remain unchanged. The redundancy of arms can improve the strength capabilities of the manipulator and can reduce the number of singular configurations. The kinematic redundancy can increase manipulator workspace by adding one or more active joints to the manipulator arms [81, 82]. In addition it allows kinematic redundancy manipulator to follow trajectories that are far from singular configurations.

The degree of redundancy depends clearly on the degree of mobility and the degree of spatiality of the analyzed mechanism; it represents the number of mobility remaining to the mechanism after immobilizing the effector element. The existence of redundant movements enables the execution of a desired movement in several ways, which further creates the possibility to optimize the operation of the robot.

Another example illustrating the influence of the degree of redundancy on parallel mechanisms with six degrees of freedom is the manipulator type 6(P) UPS which is a mechanism derived from Stewart-Gough mechanism, [83] and manipulator type 6(P) RUS [84, 85].

Position and orientation of the mobile platform are given by the **x** vector that has the expression:

$$\mathbf{x} = [\, x_E \; y_E \; z_E \; \alpha_E \; \beta_E \; \gamma_E]^T \tag{6}$$

where $\mathbf{x}_E$, $\mathbf{y}_E$ and $\mathbf{z}_E$ represent projections of the position vector $\mathbf{r}_E$ belonging to the $(CF)_E$ system on the $(CF)_O$ system and α_E, β_E, γ_E give the orientation of $(CF)_E$ system around the x, y, z axes of the $(CF)_O$ system. $(CF)_E$ and $(CF)_O$, are the notations used for the reference system of the mobile platform respectively for the reference system of the frame. The input data are entered by the vector $\mathbf{q}_1$ for the manipulator type 6(P) UPS [83] and $\mathbf{q}_2$ vector for the manipulator type 6(P) RUS [84, 85]:

$$\mathbf{q}_1 = [\rho_1 \; \rho_2 \; \rho_3 \; \rho_4 \; \rho_5 \; \rho_6 \; \delta]^T \tag{7}$$

and ρ_j is the actuating i elements length,

$$\mathbf{q}_2 = [\theta_1 \; \theta_2 \; \theta_3 \; \theta_4 \; \theta_5 \; \theta_6 \; \delta]^T \tag{8}$$

where θ_j represents the angular displacement of l_i arms.

In case of both manipulators, the position of G_1 is given by the reduntant actuator's δ displacement (G_1 is a point that belongs to the redundant joint).

6 Workspace volume

For the evaluation of a parallel manipulator performance, workspace represents an important criterion. Typically, the degree of difficulty for setting workspace grows proportionally with the increase in the number of degrees of freedom of the parallel manipulator. In the present review, the workspace volume has been considered and also, a comparative study for a series of parallel manipulators with six degrees of freedom was done. Fully working space for parallel manipulators with six degrees of freedom is a space with six dimensions, whose representation is difficult to express for human perception [86]. Therefore there are typically determined the subspaces of this space. These subspaces are three-dimensional spaces that can be defined within the spatial Cartesian coordinate system. The workspace of the robot can be briefly defined as the volume generated by a point of the effector element when the mechanism takes all possible configurations. It can be classified into two categories: reachable workspace and orientation workspace [87].

The reachable workspace is determined by the points reached by the end effector when the actuating branches go through all the possible combinations of their rotation domain. The orientation workspace is determined by the possible orientations of the end effector. It is known that an orientation of any system $\{R_1\}$, relative to the other system $\{R_2\}$ is expressed using Euler angles. The final equation of the rotation matrix related to Euler angles is obtained by three successive elementary rotations applied to $\{R_1\}$ system, by angles ϕ, θ and respectively ψ [88]. Euler angles allow the representation of the orientation workspace for most parallel manipulators as volumes with simple form. Using cylindrical coordinate representation for the orientation workspace, Merlet [89], obtains a graph of the volume of this space as a function of angle variation ψ (Fig.3), highlighting the fact that for different values assigned to the angles ϕ and θ orientation workspace volume increases as the angle value ψ approaching 0°.

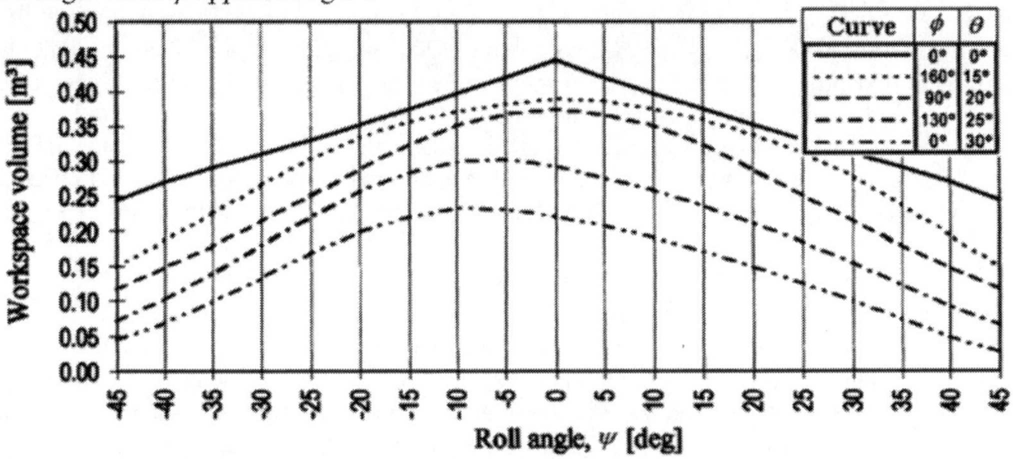

Fig. 3. The volume of the orientation workspace depending on the angle ψ [89]

Bonev and Ryu [90] propose a computational algorithm, implemented with MATLAB software application, in order to determine the influence of the characteristic position of point *C* (belonging to mobile platform), in relation to the axis of symmetry of a parallel manipulator, over the orientation workspace described. Yi Lu et al. [91] analyzed a new kinematic model of Stewart type manipulator, determining the influence of angular rotation of spherical joints. They highlight the following differences with respect to the workspace volume, compared to Stewart platform (Tab.1):

Table 1. Workspace volume, compared to Stewart platform [91].

P.M. type	e [mm]	E [mm]	θ [deg]	r_{ij} [mm]	V_w [mm³]
Yi Lu	100	250	-	500-700	71.905
Stewart	100	250	±30	500-700	20.793
Stewart	100	250	±35	500-700	33.585
Stewart	100	250	±40	500-700	48.745
Stewart	100	250	±45	500-700	58.947
Stewart	100	250	±50	500-700	77.163

In the Table 1, e and E are the distances from B_i to o and B_i to O respectively. O and o are the center point of B, m (B represent the frame end m represent mobile platform), θ is the limited rotational angle of the S joint. In the case of Stewart manipulator, it can be observed, that with increased rotation angle of the spherical joint θ within ± (30...50)°, the volume of the workspace increases as well. On the other hand, pairs of kinematic elements which make the link between mobile platform and fixed base plate have dimensions imposed by the mechanism configuration. Spherical joints that theoretically can achieve 360° rotation around an axis, actually they are restricted in terms of the angular movement and for this reason it is necessary to determine the maximum value of the rotation angle θ, for each kinematic joint, in the design stage [92].

Lim et al. [92] examine the workspace volume of a parallel manipulator with suspended mobile platform (Tab.2).

Table 2. Workspace volume of P.M. by Lim [93]

P.M. type	L [mm]	μ [deg]	γ [deg]	r [mm]	V_w [mm3]
Lim	800	100	82	300	39.334

Coppola G. et al. [94], in their study on the reconfigurable parameters of a parallel mechanism with six degrees of freedom, highlight its volume workspace (Tab.3):

Table 3. Workspace volume of P.M. by Coppola [94]

P.M. type	b [mm]	r [mm]	α [mm]	l_{min} [mm]	l_{max} [mm]	α_{min} [deg]	α_{max} [deg]	V_w [mm^3]
Coppola	85	125	32,5	150	300	45	145	26.200

In the Table 2, $l_{min}$and l_{max} are minimum and maximum values of the actuating branch length, *a, b,* and *r* are the geometrical values which correspond to the lengths of their respective vectors and α_{min} and α_{max} are the minimum and maximum values assigned to the angular movement of the joint.

It should also be pointed out that several works from the literature describe various ways to use CAD software to investigate the working space of parallel manipulators [95-97]. On the basis of such methods, an approximate measurement of the inclination angle of the mobile platform could be made.

Another problem of parallel mechanisms, in general, is the motion limitation imposed on their kinematic elements in order to avoid any collision between them. Basically, these limitations determine a restricting workspace and the analysis of positions that may cause such collisions is a prerequisite for the proper functioning of the manipulator [98].

7 Conclusions

The present work reviews the main aspects of determining the workspace and avoiding singular configuration of 6-DOFs parallel manipulators, highlighting the importance of existing redundant movements within these types of manipulators. The paper also revealed the progress made in the last twenty years in the development of of 6-DOFs parallel manipulators which increasingly find a wide scope of applications in different industrial areas such as robotics and assisted medicine.

Throughout the work, there were outlined three types of singularities encountered in the modelling of different types of parallel manipulators, and three types of redundancy. Also, an analysis was made of the dimension of the workspace for a series of parallel manipulators, highlighting a number of factors that influence its dimensions:

• the design parameters *(r, l, a, b, e)*;
• the value of rotation angle θ of spherical joint, in the case of Stewart manipulators;

- position of characteristic point C (belonging to mobile platform), in relation to the symmetry axis of parallel manipulator;
- variation of the angles, ϕ, θ and ψ ;
- the motion limitation imposed on cinematic elements, to avoid collision between them.

Knowing the characteristics of parallel mechanisms with six degrees of freedom, a new direction for future research would be the design of a novel type of parallel manipulator, with six degrees of freedom, able to perform complex milling process; moreover, once designed such type of robot, the optimum parameters necessary to achieve a larger workspace have to be inferred.

References

1. J. P. Merlet, *Parallel robots. 2nd ed.*, (2006)
2. D. Zhang, *Parallel robotic machine tools,* (2010)
3. J. P. Merlet, Tehnical Report 1645, INRIA, (1992)
4. C. Gosselin, J. Angeles, J. Mech. Transm.-T. ASME, **110**, 35-41 (1988).
5. J. Gallardo, R. Rodrıguez, M. Caudillo, J. M. Rico, Mech. Mach. Theory, **43**, 201–216 (2008)
6. R. Di Gregorio, J. Mech. Transm.-T. ASME., **126**, 850–855 (2004)
7. C. M. Gosselin, J. Angeles, J. Mech. Transm.-T ASME., **111**, 202–207 (1989)
8. Y. Zhang, C. Crane, J. Duffy, J. Robotic Syst., **15**, 299–308 (1998)
9. K. Wohlhart, Mech. Mach. Theory, **29**, 581–589 (1994)
10. C. Innocenti, V. Parenti-Castelli, Mech. Mach. Theory, **28**, 553–561 (1993)
11. K. M. Lee, D.K. Sha, *Proceedings IEEE/ICRA.*, **1**, 345–350 (1987)
12. H. S. Kim, L.W. Tsai, J. Mech. Transm.-T. ASME., **125**, 92–97 (2003)
13. C. H. Liu, S. Cheng, J. Mech. Des. ASME., **126**,1006–1016 (2004)
14. Z. Huang, Y. Fang, Mech. Mach. Theory, **31**, 1009–1018 (1996)
15. Y. Fang, Z. Huang, Mech. Mach. Theory, **32**, 789–796 (1997)
16. Z. Huang, J. Wang, Mech. Mach. Theory, **36**, 893–911 (2001)
17. S. K. Agrawal, *Proceedings of 8th World Congress on TMM,* 405-408 (1990)
18. Z. Huang, J. Wang, *Proceedings of A Symposium Commemorating of Sir Robert Stawell Ball*, (2000)
19. Z. Huang, J. Wang, Y. Fang, Mech. Mach. Theory, **37**, 229–240 (2002)
20. J. Merlet, C. Gosselin, N. Mouly, Mech. Mach. Theory, **33**, 7-20 (1998)
21. F. Bulca, J. Angeles, P.J. Zsombor-Murray, Mech. Mach. Theory, **34**, 497-512 (1999)
22. A. Kosinska, M. Galicki, K. Kedzior, J. Robotic Syst., **20**, 539-548 (2003)
23. M. Arsenault, R. Boudreau, J. Robotic Syst., **21**, 259-274 (2004)
24. A. Dunning, G. Tolou, Int. J. Mech. Sc., **2**, 157–168 (2011)
25. L.-W. Tsai, *Robot Analysis:The Mechanics of Serial and Parallel Manipulator,* (1999)
26. D. Chablat, P. Wenger, IEEE Trans. Robot. Autom., **19**, 403–410 (2003)
27. X. J. Liu, J. Wang, F. Gao, L.P. Wang, IEEE Trans. Robot. Autom., **17**, 959–968 (2001)
28. C. Reboulet, R. Pigeyre, *Proceedings of the ISRAM.*, 293–298 (1990)
29. M. Valenti, ASME. Mech. Eng., **17**, 70–75 (1995)
30. K. Cleary, T. Brooks, *Proceedings of the IEEE/ICRA.*, 708–713 (1993)
31. Z. Zhou, J. Xi, J. Mech. Des. **128**, 403–412 (2006)
32. D. Hong et al, Mech. Struct. Mach., **31**, 509–528 (2003)
33. J. Angeles, Springer, (1997)
34. V. Parenti-Castelli, R. Di Gregorio,.J. Mech. Des., **122**, 294–298 (2000)
35. D. Stewart. *Proceedings of Int.Mech.Eng.*, **15**, 371–386 (1965)
36. R. Clavel, *Proceedings of the ISR.,* (1988)

37. S. Staicu, D.C. Carp-Ciocardia, *Proceedings of the IEEE/ICRA.*, 416–412 (2003)
38. L.W. Tsai, R. Stamper, ASME. Des. Eng. Tech. Conf. (1996)
39. J. M. Hervé, F. Sparacino, *Proceedings of the ARK.*, (1992)
40. G. Gogu, *Report ROBEA MAX CNRS*, (2003)
41. T. G. Ionescu, Mech. Mach. Theory, **38** (2003)
42. G. Gogu, Mech. Mach. Theory, **40,** 1068–1097 (2005)
43. C. Gosselin, J. Angeles, J. Mech. Transm.-T. ASME., **111**, 202–207 (1989)
44. B. Dasgupta, T.S. Mruthyunjaya, Mech. Mach. Theory, **34**, 1135–1152 (1998)
45. L.W. Tsai, *Robot Analysis: The Mechanics of Serial and Parallel Manipulator*, (1999)
46. T.R. Kane, D.A. Levinson, *Dynamics, Theory and Applications,* (1985)
47. M. Sorli, C. Ferarresi, M. Kolarski, B. Borovac, M. Vucobratovic, Mech. Mach. Theory, **32**, 51–77 (1997)
48. Z. Geng, L.S. Haynes, J.D. Lee, R.L. Carroll, Robot. Autonom. Syst., **9**, 237–254 (1992)
49. L.W. Tsai, R. Stamper, ASME. Des. Eng. Tech. Conf., (1996)
50. A.V. Nguyen, B.C. Bouzgarrou, K. Charlet, A. Béakou, Mech. Mach. Theory, **93**, 65–82 (2015)
51. S. Staicu, X.J. Liu, J. Li, Springer, **58**, 217–235 (2009)
52. B. Dasgupta, T.S. Mruthyunjaya, Mech. Mach. Theory, **33**, 711–725 (1998)
53. C.L. Lin, H.Y. Jan, J.R. Lin, T.S. Hwang, Eur. J. Control, **3**, 201–212 (2008)
54. S. Parsa, R. Boudreau, J.A. Carretero, Mech. Mach. Theory, **85**, 53–63 (2015)
55. P.C. Lee, J.J. Lee, Mech. Eng., **7**, 163–187 (2012)
56. K.H. Hunt, Cambridge University Press (1978)
57. J. P. Merlet, Int. J. Robot. Res., **8**, 45–56 (1989)
58. C. Gosselin, J. Angeles, *Proceedings of IEEE T. Robotic. Autom.*, **6**, 281–290 (1990)
59. J. Craig, *Introduction to Robotics, Mechanics and Control, Third Edition*, (2005)
60. D. Zlatanov, I.A. Bonev, C. Gosselin, *Proceedings of the IEEE/ICRA,* 496–502 (2002)
61. A. Wolf, M. Shoham, J. Mech. Transm.-T. ASME., **125**, 564–572 (2003)
62. R. Penne, E. Smet, P. Klosiewicz, J. Intell. Robot. Syst., **62**, 205–216 (2011)
63. S. Staicu. Robotica, **27**, 199–207 (2009)
64. F. Hao, J.M. McCarthy, J. Robot. Syst. **15**, 43–55 (1998)
65. A. Dandurand, *The rigidity of compound spatial grids.*, **10**, 41–56 (1984)
66. A. Stoica, D. Pisla, S. Andras, B. Gherman, B.Z. Gyurka, N. Plitea, Mech. Eng., **8** ,70–79 (2013)
67. C. Gosselin, J. Angeles, IEEE T. Robotic. Autom., **6**, 281–290 (1990)
68. O. Ma, J. Angeles, *Proceedings of the IEEE ICRA.*, 1542–1547 (1991)
69. P. Ben-Horin, M. Shoham, Mech. Mach. Theory, **41**, 958–970 (2006)
70. Z. Tong, J.F He, H.Z. Jiang et al, Robotica, **30**, 305–314 (2012)
71. A. Hara, K. Sugimoto, J. Mech. Transm.-T. ASME.,**111**, 34–39 (1989)
72. M. Taniguchi, M. Ikrda, A. Inagaki, Int. J. Jpn. Soc. Precis. Eng., **26**, 35–40 (1992)
73. Y. Yun, Y. Li, Springer, **61**, 829–845 (2010)
74. J. Wang, C. Gosselin, J. Mech. Transm.-T. ASME., **126**, 109–118 (2004)
75. I. Ebrahimi, J.A. Carretero, R. Boudreau, Mech. Mach. Theory, **42**, 1007–1016 (2007)
76. S. Lee, S. Kim, *Proceedings of the IEEE/CDC.*, **2**, 1097–1102 (1993)
77. R. Boudreau, S. Nokleby, Mech. Mach. Theory, **56**, 138–155 (2012)
78. K.E. Zanganeh, J. Angeles, *Proceedings of IEEE/ICRA.*, 3043–3048 (1994)
79. H. Cheng, G.F. Liu, Y.K. Yiu, Z.H. Xiong, Z. Li, *Proceedings of the IEEE/IROS*, 171–176 (2011)
80. J. P. Merlet, *Redundant parallel manipulators*, **8**, 17–24 (1996)
81. J. Wang, C. Gosselin, J. Mech. Des., **126**, 109–118 (2004)
82. J. Kotlarski, B. Heimann, T. Ortmaier, Mech. Eng., **7**, 120–134 (2012)

83. P. Last, C. Budde, J. Hesselbach,. *Proceedings of the 2005 IEEE/CASE.*, 393–398 (2005)
84. Y. Jin, I.M. Chen, G.L. Yang, Mech. Mach. Theory, **44**, 912–922 (2009)
85. E. Ottaviano, M. Ceccarelli, *Optimal design of CAPAMAN with prescribed workspace*, 35–44, (2001)
86. D.I. Kim, W.K. Chung, Y. Youm, IEEE/ICRA., 2986-2991 (1997)
87. I.A. Bonev, J. Ryu, Mech. Mach. Theory, **36**, 15-28 (2001)
88. A. Năstase, *Mecanica Roboților*, **2**, 24-30 (2012)
89. J.P. Merlet, Mech. Mach. Theory, **29**, 1099-1113 (1994)
90. I.A. Bonev, J. Ryu, Mech. Mach. Theory, **36**, 1-13 (2001)
91. Y. Lu, X. Li, C. Zhang, Y. Liu, Robotica, 1-16 (2014)
92. H. Lim, S.H. Lee, B.R. So, B.J. Yi, Int. J. Cont. Autom. and Syst., **13**, 942-950 (2015)
93. M. Majid, Z.A. Huang, Y.L. Yao, Int. J. Adv. Manuf. Tech., **16**, 441–449 (2000)
94. G. Coppola, D. Zhanga, K. Li, Robot. Cim.-Int. Manuf., **30**, 99–106 (2014)
95. K. Arrouk, B. Bouzgarrou, G. Gogu, Appl. Mech. Mater., **162**, 131-140 (2012)
96. K. Arrouk, B. Bouzgarrou, G. Gogu, RAAD., (2010)
97. K. Arrouk, B. Bouzgarrou, G. Gogu, Springer, **5**, 605-612, (2010)
98. http://www.orbitmotionsystems.com., Operation Manual PacDrive Robot D2, Edition 2008-08

An analysis of the impact of soil mass pH on the wear process of steel

Jerzy Napiórkowski[1,*], *Magdalena* Lemecha[1] , and *Mariusz* Rosiak[2]

[1]University of Warmia and Mazury in Olsztyn, Faculty of Technical Sciences, Exploitation and Diagnostics Vehicles and Machines. St. M. Oczapowskiego 11, 10-079 Olsztyn, Poland
[2]Opole University of Technology, Faculty of Mechanical Engineering
Department of Manufacturing Engineering and Production Automation, ul. S. Mikołajczyka 5, 45-271 Opole, Poland

Abstract. The chemical properties of soil mass are among the most important factors affecting the intensity of the wear of working elements. The paper analyses the wear process of 38GSA steel, used in working elements which cut soil, depending on the corrosiveness of the chemical environment. The tests were conducted in natural soil conditions. Changes in the mass of samples depending on soil pH were measured during the tests. Based on the obtained results, it has been concluded that along with an increase in soil acidity, the steel wear intensity also increases. The highest value of wear was observed in acidic soil, followed by neutral, with the lowest value corresponding to alkaline soil.

1 Introduction

Intense wear, being a process of qualitative and quantitative physicochemical changes occurring on the friction surface, is a phenomenon accompanying the operation of working elements in soil mass [1, 2].

The process of the rational selection of the construction and technological forms of working elements, and of planning the course of operation, is possible when their wear process is known. The identification and description of wear in soil comes down to a comprehensive description of the process of friction involving: the soil - the working element - the parameters of work [3]. The soil is a complex, animate creation of nature, in which continuous processes of decomposition and synthesis occur for both mineral and organic compounds, along with their displacement [4].

Due to the multiplicity of factors affecting the course of wear for working elements in soil mass, so far there has been no formalised theory of mutual interactions between the soil and the working element. Most papers indicate that the wearing properties of soil result from its grain sizes, cohesiveness and humidity [5, 6].

Hydrogen-bearing properties belong to the phenomena involving the destruction of metals as a result of environmental impact [7]. Hydrogen-related wear has been described,

*Corresponding author : napj@uwm.edu.pl

among other things, for the surfaces of rotors and timing gears of pumps, in braking systems and rolling bearings.

The aim of the paper is the analysis of the impact of soil pH on steel destruction processes in abrasive soil mass.

2 The characteristics of soil pH

One of the most important chemical features of soil mass is its pH. Soil pH is a condition of soil solution determined by the ratio of hydrogen (H+) to hydroxide (OH-) ions, affected by the presence of acids, alkalis and salts [8]. A solution may have one of three types of pH, namely: acidic, alkaline and neutral.

Based on electrical conductance, it has been concluded that at a temperature of 22°C the product of the concentrations of hydrogen and hydroxide ions is constant and equals 10-14. This means that in one litre of distilled water there are 10-7 gram equivalents of hydrogen ions and just as many hydroxide ions. Soil pH is therefore dependent on the concentration of [H+] and [OH-] ions.

Neutral solutions: [H+] = [OH-] = 10-7 [H+],
Acidic solutions: [H+] » [OH-] = 100÷10-7 [H+],
Alkaline solutions [H+] « [OH-] = 10-7÷10-14 [H+].

In order to simplify the mathematical calculations, Sorensen introduced the pH designation - a negative logarithm of the concentration of hydrogen ions in a solution, pH = -log[H+].

The acidity of soil is caused by hydrogen ions [H+] contained in a soil solution. The acidity derived from the [H+] ions contained in a soil solution is called the active acidity. The acidity derived from adsorbed [H+] and [Al3+] ions is known as the potential acidity. Two types of negative charges have been observed in soil colloids:

a. constant charges,

b. charges dependent on the concentration of hydrogen ions.

The former are present mainly in clay minerals. They are generated by electrostatic forces originating from isomorphous replacements in the crystals of those minerals. The anions bound by those constant charges are replaceable at any values. The second type charges are not constant and depend directly on pH. They originate from several sources. The first of them are H2SiO3 and Al(OH)3 on the free surfaces of crystals and the outer surfaces of silicate minerals. The second are the carboxylic (-COOH) and phenyl groups of topsoil colloids. Each of these groups contains covalently bound hydrogen which does not dissociate at low pH values. However, as pH increases, H+ ions begin dissociating, leaving negative charges on the colloids. In this manner, hydrogen is replaced by other cations, which themselves are also replaceable [9].

Another feature of soil humidity, defined as the amount of water in the medium. During the cultivation of soils with low humidity, which are characterized by a high compactness, it consumes flank plowshares. This reduces the width of the trapezoidal and the length of the bow section of working elements. In the studied conditions, soil moisture was 8.5% ± 2.3%. With the saturation of the soil with water followed by reducing the amount of wear, but the degree of these changes it depends on the type of soil. All songs soil change their properties with the change of water content.

Another condition affecting the process of abrasive wear in abrasive soil bulk density is, which is the ratio of the weight of the soil to the volume they occupy. During the research the density was 1.54 g*cm^3.

Compactness of soil, which was 2.53 ± 0.87MPa is defined as the degree of binding of the particles.

Shear stress in the layer of soil within a range of 10-20 cm were 45kPa.

3 The methodology of the research

The tests involved samples 30 x 20 x 10 mm in size, made of 38GSA steel. This is martensitic steel which contains bainite and troostite. The chemical composition was as follows: C - 0.38%, Mn - 1.07%, Si - 1.17%, P - 0.028%, S - 0.02%, Cr - 0.18%, Cu - 0.16%, Al - 0.022%. The hardness of the steel ranged between 409 and 442 HV10. Some of the samples had their wear evaluated by placing them in a soil of diversified pH, while another part was subjected to the examinations of wear in a laboratory device (Figure 1). The soil was collected from a pine forest from a depth of approximately 20 cm. Its pH amounted to 4.2. It was measured by means of a FieldScout pH meter from Spectrum Technologies, Inc. In accordance with the quality standard PN-EN ISO 14668-2(2004), the soil was defined as loamy sand, and it consisted of: sand: 85.8%, silt: 11.2% and clay: 3%. Soil humidity was evaluated by means of the measurement of the mass of the solid phase dried at a temperature of 105°C, and it ranged between 10 and 12%. Prior to the test, the samples were subjected to washing in an ultrasonic cleaner with a washing agent based on potassium hydroxide. The value of the mass wear of the samples was determined by means of a laboratory weighing scale with an accuracy of +/- 10-4 g. The samples were placed in vessels containing 2 kg of soil and stored for 21 days. The soil was deacidified by means of burnt lime (CaO) in various quantities in order to obtain the varying values of pH. The tests of wear were conducted in a wear testing machine utilising the "spinning bowl" method. Friction tests were conducted at a load of 53 kPa, a cutting speed of 2 ms-1, and a friction path of 20 000 m. The assessment of the surface after wear was performed upon completing the whole friction path under a JSM 5310LV type JEOL scanning electron microscope in a digital system. The microscope was equipped with a detector of reflected electrons, a detector of secondary electrons, and an EDS type Ultra DR7 ThermoScentific X-ray spectrometer.

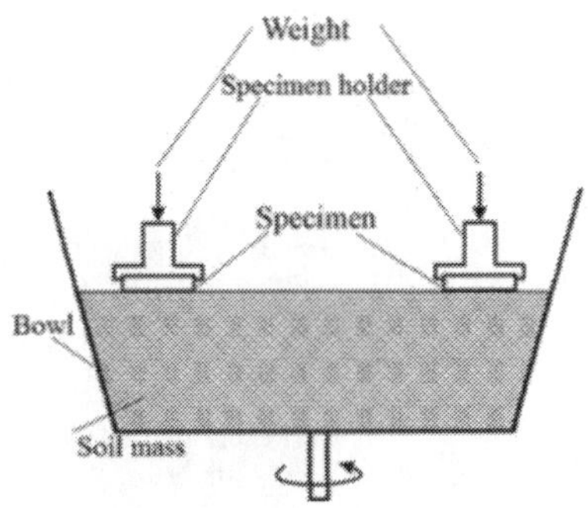

Fig. 1. A view of the wear testing machine.

4 Analysis of the test results

Figure 2 presents the relation between the loss in weight of the samples and soil pH. The elements stored in a soil having a considerable concentration of hydrogen ions (pH) are characterised by an increased loss in weight compared to the steel placed in more neutral and alkaline environments.

DOI: 10.1051/matecconf/20179405004

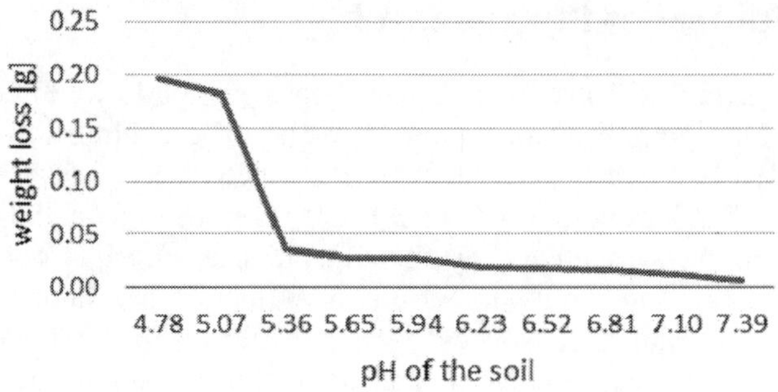

Fig. 2. The dependence of the samples' loss in weight on soil pH.

The obtained dependence is confirmed by the test results involving wear during friction in soil, presented in Figure 3.

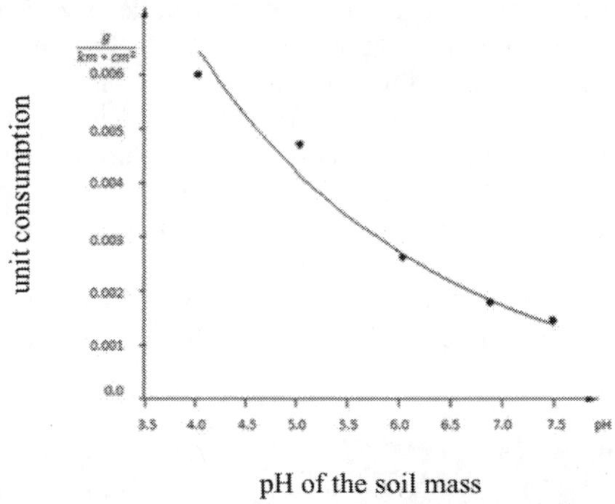

Fig. 3. The course of the intensity of wear depending on soil pH.

Fig. 4. A view of the surface after friction tests in the very acidic soil.

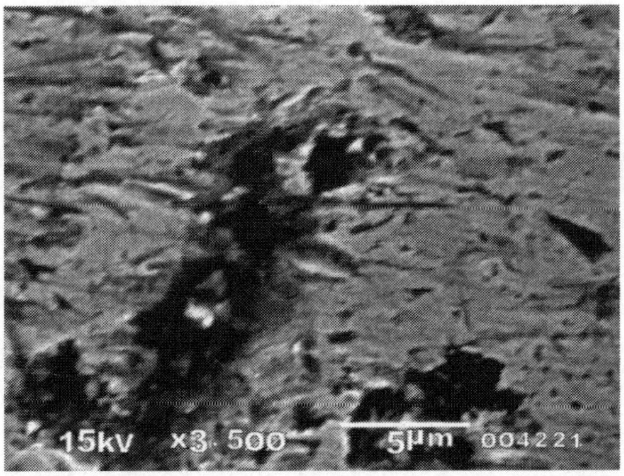

Fig. 5. A view of the surface after friction tests in the neutral soil.

An important impact of soil acidity on the course of wear was observed based on the obtained results. No statistical differences between the intensity of wear in neutral and alkaline soils were observed. The wear of 38GSA steel in the very acidic soil was over 3 times, and in the acidic soil over 2 times more intense than in the neutral soil.

The surface of wear in the very acidic soil (Figure 4, Figure 5) is characterised by the mutual interactions of chemical (the creation of oxides, deep pitting) and mechanical wear (micro-cutting and furrowing). Mechanical wear is predominant in the case of soils with neutral pH. Locally present pitting can indicate the presence, to some slight extent, of accompanying chemical wear.

The phenomena observed during wear in very acidic and acidic soils can be categorised as hydrogen-related wear. Hydrogen originates from the surrounding environment, including the increased acidity of soils, and from technological hydrogen contained in metal.

The increased saturation with hydrogen for these types of soils causes its intense adsorption on the surface of a working element, followed by the diffusion of hydrogen into the surface layer. When approaching the surface of metal, the particles of hydrogen in close proximity to it (less than 0.5 nm away) are subject to sufficient interaction with this surface, so that the phenomenon of adsorption can take place. The dispersion forces cause physical adsorption, resulting in the creation of a layer of non-localised moving particles on the surface of a solid, creating a kind of a two-dimensional liquid. The particle which will break the barrier of the activation energy may be bound to it by the forces of a specific character, size and nature, typical for covalent or ionic chemical bonds. It is a chemical reaction known as chemical adsorption or chemisorption. In the process of friction there are considerable concentrations of hydrogen present locally in the crystalline grid, along with the processes related to lowering the plasticity and decohesion of the fragments of surface layers, causing lasting mechanical damage, and as a final result intensifying the wear. This is directly related to the "promoters" of penetration.

The process of wear observed in acidic soils corresponds to classic hydrogen-related cracking wear. During the wear, hydrogen is accumulated in the places of the discontinuous structure of the surface layers, such as the boundaries of grains and phases, as well as dislocations. It remains there in a lasting manner due to the high bond energy of the "hydrogen atom – trap" complex. Cyclically repeating mechanical impulses move the subsequent portions of hydrogen atoms to the places of their storage, until reaching local concentrations which trigger nucleation and the creation of 1 – 2 μm deep local cracks in the surface layer.

5 Conclusion

The elevated hydrogen content precipitates in a reaction of the components of steel in an acidic environment. Hydrogen-related wear considerably affects the wear of soil handling elements, although this phenomenon is not predominant. It should only be considered as wear which intensifies the abrasive wear, which is the primary type of wear present in soil handling elements.

The mechanisms of hydrogen-related wear are described by the theories of the pressure of molecular and adsorptive hydrogen and the maximum triaxial stresses. In the case of wear in soil mass, the observed phenomena are explained by the theory of absorption. This states that the diffusing hydrogen is adsorbed on the tips of a fissure, reducing the surface energy of metal atoms, and being subject to tensile stresses it causes the creation and propagation of a fissure, resulting in the local displacements of material. The intensity of the diffusion of hydrogen into the material causes brittle cracking in the micro-volumes of the surface layer and its destruction by the forces of friction.

References

1. Napiórkowski J. (ed.), *Research and modeling of abrasive wear and fatigue*, (University of Warmia and Mazury in Olsztyn, 2014)
2. Bayhan, Y., Tribology International **39**, 570–574 (2006)
3. Napiórkowski, J., Drożyner, P., Kołakowski, K,; Mikolajczak, P., International Virtual Journal for Science, Technics and Innovations for the Industry **6 (11)**, 27–29 (2012)
4. Bhakat A. K., Mishra A. K., Mishra N. S., Wear **263**, 228-233 (2007)
5. Stachowiak, G. W.; Batchelor, A. W. *Engineering tribology* (3rd ed. Amsterdam, Boston: Elsevier Butterworth-Heinemann, 2005)
6. Gościański, M., Łabęcki, M., Kapcińska, D., Pirowski, Z., Journal of research and applications in agricultural engineering **52(2),** 43–51 (2007)
7. Stabryła J., Starczewski L., The impact of water-bearing soil components wear agricultural tools, Maintenance Problems 1-2006, 199-206
8. Uggla H., *Agricultural soil science* (PWN Warsaw, 1981)
9. Fred A., Soil Science Society of America Journal **35** No. 3, 420-426 (1971)

The tool profile and the localization of the revolution axis for irregular outline milling

Eugen Pămîntaş[1,*]

[1]Politehnica University of Timişoara, Materials and Manufacturing Engineering Department, 300222, Timişoara, Romania

Abstract. For the most cases of manufacturing cutting processes in today's flexible manufacturing systems, the cutting tools which are needed are almost standard. The problem of choosing the appropriate tool shape and dimension is often made by the tool management system according with the shape of mechanical part which must be manufactured. But, the accuracy of profiled surfaces, such as the helical channel of tools depends essentially on three aspects. The first one is the correct determination of geometric profile. The second is the correct execution of that profile. Finally, the third issue is related the positioning of the tool relative to the axis of the blank. The accurate determination of the active profile of the profiled disk milling is an essential constraint, but not the only one to obtain the irregular outline shapes by milling. The settings up parameters are those which supply the closeness of the part spiral shape. The paper presents an iterative analytical calculation procedure for the points from the profiled shape of the disk type of milling cutter. Also, the paper shows a set of analytical expressions for the calculus of the disk type of milling cutter localization parameters, under the condition of rigidity and strength requirement insurance.

1 Introduction

For the most cases of manufacturing cutting processes in today's flexible manufacturing systems, the cutting tools which are needed are almost standard.

The problem of choosing the appropriate tool shape and dimension is often made by the tool management system according with the shape of mechanical part which must be made out the type of workmaterial.

Much more, the tool management system, which is an intermediate link between manufacturing process, manufacturing programme planning, tools and tools storage, can graphically describe through its data base the type, dimensions, shape of tools to be used in NC production, and also can help the tool adjuster and ensures a very fast and simple design of tool and pre-setting sheets [1].

The tool components (cutting tool inserts, shafts, drifts, supporting shims, tool holders and clamping hooks etc) are defined and kept in the tool base by means of the graphic software. By the other modules the shapes of the individual tool components are designed

* Corresponding author: eugen.pamintas@upt.ro

and drawn, their dimensions are defined and the tool components are assembled into a necessary tool for particular machining operation. The prepared drawing is attached to the tool and pre-setting sheet are automatically completed with the other technologically data.

Not so simple is the problem of designing the milling tools for the mechanical parts with complex special shapes, when the determination of the tool's profile is a matter of great importance. In Romanian and foreign literature lists numerous ways of solving this problem, especially for specific cases of tools and usually with reference to the facilities offered by various software systems for programming CNC machine tools [2-4].

How the profiled shape of the disk type of milling cutter can be find in an analytically way is shown in the following.

2 Calculation methodologies

In the conceiving process of the mechanical part, different kinematical laws and rules must be taken into considerations. As a result of the analytically calculation, the characteristic shape points obtained are usually situated on different analytical unknown curves [5, 6].

Much more, even the analytic equation of the profile line can be established, in the practical achievement of the tool, it is necessary for the execution workshop to approximate the point by point profile by other easy to manufacture curve like areas of circles and straight line segments. Using these approximations a reduced number of profile points are taken into account with high accuracy if the errors of approximation are minimized by mathematical methods carried out using the computer power calculation.

2.1 The equation of helicoidally shape of the part

That is why, most of complex profiles can be considered that are generated by the movements of a curved line (C_M) (see figure 1) along a curve (C_S) with a known mathematical equation [7].

For example, to simplify the calculus method described, if it consider (C_M) a segment of circle (in cross section of the cylindrical part) and (C_S) a helicoidally curve, the parameter equations of the part cross profile, in the mobile reference system $OM_xM_yM_z$ are the following (r_{wp} – the radius of workpiece; r_{cM} – the radius of circle (C_M)):

$$\begin{aligned} M_x &= r_{cM} \cdot \sin\theta \\ M_y &= r_{wp} - r_{cM} \cdot \cos\theta \\ M_z &= 0, \end{aligned} \tag{1}$$

where t take values between $-\pi/2$ and $\pi/2$ along OM_z or OS_x axis.

The part helicoidally shape will be obtained by an axial movement of the curve (C_M), at the same time and in correlation with a rotation by φ degree around OM_z or OS_x axis.

To transfer the equations (1) from the mobile system to the stationary system $OS_1S_2S_3$ pendent of part, the following matrix are used (a rotation with the angle φ around OS_z and a translation with $P_H/2$ along OS_z axis in negative sense):

$$\begin{vmatrix} S_x \\ S_y \\ S_z \end{vmatrix} = \begin{vmatrix} \cos\varphi & \sin\varphi & 0 & 0 \\ -\sin\varphi & \cos\varphi & 0 & 0 \\ 0 & 0 & 1 & -\frac{p_H}{2\pi} \end{vmatrix} \cdot \begin{vmatrix} M_x \\ M_y \\ M_z \end{vmatrix} \tag{2}$$

 DOI: 10.1051/matecconf/20179405005

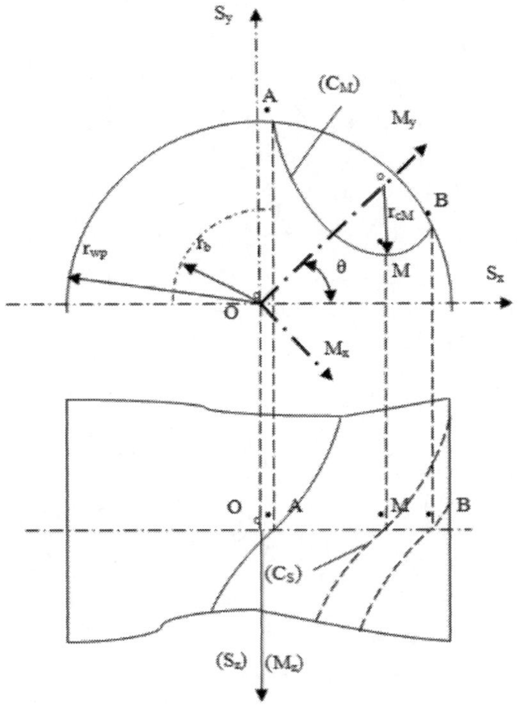

Fig. 1. Calculus scheme of the tool's profile.

The result of counting matrix (2) is:

$$S_x = r_{wp} \cdot \sin\varphi + r_{Cc} \cdot \sin(\theta - \varphi)$$
$$S_y = r_{wp} \cdot \cos\varphi - r_{cM} \cdot \cos(\theta - \varphi) \qquad (3)$$
$$S_z = -p_H \cdot \varphi / 2.$$

2.2 The equation of tool's profile

These are the coordinates of point M situated on the surface profile AB in the stationary coordinate system $S_xS_yS_z$, linked to the workpiece. To establish the tool's profile which will manufacture the part profile, the coordinates (3) must be transferred in an orthogonal coordinate system linked to the tool.

Supporting that the part will be manufactured with a disk-milling tool with $2R_{exT}$ diameter, then the position of tool coordinate system $OT_1T_2T_3$ is so as seen in figure 2. The system $OT_1T_2T_3$ is rotated with angle δ around OS_y axis identically with OT_2 axis.

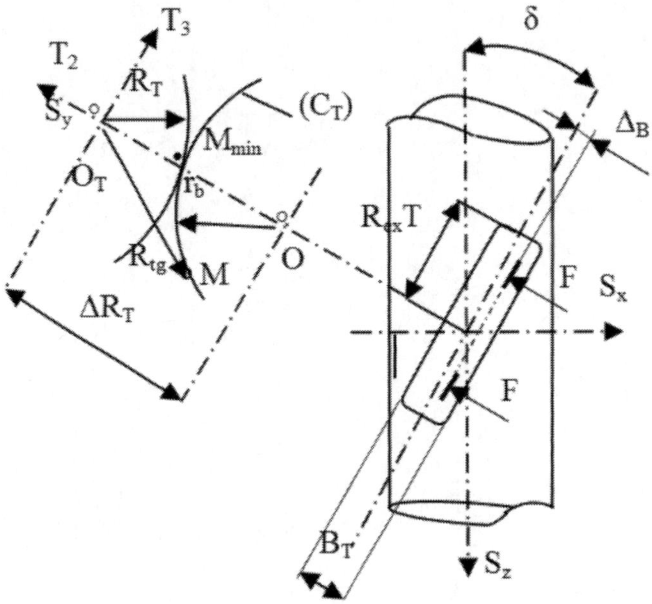

Fig. 2. The position of tool's coordinate system according with part stationary system

The origin O_T of tool's coordinate system is translated along OS_y with ΔR_T value (equation (4), where r_b is the minimum radius of profile on workpiece):

$$\Delta R_T = r_b + R_{exT} \tag{4}$$

To transfer the coordinates of point M from stationary system OS_xSySz in the tool's system of coordinates $OT_1T_2T_3$ will use the following matrix:

$$\begin{vmatrix} T_1 \\ T_2 \\ T_3 \end{vmatrix} = \begin{vmatrix} \cos\delta & 0 & -\sin\delta & 0 \\ 0 & 1 & 0 & -\Delta R_T \\ -\sin\delta & 0 & \cos\delta & 0 \end{vmatrix} \cdot \begin{vmatrix} S_x \\ S_y \\ S_z \end{vmatrix} \tag{5}$$

The results of the counting matrix (5) are:

$$\begin{aligned} T_1 &= r_{wp} \cdot \sin\varphi \cdot \cos\delta + r_{cM} \cdot \sin(\theta - \varphi) \cdot \cos\delta + p_H \cdot \varphi \cdot \sin\delta / 2\pi, \\ T_2 &= r_{wp} \cdot \cos\varphi - r_{cM} \cdot \cos(\theta - \varphi) - \Delta_{B,} \\ T_3 &= r_{wp} \cdot \sin\varphi \cdot \sin\delta - r_{cM} \cdot \sin(\theta - \varphi) \cdot \sin\delta - p_H \cdot \varphi \cdot \cos\delta / 2\pi, \end{aligned} \tag{6}$$

To determine the profile curve (C_T) of the tool we must know the value of angle θ. In this case the curve must be sectioned perpendicular on tool axis O_TT_1. The equation of sectional plane is $T_1 = \Delta_B$, where Δ_B range between – $B_T/2$ and + $B_T/2$, and its expression is the following:

$$\Delta_B = r_{wp} \cdot \sin\varphi \cdot \cos\delta + r_{cM} \cdot \sin(\theta - \varphi) \cdot \cos\delta + p_H \cdot \varphi \cdot \sin\delta / 2\pi \tag{7}$$

In this manner, the values of angle θ can be established for a desired number of profile points:

$$\theta = \varphi + \arcsin \frac{k - r_{wp} \cdot \sin\varphi \cdot \cos\delta - p_H \cdot \varphi \cdot \sin\delta / 2\pi}{r_{cM} \cdot \cos\delta}, \tag{8}$$

and then all the coordinates of M point of the tool can be determinate.

The radius of tool for different values of angle φ (not only positives but also negatives) is given by the following equation:

$$R_T = \left(\sqrt{T_2^2 + T_3^2}\right)_{\min}, \tag{9}$$

In line with the manner fully described already, knowing the axial value ΔR_T and the tool radius R_T, the profile of disk-milling tool is established to the full.

In order to carry into effect the foresaid methodology of establishing the disk-tool profile it must consider step by step the following stages:

- Determining the part profile curve and approximate it by arc segments and straight lines segments;
- Defining the curve (C_S);
- Deciding upon the characteristics of disk-milling:
  - the maximum external radius R_{exT} – by constructive and stress considerations,
  - the breadth of disk-tool B_T – at least as the part profile breadth;
- Calculation of angle θ for a few dozen points (for example it can do values from –BT/2 to BT/2 to with a step by 0,1 mm;
- Calculation of the coordinates of tool profile points T_1, T_2, T_3;
- Calculation of tool profile radius R_T giving values for angle φ as was shown before;
- Drafting the tool profile, checking the accuracy and making adjustments;
- Rough modelling of the tool and simultaneous final modelling whose result is a complete solid model of the tool. Both activities can progress simultaneously in on-line exchange of data in case of using CAD/CAM programs. Another result of final modelling of the tool is the engineering drawing and individual details such as rounding, transitions etc.

2.3 Additional results

The calculation equations of the disk-tool profile established above can be used also in other cases of complex surface part processing. Let afford some proofs on this account:

- for a profiled blanket – $r_{wp} = 0$ and the curve (C_S) is a straight line inclined or not (φ = 0);
- for a cylindrical part with a non-helicoidally profile – (C_S) is a straight line along OS_z axis => φ = 0;
- for a part which are manufactured with a ball-end milling cutter, only angles φ and θ exists in equations (6), that is why these will change into the following:

$$\begin{aligned} T_1 &= r_{wp} \cdot \sin\varphi \cdot + r_{cM} \cdot \sin(\theta - \varphi) \\ T_2 &= r_{wp} \cdot \cos\varphi - r_{cM} \cdot \cos(\theta - \varphi) - r_b \\ T_3 &= -p_H \cdot \varphi / 2\pi, \end{aligned} \quad ...(10)$$

The other curve (C_T) which must be tangent to the tool is obtained giving values to the angle φ. This fact is happening when R_T become minimum. In this case, the radius R_T is

DOI: 10.1051/matecconf/20179405005

identically with tool's radius. To minimize the value of radius R_T, an iterative process must be made by the aide of computer. In this manner the tool profile is known throughout through: the values of profile height T_2 and the profile radius R_{Tmin}.

3 The disk type milling-tool's theoretical installation parameters

In the general case, the relative of the tool as against with the formation surface is measured by six parameters, namely the linear travel parameters l_X, l_Y, l_Z and the angular parameters G, H_z and L which characterize the spatial rotation of the coordinate system $X_f Y_f Z_f$ against with the coordinate system XYZ. In the particular case of the helicoidally surfaces processing with disk-milling tools in case that the formation is made through self generation, the parameters number for the tool installation decrease at three.

As a general rule, the origin of the coordinate system $X_f Y_f Z_f$ of the tool, this being the case in fig. 3, is displaced along the reference system axes of the piece XYZ, with the values l_X, l_Y and l_Z (the index corresponds to the axis along which the origin is displaced).

The coordinate system rotation XYZ round about the axis Z_f (round about the active surface axis) with angle H_z leads at the motion which realized the surface self generation. Therefore the angle H can have a certain amount.

The rotation H_z together with rotation G, performs round about the axis Y_f, will keep it in mind to ensure for the plan $Y_f O_f Z_f$ a parallel position with the axis Z.

In that case the distance "a" will be finding between the axis Z and Z_f, and the pivoting angle L of the coordinate system $X_f Y_f Z_f$ round about the axis X_f, will be the supplement angle between the crossing axes Z and Z_f (see figure 3).

In this way the angle H_z will be the rotation angle in elevation of the coordinate system $X_f Y_f Z_f$ round about the axis Z.

It admits "a" and "H_z" the installed parameters of the tool, it results the relative displacements l_X and l_Y can be calculated with the following relations:

$$l_x = a \cdot \cos H_z, \qquad (11)$$
$$l_y = a \cdot \sin H_z.$$

It results that the axis position of the disk tools as against the piece will be determined with the helping of four parameters: A, ε (ε=π-L), H_z and l_z.

If it considers that the axis X_f of the tool is in the piece plane XOY then $l_Z = 0$ and so the axis position of the disk type milling tool can be determined by three theoretical parameters a, ε and $H_Z^{'}$

3.1 The determination of the angle $H_z^{'}$

The value of the angle $H_Z^{'}$ differs from the angle H_z through of the origins motion O and O_f of the systems of coordination along the axis Z.

Thus, if the distance between the origins is l_Z (see figure 3) the totalized angle $H_Z^{'}$ will have the following value:

 DOI: 10.1051/matecconf/20179405005

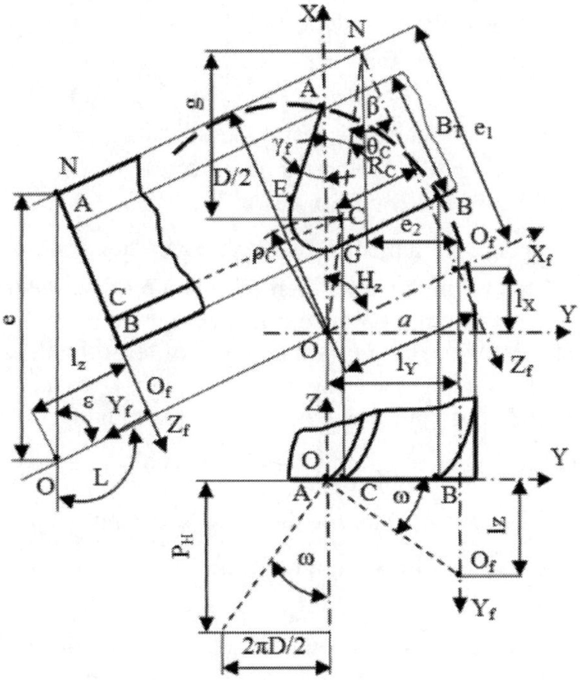

Fig. 3. The calculus scheme for the installation parameters of disk-milling tool

$$H_z' = H_z - \frac{l_z}{p}, \tag{12}$$

If the values of H_z and l_z are they defined from figure 3 as is detailed shows in [8], then H_Z' has the following equation:

$$H_z' = \arccos\frac{ctg\varepsilon}{tg\omega} + \theta_c + \frac{a}{p}\sqrt{tg^2\omega - ctg^2\varepsilon}, \tag{13}$$

and must ensure the third condition of cutting process:

$$H_z' \ge \frac{\rho_c^2 \cdot \sin 2\gamma_f}{p^2} + \gamma_f; \qquad p = \frac{p_H}{2\pi}. \tag{14}$$

3.2 The determination of the angle ε

Linked with the installation parameters of the disk-tool (a, ε and H_Z'), often for to facilitate the calculations we can use other two auxiliary parameters of calculation, namely:

- ρ_c – the primitive radius of the piece, respectively
- R_c – the primitive radius of the tool, dependent on *a*, ε and H_Z'.

In the figure 3 it represents a helicoidally surface characterized by the amounts ρ_c and R_c. The primitive radius ρ_c it is obtained from the condition that the angle of the helix line, which passes through the cylinder of radius ρ_c, it be equal with the angle ε.

From figure 3 it can see the following:

$$ctg\varepsilon = \frac{2\pi \cdot \rho_c}{p_H}, \quad \rho_c = \frac{p_H}{2\pi} ctg\varepsilon = p \cdot ctg\varepsilon, \quad p \cdot ctg\varepsilon > \rho_c \cdot \cos\gamma_f. \tag{15}$$

in accordance with the first processing condition [9].

Also, from figure 3 it deduces that:

$$R_c = a - \rho_c \cdot \cos(H_z - \theta_c), \tag{16}$$

$$a = (g + \rho_c) \cdot \cos(H_z - \theta_c). \tag{17}$$

Accordingly with these forenamed, it result the following:

- the parameter "a" is entirely determinate through equation (17);
- the parameter ε is given by the condition (15);
- the parameter $H_z^{'}$ must simultaneous satisfy the condition (14) and the equation (17).

If the inequality 4 isn't coming true, then the profile of the helicoidally flute which is manufactured will be under cutting on the point A zone;

If the equation (17) isn't carrying out, then the profile of the disk type milling tool will has the top radius R_c smaller than that maximum possible. Because, isn't permitted that the flute profile to be rounded around point A zone, the inequality (14) must be compulsory carrying out. The unfulfilled of equation (17) is admissible. Naming with $H_{zAG}^{'}$ the value of $H_z^{'}$ given by the inequality 4(1) and with $H_{zAG}^{'}$ the value of $H_z^{'}$ determinate through equation (17), the installation of tool is optimum if:

$$H_{zGE}^{'} \geq H_{zAG}^{'}. \tag{18}$$

To compute the optimum installation and localization parameters for a disk type milling tool, two parameters are necessary to be known: the angle ε and the external diameter of the disk-tool D.

It is well known that for the parameters ε and D, some range of values are admitted, from among only the optimum values will be selected

The angle ε is choosing on a value which carrying out on the limit the inequality (15).

The maximum value of the angle ε after allowing for the inequality 15) is marked with ε_{max}. In the case $\varepsilon<(\pi/2)-\omega$, the equation (17) has not a real solution, in which case the inequality (18) can't be coming true for such values of angle ε.

Such a situation is meeting when the profile of the flute has a great radius of rounding R_c, and the number of tool's tooth are small.

For a value of ε which is in accordance with the inequality:

$$ctg\varepsilon > \frac{\rho_c \cdot \cos\gamma_f}{p}, \tag{19}$$

the inequality 4, which limits the minimum value of $H_z^{'}$, will exceed surely the value $H_{zAG}^{'}$. As a consequence, the section AG of the tool's profile can be made for any value of the angle $H_z^{'}$.

Decreasing of the angle ε with a quota, such as the inequality (15) is coming true, also decrease in a large measure the possibility to carrying out the inequality (18), and finally leads to increasing the breadth of disk type milling tool B_T.

4 Conclusions

To solve the problems attached to the part shape formulation by milling is necessary to know the parameters which characterize the calculation points M. These points are emphasizing for the disk or ball-end milling tool. Using the established equations in this article, the tool profile can be entirely defined.

The optimum value of the parameters a, ε and $H_Z^{'}$ will be considered those that on the one hand ensures carrying out processing conditions of the nominal surface of the piece, and on the other hand ensures carrying out the constructive and exploitation conditions of the tool, still in the designing phase.

For this reason, often for the realization of a rationale construction tool, it prescribes the analysis of the installation parameters influence of the tool axis over its one or the other constructive parameter.

Using the established equations in this article, the tool profile can be entirely defined. The results can be used in a fully automatic process of tool manufacturing.

Once it was determined by analytical equations, the profile of tool and its installation parameters, should be checked the curve that will get the points, will make approximations with arcs or line segments, where appropriate, depending on the possibilities of programming machine tools and, very important then, it will check for any undercut areas by generating simulation profile processed.

References

1. J. Balic, *Constitution to integrated manufacturing* (DAAAM Publishing Series, Vienna, 1999)
2. D. N. Reshetov, V. T. Portman, *Accuracy of machine tools* (ASME Press, New York, 1988)
3. V. S. Mihai, *Contributions to the generation of manufacturing technology on CNC machine tools to the profiled milling cutters*. (PhD Thesis, Technical University of Cluj-Napoca, 1998)
4. L. D. Beju, D. P. Brîndaşu, N.C. Muţiu, J. Rothmund, J. Adv. Manuf. Techno. **83**, 2111 (2016)
5. Gh. Atanasiu, Gh. Munteanu, M Postolache, *Linear algebra. Analytical and differential geometry. Differential Equations - Collection of problem (*All Publishing, Bucureşti, 1998)
6. V. Balan, *Linear algebra. Analytical geometry (*Fair Partners Publishing, Bucureşti, 1999)
7. I. Grozav, E. Pămîntaş, *The theory and computer aided design of cutting tools*, ("Politehnica" Publishing, Timişoara, 1994)
8. I. Pop, E. Pămîntaş, *Proceedings of PUPR Conference*, **II**, 1989
9. I. Pop, V. Putz, V. Domilescu, *Proceedings of PUPR Conference*, **II**, 1989

Assemblability analysis of the modular fixturing systems with accurate grid pattern holes using Monte Carlo simulation

Tudor Păunescu[1,*]

[1]Transilvania University of Brasov, Department of Manufacturing Engineering, Mihai Viteazu No.5, Brașov, Romania.

Abstract. This paper focuses on the assemblability of modular fixturing systems. Two types of stochastic models are proposed: for two holes-two pins (2H2P) and for two holes-two pins-two holes (2H2P2H), both used by modular fixturing systems with multipurpose grid pattern holes. The probability of successful assemblies to engineering specifications is calculated using the Monte Carlo simulation. The weights of the assemblability criterion variables are analysed and several design recommendations are indicated.

1 Introduction

Modular fixtures (MF) is one of the most appropriate flexible fixturing techniques used by CIM systems. It is well known that MF is a workholding using standardized reusable components. After the fixture is built and the machining process is completed, the fixture can be disassembled and the elements can be used to build another fixture. MF fills the gap between the general-purpose and special-purpose fixtures.

A lot of academic and applications papers have been published in this area, much focus has been put on intelligent methods for computer aided fixture [1, 2].

The present paper focuses on modular fixturing systems (MFS) with grid pattern holes (MFSH) because offer several advantages over modular systems with T slots: more positional possibilities of the modules on the base plate, more security, repeated setups are easier, if grid pattern holes are damaged, these are repairable.

MFSHs are based on alternating dowel holes and tapped holes, or multipurpose holes. The second is very popular, because the coaxial arrangement of threaded and cylindrical locating surface allows positioning and fastening at the same time.

The MFSHs use two locating principles of modules: if the locating module has a symmetry axis one hole – one pin – one hole assembly is used; if the locating module has not a symmetry axis the assembly is based on two holes-two pins-two holes.

The pins are rigid bodies (position sleeves (fig. 1a), position screws (fig. 1c)) or elastic bodies (slotted position sleeves (fig. 1b), collets).

Is necessary to analyse the assemblability of the modules, as first step of optimizing tolerances.

* Corresponding author: t.paun@unitbv.ro

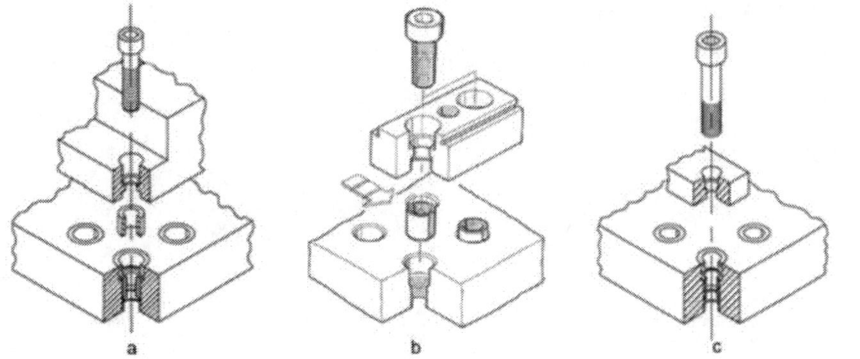

Fig. 1. Three positioning systems based on: rigid sleeve (a), slotted sleeve (b) and position screw (c) [3 - 6].

Grid pattern of MFSH for medium-size workpieces are spaced 25, 40, 50 ±0.01 mm, on every grid tooling plate or blocks. Usually, the holes and pins have 5-7 IT grades (Table 1).

Table 1. Tolerances of the MFSH.

No.	MFSH	D_1 [mm]	D_2 [mm]	D_3 [mm]	pitch [mm]
1	AMF M12 [4]	Φ16F7	$\Phi16^{0.011}_{0.00}$	Φ16F7	40±0.01
2	AMF M16 [4]	Φ22F7	$\Phi22^{0.015}_{0.002}$	Φ22F7	50±0.01
3	Halder M12 [5]	Φ12F6	Φ12k5	Φ12F6	50±0.01
4	Halder M16 [5]	Φ16F6	Φ16k5	Φ16F6	50±0.01
5	Kipp M8 [6]	Φ12H6	Φ12h6	Φ12H6	25±0.01
6	Kipp M12,16 [6]	Φ12,16F7	Φ12,16g5	Φ12,16H7	50±0.01
7	Norelem M6 [7]	Φ12E6	Φ12n6	Φ12E6	25±0.01
8	Norelem M10 [7]	Φ18E6	Φ18n6	Φ18E6	40±0.01

Notations: D_i, i=1-3, centering diameter; 1-fixture base; 2-locating component (locating screw or sleeve); 3-modular mounted component.

2 The assemblability problem of modular fixturing systems

The usual solving methods for assemblability problem are: the worst case method (WCM) and statistical methods, mainly Monte Carlo simulations (MCS). Statistical tolerance design is accepted by industry, particularly in case of modular fixturing systems if a module fails to assemble it can be replaced with another interchangeable module. Because WCM assumes extreme conditions that rarely happen, the percent of components which cannot be assembled is zero. The MCS can generate a more reasonable results than WCM, because is possible to increase the tolerances for individual dimensions, in this way manufacturing costs could be reduced.

The assemblability problem calculates the probability of successfully parts assembling when the dimensions and tolerances of the parts, mating constraints, and an assembly sequence are given. In case of modular fixture, an assembly sequence is not required because each locating module has unique assembly scheme.

The main sources of errors are: geometric errors, elastic deformations due to fixturing and cutting forces, wear and thermic deformations. In this approach, geometrical models of positioning modules are considered. Also, suppose that the locating modules has a low ratio of height to length, the geometrical models are simplified to 2D.

 DOI: 10.1051/matecconf/20179405006

2.1 Assemblability of 2H2P

The 2H2P assembly is used by the Norelem MFSH [7] for two components: rest and seating ledge (fig. 2) and riser block plug-in for locating pin.

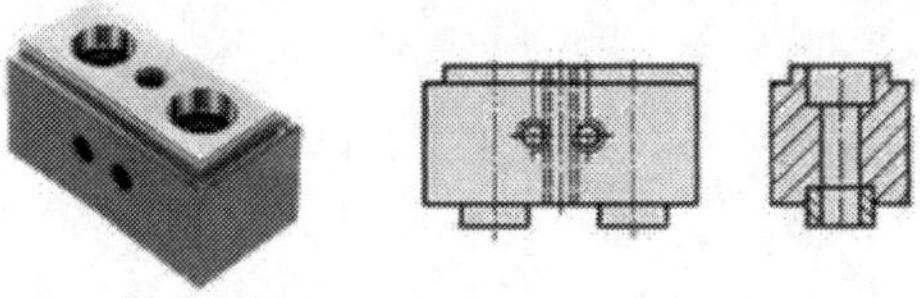

Fig. 2. Support-stop block with force fits between the sleeves and the holes of module [7].

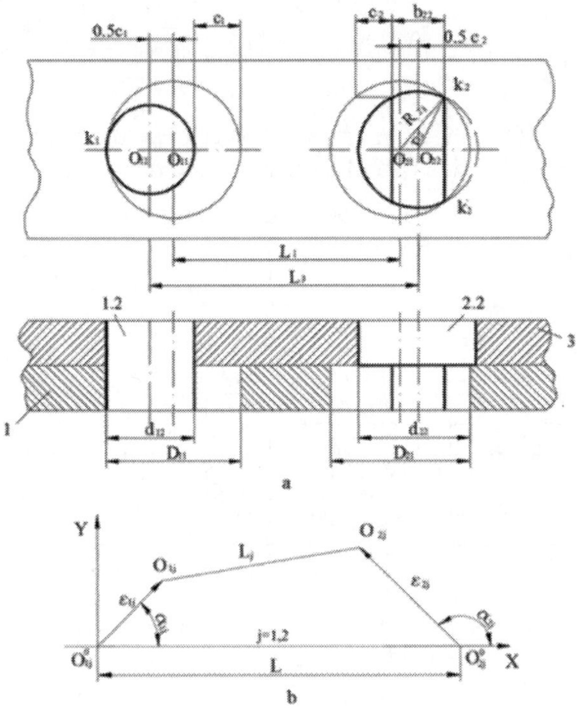

Fig. 3. a-2H2P in a limit assembly position; b-vectorial summation of centre distances.

Notations

Indexes for D_{ij}, d_{ij}, UL_j, LL_j, ul_j, ll_j, c_j, L_i, α_{ij}, ε_{ij}:

i=1-3: 1-base module; 2- rigid sleeve; 3 - mounted module;

j=1, 2: 1- for round sleeve; 2 - for diamond sleeve and corresponding hole;

L_i, i=1-3 centre distance between axes;

D_{ij}, d_{ij}, i=1-3, j=1, 2diameters of holes and sleeves respectively;

c_j, j=1, 2 clearances between the holes and sleeves;

TPp_i, i=1, 3 the positional tolerance of the centering hole;

TPs_{22} locating diamond pin symmetry tolerance;

T_{22} tolerance of the diamond pin width;

UL_j, LL_j, ul_j, ll_j, j=1, 2 upper and lower limits for holes and pins respectively;

m-minimum, M-maximum.

Basic sizes of holes and sleeves/pins are equal, and the limits are equal respectively. The same rule is applied for basic size of centre distance between the two holes.

A lot of MFSH producers use the conventional tolerancing (±0.5T) of pitch. If the ISO position tolerance is used, is necessary to make following substitution: TPp→ 0.5T.

In fig. 3a and 4a clearances between the parts are shown with some exaggeration.

Using diamond pin, clearance in one direction ($c_2$in fig.3a) is much more than radial clearance, but contact surface area is smaller than round pin. Generally, the width of a diamond pin is usually 1/8 of D[8].

The lateral clearancec_{2l}can be approximately calculated as [9]:

$$c_{2l} \approx c_{2r}\frac{D_2}{b_2} \tag{1}$$

A more accurate formula is [9]:

$$c_{2l} = \sqrt{b_2^2 + D_2^2 - d_2^2} - b_2 \tag{2}$$

The 2H2P problem with two round pins was studied by A. Shan [10]. The probability of successful 2H2P assemblies to specifications is estimated using a Monte Carlo simulation. Two inequations are used to calculate the feasibility of the 2H2P assembly:

$$DPI \geq DHI \text{ and } DPO \leq DHO \tag{3}$$

DPI is inside distance between two pins; *DPO* is outside distance between two pins; *DHI* is inside distance between two holes; *DHO* is outside distance between the holes.

In this paper, one round and one diamond locating pin are considered, the feasibility condition uses just one inequation (6). From this general condition the worst case and the assembly having two round pins result as particular cases.

During assembling the mounted module can be slightly moved, so the axis $O_{12}O_{22}$ and $O_{11}O_{21}$ are coincident. One critical position of two components has three contacts between pins and holes:K_1, $K_2$and K_2' (fig. 3a). The assemblability criterion is:

$$0.5\,(c_{1r} + c_{2l}) \geq L_2 - L_1 \tag{4}$$

For the other critical position:

$$0.5\,(c_{1r} + c_{2l}) \geq L_1 - L_2 \tag{5}$$

From 4,5and 2 inequations results the following assemblability criterion:

$$0.5\,(c_{1r} + c_{2l}) = 0.5\left(c_{1r} + \sqrt{b_2^2 + D_2^2 - d_2^2} - b_2\right) \geq |L_1 - L_2| \tag{6}$$

If the both pins are round b_2=$d_2$then$c_{2l} = c_{2r}$, from assemblability condition 6 results:

$$0.5\,(c_{1r} + c_{2r}) \geq |L_1 - L_2| \tag{7}$$

A common source ofL_1, L_3 errors are the position tolerances (TPp) of holes. Usually, the distributions are normal, the magnitude of the first vector-error $\varepsilon_l$is 0 … TPp/2. Another parameter of the vector-error is the orientation. It is assumed that α_l is a random variable with a uniform distribution from 0 to2π. The diamond head asymmetry must also be taken into consideration, when L_2 is calculated:

$$L_2 = L_3 + s\varepsilon_s,\ \ s = +1\ or - 1, \varepsilon_s = 0 \ldots 0.5TP_s \tag{8}$$

If the worst case (100% interchangeability) is considered, the following values are assigned to variables: minimum clearances, one of the centre distance has a minimum value and the other the maximum value, maximum width and asymmetry of the diamond pin, therefore the assemblability condition is:

$$0.5\,(c_{1rm} + c_{2lm}) = 0.5\left(c_{1rm} + \sqrt{b_{2M}^2 + D_{2m}^2 - d_{2M}^2} - b_{2M}\right) \geq TPp_1 + TPp_2 + 0.5TPs_2 \quad (9)$$

$$0.5\,(c_{1rm} + c_{2lm}) \approx 0.5\left(c_{1rm} + \frac{d_{2m}}{b_{2M}} c_{2rm}\right) \geq TPp_1 + TPp_2 + 0.5TPs_2 \quad (10)$$

Usually TPp=TPp$_1$=TPp$_2$ and $c_{rm} = c_{1rm} = c_{2rm}$, from condition 10 results:

$$0.5c_m\left(1 + \frac{d_{2m}}{b_{2M}}\right) \geq 2TPp + 0.5TPs_2 \quad (11)$$

If both sleeves are round:

$$c_m \geq 2TPp \quad (12)$$

2.2 Assemblability of 2H2P2H

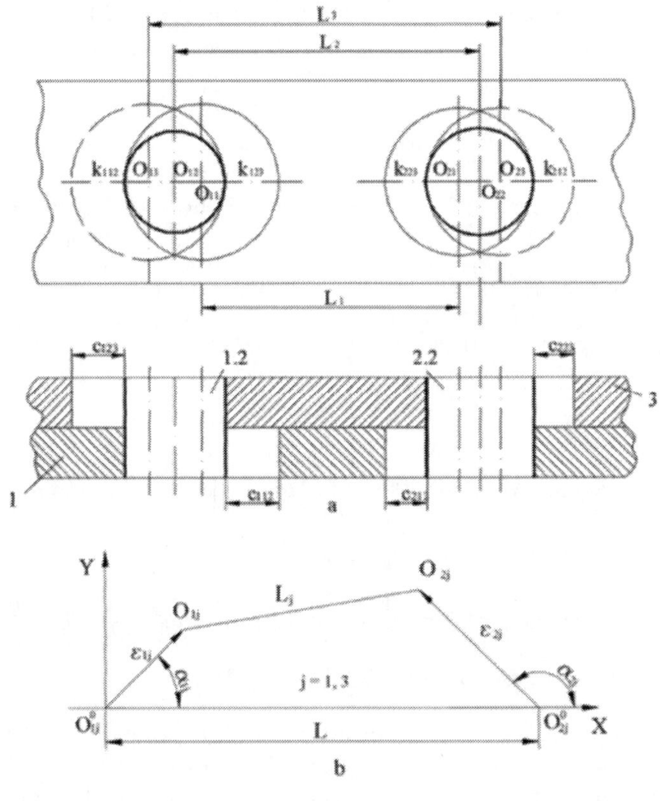

Fig. 4. a-2H2P2H in a limit assembly position; b-vectorial summation of centre distances.

In fig. 4a clearances are noted as: c_{ijk}, i=1,2; j,k=1-3, jk defines de clearance between the assembly components, for example c_{123} is clearance between sleeve/position screw 12 and mounted module 3. Unlike the 2H2P notations, because all clearances are radial, the r index is not used.

The assemblability criterion for a general 2H2P2H assembly is:

$$0.5(c_{112} + c_{123} + c_{212} + c_{223}) \geq |L_1 - L_3| \quad (13)$$

If the worst case is considered, similar to 2H2P, the following substitutions are made: minimum clearances ($c_m = c_{112} = c_{212} = c_{123} = c_{223}$), one of the centre distance has a minimum value and the other the maximum value ($|L_1 - L_3| = TPp_1 + TPp_2 = 2TPp$), so assemblability criterion is:

$$c_m \geq TPp \quad (14)$$

3 Statistical approach of MFSH assemblability problem using Monte Carlo simulation

MCS is an efficient tool of tolerance analysis to the design of mechanical assemblies [11-15]. It is applied as a pure statistical method, or in combination with other methods.

For 2H2P and 2H2P2H assemblies the input variables are: position tolerance of grid pattern and mounted module, common nominal diameter of holes and sleeves, upper and lower specification limits of holes and pins, respectively. Particularly, for diamond pin the additional input variables are: the width of head and its tolerance, the symmetry tolerance. The successful 2H2P2H assemblies using Monte Carlo simulation can be calculated with following algorithm, which simulates the 2D assembly process of modules:

- After selecting the distributions of holes and pins diameters, the random values of D_{11}, D_{21}, d_{12}, d_{22}, D_{13}, D_{23} are generated and then the clearances are calculated: $c_{i12}=D_{i2}-d_{i2}$, $c_{i23}=D_{i3}-d_{i2}$, i=1,2.
- The centre distances between axes of the two adjacent holes L_i, i=1,3 are calculated (fig. 4b). Random error-vector $\boldsymbol{\varepsilon}_{ij}$, i=1, 2, j=1,3, is generated $\varepsilon_{ij}= 0\ldots TPp_j/2$ where TPp_j is the positional tolerance of the centering holes and α_{ij} is the direction of $\boldsymbol{\varepsilon}_{ij}$. The random variable α_{ij} has a uniform distribution from 0 to 2π.
- If the assembly 2H2P2H is not feasible $0.5(c_{112} + c_{123} + c_{212} + c_{223}) < |L_1 - L_3|$it is indexed, then go to step 1.
- The calculating steps 1...3 are repeated with a large sample size, and the results are processed.

The algorithm for 2H2P follows similar steps, but clearances between locating module 3 and sleeves 2 have zero value and assemblability criterion is calculated with (11).

In the first step, this paper investigates the solutions to the assemblability problem offered by the MSFH designers (table 2).

In table 2, q is the index of MSFH interchangeability. It is difference between the clearances and variation of the centre distances:

$$q_{2H2P} = c_m - 2TPp - 0.5TPs_2 \quad (15)$$

$$q_{2H2P2H} = 2(c_m - TPp) \quad (16)$$

If $q \geq 0$ all assemblies are possible, therefore the Monte Carlo simulation is uselessly.

Monte Carlo simulations with a 10^8 sample size show that in the majority of commercial MSFH the 100% interchangeability is not used (table 2).

The estimation of sensitivities is not a direct result of MCS. In the present work, the weight of minimum clearance, position tolerances of holes, pins and holes tolerances are analysed. The sensitivity information can be obtained using the method: holding the limits of the independent variables constant except one being studied.

Table 2. Analysis of the commercial MFSH assemblability.

No.	MFSH		q	Distributions	
	Name	Type	[μm]	normal	uniform
1	AMF M12 [2]	2H2P2H	-10	1	0.99999998
2	AMF M16 [2]		-10	1	1
3,4	Halder M12, M16 [3]		-6	1	1
5	Kipp M8 [4]		-20	1	0.99925001
6	Kipp M12,16 [4]		24	1	1
7,8	Norelem M6, M10 [5]	2H2P	-1	1	0.99994901

Designers can improve the modular systems by calculating the weight that each variation contributes to resultant assembly variation.

The uniform distributions of the position tolerances of holes, pins and holes tolerances are considered. As is known the uniform distribution predicts greater variation than the normal distribution, but not as great as worst case.

Because few rejected assemblies per million will be predicted, in this research the sensitivity analysis uses 10^9 sample size.

3.1. Sensitivity analysis of 2H2P assembly

Sensitivity analysis uses a2H2P assembly with usual values as basis of comparison: basic size of holes and sleeves diameter d=12mm; pitch p=40mm; the positional tolerance of the centering holes $TPp_1=TPp_2=0.01$mm; upper and lower limits for holes UL=0.01mm, LL=0mm, T=UL-LL=0.01mm;upper and lower limits for pins ul=-0.01mm, ll=-0.02mm, t=ul-ll=0.01mm, $c_{min}=0.01$mm.The width of diamond sleeve is 10mm with 0.01mm symmetrical tolerance, diamond head asymmetry geometrical tolerance TPs=0.01mm.

Results of Monte Carlo simulations using Mathcad are: $27/10^6$ 2H2P-RPRP (RP-round pin) and $87/10^6$ 2H2P-RPDP (DP-diamond pin)) assemblies are rejected, so high quality level can be considered.

Usually the TPp= 0.01mm (see table 1), holes with TPp<0.01mm are too expensive. In this analysis 0.02mm is the maximum value of positional tolerance.

Because minimum clearance is possible to be greater or smaller than 0.01mm, both variants are investigated: c_{min}=0…0.01mmand c_{min}=0.01…0.02mm.

If T_{22}=[0.02…0.4] rejects remain constant 87 per million, so designer can assign a general tolerance for linear dimensions, medium or coarse class (ISO2768-1), and the diamond pin can be cheaper.

In fig. 5 and 6 with a horizontal marker is shown 2700 rejects per million, corresponding to tree-sigma rule (±3σ). Sensibility analysis of the 2H2P assemblies reveals:

- increasing of the centering holes positional tolerance (TPp) and decreasing of the minimum clearance (c_{min}) have great influence on probability of successful 2H2P assemblies comparatively with hole or pin tolerance (T, t) (fig.5 and fig.6);
- if minimum clearance decreases results more rejected parts of the 2H2P-RPDP than 2H2P-RPRP assembly, as a result of diamond pin asymmetry;
- if minimum clearance increases a practically 100% probability is achieved ((fig.5 and fig.6);
- increasing or decreasing of diamond width tolerance has not effect on assemblability (fig.7);
- diamond pin asymmetry is more important than diamond width tolerance (fig.7), but the effect on assemblability is less important than positional tolerance of holes and minimum clearance;

 DOI: 10.1051/matecconf/20179405006

- increasing of hole or pin tolerance has a minor impact on assemblability (fig.5 and fig.7).

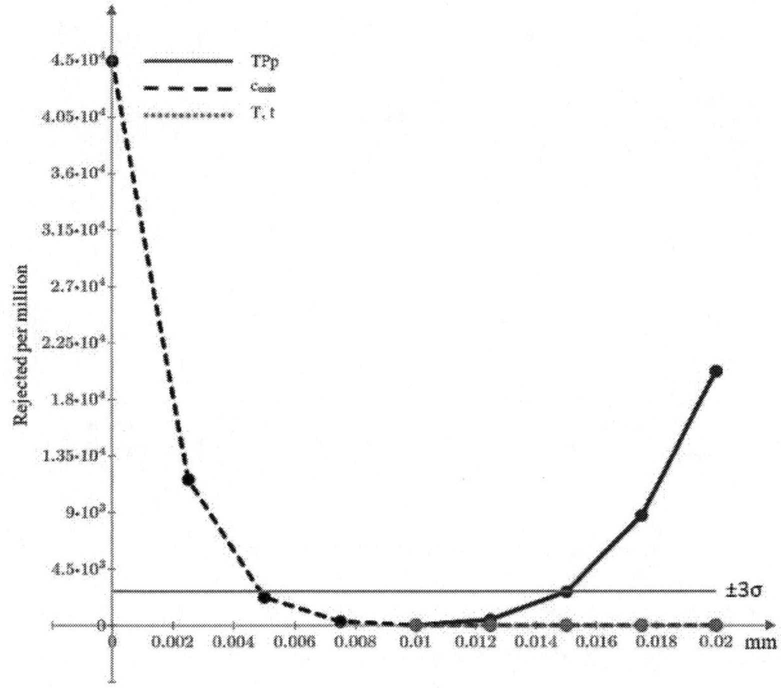

Fig. 5. Influence of positional tolerance of the centering holes, minimum clearance, hole or pin tolerance for 2H2P-RPRP.

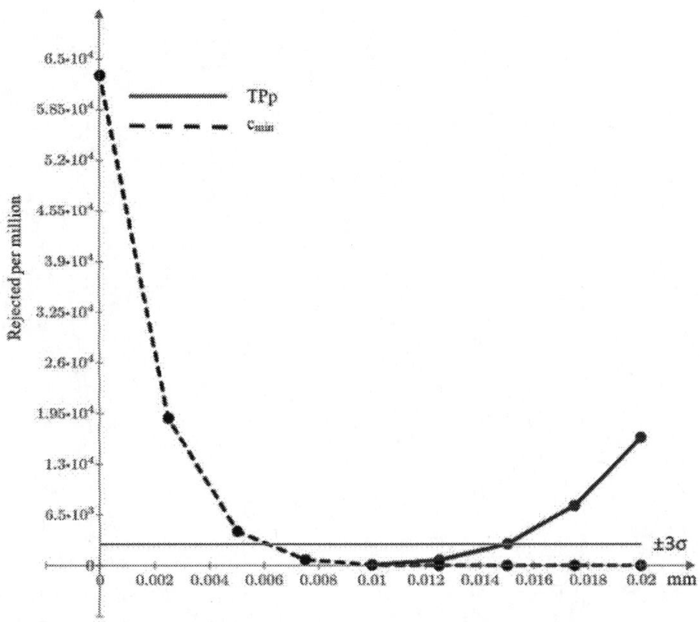

Fig. 6. Influence of positional tolerance of the centering holes and minimum clearance for 2H2P-RPDP.

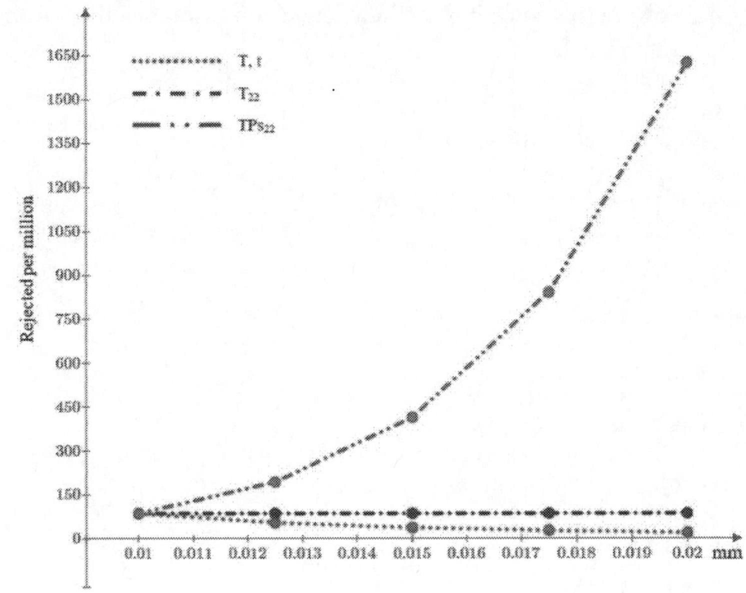

Fig. 7. Influence of hole or pin tolerance, the diamond pin width and symmetry tolerances for 2H2P-RPDP.

3.2 Sensitivity analysis of 2H2P2H assembly

The 2H2P2H assembly used as basis of comparison has the same values as 2H2P-RPRP assembly and the same sensitivity analysis method is applied.

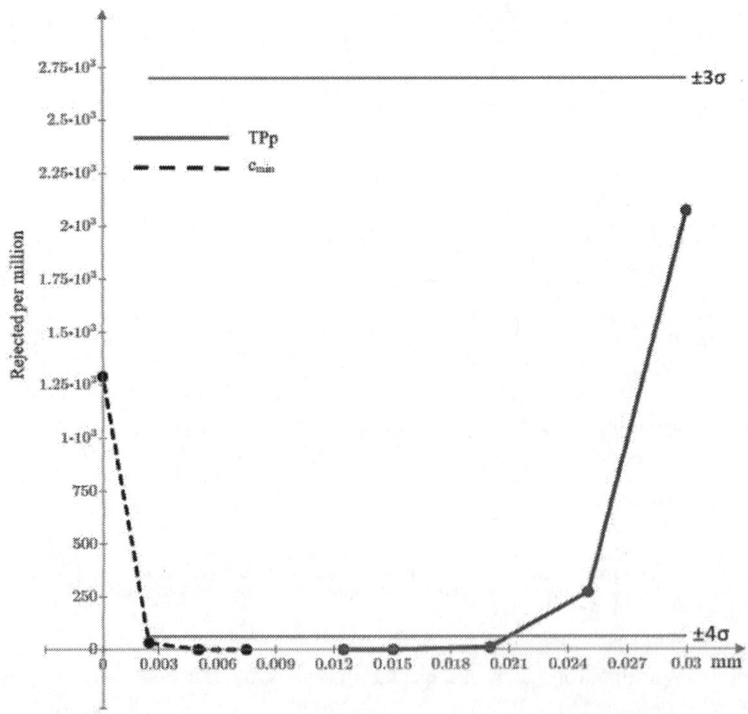

Fig. 8. Influence of the centering holes positional tolerance and minimum clearance for 2H2P2H.

Additionally, in fig. 8 and fig.9 with a horizontal marker is shown 63.4 rejects per million corresponding to four-sigma rule (±4σ).

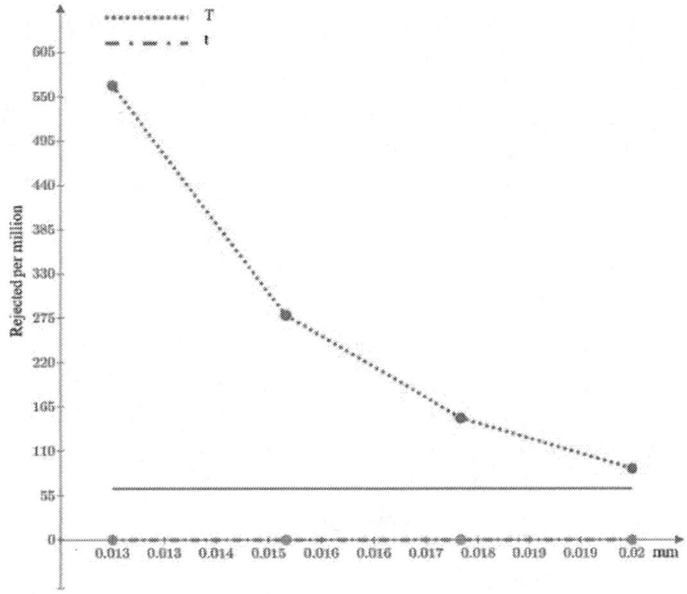

Fig. 9. Influence of hole and pin tolerance for 2H2P2H.

Analysing the 2H2P2H dimensional sensibility reveals:

- if $TPp \in [0.0125 \ldots 0.02]$hole positional tolerance increasing has minimum influence on probability of successful 2H2P2P assemblies, but if $TPp \in [0.02 \ldots 0.03]$the influence is greater(fig. 8);
- influence of minimum clearances is greater if the locational clearance fit is considered (c_{min}=0) (fig. 8);
- if hole tolerance is increased the number of rejected assemblies decreases (fig. 9);
- increasing of pin tolerance has zero impact on assemblability, because the worst case condition is satisfied (fig.9).

4 Conclusions

This paper focuses on the assemblability problem of the modular fixturing systems with accurate grid pattern holes in a statistical framework.

The two main purposes of this work are:

- to estimate the probability of successful 2H2P and 2H2P2H assemblies;
- to analyse the sensitivity of these assemblies.

Stochastic models of the 2H2P-RPRP, 2H2P-RPDP and 2H2P2H assemblies are proposed. The assembly problem of modular fixturing systems with grid pattern holes is solved using Monte Carlo method.

The weights of minimum clearance, position tolerances of holes, pins and holes tolerances, tolerance and asymmetry of the diamond pin reveals:

- comparison of the results between the 2H2P and 2H2P2H shows that the second assemblies can easier satisfy the assemblability condition;
- increasing of the centering holes positional tolerance and decreasing of the minimum clearance between pin and hole have great influence on probability of successful 2H2P assemblies comparatively with hole or pin tolerance;

- for 2H2P-RPDP assemblies, diamond pin asymmetry tolerance is more important than diamond width tolerance;
- hole or pin increasing tolerance has a minor impact on assemblability of 2H2P;
- if TP<0.02mm hole positional tolerance increasing has minimum influence on probability of successful 2H2P2P assemblies, but if TP>0.02mm the influence is greater;
- if hole tolerance is increased the number of rejected 2H2P2P assemblies decreases.

Based on this approach and results presented above, designers can optimize the modular fixturing systems with grid pattern holes for medium-size workpieces.

A future area of study is to integrate the results of the present study in an efficient method to optimize the fixturing modular systems correlated to the workpiece locating accuracy.

References

1. J. Marrs, *Machine Designers Reference* (Industrial Press Inc., New York, 2012)
2. N. P. Maniar, D. P. Vakharia, Int. J. of Scientific & Eng. Research **4**, 1 (2013)
3. Carr Lane modular fixturing systems, http://www.carrlane.com
4. Andreas Maier GmbH & Co. KG, http://www.amf.de
5. Erwin Halder KG. http://www.halder.de
6. Heinrich Kipp Werk. http://www.kipp.com
7. Norelem. http://www.norelem.fr
8. P. H. Joshi, R. O. Parmley, *Jigs and Fixtures Design Manual* (McGraw-Hill Professional Publishing, New York, 2003)
9. S. Vasii-Rosculeţ, N. Gojinetchi, *Proiectarea dispozitivelor [Fixture Design]* (Didactic and Pedagogical Publishing., Bucharest,1982)
10. A. Shan, R. N. Roth, R. J. Wilson, Int. J. Adv. Manuf. Technol. **15**, 222 (1999)
11. K. W. Chase, A. R. Parkinson, Research in Eng. Design **3**, 23 (1991)
12. J. Gao, K. W. Chase, S. P. Magleby, *Proc. of the ASME Design Eng. Tech. Conf.*, **1** (1995)
13. Y. S. Hong, T.C. Chang, Int. J. of Prod. Research **40**, 2425 (2002)
14. F. Scholz, Boeing Information & Support Services **1** (1995)
15. S. H. Huang, Int. J. Prod. Res. **42**, 4871 (2004)
16. H. Geng, *Manufacturing Engineering Handbook* (McGraw-Hill Education Publishing, New York, 2016)

Accuracy analysis of the modular fixturing systems with grid pattern threaded holes

Tudor Păunescu [1,*], and *Rodica* Păunescu[1]

[1]Transilvania University of Brasov, Department of Manufacturing Engineering, Mihai Viteazu No. 5, Braşov, Romania.

Abstract. Using a baseplate with a grid pattern threaded holes and the machine tool to accurately locate the modular locators upon the baseplate, the positional accuracy of the workpiece can be improved and results a relatively inexpensive modular fixturing systems. A stochastic mathematical model of the geometrical accuracy and a solving algorithm for these modular fixturing systems are proposed. Four types of modular fixturing systems with grid pattern threaded versus the most important classical modular-fixturing systems with multipurpose holes are analysed. Monte Carlo approach is used to calculate and to compare the locating accuracy of these modular fixturing systems.

1 Introduction

Two primary forms of modular fixturing systems (MFS) are available: with grid pattern holes and with T-slots. The first system offers several advantages over second: more positional possibilities, more security, repeated setups are easier, if grid pattern holes are damaged, these are repairable [1]. Commercial grid-pattern modular fixturing systems are based on alternating dowel and tapped holes or multipurpose holes (MFS-MH). The last system has an alignment bushing and threaded insert in each hole, so a multipurpose hole can be used as an alignment hole, mounting hole, or can have both functions.

The present paper focuses on the MFS with a particularity: baseplates have a grid pattern with threaded holes (MFS-TH) and the machine tool is used to accurately locate the modular components upon the baseplate.

The idea to use machine tool to setup modular work holdings is not a new one. Over the past decades several boring machines were equipped with a tool (fig. 1a) to accurately align the locating modules (fig. 1b).

In fig. 1c is shown an example of prismatic part located with 3-2-1 method: the machine tool plate is the primary surface, the two modules locate part on secondary surface and one locating module restricts the last degree of freedom. The screws and clamps are not shown.

* Corresponding author: t.paun@unitbv.ro

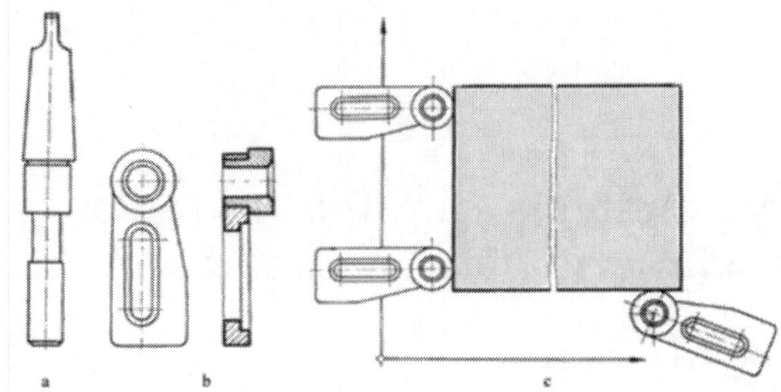

Fig. 1. The tool (a) used to accurately align the locating module (b), an example of 3-2-1 locating scheme (c) [2].

Other modular fixturing system using machine tool to setup locators is proposed by P. V. Hudimac [3].

The author claims that [3]:

- MFS-TH eliminates the need to accurately align the grid-pattern reference frame of the baseplate with the coordinate system reference frame of the machine tool.
- MFS-TH eliminates the dependence upon the manufactured precision of the baseplate, resulting an inexpensive baseplate.
- The stacking-up of modular fixturing components are free of errors accumulation.
- Locating and clamping elements can be easily assembled and disassembled.

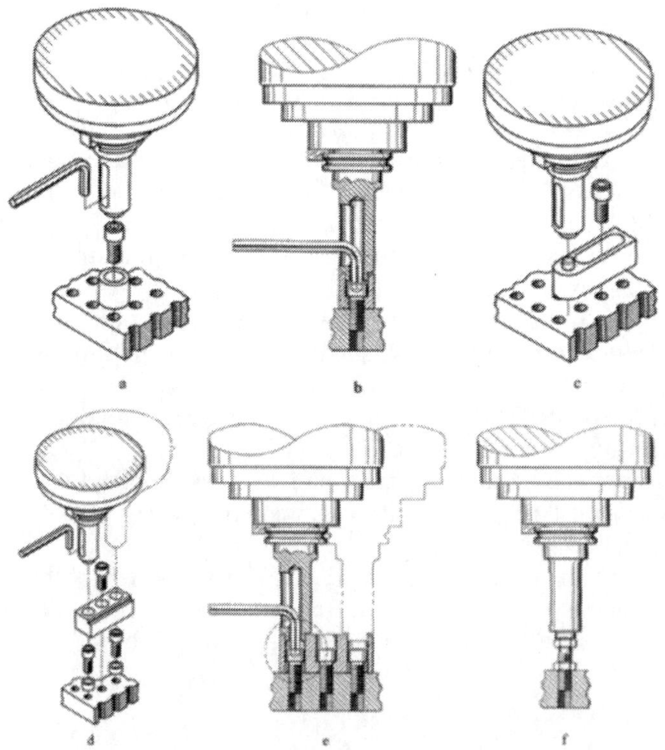

Fig. 2. How is assembled a cylindrical locating module (a-c), a single edge locator (d, e) and an adjustable support (f) [3].

As you see in fig. 2a-e, the locating tool is positioned within a machine tool spindle, the nose is accurately centered in the counterbore of the locating module and wrench is engaged in the head of screw. Wrench is turned to engage fastener into the threaded hole of the baseplate and, finally the locating module is affixed into position. In fig. 2f the end surface of tool contacts the top surface of the threaded adjustable support to set the Z-axis workpiece reference surface. In our opinion, the cylindrical locating module (fig. 2a, b, c) can be easily and accurately assembled, but is more difficult to accurately assemble the locating modules with two holes (fig. 2d and e).

Both solutions described above use cylindrical pin and hole with clearance fit, to accurately align locators. The clearance between the locating tool nose and module counter bore is a source of positional error. If the centering surfaces are conical the clearance is zero and this source of error is eliminated. For example, two types of locators with conical surfaces, and two modular fixtures are shown in fig. 3 [4].

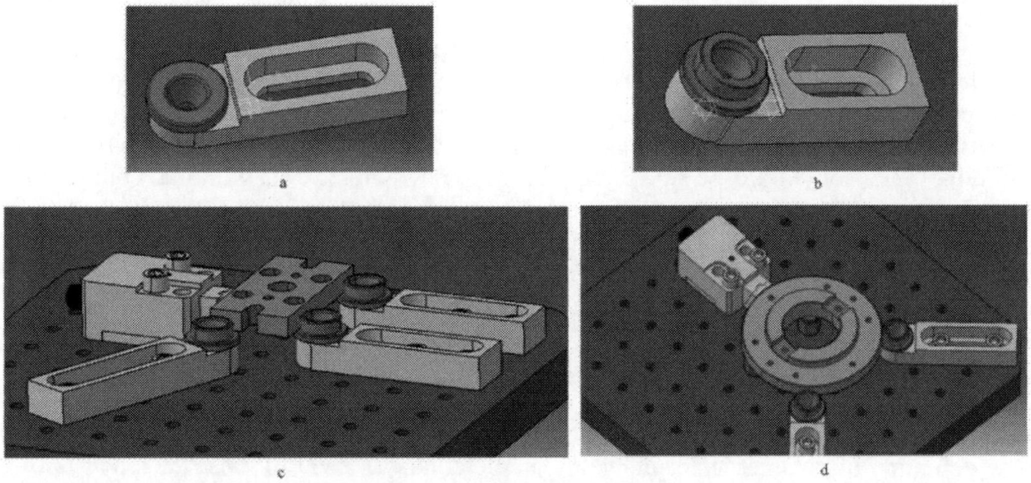

Fig. 3. Two types of modular locators (a, b) and examples of modular fixtures (c, d) [4].

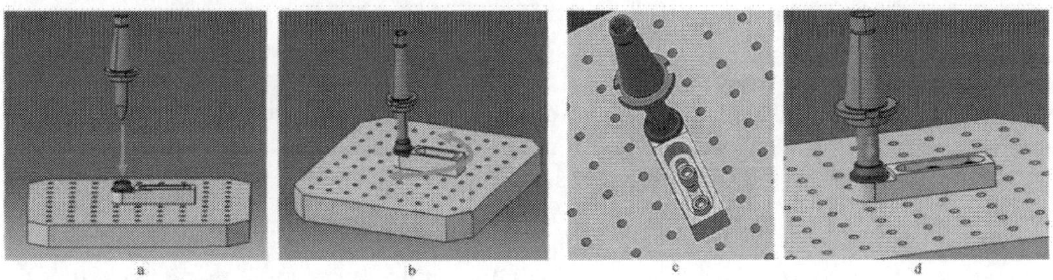

Fig. 4. Steps of locator-baseplate assembly [4].

Steps of locator-baseplate assembly, shown in fig.4, are described below:

- The modular locator is placed in an approximate position on the baseplate, the tool is lowered to programmed target. A clearance between the tool and sleeve is necessary for rotational adjustment of the locator.
- The locator is manually rotated to fit to threaded holes.
- Screws are tighten with low forces.
- The tool is lowered in the final position until the clearance between the tool and sleeve is zero, then the screws are tighten again.

2 Sources of geometrical errors and modelling

If the machine tool is used to accurately locate the modular components upon the baseplate, a number of error sources contribute to the resulting workpiece location error, which can lead to poor part quality:

- machine tool positioning inaccuracy;
- inaccuracy of the tool holder;
- inaccuracy of the locating tool;
- clearance between locating tool nose and the locator hole.

One of the most important error of the machine tool is the quasi-static geometric/kinematic error. The geometric errors may be compensated, a lot of machine tool manufacturers provide the errors mapping and a subsequent compensation in the same manner as has been well established in CMMs [5]. Now, most modern controllers have the capability of volumetric positioning error compensation.

For a 3-axis machine, there are 6 errors per axis, a total of 21 errors (6x3 plus 3 squareness errors): linear displacement errors, vertical and horizontal straightness errors, roll, pitch and yaw angular errors and squareness errors.

Machine tool accuracy has generally been described by the linear positioning accuracy and repeatability of the axes, but this is a deficient specification because other geometric effects: angular, straightness and squareness errors are ignored. A more comprehensive way to define precision would be to specify the accuracy for the full working volume of a machine tool taking into account all geometric errors [6].

Traditionally, manufacturers have ensured accuracy of parts with linear calibration of each axis of the machine tools, but this method is inadequate for ensuring accuracy of 3D parts. The conventional definition of the 3D volumetric positioning error is the root mean square of the 3 axes displacement error. This approach is correct if the dominate errors are the displacement errors.

In our days the displacement errors are reduced considerably, the dominate errors are straightness and squareness errors. ASME B5.54 and ISO 230-6 standards are working on a new definition of volumetric accuracy. A complete measurement of those 21 errors is very complex and time consuming, for those reasons the body diagonal displacement error is a good quick check of the volumetric error [7, 8].

Machine tool builders list accuracy and repeatability, but often is not given the method used. A number of standards exist but differ in their analysis procedures and in key parameter definition (ISO 230-1 2012, ASME B5.54, JIS B 6336-1986). As a result the values reported for positional accuracy and repeatability can vary depending on which standard was used [10, 11]. More, if the same workpiece is machined at different locations of the work table, different positioning errors will be obtained [12].

In our approach the following hypotheses are considered:

- errors are quasi-static and geometric;
- vertical machine centers are used to align the locators;
- because the movement on Z axis is short the straightness and squareness errors are neglected, so a 2D model is proposed, and the spatial errors are projected on XY plane;
- geometric error of the cylindrical locator axis is related to machine coordinate system (machine zero).

Errors are summed as vectors (fig. 5b), chained tip-to-tail. In statistical tolerance design each vector-error has two parameters: magnitude and angle [9]. The first parameter describes the distance of the axis relative to its true position, and the second dictates the direction. It is assumed that the random variable angle is a uniform distribution from 0 to 2π.

In fig. 5a is shown the worst aligning case of the locator (variant 1 in table 1): the locating tool is mounted in the tool holder (2), a clearance exists between the tool (3) and locator (4).

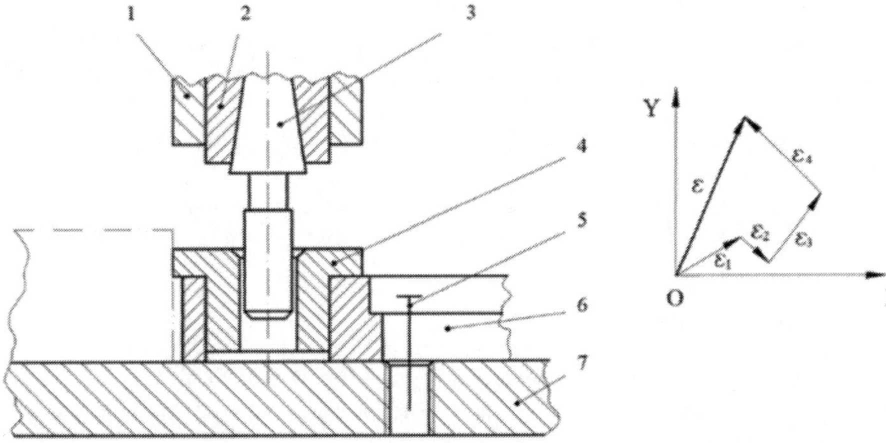

Fig. 5. Drawing section on subassembly locator-aligning tool (a), summed errors as vectors (b).

Because the centering surface can be cylindrical or conical, and the tool holder is used or not, four variants of locating schemes can be generated (table 1).

Table 1. Aligning schemas of the cylindrical modular locators.

No	Source of error	Variant			
		1	2	3	4
1	Machine tool positioning inaccuracy (ε_1)	x	x	x	x
2	Inaccuracy of the tool holder (ε_2)	x	x	-	-
3	Inaccuracy of the locating tool (ε_3)	x	x	x	x
4	Clearance between locating tool nose and the locator hole (ε_4)	x	-	x	-

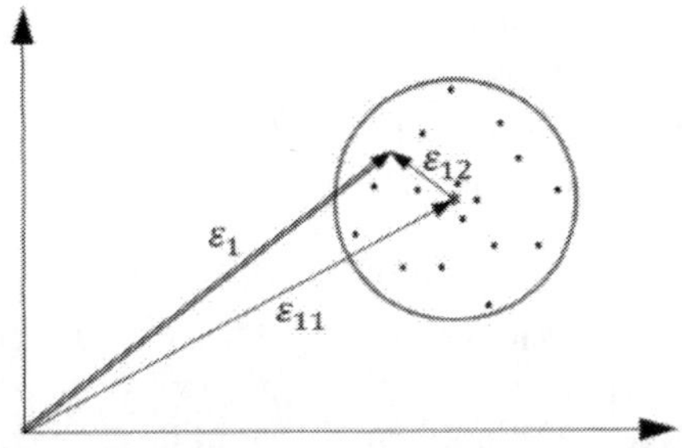

Fig. 6. Machine tool positioning accuracy (ε_{11}) and repeatability (ε_{12}) summed as vectors.

Machine tool positioning accuracy (ε_1) has two components (fig.6): positioning accuracy (ε_{11}) and repeatability (ε_{12}). Today, the positioning accuracy of medium size vertical machine centers is ±0.002…±0.005mm and generally the repeatability is half of the accuracy: ±0.001…±0.003mm.

The second source is inaccuracy of the tool holder (ε_2). The runout of tool holders is under ro_2=0.01mm, but runout of the new generation of high-accuracy tool holders is 0.003 mm.

Inaccuracy of the locating tool (ε_3) is the third source of positioning error, the runout is $ro_3 = 0.005 \ldots 0.01mm$.

If the centering surface of the locating tool is cylindrical, fits H6/g5 and H7/g6 having a minimum clearance are reasonable.

3 Monte Carlo simulation

Generally, the geometrical accuracy analysis of the MFS is a difficult task because modular locating schemes and the dimensional chains are spatial, multi-dimensional, nonlinear and often in presence of clearances.

Compared with other methods Monte Carol simulation (MCS) offers many advantages: simplicity, flexibility, it can be used for nonlinear assembly functions, other distributions than the normal may describe the fixture modules dimension better and more easily. But MCS analysis has two disadvantages: the computational effort can be large, so hundreds of thousands of samples or more are required for accurate results and MCS cannot allocate optimal tolerances.

The problem: estimating the statistical positioning accuracy of the modular locator positioned by means of machine-tool spindle.

In case of variant 1 (see table 1) the variables are:

- machine tool positioning inaccuracy (ε_1): positioning accuracy (ε_{11}) and repeatability (ε_{12}) are 2D vectors, first has a constant magnitude and the second is a random variable with a normal or uniform distribution from 0 to repeatability, both have directions uniform distributed from 0 to 2π;
- inaccuracy of the tool holder (ε_2) and inaccuracy of the locating tool (ε_3) are vectors with magnitudes $0\text{-}ro_2/2$, respectively $0\text{-}ro_3/2$, and angles have a uniform distribution from 0 to 2π;
- ε_4 is generated by the clearance between the tool (4) and locator (3) $c=D_4\text{-}d_3$; usually the diameters of the centering surfaces have normal distributions or uniform, magnitude is random variable with a uniform distribution from 0 to c/2 and the direction is random variables with a uniform distribution from 0 to 2π.

The proposed algorithm simulates the 2D positioning process:

- The error-vector ε_1 is generated as result of adding ε_{11} and ε_{12}.
- The error-vectors ε_2 and ε_3 are calculated.
- The last error-vector ε_4 is generated.
- ε_i, i=1-4 are chained heads to tails.
- The steps 1…4 are repeated, sample size is n=10^6.
- Resulting errors are statistical processed, the positioning error of the locator is calculated based on $\pm 3\sigma$.

For variant 2 $\varepsilon_4=0$, for 3 $\varepsilon_2=0$ and for 4 $\varepsilon_2=0$ and $\varepsilon_4=0$.

Results of the Monte Carlo simulations using Mathcad are shown in table 2. Three cases are considered: high, medium and low positioning accuracies. For each case, in the first row of the table 2 random variables with uniform distributions are considered, in the second row accuracy and repeatability of machine tool have uniform distributions and other ones have normal distributions.

DOI: 10.1051/matecconf/20179405007

Table 2. Results of MFS-TH Monte Carlo simulations.

No	Specifications	Positioning accuracy of the modular locator $\varepsilon_x = \varepsilon_y$, ε_r [µm]			
		Var.1	**Var.2**	**Var.3**	**Var.4**
1	High accuracy: acc=2µm, rep=1 µm, $ro_2=ro_3$=3µm, H6/g5	±12, 13.7	±6.3, 7.1	±11.4, 12.9	±4.9, 5.4
		±10.4, 11.7	±4.9, 5.5	±10.1, 11.3	±4.1, 4.7
2	Medium accuracy: acc=3µm, rep=1.5 µm, $ro_2=ro_3$=5µm, H6/g5	±14.5, 16.6	±10.2, 11.5	±13.2, 15	±7.8, 8.5
		±12.2, 13.8	±7.7, 8.5	±11.6, 13	±6.6, 7.5
3	Low accuracy: acc=5µm, rep=2.5 µm, $ro_2=ro_3$=10µm, H7/g6	±24.1, 27.4	±19.5, 21.9	±20.8, 23.7	±14.4, 15.7
		±19.2, 21.9	±14.5, 16.6	±17.6, 20	±12.1, 13.9

In every cell of the table 2 the first value is positioning accuracy on X and Y axis ($\varepsilon_x = \varepsilon_y$, fig.7), the second value is the half of positional error ($\varepsilon_{ri} = \sqrt{\varepsilon_{xi}^2 + \varepsilon_{yi}^2}, i = 1 - n$).

If 10^6 sample size is used, the stability of results is very good: for second variant, medium precision and uniform distribution the positioning error varies ±0.4%.

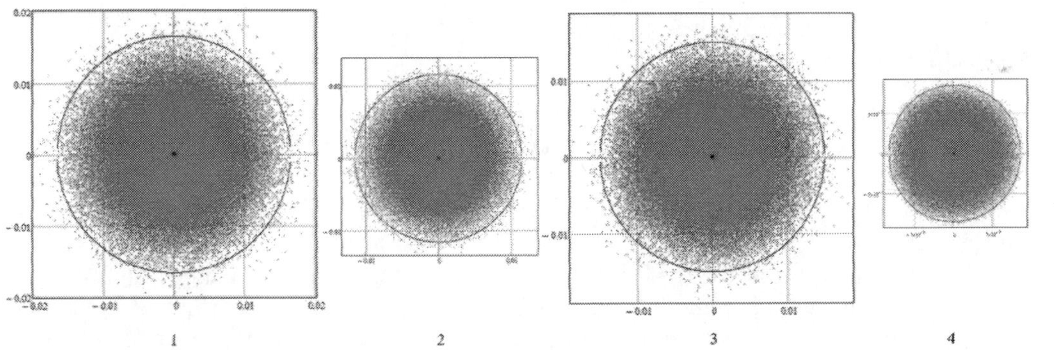

Fig. 7. Locator positioning accuracy for medium precision, uniform distributions (row 3, table 2).

First of all, influence of clearance between locating tool nose and the locator hole is greater than accuracy of machine tool, roundness of the centering tool and tool holder (variants 1, 2 and 3, 4, table 2).

If the centering surfaces are conical the clearance is zero, but the locator aligning is more difficult.

Secondly, machine tool positioning accuracy and repeatability are a significant sources of errors.

4 Comparison between MFS-TH and MFS-MH

MFS-MHs (threaded and cylindrical locating surfaces are coaxial) use cylindrical locating surface of screws/sleeves to accurately locate the modular locators. In contrast to MFS-TH the locating modules of MFS-MH are aligned outside of machine tool.

The mathematical model and solving of the MFS-MH are similar with MFS-TH, but the datum is the baseplate frame.

Table 3. Results of Monte Carlo simulations for couple of commercial MFS-MHs.

MFS-MH	Pin/Sleeve	Hole 1	Hole 2	Pitch	$\varepsilon_x = \varepsilon_y$, ε_r[μm]
Halder L12	12 k5	12 F6	12 F6	50 ± 0.01	± 18.6, 21.3
					±17.0, 19.2
KippM12	12 g5	12 F7	12 F7	50 ± 0.01	±25.3, 28.8
					±23.5, 26.3
Nabeya M12	$12^{0.021}_{0.01}$	$12^{0.009}_{0.004}$	$12^{0.009}_{0.004}$	80 ± 0,01	±10.8, 12.5
					±9.0, 10.3
AMF M12	$12^{0.011}_{0}$	12 F7	12 F7	40 ± 0.01	±19.9, 23.1
					±17.6, 20.0

Sources of errors are: position tolerance of the grid pattern, clearances between the centering hole of fixture base and screw/sleeve, and between mounted locator and screw/sleeve.

The input variables are: position tolerance of grid pattern (TPp), upper and lower specification limits of holes and screw/sleeve, respectively. Two different distributions of input data were tested: normal and uniform. The sample size is 10^6 to get a reliable measure of the output distribution. Data analysis of Monte Carlo simulations from table 2 and 3 shows that fixtures based on MFS-TH can be more accurate than MFS-MH.

5 Conclusion

MFS-THs are feasible in the job and batch productions. Because MFS-THs eliminate the need to align the grid-pattern of the baseplate with coordinate reference frame of the machine tool and the dependence upon the manufactured precision of the baseplate, fixtures assembled online are cheaper than fixtures assembled offline from modules with multipurpose holes.

Most locators of the MFS-THs are cylindrical locators, these can be quickly and accurately aligned with a centering tool mounted in the machine tool shaft. Aligning the more complex locators than cylindrical locators is time expensive and inaccurate, moreover modular blocks can't be mounted, so with these modules only a limited number of workholdings can be assembled.

In this paper several MFS-THs are analysed, and a new variant with conical centering surface is proposed.

A general stochastic model of the MFS-TH locating accuracy is proposed, and it is solved using Monte Carlo simulation.

The positional accuracy of four MFS-THs with low, medium and high precision are calculated. Monte Carlo simulations show that with components of MFS-THs can be assembled accurate modular fixtures.

Even MFS-THs currently aren't used, these MFSs can be more accurate and cheaper than MFS-MHs.

References

1. ***, *Modular Fixturing Handbook* (Carr Lane Mfg. Co. St. Louis, Missouri, 1991)
2. E. Botez, *Tehnologii de programare numerică a maşinilor unelte [Numerical Programming Technologies of the Machine Tools]* (Technical Publishing House, Bucharest, 1973)
3. P.V. Hudimac, Machine-set modular-fixturing system. Patent US5509214 (1996)
4. B. Brânzoi, *Dispozitive de prindere modulare setabile pe centre de prelucrare [Modular Fixtures Configured by Machining Centers]* (M.S. Thesis, Transilvania University of Brasov, 2015)
5. J. Gu, *43rd Proceedings of the North American Manufacturing Research* **xxx**, (2015)
6. S. Fletcher, *Laser Metrology and Machine Performance* (Euspen Ltd, Cranfield University, 2009)
7. O. Svoboda, P. Bach, *IMTS 2004 Manufacturing Conference,* **1** (2004)
8. C. Wang, *Seven International Symposium on Instrumentation and Control Technology*, ISICT08, **7129**, (2008)
9. S. H. Huang, *Int. J. Prod. Res.* **42**, 4871 (2004)
10. S. Klabunde, R.Schmidt, Modern Machine Shop **3,** (1998)
11. B. Mullany, Transactions of NAMRI/SME, **36,** 309 (2008)
12. K. Gupta, *5th International & 26th All India Manufacturing Technology, Design and Research Conference* (AIMTDR), **1** (2014)

Design of a robotized workstation making use of the integration of CAD models and Robotic Simulation software as way of pairing and comparing real and virtual environments

Karol Velíšek[1], *Radovan* Holubek[1,*], *Daynier Rolando* Delgado Sobrino[1], *Roman* Ružarovský[1], and *Nina* Vetríková[1]

[1] Slovak University of Technology in Bratislava, Faculty of Materials Science and Technology in Trnava, Institute of Production Technologies, 91701 Trnava, Slovakia.

Abstract. Over the last years, there has been an increasing tendency and pressure on the faster implementation robotic devices and systems in manufacturing. Such transition involves several disciplines starting with the prototyping of CAD models itself. The paper addresses the creation of CAD models and is mainly aimed at their integration in a given simulation environment according to the conception and guidelines of Industry 4.0, where the part itself becomes the entity carrying most of the needed information at any time of a production process. The creation of such CAD models is key for the further and better customization of simulations. In other to better exemplify all this, the paper describes the whole process of "virtual to real life implementation" of a given robotized workplace needed to be developed at the Institute. The design of such robotized workplace included the use of an ABB IRB 120 robot and several other devices which were all designed, simulated and analyzed in a virtual environment before the final development and implementation. This paper helped demonstrating the importance of having exactly the same model (real and virtual) with respect to the success of the offline simulations.

1 Introduction

Since the very beginning and course of the 3rd Industrial Revolution, the use of PLCs, Robotics and automation in general have played an outstanding role in the development of manufacturing. They have all become an irreplaceable asset that has survived all time boundaries and keep being of vital importance today. Robotics and automation have not only helped increasing productivity and taken manufacturing to another level, but they have also had a significant role in cost reduction and have contributed to make work safer than ever. However, everything has not been that easy, together with all these benefits companies have also seen themselves immersed in the constant training of their workers

*Corresponding author: radovan.holubek@stuba.sk

and redesign of their production plants in order to really meet all the possibilities robotization and automation offer.

All what is mentioned above has not only impacted production giants but also small medium enterprises and educational institutions like the Institute of Production Technologies itself.

This institution shares a great responsibility in the training of national specialist and students because of its strategic positioning in the canter and surroundings of the bigger manufacturing hubs of the country.

It has been proved that the most effective way to deal with new technological challenges and new tendencies, e.g. Industry 4.0 nowadays, is the training and preparation of the future workforce since the very base.

This specially applies not only to Robotics and Automation but also to their linkage to CAD designs software which are all very important and complex disciplines whose the Slovak industry so badly asks for, needs and the present papers moves around.

Being aware of all above mentioned elements, the present paper introduces the design of a robotized workstation.

Such design has been conceived from the very CAD design and has been then simulated offline in RobotStudio before the real implementation, this has offered the possibilities of pairing and comparing both the virtual and real environments and somehow qualitative study the effectiveness of the models before the real implementation in one of the teaching laboratories of the institute.

The rest of this paper is divided in: Section 2: Design of the workstation in CATIA V5, section 3: Modelling of the workstation in the RobotStudio environment, Section 4: Construction of the designed workstation and analysis of the virtual and real environments, Section 5: Conclusions and further research, Section 6: Acknowledgement and finally a section of references.

2 Design of the workstation in Catia V5

This section shows the chronological steps in the creation of the workstation. The whole conception started from considering the robot type (ABB IRB 120), the parts to manipulate and the pallets to be used.

As for the robot type it was mainly useful to take into account its weight, dimensions, movements angles among a few other technical information available at the manufacturer' website, see [1].

Based on the previous information, the base frame of the workstation was given the dimensions shown below. It was important to create a stable frame able to deal with the full-speed movements and acceleration of the robot in the worst case scenarios.

Most of the profiles used were made of aluminum and can be found in [2].

- Dimensions of the base frame: 1300x1000x1000 mm, it was made of 40x40 mm aluminum profiles.

As it can be also seen in Fig. 1, a wooden table was inserted into the design so as to secure a much better stiffness of it.

At the same time it will be also used for placing the robot controller, the wooden material does better avoids shakings who may well have a negative effect on the structure and integrity of the Controller.

Fig. 1. Design of the base frame.

In a similar way the work table was also created, see Fig. 2.

- Dimensions of the profiles in the work table: 1400x1200 mm. The profiles were of 40x80 mm and were cut (in length) at 1400mm, while 15 of those profiles were fixed together giving the needed width of 1200 mm.

Fig. 2. Design of the work table.

A better fixation of the robot on the above shown table is to be achieved by using the steel base plate under its base.

This has been also design as shown in Fig. 3.

Dimensions of the steel plate: 370x400x10 mm.

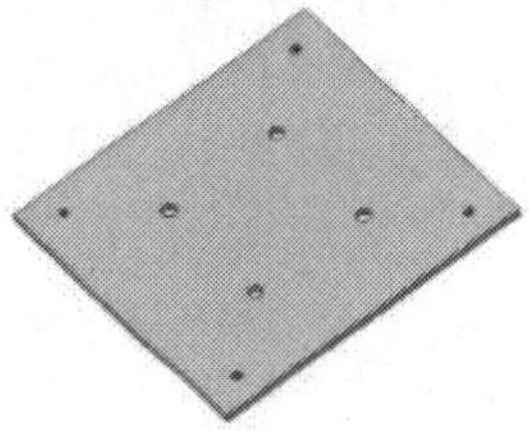

Fig. 3. Design of the steel base plate for the base of the robot.

Fig. 4 shows all of the previous element together in what would be the final workstation, still it does not include a series of safety elements [3].

Fig. 4. Designed workstation without safety elements.

As for fulfilling some of the safety regulations of the robotized workstation to be located at a student's laboratory, the following safety barriers were also designed and added to the surroundings of the robot, see Fig. 5.

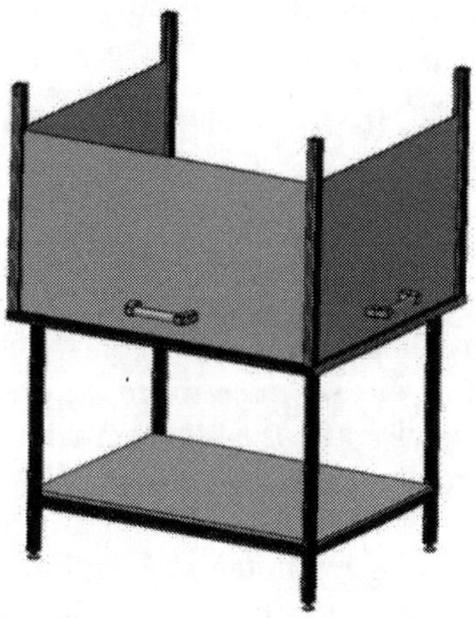

Fig. 5. Designed workstation with safety barriers.

3 Modelling of the workstation in the RobotStudio environment

In this section, all of the previous CAD models have been imported into the RobotStudio environment. Despite existing many options for file conversion, in this case, the STEP files coming from CATIA V5 were converted to SAT files using a CAD exchanger tool. With all this, the main idea was to look for the better layout configurations and prove the kinematics and reach of the robot to execute given tasks which was key prior to the definite construction of the workstation [4]. The following Fig. 6 shows the virtual environment of the workstation while executing a given operation in the student's laboratory of the institute. Based on this off-line analyses and simulations, the final workstation is to be finally constructed and compared [5]. Notice that after being all file types imported, it was necessary to place the robot in the "right "position on the previously designed steel base plate, for this it was necessary to make an analysis of the reach of the robot and its kinematics.

 DOI: 10.1051/matecconf/20179405008

Together with all this, several targets were created off-line so as to verify the positioning and visualize the functioning of the robot through simulation.

From all this process, students will contribute and continue to develop an existing database of CAD and SAT files that will be made available once appropriated and will have a positive impact on the teaching process [6, 7].

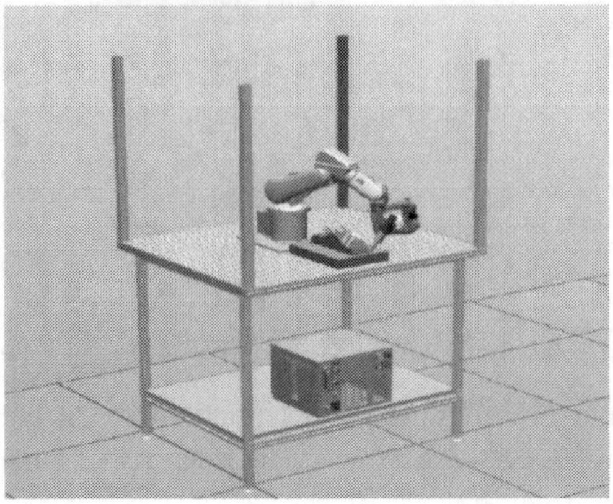

Fig. 6. Designed workstation in the RobotStudio environment at an institute's laboratory.

4 Construction of the designed workstation and analysis of the virtual and real environments

Taking into account all of the design elements mentioned above, also the off-line simulations and analyses made in RobotStudio, the physical version of the designed workstation is shown next in Fig. 7.

Notice that, as somehow already mentioned before, this is an educational workstation that besides serving research purposes leans mainly towards an enhancement in the teaching process in the laboratories of the faculty.

Within this workstation students are able to simultaneously play off-line simulations and execute those same movements in the real station, we can also make the students send their assignments via Ethernet and directly see their results in the workstation and in a similar way input their off-line programs into the USB ports of the robot controller [8].

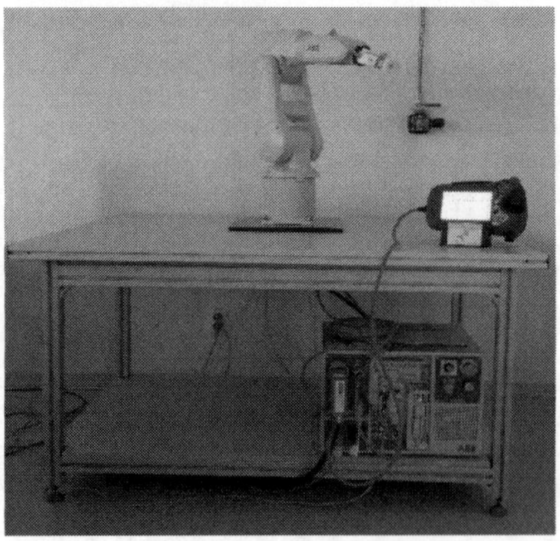

Fig. 7. Constructed workstation located at an institute' laboratory.

5 Conclusions and further research

The present paper presented in a synthetized way the design, off-line simulation, tests and construction of a robotized workstation at one of the laboratories of the Institute of Production Technologies.

The whole process started with the CAD designs of the components of the workstation, then these were imported, simulated and tested off-line in the RobotStudio software. Once positioning, reach and other issues were analyzed the modelled workstation was constructed and finally compared with the models by executing the same movements created off-line.

The whole idea will positively impact the teaching process since it allows students to create off-line models and then simultaneously compare them with the real workstation. All this will also contribute to the creation of an educational database of files that will be used for exercises and much more within the teaching activities.

Further research is to encompass the construction and inclusion of many other devices and elements pertaining to the same station which have been already modelled and taken into account for this initial conception.

This work was supported by KEGA 027STU-4/2014: Establishment virtual laboratory of robotics and manipulation techniques.

References

1. ABB, Technical data for the IRB 120 industrial robot. Available from: http://new.abb.com/products/robotics/industrial-robots/irb-120/irb-120-data (2016)
2. FM Systeme Kompatible Profilreihen & System-Engineering, Germany (2013)
3. J. Johnson, M. Kwoka, H. Houayek, I. Walker, K. Green, Design, construction, and testing of a novel robotic workstation. In *Proceedings fourth international conference on computational intelligence, robotics, and autonomous systems* (CIRAS). Palmerston North, New Zealand, 155–160, (2007)
4. P. Neto, J.N. Pires, A.P. Moreira, CAD-based off-line robot programming. Robotics Automation and Mechatronics (RAM), *IEEE Conference*, 516-521, (2010)
5. J. Zhang, X. Fang, L. Qi, Robot. Comput. Integr. Manuf., **38**, 9–15 (2016)
6. S. Hao Deng., Z. Hua Cai., D. Dan Fang., H. Lin Liao., G. Montavon, Surface and Coatings Technology, **206**, 19–20, 3875-3882 (2012)
7. R. Holubek, D.R. Delgado Sobrino, P. Košťál, R. Ružarovský, Applied Mechanics and Materials: Novel Trends in Production Devices and Systems II, **693**, 62-67, (2014)
8. R. Holubek, K. Velíšek, M. Šimúnová, N. Vetríková, D. R. Delgado Sobrino: On the possibilities and the advantages of robot off-line programming: Teaching of off-line programming methods as a tool to further industrial implementation utilization. In *3rd International conference on Advances in mechanical, aeronautical and production techniques - MAPT*. 1. Issue. New York: Institute of Research Engineers and Doctors, (2015)

ERP processes automation in corporate environments

Victor Antonoaie[1], *Adrian* Irimeş [2], and *Lucia-Antoneta* Chicoş[3,*]

[1] Transilvania University of Braşov, Department of Management and Economic Informatics, Colina Universităţii 1, Braşov, Romania
[2] IBM Romania, Turnului 5, Braşov, Romania
[3] Transilvania University of Braşov, Department of Manufacturing Engineering, Mihai Viteazul 5, Braşov, Romania

Abstract. The automation processes are used in organizations to speed up analyses processes and reduce manual labour. Robotic Automation of IT processes implemented in a modern corporate workspace provides an excellent tool for assisting professionals in making decisions, saving resources and serving as a know-how repository. This study presents the newest trends in process automation, its benefits such as security, ease of use, reduction of overall process duration, and provide examples of SAP-ERP projects where this technology was implemented and meaningful impact was obtained.

1 Automation principles

The IT industry has, for decades, focused on process automation and ways to increase IT productivity and efficiency with aim to reduce operator errors. Due to market condition, the need to provide information and faster speed to market, the focus is shifting to enhance business productivity. The goal for these improvements is to help capture new markets and gain a competitive edge in today's business and technology race. In order to achieve these goals, the company needs to drive down costs and improve efficiency, to free funds and resource to make strategic investments. Furthermore, making sure that the IT specialists are focusing on innovation in the base areas of the business, working through complex problems, and improving productivity is of the highest importance [1]. Thus, automation is becoming ever more critical.

Until now, traditionally, the automation process relies on scripts that combines computer commands that are designed for a single platform or for a specific product. In order to achieve this, traditional process automation involves design and implementation of multiple instances of the same product. This type of automation is now becoming obsolete and modern methods are required. Now, solutions that use modern analytics-based automation are becoming the building block of the next generation enterprise IT system. One example is the IBM Enterprise IT Automation Services, which is seen deployed in various organizations, helping them to become industry leaders.

* Corresponding author: l.chicos@unitbv.ro

The standard IT setup of today's enterprise includes technologies from an array of vendors as well as various sourcing strategies and delivery models that span several platforms and products. This modern and complex IT environment, makes fixing operations issues, even small ones, a challenge. Even trivial tasks and problems caused by the simplest issues, such as user password management, grow out of control due to an error file that is too large to use, and due to this the time to identify and fix an error takes many hours. These types of events can cause interruption or in time severely degrade overall performance, and can as a result have an impact on the enterprise business revenue.

The topic of this paper is to present the newest trends in process automation, its benefits (security, ease of use, reduction of overall process duration) and provide examples of ERP-SAP projects where this technology was implemented and meaningful impact was obtained.

2 Implications and uses of automation technology in the IT sector

2.1 What is Enterprise IT Automation services?

Regardless of how good and solid the enterprise IT infrastructure is, how thorough and proactive the maintenance plans it has employed, problems with inevitably occur. Companies are dependent on IT staff to manage the errors, which itself is a risk. Well established processes and trimmed staffing models no not fully eliminate the time wasted as IT technicians respond to incidents and requests. Considering the criticality of certain incidents types the failure to solve them in time can have a significant business impact. To minimize the delay between the time of the failure and the action taken to fix the problem is to have in place autonomous software solutions, robots that monitor the enterprise IT landscape and that takes actions when something breaks.

Enterprise IT Automation Services provides the ability to fix problems immediately, as they occur, reducing the Mean Time to Respond (MTTR) from days and hours to minutes. By doing so, it prevents severity 3 and 4 events from becoming severity 1 events.

The enterprise IT infrastructure is continuously being analyzed and logged by Enterprise IT Automations Services, tracking and logging everything and so providing granular, transparent and deep level details. The created data can be then analyzed further in order to identify patterns and provide starting points for continuous improvements.

2.2 What are the benefits of Enterprise IT Automation services?

Automation service solutions [2] have been deployed consistently trough the clients' landscape worlds wide and demonstrated value to clients in areas like:

• Service level and quality improvements - Virtual engineers are autonomous dispatched to address and fix issue as they occur. Enterprise IT automation services dramatically reduce the MTTR by removing the dependency on human engineers to handle a large volume of the issues signalized. The risk for human error is also mitigated for the problems where virtual engineers are programmed to fix.

• IT staff productivity and efficiency improvements - Provides enterprise with the ability to do more work with less resources. The Return of Investments (ROI) is calculated considering the existing baseline and the potential gain by automation. Parallel to this cost benefits arisen with improved service levels and greater availability. Another indirect benefit results from freeing IT operations resources. Technicians are no longer distracted by low service interruptions and they can concentrate on fixing complex problems more effectively.

• Continual services improvements - Efficiencies and improvements vary from one implementation to another, based on the baselines and penetration rates. The persistent pattern, however, across all is an improvement in the form of reduced response and resolution time, improved quality of service, and a reduction in outages.

2.3 How does Enterprise IT Automation services work?

Enterprise IT Automation Services includes several key components [1]:
• The virtual engineer is a software component that uses algorithms to analyze and automatically take action on issues that can impact the business process, without any form of human intervention. The virtual engineer continuously monitors the system for problems for which it is programmed for, and when it spots such an issue, it goes through a series of well-defined programmed actions (in a logical preset order) that are specific to the situation in order to remediate the issue.
• Patterns describe the behaviour that virtual engineers must follow to respond to an incident. There are standard patterns, client-specific patterns, and new patterns that are developed based on insights discovered by analytics.
• Analytics helps to identify what can be automated in further deployments, monitor the system behaviour of the automation, learning from company global experience, and so on. Analytics is used to measure the effectiveness of automation and patterns and to provide information that can be used for new development, changes to the existing systems, or determining the root cause of problems.
• Enterprise IT Automation Services is deployed as a service on multiple platforms like the IBM SoftLayer® cloud infrastructure. Therefore, it requires minimal effort and infrastructure commitments on an organization's IT environment.

Fig. 1 provides a high level view of how IBM Enterprise IT Automation Services integrates with the IT environment of an organization enabling the virtual engineer to detect, diagnose and either fix the problem or escalate it to a human engineer for resolution.

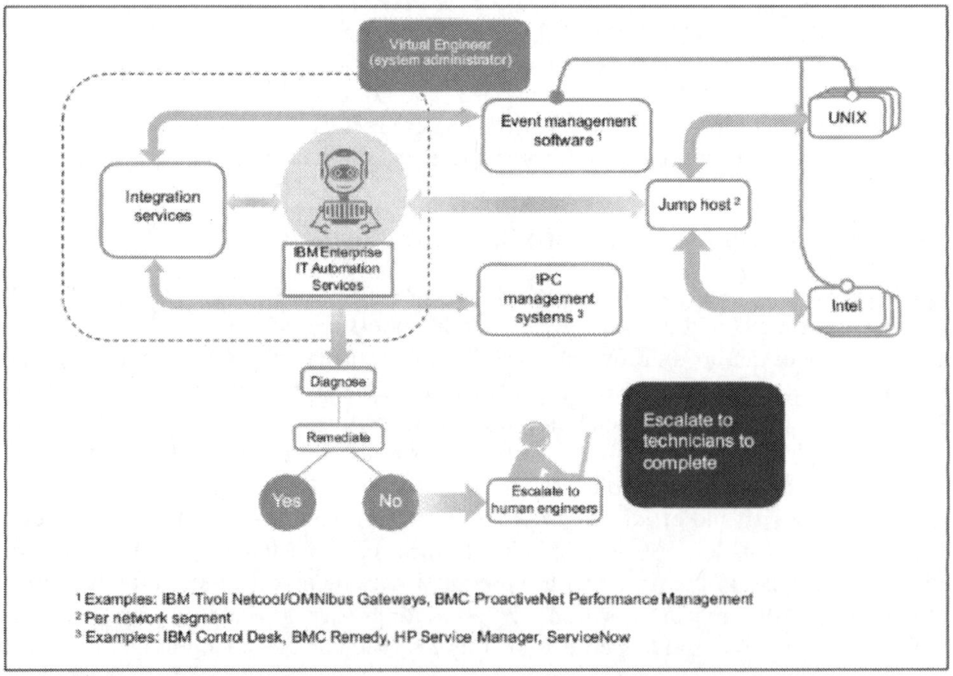

Fig. 1. Integration of IT Automation Services with the company IT environment [1].

2.4 How are Enterprise IT Automation Services capabilities deployed in an enterprise IT environment?

Enterprise IT Automation Services is deployed in two phases:
• Enablement

At first a onetime setup implementation is needed for the automation framework. Enterprise IT Automation Services slots into the enterprise IT operations as a new step in the support flows of IT technician's engagements. Enterprise IT Automation Services integrates into an organization's environment without requiring any changes to tools or previous automation investments. This is a simply and interface with key IT service management systems to automate use cases across the organization's IT infrastructure platforms, database, middleware and groupware applications, storage devices, network devices, and some business applications.
• Continual service improvement

A continual service improvement process based on analytics is instated to propose new automation use cases and updates to those deployed already by using:
- Trends in the behaviour of active (deployed) automation;
- New trends in the organization's environment, for example, new type of incidents, service requests and so on;
- Lessons learned from global implementations.

Are there any limits to what can be automated? Based on a clear set of questions, can be identified a system suitable for automation:
- Is the system on which automation should run reachable across the network?
- Is it possible to interact with the system programmatically by using APIs or shells?
- Is it possible to describe what have to be done on the system as a Standard Operating Procedure (SOP)?

If the answer is "yes" to all three questions, automation is applicable.

3 The support system for an ERP implementation. Case Study – SAP implementation for the financial branch of an automotive company

In this paper, the case study is structured into six main parts to better present the scale and complexity of an automation process implementation [3] into a live SAP ERP support system:
• Part 1. Scale of the implementation, integration with other applications, incident management systems, and general system description.
• Part 2. Distinguishing between the repetitive tasks, which can be automated, and difficult incidents which require a differential analysis and resolution. Automation advantages [4].
• Part 3. Designing an automation process guideline to perform the abovementioned tasks - workflow. Complete tasks versus partial automation input.
• Part 4. Testing and implementation phase. Troubleshooting and process adjustment. Implementation in a live system.
• Part 5. Post implementation analysis – time and effort saved through automation.

3.1 Part 1 - System implementation and integration

To be able to fully understand the need for an automation process [5] in an extended support system for an ERP (Enterprise Resource Planning) implementation we must first comprehend the scale and complexity of such an enterprise [6].

The case study selected by the authors focuses on a support system for a large ERP Implementation for two Banking Systems part of an automotive consortium. The SAP (Systems, Applications and Products in Data Processing) system installed for the client is part of a larger application network which handles all the functions and aspects of the company. As far as the user base is concerned, the SAP implementation is used by over 2000 individual user and several 'technical users' (a technical user is a user account which is used for communication purposes between application as well as for specialized automated tasks and jobs which require a certain set of authorizations to ease the role assignment process and ensure clear usage parameters). The system functions on a 24/7 schedule, with no downtime, with unrestricted user access. Automated processes, tasks, file transfers and other activities, part of the IT workflow [7] , which do not require direct input are scheduled, in general, outside working hours to maximize resource use (as with any application system, the server resources are limited. Therefore, some activities need to be scheduled by taking into account the overall resource use to minimize the risk of failure due to insufficient memory or lengthy run times) and to ensure minimal potential interference from other user related activities.

The system issues can be sorted by the way in which they are reported:

• issues signalled by the user: a problem encountered by a user in any part of the system, ranging from a locked transaction, errors in performing a task to system dumps. These issues are signalled to the support team through an incident management system. The support team must resolve the issue in a time frame established by the severity of the problem (from 2h in the case of a critical incident to 3 days in case of a low impact issues).

• issues triggered by failed SAP monitor: the support team sets monitor applications for key jobs [8] and processes running in the system. When one of them fails, the monitor automatically creates an incident and the support team is notified

• issues detected by the support team as part of regular checks: the team carries out checks on key parts of the system at scheduled intervals to monitor complex activities which cannot be covered by a monitor (because they require in depth analysis).

• incidents reported by other support teams: a support team from another linked application signals an issue with a cross-application process which requires the intervention of the SAP support team to fix.

3.2 Part 2 - Establishing tasks for processing by automated robot

As we have described in the above subchapter, the incidents received by the SAP-ERP support team can be very different, ranging from basic restarts of tasks to complicated program code issues. Therefore, the first step in any automation implementation must be to distinguish which of these tasks can be subjected to processing by an automated robot and which can be handled directly by the human operators.

Most of the incidents reported directly by users generally concern difficult issues, from all aspects of the application and are in most cases different, both in cause and in resolution. In this case there is no suitable way in which to automate the resolution process. The effort involved in creating the framework for each individual situation would be highly time and resource consuming and the expected savings would not be enough sufficient to merit the investment [9].

DOI: 10.1051/matecconf/20179406001

On the other hand, incidents which are repeatable and which have a clear, well established resolution process are prime candidates for robotic automation. In this category we can include failed jobs which require a simple restart, incidents reported by application monitoring which occur from various reasons, issues with data files not received by an application caused by delays etc.

The clear definition of how an incident process suitable for automation, in the general consensus of the automation guidelines [10] is: an incident which has a clearly established cause, which contains clear and complete information, which is repeating frequently enough to warrant an automated solution and which can be resolved entirely (or the automation robot would ensure a sufficient contribution to the solution workflow) by an automated robot working on a designed framework and within the set parameters of the workflow.

Therefore, the main concern when deciding to implement an automated robotic solution in a support system is to have enough such incidents to sustain the investment. More clearly put, the time and effort saved by the robot must be greater than the cost of the investment.

The main advantages of implementing an automated robot for such actions as described above would be (Fig.2):

• An Automation Robot is *consistent* throughout the entire process, which means consistent 24x7 process execution depending on the design set by the users. Moreover, the set process steps are followed correctly every time, without fault and at the same time, the results are predictable with no errors;

• A process automated by a robot has the advantage of *availability*: it can function on a 24x7 schedule, without needing vacations, without low performance periods and can be constantly monitored through specialized tools and reports;

• *Security* is a key feature of automated processes because: the robot only performs what it is programmed to do as part of a workflow. It only has access to the systems for which is programmed to work on and has a full audit trail of all actions. Moreover, it does not mine or harvest data and the risk of problems/failures is reduced because it has key control points in the process definition;

• *Efficiency* is perhaps one of the most valuable features of automation: robots performs tasks quicker and with better cost efficiency than human resources while also handling peak periods without needing resource reassignment.

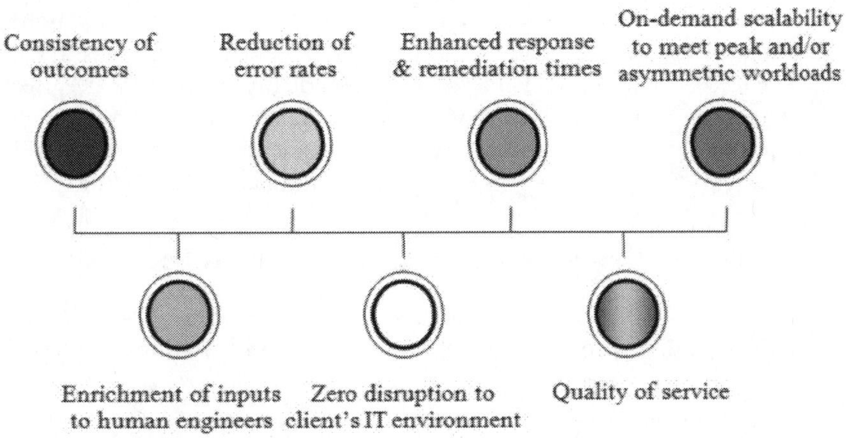

Fig. 2. Advantages of automation.

3.3 Part 3 - Designing automation process guideline

Any automation process must start from an actual activity, performed by a human user within certain set parameters, such as: duration, complexity, number of repetitions within a time frame and costs.

After finding such an activity, like the ones presented in the subchapters above, the actual guideline process can begin. The first step is always to deconstruct the overall activity into steps arranged in the logical and chronological order. Then each individual step must be analyzed separately to determine the actions involved their duration and prerequisites as well as the effects.

Once each individual action is analyzed, measured and the robotic actions are determined, the individual action times are added up. The result is the overall robotic activity duration which is then compared to the time a human user would need to perform it. Nevertheless, this is not the only time consideration taken into account. For example, a human user would perform a system check three times during a normal work day of 8 hours, as these checks are only a small part of his or hers daily activities. On average, such a check would take approximately 10 minutes per activity which would generate a total of 0.5 hours per day. This is 6.25% of the workday dedicated to these checks. By comparison, an automated robot can do the same checks in less than 1 minute, ten times faster. Moreover, the robot can perform this task every hour, not only during working times but 24x7. This can prove very useful, as the more frequent the checks, the quicker a problem can be detected and resolved.

An automated robot [11] can perform any task he is designed to, as long as the initial programming includes all possible variables and it has clear and correct instructions. This is one of the most important parts of the activity workflow.

Nevertheless, there are instances where a process can be categorized and broken down into clear fixed steps only to a certain extent. As most SAP installations in companies contain large portions of customized code and programs, the number o variables involved in some processes can be too large to analyze and create scenarios on.

But this does not exclude automation completely from these workflows. If a problem resolution contains at least one initial standard step, that step can be automated. At first it might not seem as a necessary measure, even a time/effort saving of 5 minutes per incident can lead to a significant reduction of average solution times when looking at the entire company. The best example would be the automation for reading the information in a customer incident opened to the support team. The robot reads the text and searches for key words such as the target system, the client, transaction code used, error message code, etc., designed for each type of incident. If the robot does not find any of the key word in the message, it instantly e-mails the user and requests the missing information. If all the information required is present, the robot adds a line in the incident stating this fact which signals the human user that he can begin working on finding a solution to the issue.

3.4 Part 4 - Testing and implementation of automation robot

As with any new functionality to be implemented into a live production system, thorough testing and troubleshooting is required before the go live. Most SAP installations involve a three tier system, with a development system (where the programs are designed), an integration/testing system (which is an identical copy of the production system – it is here where programs are tested for functionality and potential issues) and a production system (a live system, where the users operate – and from where incidents will originate).

An automation robot implementation is no different from any other developed program or feature, therefore the first step in the installation is to apply the robots to the testing system. As this system is an identical copy of the production system, developers can generate the exact circumstances of a system issue and see how the robot performs.

It is also in this test phase where most of the adjustments can be made to the process and workflow, as even the best design needs real world issues testing to ensure complete functionality. This is also the stage where developers check if the robot is authorized to perform the designed tasks and can interact correctly with all the desired interfaces.

Successful test results in this pre-production system is mandatory before the automation robots can be implemented in the live system, as any potential flaw there could cause significant business impact.

When the automation robot receives the testing green light from all parties involved, specifically the developers, the business test team and the system administrator, in the next step is to install the automated solution in the production system.

The automation robot functions from a separate server and is not installed in the system. The robot functions in the system in much the same way in which a human user would. It would access it using preset credentials, follow SAP specific transaction patterns and perform general user actions.

3.5 Part 5 - Post implementation analysis

Even though the save in effort (FTE - Full Time Equivalent - the number of hours a human employee would work on a regular basis) and time is established in the early stages of the automation design, the real results can only be seen after the robot has been implemented in the production system and has been functioning for a longer period of time.

Based on the results of similar implementations in five different companies from various industries, the benefits of robotic automation implementation can be seen in table 1 and following charts (Fig.3 – 6).

Table 1. Analysis of automation by industry.

Company Type/Indicator	Telecommunication	Energy Distribution	Financial Services	Technology Company	Large Retailer
Total FTE before automation	450	400	800	360	680
Number of processes automated	15	12	32	14	10
Number of robots	160	120	180	130	168
FTE Saved	300	250	360	210	342
Projected ROI % (after 3 years)	80	65	85	72	59

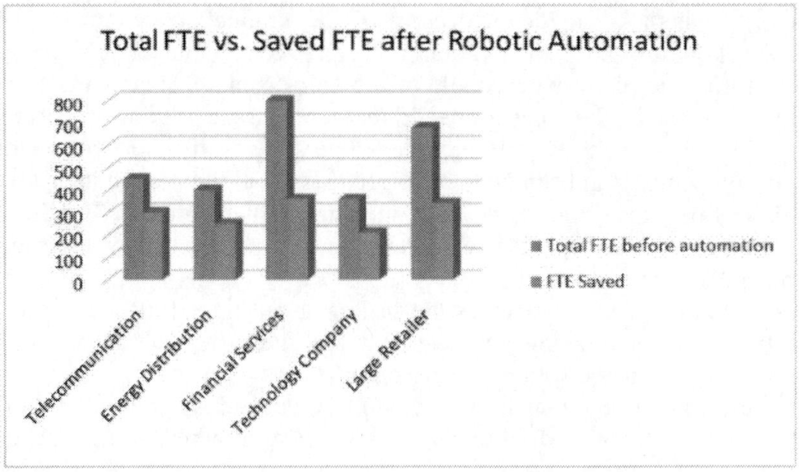

Fig. 3. Total FTE vs. saved FTE.

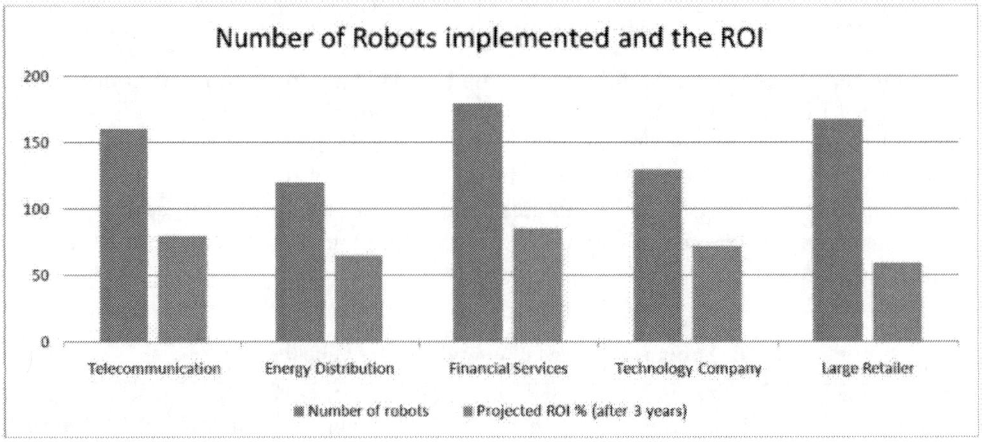

Fig. 4. Number of robots implemented and ROI.

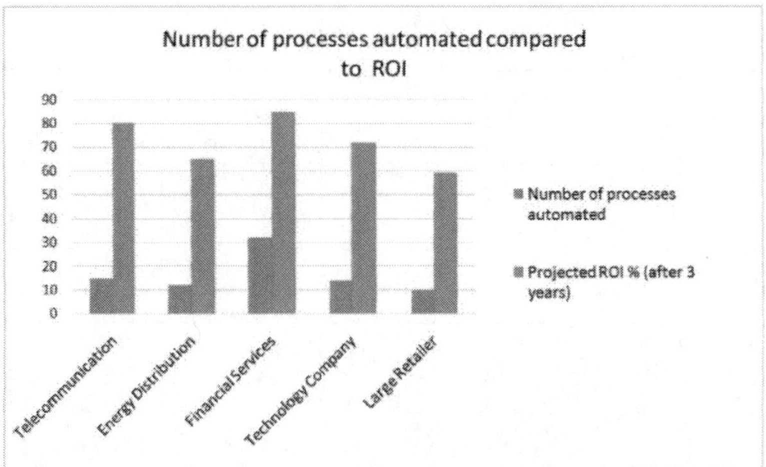

Fig. 5. Number of processes automated compared to ROI.

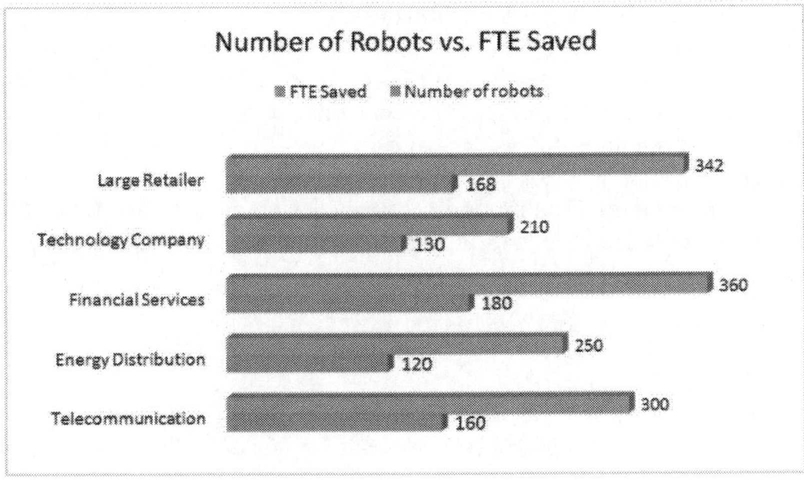

Fig. 6. Number of robots vs. FTE saved.

As it can be clearly deduced from the above charts, the advantages of processes automation are evident. There is a clear decrease in the FTE (Full Time Equivalent) after the automation robots took over certain task. Moreover, the more processes automated the greater the number of FTE saved. On top of that, the projected ROI (Return of Investment) increases proportionally to the number of company processes automated.

Automating processes with robots [12] is only one of the aspects of overall IT process optimization. To complement the work of automated robots in an IT company the future solution is the deployment of cognitive assistant machines [13] in these systems.

The cognitive assist is based on a super computer with a high capacity for assimilating and disseminating information. The processing power provided by this machine [14] is used to create a program which can answer real language questions based on previously registered information.

The application of such a machine in a SAP support system is mainly targeted on uploading operational manuals, problem resolutions, system designs and any other system related documentation. The purpose of this action is to create a large enough database to, potentially, allow for this cognitive assistant to provide solutions to issues based on similar problems faced in the past and system characteristics.

Even though such a task could take a long time to accomplish (taking into account the sheer volume of data needed to be inserted into the system) the results could prove extremely useful as the assistant could provide instant solutions to issue which otherwise would take hours/days to analyze. Moreover, the live database so created could provide a much more efficient way of storing and finding needed information compared to general database queries.

4 Conclusions

As it has presented in the above chapters of this papers, Robotic Automation of IT Processes is an inevitable step in the development of this ever-changing industry, as more and more companies have entered this field, both as providers of automated solutions as well as its recipients.

The advantages of such an implementation in a SAP-ERP system are clear and strong: reduction of response time, a cut in man hours spent on repetitive tasks, reduction of costs and avoidance of potential faults through human failure.

An automated robot is an excellent tool for assisting human operators in repetitive tasks as well as performing individual tasks. Moreover, when coupled with a well designed data base and a cognitive assistant, it can also serve as a live search data repository and perhaps event provide analytic solutions to system issues.

Even though the initial costs of automating suitable processes with robots are significant, both in terms of funding and effort in designing the workflow, the potential savings for the company are clear and failing to join the automation revolution could cost companies their business advantage in the future.

References

1. F. Bouz, H. Dave, *IBM Enterprise IT Automation Services - Dynamic Automation in the Age of Cognitive, Dynamic Learning and Self-healing Systems* (IBM Redbooks, NY, 2016)
2. M. A. Rochester, 5 Things to Know About IBM Enterprise IT Automation Services in the Age of Cognitive, Dynamic Learning and Self-healing Systems, https://www.ibm.com/developerworks/community/blogs/5things/entry/IBM_Dynamic_Automation_in_the_Age_of_Cognitive_Dynamic_Learning_and_Self_healing_Systems?lang=en, accessed 02.06.2016
3. D. Krapohl, Integrated Approach to Organizational Development, http://www.augmentedintel.com/wordpress/index.php/integrated-approach-to-organizational-transformation/, accessed 10.06.2016
4. Process Automation. Streamline operations and increase productivity, http://www-01.ibm.com/software/info/bpm/process-automation/, accessed 14.05.2016
5. B. Violino, Robotic process automation: The new IT job killer?,http://www.infoworld.com/article/2898108/robotics/robotic-process-automation-new-it-job-killer.html, accessed 2016
6. G. Nizri, IT Process Automation is the Way of the Future – Don't Get Left Behind! http://ayehu.com/it-process-automation-is-the-way-of-the-future-dont-get-left-behind/, accessed 10.06.2016
7. D. Georgakopoulos, M. Hornick, A. Sheth, *An overview of workflow management: From process modeling to workflow automation infrastructure* (Kluwer Academic Publishers, Boston, 1995)
8. A. Kento, *Efect of Job Size Characteristics on Job Scheduling Performance. Job Scheduling Strategies for Parallel Processing*, (Springer Berlin Heidelberg, 2000)
9. S.M.Merritt, J. L. Unnerstall, D. Lee, K. Huber, Hum Factors **57**, 740 (2015)
10. B. Lydon, Industry 4.0 for Process Automation – Process Sensors 4.0 Roadmap, http://www.automation.com/portals/process-automation/industry-40-for-process-automation-process-sensors-40-roadmap, accessed 30.05.2016
11. What is Robotic Process Automation?, http://www.irpanetwork.com/what-is-robotic-process-automation/, accessed 28.06.2016
12. D.Williams, IT Operations Run Book Automation, https://www.gartner.com/doc/492859/it-operations-run-book-automation, accessed 28.06.2016
13. D. A. Ferrucci, IBM J RES DEV **56**, 1:1 (2012)
14. T. Winograd, R. Flores, *Understanding Computers and Cognition* (Ablex Publishing, New Jersey,1990)

A multi-criteria decision making approach for supplier selection in the flexible packaging industry

Ciprian Cristea[1,*], and *Maria* Cristea[1]

[1]Faculty of Electrical Engineering, Technical University of Cluj-Napoca, Romania

Abstract. The supplier selection problem represents one of the most important components of the supply chain management. This article presents a multiple criteria decision making analysis contributing to the selection of the most convenient supplier in the flexible packaging industry. Due to the fact that in today's supply chain management, the performance of potential suppliers is evaluated against multiple criteria rather than taking into account only the cost factor, the appropriate criteria in the supplier selection have been identified, and the considered variants are assessed, considering the decision maker's preferences and existing constraints. The variants are ranked in terms of their suitability for selecting a supplier with the use of Electre III method. The results obtained from the simulation experiment suggest that this methodology is a feasible decision support model.

1 Introduction

In order to be up-to-date and feasible on the contemporary global market, characterized by issues like intense competition, high consumer expectations for quality products and services, low profit margins and short lead-times, enterprises need to take the advantage of any opportunity to improve their business processes [1]. Firms realize the significance of supply chain management in their organizations due to the fact that their success is not achievable without an efficient one which plays an important role in the attainment of sustained competitive advantage on the market. Supply chain management represents the overview of the product's lifecycle: from the supply which focuses on obtaining raw materials, to the manufacturing of the product that focuses on transforming the obtained raw materials into finished products, to their distribution to customers through distributors, warehouses and retailers [2]. Supplier selection occupies an important position in the supply chain management, and is a research area that continually receives a relatively large amount of attention in the literature, from both practitioners and academics.

Supplier selection is the process by which companies identify, evaluate and choose the suppliers to become part of their supply chain. The main objective of the supplier selection process is to maximize overall value to the purchaser, reduce acquisition risk, and build a close and long term relationships between buyers and suppliers [3]. The selection of the

* Corresponding author: ciprian.cristea@emd.utcluj.ro

right suppliers is essential for enhancing the enterprises' competitiveness and to promptly react to market requirements and innovation process [4]. Likewise, selecting the right suppliers can decrease operational costs and delivery time [5]. In today's financial and production environment, manufacturing firms depend very much on suppliers for providing materials and components used in finished products, whereas the costs of raw materials and component parts constitute approximately 50% to 70% of the production costs [6]. Previous studies revealed that cost is not the only criterion to be considered when enterprises select the preferred suppliers. The selection of appropriate suppliers implies consideration of multiple qualitative and quantitative criteria. Thus, in [7] quality is considered the critical criterion, followed by deliverability, price, manufacturing capability, service, management, and technology.

Supplier selection problem received a lot of attention recently. Multiple studies using different analysis methods have been performed. Some of the methods mentioned in literature are: analytical hierarchy process (AHP) [8], analytic network process (ANP) [9], data envelopment analysis (DEA) [10], Elimination and Choice Expressing Reality (ELECTRE) [11], Preference Ranking Organisation Method for Enrichment Analysis (PROMETHEE) [12], Techniques for Order Performance by Similarity to Ideal Solution (TOPSIS) [13], game theory [14], linear programming [15], non-linear programming [16], multi-objective programming [17], goal programming [18], stochastic programming [19], mixed integer programming [20] and mixed integer nonlinear programming [21], genetic algorithm [22], grey system theory [23], neural network [24], Bayesian networks [25], decision tree [26], case based reasoning [27], ant colony algorithm [28], and their hybrids.

The rest of the paper is organized as follows: in the second section the methodology of multiple criteria decision making is presented. The third section describes the case study based action research for the supplier performance evaluation within a large company performing in the field of flexible packaging production. The variants (suppliers), evaluation criteria and decision making's preferences are detailed in this section. Furthermore, the results of the computational experiments are presented in the fourth section and, finally, the last section summarizes the conclusions.

2 Methodology

Selecting the suitable supplier considerably decreases purchasing costs, improves competitiveness in the market, enhances supply chain performance and improves end-user satisfaction. Due to the strategic importance, various decision makers from departments other than purchasing such as finance, production, and marketing are frequently involved in the decision making process for the supplier selection process [1]. The process of selecting the supplier implies the evaluation of various suppliers based on diverse quantitative and qualitative criteria. Therefore, choosing the right suppliers is considered a multiple criteria decision-making problem.

Multi-criteria decision-making is probably the most widely used decision methodology in the business and engineering fields, being able to improve the quality of decisions by making the process more efficient, rational and explicit [29]. The decision processes may be divided into two stages. The first stage is evaluation. The assessment criteria used to measure the performance of diverse alternatives of the problem are established. The alternatives of the problem are assessed through the pre-determined criteria. This stage requires the knowledge of experts to qualitatively and/or quantitatively evaluate each alternative under each defined criterion. The outcome of this stage is frequently a matrix, called the decision matrix, composed of the alternatives as its rows and the criteria as its columns. The performances of alternatives under each criterion are found in its cells. The second stage of the decision-making process consists of the ranking and selecting the alternatives based on their performances [30].

The ELECTRE III method was chosen because it can develop the analysis better than other multiple criteria decision making methods; likewise for its capability to deal with imprecise, inaccurate, uncertain data, inevitable to complex processes in complicated human decision [31]. The ELECTRE III method is based on the application of the outranking relation, in order to rank a finite set of alternatives from the best to the worst, based on using a set of evaluation criteria, and on the preferential information submitted by the decision makers [32].

The computational algorithm of the ELECTRE III method includes the following three phases [32-34]:

a. Establishing a set of alternatives (A) and a family of criteria (F) combined with the construction of a performance matrix. In this phase, the preferential information is given by the decision makers in the form of criteria weights (w) and the indifference (q), preference (p) and veto (v) thresholds.

b. Construction of the outranking relation ($S(a,b)$) is a result of computation procedure, which indicates the extent to which "alternative (a) outranks alternative (b)" overall. The outranking relation is built on the basis of concordance and discordance tests. The concordance index indicates the truth of the statement "alternative (a) outranks alternative (b)". The value of concordance index is between 0 and 1. The global concordance indicator is determined as the weighted sum of concordance indexes for all criteria. The discordance test evaluates the strength of evidence against the statement "alternative (a) outranks alternative (b)", which can be repealed if the difference of performance between alternative (a) and alternative (b), on any criterion is higher than the veto threshold. The value of this index is between 0 and 1. The outranking relation is expressed by the degree of credibility ($d(a,b)$). It is an equivalent of the global concordance indicator ($C(a, b)$) determined in the concordance test weakened by the discordance indexes ($D_j(a,b)$) computed in discordance test. The values of the credibility degree range between 0 and 1.

c. Exploitation of the outranking relation ($S(a,b)$). Based on the calculation of the credibility degree the method establishes two preliminary rankings using the distillation procedure. Using this procedure a descending and an ascending preorder can be acquired. The final ranking is presented either in the form of the ranking matrix or the outranking graph, as the results of intersection of the above mentioned complete preorders.

3 Case study

In order to illustrate the implementation of the proposed decision making method to the supplier selection problem, a case study conducted in collaboration with the management of a large leading firm performing in the field of flexible packaging production is presented. The company has manufacturing facilities in Romania and other facilities in Europe, being capable of timely responding to market needs in terms of quality, flexibility and production standards. To ensure a strategic cultural and operating closeness to the global end users, the company's subsidiaries are based near their customers. The firm exports its products to most of the major markets across the world.

This case study deals with supplier selection among a set of seven candidates (Alpha, Beta, Gamma, Delta, Epsilon, Zeta and Eta), which can provide materials according to a set of requirements indicated by the company. Most of the case firm's suppliers are in foreign countries. The evaluation was performed by three decision-makers, managers representing three different corporate functions (sales, engineering and finance departments), whose various point of views have to be comprised in the model. The decision makers were interviewed for identifying the adequate criteria, weights, and the rates of the evaluated suppliers. The weights and the value of thresholds are depicted in table 1.

Table 1. Thresholds and criteria weights.

Criteria \ Preferences	Indifference *q*	Preference *p*	Veto *v*	Weights
C_{I}	0.2	0.4	0.6	0.2
C_{II}	0.1	0.2	0.4	0.1
C_{III}	0.2	0.3	0.5	0.08
C_{IV}	0.15	0.2	0.5	0.05
C_{V}	0.1	0.2	0.4	0.05
C_{VI}	0.15	0.2	0.6	0.03
C_{VII}	0.15	0.2	0.6	0.01
C_{VIII}	0.1	0.15	0.4	0.03
C_{IX}	0.1	0.2	0.5	0.03
C_{X}	0.1	0.2	0.3	0.01
C_{XI}	0.1	0.3	0.6	0.04
C_{XII}	0.15	0.3	0.5	0.03
C_{XIII}	0.1	0.3	0.6	0.02
C_{XIV}	0.1	0.3	0.5	0.01
C_{XV}	0.1	0.2	0.6	0.02
C_{XVI}	0.1	0.3	0.6	0.05
C_{XVII}	0.05	0.3	0.7	0.05
C_{XVIII}	0.1	0.2	0.5	0.04
C_{XIX}	0.1	0.15	0.4	0.03
C_{XX}	0.1	0.2	0.3	0.01
C_{XXI}	0.1	0.3	0.7	0.03
C_{XXII}	0.1	0.2	0.5	0.04
C_{XXIII}	0.1	0.2	0.5	0.04

The decision makers have proposed 23 criteria (C_I to C_{XXIII}) in order to perform a complete evaluation of the suppliers. The determined criteria are presented as follows.

Price (C_I – [Euro/kg]) has a crucial role in supporting managers to make a decision because the cost of raw materials and component parts represent approximately 50% to 70% of the production costs. In a competitive manufacturing environment many companies are forced to adopt target costing for achieving consistent profitability. *Quality* (C_{II} – [points]) of the raw materials delivered by the supplier must conform to a set of technical properties (dynamic COF (coefficient of friction), heat seal range, seal strength, oxygen permeability, water permeability, corona treatment level, optical density, haze) request by the firm.

Payment terms (C_{III} – [days]) affect the firms' cash-flow. The shorter the term for the company s to pay its suppliers, the less money has on hand, fact which is not favorable for the cash-flow. *Response time* (C_{IV} – [points]) is increasingly appreciated taking into account the short deadlines, better production processes, reduced time-to-market, multiple marketing campaigns opportunities. *Standard delivery period* (C_V – [days]) represents the normal time between the placement of an order and receipt of the delivery. Due to aggressive marketing strategies, in emergency situations, the supplier should be able to deliver sooner than standard delivery period. *External audit results* (C_{VI} – [points]) could influence the customer's purchasing decisions. In order to ensure compliance to requirements, an audit can apply to a product, process, or to an entire organization.

Implemented ISO standards (C_{VII} – [points]) show that the supplier has a quality management system that meets the requirements and it has the ability to supply consistent, conforming products or services. *Production capacity* (C_{VIII} – [points]) and *available production capacity* (C_{IX} – [points]) are very important especially when the demand for the buyer's product is tied to unforeseen market events. *Reputation* (C_X – [points]) takes into account the performance history of the supplier, its size and feedback from its clients, previous global experience. *Geographical location* (C_{XI} – [points]) should be assessed carefully in order to eliminate the risk of disruption in the supply chain. *Product diversity* (C_{XII} – [points]) measures the ability of the suppliers to manufacture different products. *Flexibility* (C_{XIII} – [points]) refers to the ability of the supplier to adapt production resources in order to meet the changing demands of customers. *Response to customer complaints* (C_{XIV} – [points]) should be quick and efficient so as to keep customers happy and is critical to the success of any business. *Customs duty and procedures* (C_{XV} – [points]) depends on the location the goods came from. *Technology* (C_{XVI} – [points]) of suppliers was assessed based on the type of processes, machinery and equipment. Likewise, access to appropriate manufacturing technologies and patents were evaluated.

On-time delivery (C_{XVII} – [points]) refers to the supplier's commitment to maintain the delivery schedule determined by the firm. *Research & development* (C_{XVIII} – [points]) enables a supplier to improve its current technologies and products, and can boost its competitive advantage. *Reliability of delivery* (C_{XIX} – [points]) indicates the supplier's ability to predictably supply the requested products as promised. *Financial strength* (C_{XX} – [points]) is very important because a supplier should have adequate funds to deliver the products to the customer. The financial strength is evaluated based on the financial ratio, such as activity, profitability, liquidity and solvency. *Warranty terms* (C_{XXI} – [points]) represent a type of guarantee that a supplier makes based on terms and situations when replacements or repairs will be needed if the product does not have the properties as originally described for a set period of time. *Communication* (C_{XXII} – [points]) is essential in developing more collaborative and trusting relationships, especially when dedicated contact persons from the supplier for specific projects are assigned. *Quality of the human resources* (C_{XXIII} – [points]) depends on supplier employee's attitudes, competences and skills. A high quality of human resources represents a source of sustained competitive advantages.

Table 2 presents the performance of all the alternatives to be assessed with respect to the identified criteria that meet managers' expectations.

Table 2. Performance matrix.

Criteria \ Alternatives	Alpha	Beta	Gamma	Delta	Epsilon	Zeta	Eta
C_I	1.65	1.6	1.65	1.68	1.55	1.55	1.5
C_{II}	99	80	90	75	85	50	70
C_{III}	120	90	90	90	90	60	30
C_{IV}	80	70	60	65	90	90	50
C_V	60	30	45	45	45	21	70
C_{VI}	49.5	42.5	45	40	35	30	35
C_{VII}	33	26.67	30	30	26.67	20	23.33
C_{VIII}	24.75	17.5	17.5	22.5	17.5	7.5	20
C_{IX}	80	60	50	60	50	30	50
C_X	19.8	14	16	16	14	10	12
C_{XI}	90	99	70	60	60	99	50
C_{XII}	24.75	17.5	20	22.5	20	5	12.5
C_{XIII}	10	16	10	10	14	18	10
C_{XIV}	30	23.33	26.67	23.33	20	16.67	16.67
C_{XV}	33	33	33	26.67	16.67	33	16.67
C_{XVI}	39.6	34	36	38	34	16	28

DOI: 10.1051/matecconf/20179406002

C XVII	18	18	17	16	14	16	12
C XVIII	28.29	20	25.71	28.29	22.86	2.86	14.29
C XIX	9.9	6	6	8	8	5	6
C XX	12.38	10	11.25	12.38	8.75	6.25	11.25
C XXI	49.5	40	45	45	35	25	30
C XXII	66	60	60	53.33	66	46.67	40
C XXIII	22	20	20	20	20	17.78	17.78

4 Results

The application of ELECTRE III method to the supplier selection problem at a large company implies a significant amount of computation that, without the use of a specialized software package, is hardly imaginable. The computational experiments were performed using a software developed by the authors of this paper based on the ELECTRE III method algorithm. The software is a Windows Forms application, built using Microsoft .NET Framework and Visual Studio 2015. The language used is C#, version 4.5, which is a powerful object-oriented programming language developed by Microsoft.

The user interface uses Windows Forms, which offer a lot of flexibility in gathering and displaying data, making use of complex controls such as the Data Grid View Control and the Tab Control. The first control displays and manipulates data displayed in a grid. It allows defining cells and adding rows. The second control defines a list of tabs, which are control containers, which can be brought into and out of view using a navigation bar. This simplifies the user interface and increases usability. The business logic is implemented using both user-defined data structures and existing data structures, such as the Dictionary, which stores key-value pairs, such as the column-name – column-value within a record of data. These data structures, coupled with the powerful concepts of object-oriented programming, like inheritance and polymorphism, help keep complexity small, while still remaining expressive. Data is saved in binary format on disk, using a process called serialization. It is fast and easy to import and export, while it remains inaccessible by other applications, thus improving integrity.

The developed software starts with a specified set of alternatives assessed on a well-founded family of pseudo-criteria and aggregates these preferential data into a fuzzy outranking relation. Once the application is started, in order to create the supplier selection problem at the company performing in the field of flexible packaging production, criteria and alternatives are defined via File from menu bar. For every criterion, the direction of the preference has to be specified. The direction of the preference indicate if the criterion is diminishing, like for example the price of raw materials and component parts (the score of the criterion has to be minimized), or increasing, like for example the quality of the raw materials (the score of the criterion has to be maximized). Once the alternatives and criteria have been defined, the performance of the alternatives is entered by selecting Input/Performance Matrix from the menu bar. The user has to define the indifference, preference and veto threshold for each criterion via Input/Thresholds from menu bar. Table 1 and 2 summarize data that was loaded in the application. Having established the alternatives, criteria, performances and thresholds, the user may proceed to the calculation of data by selecting Action/Compute from the menu bar. In order to establish the sequence of alternatives using the processes assigned to the ELECTRE III, three types of complementary results (the concordance matrix, the discordance matrix and the credibility matrix) are returned by the application. Figure 1 shows the credibility matrix calculated by the application.

 DOI: 10.1051/matecconf/20179406002

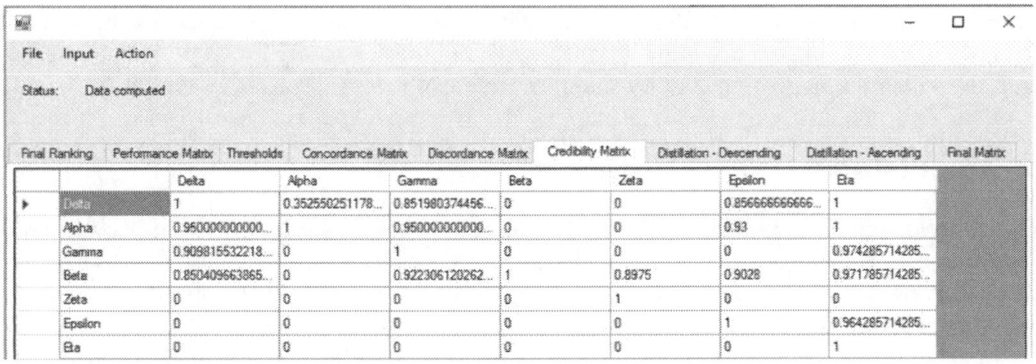

	Delta	Alpha	Gamma	Beta	Zeta	Epsilon	Eta
Delta	1	0.352550251178...	0.851980374456...	0	0	0.856666666666...	1
Alpha	0.950000000000...	1	0.950000000000...	0	0	0.93	1
Gamma	0.909815532218...	0	1	0	0	0	0.974285714285...
Beta	0.850409663865...	0	0.922306120262...	1	0.8975	0.9028	0.971785714285...
Zeta	0	0	0	0	1	0	0
Epsilon	0	0	0	0	0	1	0.964285714285...
Eta	0	0	0	0	0	0	1

Fig. 1. Credibility matrix.

The credibility index aggregates the total concordance and discordance indices, indicating whether the outranking hypothesis is true or not. It can be noticed that the diagonal of the credibility matrix is the unity because the alternative is compared to itself. Taking, for instance, the example of alternatives Gamma and Alpha, there are no arguments that Gamma outranks Alpha overall ($S_{(Gamma,\ Alpha)}=0$), while there are some high chances that supplier Alpha outranks supplier Gamma ($S_{(Alpha,\ Gamma)}=0.95$). The next step is to explore this matrix. In order to rank the alternatives, the application is exploiting the outranking degrees by the descending and ascending distillation procedures, which lead to two complete pre-classifications. The descending distillation is shown in fig. 2 (a).

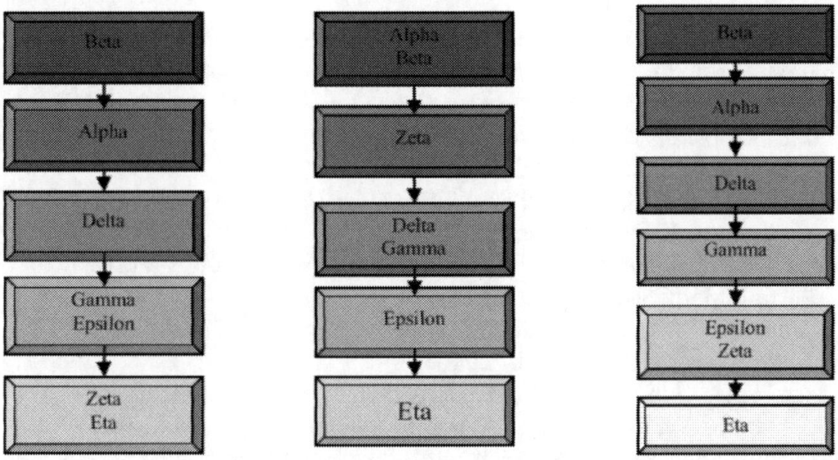

(a) (b) (c)

Fig. 2. Descending distillation (a), ascending distillation (b), final ranking (c).

The descending distillation consists of selecting the best ranked alternatives initially and ending with the worst. The best alternative (Beta) is taken out from the credibility matrix, and then the above step is gradually repeated for the remaining alternatives. The procedure ends when each alternative is assigned a rank. The result of descending distillation showed preference for Beta, followed by Alpha, Delta, Gamma, Epsilon, Zeta and Eta. Regarding the supplier Gamma and Epsilon, as well as Zeta and Eta, there was no preference between them. The ascending distillation, illustrated in fig. 2 (b), uses the same procedure as the descending one, but with successive removal of the worst alternative. The ascending distillation showed that Beta and Alpha were ranked as the best options, followed by Zeta, Gamma, Delta, Epsilon and Eta. Regarding the supplier Delta and Gamma, there was no preference between them. The final ranking is obtained by calculating the intersection of the two pre-classifications, as depicted in fig. 2 (c). It is noticed that Beta was preferred over the other suppliers, being the most adequate solution. Regarding variants Epsilon and Zeta, there was no preference between them. At the other extreme, Eta is positioned at the bottom of the ranking, which denotes that it represents the least preferred supplier.

5 Conclusions

The supplier selection process is one of the most important decision making problems in the supply chain management. The selecting process is critical for enhancing the company's competitiveness, and requires the assessment of different alternative suppliers based on different criteria.

In this paper, a practical application of a multiple criteria decision making methods, namely the ELECTRE III method, is proposed to deal both with the quantitative and qualitative criteria and to select the suitable supplier effectively. The proposed decision making model has been applied to a supplier selection process of a company working in the field of flexible packaging production. Seven suppliers have been identified and assessed according to twenty three criteria that meet stakeholders' expectations. By running the developed application, the final ranking of suppliers was obtained. The results reveal that ELECTRE III method may be a very useful tool to support managers in solving supplier selection problems.

References

1. A. Sanayei, S.F. Mousavi, A. Yazdankhah, Expert Syst. Appl., **37**, 7, (2010)
2. S.I. Omurca, Appl. Soft Comput., **13**, 8, (2013)
3. C.-T. Chena, C.-T. Lin, S.-F. Huangb, Int. J. Prod. Econ., **102**, 13, (2006)
4. G. Bruno, E. Esposito, A. Genovese, M. Simpson, Expert Syst. Appl., **43**, 15, (2016)
5. Z.H. Che, H.S. Wang, Comput. Ind. Eng., **55**, 24, (2008)
6. D.M. Lee, P.R. Drake, Int. J. Prod. Res., **48**, 32, (2010)
7. W. Ho, X. Xiaowei, K.D. Prasanta, Eur. J. Oper. Res., **202**, 9, (2010)
8. T. Peri´c, Z. Babi´c, I. Veˇza, Int. J. Comput. Integr. Manuf., **26**, 14, (2013)
9. R.H. Lin, Appl. Math. Model., **33**, 7, (2009)
10. A. Hadi-Vencheh, M. Niazi-Motlagh, Int. J. Comput. Integr. Manuf., **24**, 9, (2011)
11. M. Sevkli, Int. J. Prod. Res., **48**, 13, (2010)
12. Y.-H. Chen, T.-C. Wang, C.-Y. Wu, Expert Syst. Appl., **38**, 7, (2011)
13. M.N. Mokhtarian, A. Hadi-Vencheh, Appl. Soft Comput., **12**, 10, (2012)
14. M. Agha Mohammad Ali Kermani et al., Int. J. Comput. Integr. Manuf., **25**, 9, (2012)
15. C.-T. Lin, C.-B. Chen, Y.-C. Ting, Expert Syst. Appl., **38**, 6, (2011)
16. B.-M. Hsu, C.-Y. Chiang, M.-H. Shu, Expert Syst. Appl., **37**, 9, (2010)
17. M. Yu, M. Goh, H. Lin, Eur. J. Oper. Res., **219**, 7, (2012)
18. T.J. Kull, S. Talluri, IEEE Trans. Eng. Manag., **55**, 11, (2008)
19. L. Li, Z.B. Zabinsky, Int. J. Prod. Econ., **134**, 13 (2011)
20. C. Basnet, J.M.Y. Leung, Comput. Oper. Res., **32**, 14, (2005)
21. S.H. Ghodsypour, C. O'Brien, Int. J. Prod. Econ., **73**, 13, (2001)
22. A.F. Guneri, T. Ertay, A. Yücel, Expert Syst. Appl., **38**, 11, (2011)
23. M. Tseng, Appl. Soft Comput. J., **11**, 10, (2011)
24. C.C. Lee, C. Ouyang, Expert Syst. Appl., **36**, 10, (2009)
25. L. Ferreira, D. Borenstein, Expert Syst. Appl., 39, 11, (2012)
26. X. Guo, Z. Yuan, B. Tian, Expert Syst. Appl., **36**, 8, (2009)
27. F. Faez, S.H. Ghodsypour, C. O'Brien, Int. J. Prod. Econ., **121**, 14, (2009)
28. Y.L. Tsai, Y.J. Yang, C.-H. Lin, Expert Syst. Appl., **37**, 9, (2010)

29. X. Wang, E. Triantaphyllou, Omega, **36**, 19, (2008)
30. G.A. Montazer, H.Q. Saremi, M. Ramezani, Expert Syst. Appl., **36**, 11, (2009)
31. C.E.D. Infante et al., J. Clean. Prod., **52**, 12, (2013)
32. M.K. Janiaka, J. Żakb, *Transport. Res. Proc.*, **3**, 10, (2014)
33. T.M. Mro´z, Appl. Therm. Eng., **28**, 9, (2008)
34. M. Cristea, C. Cristea, *Proceedings of the Annual Sessions of Scientific Papers "IMT Oradea"*, 6, (2016)

A multi-criteria decision making approach for the selection of a flexible packaging equipment

Ciprian Cristea[1,*], and *Maria* Cristea[1]

[1]Technical University of Cluj-Napoca, Faculty of Electrical Engineering, Romania

Abstract. Flexible packaging is one of the fastest growing segments of the packaging industry, combining paper, plastic film and aluminum foil to deliver a broad array of products for food and beverage, personal care products, and pharmaceutical industries. In order to preserve the quality and safety of the product contained in them, there are currently a variety of flexible packaging equipment options available. However their relative costs and performance differ. This study proposes an application of a ranking methodology to assess a selection of the most suitable flexible packaging equipment. The adequate criteria in the selection of equipment have been identified, and the considered options are assessed, considering the decision maker's preferences and existing constraints. The options are ranked in terms of their suitability for selecting equipment by using the Electre III method. The results obtained from the simulation experiment highlight the effectiveness of the model in outranking different options in the process of equipment selection.

1 Introduction

In the contemporary competitive manufacturing environment, to cope with increasingly high consumer expectations regarding quality products and services, shortening of product life cycles, low profit margins, short lead-times and rapid innovation, organizations have to invest and improve their production facilities. Technology is one of the most important resources of companies that have to take into account any opportunity to optimize their manufacturing processes. In order to enhance their production and productivity, practitioners and academics have come to the same conclusion: the firms have to invest in new technology and acquire latest pieces of equipment. Despite the fact that innovation and investment in new technology involve high costs, these represent some of the few sources of growth, being a positive relationship between the development of an enterprise and the equipment investment rates [1]. The adequate equipment selection is time consuming and involves a complex decision. In this process, multiple decision makers, with various expertise and perspectives, are frequently implied and have to deal with ambiguity and uncertainty, such as undifferentiated alternatives, incomplete information and inadequate understanding [2].

The global packaging market stood at $812 billion in 2014, while sales from packaging industry are projected to increase by 3.3 percent in real terms for 2015 to $839 billion, and

* Corresponding author: ciprian.cristea@emd.utcluj.ro

an annual growth of 3.5% per year to 2020 is forecasted, with sales to reach $997 billion by 2020 [3]. In order to cope with the growing demand and the huge interest in new types of packaging and the latest film technology, the companies have to invest in the latest equipment technologies for further developments in different sectors.

Flexible packaging is probably the most economical method to package, preserve and distribute a broad array of products for food and beverage, personal care products, pharmaceuticals and other products that need extended shelf life. In the contemporary retail market, more than two-thirds of the purchasing decisions are made in the store, when standing in front of the shelf, frequently preferring this to other forms of communication. The consumers decide within seconds whether they will purchase a product, so packaging has to absolve also persuasive functions, transfer emotions and sensual impressions [4].

The selection of new equipment is a time-consuming and complex process that necessitates experience and advanced knowledge. For an adequate assessment the decision maker may require a large amount of data to be analyzed and many factors to be taken into account. In order to select the most appropriate equipment among others, the decision-maker should be an expert or at least be very familiar with its specifications [5]. The middle and upper management have a ninety-four percent involvement in authorization for final selection, while the remaining six percent belongs to engineering staff [6].

The equipment selection problem received a lot of attention recently. Multiple studies have been performed dealing with this subject, such as analytical hierarchy process (AHP) [7], artificial neural network [8], Elimination and Choice Expressing Reality (ELECTRE) [9], multiobjective optimization on the basis of ratio analysis (MOORA) [10], fuzzy AHP [11], combination of AHP and Preference Ranking Organisation Method for Enrichment Analysis (PROMETHEE) [12], fuzzy Techniques for Order Performance by Similarity to Ideal Solution (TOPSIS) [13], fuzzy AHP and Grey Relational Analysis (GRA) [14], VIsekriterijumsko KOmpromisno Rangiranje (VIKTOR) [15], Step-wise Weight Assessment Ratio Analysis (SWARA) and COmplex PRoportional ASsessment of alternatives with Grey relations (COPRAS-G) [16].

The rest of the paper is organized as follows: in the second section the methodology of multiple criteria decision making is presented. The third section describes the case study based action research for the equipment performance evaluation within a firm working in the field of flexible packaging production. The alternatives (equipment), evaluation criteria and decision making's preferences are described in this section. The results of the computational experiments are presented in the fourth section and, finally, the last section summarizes the conclusions.

2 Methodology

The multiple criteria decision making methods are used for solving complex problems featuring high unpredictability, various forms of data and information, conflicting objectives, multi interests and perspectives, and the accounting for complex and evolving socio-economic and biophysical systems [17]. The selection of optimal equipment from a finite number of alternative equipment on the basis of many different and conflicting attributes is a multi-criteria decision making problem. In order to select the optimum equipment, practitioners and academics have found that the price of the equipment shouldn't be the overriding determinant. The analysis of several alternative solutions involves the assessment based on diverse quantitative and qualitative criteria. Frequently, these goals are in conflict, so that solutions differ from each other in many features. Therefore, there is no single solution that optimizes all criteria simultaneously, and there is no exclusive criterion that describes the consequences of each alternative solution correspondingly. Multiple criteria decision making methods use mathematical models that assess alternative scenarios of equipment, taking into account both their objective characteristics and the preferences of the decision makers [18].

Among the existing methods, the ELECTRE III was chosen for its ability to deal with inaccurate, imprecise, uncertain data, inevitable to complex processes in human decision [19]. This method is based on the application of the outranking relation, in ranking a finite set of alternatives from the best to the worst, on the basis of using a set of evaluation criteria [20].

The ELECTRE III method algorithm comprises the following three phases [21-23]:
1) *Establishing a set of data* – a set of alternatives and a family of criteria used for comparison are defined, followed by the performance matrix development. In order to produce outranking relations with an allowance for data uncertainly, the indifference (q_i), preference (p_i), and veto (v_i) threshold, as well as criteria weight indexes (w_i) are defined.
2) *Construction of the outranking relations* – the possible alternative solutions are pairwise compared and each pairwise comparison is characterized by an outranking relation. The outranking relation is built on the basis of the concordance and the discordance matrices. The concordance index for each pair of alternatives *A* and *B* is determined by comparing the performances of both alternatives for every criterion. The concordance index shows the truth of the statement "*A* outranks *B*". The value of concordance index is between 0 and 1, the full truth of the assertion being indicated by a value of 1, while a value of 0 highlights that the statement is false. The concordance index is defined as follows [24]:

$$C(A,B) = \frac{1}{W} \cdot \sum_{i=1}^{n} w_i \cdot c_i(A,B) \tag{1}$$

where:

W – sum of all weights of criteria ($W = \sum_{i=1}^{n} w_i$);

w – weight of criterion i;
n – number of criteria;
$c_i(A,B)$ – concordance index over alternatives A and B with respect to the criterion i;

$$c_i(A,B) = \begin{cases} 1, \text{if } q_i(f_i(A)) \geq f_i(B) - f_i(A) \\ \dfrac{p_i(f_i(A)) + f_i(A) - f_i(B)}{p_i(f_i(A)) - q_i(f_i(A))}, \text{if } p_i(f_i(A)) > f_i(B) - f_i(A) > q_i(f_i(A)) \\ 0, \text{if } p_i(f_i(A)) \leq f_i(B) - f_i(A) \end{cases} \tag{2}$$

where:
$f_i(A)$ – performance of alternative A based on the criterion i;
$f_i(B)$ – performance of alternative B based on the criterion i;
q_i – indifference threshold for the criterion i;
p_i – preference threshold for the criterion i.

The concordance index for each pair of alternatives *A* and *B* is determined by comparing the performances of both alternatives for every criterion.

The discordance index for each pair of alternatives *A* and *B* is determined by comparing the performances of both alternatives for every criterion. The strength of the evidence against the statement "*A* outranks *B*" is evaluated by the discordance index. If the difference of performances between the alternative *A* and *B*, on any criterion i, is higher than the veto threshold v_i, it can be overruled. The discordance index for each criterion i is given by equation (3) [25].

$$D_i(A,B)=\begin{cases}0, \text{if } p_i(f_i(A))\geq f_i(B)-f_i(A)\\ \dfrac{f_i(B)-f_i(A)-p_i(f_i(A))}{v_i(f_i(A))-p_i(f_i(A))}, \text{if } v_i(f_i(A))>f_i(B)-f_i(A)>p_i(f_i(A))\\ 1, \text{if } v_i(f_i(A))\leq f_i(B)-f_i(A)\end{cases} \quad (3)$$

where:
v_i – veto threshold for the criterion *i*;

The outranking relation is expressed by the degree of credibility. The credibility index *S(A,B)* is defined as follows [22]:

$$S(A,B)=\begin{cases}C(A,B), \text{if } C(A,B)\geq D_i(A,B)\ \forall i\\ C(A,B)\cdot\prod_{i\in J(A,B)}\dfrac{(1-D_i(A,B))}{(1-C(A,B))}, \text{otherwise}\end{cases} \quad (4)$$

where:
J(A,B) – the set of criteria for which *$D_i(A,B)>C(A,B)$*.

The credibility matrix is given by the degrees of credibility.

3) *Exploitation of the outranking relations* – in order to rank the alternatives, a distillation procedure is used. A descending and an ascending preorder can be obtained by using this procedure. The final ranking is acquired by combining these preorders. The ranking matrix contains the aggregation of the above mentioned preorders. There are the following cases [22], [26]:

a) *If alternative A is higher ranked than alternative B in both distillations or A is better than B in one distillation and has the same position in the other one, then variant A is better than variant B: A P+ B;*

b) *If variant A is higher ranked than variant B in one distillation but B is better ranked than A the other distillation, then A is incomparable to B: A R B;*

c) *If alternative A has the same position in the ranking system than alternative B in both distillations, then variant A is indifferent to variant B: A I B;*

d) *If variant A is lower ranked than variant B in both distillations or A is lower ranked than B in one distillation and has the same rank in the other distillation, then A is worst than B: A P – B.*

The final ranking is achieved by summing the number P+ for each alternative.

3 Case study

In order to demonstrate and validate the application of the above mentioned multiple criteria decision making method for solving the equipment selection problem, a case study conducted in collaboration with the management of a leading company working in the field of flexible packaging production is presented.

This case study deals with the equipment selection among a set of seven candidates (Sigma, Tau, Upsilon, Phi, Chi, Psi and Omega), according to a set of requirements indicated by the decision makers. The assessment was performed by managers representing different corporate functions (engineering, manufacturing, sales, and finance departments) whose different points of view have to be comprised in the model. In order to identify the suitable criteria, weights of criteria, and the rates of the evaluated equipment, the decision makers were interviewed. The weights of the criteria and the value of thresholds are presented in table 1.

Table 1. Thresholds and criteria weights.

Criteria \ Preferences	Indifference *q*	Preference *p*	Veto *v*	Weights
C_{I}	0.2	0.4	0.6	0.14
C_{II}	0.15	0.5	0.65	0.08
C_{III}	0.1	0.3	0.4	0.06
C_{IV}	0.2	0.3	0.5	0.05
C_{V}	0.1	0.3	0.6	0.02
C_{VI}	0.1	0.3	0.5	0.03
C_{VII}	0.15	0.25	0.35	0.02
C_{VIII}	0.1	0.2	0.6	0.02
C_{IX}	0.2	0.3	0.7	0.03
C_{X}	0.1	0.15	0.3	0.05
C_{XI}	0.2	0.4	0.6	0.02
C_{XII}	0.2	0.3	0.5	0.02
C_{XIII}	0.1	0.15	0.35	0.03
C_{XIV}	0.2	0.4	0.7	0.02
C_{XV}	0.1	0.2	0.6	0.04
C_{XVI}	0.3	0.45	0.6	0.02
C_{XVII}	0.15	0.25	0.4	0.01
C_{XVIII}	0.2	0.3	0.4	0.01
C_{XIX}	0.2	0.4	0.6	0.03
C_{XX}	0.1	0.3	0.6	0.04
C_{XXI}	0.1	0.2	0.45	0.03
C_{XXII}	0.2	0.3	0.4	0.02
C_{XXIII}	0.2	0.4	0.6	0.03
C_{XXIV}	0.15	0.25	0.4	0.01
C_{XXV}	0.2	0.35	0.5	0.01
C_{XXVI}	0.1	0.3	0.5	0.01
C_{XXVII}	0.15	0.4	0.6	0.03
C_{XXVIII}	0.1	0.2	0.3	0.01
C_{XXIX}	0.3	0.5	0.7	0.02
C_{XXX}	0.2	0.4	0.6	0.02
C_{XXXI}	0.2	0.3	0.5	0.03
C_{XXXII}	0.1	0.2	0.6	0.04

The decision makers have proposed thirty-two criteria (C_I to C_{XXXII}) in order to perform a complete evaluation of the equipment. The determined criteria are presented as follows: *price* (C_I – [Euro]), *quality* (C_{II} – [points]), *maximum speed* (C_{III} – [lm/min]), *maximum print width* (C_{IV} – [mm]), *flexographic printing repeat* (C_V – [points]), *maximum web width and web dimensions* (C_{VI} – [points]), *material diversity* (C_{VII} – [points]), *number of printing decks* (C_{VIII} – [points]), *drying system and UV curing system* (C_{IX} – [points]), *change over time* (C_X – [points]), *central impression drum cylinder or in line cylinders* (C_{XI} – [points]), *temperature controller* (C_{XII} – [points]), *transmission type and drive system* (C_{XIII} – [points]), *computerized setting of print decks* (C_{XIV} – [points]), *automatic module register* (C_{XV} – [points]), *production controls* and job chart (C_{XVI} – [points]), *reverse printing* (C_{XVII} – [points]), *turret unwinder and rewinder capacity* (C_{XVIII} – [points]), *automatic pressure module* (C_{XIX} – [points]), *automatic defect detection system* (C_{XX} – [points]), *delivery terms* (C_{XXI} – [points]), *delivery time* (C_{XXII} – [months]), *terms of payment* (C_{XXIII} – [points]), *assembly and start-up of the machine* (C_{XXIV} – [points]), *machine acceptance tests* (C_{XXV} – [points]), *service project manager* (C_{XXVI} – [points]), *warranty and after warranty service*

 DOI: 10.1051/matecconf/20179406003

(C_{XXVII} – [points]), *corona treatment systems* (C_{XXVIII} – [points]), *job types and runs* (C_{XXIX} – [points]), *plate-cylinder assembly* (C_{XXX} – [points]), *video web inspection* (C_{XXXI} – [points]), *ink management and doctor blade chamber* (C_{XXXII} – [points]).

Table 2 indicates the performance of all the alternatives to be evaluated with respect to identified criteria that meet managers' expectations.

Table 2. Performance matrix.

Alternatives / Criteria	Sigma	Tau	Upsilon	Phi	Chi	Psi	Omega
C I	1800000	900000	1600000	1650000	1400000	1700000	1300000
C II	90	50	70	95	80	85	75
C III	400	300	200	360	400	350	380
C IV	1270	1100	850	660	1200	1250	1270
C V	53.33	53.33	56.67	56.67	53.33	53.33	50
C VI	72	56	40	48	64	72	68
C VII	45	40	45	47.5	42.5	45	42.5
C VIII	8	8	8	8	8	8	8
C IX	32.73	27.27	21.82	25.45	29.09	32.73	30.91
C X	23.75	17.5	20	23.75	17.5	22.5	22.5
C XI	47.5	40	30	35	45	45	45
C XII	11.25	9.38	8.75	10.63	10	10.63	10.63
C XIII	15.83	12.5	13.33	15	14.17	15	15
C XIV	38	28	30	34	30	34	34
C XV	60	46.67	46.67	53.33	53.33	56.67	56.67
C XVI	21.25	17.5	17.5	21.25	20	20	20
C XVII	50	50	80	90	50	60	50
C XVIII	40	35.56	31.11	37.78	35.56	40	40
C XIX	28.33	23.33	23.33	25	25	26.67	28.33
C XX	9	7	7.5	8.5	8	9	8.5
C XXI	17.5	20	22.5	18.75	21.25	20	20
C XXII	8	6	6	7	5	7	7
C XXIII	14	14	18	15	18	15	14
C XXIV	30.91	25.45	21.82	30.91	27.27	29.09	32.73
C XXV	48.39	48.39	45.16	54.84	45.16	51.61	51.61
C XXVI	11.43	8.57	8.57	12.86	10	10.71	10.71
C XXVII	10	8.13	7.5	10.63	8.75	9.38	10
C XXVIII	9.47	7.37	7.37	8.95	7.37	8.95	9.47
C XXIX	21.25	17.5	20	22.5	18.75	20	20
C XXX	20	16.67	15.56	20	18.89	20	18.89
C XXXI	15.83	11.67	12.5	14.17	12.5	14.17	15
C XXXII	19	14	15	16	15	17	17

4 Results

The implementation of ELECTRE III method to the equipment selection problem at a leading company working in the field of flexible packaging production is, practically speaking, not feasible in the absence of specialized software. This problem and the complexity of the decision making process are amplified by the large number of criteria a high performance equipment has to meet for a proper assessment. To address this difficulty, we have developed a software, executable within the Microsoft Windows operation system, based on the ELECTRE III method algorithm, capable of performing the computational experiments. The decision support system has been developed as a Windows Forms application, and has been implemented using .NET and Visual Studio 2015. The language used is C#, which provides full support for object-oriented programming, allowing for the key concepts to be expressively and concisely represented. It uses inheritance, containment, encapsulation and polymorphism in order to provide much greater flexibility. Further development could consist in creating an automated test framework to ease the effort of making changes and adding new features. Another improvement could be to isolate the business logic into a framework which could power separate front ends, such as mobile devices, web applications, or other applications which might import the functionality either as a package or a DLL.

In order to indicate which the most appropriate flexible packaging equipment solution is, the user of the application has to define a set of alternatives and a set of comprehensive criteria with which to compare them by selecting File/New from the menu bar. Each criterion has an increasing preference direction if the objective is to maximize the criterion, like for example the maximum speed, or decreasing preference direction if the goal is to minimize the criterion, like for instance the change over time. The performance of the alternatives to be assessed with respect to identified criteria is depicted via Input/Performance Matrix from the menu bar. The preference parameters for each criterion are depicted via Input/Thresholds from menu bar. The preference parameters and the performance of the alternatives were entered in the application according to the experts' options shown in tables 1-2. Having established the input data, we may continue to the calculation of the rankings by selecting Action/Compute from the menu bar. Following the performed computational experiments, intermediate results (the concordance matrix, the discordance matrix and the credibility matrix) were acquired, accessible via Action from the menu bar. Figure 1 presents the concordance matrix computed by the application.

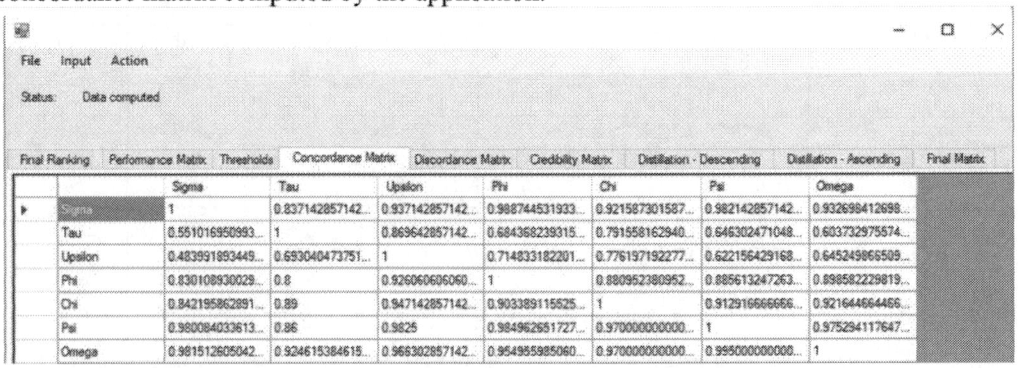

	Sigma	Tau	Upsilon	Phi	Chi	Psi	Omega
Sigma	1	0.837142857142...	0.937142857142...	0.988744531933...	0.921587301587...	0.982142857142...	0.932698412698...
Tau	0.551016950993...	1	0.869642857142...	0.684368239315...	0.791558162940...	0.646302471048...	0.603732975574...
Upsilon	0.483991893449...	0.693040473751...	1	0.714833182201...	0.776197192277...	0.622156429168...	0.645249866509...
Phi	0.830108930029...	0.8	0.926060606060...	1	0.880952380952...	0.885613247263...	0.898582229819...
Chi	0.842195862891...	0.89	0.947142857142...	0.903389115525...	1	0.912916666666...	0.921644664466...
Psi	0.980084033613...	0.86	0.9825	0.984962651727...	0.970000000000...	1	0.975294117647...
Omega	0.981512605042...	0.924615384615...	0.968302857142...	0.954955985060...	0.970000000000...	0.995000000000...	1

Fig. 1. Concordance matrix.

Analyzing Figure 1, one may state that there are few arguments that alternative Upsilon is at least as good as alternative Sigma ($C_{(Upsilon,\ Sigma)}$=0.48), while there are some high chances that equipment Sigma is at least as good as equipment Upsilon ($C_{(Sigma,\ Upsilon)}$=0.93). Following the performed computational experiments, two pre-classifications were obtained. The results from the partial pre-classifications are aggregated into the ranking matrix, depicted in Figure 2. The final ranking of the alternatives is acquired by summing up the number of P+ from the ranking matrix, for each variant.

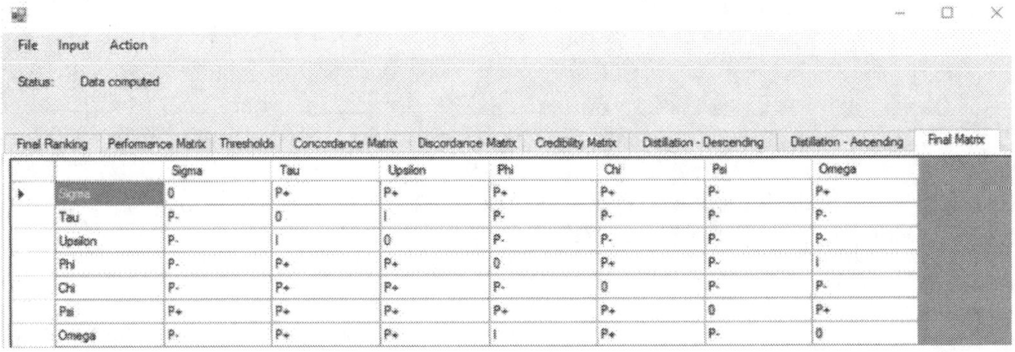

	Sigma	Tau	Upsilon	Phi	Chi	Psi	Omega
Sigma	0	P+	P+	P+	P+	P-	P+
Tau	P-	0	I	P-	P-	P-	P-
Upsilon	P-	I	0	P-	P-	P-	P-
Phi	P-	P+	P+	0	P+	P-	I
Chi	P-	P+	P+	P-	0	P-	P-
Psi	P+	P+	P+	P+	P+	0	P+
Omega	P-	P+	P+	I	P+	P-	0

Fig. 2. Ranking matrix.

It can be noticed that equipment Psi got preference over the others, followed by alternative Sigma. The equipment at the third best position were Phi and Omega, being no preference between them, followed by alternative Chi. Finally, the variants Tau and Upsilon got the worst evolution according to the criteria analyzed.

5 Conclusions

The growth of the flexible packaging industry has led to an increased need for having and using proper selection of new equipment. The selection process is vital for optimizing performance and productivity, and necessitates the evaluation of various equipment variants based on different criteria.

A practical application of a multiple criteria decision making methods, namely the ELECTRE III method, is proposed in this study to deal with the decision making situation of the equipment selection in industrial environments considering both quantitative and qualitative attributes. The application proposed in this paper helps the decision makers from a company working in the field of flexible packaging production in ranking equipment. Seven equipment have been identified and evaluated according to thirty two criteria that met decision maker's expectations. The final ranking of the equipment was achieved by running the software with the collected data. The results reveal that ELECTRE III method may be a very useful tool to support managers from the flexible packaging industry in facing the problem of selecting suitable equipment for their particular applications.

References

1. S. Vranakis, P. Chatzoglou, JIEM **7**, 24 (2014)
2. B. S. Ahn, K. S. Park, C. H. Han, J. K. Kim, Eur. J. Oper. Res. **125**, 9 (2000)
3. Smithers Pira, The Future of Global Packaging to 2020, (2015)
4. D. Aloini, R. Dulmin, V. Mininno, Expert Syst. Appl. **41**, 9 (2014)
5. Z. Ayag, R.G. Ozdemir, Int. J. Prod. Econ. **140**, 7 (2012)
6. W. Gerrard, Proceedings of 4th National Conference on Production Research **7**, (1988)
7. S. Chakraborty, D. Banik, Int. J. Adv. Manuf. Tech. **28**, 9 (2006)
8. M. Alberti, J. Ciurana, C. Rodriguez, T. Ozel, J. Intell. Manuf. **22**, 15 (2011)
9. C.M. Balaji, et al., Proceedings of the IEEE international conference on automation science and engineering, **6** (2009)
10. S. Chakraborty, Int. J. Adv. Manuf. Tech. **54**, 12 (2011)
11. O. Duran, J. Aguilo, Expert Syst. Appl. **34**, 8 (2008)
12. M. Dagˇdeviren, J. Intell. Manuf. **19**, 10 (2008)

13. M. Yurdakul, Y.T. Ic, J. Mater. Process. Tech. **209**, 8 (2009)
14. A. Samvedi, V. Jain, F.T.S. Chan, Int. J. Prod. Res. **50**, 11 (2011)
15. P. Chatterjee, V.M. Athawale, S. Chakraborty, Robot. CIM-INT Manuf. **26**, 7 (2010)
16. M. H. Aghdaie, S. Hashemkhani Zolfani,E.K. Zavadskas, Eng. Econ. **24**, 13 (2013)
17. J.J. Wang, Y.Y. Jing, C.F. Zhang, J.H. Zhao, Renew. Sust. Energ. Rev. **13**, 16 (2009)
18. A. Papadopoulos, A. Karagiannidis, Omega **36**, 11 (2008)
19. C.E.D. Infante et all., J. Clean. Prod. **52**, 12 (2013)
20. M.K. Janiaka, J. Żakb, Transport. Res. Proced. **3**, 10 (2014)
21. J. Zak, S. Weglinski, Transport. Res. Proced **3**, 10 (2014)
22. C. Giannoulis, A. Ishizaka, Decis. Support Syst. **48**, 10 (2010)
23. M. Cristea, C. Cristea, Proceedings of the Annual Sessions of Scientific Papers "IMT Oradea", **6** (2016)
24. M.S. Pishvaee, R.Z. Farahani, W. Dullaert, Comput. Oper. Res. **37**, 13 (2010)
25. J. Figueira, et al., *Multiple Criteria Decision Analysis: State of the Art Surveys*, (Springer, Berlin, 2005)
26. G.-H. Tzeng, J.-J. Huang, *Multiple Attribute Decision Making: Methods and Applications*, (CRC Press, Boca Raton, 2011)

 DOI: 10.1051/matecconf/20179406004

Problems of integration of a manufacturing system with the business area of a company on the example of the Integrated Manufacturing Systems Laboratory

Grzegorz Ćwikła[1,*] and *Krzysztof* Foit[1]

[1]Silesian University of Technology, Faculty of Mechanical Engineering, Institute of Engineering Processes Automation and Integrated Manufacturing Systems, Konarskiego 18A, PL44-100, Poland

Abstract. The paper presents issues related to data acquisition and communication between the manufacturing system and the business area of a company, understood as the transmission of real-time information about the status of the production system (the state of equipment, production tasks, workers, the flow of materials, work in progress and products), and, production orders, which is the basis of the Industry 4.0 concept. Data can be acquired from various sensors, PLCs and automatic identification systems, through interfaces (e.g. IO-Link, OPC, MTConnect), providing translation of data between different producer's equipment and software. The article presents assumptions, concept, interfaces, as well as hardware and software equipment of the Integrated Manufacturing Systems Laboratory, allowing research and teaching of integration of production and business areas. Laboratory workstations equipped with PLCs (3 different producers), sensors (e.g. distance, proximity, colour, position), actuators, HMI panels, hardware interfaces, RFID readers, barcode readers and vision systems, interfacing through OPC, IO-Link or Profinet with SCADA and other IT systems have been presented. An example of systems integration has also been presented.

1 Introduction

Issues related with the smooth flow of information in the enterprise are particularly important in the contemporary market situation, when companies are forced to compete with often global-scale competitors. Companies are looking for more sophisticated opportunities to optimize production and business processes, because the industry is running out of easy ways to increase efficiency by reducing operating costs, therefore it is necessary to ensure the continuous optimization of the functioning of the company, based on collecting and processing information about its functioning in real time, which is sometimes defined as an Industry 4.0 concept/initiative or Smart Manufacturing [1].

* Corresponding author: grzegorz.cwikla@polsl.pl

Efficient, real-time feed of information about the dynamically changing state of machinery and equipment, production orders, flow of materials and semi-finished products, tasks performed by employees, quality of products and processes, and other necessary data can allow for the making of proper, optimised decisions, regarding i.a. dispatching of production tasks, improved use of available material, hardware (machines, equipment) and human resources, planning of equipment maintenance and investment in new machines [2].

The issue of acquisition of data from production systems is an interdisciplinary problem, requiring a combination of experience in areas such as industrial automation, computer science, electronics, communication and management. Industrial data acquisition for the needs of company management support is a broad issue, requiring an individual approach to the specific conditions and needs of a particular user [3,4]. Universal methodology, supporting design of integrated information acquisition systems, called MIAS (Manufacturing Information Acquisition System), enabling the systematic design of a production data acquisition system, aimed at company management support, has been proposed in [5]. The main factors hindering the development of a universal methodology for data acquisition production are a huge diversity of existing production systems associated with the variety of technological processes, differences in the scale of production, specificity and organizational culture of companies, as well as a multitude of IT solutions, supporting the different areas of a company. IT and control systems used in companies are often a patchwork of solutions from different periods of time, from different suppliers, that are frequently not compatible with one another. Issues of data acquisition from different production systems, including all stages (i.e. accessing raw data, pre-processing, archiving) and technical means (data sources, communication and interfacing) of data acquisition have been discussed in [6].

All previously mentioned factors are the cause, that data acquisition and integration of manufacturing systems, with a business area, should become the subject of research and development of new solutions that could improve the management of processes taking place in companies. For this purpose the Integrated Manufacturing Systems Laboratory in the New Technologies Centre of the Silesian University of Technology, was established and equipped with modern automation devices and advanced software, allowing research and development on the field of the acquisition of production data and integration of production systems with the business area of companies. It allows the conducting of research, and the development of control systems, as well as for lectures and workshops for students specialising in industrial automation, or production engineering and management.

2 Concept of the Integrated Manufacturing Systems Laboratory

The problem of the integration of the production system with the business area, being an important part of the Industry 4.0 concept [1] requiring full interconnectivity of machines, systems, processes and products in "smart" networks, can be divided into several specific tasks, such as:

- collecting data through sensors and other devices,
- acquisition of data from non-automated or partially automated production systems, requiring additional data sources, e.g. automatic identification systems,
- generating data in controllers (PLC - Programmable Logic Controller, CNC, etc.),
- supervisory control in HMI (Human-Machine Interface) and SCADA (Supervisory Control And Data Acquisition) systems,
- communication and interfacing in heterogeneous production system environments,
- data archiving and processing in both the middle-layer and business-layer systems (e.g. MES - Manufacturing Execution System, ERP - Enterprise Resource Planning).

These subjects can be divided between two main areas: hardware (automated machines, actuators, semi-automatic data sources, control systems, hardware layer of communication interfaces) and software (standards, interfaces and formats of communication, industrial databases, MES systems and ERP software) [3].

In accordance with these objectives, a laboratory equipped with a set of hardware components, responsible for generating data (a small production system with sensors and automatic identification subsystems - vision system, barcodes and RFID - Radio-frequency Identification), industrial control and data processing systems (PLCs and other hardware), communication interfaces, as well as software (OPC - OLE for process control), HMI/SCADA, industrial databases, and Manufacturing Execution Systems and Enterprise Resources Planning) has been established.

2.1 The specificity of communication in production systems

All hardware and software elements of the Laboratory are connected by communication infrastructure, based on the Ethernet hardware layer. Ethernet network connects components that communicate using different protocols, including among others TCP/IP, Profinet, Modbus IP. Some elements are connected using other interfaces, such as RS-232, RS-485 (Modbus RTU), Profibus, CANbus and CC Link. There is also a set of modern IO-Link devices (IO-Link masters, sensors and actuators) - IO-Link is a relatively new communication standard gaining popularity in European industry, allowing the creation of complex control systems with remote input/output islands, based on standard 3 or 4-wire cables, which allows the modernization of existing control system by modern sensors (smart sensors) and actuators [7]. The set of communication interfaces can be easily extended, allowing for testing different solutions used in companies.

The diversity of control and communication systems in the Laboratory allows the simulation of real situations commonly found in the industry - co-existence of hardware and software solutions from different periods of time and from different suppliers. Nowadays, communications is increasingly being dominated by solutions based on Ethernet and is compatible with international standards (e.g. Profinet), displacing older and proprietary communication solutions. Often, however, the older solutions (interfaces, operating systems), even over 20 years old are still used, and the company does not plan to abandon them, forcing integrators to create communication systems including those obsolete hardware components for a task of acquiring complete set of data on technological processes and equipment status [6].

2.2 Acquisition of data from non-automated production systems

Designers creating a production data acquisition system should not forget that usually not all parts of the production process are automated (automated parts of company operation are easier object of data acquisition, equipped with wide selection of data sources, requiring often only the establishment of appropriate interfaces), so in order to obtain a full picture of the situation, solutions allowing data acquisition from non-automated parts of the process should be developed as well [8].

In non-automated (or partially automated) production systems, there is a need to replace previously used manual methods of data acquisition (such as verbal communication and reports written in logbooks) with semi-automatic solutions, speeding up data acquisition and allowing for the immediate pre-validation. Such solutions may include automatic identification systems (barcodes, RFID), as well as methods based on image recognition systems, dedicated HMI stations with appropriate programs and other solutions, depending on the specifics of the processes subject to reporting [8]. Location of objects (products, lots of WIP, mobile machinery, means of transportation, etc.) and workers can be acquired using a relatively new class of systems - Real-Time Locating Systems (RTLs) [9].

In the Laboratory of Integrated Manufacturing Systems semi-automatic (or assisted-manual) data acquisition solutions are also represented [6], including 1D and 2D barcode technologies, RFID systems operating in different frequency bands, vision systems (advanced smart camera and vision sensors), as well as HMI touch panels, that can be programmed to support data acquisition tasks.

2.3 Organization chart of the Laboratory

Based on the above assumptions, the organizational structure of the Integrated Manufacturing Systems Laboratory, including a set of specialised workstations and additional hardware and software modules has been proposed (Figure 1).

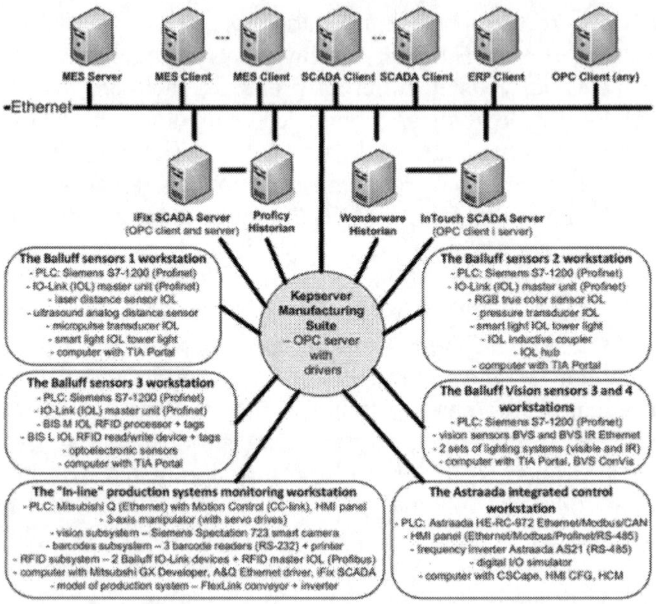

Fig. 1. Organization chart of the Integrated Manufacturing Systems Laboratory

Main workstations, subsystems and software modules of the Laboratory are as follows:

- The "In-line" production systems monitoring workstation
- The Balluff sensors 1workstation – IO-Link distance and position sensors
- The Balluff sensors 2 workstation – IO-Link pressure and colour sensors
- The Balluff sensors 3workstation – IO-Link RFID system, other various sensors
- The Balluff vision sensors 4 and 5 workstations – vision sensors and lighting
- The Astraada integrated control workstation – PLC, HMI, inverter, motor, sensors
- Software tools allowing programming of devices (PLCs, HMIs)
- The OPC server package, including drivers for various PLCs and devices (KEPServerEX Manufacturing Suite)
- SCADA systems (Proficy iFix, Wonderware InTouch, Citect SCADA)
- Industrial Databases (Proficy Historian, Wonderware Historian)
- MES software – (Proficy Plant Applications, Wonderware MES)

3 Main control and data acquisition workstations

The Laboratory should provide opportunity to get acquainted with the widest range of hardware and software currently encountered in the industry. The workstations can be regarded as independent, allowing learning of the PLCs and I/O systems programming. They can also be considered as a part of an integrated system of data acquisition and supervisory control, acting as data sources and objects of supervisory control or management.

3.1 The "In-line" production systems monitoring workstation

It is the most complex and at the same time, the oldest workstation, forming the nucleus of the Laboratory, described previously in [10], partially rebuilt (exchange of TURCK RFID subsystem into Balluff IO-Link RFID system, the installation of various sensors and IO-Link Smart Light programmable tower light). It is an integrated system that allows for the monitoring of the production process (the miniature modular production system to be monitored can be set inside, under the frame of the "In-line" workstation) with an advanced vision system (Siemens smart camera), RFID tags and bar codes, with the support of conventional sensors for industrial automation. The main function modules of the "In-line" workstation are three semi-automated data acquisition subsystems, a chassis acting as the supporting structure for the other elements (with 3-axis manipulator carrying Smart camera and one RFID heads), an advanced control system (based on Mitsubishi's Q series PLC and HMI panel), and mechatronic drives.

3.2 The Balluff sensors 1, 2 and 3 IO-Link workstations

These are 3 similar workstations built on the basis of Balluff's IO-Link elements and Siemens S7-1200 PLCs, where the emphasis is on the use of modern methods of communication between PLC and sensors / actuators. It has been decided to use a new, communication standard called IO-Link [7], currently gaining popularity, which allows two-way communication between the IO-Link master unit and slave devices – intelligent sensors (smart sensors) and actuators, using typical 3-wire cables. The IO-Link master unit can be connected to PLC or other higher-level device using Profinet, Profibus or other interfaces, which allows the easy creation of remote input/output islands. IO-Link is not regarded as an industrial network, because the I/O sockets of IO-Link master unit cannot be connected to the other master units on the other levels of the hierarchy (it is still possible to have more master units connected by Profinet or Profibus). IO-Link allows for the shortening of the distance between places where analogue data is collected and digitized – A/D conversion can take place directly in smart sensors or in special analogue sockets of the IO-Link master, preventing distortion of the signal. Typically, it is possible to connect IO-Link smart sensors / actuators or standard discrete sensors to IO-Link master sockets and full diagnostics of master - sensors - actuators connections is also possible. Thanks to bi-directional communication, smart IO-link devices allow remote configuration and diagnostics without physical access to devices, which is very convenient when they are not easily accessible. IO-Link makes it easy to modernize the control system through the exchange of sensors and leaving the existing wiring in place. A diagram of the possible connections of devices in the IO-Link is shown in Figure 2.

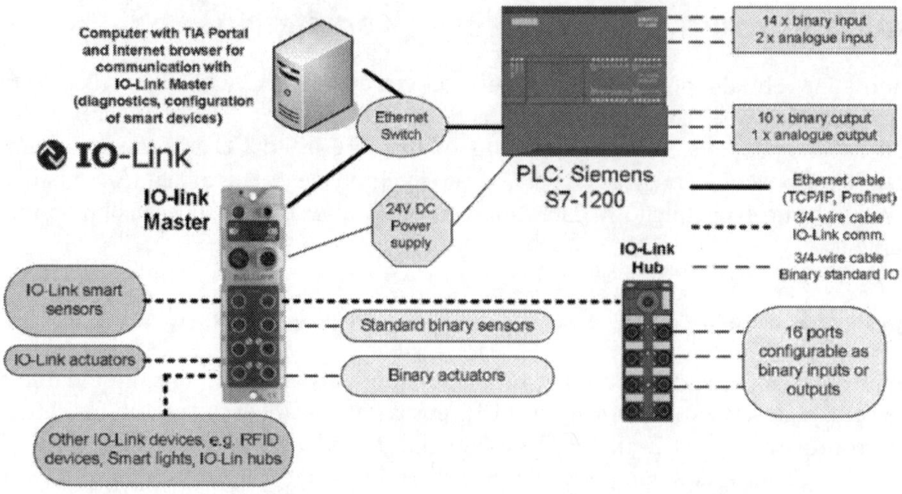

Fig. 2. Possible connections of IO-Link devices in the Laboratory's workstations

As mentioned before, these workstations are built on the basis of Siemens S7-1200 PLCs and Balluff's IO-Link master units, allowing for direct connection of up to 16 slave devices that communicate using the IO-Link protocol or standard discrete I/O signal. An interesting device is the IO-Link hub (installed in workstations 2 and 3), that allows the connection of up to 16 (2 pins in each of 8 sockets) standard discrete sensors or actuators (in this case it is not possible to connect smart sensors). Another interesting solution is the inductive coupler used in the Balluff sensors 2 workstation, allowing wireless transmission for a distance of up to 5 mm for both the IO-Link or SIO (Standard Input-Output) signal, as well as supplying power for the connected devices. This allows for easy communication with moving machine parts, replacing unreliable mechanical connections.

The other devices available in the Balluff sensors 1, 2 and 3 workstations are a variety of sensors (listed in Figure 1) and RFID devices, as well as output devices - Smart Light tower lights. The Balluff sensors 1 workstation, together with the specified devices is shown in Figure 3 (left). The workstation's main components are Simatic S7-1200 PLC (1) and IO-Link master unit BNI005H (2), allowing the connection of up to 8 intelligent IO-Link sensors or actuators (connecting classic sensors is also possible), both interfaced using Profinet. PLC is equipped with a simulator and an ultrasound distance sensor (3), while remaining sensors, optoelectronic (laser) distance sensor BOD0012 (4) and Micropulse position sensor BTL1F5N (5) are connected using the IO-Link standard. The Smart Light BNI0086 (LED programmable tower light) (6) is currently the only output device (Figure 3, left).

3.3 The Balluff vision sensors 4 and 5 workstations

The main components of these workstations are Balluff's vision sensors (BVS 001M and BVS 001H), the first of which requires illumination of the observed objects in visible light, and the second one requires infrared lighting), that differs from the conventional "full-size" or smart camera vision systems by the simplicity of its installation, application and programming, but they are slightly less capable (smaller resolution and configuration flexibility, the lack of a possibility to exchange lenses). It allows for 1D/2D barcodes reading, feature presence/absence and inspection, pattern detection, position detection, dimension verification, edges counting, etc.. All configured inspections compensate for the parts rotation, minimizing the need to carefully position parts, while reducing setup costs [11]. The vision sensors communicate with the surroundings (Siemens PLC and a computer used for programming by ConVis configurator) via Ethernet and RS-232 interfaces. The outputs of the sensor signals can be discrete signal (success or failure of inspection), the numerical values of the parameters (the coordinates of objects, dimensions and distances), and strings read from the barcodes. An important element of the vision sensor workstations are sets of LED light sources, operating in the visible light or infrared – spotlight, linear and backlight. The vision sensors workstations enables research and teaching on the use of vision systems in production processes - the selection of suitable sensors, optical characteristics and lighting, adjusting the position of the sensor in the analyzed scene, configuration procedures, object recognition, as well as the programming of industrial controllers for use with vision sensors.

3.4 The Astraada integrated control workstation

The Astraada workstation (Figure 3, right) allows research and teaching on a number of issues related to the control of technological processes. The main elements are Astraada HE-RC972 PLC (7) with Ethernet (there are various protocols available, including Modbus IP, FTP and HTTP), RS232 and CAN interfaces, colour touch screen HMI (8) Astraada AS43TFT0725 (Ethernet, USB, 3xRS232/485 interfaces) and the Astraada AS21DRV20C4 frequency inverter (12), connected with the AC motor (13). The PLC communicates with the HMI panel via Modbus IP, and the HMI with a frequency inverter via RS485 (Modbus RTU), enabling visualization of the state and setting the parameters of the inverter. Inductive (10) and capacitive (11) sensors are also available, as well as an I/O simulator (9).

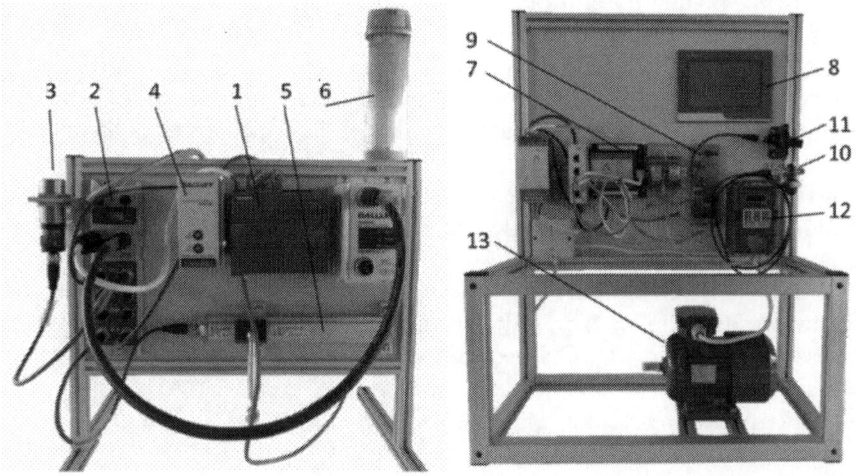

Fig. 3. The Balluff sensors 1 (left) and the Astraada integrated control (right) workstations

The range of possible exercises includes programming the controller, HMI and inverter and the cooperation of these devices using different media and protocols. The Astraada workstation is naturally integrated with the network of the Laboratory and higher level software, in particular KEPServerEX OPC server, that can communicate with the PLC and the HMI, allowing interaction with the hardware for any OPC client applications.

4 Software used for the data acquisition and production systems integration

The software installed in the Laboratory, combined with the hardware allows a complete representation of issues related to the Industry 4.0 concept, and thus the subject of integration of equipment, staff and circulation of materials in a thoughtful, orderly system, capable of delivering real-time information about the status of machines and equipment, execution of orders, crew activity, and product quality and genealogy. IP addresses of computers and other devices installed in the Laboratory belong to a common address pool, so all devices can communicate with each other, which makes PLC programming from any computer in the Laboratory or access to any data possible.

There are 3 main types of software installed in the Laboratory:

- software tools necessary for configuring and programming of devices (PLCs, HMIs, frequency inverters, vision sensors, IO-link masters and smart sensors),
- software interfacing between hardware and IT systems – OPC server and drivers, dedicated drivers,
- IT systems exchanging, archiving and consuming data from hardware sources – SCADA, Historian, MES, ERP, etc.

A study of the situation in the industry [12] shows that many of the problems with the integration of systems are associated with a variety of operating systems and security software used on computers, which resulted in the installation of different operating systems, anti-viruses and firewalls on the computers used in the Laboratory.

4.1 The OPC server

OPC was chosen as the main standard integrating hardware layer of the Laboratory workstations with SCADA, archiving and production management software due to its versatility, growing popularity and other advantages [13]. KEPServerEX OPC server has been selected as the main hardware-software interface. It is equipped with a wide range of drivers (KEPServer Manufacturing Suite) enabling communication with different devices. The OPC server communicates with devices from different manufacturers using dedicated drivers, and a variety of transmission media and protocols, while it shares the collected data in a standardized form (OPC protocol, SNMP, ODBC, XML, web services), allowing access to current and historical process data, alarms, for any compatible client software.

As a result, there is no need to create special communication drivers for each program that requires connection to specific hardware, moreover, the data is retrieved from the PLC to the OPC server only once, where it can be made available to many clients (reducing the communication load of PLCs). The OPC server can be the central point of integration of different areas of a company. OPC server can exchange data with other OPC servers, it is possible to apply redundant architecture, furthermore, some SCADA systems can be configured not only as an OPC client, but also an OPC server.

4.2 HMI/SCADA, Historian and MES software

The software installed in the Laboratory, communicating with the hardware by the OPC protocol includes: SCADA systems used for supervisory control and data acquisition from individual workstations, Historian industrial database useful for storing large amounts of data, and software belonging to higher level IT systems – production and business management level systems – Manufacturing Execution Systems (MES) and ERP. A common feature is the ability to use the OPC server as a source of data, although it is also possible to use dedicated drivers, and in the case of MES systems, the preferred data source is the Historian industrial database. The distinguishing features of the Historian, from typical SQL databases, include high performance of mass simple process data storage, ease of configuration, reliability, zero-downtime maintenance, automatic data compression, redundancy and the ability to pre-process the data (data integration - calculation of the synthetic data based on the values of many simple process variables) [14].

The Integrated Manufacturing Systems Laboratory uses integrated software packages, forming complete platforms, e.g. GE Intelligent Platforms (formerly Proficy) or Invensys Wonderware, which are complete solutions, including, inter alia, SCADA systems (GE iFix, Wonderware InTouch), industrial database (GE Historian, Wonderware Historian) and MES (GE Production Performance - formerly Proficy Plant Applications, Wonderware MES) [15]. This provides the opportunity to get acquainted with the solutions used in the largest enterprises [16].

5 Example of the integration of workstations and IT systems

The first task during the exemplary application was to get acquainted with the specifics of the communication protocols of existing equipment and checking the availability of OPC drivers. The next steps were: creating demonstration programs for each PLC and workstation, installing the appropriate OPC drivers in KEPServer, the configuring PLC - OPC server connection, creating a database containing tags retrieved from the physical addresses in the PLCs, and validation of the communication using the OPC Quick Client application. The last stage of integration was gaining access to the data collected from the OPC client application. In this case HMI/SCADA iFIX was used, in which firstly the OPC communication driver had to be defined, then the database of tags was created, and finally SCADA applications using collected data from selected workstations were designed.

In order to provide data that can be accessed through OPC, demonstration programs have been developed for all workstation PLCs, using appropriate development environments (Mitsubishi GX IEC Developer, Simatic TIA Portal Step7, Cscape). These programs use various logic blocks, including Boolean operations, counters, timers, and other functions in order to generate different types of data, like Boolean, integer, real and strings, based on the state of inputs connected to simulators, sensors, or, in the case of the "in-line" workstation, data acquired from barcode and RFID readers.

5.1 OPC server configuration and testing

The main steps of setting up KEPServer OPC server are: adding of a channel (at this stage it is possible to choose which communication driver will be used), adding a device to the channel (defining device network address, protocol, communication settings and other parameters), and finally, adding tags (variables) to the devices.

Three separate channels have been defined because every workstation's PLC required another communication driver. The Mitsubishi Ethernet, Modbus TCP/IP Ethernet and Siemens TCP/IP Ethernet have been used, respectively, for workstations.

At the second stage PLCs have been added to the channels. Thanks to the local Ethernet network and its interfaces, all PLCs and PC with an OPC server in the same network segment, are available in uniformed TCP/IP address space, it is possible to exchange data with all PLCs and other networked devices (like IO-Link master units) in the Laboratory.

The most difficult part of the OPC set-up procedure is the defining of tags, because each PLC and protocol has its own methodology of addressing and accessing of different types of data, inputs, outputs, registers, variables and function blocks. In the case of relatively widespread Modbus protocol, it is possible to translate internal PLC addresses with the Modbus addressing translation table, that has to be combined with knowledge about the structure of a specific PLC (number of discrete/analogue inputs and outputs, flags, registers, system variables etc.). Mitsubishi and Siemens PLCs are both different and it is necessary to find technical data and addressing convention for each. KEPServer's help files describe the basics of addressing methods. The final effect of the OPC server configuration (Figure

4) is a set of channels with defined devices and tag databases for each PLC device, accessing data changed in the PLC programs.

Validity of KEPServer configuration and communication can be verified using OPC Quick Client, available in the KEPServer set of programs. The status of communication with PLCs can be displayed, data quality on all defined tags should be "good", and value of variable in Quick Client should change together with changes in the PLCs.

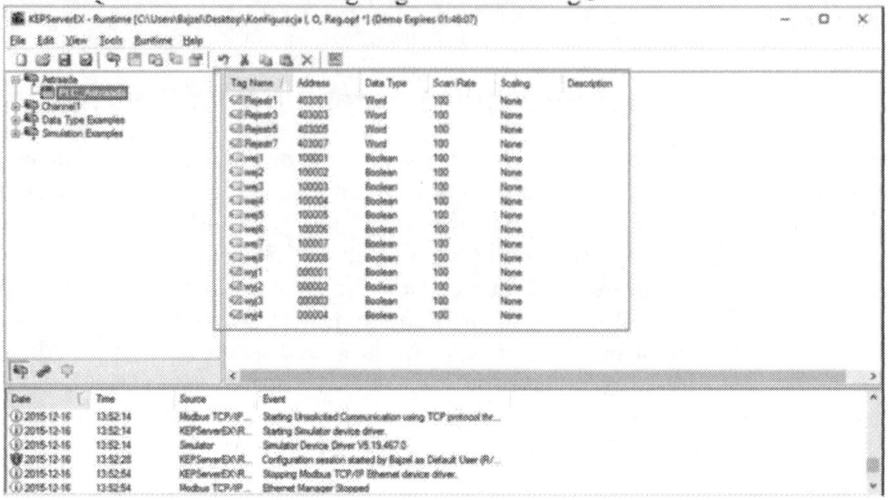

Fig. 4. KEPServer configuration for the Astraada workstation

5.2 HMI/SCADA iFIX applications

In order to check the last stage of integration, GE Intellution Proficy iFIX applications, displaying data acquired from workstations, have been developed. It required configuring the OPC driver in the iFIX configuration module, adding tags to the iFIX tag database, and then designing HMI screens displaying data from PLCs in various forms, like numerical display, bit lamps, character string displays, animated objects (Dynamo) and charts, etc.

6 Conclusions

The Integrated Manufacturing Systems Laboratory provides a unique opportunity to get acquainted and conduct exercises or research associated with the issue of the integration of manufacturing systems with software supporting company management at different levels. Previous solutions typically focused on one of these aspects, either the issue of designing, configuring and programming of automatic control systems, or only enterprise management support software (MES, ERP). Data processed in software supporting the management installed in the Laboratory are the actual, non-simulated data from workstations hardware, that can be configured and programmed for specific applications, making realistic presentation of these issues easier.

The main problems encountered during the operation of the Laboratory are the high degree of complexity and the variety of hardware and software solutions represented in it, which is a problem assuming the limited time of workshops, and existing conflicts between the presentation of issues of systems integration and programming of devices – presentation of issues of systems integration is possible if workstations are programmed with prepared demos, which often are overwritten during the PLC programming workshops.

The advantage of the OPC technology is that data acquired once from PLC can be reused many times in client applications. Experiments have shown that, thanks to the OPC, data can be obtained from different devices, but various difficulties have also been encountered. Access to simple data, such as the states of PLC's inputs, outputs and registers is relatively easy, but it is not always possible to obtain more complex data stored in function blocks specific for PLC manufacturers (e.g. to get data from optimized function blocks, dedicated Siemens software has to be used). Moreover, equipment from different producers and periods of time is often visible in the shopfloor - access to older devices, for which there is no documentation and source software can be a problem because of a lack of interfaces or knowledge on the meaning of data in the controller's memory - it has to be recovered through reverse engineering.

The Laboratory allows students to get familiar with the issues of integration of the various systems of control and management, very important in the context of the need of improving the efficiency of business management, and implementation of the Industry 4.0 concept.

The Integrated Manufacturing Systems Laboratory has been partially equipped thanks to cooperation from Balluff Sp. z o.o. and Balluff GmbH.

References

1. S. Weyer, M. Schmitt, M. Ohmer, D. Gorecky, IFAC-PapersOnLine **48-3**, 579 (2015)
2. K. Herbuś, P. Ociepka, IOP MSEng, **95**, 012084, (2015)
3. A. Gwiazda, A. Sękala, Z. Monica, W. Banaś, Adv. Mat. Res., **1036**, 1023 (2014)
4. A. Sękala, G. Ćwikła, G. Kost, IOP MSEng, **95**, 012097 (2015)
5. G. Ćwikła, AppMechMat, **474**, 27 (2014)
6. G. Ćwikła, AdvMatRes, **837**, 618 (2014)
7. http://www.io-link.com/share/Downloads/At-a-glance/IO-Link_System_Description_engl_2013.pdf (2013)
8. G. Ćwikła, D. Krenczyk, A. Kampa, G. Gołda, IOP MatScEng, **95**, 012153, (2015)
9. N. Pirzada, M. Yunus Nayan, F. Subhanc, M. Fadzil Hassan, M.A. Khan, AASRI Procedia **5**, 92 (2013)
10. G. Ćwikła, Adv. Mat. Res., **837**, 334 (2014)
11. http://www.balluff-ua.com/pdf/bvs_manual_en.pdf (2015)
12. L. Monostori, Procedia CIRP **17**, 9 (2014)
13. G.G. Florea, L.A. Ochean, ProcIFAC InfContrProbManuf,1844 (2012)
14. B.R. Mehta, Y.J. Reddy, Industrial Process Automation Systems. Design and Implementation (Elsevier, 2015)
15. M. Bączkowicz, A. Gwiazda, IOP MatSciEng, **95** (2015)
16. K. Herbuś, G. Kost, D. Reclik, J. Świder, AdvMatRes, **837**, 582 (2014)

ERP system for 3D printing industry

Bogdan Deaky[1,2*], and *Luminița* Pârv[1]

[1]Transilvania University of Brașov, Department of Manufacturing Engineering, Mihai Viteazu No.5, Brașov, Romania
[2] Bluemind Software, Str. Iancu de Hunedoara 45/A/17, 530194, Miercurea-Ciuc, Harghita, Romania

Abstract. GOCREATE is an original cloud-based production management and optimization service which helps 3D printing service providers to use their resources better. The proposed Enterprise Resource Planning system can significantly increase income through improved productivity. With GOCREATE, the 3D printing service providers get a much higher production efficiency at a much lower licensing cost, to increase their competitiveness in the fast growing 3D printing market.

1 Introduction

Enterprise Resource Planning (ERP) packages are information systems that automate the business processes of organizations thereby improving their operational efficiency substantially. ERP projects that involve customization are often affected by inaccurate estimation of efforts [1]. Complex information systems like the ERP integrate the data of all business areas within the organization. The implementation of ERP is a difficult process as it involves different types of end users [2, 3].

Enterprise resource planning systems are extensive software systems that integrate a number of business processes, such as manufacturing, supply chain, sales, finance, human resources, budgeting, and customer service activities.

Various studies have asserted that ERP systems can increase competitive advantage in the Information Technology (IT) era. Particularly in the e-business era, there has been a global trend to integrate business processes based on the company's strategic implementation [4, 5]. A review of the potential benefits from ERP implementation can be classified into tangible and intangible benefits [4].

Tangible benefits include reduction of inventory, reduction of personnel, increased productivity, improvements in orders management, quicker closing of financial cycles, increase of 'on-time' deliveries, improved customer service, reduction in IT and procurement costs, improvement of cash flow management, increase of revenue and profits, reduction in transportation and logistics costs and reduction in the need for system maintenance [1-4].

Intangible benefits refer to the increased visibility of corporate data, new or improved business processes, improved responsive- ness to customers, improved communications, unanticipated reduction in costs, better integration between systems, standardization of

* Corresponding author: bogdan@bluemind-software.ro

computing platforms, increased flexibility, global sharing of information, improved business performance and improved visibility into the Supply Chain Management process [5,6].

2 GOCREATE – General presentation

GOCREATE is a cloud-based production management and optimization application which helps 3D printing service providers' use their resources better to significantly increase the income through improved productivity.

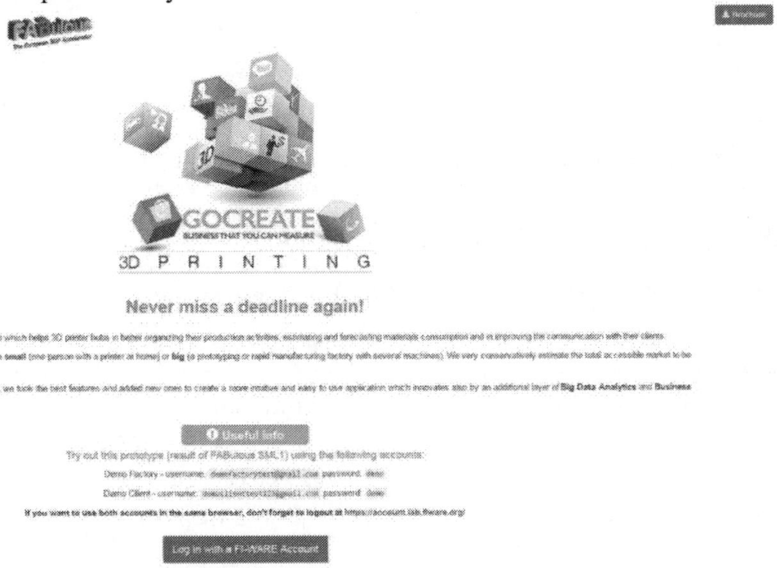

Fig. 1. GOCREATE Login page.

GOCREATE (Figure 1.) is a cloud-based application offered as Software as a Service (SaaS). It helps 3D printer hubs (small and big) with automatic scheduling, material consumption tracking and forecasting and by improving the communication with their clients.

2.1 Information flow

GOCREATE is built with FIWARE technology. The FIWARE mission statement is "to build an open sustainable ecosystem around public, royalty-free and implementation-driven software platform standards that will ease the development of new Smart Applications in multiple sectors", For more please refer to [x]

[x] FIWARE, "About us", https://www.fiware.org/about-us/

The information flow between the various GEs (Generic Enablers) and our own systems in GOCREATE is shown in Figure 2.

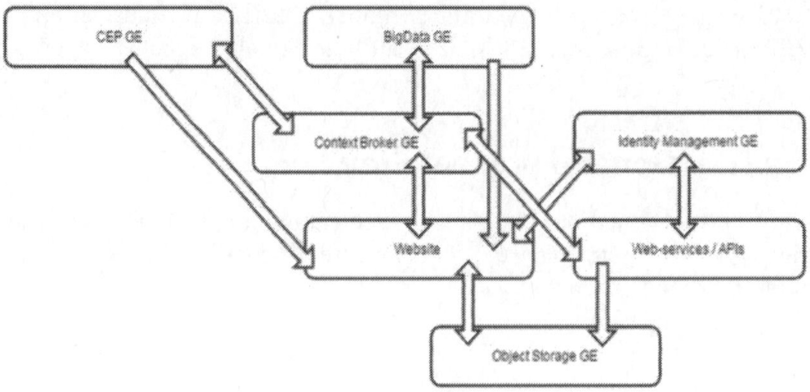

Fig. 2. Information flow diagram (GE-Generic Enabler).

2.2 FIWARE technology usage

The FIWARE Enablers used by GOCREATE are presented in Table 1 [7].

Table 1. FIWARE tools.

FIWARE tools	Function	Observation
Keyrock	- implementation of the FIWARE Identity Management Generic Enabler - account management and authentication and token/user validation for security	Keyrock Open Specification
Orion	- stores all the context data - used everywhere where context data is involved	exception: payment transactions
Object Storage	- stores the files (binary STL or other formats) in the FIWARE cloud	
Wirecloud	- generate customizable dashboards (for our own use)	
CEP	- generate alerts based on complex events.	recognizing started but failed payments
Cosmos and SpagoBI GE	- identify the trends and create valuable reports	Big Data
EXTRA: 3D-UI-XML GE	- displaying the uploaded 3D models	STL files

3 GOCREATE – ERP system functions

Some individual characteristics of information system users have been shown to be associated with different levels of information system usage.

GOCREATE 3D Printing-provides the following facilities for the Customer (Table 2):

There were two Service Maturity Levels in the FABulous accelerator's 2nd open call. Bluemind Software was selected to complete both SLM1 and SML2. At the moment of writing this paper, it is completing SML2.

Table 2. Facilities provided by the system for the customer.

Facilities for the customer	**SML1**	**SML2**
– viewing the Factory data/capabilities and its Machines	✓	✓
– sending Orders directly	✓	✓
– sending Orders directly via third parties	x	✓
– checking order status	✓	✓
– checking personal order history	✓	✓
– having a dialog with the Factory	x	✓

GOCREATE provides the following facilities for the Factory (Table 3).

Table 3. Facilities provided by the system for the factory.

Facilities for the factory	**SML1**	**SML2**
- defining the Factory data	✓	✓
- defining the available Machines and Materials and their associations	✓	✓
- factory presentation page	✓	✓
- receiving Orders and setting Order status	✓	✓
- saving the objects to be printed and adding new parameters to the Order model for additional optimizations	x	✓
- having a dialog with the Customers	✓	✓
- scheduling recurring events and customer orders to machines	✓	✓
- schedule visualization	✓	✓
- alerts (related to deadlines, machine load limits etc.)	x	✓
- visualizing Order statuses and Order history	✓	✓
- connecting to other systems which facilitate other aspects of the business - by providing interfaces as web-services/APIs, to having orders placed	x	✓
- information offer as open data	✓	✓
- computation of monthly material consumption and forecasts	x	✓
- adding other operations to the manufacturing process (e.g.	x	✓

	finishing operations)		
-	customizable dashboards	x	✓
-	access to selected relevant Big Data Analysis and BI process results	✓	✓

GOCREATE interface samples are shown below.
One of the useful features of GOCREATE is an automatically generated and updated public webpage, where the 3D printing service providers can start receiving orders.

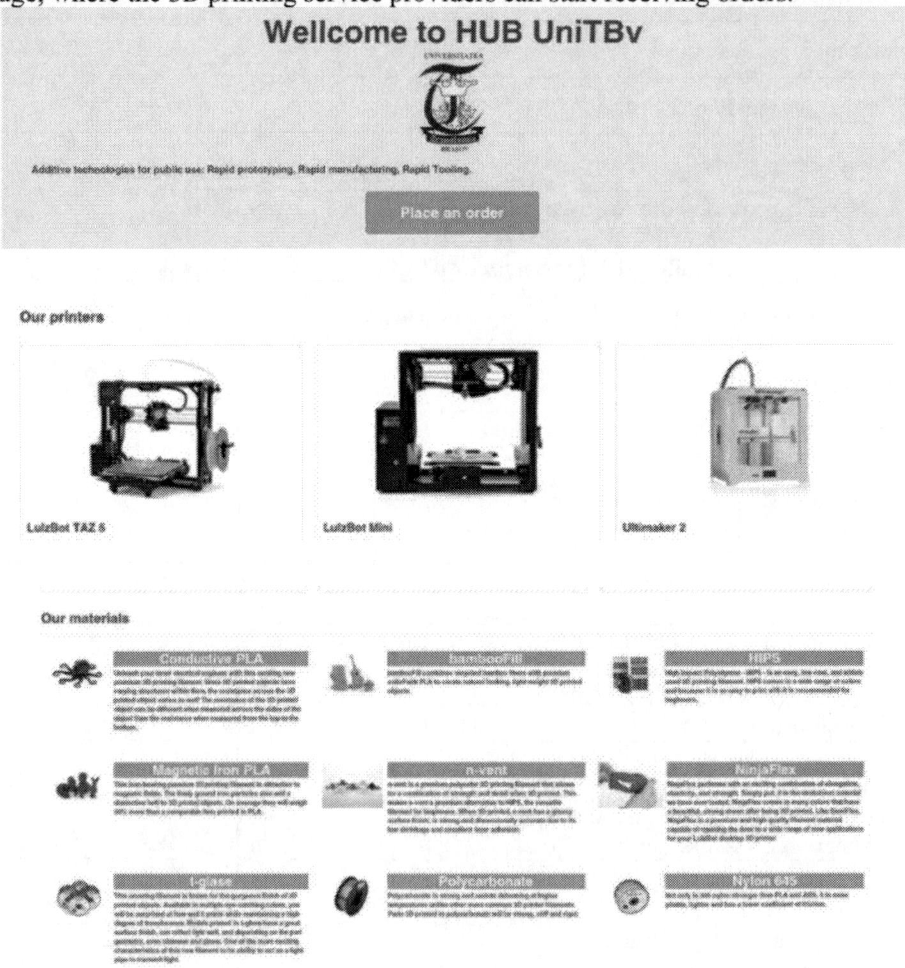

Fig. 3. Public Page

Orders can be submitted by clients from the public page, but also by the factory. This way, the Factory won't be stuck when a client places an order by phone.

 DOI: 10.1051/matecconf/20179406005

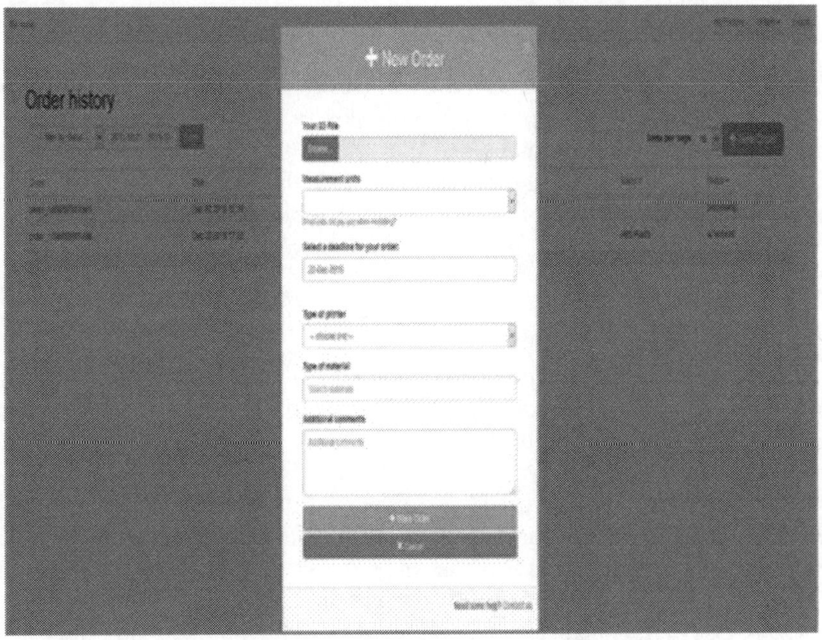

Fig. 4. A new order submission.

One of the important differentiators of GOCREATE is the automation of the order scheduling process. It not only saves time but it also very easy to use (maximum one click).

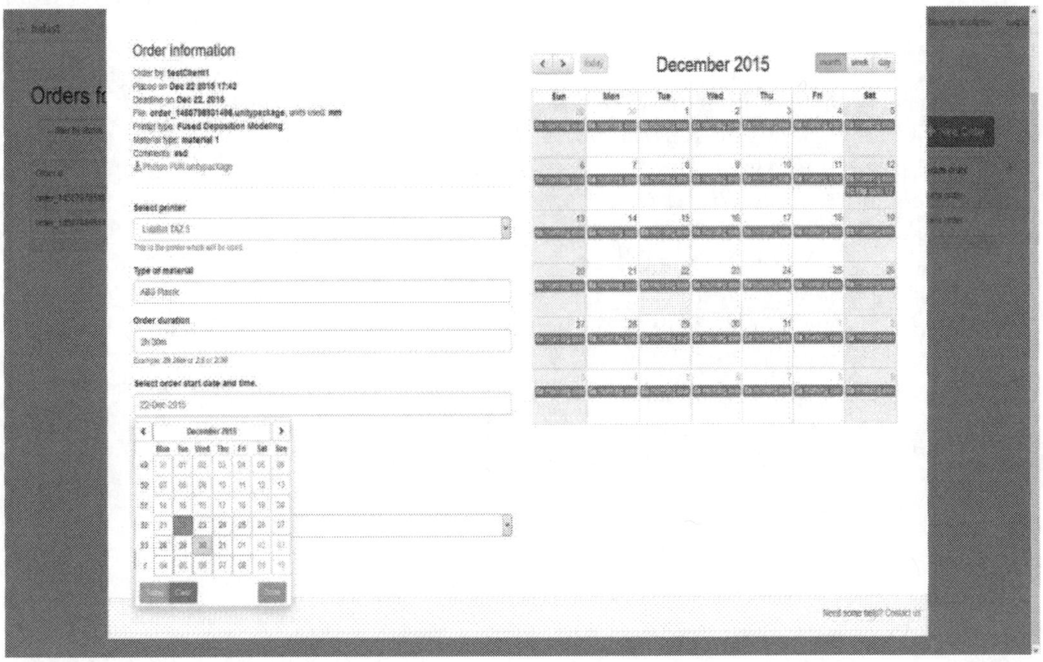

Fig. 5. Automatic order placement in the schedule of the appropriate 3D printer.

4 Conclusion

Enterprise resource planning systems are different from other innovations of IT because of the socio-technical challenges due to the complexity involved in the implementation process and the different types of end users [9].

In general terms, GOCREATE ERP aims to provide the following benefits:

- give a global, real-time view of data that can enable companies to address concerns proactively and drive improvements;
- improve financial compliance with regulatory standards and reduces risk;
- automates core business operations such as lead-to-cash, order-to-fulfilment and procure-to-pay processes;
- enhances customer service by simple embedded dialog elements.

GOCREATE is the first step in addressing the whole 3D printing industry. Besides the needs of 3D Printing service providers, it already takes into account the needs of the 3D printer and material manufacturers.

GOCREATE addresses a geographically dispersed market. The sales strategy aims to get GOCREATE into the market through partners who already sell to our clients. Most of them are 3D printer and material manufacturers.

The advantages offered by GOCREATE for partners are:

- their sales are boosted by added value and features like automatic material ordering;
- they gain a new direct advertising and sales channel, using the various forms of communication of the new market;
- they receive access to market information based on real-time and accurate data.

We hereby acknowledge Bluemind Software for allowing us to publish the above information and the FABulous accelerator and the RICAP mentoring program for business development and internationalization for making it all possible.

References

1. T. Chandrakumar, S. Parthasarathy, Comp. Stand. Int. **47**, 100 (2016)
2. C. A. Rajan. R Baral, IIMB Manag. Rev. **27,** 105 (2015)
3. K. I. Weinrich, N. Ahmad, Int. J. Ent. Inf. Sys. **5(1)**, 55 (2009)
4. Y. C. Shen, P.s. Chen, Wang, C.H. Comp. Ind. **75**, 127 (2016)
5. K. Bokovec, T. Damij, T. Rajkovic, Comp. Ind. **73**, 93 (2015)
6. D. Jain, Y. Sharma, Annual Res. J. SCMS **4**, 140 (2016). Available at SSRN:http://ssrn.com/abstract=2755457
7. FIWARE Catalogue, FIWARE (Core Platform of the Future Internet), http://catalogue.fiware.org , Accessed July 2016
8. R. Buican, Gh. Oancea, C. Lancea, A. Pop, App. Mech. Mat. **760**, 515 (2015)
9. J. Sujanova, P. Gabris, M. Licko, P. Pavlenda, R. Stasiak-Betlejewska, *Proc. 13th Eu. Conf. Knowl. Manag.*, **1-2**, 1135 (2012)

Model for determining and optimizing delivery performance in industrial systems

Flavia Fechete[1,*], *and Anişor* Nedelcu[1]

[1]Transilvania University of Braşov, Department of Manufacturing Engineering, Mihai Viteazu No.5, Braşov, Romania.

Abstract. Performance means achieving organizational objectives regardless of their nature and variety, and even overcoming them. Improving performance is one of the major goals of any company. Achieving the global performance means not only obtaining the economic performance, it is a must to take into account other functions like: function of quality, delivery, costs and even the employees satisfaction. This paper aims to improve the delivery performance of an industrial system due to their very low results. The delivery performance took into account all categories of performance indicators, such as on time delivery, backlog efficiency or transport efficiency. The research was focused on optimizing the delivery performance of the industrial system, using linear programming. Modeling the delivery function using linear programming led to obtaining precise quantities to be produced and delivered each month by the industrial system in order to minimize their transport cost, satisfying their customers orders and to control their stock. The optimization led to a substantial improvement in all four performance indicators that concern deliveries.

1 Introduction

Industrial systems performance plays a crucial role for their survival and maintaining on the market, given the fierce competition which led to the disappearance from the market of those who are not progressing and therefore do not obtain high performance.

The current economic environment is no longer confined to strictly economic performance targets, it also requires a global vision of what the entity's performance really means.

Therefore it is no longer sufficient an exclusively economic approach to assess the financial performance of an organization, it requires a global vision that may relate to product and processes quality, entity's costs, on time delivery of their products or even to employees satisfaction. Delivery performance requires a special attention because it is closely related to customer satisfaction, the size of future orders, customer loyalty, and it may have a major implication on the revenue and expenses budget of the company.

* Corresponding author: fechete_flavia@yahoo.com.

In this regard it is imperative to meet the precise quantities to be produced so that delivery is made on time, according to the customer orders, and at the same time the delivery fits in the planned transport costs.

The present work aims to help the company to improve decision management system, to create a decision model that will contribute to improvement. The major objective of this paper is to create a model for the delivery performance of an industrial system, a model that will help the managers to make the best decisions on production optimization.

For this purpose, there will be used the linear programming to obtain the precise quantities to be produced and delivered each month by the industrial system in order to minimize their transport cost, satisfying their custormers orders and to control their stock.

Modeling and optimization techniques are used not only in the traditional supply chain management, but also in other functional areas of organizations as finance, accounting, marketing [1].

The problem of the delivery performance lies in the attention of many other researchers [2 - 4], though the delivery optimization focuses particularly on cost reduction, without taking into account other important factors.

In this regard, it was made a brief introduction to the theory of mathematical programming, following with a case study in order to develop the linear programming used to determine an optimal production program. The case study was conducted on an industrial system that produces and sells outer rings for cam follower rollers. The research was focused on delivering the products manufactured by the organization, it was intended to determine the optimal production program that satisfies customers orders and also minimizes the transport cost. Thus, it was determined the objective function and its constraints and using the linear programming module of the software Quantitative Management (QM), we were able to determine the amount of each product that must be manufactured in order to minimize the costs of delivery.

2 Method

The decision process involves evaluating several decisional alternatives based on economic indicators, in order to choose the optimal variant. For choosing the optimal decision it is necessary to hierarchy available variants, depending on the desired criteria. One of the most used methods in this regard it is economic modeling by linear programming.

Linear programming problem is within the general mathematical programming models and it is characterized by the fact that both the objective function and the constraints are expressed mathematically by linear functions [5].

The linear programming problem contains restrictions, as well as a criteria of "performance" to assess the effectiveness of each activity. Depending on the purpose, we can choose as a criterion of efficiency an indicator measuring the effort, one that measures the result or an indicator expressed as a ratio of result and effort (or effort and result) [6].

Maximum efficiency means minimizing the effort and maximizing the outcome and the optimal concept is defined in this case as a program that minimizes or maximizes an objective function and simultaneously satisfies all technical and economic restrictions [7].

In order to determine the optimal production program that satisfies customers orders and also minimizes the transport cost there will be used the following objective function [1, 6, 8]:

$$min\left[\sum_{i=2}^{n}((c_i \times x_i + c_i^s \times x_i^s + d \times s_i) + c_1 \times x_1 + c_1^s \times x_1^s)\right], \quad (1)$$

With the following constraints:

$$\sum_{k=1}^{i}(x_k + x_k^s) - s_{i+1} = \sum_{k=1}^{i} R_k , \tag{2}$$

$$i=1,2, \ldots, n-1 \tag{3}$$

$$x_n + x_n^s + s_n = R_n \tag{4}$$

$$0 \leq x_i \leq y_i,\ i=1,2, \ldots, n \tag{5}$$

$$0 \leq x_i^s \leq y_i^s,\ i=1,2, \ldots, n \tag{6}$$

$$s_i \geq 0,\ i=1,2, \ldots, n, \text{ where} \tag{7}$$

n – number of months;
$y_i, i=1,2,\ldots,n$ - monthly production volume achievable on normal program;
y_i^s , $i=1,2,\ldots,n$ - monthly production volume achievable overtime;
$c_i, i=1,2,\ldots,n$- monthly cost of production on normal program;
c_i^s , $i=1,2,\ldots,n$ - monthly cost of production overtime;
d - unit cost of storage;
x_i, $i=1,2,\ldots,n$ – quantities that must be monthly produced on normal program;
x_i^s, $i=1,2,\ldots,n$ - quantities that must be monthly produced overtime;
s_i, $i=1,2,\ldots,n$ – quantities that must be monthly stored.

In order to achieve the minimum costs of production and storage, variables x_i, $x_i^s s_i$ will be determined.

Determining the optimal production program is one of the major concerns of managers because it is closely related to the economic performance of the organization, deciding the success of the organization [9].

This paper presents below a study case of modeling and optimization the delivery performance of an industrial system by using linear programming. There is considered an industrial system that must sell their product during a period of 12 months, based on the customer orders. The paper objective is to analyze the current situation of delivery and to determine an optimal production program by using linear programming in order that shipping costs to be minimal.

3 Results

The delivery area is one of the most important aspect for each company, it may influence the customer behavior, the number of orders and the sales. The industrial system shows a major interest on this particular area, desiring to satisfy customer needs and minimizing their transportation expenses.

In order to identify the optimizing area, there will be first analyzed the main key indicators monitored in the delivery, as we proposed below. These indicators are:

A. On-Time Delivery - OTD

This indicator measures the percentage of all orders delivered by the requested delivery date, as indicated in the PO/contract during a defined period of time.

$$OTD = \frac{Nrct}{Nrcl} \times 100, \text{ where} \tag{8}$$

OTD = On-Time Delivery;
Nrc_t = Number of orders delivered at requested date;
Nrc_l= Number of orders.

B. Supplier Lead-Time Variability - SLT

This indicator is the average of the absolute percentage differences (APD) between the supplier's forecasted lead time and the actual lead time for each order placed with the supplier. This indicator can be calculated for any supplier that supplies products to the requesting facility. It can be measured over any time period, but one year is typical; usually measured in days.

$$SLT = \frac{\Sigma APD}{Nrc} \tag{9}$$

$$APD = \frac{Iprg - Trl}{Trl}, \text{ where} \tag{10}$$

SLT = Supplier Lead-Time Variability;
APD = Absolute percentage differences between the expected time delay in delivery of the supplier and real time delivery;
Nr_c = Number of orders;
I_{prg} = Expected delay;
T_{rl} = Real time delivery.

C. Backlog efficiency - BE

This indicator measures the performance of the backlog comparing 2 consecutive months.

$$BE = \frac{Rlc}{Rt} \times 100, \text{ where} \tag{11}$$

BE = Backlog efficiency;
R_{lc}= Total value of backlog delivered this month;;
R_t= Total value of backlog this month;.

D. Transport efficency - TE

$$TE = \frac{Crt - Cpt}{Crt} \times 100, \text{ where} \tag{12}$$

TE = Transport efficency;
C_{rt} = Actual cost per transport;
C_{pt} = Forecasted cost per transport.

Table 1 presents the results obtained from the industrial system for each performance indicator.

Table 1. Data used to calculate delivery indicators.

Month	Number of orders delivered at requested date	Number of orders	Forecasted cost per transport (euro)	Actual cost per transport (euro)	Backlog (Pieces)	Real time delivery (days)	Expected time delay in delivery (days)
Jan	252	357	3500	14645,1	235805	6,5	5,5
Feb	211	416	3500	18375,8	219434	6,8	6,2
Mar	253	495	3500	7690,0	116117	6,9	7,1
April	238	431	3500	7924,9	37888	7,0	8,9
May	260	408	3500	6867,8	91333	6,5	6,3

June	294	404	3500	3427,2	48175	6,3	6,1
July	338	390	3500	2455,8	36523	6,0	5,9
Aug	346	370	3500	6873,0	63181	5,9	5,4
Sept	346	393	3500	6590,1	30351	5,7	4,9
Oct	345	383	3500	2844,8	52786	5,9	6,1
Nov	386	418	3500	2867,2	49517	5,5	4,7
Dec	274	290	3500	1646,4	12379	13,0	12,3

Using the data from Table 1 the delivery indicators were calculated; the evolution of the four performance indicators over the period of 12 months, it is presented in Table 2.

Table 2 Indicators evolution during months January-December.

Month	OTD (%)	SLT (%)	BE (%)	TE (%)
January	70,588	94,714	90,75	-318,43
February	50,721	97,473	91,40	-425,02
March	51,111	99,408	95,76	-119,71
April	55,220	95,009	98,49	-126,42
May	63,725	99,202	96,68	-96,22
June	72,772	99,295	98,16	2,08
July	86,667	99,374	98,64	29,83
August	93,514	97,492	97,54	-96,37
September	88,041	95,761	98,76	-88,29
October	90,078	99,328	98,24	18,72
November	92,344	95,981	98,21	18,08
December	94,483	98,055	99,45	52,96

As it can be seen, the risk is on having an efficient transport, namely, that the delay in delivery of the ordered pieces increases the transportation expenses for the backlog products. Indicators evolution and graphic analysis is provided in Fig. 1.

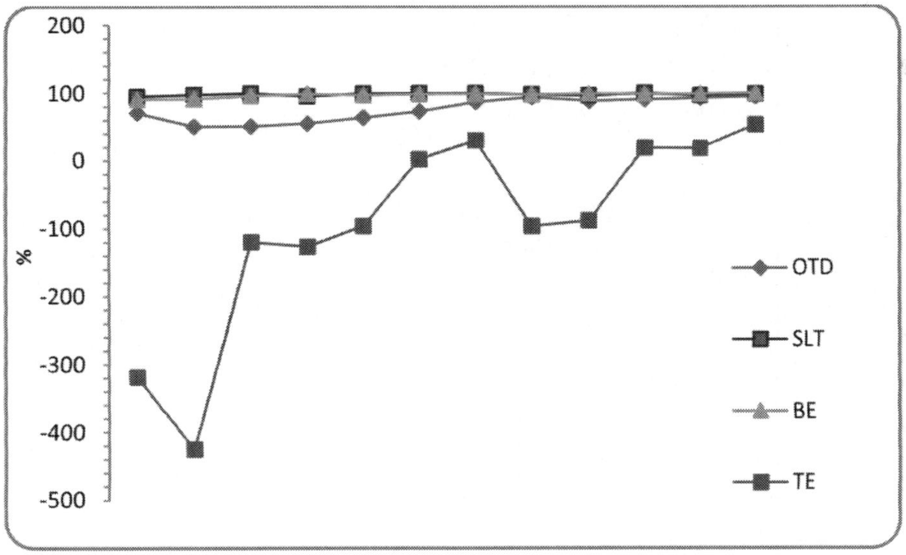

Fig.1. Delivery indicators: OTD, SLT, BE and TE.

This paper aims to optimize the delivery area of the analyzed industrial company, which is a major problem for this company, owed to backlog in delivery and significant transportation costs because undelivered parts incur much higher costs.

Thus, knowing the amount of pieces that will be delivered for the next 12 months, according to the customer orders, the production capacity for each month, and the cost per piece, there will be determined the optimal production by using linear programming in order that shipping costs to be minimal.

The industrial system spends 25 euro/min for getting a piece of the outer ring and the cost of delivery of the backlog on a special transport is 5 euro/piece.

Annotations:
$x_1, x_2, x_3, \ldots, x_{12}$ = The quantity of parts to be delivered in 12 months.

It will thus obtain the following restrictions:

$$\begin{cases} x_i > 0, \quad i=\overline{1,12} \\ x_1 \geq 2\,548\,556 \\ x_2+(x_1-2\,548\,556) \geq 2\,550\,855 \\ x_3+(x_1+x_2-5\,099\,411) \geq 2\,735\,389 \\ x_4+(x_1+x_2+x_3-7\,834\,800) \geq 2\,503\,787 \\ x_5+(x_1+\ldots+x_4-10\,338\,587) \geq 2\,750\,643 \\ x_6+(x_1+\ldots+x_5-13\,089\,230) \geq 2\,624\,632 \\ x_7+(x_1+\ldots+x_6-15\,713\,862) \geq 2\,682\,187 \\ x_8+(x_1+\ldots+x_7-18\,396\,049) \geq 2\,563\,748 \\ x_9+(x_1+\ldots+x_8-20\,959\,797) \geq 2\,451\,276 \\ x_{10}+(x_1+\ldots+x_9-23\,411\,073) \geq 2\,992\,994 \\ x_{11}+(x_1+\ldots+x_{10}-26\,404\,067) \geq 2\,761\,731 \\ x_{12}+(x_1+\ldots+x_{11}-29\,165\,798) \geq 2\,258\,991 \\ x_1 \leq 2\,700\,000 \\ x_2 \leq 2\,500\,000 \\ x_3 \leq 2\,700\,000 \\ x_4 \leq 2\,800\,000 \\ x_5 \leq 2\,800\,000 \\ x_6 \leq 2\,700\,000 \\ x_7 \leq 2\,800\,000 \\ x_8 \leq 3\,000\,000 \\ x_9 \leq 2\,800\,000 \\ x_{10} \leq 3\,000\,000 \\ x_{11} \leq 2\,800\,000 \\ x_{12} \leq 2\,100\,000 \end{cases} \tag{13}$$

The data used to optimize the delivery performance are presented in Table 3:

Table 3. Required data for the optimization process.

Month	Customer orders (pieces)	Production capacity (pieces)	Expenses (euro/min)
January	2548556	2 700 000	25
February	2550855	2 500 000	25
March	2735389	2 700 000	25
April	2503787	2 800 000	25
May	2750643	2 800 000	25
June	2624632	2 700 000	25
July	2682187	2 800 000	25
August	2563748	3 000 000	25
September	2451276	2 800 000	25
October	2992994	3 000 000	25
November	2761731	2 800 000	25
December	2258991	2 100 000	25

Considering the cost of production and transport, the objective function becomes:

$$f(x) = 25\,x_1 + 25\,x_2 + \ldots + 25\,x_{12} + 5\times(x_1 - 2\,548\,556) + 5\times(x_1 + x_2 - 5\,099\,411) + 5\times(x_1 + x_2 + x_3 - 7\,834\,800) + 5\times(x_1 + \cdots + x_4 - 10\,338\,587) + 5\times(x_1 + \cdots + x_5 - 13\,089\,230) + 5\times(x_1 + \cdots + x_6 - 15\,713\,862) + 5\times(x_1 + \cdots + x_7 - 18\,396\,049) + 5\times(x_1 + \cdots + x_8 - 20\,959\,797) + 5\times(x_1 + \cdots + x_9 - 23\,411\,073) + 5\times(x_1 + \cdots + x_{10} - 26\,404\,067) + 5\times(x_1 + \cdots + x_{11} - 29\,165\,798). \quad (14)$$

The minimum of the function:

$$f(x) = 25\,x_1 + 25\,x_2 + \ldots + 25\,x_{12} + 55\,x_1 + 50\,x_2 + 45\,x_3 + 40\,x_4 + 35\,x_5 + 30\,x_6 + 25\,x_7 + 20x_8 + 15\,x_9 + 10\,x_{10} + 5\,x_{11} - 172\,961\,230. \quad (15)$$

The final objective function will be:

$$\min f(x) = 80\,x_1 + 75\,x_2 + 70\,x_3 + 65\,x_4 + 60\,x_5 + 55\,x_6 + 50\,x_7 + 45\,x_8 + 40\,x_9 + 35\,x_{10} + 30\,x_{11} + 25x_{12} - 172\,961\,230. \quad (16)$$

To simplify calculations, data were inserted in the software Quantitative Management (QM) as shown in Fig. 2.

	X1	X2	X3	X4	X5	X6	X7	X8	X9	X10	X11	X12		RHS
Minimize	80,	75,	70,	65,	60,	55,	50,	45,	40,	35,	30,	25		
Constraint 1	1,	0,	0,	0,	0,	0,	0,	0,	0,	0,	0,	0,	>=	2.548.556
Constraint 2	1,	1,	0,	0,	0,	0,	0,	0,	0,	0,	0,	0,	>=	5.099.411
Constraint 3	1,	1,	1,	0,	0,	0,	0,	0,	0,	0,	0,	0,	>=	7.834.800
Constraint 4	1,	1,	1,	1,	0,	0,	0,	0,	0,	0,	0,	0,	>=	10.338.590
Constraint 5	1,	1,	1,	1,	1,	0,	0,	0,	0,	0,	0,	0,	>=	13.089.230
Constraint 6	1,	1,	1,	1,	1,	1,	0,	0,	0,	0,	0,	0,	>=	15.713.860
Constraint 7	1,	1,	1,	1,	1,	1,	1,	0,	0,	0,	0,	0,	>=	18.396.050
Constraint 8	1,	1,	1,	1,	1,	1,	1,	1,	0,	0,	0,	0,	>=	20.959.800
Constraint 9	1,	1,	1,	1,	1,	1,	1,	1,	1,	0,	0,	0,	>=	23.411.070
Constraint 10	1,	1,	1,	1,	1,	1,	1,	1,	1,	1,	0,	0,	>=	26.404.070
Constraint 11	1,	1,	1,	1,	1,	1,	1,	1,	1,	1,	1,	0,	>=	29.165.800
Constraint 12	1,	1,	1,	1,	1,	1,	1,	1,	1,	1,	1,	1,	=	31.424.790
Constraint 13	1,	0,	0,	0,	0,	0,	0,	0,	0,	0,	0,	0,	<=	2.700.000
Constraint 14	0,	1,	0,	0,	0,	0,	0,	0,	0,	0,	0,	0,	<=	2.500.000
Constraint 15	0,	0,	1,	0,	0,	0,	0,	0,	0,	0,	0,	0,	<=	2.700.000
Constraint 16	0,	0,	0,	1,	0,	0,	0,	0,	0,	0,	0,	0,	<=	2.800.000
Constraint 17	0,	0,	0,	0,	1,	0,	0,	0,	0,	0,	0,	0,	<=	2.800.000
Constraint 18	0,	0,	0,	0,	0,	1,	0,	0,	0,	0,	0,	0,	<=	2.700.000
Constraint 19	0,	0,	0,	0,	0,	0,	1,	0,	0,	0,	0,	0,	<=	2.800.000
Constraint 20	0,	0,	0,	0,	0,	0,	0,	1,	0,	0,	0,	0,	<=	3.000.000
Constraint 21	0,	0,	0,	0,	0,	0,	0,	0,	1,	0,	0,	0,	<=	2.800.000
Constraint 22	0,	0,	0,	0,	0,	0,	0,	0,	0,	1,	0,	0,	<=	3.000.000
Constraint 23	0,	0,	0,	0,	0,	0,	0,	0,	0,	0,	1,	0,	<=	2.800.000
Constraint 24	0,	0,	0,	0,	0,	0,	0,	0,	0,	0,	0,	1,	<=	2.100.000

Fig. 2. Objective function and restrictions introduced in module Linear Programming of QM.

By solving the objective function in the QM program, it will obtain the results presented in Fig.3 and in Fig.4.

	X1	X2	X3	X4	X5	X6	X7	X8	X9	X10	X11	X12		RHS
Minimize	80,	75,	70,	65,	60,	55,	50,	45,	40,	35,	30,	25,		
Constraint 1	1,	0,	0,	0,	0,	0,	0,	0,	0,	0,	0,	0,	>=	2.548.556,
Constraint 2	1,	1,	0,	0,	0,	0,	0,	0,	0,	0,	0,	0,	>=	5.099.411,
Constraint 3	1,	1,	1,	0,	0,	0,	0,	0,	0,	0,	0,	0,	>=	7.834.800,
Constraint 4	1,	1,	1,	1,	0,	0,	0,	0,	0,	0,	0,	0,	>=	10.338.590,
Constraint 5	1,	1,	1,	1,	1,	0,	0,	0,	0,	0,	0,	0,	>=	13.089.230,
Constraint 6	1,	1,	1,	1,	1,	1,	0,	0,	0,	0,	0,	0,	>=	15.713.860,
Constraint 7	1,	1,	1,	1,	1,	1,	1,	0,	0,	0,	0,	0,	>=	18.396.050,
Constraint 8	1,	1,	1,	1,	1,	1,	1,	1,	0,	0,	0,	0,	>=	20.959.800,
Constraint 9	1,	1,	1,	1,	1,	1,	1,	1,	1,	0,	0,	0,	>=	23.411.070,
Constraint 10	1,	1,	1,	1,	1,	1,	1,	1,	1,	1,	0,	0,	>=	26.404.070,
Constraint 11	1,	1,	1,	1,	1,	1,	1,	1,	1,	1,	1,	0,	>=	29.165.800,
Constraint 12	1,	1,	1,	1,	1,	1,	1,	1,	1,	1,	1,	1,	=	31.424.790,
Constraint 13	1,	0,	0,	0,	0,	0,	0,	0,	0,	0,	0,	0,	<=	2.700.000,
Constraint 14	0,	1,	0,	0,	0,	0,	0,	0,	0,	0,	0,	0,	<=	2.500.000,
Constraint 15	0,	0,	1,	0,	0,	0,	0,	0,	0,	0,	0,	0,	<=	2.700.000,
Constraint 16	0,	0,	0,	1,	0,	0,	0,	0,	0,	0,	0,	0,	<=	2.800.000,
Constraint 17	0,	0,	0,	0,	1,	0,	0,	0,	0,	0,	0,	0,	<=	2.800.000,
Constraint 18	0,	0,	0,	0,	0,	1,	0,	0,	0,	0,	0,	0,	<=	2.700.000,
Constraint 19	0,	0,	0,	0,	0,	0,	1,	0,	0,	0,	0,	0,	<=	2.800.000,
Constraint 20	0,	0,	0,	0,	0,	0,	0,	1,	0,	0,	0,	0,	<=	3.000.000,
Constraint 21	0,	0,	0,	0,	0,	0,	0,	0,	1,	0,	0,	0,	<=	2.800.000,
Constraint 22	0,	0,	0,	0,	0,	0,	0,	0,	0,	1,	0,	0,	<=	3.000.000,
Constraint 23	0,	0,	0,	0,	0,	0,	0,	0,	0,	0,	1,	0,	<=	2.800.000,
Constraint 24	0,	0,	0,	0,	0,	0,	0,	0,	0,	0,	0,	1,	<=	2.100.000,
Solution->	2.634.800,	2.500.000,	2.700.000,	2.503.790,	2.750.640,	2.624.630,	2.682.190,	2.563.750,	2.564.990,	3.000.000,	2.800.000,	2.100.000,		$1.653.001.250,

Fig. 3. The optimized solutions obtained by solving the objective function.

 DOI: 10.1051/matecconf/20179406006

As it can be seen in Fig. 4, to obtain optimal costs, the industrial company should deliver, respecting customer requirements and production capabilities, the following quantities: in January 2 634 800 pieces, in February 2 500 000, in March: 2 700 000 pieces, in April 2503790 million pieces, May: 2 750 640 units, in June 2 624 630 units, in July 2 682 190 units, in August: 2 563 750 pieces, in September 2 564 990, October 3 000 000 units, in November: 2 800 000 units, and at the end of the year, in December, 2 100 000 pieces.

Variable	Value	Reduced	Original	Lower	Upper
X1	2.634.800,	0,	80,	75,	Infinity
X2	2.500.000,	0,	75,	-Infinity	80,
X3	2.700.000,	0,	70,	-Infinity	80,
X4	2.503.790,	0,	65,	60,	80,
X5	2.750.640,	0,	60,	55,	65,
X6	2.624.630,	0,	55,	50,	60,
X7	2.682.190,	0,	50,	45,	55,
X8	2.563.750,	0,	45,	40,	50,
X9	2.564.990,	0,	40,	35,	45,
X10	3.000.000,	0,	35,	-Infinity	40,
X11	2.800.000,	0,	30,	-Infinity	40,
X12	2.100.000,	0,	25,	-Infinity	40,

Fig. 4. The optimal quantities to be delivered by the industrial company.

After applying the new optimized variants for delivery it can be seen that the optimization led to a substantial improvement in all performance indicators that concern deliveries.

After optimization, the performance indicators OTD, SLT, BE and TE have values between 94 and 100 % (Fig.5), which means that deliveries are made according to the clients needs, without additional expenditure of special transport.

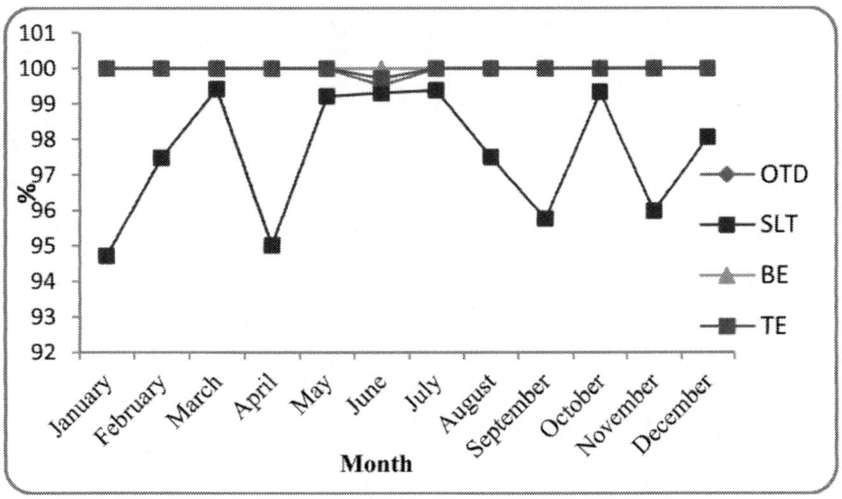

Fig. 5. Delivery indicators after optimization.

4 Conclusions

The main objective of business process modeling is to optimize decision-making process, as a determining factor in the transition from experience, intuition complex, to a scientific approach based on information-reasoning. Economic modeling process enables the manager making the best decision, giving it multiple ways to make an efficient agreement between human, material, financial with the main objectives of the organization.

This paper objective is to help managers to improve the delivery performance of the organization by determining and implementing an optimal production program. Modeling a manufacturing program using the QM software led to obtaining precise quantities to be produced and delivered each month by the industrial system in order to minimize their transport cost, satisfying their custormers orders and to control their stock.

Future research will focus on multi-criteria modeling methods, namely the multidimensional linear programming that allows to use the objective function as a utility function that will combine both criteria of maximizing profits and minimizing costs.

References

1. D. Lixandroiu, *Economic processes modeling* (Transilvania University Publishing House, Braşov, 2012).
2. A. Guiffrida, R. Nagi, Int. J. Production Economics **102,** 22–36 (2006).
3. A. Gunasekarana, C.Patelb, R.E.McGaugheyc, Int. J. Production Economics **87,** 333–347 (2004).
4. H. Shina, W.C.Bentonb, M.Junc, Comp & Op R **36**, 2462 – 2471 (2009).
5. M. Beldiga, Sci.Jur.S.Univ.Mold **7,** 34-39 (2012).
6. A .Nedelcu, *Technology and flexible manufacturing systems* (Lux Libris Publishing House, Brasov, 2000).
7. C. Anghelache, *Modeling and simulation of decision-making processes* - lecture notes (Artifex, Bucuresti, 2003).
8. M. Andrasiu, *Multi-criterion Decision Methods*, (Tehnica Publishing House, Bucuresti, 1986).
9. A. Rusu, *Operational Research* (Iasi, 2007).

MATEC Web of Conferences **94**, 06007 (2017) DOI: 10.1051/matecconf/20179406007
CoSME'16

Decision models in designing flexible production systems

Adriana Florescu[1,*], and *Sorin Adrian* Barabaş[1]

[1]*Transilvania* University of Brasov, Faculty of Technological Engineering and Industrial Management, Romania

Abstract. Flexible production system is a complex whole that raise some issues in terms of its design and in relation to the conditions for implementing it. To implement a flexible production system configuration must be found that satisfies both economic and system performance requirements. The configuration which best meet the objectives of introducing a flexible production system must be sought in the set of alternatives defined and evaluated. In this paper we present a methodology of realising the configuration and complex evaluation of the analyzed system. It will be developed models which generate new alternative configurations, optimization and evaluation models of the performance of the flexible production system. This will create a framework for interactive decision support, user-oriented that can be used by management to solve this selection problem. The applicative character of the study consist in tracking of the technological process in real time using the developed software package on the designed system, based on mathematical models for configuration and optimization of the system.

1 Introduction

The researches performed within the scientific paper, will be directed to the study of Flexible Manufacturing Systems (FMS), in order to know their behaviour and their performances very well, and if it is possible, before their physical manufacturing, and in order to establish on scientific bases dimensioning models, representation and simulation of FMS.

By introducing flexible production systems there can be obtained important advantages concerning the functioning of the organization and other economic advantages concerning: the possibility to change the reference (a new system adaptation) almost without losing time, for new equipment and disconnecting the attendance staff reported to the work pace of the machine, lead to an important increase of the use degree of the machine; fabrication in batches sizes, according to the necessary amount and at the time required by mounting (the possibility to deliver the pieces manufactured "Just in Time"), make possible an important decrease of the pieces volume in the workshop and warehouses (production decrease in processing and stocks, with a rate of 50-60 %) and it has influence upon a quick reaction capacity towards the market requirements oscillations, at the reference change;

* Corresponding author: fota.a@unitbv.ro

production time decrease, often with more than 50%, leads to delivery time decrease and thus to creating decisive competition advantages; improving quality by strictly respecting the stages of the production process; realizing a production process easily to be controlled in its essential points; work force qualification increase by its movement towards preponderantly intellectual activities; better use of the capital by its distribution between the buildings and the equipment of the process.

By the possibility of the optimal use both of the system capacity and the staff and their high availability, which can be reached today, of machines and plants, on the one hand and of almost unlimited enlargement, replacement and change flexibility, on the other hand, result important productivity increase on long terms of systems use (8-10 times reported to classical systems).

Taking into account the actual stage of the research and technical scientific achievements in the domain of flexible systems the following conclusions can be stated. Speciality literature [1, 2] offers disparate information concerning the construction and the implementing of flexible production systems:

- In their evolution of about three decades, there can be noticed efficient flexible systems implementing for cutting processes, but also many failures, determined most of the times by an over evaluation of technical solutions, by the impossibility of managing efficiently these systems which are extremely complex, by huge investment that they need. It is very difficult a correct substantiation of the investment costs, which most of the times exceed the expectancies;
- A difficulty which tempers the rhythm of FMS implementing is caused by the impact generated on the work market. The decrease sometimes dramatically of the work force necessary number, simultaneously with a qualification rise, by its movement towards intense intellectual activities, with a lot of stress, generates social problems and integration problems in the new environment;
- For now there isn't a unitary international definition for flexible manufacturing systems. The definition of the FMS concept differs from a country to the other and from an investigator to the other too. After gaining experience in their design and in their use, there was noticed a clear tendency to get different notions close to each other, as well as definitions on which the concept of flexible production systems are based;
- In most of the speciality works "flexibility" is often used as a "generic collocation" and it is not used in general, as a complex concept. Thus, in a very huge number of works „flexibility" is treated partially, superficially, in a manner which is confuse and sometimes wrong. There still are many opinions concerning the quantification way of the system flexibility. It is difficult to anticipate, in the concept stage, the optimal flexibility degree, as this is one of the major causes of the inefficiency concerning FMS exploitation. In most of the speciality works there is no distinction between the automation concept and the flexibility one. In reality they are different concepts;
- Actual flexible production systems are not always approached globally, interdisciplinary. Due to their complexity, these systems need a synthesis approach and implicitly the adequate expert. There is not presented a clear methodology, accepted by everybody in order to determine the structure of the flexible production systems, according to the system's inputs and outputs and to the interdependences between them and the exterior environment.

Advanced production systems development requires profound and continues changes which will affect both the technical structure, through increased automation, robotics and information technology, and also the methods and techniques of management and organization of such systems. To ensure their competitiveness, companies will have to rethink a new strategy design and implementation of these new and efficient systems.

2 Modeling in Flexible Production Systems

As far as the functioning modelling of the flexible production systems is concerned, there still are a lot of attempts to define and analyse the performance indicators of the system [3]. From the point of view of global characteristics, flexible production systems are superior to production systems which they replace. The huge number of the diverse configurations of flexible production systems showed that it is possible the flexible automation in the domain of individual fabrication, of small series and medium one, both from technical and economic point of view.

2.1 Models applied in the design of FMS

The decision models encountered in the literature [4, 5] that allow selecting optimal alternative configuration of a flexible production system have different characteristics depending on conditions requirements for sizing and configuration of FMS. When designing any flexible manufacturing system it is taken into account the following: the behaviour in time of FMS, degree of flexibility, decision variables of FMS, certainty of data programming / planning.

The behaviour in time of a FMS is analyzed by applying two types of models: dynamic and static model. Dynamic models are applied generally in process modeling of waiting at the workstation in the system, traceability of materials and flexible manufacturing system simulation as a large, complex dynamic system. Thus, basic models come from queuing theory and dynamic of large systems. Static models include in the analysis i.e. calculation of productivity of the workstation. Productivity of the entire flexible manufacturing system is determined by the productivity of the "bottleneck" workstation. Linear programming models are used.

The degree of the flexibility of the system requires the use of the following two types of decision models: models with one period, where the analysis focuses on the representative period for which it can be designed a acceptable configuration of FMS; multi-period models used to establish a development strategy and step by step adjustment of the system configuration depending on the required production changes. The lifetime of FMS is divided into sub-periods, each with its own program of production.

Decision variables in a flexible production system are based on models that trace products association to routes and operations to machines and models for selecting the equipment that will be integrated into FMS. The models which aim associating products at routes and operations to machines are used to determine the sequence of operations (technological plan itineraries / routes) consistent with planned production. The models that allow the selection of equipment to be integrated in FMS, allow the determination of different equipment: machines, working posts, workstations, loading / unloading station, industrial logistics / vehicles to be included in FMS. Models for determining the type and number of pallets or devices determine the right level of pallets. The number of pallets circulating in FMS determines how the available capacity is used.

The certainty degree of the planning data requires the use of deterministic and stochastic models. The deterministic models take into account that the information received from the outside about outputs performed in FMS are considered to be certain. The stochastic models are premised on the fact that the desired outputs are the result of a forecasting process which is, in many cases, the topic of greatest uncertainty.

Modelling and simulation of a flexible manufacturing system involves creating a model of the system and implement it in a programming language, aimed to checking the functioning of the system and assessing its performance.

2.2 A model for optimal configuration of a Flexible Manufacturing System

On the basis of a flexible manufacturing system design should stand current task production, characterized by a predetermined typology, [6]. In order to choose the type and number of workstations in a flexible manufacturing system will be developed methodology for determining the degrees of consistency, affinity and differentiation and will make their quantitative analysis. As a result of this analysis, conclusions as follows are coming out, regarding time and working station number of flexible manufacturing systems selection as well as machine number of each working station. These conclusions were be used in this paper in view of drawing up the graphic model.

The affinity degree is determined by the equation (1):

$$AD = m_g\{TO_i(I_k)\}, 1 \le k \le r, 1 \le i \le q \tag{1}$$

where: TO_i are the technological operations, I_k - generalized item and m_g - different commune elementary geometrical components of all generalized manufacturing task. This expresses reducing of total number of workstation (q_{total}) of the FMS, increase of the processing process in flexible manufacturing system concentration degree, by the group technology.

FMS results as a less flexible structure, objective expected in design by: $min\ q_{total}$ $\{m_i, 1 \le i \le q\}$; m_i - machines in system; n_g - representing the total number of commune and not commune geometrical elements - in equation (2), which defines the generalized item (I_k) and then, the typological nucleus of generalized manufacturing task; it results that difference expresses (n_g-m_g) the not-commune number geometrical elements (completely different, specific) of different items included in manufacturing task, which distinguish them and technologic by different complementary technological operations (TO_c) and additional technological operations (TO_a). This difference expresses a certain indicator of the processing process division degree in flexible manufacturing system which can't be made by group technology, but only by different TO_i (TO_c) technologies.

Consequently the difference (equation 2) - degree of differentiation (DD) needs just as much (m_i) of different types:

$$DD = (CD - AD) = (n_g - m_g)TO_c \tag{2}$$

$$q_{different} = (n_g - m_g)TO_i \tag{3}$$

The number (m_i) on which the different TO_i types for some specific item included in generalized manufacturing task are executed. But, min $m_g TO_i$ expresses on the one hand different (m_g) TO_i, representing commune (m_g) TO_c, consecutively executable by group technology. Consequently, flexible manufacturing systems should include:

$$q_{commune} = \min m_g\ (TO_i) \tag{4}$$

which will execute the commune geometrical elements (m_g) and commune technological operations (TO_i) for all items included in generalized manufacturing task.

From the technological point of view flexible manufacturing system is consisting of:

$$q_{tot} = q_{com} + q_{dif} = \min m_g(TO_i) + (n_g - m_g)(TO_i) \tag{5}$$

 DOI: 10.1051/matecconf/20179406007

Table 1 presents the values of affinity degrees (*AD*), consistency degrees (*CD*) and the difference (*CD-AD*), calculated (in %) according to the previous methodology for designed FMS in which considered six family parts ($FP_1...FP_6$), [6]:

$$CD = \frac{n_{gk}}{n_{g1}} \tag{6}$$

$$AD = \frac{m_{gk}}{n_{g1}} \tag{7}$$

$$DD = \frac{n_{gk}-m_{gk}}{n_{g1}} \tag{8}$$

Table 1. Analysis of affinity, consistency and differentiation degrees

FP_k	n_{gk}	m_{gk}	CD [%]	AD [%]	DD [%]
FP_1	50	47	100	94	6
FP_2	48	42	96	84	12
FP_3	30	28	60	56	4
FP_4	27	27	54	54	0
FP_5	29	28	58	56	2
FP_6	27	25	54	50	4

Based on consistency degrees, affinity and differentiation degrees determined by the methodology developed it is presented in Figure 1 the general structure of FMS analyzed for the six families of cylindrical parts.

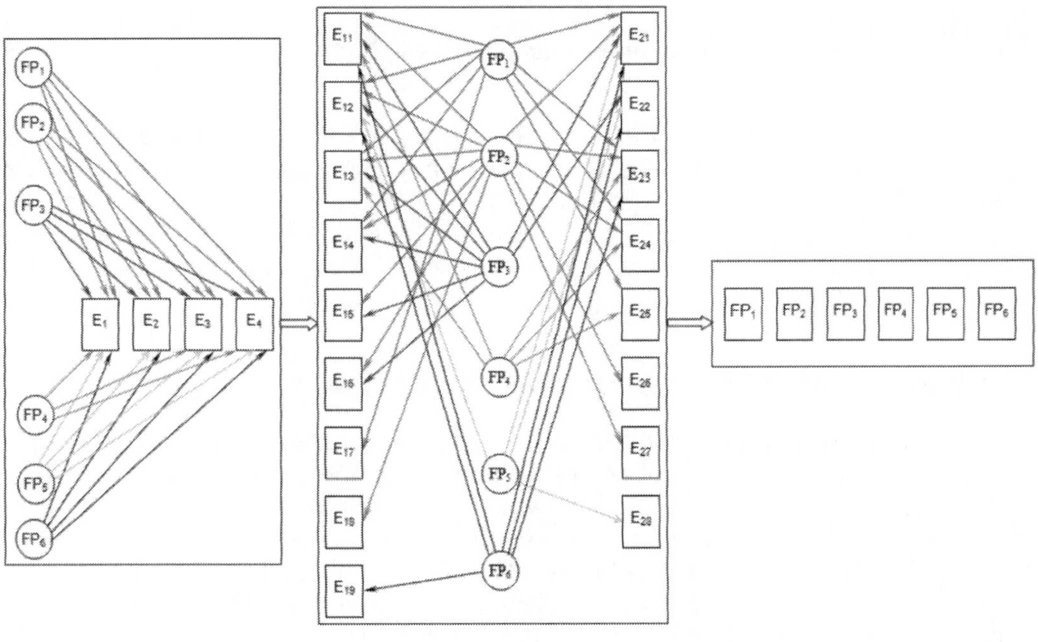

Fig. 1. The structural scheme of the system

Determining the number and type of machine-tools, technological equipment, workstations, stations of measuring - control and their arrangement in system was based on the following factors: total production volume expressed as number of items processed and total unit cost / piece, material flow, technological itineraries and other components of the FMS.

2.2.1 Establishing the optimal configuration and simulation of the system

Configure a handling system depends mainly on: typological diversity of the task of production, the type of objects handled, the amount of products handled, transport distances, type of production system serviced, the pace imposed of the material flow, diagram routes material flows, routes transport. The basic configurations of a FMS commonly found [1, 6] are in a straight line (open), in the loop (closed), in large scale open expanses and centred on robot cells. A major influence on the determination of the structure of this FMS had the degree of affinity or differentiation. The basis for establishing the system configuration is the relationship systemic between information technological flow - the technology of group on entry and task manufacturing generalized from out of the system, [6].

In Figure 2 is shown the system designed, on which was implemented and validate the optimization model previously developed. To study the behaviour of a real system for purposes of control and management of the system was developed a simulation program using the programming environment Cx-Supervisor Developer.

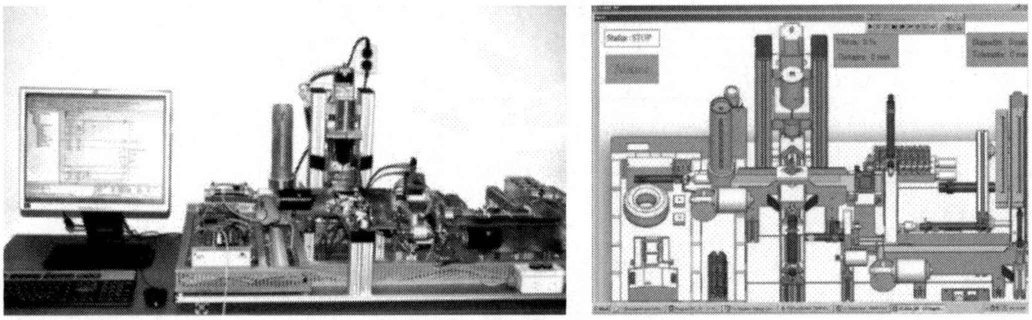

Fig. 2. The structuring and simulation of FMS. The sequence of the application - simulation software

This allows for modification of the technological flux based on relevant factors: number and type of workstations and tools-machines from the system, number and type of vehicles (conveyors) and speed of transport, logistics subsystem type used, the type of parts and their technological route, duration of the processing cycle of each type of piece to each module of the flexible system, the number of parts simultaneously in processing and batch to batch volume.

3 The model of decision in route optimization

Mathematical optimization models used in the configuration of flexible production systems aimed at reducing the number of different alternative configurations and thus to exclude those suboptimal configurations, in the previous stage of the system designing. This decision problem applies to complex systems there are different routes that can be used alternatively and / or in parallel for manufacturing a typological diversity of parts.

Different mathematical optimization approaches proposed in the literature [5 - 9], covering the wide range of different variables of decision. Decision-making process in determining the optimal alternative configuration of the FMS is to establish the first performance evaluation procedures that allow rapid a rough estimate of the performance of FMS. Then, the process continues progressively to leave some configurations that will be analyzed in detail. Thus, it is possible to eliminate a significance number of alternatives, earlier with a small computational effort. In general, the procedure is based on the classical model of closed queuing networks systems (the CQN model) [10].

3.1 The optimization model

The model of decision for route optimization associates the share of production for different types of items on different routes. To develop optimization model can be considered following inputs:
- are defined several routes of transport available (R_k): $i = (1, ..., R_k)$;
- the share of production is considered variable: $Q_{ki} = var$

The mathematical model of distribution of the total volume of production (total share of production – Q_{tk}) on alternative routes (R_k) is shown in equation (1):

$$Q_{tk} = \sum_{i=1}^{R_k} Q_{ki} \,, k = 1, \dots, n \tag{9}$$

Q_{ki} share of production is calculated based on the ratio of the average number of parts / items (P_k) - k per unit time manufactured on processing routes (R_k) and number of parts - k manufactured per unit time. Thus, by variable Q_{ki} are optimizing the most significant indicators of performance such as the productivity and waiting times in the system.

A frequent optimization model in the literature [5, 10] is the linear optimization model for determining the optimum distribution of the productivity of individual parts of transport routes.

It is considered that the system works with a large number of parts (default paddles) and the objective optimization is to maximize productivity by associating parts on various routes. It can be determined the degree of load (w) in each workstation (q) for each of the routes (i), according to the equation (9), in which: TO_{pqi} - average number of operations from workstation (q) for parts (k) processed on routes (i); $\bar{T}_{pqi}$ - average time for processing workpieces (p) from the station (q) moving on the route (i).

$$w_{kqi} = TO_{kqi} \cdot \bar{T}_{kqi} \tag{10}$$

$$q = 1, 2, \dots, m; \; k = 1, 2, \dots P_k; \; i = 1, 2, \dots, R_k$$

The decision model is represented by the objective function (9). Decision variables [10] are the number of parts processed per unit of time and moving on the alternative routes (x_{ki}).

$$\max Q(x) = \sum_{k=1}^{P_k} \sum_{i=1}^{R_k} x_{ki} \tag{11}$$

The objective function describes the maximize productivity of the entire flexible production system, which is the sum of the average number of parts that are processed per unit time in the system. The variables in the model are: $Q(x)$ – productivity of FMS and x_{ki}.

Are imposed the conditions, where N_q is number of machines / workstation in system:

$$Q_{tk} \sum_{p=1}^{P_k} \sum_{i=1}^{R_k} x_{pk} = \sum_{i=1}^{R_k} x_{ki} \qquad k = 1, 2, \dots, P_k \tag{12}$$

$$\sum_{k=1}^{P_k} \sum_{i=1}^{R_k} x_{ki} \cdot w_{kqi} \; \le \; N_q \qquad q = 1,2, \dots, m \tag{13}$$

The linear optimization model is developed and it can be solved with a standard algorithm. Some of the results obtained and validated on designed system are presented in Table 2.

The objective function from equation (9) expresses maximize productivity whole system and represents the average number of parts that are processed per time unit in the system. Productivity of the system includes the sum of all variables, according to equation (9).

The share of production of each type of piece was obtained by dividing the parts productivity at the system productivity.

Table 2. Results. The optimal solution

Part P_k	Route R_k	Decision variables	Q_{ki}	Q_{ik}
P1	1	x_{11}	0,0934	0,18
P2	2	x_{12}	0,0866	
P3	1	x_{21}		0,13
P4	1	x_{31}		0,12
P5	1	x_{41}	0,0729	0,14
P6	2	x_{42}	0,0671	0,12
P7	1	x_{51}	0,0818	0,16
P8	2	x_{52}	0,0782	
P9	1	x_{61}		0,15

4 Conclusions

Based on research conducted were developed graphic-analytical dynamic model to configure the FMS and the optimization system model. A flexible system was designed and optimized, including industrial logistics subsystem and subsystem management process. Was performed simulation of functioning of the system and optimization in real-time environment and by implementing a monitoring software and SCADA control, using programming environment Cx-Supervisor Developer.

The applicative character of the research theme is demonstrated simulating the flexible manufacturing system's performance for cylindrical parts processing. After simulating the functioning of the flexible production system, the validity of the models will be confirmed, as well as their utility in the design and the management of the processes in real time and environment. The implementation of flexible production system in an integrated production system, is an important result, which leads to optimization of the material flows and informational of enterprise.

References

1. M. P. Groover, *Automation, Production Systems and Computer-Integrated Manufacturing*, 3rd Edition, Prentice Hall, Pearson (1987)
2. K. H. Han, Int. J. Math. Comp. Sim., **6,** 4 (2012)
3. D. Catrina, *Flexible Manufacturing Systems*, MatrixRom Publish House, Bucharest (2008)
4. B.S.P. Reddy, C.S.P. Rao, Int. J. Sim. Mod., **10** (2011)
5. I. Bojan, *Flexible Production Systems,* Dacia Publish House, Cluj-Napoca (1999)
6. A. Fota, *Designing of the machines systems. Modeling and simulation,* Transilvania University of Brasov Publish House (2004)
7. M. Anderson, G. Olsson, *Proceedings of 1998 Winter Simulation Conference*, Washington, DC, pp. 935-941 (1998)
8. G. A. Chang, W. R. Peterson, *122nd Anual Conference& Exposition*, June 14-17, Seatle, WA, American Society for Engineering Education (2015)
9. F. Zammori, M. Braglia, M. Frosolini, Int. J. Ind. Eng. Comp., **2**, p. 593-616 (2011)
10. G. Secco-Suardo, *Optimization of a Closed Network of Queues*, Electronic Systems Laboratory, MIT, Cambridge, Massachusetts (1978)

MATEC Web of Conferences **94**, 06008 (2017) DOI: 10.1051/matecconf/20179406008
CoSME'16

Online measurement of work safety culture – statement of research

Izabela Gabryelewicz[1,*], *Patryk* Krupa[1], *Joanna* Sadłowska – Wrzesińska[2]

[1]University of Zielona Góra, ul. Szafrana 4, 65-516 Zielona Góra Poland
[2]Poznan University of Technology, ul. Strzelecka 11, Poznań 60-965, Poland

Abstract. The article presents a developed Analytic Platform - Direct Safety, which can be helpful, especially for small enterprises, in determining own policy in the field of occupational safety. Small companies need indicators, on the basis of which they could anticipate and undertake actions aimed at weak sides of the occupational safety. The developed application provides means for analysing a series of aspects, variants and compilations between particular data. One can conduct an analysis anytime, arbitrarily changing selected criteria. The developed tool allows a complex analysis of collected data, considering all values obtained during the analysis of the company's occupational safety state. The article describes the possibilities of the developed Analytic Platform - Direct Safety, on the basis of an analysis of results of the research in the field of measuring the level of the occupational safety culture conducted at a small manufacturing company

1 Introduction

Each company is an enterprise aimed at supplying particular goods or services [1]. Enterprises are characterised by significant functional and organisational complexities and a large diversification of aims (considering that the most essential aim is profit).

Processes taking place in an enterprise as well as its structure should correspond to features of an intelligent enterprise in order to provide a proper and effective operation [2] (Table 1).

An analysis of the state and functioning of enterprises is multidimensional as enterprises (in major part) have a fractal and hierarchic character. In order to properly function, an enterprise should be:

- safe (lack of accidents and injuries),
- healthy (lack of occupational diseases),
- efficient (high quality production and fully utilised technical and HR infrastructure).

In a general meaning, safety, in its widest meaning, has the largest influence on a proper functioning of an enterprise. Providing safety in a manufacturing enterprise is a complex process and depends on numerous systems. Despite systems of various levels of complexity and automation interpenetrate themselves, the final link of a safety system is a human, at the same time being an addressee of results of the operation of such system [3].

*Corresponding author: i.gabryelewicz@ibem.uz.zgora.pl

Table 1. Comparison of traditional (bureaucratic) and intelligent organisations [2].

Traditional organisation (bureaucratic)	Modern organisation (based on knowledge)
Work fully utilising knowledge of a small part of employees	Work based on knowledge of all employees
Meaningless repetitions of actions and procedures	Work based on innovations
Individual or team work	Team work
Work consisting in serving functions	Project work
Use of individual skills	Use of skills of teams
Strong position of managers, often striving towards autocratic management	Lack of typical managerial posts – manager as coach and initiator
Orientation to company's interior to maintain internal balance	Orientation on customers
Coordination from higher level	Coordination by team members

Figure 1 presents elements of an enterprise on which safety depends. Humans have an influence on a proper functioning of particular work safety elements. A human is a superior factor in functioning of each enterprise and a proper operation of each company depends on his/her knowledge, attitude and views [4].

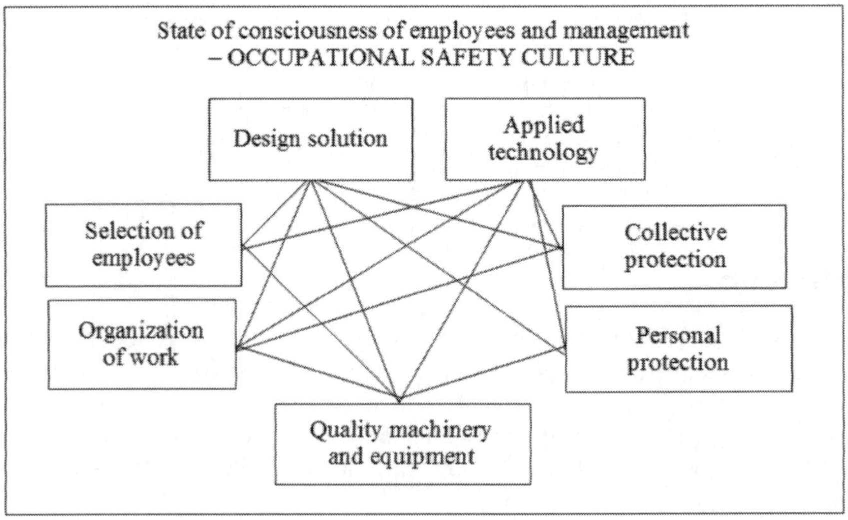

Fig. 1. Relations between elements having an influence on occupational safety state (own elaboration)

Such understanding of the human's function can be specified as occupational safety culture. Measuring occupational safety culture as a determinant of occupational safety is becoming more essential and more popular. There are developed various surveys, control lists, which measure the level of occupational safety culture. The level of occupational safety culture is a symptom thanks to which we are able to predict consequences of activities. Thus, measurements of occupational safety culture should be treated as a quantitative symptom model (cause and effect). A low level of occupational safety culture should be regarded as a symptom of low occupational safety, what results in an increased number and seriousness of work accidents [5]. As a result, such state can cause a higher number of occupational diseases of employees, a greater staff turnover, a lower quality of products or services but also negative effects on the environment quality.

DOI: 10.1051/matecconf/20179406008

When searching for reasons of a low level of occupational safety, some features characterising an enterprise do not have a character of reasons when regarded individually. However, when other circumstances occur, they can lead to the risk of employees' loss of life or health, a lower production quality or a negative influence on the environment. Thus, we have to assume that abnormalities in enterprise functioning occur as a result of combined acts and coincidences, i.e. an interference of symptoms, what causes the effect of negative synergy.

2 Construction of analytical platform – used tool and technologies

In most of cases, collection of data is only a beginning of studies. The most serious problems occur during their analysis. Analysing numerous collected data, considering various correlations between them, is very difficult or even impossible without proper tools [6-8]. In response to this problem, there was developed the Analytical Platform, which enables multidimensional and multi-aspect analysis of collected data. Such tool is indispensable due to the necessity of fast decision making. The Analytical Platform was developed with a simple operation in mind and due to often limited financial resources, especially in case of small companies, only free software was used.

The application was developed using various programming languages, including: HTML, PHP, SQL, CSS. The application was developed using public free software such as: operating system (Linux Mint) with server software (Apache –www server, MySQL – database, PHP –PHP language parser) and Bluefish –HTML and PHP codes editor.

The platform enables online and offline studies, as well as those in the traditional paper form. Requirements of the developed application depend on the way of conducting research. Requirements for offline work using a single computer without an access to the Internet:
• operating system: Windows, Linux, MacOS,
• server applications (Apache, PHP, MySQL),
• installed Analytical Platform.

In case of launching online surveys, a device with an access to the Internet, should meet the following requirements:
• any operating system with a web browser,
• an access to a network in which the software is installed.

Figure 2 presents the system construction at completing an online survey. Figure 3 presents the system construction at completing a paper survey. In each option, the system includes the following elements:
1. Survey management module – tool for:
• defining survey templates,
• management of survey programs.
2. Completing survey data – module for completing surveys by respondents
3. Analytical Platform – analytical module for:
• generating automatic summary reports,
• creating partial reports (from selected areas),
• complete processing of survey results.
4. Database – including:
• information concerning system users and administrators,
• results of respondents' answers.

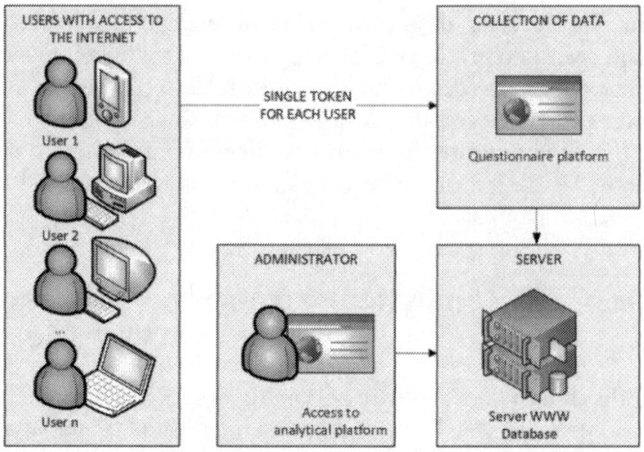

Fig. 2. System construction at completing surveys online.

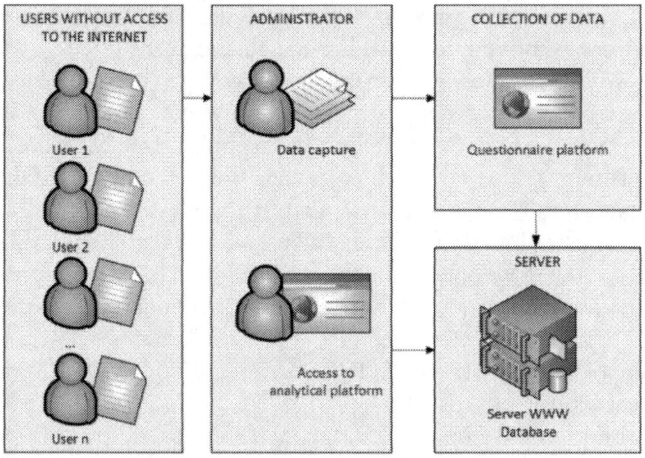

Fig. 3. System construction at completing surveys in paper form.

3 Analytical platform – user interface – completing survey online

In order to have an access to complete a survey, one should log in using TOKEN. Tokens are generated by the administrator. A user is redirected to the website where he/she can complete a survey. Generated tokens include encoded information, which allow e.g. company identification, its branch and token status. Status includes information about:

- activity or use of token,
- company for which it was developed,
- generation date,
- use date.

Generation of tokens takes place in the administration panel, when one can determine:

- number of created tokens,
- length (number of characters),
- company for which a particular group of tokens is addressed,
- company branch.

 DOI: 10.1051/matecconf/20179406008

Figure 4 presents the overview of a snapshot including social and demographic data, necessary for further analysis of obtained data. Figure 5 presents a fragment of a survey, used for studying the level of occupational safety culture of an enterprise.

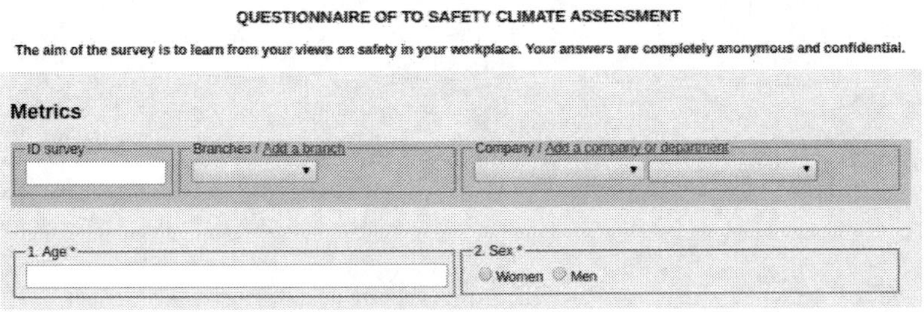

Fig. 4. Survey snapshot with social and demographical data (own elaboration)

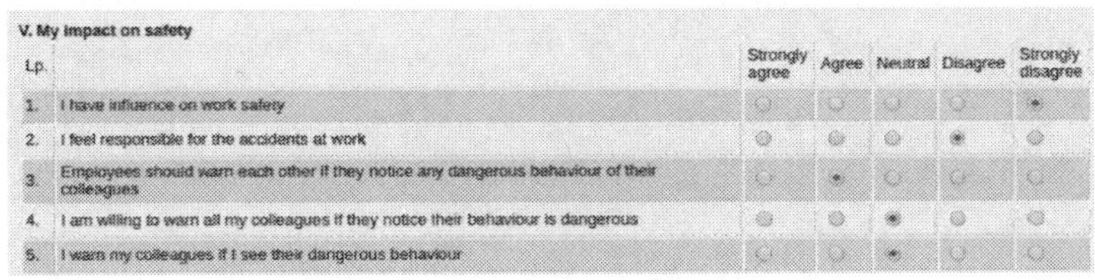

Fig. 5. Part of survey for measuring the level of occupational safety culture of an enterprise (own elaboration).

After completing a survey form, we have possibilities of analysing obtained data.

4 Analytical platform – user interface – data analysis

Occupational safety culture module of the Analytical Platform includes a tabular summary of social and demographical data (Figure 6). Such summary is helpful when selecting data for analysing dependences between various social and demographic factors and the level of occupational safety culture, considering particular designata of safety culture. Figure 7 presents a general overview and elements of the interface for analysing results of data collected during survey studies.

NUMBER OF RESPONDENTS / EDUCATION

	Generally		Women		Men	
Education	Count [1215]	% [100]	Count [411]	% [33.83]	Count [804]	% [66.17]
Higher	229	18.85	69	5.68	160	13.17
Secondary	540	44.44	173	14.24	367	30.21
Vocational	361	29.71	131	10.78	230	18.93
Primary	67	5.51	32	2.63	35	2.88

NUMBER OF RESPONDENTS / SENIORITY

	Generally		Women		Men	
Seniority	Count [1215]	% [100]	Count [411]	% [33.83]	Count [804]	% [66.17]
to 2 years	80	6.58	32	2.63	48	3.95
3 - 5 years	131	10.78	34	2.80	97	7.98
6 - 10 years	198	16.30	66	5.43	132	10.86
11 - 15 years	204	16.79	71	5.84	133	10.95
16 - 20 years	176	14.49	57	4.69	119	9.79
over 20 years	418	34.40	149	12.26	269	22.14

Fig. 6. View on table with collective summary of social and demographic data.

 DOI: 10.1051/matecconf/20179406008

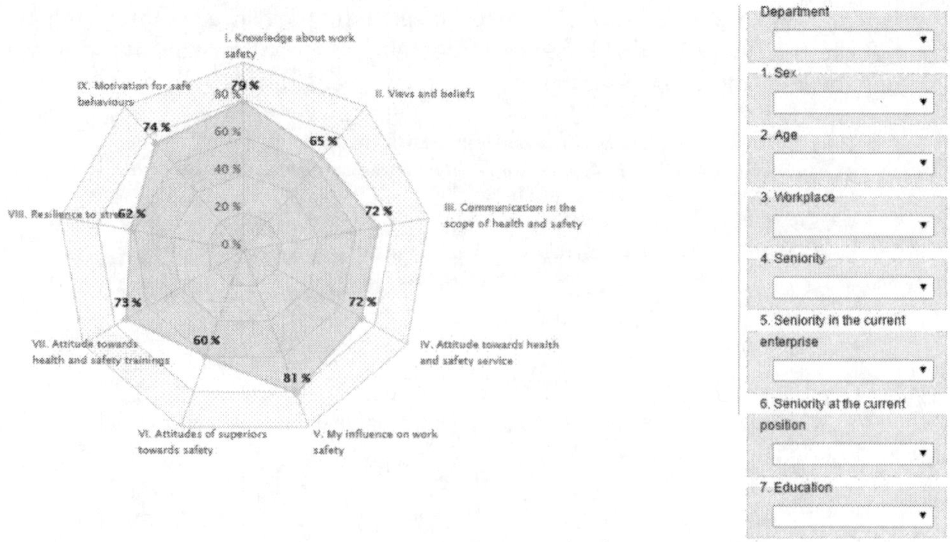

Fig. 7. View on and elements of the interface for analysing research results.

The platform enables a simple analysis of data in accordance to seven basic categories: sex, age, position, general job seniority, job seniority at current workplace, job seniority at current position and education. The Analytical Platform enables analysing of collected data in by single categories. One can check the level of safety culture among employees in accordance to their age (Figure 8) or their education (Figure 10).

Use of indicators (Figure 9 and Figure 11) enables tracing of trends of the level of occupational safety culture depending on a selected analysed social and demographical category.

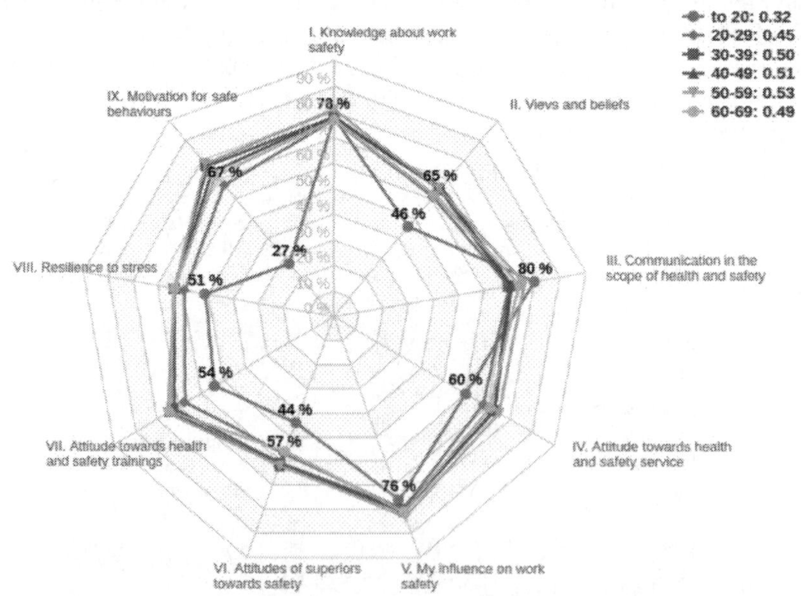

Fig. 8. Generated Safety Culture Net – depending on employee's age (own elaboration).

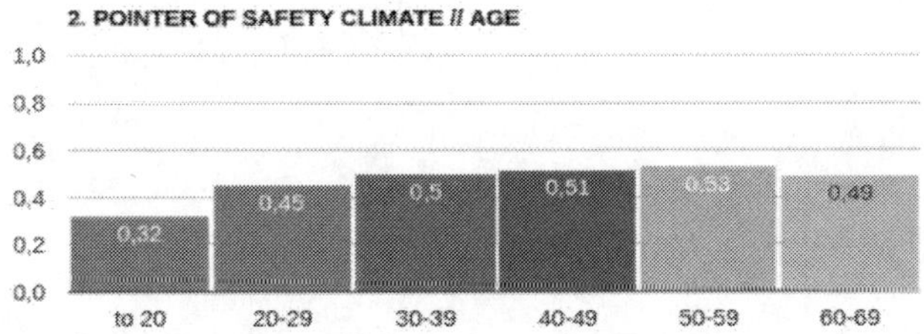

Fig. 9. Generated bar graph depicting the level of safety culture depending on the employee's age in indicative terms (own elaboration).

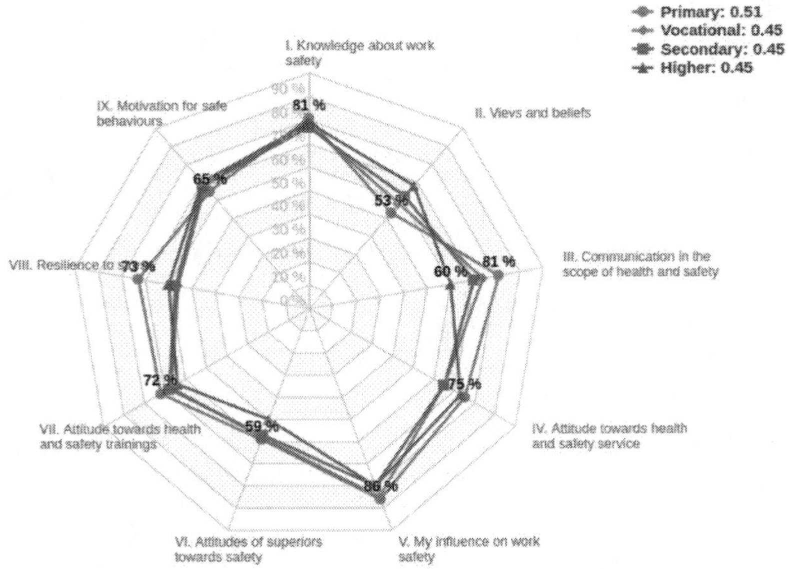

Fig. 10. Generated Safety Culture Net – depending on employee's education (own elaboration).

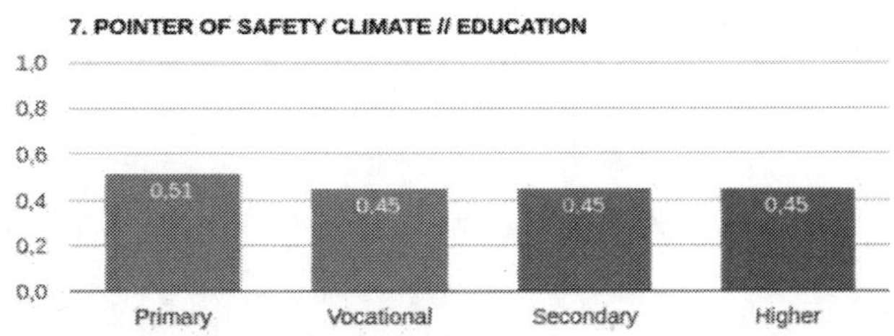

Fig. 11. Generated bar graph depicting the level of safety culture depending on the employee's education in indicative terms (own elaboration).

The Analytical Platform enables also an analysis of data in accordance to selected criteria, which can be freely configured. One can check the level of safety culture among women aged over 29 with secondary education. (Figure 12). Changing the criteria on a side panel, one can analyse the level of safety culture among the youngest employees (regardless sex), with primary education (Figure 13).

 DOI: 10.1051/matecconf/20179406008

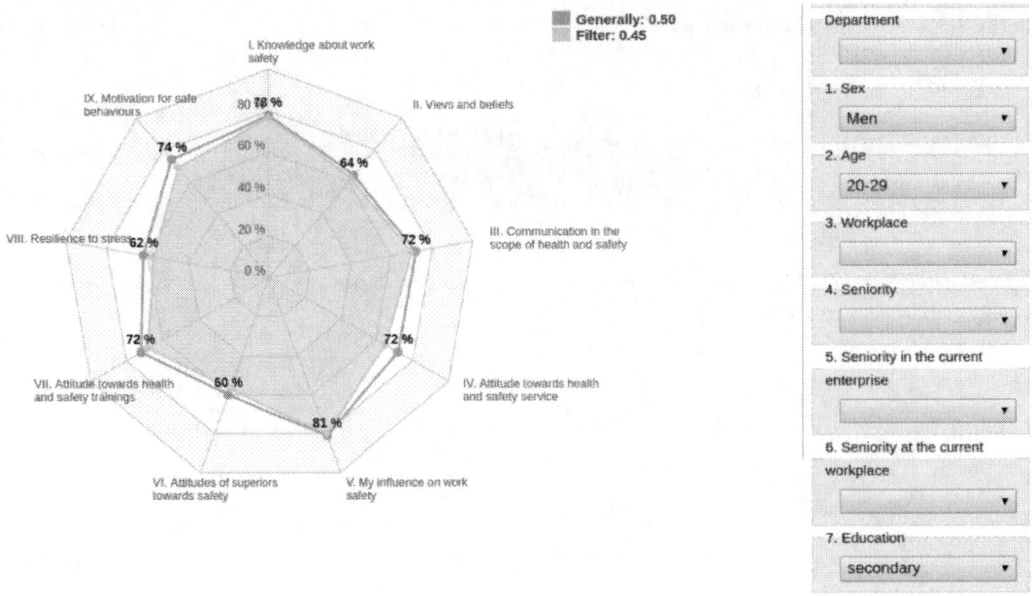

Fig. 12. Generated safety culture net – multi-aspect analysis of criteria (own elaboration).

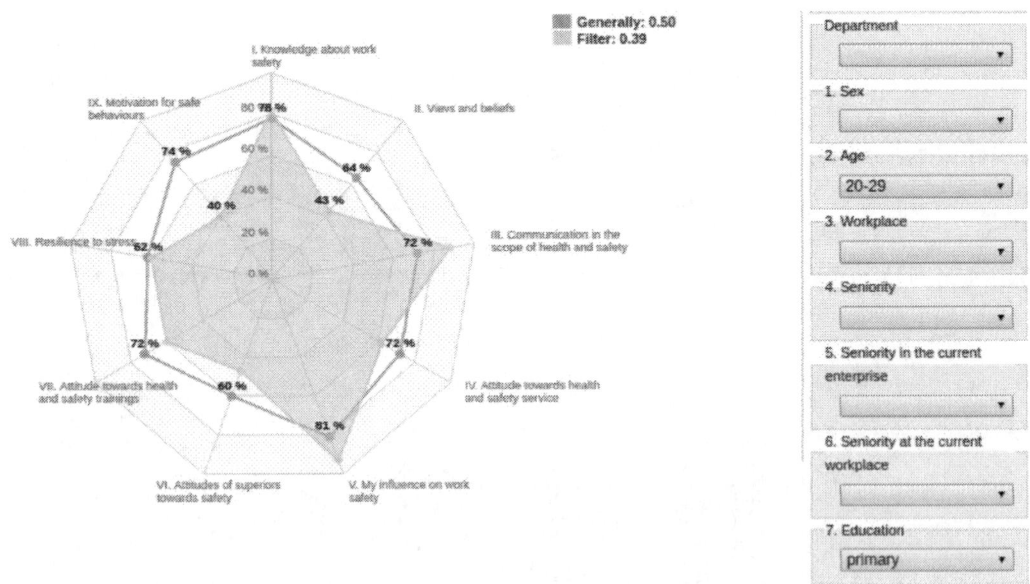

Fig. 13. Generated safety culture net – multi-aspect analysis of criteria (own elaboration).

5 Analytical platform functionality

The Analytical Platform was developed using free of charge tools and technologies. It has a friendly and intuitive interface. Other features of the Analytical Platform include:

• It supports studies realised at the same time at various work places, regardless locations. It enables obtaining data without the necessity of visiting plants, what often relates with expensive trips and spent time. Remote data collection facilitates the process of data collecting and enables collection of a larger amount of data during a shorter period of time.

• It supports cyclically conducted research, regularly repeated. It is possible to consider changes in the group of respondents (staff turnover) in consecutive cycles of studies. It allows defining of new versions of surveys, what is a fundamental research process, after conducting pilot studies.
• The database includes all versions of surveys (before pilot studies and modified after them). The database stores also results of all survey studies.
• The Analytical Platform provides maintenance of confidentiality and anonymity.
• Survey results collected in the database can be used for further analyses, e.g. for statistical aims.
• It enables creation of surveying state (e.g. the number of completed surveys, arranged in accordance to particular branches).

The above mentioned features of the Analytical Platform can be specially useful for comparative studies, when it is necessary to compare results from various analysed companies (e.g. due to a branch, size, social and demographic structure). Stored data can be useful for studies in the time context. When there is a need to study changes taking place in an enterprise:
• in the course of time,
• after introducing particular activities in the field of occupational safety,
• changes of the management,
• other changes.

After that, it would be necessary to check how such changes influence the level of occupational safety culture.

6 Conclusions

Research works aimed at decreasing threats are not able to replace indispensable investments in proper functioning of an enterprise [9]. However, even at small resources (often observed at small companies), a multidimensional analysis of the level of safety culture can have an influence on decreasing threats, mainly by shaping proper attitudes of employees and increasing their awareness. Shaping should be preceded by diagnostics in order to be efficient. Purposeful activities aimed at a particular group of employees (considering social and demographical features) will bring better results than general activities.

The presented Analytical Platform is a tool supporting the evaluation of the level of occupation safety culture. In case of collecting large amounts of data, modern technologies are a rational solution to improve work as well as to enable multidimensional analyses. They allow one to determine various correlations, which, using traditional analysis methods, would be impossible (or very difficult due to their time-consuming character) to be determined.

The possibility of comparing study results by an enterprise type or conduction time corresponds to the assumption that abnormalities in an enterprise functioning are not a constant category but they are changing depending on the environment, time, and social and demographic structure of employees and the management.

The most essential activity is obtaining information on the enterprise state. This requires a development or utilisation of proper tools and methods which are objectively oriented, in accordance to the specificity of the enterprise activity and environment in which it functions. Tools and methods and their form should be adjusted to the perceptive, decisive and executive possibilities of a user of a diagnosed enterprise.

The basic aim of developing the Analytical Platform was the necessity of having a tool for analysing a large amount of data and the possibility to determine multidimensional correlations between particular data.

 DOI: 10.1051/matecconf/20179406008

References

1. G. W., Werner, Praktyczny Poradnik Konserwacji Maszyn i Urządzeń (Practical Guide to Conservation of Machines and Devices), (WEKA Sp. Z o.o. Wydawnictwo Informacji Zawodowej, Czerwiec 2000)
2. M. Kłak, Zarządzanie wiedzą we współczesnym przedsiębiorstwie (Knowledge management in a modern enterprise), Wydawnictwo Wyższej Szkoły Ekonomii i Prawa im. Prof.EdwardaLipińskiego w Kielcach, 206 (2010)
3. M. Butlewski, M.Slawinska, Ergonomic method for the implementation of occupational safety systems, Occupational Safety and Hygiene Ii, 621, (2014)
4. S. Legutko, Trendy rozwoju utrzymania ruchu urządzeń i maszyn (Development trends in machins operation maintenance), Eksploatacja i Niezawodność **2**, 8 (2009)
5. I. Gabryelewicz, P. Krupa, J. Sadłowska-Wrzesińska, IT tool for aiding assessment of safety culture level in an organization, in: J. Grabara and S. Kot (ED): IT Systems and Decisions in Business and Industry Practice, **77** (2015)
6. I. Gabryelewicz, J. Sadłowska-Wrzesińska, E. Kowal, A.Kowal, AHFE, Elsevier.B.V., 962, (2015)
7. J. Sadłowska-Wrzesińska, I. Gabryelewicz, Springer International, CCIS **529**, 350, (2015)
8. J. Sadłowska-Wrzesińska, K. Grupka, I. Gabryelewicz, J. Grabara and S. Kot (ED): ITSDBIP, **186** (2015)
9. H. Bałuch, 57, Zeszyt nr 158, Railroading Problems, **57**, 158 (2013)

Managerial instrument for didactic staff structure optimization for Distance Learning

Cristina Gavrus[1,*], and *Georgiana* Limbasan[1]

[1]Transilvania University of Brasov, Department of Engineering and Industrial Management, Mihai Viteazul Street, No.5, Brasov, 500174, Romania

Abstract. Distance learning is a modern system for providing educational services and is relatively new in Romania, if related to the date of its emergence in Europe. More and more active working people are interested in this form of education, paying of course a special attention to its quality. It is quite difficult to appraise the quality of educational programs but several instruments and criteria have been developed over time. The present paper proposes an original mathematical instrument that is aiming at human resources, this type of resources being considered extremely important in case of providing educational service. The number of teachers is crucial for a distance learning program study, because the didactic staff must cover a number of didactic classes that take place on weekends. Concretely, this paper is focused on finding an algorithm that allows the didactic staff structure optimization. For accomplishing this objective, two managerial instruments were use. One of them is mathematical linear programing technique, that develops a mathematical model for didactic staff structure and the other one is *WinQSB* software package that tests the mathematical model.

1 Introduction

Distance learning is a modern system for providing educational services and is relatively new in Romania, if related to the date of its emergence in Europe. Over the last decade scientific and technological progress allowed a spectacular development of this form of education, as new communication tools were available for teachers as well as for students. As competition is continuously growing on educational market, people interested in distance education are of course paying a special attention to the quality of such programs. The quality standards elaborated by Romanian Agency for Quality Assurance (ARACIS) [1] are of course very important by they must be applied under the condition of resource optimization. The four categories of resources that are discussed in case of evaluation of the quality of any distance learning program study are: material resources, financial resources, human resources and informational resources. All these categories of resources have their importance and role, but when talking about the educational system human performance

*Corresponding author: cristina.gavrus@unitbv.ro

must be the core, because it directly influences the quality of didactic activities. The specialized literature is poor as regards the modalities of human resource optimization with a view to ensuring a high quality and efficient educational process. Usually, universities allocate human resources (didactic staff) to different didactic activities on the basis on human experience.

2 Distance learning in Romania

In Romania, the first distance learning study program was set up at the Academy of Economic Sciences of Bucharest in 1998. The Romanian distance learning system is a mixture between the American distance learning system that does not require the physical presence of the instructor, being based only on Internet communication tools [2] and European school that involves the physical presence of a teacher, of course combined with modern Internet communication tools [3]. In accordance with the founder of distance education in Romania, this is a modern form of education that supplies educational services for a wide variety of students from different fields. Distance learning is also helping students to learn how to learn in a modern environment supported by modern communication tools and is encouraging them to see the study activity as a continuous process [4]. In a general approach, distance education in Romania can be seen as a study system that makes available for students educational resources based on the following characteristics [4,5]: separation in time and space of students by the source that provides educational resources (the university that coordinates the distance learning study program); replacing the teaching activities (courses) by individual study; transmitting the learning resources by an environment specific for this form of education, which is the on-line electronic platform; student – teacher periodic meetings for deploying the applicative activities, such as seminars, laboratories, projects; access to a tutor for each student for coordinating the learning process and knowledge assimilation through evaluation tests; administrative assistance for every student, which is ensured by a special department that organizes the entire activity and the interface with teachers. The activities deployed within the Romanian distance learning system are: tutorial activities and homework, which are the equivalent of seminar classes of a traditional full time learning system, aided applicative activities, which are the equivalent of laboratories and project classes and individual studies, which are the equivalent of courses.

2.1 Distance learning resources

As specified before, distance learning must be based on four major categories of resources, as presented below [1, 5]. Table 1 identifies and briefly explains these categories of resources. The order of their appearance in the table does not take into consideration the importance of the resources. Each category of resources is as important as the others and they support each other. We cannot talk about a high quality distance learning without any of these categories. However, the present paper strongly highlights the importance of human resources, which are represented by didactic staff, because the human interaction is crucial for a well deployment of activities that take place in case of a distance learning study program.

Table 1. Distance learning resources.

Resource	**Explanations**
Material resources	They include proper spaces for deploying didactic activities, such as classrooms, rooms for seminars and projects, laboratories equipped with specific machine, tools and devices, Internet connected computer networks, multimedia software and hardware equipment and educational resources specific for distance learning education, such as: printed and CD courses written in distance learning technology, books, tutorials guides and others.
Financial resources	This category of resources must demonstrate that the university is capable of ensuring high quality educational services. The financial resources of a distance learning study program come from student's fee and they must fulfil certain conditions: they allow all the students to finalize studies and graduate; the financial policy of the institution clearly specifies all the expenditure and revenue; the financial resources take into consideration unpredictable situations that might appear. The fee paid by students is the value of all the services offered to students and must contain measurable elements.
Human resources	They are represented by the didactic staff involved in didactic activities; there are two categories of teachers: subject coordinators, whose role is to elaborate the educational resources and form the work team for a certain subject and tutors, who guide student during application classes, evaluate them by homework and ensure the interface among students, coordinators and educational resources.
Informational resources	The most important informational resources are represented by the educational resources and the electronic platform. The educational resources must be conceived in such a fashion that they replace the physical presence of an instructor in teaching – learning process. These resources must fulfil certain conditions, the most important are as follows: to be designed in accordance with the subject objectives; to present the competences gained by students after reading the didactic material; the material content must be divided into modules and study units and must be based on examples and auto appraising tests. The electronic platform is another important communication tool for students, subject coordinators and tutors and also, the educational resources are distributed by this platform. The electronic platform must be flexible and easy to adjust to requirements of the teaching – learning process and must make communication more efficient.

2.2 Quality management of distance learning in Romania

In order to ensuring the quality of educational process by distance learning system, the Romanian Agency for Quality Assurance in Higher Education (ARACIS) [1] have been elaborated specific standards, performance indicators and a certain methodology, regarding the accreditation of the distance learning study programs and periodical evaluation. Quality management is part of an integrated system that also contains: institutional capability and educational efficiency. The university coordinating distance learning study programs must have specific structures, policies, strategies and procedures for quality management and assurance of didactic activities [5]. The most important standards and performance indicators are [5]: existence of a set of rules regarding the initiation, approval, monitoring and periodical evaluation of study programs and activities; existence of certain procedures for periodical evaluation of didactic staff; transparency of information regarding the activities deployed within a distance learning study program and regarding the certificates, diploma and qualifications awarded to graduates. An extremely important aspect of quality management in case of distance learning system refers to an optimal use of resources, the present paper proposing, as specified in the abstract, an original model for human resources optimization.

3 Aspects regarding linear programming

As mentioned before, quality of didactic activities deployed within distance learning programs strongly depends on resource use optimization. When talking about teaching – learning process, human resources, represented by didactic staff is crucial. A series of disciplines have been developed for resource optimization, the management science being one of them [6]. Management science is in fact an interdisciplinary science that is aiming at rational approach of the managerial process of decision making, based on scientific methodologies [6, 7]. Within this science, certain quantitative techniques and methods have been developed. The method used for building a managerial instrument that allows the optimization of teaching staff structure for distance education programs is linear programming method, which is a managerial and mathematical tool for optimizing a function dependent on one or more variables and taking into consideration certain restrictions [7-9,10]. Any linear programing problem is a quantitative approach of a phenomenon that is based on a mathematical model. The mathematical model is obtained as a result of a modelling process that allows passing from the real model to the mathematical model [11]. This model contains three components [6, 11, 12]:

• **Objective function**, which is aiming at maximization or minimization of an expression and is illustrated by the equations (1):

$$[\max] f = \sum c_i \cdot x_i \quad \text{or} \tag{1}$$

$$[\min] f = \sum c_i \cdot x_i$$

Where x_i represents the variables of the objective function and c_i represents the coefficients of the objective function's variables.

• **Restrictions**, which represent certain conditions that the variables of the mathematical model must fulfil and which indicate the range of values for the objective function. Generally, restrictions might be highlighted by equation system (2).

$$\begin{cases} a_{11} \cdot x_1 + a_{12} \cdot x_2 + ... + a_{1n} \cdot x_n \le b_1 \\ a_{21} \cdot x_1 + a_{22} \cdot x_2 + ... + a_{2n} \cdot x_n \le b_2 \\ ... \\ a_{j1} \cdot x_1 + a_{j2} \cdot x_2 + ... + a_{jn} \cdot x_n \le b_j \\ a_{j+11} \cdot x_1 + a_{j+12} \cdot x_2 + ... + a_{j+1n} \cdot x_n \le b_{j+1} \\ ... \\ a_{m1} \cdot x_1 + a_{m2} \cdot x_2 + ... + a_{mn} \cdot x_n \le b_m \end{cases} \tag{2}$$

• **Non negativity restriction**, which represent other conditions imposed on the mathematical model's variables and that can be represented by equation (3).

$$x_i \geq 0, \quad i = 1...n \tag{3}$$

3.1 Mathematical model based on linear programming for structure staff optimization in case of a distance learning program study

As mentioned in section 2 of the present paper, resource optimization is important for a distance learning program but this must be done under the conditions of taking into consideration the quality standards. As specified before, distance learning study programs must fulfil certain conditions and for that the teaching – learning process to achieve its goals, it must be deployed in an efficient way at all levels. Mainly, this might be ensured if the following conditions are fulfilled: all the specifications of the quality standards are met; all the specifications regarding developing the development of the payroll are met; taking into consideration certain economic indicators. The concrete mathematical model regarding structure staff optimization in case of a distance learning program study is developed taking into consideration the three components of a linear programming problem briefly described in section 3, as follows.

• **Objective function**

Having in mind that there are more variants for developing the payroll, by considering the quality standards, it is necessary the use of an additional criterion for determining an optimal form of didactic staff structure. This criterion might be considered as the objective function of the mathematical model. For developing this model, the following notations are considered: S_P^h, S_C^h, S_L^h, S_A^h represent the hourly salary of professors, associate professors, lecturers and professor assistant; n_{dh}^P, n_{dh}^C, n_{dh}^L, n_{dh}^A represent the number of didactic hours deployed by professors, associate professors, lecturers and professor assistants. Considering these notations, the objective function might be:

$$Min\left(n_{dh}^P \cdot S_P^h + n_{dh}^C \cdot S_C^h + n_{dh}^L \cdot S_L^h + n_{dh}^A \cdot S_A^h\right). \tag{4}$$

• **Restrictions**

The restrictions of the mathematical model refer to the most important quality standards and that can be measured and are further developed.

- The restrictions regarding the tutorial activities (AT), homework (TC) and aided applicative activities (AA). The didactic materials, considered as educational resources, destined to individual study can be elaborated by teachers entitled to deploy tutorial activities. In order to fulfil the quality standard, it is considered that this kind of activities can be deployed by professors and associate professors. The following notations are considered: N_P - the number of professors; N_C - the number of associate professors; N_L - the number of lecturers; N_A The restriction related to the right of elaborating the didactic materials, as educational resources. The educational resources destined to individual study of students can be elaborated only by didactic staff entitled to deploy teaching activities. Considering the following notations: N_P, N_C, N_L, N_A - the number of professors assistants involved in the entire distance learning study program; n_{d1} - total number of subjects having tutorial activities and homework; TC – medium number of homework activities per subject (it is considered that this kind of activities can be deployed by lecturers and professor assistants); n_{d2} - total number of subjects having aided applicative activities. Considering these notations, these restrictions can be expressed as follows:

$$N_P + N_C \leq n_{d1}, \tag{5}$$

$$N_L + N_A \leq X, \tag{6}$$

where X will be further determined.

For determining the value of X the following notations are considered: H_P^{conv}, H_C^{conv}, H_L^{conv}, H_A^{conv} represent the maximum number of conventional didactic hours weekly deployed by a professor, associate professor, lecturer and professor assistant, considering full time learning; k_1 - the reducing coefficient of legal norms regarding distance learning in comparison with full time learning (for example $k_1 = 0.3$); mainly this coefficient has in mind the same quality level for both distance learning and full time learning, considering that the same didactic staff in involved in both learning systems; N – total number of study years for a distance learning study program; n_{sem}^{year} represents the number of semesters of a study year; n_{week}^{sem} represents the number of weeks of each semester; L_L, L_A - total number of didactic hours allowed for lecturers and professor assistants related to the entire study program. These numbers might be considered equal, being no significant differences between them. Therefore, the total number of didactic hours deployed by a lecturer for N study years is:

$$L_L = H_L^{conv} \cdot k_1 \cdot n_{sem}^{year} \cdot N \cdot n_{week}^{sem}, \tag{7}$$

or

$$L_L = H_L^{conv} \cdot k_1 \cdot K, \tag{8}$$

where:

$$K = n_{week}^{sem} \cdot n_{sem}^{year} \cdot N. \tag{8.a}$$

On the other hand the total number of didactic hours related to homework activities and aided applicative activities (N_{TC+AA}) for the entire study program can be expressed as follows:

$$N_{TC+AA} = 0,4 \cdot n_{d1} \cdot (AT + TC) + n_{d2} \cdot AA \cdot \frac{N_s}{25}, \tag{9}$$

considering that N_s is the total number of students of the study program and a group of students has 25 students.

On the basis of aspects previously presented, the value of X from equation (6) is:

$$X = \frac{N_{TC+AA}}{H_L^{conv} \cdot k_1 \cdot K}. \tag{10}$$

The equation (6) becomes:

$$N_L + N_A \le \frac{0.4 \cdot n_{d1} \cdot (AT + TC) + 0.04 \cdot n_{d2} \cdot AA \cdot N_s}{H_L^{conv} \cdot k_1 \cdot K} \tag{11}$$

- Restrictions regarding the legal norms related to the didactic degree of teachers. These restrictions can be expressed as follows:

$$\frac{n_{dh}^P}{K \cdot N_P} \le H_P^{conv} \cdot k_1, \tag{12}$$

$$\frac{n_{dh}^C}{K \cdot N_C} \le H_C^{conv} \cdot k_1, \tag{13}$$

$$\frac{n_{dh}^L}{K \cdot N_L} \le H_L^{conv} \cdot k_1, \tag{14}$$

$$\frac{n_{dh}^A}{K \cdot N_A} \le H_A^{conv} \cdot k_1. \tag{15}$$

These equations allow expressing the restrictions (5) and (11) in a different way:

$$\frac{1}{H_P^{conv}} \cdot n_{dh}^P + \frac{1}{H_C^{conv}} \cdot n_{dh}^C \le K \cdot k_1 \cdot n_{d1}, \tag{16}$$

$$n_{dh}^L + n_{dh}^A \le 0.4 \cdot n_{d1} \cdot (AT + TC) + 0.04 \cdot n_{d2} \cdot AA \cdot N_s \tag{17}$$

- Restrictions regarding the maximum number of subjects allowed for one teacher. Considering that n_s is the total number of subjects of the study program, these restrictions can be expressed as follows:

$$\frac{n_s}{N_P + N_C + N_L + N_A} \ge 1, \tag{18}$$

$$\frac{n_s}{N_P + N_C + N_L + N_A} \le 3. \tag{19}$$

Taking into consideration the equations (12) – (15), restrictions (18) and (19) become:

$$\frac{1}{H_P^{conv}} \cdot n_{dh}^P + \frac{1}{H_C^{conv}} \cdot n_{dh}^C + \frac{1}{H_L^{conv}} \cdot n_{dh}^L + \frac{1}{H_A^{conv}} \cdot n_{dh}^A \le K \cdot k_1 \cdot n_s, \tag{20}$$

$$\frac{1}{H_P^{conv}} \cdot n_{dh}^P + \frac{1}{H_C^{conv}} \cdot n_{dh}^C + \frac{1}{H_L^{conv}} \cdot n_{dh}^L + \frac{1}{H_A^{conv}} \cdot n_{dh}^A \ge \frac{1}{3} \cdot n_s \cdot K \cdot k_1, \tag{21}$$

- Restrictions regarding the types of activities deployed considering the didactic degree can be expressed as follows:

For professors and associate professors:

$$n_{dh}^{P} + n_{dh}^{C} \geq AT \cdot n_{s1} \tag{22}$$

For lecturers and professor assistants it is obtained an equation identical to (17).
As it can be observed, the ensemble of equations that contains the objective function (4) and the restrictions (16), (17), (20) – (22) forms the linear programing model (23):

$$\begin{cases} Min\ (n_{dh}^{P} \cdot S_{P}^{h} + n_{dh}^{C} \cdot S_{C}^{h} + n_{dh}^{L} \cdot S_{L}^{h} + n_{dh}^{A} \cdot S_{A}^{h} \\ \frac{1}{H_{P}^{conv}} \cdot n_{dh}^{P} + \frac{1}{H_{C}^{conv}} \cdot n_{dh}^{C} \leq K \cdot k_{1} \cdot n_{s1} \\ n_{dh}^{L} + n_{dh}^{A} \leq 0.4 \cdot n_{s1} \cdot (AT + TC) + 0.04 \cdot n_{s2} \cdot AA \cdot N_{s} \\ \frac{1}{H_{P}^{conv}} \cdot n_{dh}^{P} + \frac{1}{H_{C}^{conv}} \cdot n_{dh}^{C} + \frac{1}{H_{L}^{conv}} \cdot n_{dh}^{L} + \frac{1}{H_{A}^{conv}} \cdot n_{dh}^{A} \leq K \cdot k_{1} \cdot n_{s} \\ \frac{1}{H_{P}^{conv}} \cdot n_{dh}^{P} + \frac{1}{H_{C}^{conv}} \cdot n_{dh}^{C} + \frac{1}{H_{L}^{conv}} \cdot n_{dh}^{L} + \frac{1}{H_{A}^{conv}} \cdot n_{dh}^{A} \geq \frac{1}{3} \cdot K \cdot k_{1} \cdot n_{s} \\ n_{dh}^{P} + n_{dh}^{C} \geq AT \cdot n_{s1} \\ n_{dh}^{P}, n_{dh}^{C}, n_{dh}^{L}, n_{dh}^{A} \geq 0. \end{cases} \tag{23}$$

This model considers that the variables that must be optimized are n_{dh}^{P}, n_{dh}^{C}, n_{dh}^{L} and n_{dh}^{A}.

3.2 Application

For verifying the mathematical model expressed by equation system (23) it is used *Linear and Integer Programming* module of *WinQSB* software package. This module works on the basis on Simplex algorithm, which is supported by clear iterations and surely leads to the optimal solution [13, 14]. For using this module, three major steps were necessary: indicating general data, such as introducing the problem title, indicating if it is maximization or minimization problem and the number of constraints; introducing the mathematical model; solving the problem [15]. In order to introduce the mathematical model (23) in WinQSB software package, we consider the following elements: the hourly salary for a professor, associate professor, lecturer and professor assistant are: $S_{P}^{h} = 69$ lei/hour; $S_{C}^{h} = 48$ lei/hour, $S_{L}^{h} = 35$ lei/hour; $S_{A}^{h} = 27$ lei/hour; the number of conventional hours per week for every didactic degree are: $O_{P}^{conv} = 9$ hours/week, $O_{C}^{conv} = 10$ hours/week, $O_{L}^{conv} = 12$ hours/week, $O_{A}^{conv} = 13$ hours/week; total number of students for a new distance learning study program is 75; the total number of subjects for the entire study program is 50; the total number of subjects having tutorial and homework activities is 20 and the total number of subjects having aided applicative subjects is 50. Another aspect to be considered is that the study program has 4 study years, each year has two semesters and each semester has 14 weeks. For introducing the model (23) into WinQSB software package, the following notations are considered: x_1, x_2, x_3, x_4 – the number of didactic hours for a professor, associate professor, lecturer and assistant for the entire study program. Figure 1 presents the WinQSB mathematical model, where the numerical values of the coefficients of the objective function and constrains can be seen.

Variable	X1	X2	X3	X4	Direction	R.H.S.
Minimize	69	48	35	27		
C1	0.11	0.1			<=	672
C2			1	1	<=	3318
C3	0.11	0.1	0.083	0.076	<=	1882
C4	0.11	0.1	0.083	0.076	>=	627
C5	1	1			>=	210
C6	1				>=	0
C7		1			>=	0
C8			1		>=	0
C9				1	>=	0

Fig. 1. WinQSB mathematical model.

Appealing Solve the Problem function the software package displays a table containing the optimal solution. Considering the issue of human resource optimization in case of a distance learning study program, the optimal solution refers to the numbers of didactic hours necessary for each didactic degree. The optimal solution also indicates the value for the objective function, which in this case is the minimum of the salary costs for the entire study program. Figure 2 shows the optimal solution for the mathematical model (23).

Decision Variable	Solution Value	Unit Cost or Profit	Total Contribution	Reduced Cost
X1	0	69,00	0	16,20
X2	3.748,32	48,00	179.9191,40	0
X3	0	35,00	0	4,64
X4	3.318,00	27,00	89.586,00	0

Fig.2. The optimal solution.

The optimal solution of the problem highlighted in Figure 2 refers to the total number of didactic hours necessary for every didactic degree for a distance learning study program, but considering the quality standards, elaborated by the Romanian Agency for Quality Assurance in Higher Education. As it can be seen from figure 2, the number of didactic hours for each didactic degree category is: 3716 hours for associate professors and 3360 hours for professor assistant. These values of the variables ensure the minimum value of the objective function that refers to the salary costs minimization. For determining the optimal number of teachers, by didactic degree, we must consider the number of conventional didactic hours for each didactic degree (H_C^{SP} - total numbers of conventional didactic hours for associate professors for the entire study program, H_A^{SP} - total number of conventional didactic hours for professor assistants for the entire study program) :

$$H_C^{SP} = H_C^{conv} \cdot K \cdot k_1 = 10 \cdot 112 \cdot 0.3 = 336 \text{ hours} \tag{24}$$

$$H_A^{SP} = H_A^{conv} \cdot K \cdot k_1 = 13 \cdot 112 \cdot 0.3 = 436.8 \text{ hours} \tag{25}$$

The total number of associate professors and professor assistants can be determined in the following way:

$$N_C = \frac{n_{dh}^C}{H_C^{SP}} = \frac{3748}{336} = 11.15 \cong 11 \text{ associate professors,} \tag{26}$$

$$N_A = \frac{n_{dh}^A}{H_A^{SP}} = \frac{3318}{436.8} = 7.59 \cong 8 \text{ professor assistants} \tag{27}$$

4 Conclusions

The present paper approached the issue of human resource optimization in case of a distance learning study program. In authors' opinion from the four categories of resources that support the didactic activities of a study program (material resources, financial resources, human resources and informational resources) human resources are considered crucial, because the deployment of the activities related to teaching – learning process involves people and people communication, by all means: face to face communication, e-mail communication, communication via virtual electronic platforms. The authors proposed an original instrument that allows the optimization of a distance learning study program from the point of view of human resources using as instruments linear mathematical programming, based on simplex algorithm and WinQSB software package.

The optimization problem approached by the present paper is aiming at deploying a high quality educational process but also taking into consideration a high economic efficiency. The mathematical model proposed is in fact a frame model that highlights an original approach of the issue discussed. Depending on the concrete conditions existing in universities, the mathematical model presented can be corrected / completed with new constraints.

As it is known, every university from Romania takes advantage of a certain degree of university autonomy or independence, meaning that all universities apply the quality standards elaborated by the Romanian Agency for Quality Assurance in Higher Education using the internal norms. The managerial instrument elaborated as a result of researches made by the authors of this paper takes into consideration, of course the quality standard but also specific conditions and norms of the university where the authors deploy their activity, which is Transilvania University of Brasov. As specified before, this mathematical model is a general and also a frame model that can be successfully applied by any university or institution providing higher educational services, in case of full time study programs as well as in case of distance learning study programs.

The managerial instrument that allows the optimization of the didactic staff structure can be, no doubt, further developed. Some changes or improvements that can be added refer to modifying certain restrictions or adding some new restrictions, as the quality standards or the internal conditions from universities changes or even considering different objective functions. The objective function that stands on the basis of the mathematical model developed within this paper refers to the minimization of the salary costs of the didactic staff involved in distance learning teaching activities. Of course, costs are very important but not always the efficiency of certain activities means as lowest costs as possible. In case of didactic staff structure optimization, other optimization criteria might be considered. Such criteria could be: maximization of student's number, minimization of the costs involved in educational resource distribution, minimization of the number of face to face activities, or even minimization of the number of teachers. These aspects represent potential starting points in further researches of the authors of the present paper, in the field of higher education optimization.

References

1. The Romanian Agency for Quality Assurance in Higher Education – Quality Evaluation Activities Guide for University Study Programs and for Higher Education Institutions, 2015., available at: http://www.aracis.ro/fileadmin/ARACIS/Legislatie_-_Proceduri/Part_I_-_STUDY_PROGRAMMES_ACCREDITATION.pdf
2. J. Holden, Ph. Westfall, *An Instruction Media Selection Guide for Distance Learning – Implication for Blended Learning*, (USDLA, 2010).
3. H. Fry, Katteridge, S. Marshall, *A Handbook for Teaching / Learning in Higher Education. Enhancing Academy Practice.*(Fourth Edition. Routledge Taylor / Francis 2015)
4. http://www.idifr.ase.ro/despre-id, accessed on June 27, 2016.
5. http://www.aracis.ro/fileadmin/ARACIS/Legislatie_-_Proceduri/Partea_a_V-a_-_Ghid_evaluare-ID.pdf accessed on June 27, 2016.
6. D.R. Anderson, D.J. Sweeney, T.A. Williams, J.D. Camm, J.J. Cochran, M.J. Fry, J.W. Oklmann, *An Introduction to Management Science, Quantitative Approaches to Decision Making*, (Cengage Learning, 2015).
7. B. Rander, R. Stair, M. Hanna, *Quantitative Analysis for Management*, (Global Edition, Pearson, 2014).
8. W.L. Winston, C.S. Albright, *Practical Management Science*, (Cengage Learning, 2015).
9. D. Lixandroiu, *Decision Modelling in Economics*(Economica Publishing House, 2014).
10. C. Ratiu-Suciu, *Modelling and Simulation of Economic Processes* (Economica Publishing House, 2009).
11. V. Marascu-Klein, *Modelling and simulation of Production systems*, (Lux Libris, 1997).
12. J. Chaouachi, M. S. Ghorbel, *Linear Programming*, (ISTE Press LTD – Elsevier Inc, (2016).
13. F.A. Ficken, *The Simplex Method of Linear Programming*, (Dover, 2015).
14. P.-Q. Pan, *Linear Programming Computation*, (Springer, 2013).
15. Y.L. Chang, *WinQSB*, (John Wiley & Sons, Inc, 2003).

MATEC Web of Conferences **94**, 06010 (2017) DOI: 10.1051/matecconf/20179406010
CoSME'16

A real options and joint ventures perspective for strategic decision-making process in the case of dynamic industrial environments

Ana - Maria German[1,*], *Flavius Aurelian* Sârbu[1], and *Mircea* Boşcoianu[2]

[1] Transilvania University, Department of Engineering and Industrial Management, Mihai Viteazul Street, no. 5, Braşov, România
[2] Henri Coandă Air Force Academy, Department of Fundamental Sciences and Management, Mihai Viteazul Street, no. 160, Braşov, România

Abstract. Strategic decision – making process represents a challenge in a dynamic industrial environment. Nowadays, many instruments and techniques are used in the process of decision-making. The perspective regarding real options analysis (ROA), joint ventures (JVs) and strategic decision-making process (SDMP) is very important to be analyzed because it represents a key source to obtain competitive advantage in a turbulent environment characterized by competitiveness, dynamism and a proactive thinking. The aim of the paper is to offer an interesting framework that emphasizes the relationship between real options (ROs), JVs, transaction cost (TC), bargaining models (BM), agency theory (AT) and SDMP. Also, the paper analyzes a real case study and it suggests a business model that includes elements that conduce to an innovative management. The findings are very useful for decision-makers from organizations, researchers and students.

1 Introduction

The topic of uncertainty has received more and more attention in strategic management field and there has been interest in how industrial organizations might manage strategic decision - making under uncertainty. The aim of the paper is to provide a comprehensive and extension framework that emphasizes the relationship between real options (ROs), joint ventures (JVs), transaction cost (TC), bargaining models (BM), agency theory (AT) and strategic decision-making process (SDMP). In a dynamic industrial environment, the issues that are connected to the SDMP require a new thinking model. The goal of a new business model is to create and manage mechanisms that focus on the development of new opportunities by optimizing the existing knowledge, resources, capabilities and processes.

Additionally, the paper analyzes a real case study and it suggests a business model that includes elements that conduce to an innovative management. Based on the model proposed, the industrial organizations should reformulate their strategies and investors should develop new opportunities to gain competitive advantage and obtain performance.

* Corresponding author: german_ana_maria@yahoo.com

The use of these instruments will help managers to take the right decisions in their organizations. Hence, strategic decision makers must understand how to preserve the opportunity to act in turbulent industrial environments. The framework and the model represent the first step in developing a paradigm for building the decision making process based on the existing resources, the development of the new opportunities and joint ventures, the agility in reconfiguring strategies based on combination of resources and capabilities in relation with the perspective offered by real option analysis.

2 Real options analysis (ROA)

An option is "the act of choosing, the power of choice, or the freedom of alternatives" [1]. A real option is "the investment in physical assets, human competence, and organizational capabilities that provide the opportunity to respond to future contingent events" [2]. ROA represents an important tool in strategic management for many reasons. Firstly, ROA offers unique and precious predictions on industrial organization' decisions for different types of strategic choices under uncertainty. In this case, industrial organizations can "unlock value at the joint venture termination stage and an important role exists for joint ventures that are transitional investments by design" [3]. ROA "emphasizes dynamic efficiency gains, downside risk reduction, and the organization's ability to seize upside opportunities over time by shifting value chain activities across borders in response to different uncertainties" [4]. Secondly, the theory emphasizes that maintaining flexibility under uncertainty has option value for many investments [4, 5]. Thirdly, ROA offers a new perspective on organization' resource allocation processes by informing the strategic decision makers.

The critics of the literature suggest that RO are difficult to value with certainty and the methodology is complicated but Mun argues that " traditional approaches assume a statistic decision-making ability, while real options assume a dynamic series of future decisions where management has the flexibility to adapt given changes in the business environment" [6]. The volatility is often the variable that has the most important impact on the option value. As has been shown [7], there are three approaches to estimate the volatility: twin security information, Monte Carlo simulation and closed-form expression [8]. On one hand, ROA represents only an academic tool and it is not practical in the business and on the other hand ROA "ends up choosing the highest-risk projects as the higher the volatility, the higher the option value" [9].

If the uncertainty surrounding an investment is greater, the possibility that the option will be exercised, thus making its value higher is greater. In the literature, there are different types of options that influence the investment decisions: option to wait to invest, option to abandon, option to switch, option to contract (expand), option to grow, option to stage, call options, put options, timing options or portofolio of options. The real option to wait to invest indicates to wait until further information reduces the market uncertainty. The option to abandon refers to dispose of a project that is not profitable. The option to switch changes the input or the output parameters. The option to contract (expand) is influenced by the market conditions. The option to grow refers to the future opportunities and the option to stage breaks up the investment into conditional steps. [1]. A call option is "an option in which the buyer has the right, but not the obligation, to buy a specified quantity of a security at a specified price named strike price within a fixed period of time but until its expiration [10, 1]. A put option is "an option in which the buyer has the right but not the obligation to sell a specified quantity of a security at a specified strike price within a fixed period of time but until its expiration [10, 1, 11]. Timing option can be used when there are "large investment projects that show a large probability of realizing a negative net present value" [12]. Portofolio options can be very useful for joint ventures (JV). RO valuation method of investment projects is the extension of the financial options theory on real property. The two known option calculators are Black & Scholes option valuation model and the binomial option valuation model.

3 Joint ventures (JV)

JV are defined as "equity-based collaborative arrangements whereby two or more organizations each contribute resources, including equity, for the joint pursuit of economic goals" [13, 3]. According to the ROA, JVs are "analogous to call options in the sense that they provide their partners the right (not the obligation) to buy the developed technology (underlying asset) at a specific price (exercise price) at or before a specific date (expiration date)" [14]. In addition, "JVs provide the industrial organization with the option to access certain technological capabilities, acting as a platform to profit from future technological opportunities while avoiding full resources commitment, i.e. maintaining flexibility" [14].

One of the advantages of ROA is that it offers a dynamic perspective which can explain the JV's stages, from formation to exercise [15]. The options embedded in JVs will have an impact on the distribution of the equity stakes because an investor who aims to capture the dynamic RO part will invest in a smaller share of the JV because this way (s)he limits the downside risk while preserving the opportunity to capture the upside potential [4]. The value of the dynamic option part represents "a function of the same five factors that determine the value of financial options," (the value of the underlying asset, the strike price, the time to maturity, the risk-free rate and the uncertainty surrounding the underlying asset). [16, 3]. Hence, the value of the options fixed in JVs is connected to market, uncertainty and partner-related.

4 Strategic decisions-making process (SDMP)

Strategic decisions involve investments in technology and development activities that must contain resources to future benefits in a dynamic and uncertain environment. The strategic decision - making process required for investment management is correlated to the resource-based view and to its management. Resources are "the foundation for strategy and unique bundle of resources generate competitive advantage leading to wealth creation" [17]. The studies based on strategy and finance have suggested that ROA represents a powerful valuation tool as well as a "systematic strategy framework to evaluate and structure resource investments under uncertainty and that successful use of RO can lead to the benefits of downside risk reduction and upside potential enhancement" [3, 18, 2]. RO investment decisions involves the mapping investments to options, the timing investment and the structuring investment. The implementation of RO refers to the levels of analysis (i.e. managerial/organizational dimensions) and to the implementation stages (i.e. creation, evaluation, maintenance, exercise). The performance outcomes of ROA are observed in measures (e.g. traditional risk measures, Tobin`s Q etc.) and in customized measures (e.g. downside risk, economic exposures etc.) [3].

Alternative approaches that are connected to the decisions, including the distribution of JV equity shares, are transaction costs economics (TCE), the bargaining perspective (BP) and the agency theory (AT). TCE describes the organization as "an efficiency-inducing administrative instrument that facilitates exchange between economic actors" [19]. There are two assumptions underlying the TCE perspective. In the first assumption, "individuals within an organization are assumed to be boundedly rationale - they are limited in their ability to plan for the future and to accurately predict and plan for the various contingencies that may arise" – [19] and the second assumption refers to the opportunism - some economic actors are "self-interest seeking with guile" [20] or subject to "frailties of motive"

 DOI: 10.1051/matecconf/20179406010

[21]. The ownership distribution of a JV is "the outcome of negotiations in which relative power is a deciding factor because the partner's level of ownership to be positively is related to its advertising intensity, its provision of market access and the amount of technology that it contributes and to be negatively, it is related to its number of competitors" [3]. The ownership structure of industrial organizations influence agency costs. Agency theory relates to "situations in which one individual (called the agent) is engaged by another individual (called the principal) to act on his/her behalf based upon a designated schedule [22]. The AT "posits different organizational, behavioral, economical and controlling roles, and it is a useful framework which can be extricated in promulgation of the management control systems" [22].

5 Project valuation using ROA (case study)

Manufactories have to adapt to the dynamic business environment. The decision-making process should be based on combinations of options, resources and capabilities.
Problem:

An aeronautical manufactory, through a JV, decide to develop a new product to improve the performance of its airplanes. The present value underlying assets is 100 TEUR. Volatility of the logarithmic returns on the project future cash flows is 35% and the risk-free rate on a riskless asset for the next five years is 3%.

Using real options super lattice software to take the best decision, the analyze follows some steps.

Step I. Determinate the underlying asset lattice

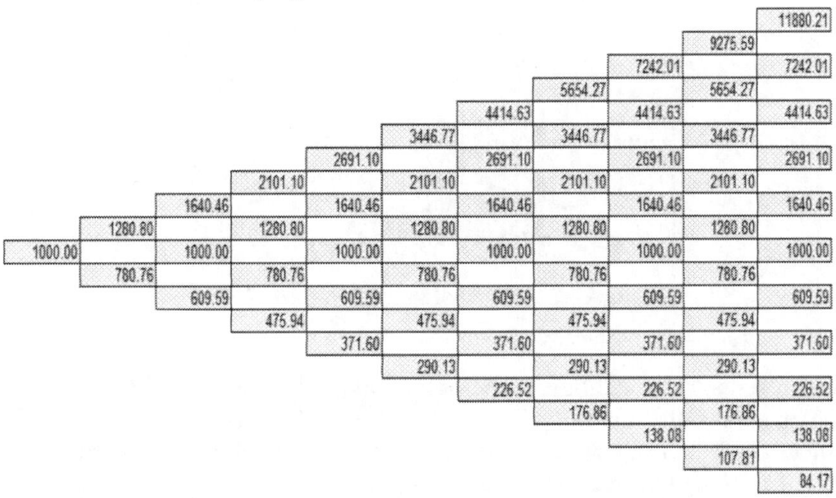

Fig. 1. Underlying asset lattice.

The figure represents the lattice evolution of the underlying with a volatility of 35%. Using the binominal approach, is calculated the value of the contraction options following a ten time-steps. All the calculations and steps are based on the up factor, down factor and risk-neutral probability analysis.

Step II. Determinate the sensitivity

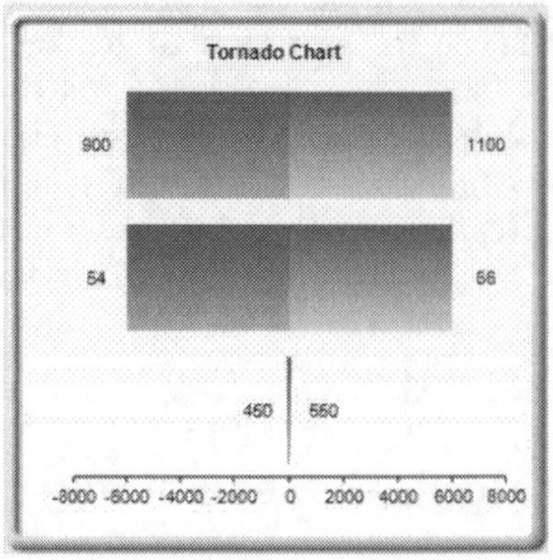

Fig. 2. Tornado Chart – determination of the sensitivity.

This analysis runs a quick static sensitivity of each input variable of the model at a time and lists the input variables with the high test impact to the lowest (according to real options super lattice software). The lattices result a good convergence. The figure represents the Tornado Chart with 35% volatility.

Step III. Simulation process

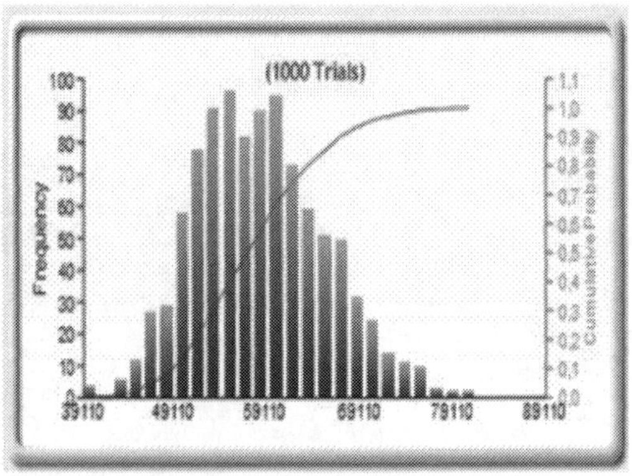

Fig. 3. Simulation process with volatility 35%.

The figure represents the simulation of the option taking into account the present value, the implementation costs, the maturity, the risk - free rate and the volatility. The figure illustrates that at 1000 trials with volatility 35%, the mean is 60.636,51 the standard deviation is 8.447,16 and the range is 50.814,05.

Step IV. Creating a business model

The industrial management represents an important topic that should be taken into consideration. In this context, it is necessary to develop a model that could be used by enterprises to obtain competitive advantage and performance.

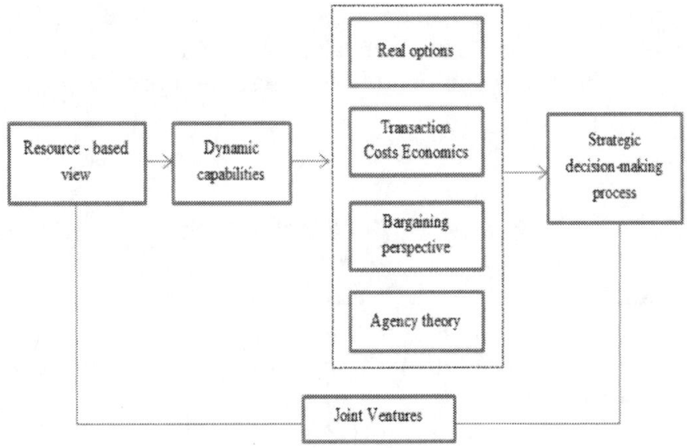

Fig. 4. Looking ahead: a general business model.

The model is focused on the resource-based view, dynamic capabilities, real options, transaction costs economics, bargaining perspective and agency theory. All this elements have a joint ventures foundation. The model emphasizes the importance of different tools in the process of elaborating the strategy. All this elements represent a source to gain competitive advantage and to obtain business performance.

It can be seen from the figures the importance of using ROA in high volatile environments and their role on protection against high uncertainty elements, especially in major investments decisions. The case study presented suggests that ROA can be used in decision-making managerial process for industrial organizations. The results obtained after the analysis suggest that it is a great choice to implement the project based on the development of a new product and the joint ventures.

6 Conclusions

In a turbulent business environment the technological progress is not sufficient. The analysis of the ways to obtain the performance of industrial organizations should be focused on the active selection of different variables/elements and on the mixing of different mechanisms that generate sustainable competitive advantage. The integration of ROs and JVs in the process of strategic decision – making, together with other paradigms that contain flexibility elements such as: TCE, BP, AT, represents an interesting instrument to gain competitive advantage. The way toward a better understanding of this elements conduce to performance, efficiency, effectiveness, competiveness and innovation.

The paper offers valuable information and provides for managers an integrative perspective of the ROA and JVs. The findings suggested in this paper show that real options and joint ventures can represent a source of competitive advantage for industrial corporations if they are put into practice by managers. By using ROA, industrial organizations can gain competitive advantage by choosing the optimal strategy. The value of strategic options is offered by variables and elements that are influenced by competition, demand and a good management. The case study presented is a strong proof that suggests that ROA should be used in industrial companies. The business model offered by the authors emphasizes precious elements such as: resource-based view, dynamic capabilities, real options, transaction costs economics, the bargaining perspective and agency theory. This elements represent the keys to an efficient decision - making process and to business performance. In addition, the framework, the case study and the business model represent a tool that is useful for managers but also, for student and researchers.

Further studies will focus on the development of a hybrid model necessary to an efficient management and an assessment procedure of the model because knowledge brings innovation and innovation brings competitive advantage and performance.

References

1. M. Branch, *Real options in practice* (John Wiley & Sons, Inc., Hoboken, New Jersey, 2004)
2. B. Kogut, N. Kulatilaka, Org. Scien. J. **12(6),** 744-758 (2001)
3. J. Reuer, T. Tong, *Real options Theory* (Elsevier, U.K., 2007)
4. J. J. Reuer, *How real are real options? The case of international joint ventures* (Oxford Blackwell Publishing, 2002)
5. J.C. Hull, *Options, futures, and other derivative securities* (Prentice-Hall, 2003)
6. J. Mun, *Real options analysis: Tools and techniques for valuing strategic investments and decisions* (John Wiley & Sons, Inc. USA, 2002)
7. L.T. Miller, C.S. Park, The Engin. Ec., **47(2),** 105-150 (2002)
8. A.M. German, M. Boscoianu, *Gaining competitive advantage in a global business environment - the case of SMEs in emerging markets* (Strategica Int. Conf., Tritonic 2015)
9. H. B. Nembhard, M. Aktan, *Real Options in Engineeing Design, Operations and Management* (Taylor & Francis Group CRC Press, 2010)
10. A. Sullivan, S. Sheffrin, M. Steven, *Economics: Principles in action. Upper Saddle River* (Pearson Prentice Hall, 2003)
11. N. Fernandes, *Finance for Executives: A Practical Guide for Managers* (NPV Publishing, 2014)
12. S. Orsag, Izvedenice, HUFA, Zagreb (2006)
13. X. Martin, R. Salomon, Organiz. Scien., **14(3),** 297–311 (2003)
14. I. Estrada, G. Fuente, N. Martin-Cruz, Res. Policy, **39(9),** 1185–1197 (2010)
15. I.R. Cuypers, X. Martin, Adv. in Strat. Mn., **24**, 107-148 (2007)
16. A. Seth, S.M. Kim, *Valuation of international joint ventures: A real options approach.* (Quorum Books, 2001)
17. C. Brush, P. Greene, M. Hart, Acad. of Mn. Execut., **15(1),** 64-78 (2001)
18. L. Trigeorgis, *Real options: Managerial flexibility and strategy in resource allocation.* (MIT Press, 1996)
19. M.J. Leiblein, J. Mn., **29(6)** 937–960 (2003)
20. O.E. Williamson, *Markets and hierarchies: Analysis and antitrust implications* (New York, Free Press, 1975)
21. H. Simon, *Models of bounded rationality: Behavioral economics and business organization* (Cambridge, MA: MIT Press, 1982)
22. M. Namazi, J. of Accoun. and Taxat., **5(2),** 38-47 (2013)

MATEC Web of Conferences **94**, 06011 (2017) DOI: 10.1051/matecconf/20179406011
CoSME'16

The use of working environment factors as criteria in assessing the capacity to carry out processes

Adam Górny[1,*]

[1]Poznan University of Technology, Faculty of Management Engineering, 11 Strzelecka St., 60-965 Poznan, Poland

Abstract. One resource indispensable for securing the capacity to carry out processes is an environment made up of social, psychological, physical and other factors. The nature of such environmental factors and the manner in which they affect the capacity to carry out processes and produce goods and/or services translates into a company's ability to meet the expectations of stakeholders, including the organization's customers. In order to define the preconditions for the efficient implementation of processes, it is necessary, among others, to ascertain the manner in which such preconditions influence process operation. One of the criteria that should be applied in such an assessment is the impact of working environment factors (temperature, lighting, noise, toxicity, dust, electromagnetic fields) on the ability to carry out work processes effectively. The paper accounts for working environment factors seen as parameters for the assessment of processes and as criteria for ascertaining the ability to perform work. The impact of the working environment on the capacity to conduct processes was assessed by means of a relationship diagram.

1 Introduction

The acquisition of the capacity to carry out production processes ties irretrievably to ensuring a proper working environment for the labor force. Its pivotal part is the physical working environment whose specifications determine workers' ability to perform work and the organization's ability to ensure safety for persons responsible for work performance [1].

In assessing an enterprise's capacity to carry out tasks, particular emphasis should be placed on the processes employed in transforming stakeholder expectations into outcomes resulting from the provision of services, the manufacturing of products or decision making [2, 3]. The ability to carry out internal and other processes defines the ability to perform such tasks.

To conduct processes, an enterprise needs to ensure access to appropriate resources which will allow it to carry them out without unnecessary hindrances and limitations [2, 4]. Such resources are technical, organizational, human, financial and other. Frequently, being in their possession or having the ability to acquire them is a precondition for attaining a

* Corresponding author: adam.gorny@put.poznan.pl

company's objectives. Such resources include the ability to adopt solutions that increase the capacity to perform work. One such solution is the establishment of a working environment that will not disrupt work performance [5]. The acquisition of resources should be preceded by an analysis of the existing state of working conditions.

One measure of equal importance for conducting the process is to identify criteria for evaluating an organization's performance. This is particularly vital for carrying out tasks that constitute a process, that are subject to systemic standards (e.g. the standards of a quality management system), and that require comparing work outcomes with any preliminary precepts and targets [6].

2 Process description and process performance conditions

Processes should be defined as a collection of mutually interrelated actions that influence one another and that are undertaken to transform inputs into outputs [6]. A simplified process-based task performance diagram is provided in Figure 1. To ensure that the transformation of inputs into outputs proceeds effectively, it is critical to consider the impact of conditions on process performance and minimize any existing or potential disruptions.

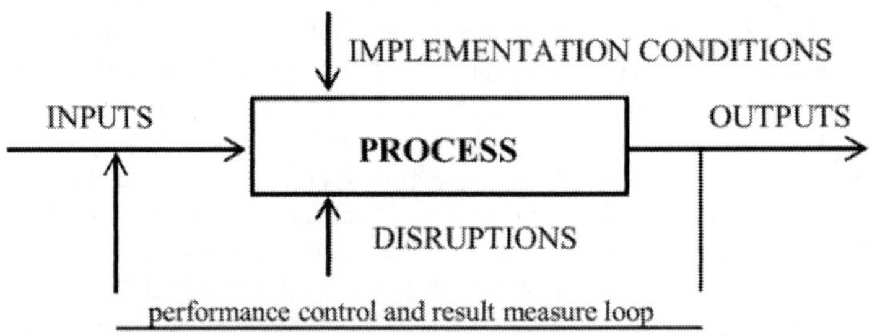

Fig. 1. Simplified process performance diagram.

A precondition for the effective performance of tasks in a process-based system is to [3, 7]:

- Identify, in advance, all processes found in an organization, arrange them into a sequence, recognize their mutual impacts, define input and output requirements and identify the processes that are particularly critical for the proper performance of tasks,
- Define the principles and methods for overseeing process performance, and
- Implement actions necessary for the achievement of planned outcomes and continuous process improvements.

The systemic approach employed in pursuance with ISO 9001 [6] may be associated with the view of an enterprise as a sum total of the effects of its processes. A process that underpins a given approach may be seen as a way to carry out tasks. Such tasks are defined in terms of the mutual relationships among them with a focus on organizational units, workstations and functions. Here, a particular role needs to be ascribed to [3, 8, 9]:

- The conditions in which specific tasks are conducted and which are central to an organization's ability to acquire and transform resources that are critical for the achievement of specific outcomes,
- Any disruptions that adversely affect the ability to carry out a process defined as a transformation of inputs into outputs. Every process disruption compromises an organization's ability to generate specific benefits. As such, it should be eliminated at the level of the components and specifications of a given process.

 DOI: 10.1051/matecconf/20179406011

The performance of processes and the interrelations among them are influenced substantially by the environment in which the processes are conducted. Success in their improvement depends on the impacts of all component elements that affect process performance [10].

As a prerequisite for its ability to improve its processes, an organization should [2-4, 10, 11]:

- Plan its actions and specifically identify its objectives and secure the required resources, including the environment in which the planned actions are to be taken,
- Make proper organizational arrangements such as the assignment of tasks and responsibilities and the acquisition of resources that are adequate to tasks and that enable the organization to respond to changes in the conditions in which tasks are performed,
- Lead and motivate the entities involved in the performance of specific tasks,
- Monitor the process by acquiring information on the effectiveness of conducting the planned actions.

In order to be capable of performing processes, an organization needs to secure the appropriate resources, including conditions for process performance. A key role among such resources is played by the working environment [12].

In an employee-friendly organization, processes are performed under strictly specified conditions which secure the satisfaction of worker needs and expectations and ensure working comfort and worker well-being. Another prerequisite for effectiveness is to abandon the policy of limiting controls to external procedures. Every employee should care for a specified section of their process thereby contributing to increases in work efficiency and ensuring that the tasks at hand are performed professionally [12, 13].

3 Working environment factors as process performance determinants

3.1 Impact of working environment factors on the ability to conduct processes

The working environment parameters defined for a workstation may be used as indicators in evaluating process performance conditions and as criteria for verifying workers' ability to operate and perform work in a specific working environment [1, 13]. The working environment may be described in terms of the level of negative and positive impacts on the worker.

Any negative or positive impacts exerted by working environment factors are a function of the conformity of working environment characteristics, as measured by means of specific indicators, with the criteria prescribed to achieve welfare in the working environment and ensure the ability of human operators to work efficiently. Where no negative impacts take place, i.e. where no such factors exist as would potentially disrupt the proper operation of processes, the broadly-defined capacity to perform occupational tasks in a proper manner is greatly increased. Such a capacity produces the following key benefits at enterprise level [1, 7, 13-16]:

- A reduction in losses caused by occupational accidents and diseases,
- Increased levels of employee motivation, teamwork and morale,
- A more efficient spending of funds earmarked to improve occupational safety,
- Greater efficiency in the performance of work,
- The ability of fewer workers to complete work within a shorter time,
- Greater efficiency achieved with the use of specific work methods,

- Unscheduled workloads (costs) reduced by effectively planning to achieve work continuity,
- Prevention of work disruptions and successful retention of most valuable workers,
- Reduced insurance expenses,
- Improved enterprise image and reputation,
- Lower risk of civil liability lawsuits.

All of the above benefits contribute to boosting the enterprise's bottom line.

Such impacts are identified with reference to the indicators and parameters used in assessing elements of the working environment. Depending on their level, individual impacts may be viewed as deleterious (hazardous) or untoward [17]. The most critical of them are such working environment factors as temperature (characteristics of the thermal environment), workstation and workplace lighting, noise and vibrations which affect workers, toxicity resulting from the use of chemicals and their mixtures in the work process, non-toxic dust, electromagnetic fields and, e.g. laser radiation levels. Their effect may be assessed by the measuring methods that help identify the actual impact of a factor on employee health and safety and on the presentation of subjective assessments of risks by workers. In the course of such evaluations, one should bear in mind that working environment factors affect [17-19]:

- Process performance, described in terms of e.g. process efficiency and effectiveness,
- The alignment of process outcomes with stakeholder expectations.

Table 1 describes the deleterious and untoward factors found in the working environment that are accounted for in assessing the capacity to conduct processes in a manufacturing company. These include the parameters applied in the assessment of [17]:

a) The thermal environment, measured by:
   - The predicted mean vote (PMV) indicator which describes the thermal comfort of a person remaining in a moderate environment on a seven-point numerical scale (from -3.0 to +3.0),
   - The conformity of the wet bulb globe temperature (WBGT) [°C] with the admissible values corresponding to the ability to perform work in a hot environment,
   - The chilling temperature indicator t_{wc} [°C] used to describe cold environments,
b) Illumination, and specifically:
   - The mean level of illumination E_{ave} [lx] relative to the required operating illumination,
   - Illumination uniformity relative to the minimum allowable illumination uniformity,
   - Light color and luminance depending on light source, measured against the requirements of the EN 12464-1:2011 standard,
c) Noise loads, described in terms of:
   - The equivalent noise level L_{Aeq},
   - The maximum noise level L_{Amax},
   - The peak noise level L_{Cpeak}.

During the evaluation, reference is made to device measurements used to formulate an objective assessment as well as subjective assessments of working conditions by workers employed at specific workstations.

The impact of environmental factors on the ability to carry out work was assessed exclusively on the basis of the disrupting factors found at the assessed workstation at the time at which the analyzed processes were performed.

Table 1. Description of the acceptability of exposure to selected deleterious and untoward factors found in the working environment.

Deleterious (untoward) factor	Measurement indicator (device-based assessment)	Working conditions evaluation (subjective evaluation by employees)
Thermal working environment (temperature and other parameters contributing to temperature impact)	· Thermal comfort $-0.7 \leq PMV \leq 0.5$ · Acceptability of moderate environment $-2.0 \leq PMV \leq 2.0$ · Acceptability of hot environment $WBGT \leq WBGT_{(accept.)}$ · Acceptability of cold environment $t_{wc} > -24°C$	Desired state - Satisfactory thermal conditions (considering the employee's protective clothing and work organization) Undesired state - Overly hot working environment (body overheating), - Overly cold working environment (body overcooling)
Lighting	· Average luminosity relative to the required operating light $E_{ave} \geq E_{(operating)}$ · Uniformity of lighting $d \geq d_{min}$ · Color of light and other evaluation parameters (such as luminance distribution) determining the ability to perform work, compliant with the EN 12464-1:2011 standard	Desired state - Ability to observe work details in performing eyesight-intensive work Undesired state - Lighting conditions (luminosity, uniformity, luminance distribution) hindering the observations of work details in performing eyesight-intensive work, - Light distribution causing the blinding of workers
Noise	$L_{Aeq} < 85$ dB $L_{Amax} < 115$ dB $L_{Cpeak} < 135$ dB	Desired state - No negative impact of noise on ability to perform work, - Acoustic comfort ensured at workstation Undesired state - Harmful effects on employee hearing organs, - Work performance hampered by negative impact of noise on the ability to communicate, - Inability to focus on task

Source: author's work based on EN 12464-1:2011, EN ISO 7730:2005, EN 27243:1993, PN-N-01307:1994 and Polish laws.

3.2 Evaluation of the impact of working environment factors on work performance

The study was conducted in a provider of gas installation services. The processes during which exposures to the disrupting factors referred to in Table 1 occurred were:

- Gas line preparation, and

– Gas line joining.

The above processes lie at the core of the operations of the assessed enterprise. The effectiveness with which they are performed is pivotal to the enterprise's market standing.

Irregularities were identified based on the activities which the workers associated with excessive exposures to working environment factors. By accounting for working environment factors, it became possible to assess their impact on the ability to carry out work tasks and conduct processes. This led ultimately to identifying circumstances in which the stresses to which the workers were subjected became excessive, forcing them to operate in improper working conditions.

In defining the existing links, use was made of the relationship diagram, which falls into the category of new quality management tools [4, 20, 21]. The relationship diagram may be used to collect and process data on the conditions affecting the performance of tasks. The diagram relies on procedures suited to the accomplishment of specific goals, including process design, by finding areas that may jeopardize process performance effectiveness [4, 20].

The study was conducted by an in-house team appointed by the company's management, comprised of technical and executive department engineers and occupational health and safety services. An auxiliary role of overseeing the investigation was entrusted to an external expert. The relationship diagram was created by:

– Describing the issue at hand, which became the central node of the diagram,
– Identifying the causes of the issue, which became the remaining nodes of the diagram,
– Identifying mutual links among such causes,
– Identifying the key causes, i.e. the parts of the diagram to or from which the greatest number of links are connected.

The parts of the diagram to or from which the greatest number of links are connected were the starting point for assessing the need to take specific improvement measures [17, 22]. The analysis of the circumstances leading to risks and strenuousness is presented in Figures 2-4.

The circumstances that caused specific irregularities are described in Table 2.

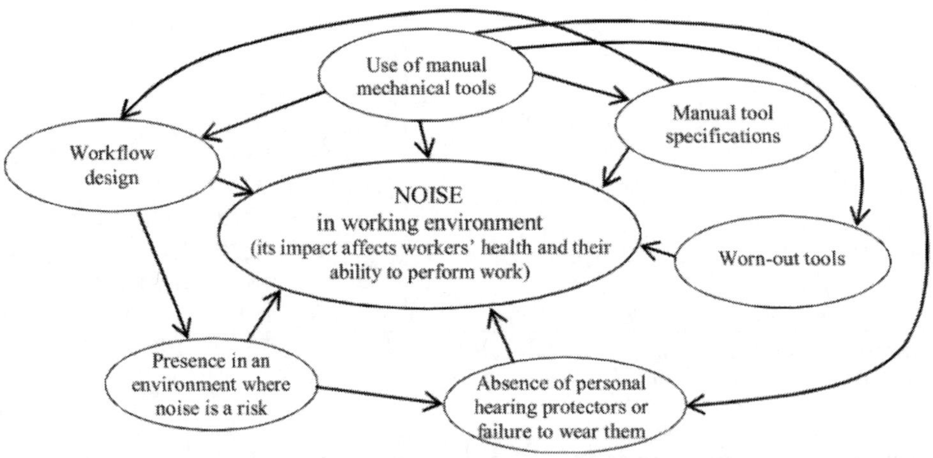

Fig. 2. Relationship diagram showing sources of noise exposures in working environment.

 DOI: 10.1051/matecconf/20179406011

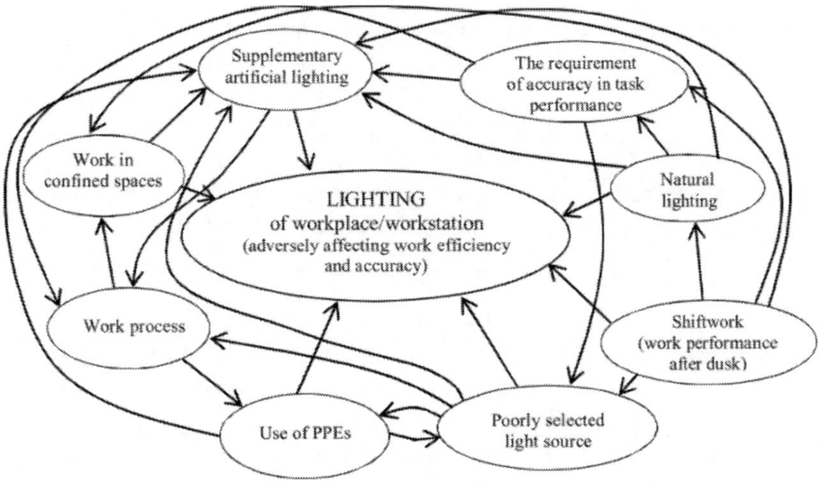

Fig. 3. Relationship diagram showing the risk of strenuous factors arising that are caused by failures to ensure efficiency and accuracy as a result of inadequate lighting in the workplace/workstation.

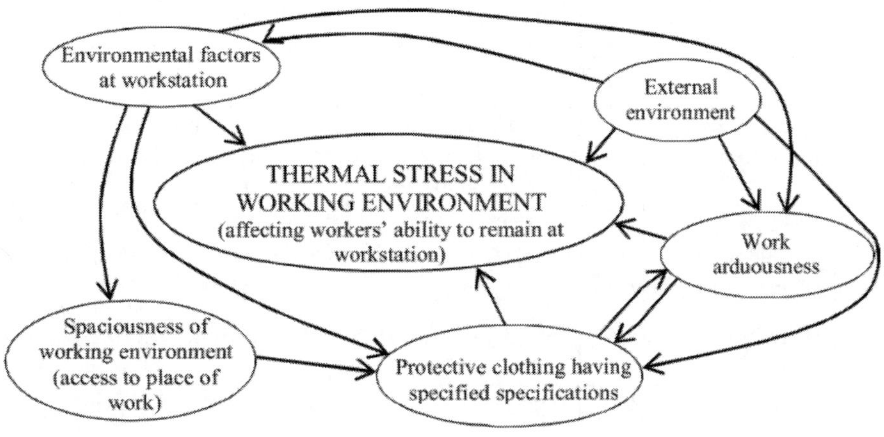

Fig. 4. Relationship diagram showing the causes of thermal discomfort in working environment.

Table 2. Description of irregularities and their causes.

Issue	Cause of issue	Description of issue
NOISE in the working environment (its impact affects workers' health and	Workflow design	Employee unable to move equipment away from his/her ear
	Use of manual mechanical tools	Due to work parameters and work methods applied at workstation, tools available to employee are a key significant noise source
	Manual tool specifications	Device parameters such as rotational torque and impact driver strength of drill or grinder) should be recognized as factors affecting noise generation
	Worn-out tools	Worn-out tools are an extra noise source in the

		performance of work
	Noise exposures	Since workers have no choice but to remain in the working environment (which is intrinsically noisy), they are exposed to noise regardless of the specific tasks they are performing at any given time
	Absence of personal hearing protectors or failure to wear them	Due to the specific nature of the work, personal hearing protectors may be the only practicable way to protect workers against noise; if workers do not wear them, this opportunity to protect them will be wasted
LIGHTING of workplace and workstation (affecting work efficiency and accuracy)	Natural lighting	The use of natural light is critical for ensuring proper workstation lighting
	Shiftwork	Work in conditions deprived of natural light (sunlight) causes significant stress, especially where supplementary artificial lighting cannot be used
	Supplementary artificial lighting	If poorly suited to working conditions and if no natural light is available, it results in inadequate illumination of the workstation
	Work in confined spaces	Makes it impossible to employ proper artificial lighting while making natural light insufficient; the issue should be recognized as particularly pressing alongside the use of proper PPEs
	Work process	Significantly affects workstation lighting, particularly if improper PPEs are worn
	Use of PPEs	
	Poorly selected light source	Results in inadequately lit work space and workstation, affecting in particular the accuracy of the performance of individual tasks
THERMAL STRESS IN WORKING ENVIRONMENT (affecting workers' ability to remain at workstation)	External environment	Impact of external environment and the inability to influence it results in significant thermal stress
	Environmental factors at workstation	Working conditions are a function of the external environment – they mirror the specific factors which result in work stress
	Work arduousness	The amount of metabolic energy tied to the arduousness of work is a critical criterion that defines thermal stress levels
	Protective clothing	Poorly selected protective clothing (i.e. one with inadequate specifications) may significantly affect thermal stress levels
	Spaciousness of working environment	The amount of space available at a workstation affects the ability of a worker to wear well-selected protective clothing

The analysis that was conducted as well as the use of the relationship diagram helped identify the factors that significantly contribute to process disruptions. To ensure that an organization's processes are conducted efficiently, it is vital to establish a thorough monitoring system and take measures that will lead to improvements. These include with respect to:

a) Workstation noise: the adoption of rules governing the use of manual mechanical tools with a view to reducing the negative impact of noise on the workers,

b) Workstation spaciousness and lighting, which affect work performance efficiency and accuracy: providing workers with a supplementary portable source of artificial light which they can use at any site during the performance of work,
c) Thermal stress experienced by workers in a working environment: using protective clothing with properly identified and selected specifications; such clothing should not hinder work performance in any adverse external conditions regardless of the degree of spatial confinement at the workstation.

As part of the detailed analysis aimed at identifying the measures that need to be taken, the organization should [4, 17-19, 21]:

- Take into account such circumstances (factors) as are likely to contribute to causing the targeted issue,
- Consider factors that might potentially hinder the organization's ability to adopt the selected solutions,
- Identify the measures that are necessary and practicable,
- Select priority measures which will make the biggest contribution to the achievement of desirable effects,
- Identify special measures associated with the innovative solutions in place,
- Define the conditions for the implementation of the project, including the selection of contractors, the planning of actions, the identification of adequate resources and the allocation of resources that are necessary of project completion.

Notably, however, none of the above matters fall within the scope of this paper.

4 Conclusions

The presumption made in the study in question was that working conditions significantly determine a man's ability to operate in a working environment and perform tasks. If such tasks are presented as a process, working environment factors need to be seen as prerequisites for work performance, whereas ensuring proper working conditions should be recognized as a disrupting factor that adversely affects process performance and the capacity to achieve the desired outcomes [18, 23].

By analyzing the issue in view of process performance described in reference to working environment factors, emphasis may be placed on the significance of working environment parameters for assessing the ability of human operators to work in specific conditions which determine their safety and their ability to perform work efficiently. Such factors are described in terms of the requirements laid down in relevant standards as well as subjective assessments by workers. Both are used in evaluating workers' ability to perform their tasks.

The relationship diagram has helped examine the issue and, on that basis, identify advisable improvement measures. The diagram has made it possible to correlate "causes" with "effects", and, as a consequence, identify links connecting individual causes. The study has found that:

- The central measure required to ensure proper thermal conditions in the working environment is to select functionally appropriate clothing,
- To ensure that workstation illumination is adequately suited to the nature of the work performed, workers should be equipped with a portable light source,
- To reduce the adverse impact of noise, it is crucial to train the workers and establish rules for the use of manual mechanical tools.

The above measures should be seen as the most critical ones for improving the current conditions. The analysis also makes it possible to identify other actions and assess their usefulness for the achievement of the desired improvements. To that end, reference should be made to the number of links found in the aforesaid relationship diagrams. The links show both the sources of problems and ways to reduce their impact.

The improvement measures adopted on that basis and aimed at eliminating or mitigating disruptions should be seen as a way to increase the capacity to carry out work processes.

By applying the above methodology to workstations exposed to other disruptions in the form of deleterious, untoward and hazardous factors having a potentially harmful effect on human health, it is possible to identify the measures that should be taken. Such measures will help improve work efficiency and effectiveness and, in effect, satisfy the expectations of the concerned enterprise's stakeholders, who include its employees. The improvement measures that allow the enterprise to enhance the conditions in which its processes are performed and reduce the disruptions to which its processes are exposed will ultimately help, inter alia, to:

- Improve work efficiency by e.g. reducing the time needed to complete tasks,
- Boost the effectiveness of task performance,
- Reduce loads having adverse impacts on workers,
- Reduce the biological cost resulting from the performance of work,
- Improve the quality of work,
- Reduce the time needed to perform tasks,
- Improve work productivity.

While no other factors having deleterious effects on the workers were observed during the study, the insights obtained suggest that the methodology described above can be successfully employed to assess the causes of other issues. Another consideration that may prove to be of significance is the possible interactions among individual deleterious factors that may produce a synergy effect.

References

1. A. Górny, The Elements of Work Environment in the Improvement Process of Quality Management System Structure, In: W. Karwowski, G. Salvendy (Eds.), *Advances in Human Factors, Ergonomics and Safety in Manufacturing and Service Industries*, 599 (AHFE 2011)
2. J. Brilman, *Modern management concepts and methods* (PWE, Warszawa 2002)
3. EN ISO 9004:2009, Managing for the sustained success of an organization. A quality management approach (ISO 9004:2009), European Committee for Standardization, Brussels
4. A. Hamrol, *Quality management with examples* (Wydawnictwo Naukowe PWN, Warszawa 2005)
5. A.W. Brown, Information & Management, **15(3)**, 125 (1998)
6. EN ISO 9001:2015, Quality management systems. Requirements (ISO 9001:2015), European Committee for Standardization, Brussels
7. A. Kawecka-Endler, B. Mrugalska, Human-Computer Interaction: Applications and Services, III, **8512**, 700 (2014)
8. A. Górny, Management Systems in Production Engineering, **3(14)**, 106 (2014)
9. E. Sgourou, P. Katsakiori, I. Papaioannou, S. Goutsos, E. Adamides, Procedia Eng., **45**, 185 (2012)
10. J. Veldman, G. Gaalman, Int. J. Prod. Econ., **149**, 202 (2014)
11. M. Kolosowski, P. Chwastyk, Procedia Engineering, **69**, 222 (2014)
12. S. Murray and M. S. Thimgan, *Human Fatigue Risk Management* (Elsevier, Amsterdam 2016)
13. R. Chaib, I. Verzea, I. Cozminca, M. Benidir, Safety Sci., **76**, 145 (2015)

14. A. Górny, Ergonomics aspects of CSR in system shaping the quality of work environment, In: P. Vink (Ed.), *Advances in social and organizational factors*, 541 (CRC Press, Boca Raton 2012)
15. A. Hedlund, M. Åteg, I.-M. Andersson, G. Rosén, J. Safety Res., **41(2)**, 145 (2010)
16. J. Ruiz, A. Coduras, J. of Business Research, **68(7)**, 1466 (2015)
17. A. Górny, *Occupational risk management* (Wydawnictwo Politechniki Poznańskiej, Poznań 2011)
18. BS OHSAS 18001, Occupational health and safety management systems. Specification (BSi, London 2007)
19. A. Górny, Modern Management Review, **22(4)**, 73 (2015)
20. A. Hamrol, W. Mantura, *Quality management. Theory and practice* (Wydawnictwo Naukowe PWN, Warszawa 2005)
21. B. Gajdzik, A. Wyciślik, *Quality, environment and safety in enterprise management* (Wydawnictwo Politechniki Śląskiej, Gliwice 2008)
22. K. Russ, Safety and Health at Work, **1(1)**, 11 (2010)
23. G. Costa, Safety and Health at Work, **1(2)**, 112 (2010)

MATEC Web of Conferences **94**, 06012 (2017) DOI: 10.1051/matecconf/20179406012
CoSME'16

Estimation of the resource buffers in the assembly process of a shearer machine in the CPPM method

Aleksander Gwiazda [1,*], *Iwona* Paprocka [1], and *Magdalena* Bączkowicz [1]

[1] Silesian University of Technology, Faculty of Mechanical Engineering, Konarskiego 18A Str., 44-100 Gliwice, Poland

Abstract. Dynamic development of scheduling systems allows significantly improving currently realized tasks. Critical Chain Project Management (CCPM) is one of the methods of project management basing on network planning. In this method is utilized the concept of a critical chain derived from the Theory of Constraints. This method allows avoiding losses considered project time and resources. It results in quicker project implementation (20-30%), and in reducing the risk level considered with tasks realization. The projects are cheaper, and the risk of cost overruns is significantly reduced. Factors that distinguish CCPM method from traditional network planning methods are: the balance of resources and the introduction of buffers. Moreover in the CCPM method key elements are: times of tasks that are reduced from traditional estimates to realistic ones. Activities associated with the task start as late as possible in accordance with the ALAP principle (As Late As Possible). This work presents the process of managing the assembly of a shearer machine taking into account the process of safety buffers utilization and the whole project optimization. It is presented the estimation of buffers capacity to obtain the improvement of project realization task.

1 Introduction

Critical Chain Project Management (CCPM) method is one of the network methods for planning, executing and managing projects in single and multi-project environments. It is distinguished from the other network methods. In particular, it differs in terms of: the scheduling procedure, the project constraints that are taken into account and the introduction of time stocks allowing avoid delays of tasks. Each project has some limitations, which are: costs, time and resources. So each project should be scheduled to achieve its objectives, taking into consideration these limitations. For this reason, it is important to appropriate manage the project already in its first phase, that means in the planning phase. Unfortunately, in many cases, the used project management system leads to the fact that the projects are not realized on time, with the ever increasing costs [1, 2]. This is why new approaches to project management are under consideration.

[*] Corresponding author : aleksander.gwiazda@polsl.pl

In order to avoid errors in the planning phase are being sought techniques that are appropriate to manage the project. Application of the proper project management methodology leads to increasing the likelihood of success and achieving significantly better results at a lower cost, less effort and in shorter time. The important factors that should be considered at the planning phase are: resources flexibility, unpredictability and uncertainty. CCPM is the method, which combines the most important features of good project management. It is a reliable technology for project management. One of the most important aspects of CCPM is the reasonable analysis of the project realization process. It is known that the time to complete a task is a statistically distributed. This means that the execution of the task may take a very short time, e.g. 3 days, or could last a very long time, e.g. 8 days. So time of a given task realization should be therefore appropriately chosen taking into account the factor of uncertainty in the given project management. The implementation of the tool such as the CCPM method allows shortening the project duration basing on the appropriate estimation of duration of all performed tasks. It is also possible to specify the spare time (safety buffers) [3 - 5].

2 Safety buffers in the CCPM method

CCPM is characterized by a different way of project scheduling. It is scheduled according to the ALAP (As Late As Possible) principle. In the other methods schedules are created according to the ASAP (As Soon As Possible) principle. In CCPM uncertainty and deviation from the schedule are natural factors affecting the duration of the project. The focus should be on managing the uncertainty and deviations that occur during the project and not to avoid them. The Critical Chain method application does not means the resignation from such activities as: project initiating, project objectives determining, determining the outputs of the project, determining the project stakeholders, defining project foredesign and limitations, determining the work division and finally the elaboration of the project diagram. The next stage is the elaboration of the schedule.

The schedule created according to the outlined guidelines will increase the likelihood that the project will be realized at the given time or even earlier [6]. It is elaborated in some previously defined steps [7]:

- elaboration of the net of dependences of planned tasks,
- estimation of aggressive times for particular project tasks,
- schedule elaboration,
- identification of the Critical Path,
- determining the Critical Chain (the Critical Path including the resources balancing),
- calculation of the size and placement of safety buffers,
- determining a new term of project realization.

A very important aspect of the project management resulting from the use of the CCPM method is its control. Safety buffers allow continuous monitoring of the entire project. This monitoring of safety buffers comes to continuous determining the level of their use. The main objective of the CCPM method is that the project buffer (main safety buffer) has not been used completely or has been used in full, but never more. If this objective is reached, it means that the project will be completed in the scheduled time (at the end of the project buffer or even earlier). In addition to control the project it is needed to determine the stage of its completion. For each project, the time values defining the progress of the project and the use of buffers are selected individually. They are dependent on the limitations and assumptions of the project. They should not be changed, however, during its lifetime. Buffer management ensures the effective risk management in the project [8, 9].

According to Goldratt it could be distinguished three basic types of safety buffers [10]. This safety buffers are designated to protect the realistic project deadline: To these buffers include:

- project buffer: a unique and single buffer to protect the project deadline,
- feeding buffer: multiple buffers to protect parts of the critical chain,
- resource buffer: multiple artificial buffers that act as warning signals to assure the availability of resources.

The buffer management according to the CCPM method is determined as such utilization of the project buffer and feeding one to increase the safety time of the project baseline schedule. It guarantees, with a high probability, that the project will be timely completed. The size of buffers is determined according to the properties of the path or chain feeding by these buffers. It could be the length of the path, its total variance, its average resource use or the number of tasks it contains. In this paper is given an example of safety buffers utilization for the technical project management. This project relates to special, piece production system of a mining shearer loader. It is a coal cutting machine which is used in the hard coal industry (Figure 1).

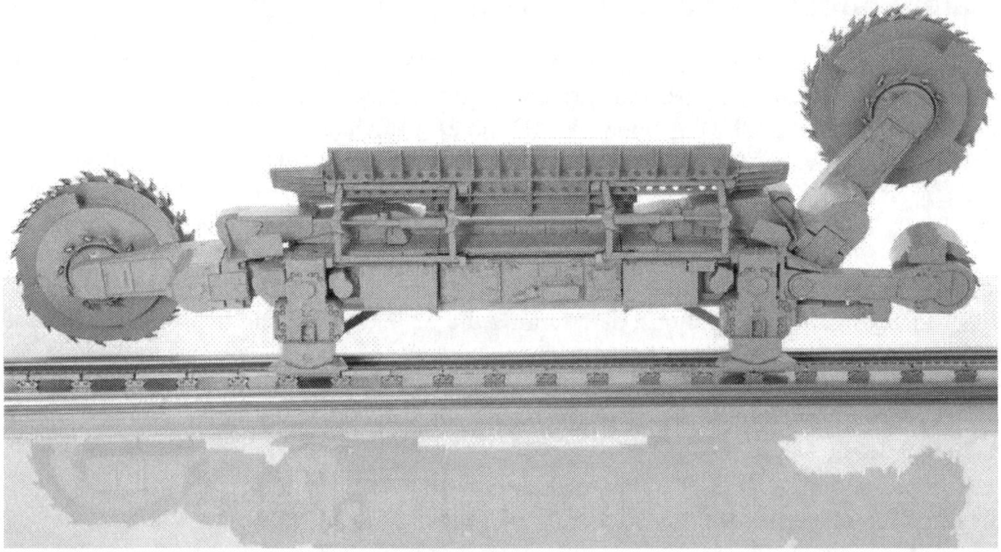

Fig. 1. Example of a mining shearer loader [11].

Shearer loader is a part of a mining complex [12, 13]. Such complex is composed of three main components: a shearer loader, an armored conveyor (being the trace of the shearer – white color in the figure), and a line of mining powered roof supports (nonvisible in the figure). The cost of a mining complex prepared for specific mining conditions is equal to some millions euro. It decides that investigation for more flexible and less costs consuming method of productions and it organization is required.

The unique designing and producing system of each shearer loader, basic of course on a number of universal modules) was the main cause deciding in selecting the CCPM method.

3 Shearer loader CCPM

The network of dependences of the analyzed production project is presented in Table 1. It consists of 17 tasks that are ordered taking into account the resources conflict (lack of welders) that mainly influence the time of this project realization.

Table 1. Network of dependences of realized tasks including the resources conflict.

Task No	Task code	Task type	Task name (in parenthesis number of subassemblies)	Preceding task No	Task duration [days]	Worker symbol
1.	**A**	**MANUFACTURING OF**	the ranging arm	J	15	a
2.	**B**		the loader	-	35	b
3.	**C**		the cutting system	-	35	g
4.	**D**		the cutting head	-	20	h
5.	**E**		the transmission of the feed drive (2)	L	45	e
6.	**F**		the drive system	-	45	c
7.	**G**		the electric drive (2)	F	75	b
8.	**H**		the transforming-converting unit BTP-1	-	70	f
9.	**I**		the block of electric devices BAE-18 (2)	-	15	g
10.	**J**	**ASSEMBLY OF**	the cutting head and the cutting system	C, D	30	a
11.	**K**		the ranging arm with the loader and subassembly obtained in the task **10**	A, B	30	h
12.	**L**		the drive system with the electric drive (2)	F, G	20	e
13.	**Ł**		the transmission of the feed drive and the subassembly obtained in the task **12**	E	35	d
14.	**M**		the subassembly obtained in the task **13** and the transforming-converting unit BTP-1	Ł, H	25	c
15.	**N**		the subassembly obtained in the task **13** and the block of electric devices BAE-18	M, I	15	c
16.	**O**		**14 + 15**	N	15	d
17.	**P**		**11 + 16**	K, O	10	e

The network of dependencies between particular tasks was used to determine the critical path of the realized project. On the basis of the determined critical path and resources conflicts between some tasks it was determined the critical chain. The resources being in conflict are critical for this project because they are common (more precisely welders that work in the conflicted tasks). Using the elaborated schedule including the relationships

 DOI: 10.1051/matecconf/20179406012

between the tasks themselves and between tasks and feeding them resources as well as taking into account their duration it was also determined the duration of the entire project, which is 230 days.

Figure 2 shows the introduction of three resource buffers (BZi) that has been estimated on the basis of historical data using statistical analysis. They are marked in orange. In this figure tasks that are non-critical are market in blue and the ones that are critical are marked in red. According to this scheme the balanced resources are marked in green. The resource buffers were introduced in places where resources conflicts were previously detected. These buffers are located between the following tasks: L - E, M - N, J - A. The unit on the diagram below is equal to 1 week.

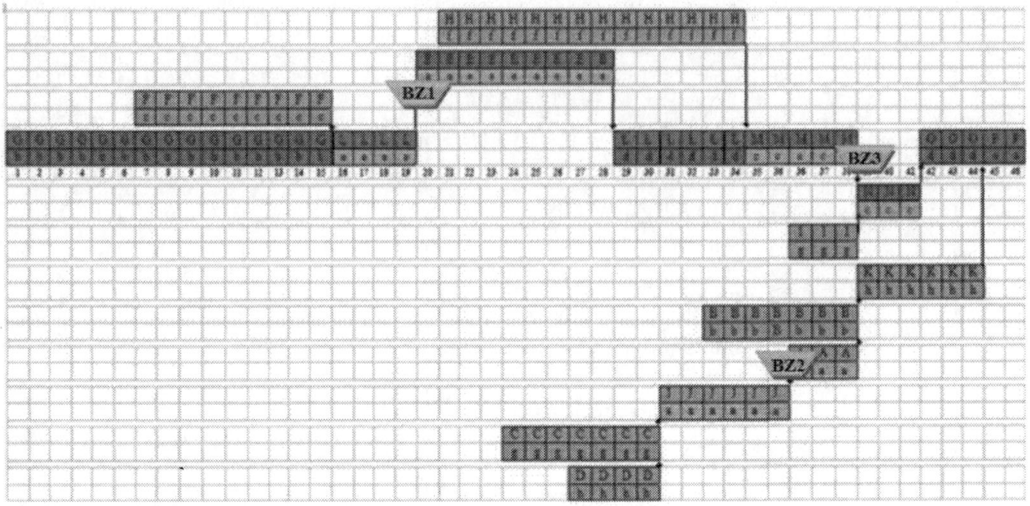

Fig. 2. Critical chain with resources buffers.

To determine the size of resource buffers (welders' availability) it was used the statistical analysis of historical data. These data concern the maximum use of resources which employees are assigned respectively to the task. In Table 2 are collected historical data on the time availability of the employee/welder "e" for the seven previously realized planning periods in the analyzed company. This table also shows the number of observations L_i and B_1, ... B_{Li}, which means employee uptime during the planning period. As the unit of time are taken days. For the planning period, in the analyzed company, it was taken last 3 quarters.

Table 2. Availability of the welder "e" ($B_1 \ldots B_{Li}$) in seven preceding projects.

i	L_i	B_1	B_2	B_3	B_4	B_5	B_6	B_7	B_8	B_9	B_{10}
1	10	43	40	41	41	39	36	31	45	30	39
2	10	32	42	37	35	27	21	40	32	36	20
3	10	38	43	36	21	38	26	20	25	38	40
4	10	37	41	35	28	22	44	37	33	19	37
5	10	28	41	40	42	39	34	24	45	38	43
6	10	30	42	43	23	42	41	22	25	29	27
7	10	35	40	25	44	27	23	26	29	25	28

To conduct the statistical analysis of the welder "e" availability it was used the program Statistica. It was realized the analysis of data matching to particular, basic statistical distributions using the Kolmogorov-Smirnov test. It proved the correlation with the normal distribution. Next there was elaborated histogram of analyzed data (ordering each of B_i

columns). Exemplar histograms, obtained in the Statistica program, are presented in Figure 3.

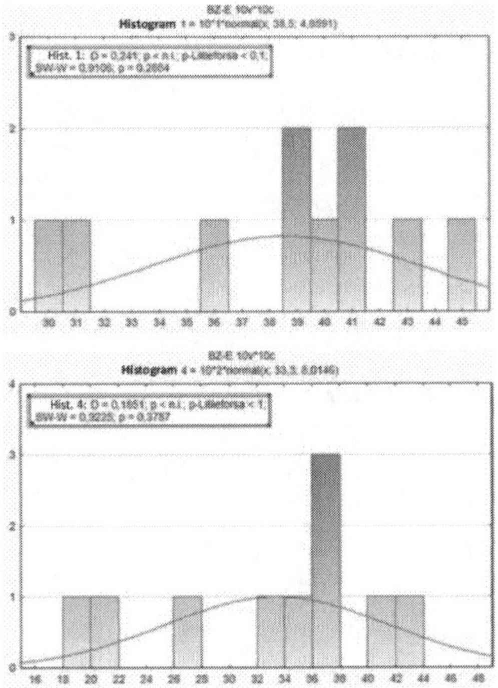

Fig. 3. Exemplar histograms chosen for analyzed variables.

The realized analysis allow elaborating, using the Statistica program, the prognosis of the welder "e" availability in the next prognosis period, taking into account the normal distribution of analyzed data. After analysis of trends and distribution parameters one obtained results presented in Table 3.

Table 3. Forecasted value for the next planning period.

Curve equation	**Results**
$\mu = 0{,}018\ x^2 - 0{,}855\ x + 37{,}02$	**31,323**
$\sigma = -0{,}209\ x^2 + 1{,}907\ x + 3{,}98$	5,86

This means that the resource "e" (welder) will be available to perform the tasks L and E on average for 32 days. To finally determine this resource buffer it must be also considered the downtime of the employee "e". This analysis was conducted analogously basic on the historical observation the previous planning periods. The result shows that the resource "e", in the next planning period will not be available to perform the tasks L and E on average for 10 days.

Estimation of this resource buffer was based on the historical data gathered during previous seven planning periods. The forecast showed that in the next planning period the availability of the analyzed resource, which the employee marked with an "e", will be 32 days. On the other hand, his unavailability is 10 days. Hence it could be concluded that 10 days is the time that is needed as a spare time to perform the tasks L and E. Assuming that the welder "e" will be available on the first day of this task realization, and the time of availability of the resource "e" is equal to the time of project tasks realization, one could estimate the size of this resource buffer. Therefore the duration of task L being realized by the resource "e" is 10 days. This means that the remaining time which may have the employee "e" is 22 days (32 - 10 = 22 days). It is the time that could be attributed to the task E, which also is realized by the employee "e". In summary, the duration of the task L is estimated to be 10 days. After this task the BZ1buffer was introduced, sized to 10 days. Next 22 days is the estimated time to perform the task E. The location of the BZ1 buffer (in violet) is shown in Figure 4. It was elaborated in the Gantt Project program.

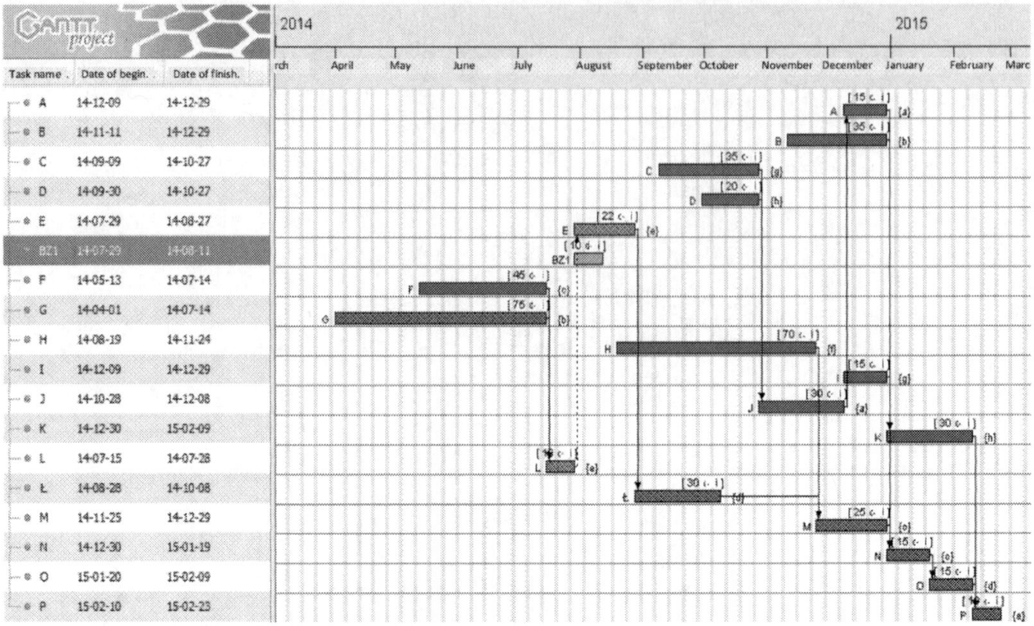

Fig. 4. Location of the BZ1 buffer (d. – days).

The other resource buffers (BZ2 and BZ3) were determined analogously. The next type of safety buffers that was introduced in this project was the feeding buffer BZA (Figure 5).

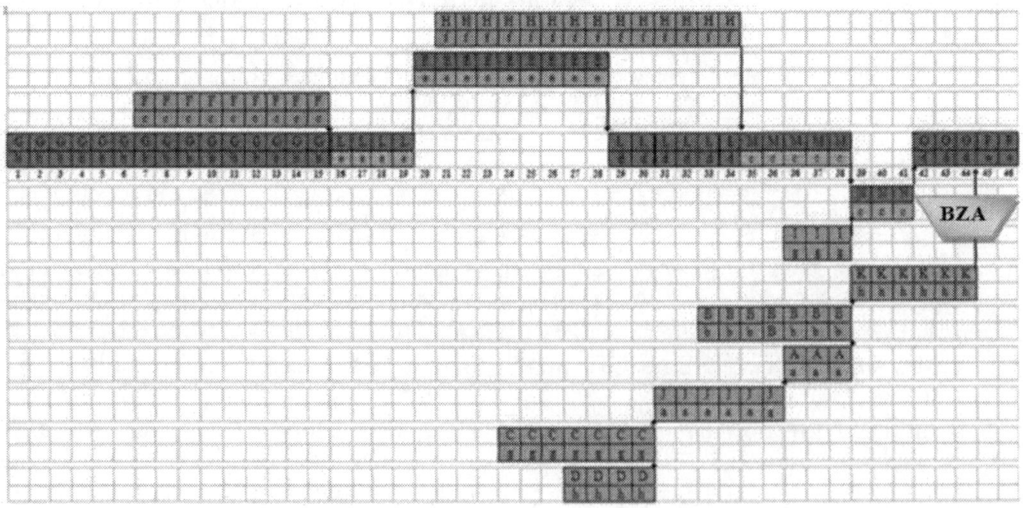

Fig. 5. Location of the BZA buffer.

The main task of the feeding buffer is to protect the critical chain. For the analyzed case it was determined the location of the feeding buffer BZA at the end of non-critical tasks. Its size was determined basing on safe and aggressive estimates. The protected is the next string of project tasks: K - B - A - J - C - D. After statistical analysis it was elaborated the

Gannt diagram including the BZA buffer (9 days). It is presented in green in the diagram below (Figure 6).

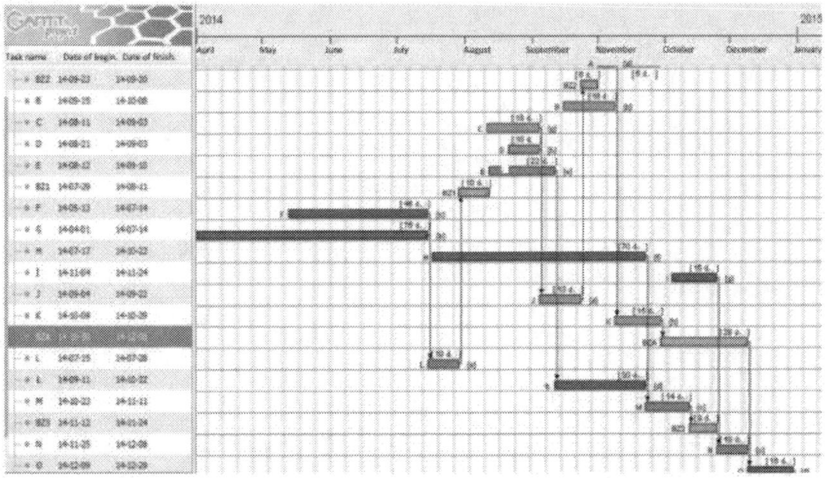

Fig. 6. Gantt diagram with the BZA buffer (d. – days).

The conducted works according estimating the duration of analyzed tasks and entering the BZA buffer allowed shortening the duration of the analyzed, non-critical tasks: K - B - A - J - C - D. Duration of the tasks K, B, C and D was shortened by 51 days. The size of the BZA buffer is 11 days. Finally it was introduced the project buffer BP, what is presented in Figure 7.

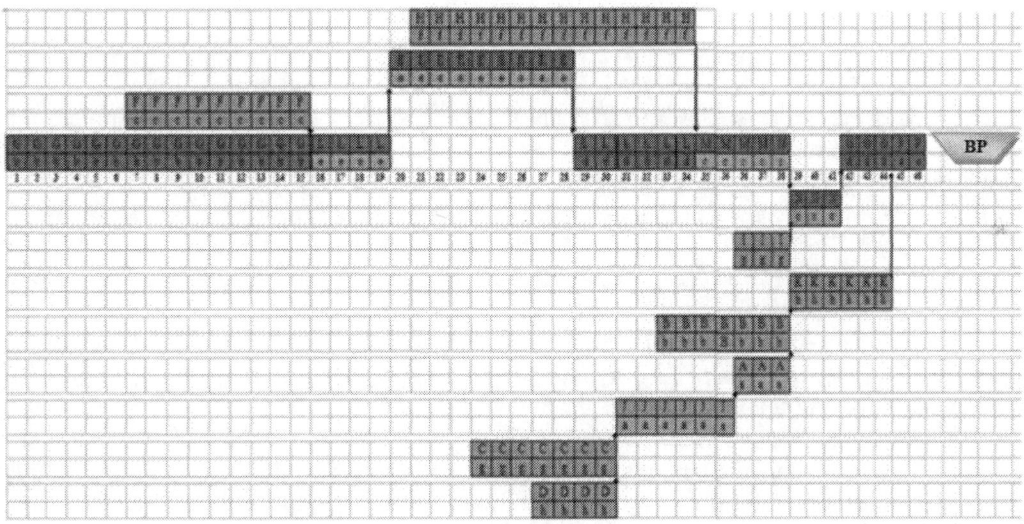

Fig. 7. Location of the PB buffer.

The project buffer is designed to protect the entire project against any delays that could occur during critical tasks realization. The buffer size was evaluated by two methods selected to compare the results. The first method is the half of the chain. The second method, used to calculate the buffer size of the project, was the Root Squared Error method. It was determined that the size of the project buffer PB should be equal to 29 days. The Gantt diagram, including all buffers analyzed in the project, is presented in Figure 8. The BZAi buffers represent feeding buffers designated for individual tasks.

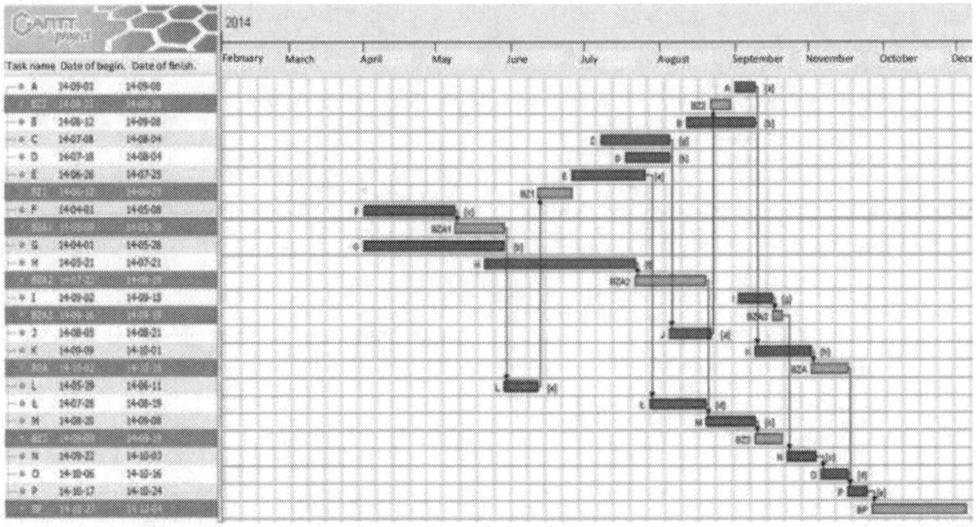

Fig. 8. Gantt diagram with the PB buffer.

The introduced changes resulted in the fact that the project completion date has to be changed. The changings are discussed in conclusion. It must be stated that the size of the BR buffer, estimated by the method of the half of the chain was 29 days and the Root Squared Error Method was 32 days. But taking into account the accuracy of calculation it was chosen the term of 29 days.

4 Conclusions

In the presented paper it was analyzed the problem of a production process scheduling. This process could be elaborated using different approaches [15 - 17]. It must be stressed that this process is strongly related with the ordered material and adopted welding condition and parameters [18, 19].

Due to conducted analysis it was elaborated the realistic production schedule for the analyzed shearer loader. Three resource buffers BZ1 (for tasks L, E), BZ2 (for tasks J, A) and BZ3 (for tasks M, N) were introduced. The buffer sizes and durations of tasks were estimated on the basis of statistical analysis. Historical data from surveys of employees allowed verifying the availability and unavailability of staff for the modified tasks. Statistical analysis concerning forecasting the parameters for describing the distribution of time availability and unavailability of these resources for the next planning period showed that it is possible to reduce the total time of analyzed task by 75 days. At the same time the sum of the resource buffers amounted to 25 days. The introduction of these buffers caused the change in the deadline for project completion. The calculation of the duration of non-critical tasks and the size of the feeding buffer BZA for this sequence of task was based on the parameters estimation of the normal distribution that governs these tasks realization. The modified time of these tasks was shortened by 51 days and the entered buffer size was 11 days. The introduction of this buffer did not result in changes in the date of the project completion. Additionally three feeding buffers for single tasks (BZA1, BZA2 and BZA3) were introduced. The tasks, to which these buffers have been allocated, were analyzed whether they are governed by the normal distribution. Basing on the calculation of parameters for the normal distribution (average value and the standard deviation) it was set the sizes of particular buffers. The introduction of these buffers resulted in a total reduction of tasks duration by 48 days. These changes did not affect the project completion. Finally it was introduced the project Buffer BP. It was assessed by two different methods: of half of a chain and method of Root Squared Error. As the result of calculations it was obtained two different project buffer sizes, respectively 29 days and 36 days. Due to the fact that in the Root Squared Error method the results were obtained with a relatively low probability it was decided to select the result obtained with the first method. The introduction of the project buffer and the task with newly determined duration times, in the critical chain, caused the change of the project completion date. The project was shortened by 56 days.

Generally the result of network planning using the CCPM method for the project of mining shearer loader was shortening the term of the project completion. If the project will be performed without the use of buffers, it is estimated that it will be completed in 123 day. However, in the case of utilization of all buffers, it is estimated that the project will end no later than in the 172 day. The analysis of previous projects and presented results showed a significant reduction of the project cost as well as the project duration. It is mainly related with the fact of eliminating the critical situations related with resource conflicts and more realistic predictions of the project realization terms. The proposed method is the most appropriate for projects with more restricted availability of resources. In the case of the analyzed project it gave a reduction of the project cost by 8%. The project duration, in comparison to the longest previous project, was about 14%.

References

1. L, P. Leach, *Critical Chain Project Management* (Artech House Publishers, London, 2004)
2. D. Updegrove, *The Critical Chain Implementation Handbook* (CreateSpace Independent Publishing Platform, North Charleston, 2014)
3. Ch. Schwindt, J. Zimmermann (eds.), *Handbook on Project Management and Scheduling* (Springer Verlag, Heidelberg, 2015)
4. G. K. Rand, Int. J. of Proj. Man. **18/3.** 173 (2000)
5. Zh. Zheng, Z. Guo, Y. Zhu, X. Zhang, J. of Neurocomp. **143,** 282 (2014)
6. Z. A. Banaszak, M. B. Zaremba, J. of Intell. Man. **17/6**, 641 (2006)
7. A. Hutanu, G. Prostean, A. Badea, Procedia – Soc. and Behav. Sc. **197,** 1416 (2015)
8. F, Marle,L.-A. Vidal, *Managing Complex, High Risk Projects: A Guide to Basic and Advanced Project Management* (Springer Verlag, Heidelberg, 2016)
9. O. I. Tukel, W.O. Rom, S. D. Eksioglu, Eur. J. of Op. Res. **172/2**, 401-416 (2006)
10. E. M. Goldratt, *Critical Chain* (The North River Press, Great Barrington, 2002)
11. http://prototypen.de/objetpolyjet-stl-140.html
12. A. Gwiazda, W. Banaś, A. Sękala, K. Foit, P. Hryniewicz, G. Kost, IOP Conf. Ser.: Mater. Sci. Eng. **95,** doi:10.1088/1757-899X/95/1/012104 (2015)
13. A. Gwiazda, App. Mech. and Mat. **474,** 147 (2014)
14. http://www.cat.com/en_ZA/products/new/equipment/underground-longwall/shearers/18346615.html
15. G. Ćwikła, D. Krenczyk, A. Kampa, G. Gołda, IOP Conf. Ser.: Mater. Sci. Eng. **95,** (2015)
16. I. Paprocka, W. Kempa, K. Kalinowski, C. Grabowik, IOP Conf. Ser.: Mater. Sci. Eng. **95,** (2015)
17. K. Foit, A. Gwiazda, W. Banaś, A. Sękala, P. Hryniewicz, IOP Conf. Ser. Mater. Sci. Eng. 95, (2015)
18. S. Topolska, J. Łabanowski, Mat in Tech. **49/4**, 481 (2015)
19. J. Łabanowski, A. Świerczyńska, S. Topolska, Pol. Mar. Res. **21/4**, 108 (2014)

MATEC Web of Conferences **94**, 06013 (2017) DOI: 10.1051/matecconf/20179406013
CoSME'16

Towards an information extraction and knowledge formation framework based on Shannon entropy

Dragoș Iliescu[1,*], and *Marian* Gheorghe[1]

[1]University Politehnica of Bucharest, Faculty of Engineering and Management of Technological Systems, 313 Splaiul Independentei, 060042 Bucharest, Romania

Abstract. Information quantity subject is approached in this paperwork, considering the specific domain of nonconforming product management as information source. This work represents a case study. Raw data were gathered from a heavy industrial works company, information extraction and knowledge formation being considered herein. Involved method for information quantity estimation is based on Shannon entropy formula. Information and entropy spectrum are decomposed and analysed for extraction of specific information and knowledge-that formation. The result of the entropy analysis point out the information needed to be acquired by the involved organisation, this being presented as a specific knowledge type.

1 Introduction

Information acquisition and knowledge formation represents a complex domain for study and research. A specific system is considered to control the information for extraction, verification and storage. This system refers to information management domain.

Related knowledge formation is then expected, based on managed information. Knowledge formation depends with the quality of managed information as only true justified information can be transformed into knowledge [1 - 3].

The 2015 version of the ISO 9001 standard express in clause 7.1.6 a specific requirement regarding the identification and management of organisational knowledge [4]: "The organization shall determine the knowledge necessary for the operation of its processes and to achieve conformity of products and services. The knowledge shall be maintained and made available to the extent necessary."

The above clause raises several challenges for organisations: how to identify needed or missing knowledge, how to segregate critical important knowledge and how to spread them across organisation. What information source to consider and which particular technology need to be implemented to control the transformation of information to knowledge [5].

* Corresponding author: dragosiliescu2005@gmail.com

Information sources for above objectives are diverse. The international standards ISO 9001:2015 and ISO 9004:2010 recommend that own organisation experience to be considered for above objective, along with certain internal and external sources [4, 6].

Description of a framework to evaluate the quantity of extracted information and how the collected information can be evaluated for knowledge formation represents the main topic of present work.

A case study will be described to exemplify the exposed concept. Information describing the nonconformities occurred in the period of 2014 and 2015 were gathered from a heavy industrial equipment manufacturing company.

Paper is structured as follows: section two defines the categories of identified nonconformities, section three operate the gathered data for information quantity based on Shannon formula and offer an interpretation of obtained results, section four presents the conclusion of the work as well as further possible development.

2 Nonconformities as information source

The records of identified nonconformities during the manufacturing cycle represent the data source of present work. Deviation is herein considered as any individual lack of compliance against a specified requirement. Nonconformity can be determined by one or a group of deviations [7] recorded in a specified moment of time within the manufacturing cycle.

The 2015 edition of ISO 9001 requests in clause 10.2.1 organisations to perform a series of specific activities to control and to correct the identified nonconformities. In clause 10.2.2 it is requested a record log to be maintained by organisation, describing the identified nonconformities by category and nature. Results of applied corrections [4] should also to be recorded. Data gathered for the study herein were extracted from this type of data log within the period of 2014 and 2015.

A total amount of 23 deviation categories, noted $x_1 \ldots x_{23}$ (Table 1), have been identified, 278 occurrences for 2014 and 346 in 2015 were recorded.

Table 1. Deviation categories.

Deviation category designation			
x_1	Shape and dimensions in finished condition state	x_{13}	Deviations found by magnetic particle inspection
x_2	Material cleanliness	x_{14}	Wrong or skipped process
x_3	Shape and dimensions in rough condition state	x_{15}	Process parameters
x_4	Product hardness	x_{16}	Defects in weld material
x_5	Deviations found by ultrasonic inspection	x_{17}	Other material characteristics
x_6	Surface condition	x_{18}	Product chemical composition
x_7	Cracks	x_{19}	Deviations found by dye-penetrant inspection
x_8	Deviations in mechanical test results	x_{20}	Deviations found by penetrant radiation inspection
x_9	Deviations found by visual inspection	x_{21}	Shape and dimensions for heat treatment
x_{10}	Shape and dimensions in rough-machined condition state	x_{22}	Microstructure
x_{11}	Chemical composition on liquid steel	x_{23}	Process documentation readability
x_{12}	Weight		

3 Estimation of information quantity

Shannon formula for entropy is employed to measure the information quantity [8, 9]: entropy of a variable X (noted as $H(X)$), with discrete values of $\{x_i \mid i=1...n\}$ and corresponding probability distribution of $\{p_i \mid i=1...n, p_i \geq 0\}$ is shown by eq. 1.

$$H(X) = -\sum_i p_i \log_2(p_i) \quad (1)$$

Applied for current case, variable X represents the likelihood of nonconformity to occur. The probability p_i (related with x_i deviation category) is approximated by eq. 2 [8, 9], where ω_i represents the occurrence of deviation category x_i and ω represents total occurrence counted for variable X.

$$p_i = \frac{\omega_i}{\omega}; \sum_i p_i = 1; \; i = 1 \ldots n \quad (2)$$

For the case of equiprobability ($p_1 = p_2 = p_3 = \ldots = p_n$), $H(X)$ gets the maximum value [8, 9] based on formula $H(X)_{Max} = log_2(n)$, where n represents the number of X's discrete values. The meaning of maximum entropy principle is related here with the information quantity to acquire: depending upon the number of discrete values recorded for variable X, a specified number of bits are needed to code that particular information [8 - 10]. It is considered herein that each x_i term can take only two values: yes or no, so that the entropy to be measured in bits. For the current case study, 5 bits (upper of 4.52 in $log_2(23) = 4.52$) codes the embedded information.

3.1 2014 versus 2015 entropy comparison

Gathered data were analysed, based on the eq. 2, for probability estimation. Results for each x_i term are depicted in Table 2 and Figure 1. The term of $p_i \log_2(p_i)$ reflects the effect on entropy of each individual term x_i.

Table 2. Calculation for occurrence probability as 2014 and 2015.

	2014		2015			2014		2015	
	ω_i	$p_i\log_2(p_i)$	ω_i	$p_i\log_2(p_i)$		$\omega_{i\,2014}$	$p_i\log_2(p_i)$	ω_i	$p_i\log_2(p_i)$
x_1	67	0.495	86	0.499	x_{13}	6	0.119	2	0.043
x_2	31	0.353	13	0.178	x_{14}	5	0.104	17	0.214
x_3	25	0.313	44	0.378	x_{15}	4	0.088	1	0.024
x_4	20	0.273	61	0.441	x_{16}	2	0.051	3	0.059
x_5	18	0.256	15	0.196	x_{17}	1	0.029	-	-
x_6	18	0.256	6	0.101	x_{18}	1	0.029	-	-
x_7	16	0.237	12	0.168	x_{19}	1	0.029	2	0.043
x_8	16	0.237	26	0.281	x_{20}	1	0.029	1	0.024
x_9	16	0.237	20	0.238	x_{21}	-	-	1	0.024
x_{10}	14	0.217	28	0.294	x_{22}	-	-	1	0.024
x_{11}	9	0.160	6	0.101	x_{23}	-	-	1	0.024
x_{12}	7	0.134	-	-					

 DOI: 10.1051/matecconf/20179406013

The 2014 data are sorted in descendent probability order base (Figure 1). The 2015 results preserved the general trend developed by 2014 records, but some particular variations are noted among the terms and this variation will be further analysed. The change in entropy for the variable X is constraint by the relation of $H(X) \geq 0$, but the change in probability distribution for each of x_i terms, 2015 versus 2014, is unconstrained.

Estimation of entropy for 2014 (eq. 1 based) shows a value of 3.647, versus the entropy of 2015 which shows a value of 3.357. In both cases, a number of 4 bits are needed to code the embedded information. As consequence, even there is a difference between the two values, no significance has been associate to the entropy change.

Variation in probability distribution is shown in Figure 2. The results offer an image over the entropy spectrum, "as before" compared with "as after".

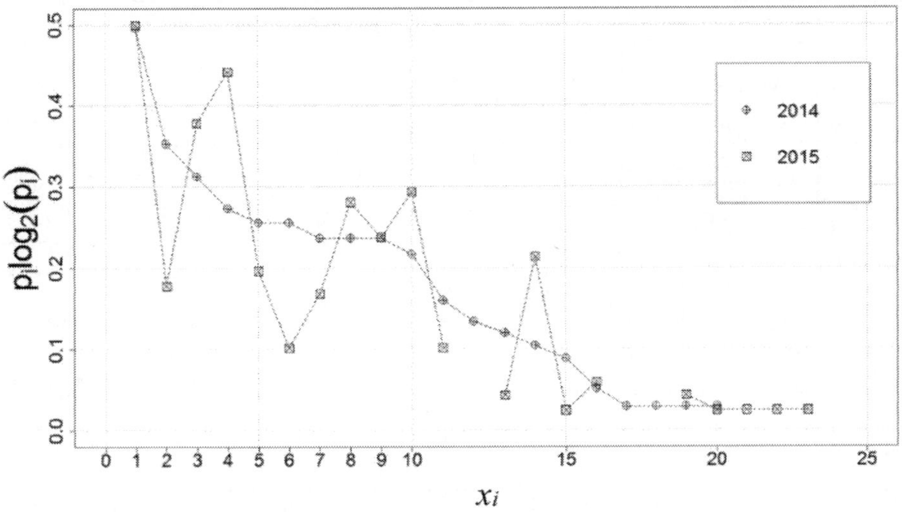

Fig. 1. Individual effect on entropy for each x_i term.

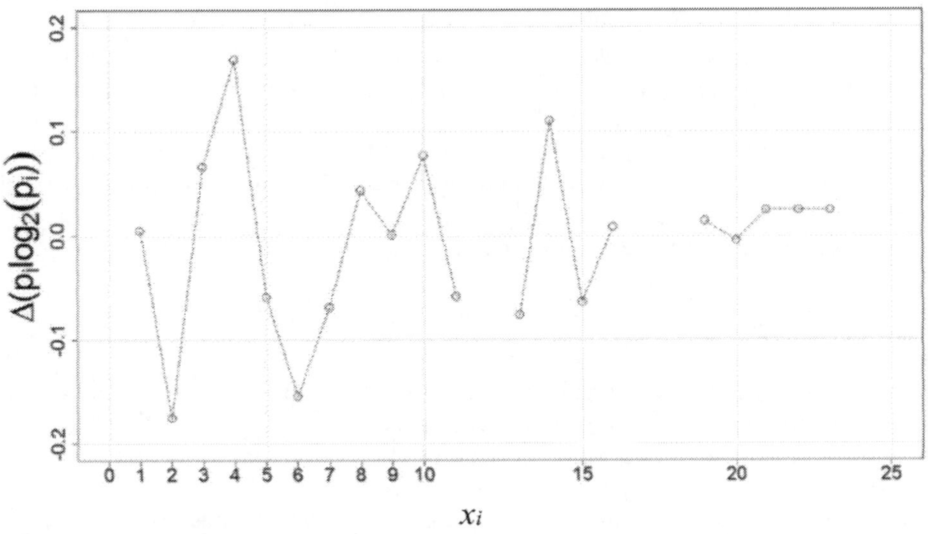

Fig. 2. Difference 2015 vs. 2014 in effect on entropy.

3.2 Information extraction and knowledge formation

Data gathered to estimate the information quantity in regard with the domain of nonconforming product management show a non-significant difference between the entropies estimated for 2014 ($H(X_{2014}) = 3.647$), versus 2015 ($H(X_{2015}) = 3.357$).

Some of the x_i terms show a significant effect on the entropy variation. Positive values in Figure 2 reflect the terms impacting the information quantity. Terms 4, 14 and 10 (Figure 2) are candidates for the terms which significantly determine a positive variation in entropy. This case is referenced as Case 1 in Table 3. Other terms show influence near zero, or there are lack of records (Case 2), and several other terms show a negative influence (Case 3).

Table 3. Extracted information and acquired knowledge.

Case	Extracted information	Knowledge definition description
1	These terms can be treated for informativeness significance:	Terms were segregated and sorted as informativeness significance order base. The three most informativeness were selected.
a	Product hardness (x_4).	
b	Wrong or skipped process (x_{14}).	
c	Shape and dimensions as rough-machined condition state (x_{10}).	
2	These terms (x_1, x_9, x_{12}, x_{16}, x_{17}, x_{18}, x_{20}...x_{23}) should be potentially considered for control improvement by:	Terms with no induced entropy change were selected to represent those domains were, potentially, an improved process control should be expected as necessary.
a	Process parameters control.	
b	Output measuring technique, capability, etc.	
c	Clear (noise reduction) communication channel	
3	These terms (x_2, x_5...x_7, x_{11}, x_{13}, and x_{15}) should not be considered for informativeness purposes.	Terms selected for being part in indistinguishable group should be confirmed for past effective corrective actions.

For those cases where the induced variation on entropy exists, either positive or negative, the meaning can be related with a successful transmission from the observed process to the observer. Only the terms inducing the positive entropy change will be counted as informativeness. These cases will transfer the information to the observer.

Case of the terms having an insignificant effect, for example terms 1, 9 and 16 in Figure 2, can be interpreted as being not for interest in the described context. However, they still should be counted as domains potentially requesting process control capability improvement, either in terms of process parameters control or in terms of communication by decreasing the noise over the communication channel.

4 Conclusions

Domain of nonconforming product management has been considered as information source, the main topic in present research was oriented on estimation the information quantity, how information extraction can be possible and how information to be transformed into knowledge.

The entropy concept was used to evaluate the occurred changes into the current manufacturing states. Comparison of collected data was main engine in measuring the change in entropy spectrum, and finally, getting the main factors to be considered for contributions in information quantity.

 DOI: 10.1051/matecconf/20179406013

Regarding the cases where improvements in process control capability are needed, the term of shape and dimensions in finished condition state was identified. This identification represents the second information type revealed by the analysis herein.

The third type is represented by the identification of that group of terms that should not be counted for informativeness. The last identified information showed a particular aspect. If the process under observation has recorded an improvement action in the past, then the results can be interpreted as a confirmation of the improvement effectiveness, this information reclaiming further investigations. All above acquired knowledge are part of the knowledge-that domain.

Returned benefits, as coming out from the three cases depicted above, refer to the selection of appropriate knowledge to be acquired, based on minimum collection of data. Overall, gathered data are discrete, and not continuous, cost of collecting being minimal.

Several limitations can be described into the present work. The entropy formula considers a noiseless communication channel. Other way said, there are no expected modifications in the data content, or meaning, during transmission within the information system. This limitation can eventually affect all the cases depicted in the paperwork, mostly Cases 2 or 3, as the change in entropy seems to be minimal as a result of a blocked communication channel, thus not reflecting actual state of the process. Another implication refers to the state of correlation that potentially exists among the considered terms, but the used entropy formula presumed the existence of independent discrete terms.

References

1. M. Carrara, D. Chiffi, Recent Trends in Philosophical Logic **41**, 31 (2014)
2. D. Davidson, *Truth and Interpretation. Perspectives on the Philosophy of Donald Davidson* (Blackwell, Cambridge, 1986)
3. E. Gettier, Analysis **23**, 121-3 (1986) http://fitelson.org/proseminar/gettier.pdf (accessed on 27.01.2016)
4. ***, ISO 9001:2015 Quality management systems - Requirements (2015)
5. ***, http://www.cavendishscott.com/ articles-news/iso-90012015/what-is-organization al-knowledge-in-iso-90012015 (accessed on 21.03.2016)
6. ***, ISO 9004:2010 Managing for sustained success of an organization. A quality management aproach (2010)
7. ***, ISO 9000:2015 Quality management systems. Fundamentals and vocabulary (2015)
8. E. C. Shannon, Bell Syst. Tech. J. **27**, 379 (1948)
9. A. Caticha, *Proc. 27th Int. Conf, Workshop on Bayesian Inference and Maximum Entropy in Sci. Eng.* (Saratoga Springs, New York, 2007)
10. K. Conrad, http://www.math.uconn.edu /kconrad/ blurbs/analysis/entropypost.pdf (accessed on 17.02.2016)

Investigations regarding the lowering of specific intellectual property risks identified in the production process

Ramona Pakocs[1,*], *Nouraş Barbu* Lupulescu[1]

[1]Transilvania University of Brasov, Department of Manufacturing Engineering, Mihai Viteazu No.5, Braşov, Romania

Abstract. The main purpose of this research is to decrease the emergence of specific intellectual property risks within the production process as well as increasing risk management performance of IP by preventing them. In order to achieve this, previous studies regarding the main specific intellectual property risks from industrial companies were analyzed together with their managerial methods as well as the possibility of reducing their emergence. As a result of the research conducted were identified five types of intellectual property risks that have a high potential of emergence in the production process, namely: the risk of production of goods in violation of IP rights; the know-how, production knowledge and trade secret disclosure risk; the technological risk of unprotected utility models; the technological risk of unprotected integrated circuits topographies and finally the risk of product counterfeit. In order to achieve the main purpose of our investigation, we have proposed new formulas for estimating the specific intellectual property risks identified in the production process. Their purpose was to minimalize the risk's negative effects on industrial companies and to increase the managerial performance from the intellectual property domain through a new type of management appropriately named: intellectual property management. The research is finalized with a case study regarding the lapse of rights of a patented invention. Based on a case analysis, it was proved that the exploitation of an invention without a contract represents a counterfeit.

1 Introduction

Managing specific intellectual property risks represents nowadays a really delicate and complex problem due to the continuous development of new technologies.
Acknowledging the importance of the probability of apparition of specific production risks and the interest toward preventing them should be a part of the priority objectives of each industrial company's management plan.

Based on previous national and international investigations from specialty literature [1, 2], as well as from certain intellectual property analysis and case studies [3-9], it can be

* Corresponding author: ramona.pakocs@unitbv.ro

observed that the specific intellectual property production risks are acknowledged at a managerial level only after they were already present. This problem is mainly encountered in industrial companies and its consequences can be extremely negative over the financial situation and can increase the bankruptcy risk.

An investigation based on the questionnaire method was thus realized in order to identify the most common specific intellectual property production risks from certain industrial companies.

A long national and international investigation [3, 4], that has interviewed 258 managers who lead companies specialized in production activities has led to important results regarding the categories of specific intellectual property production risks. As such, the study has emphasized the most threatening risks, as follows (Figure 1):

Fig. 1. The most threatening IP production risks.

Where:
- *RpgvIP* = risk of production of goods in violation of IP rights;
- *DRk-h* = disclosure risk of know-how, knowledge of production, trade secret;
- *TRum* = technological risk of unprotected utility models;
- *TRtic* = technological risk of unprotected topographies of integrated circuits;
- *Rcp* = risk of counterfeit of products.

Based on some previous studies [3-6], a series of specific intellectual property risks were identified in the product realization process from within the Quality Management System, namely: risk of production of goods in violation of IP rights; disclosure risk of know-how, knowledge of production, trade secret; technological risk of unprotected utility models; technological risk of unprotected topographies of integrated circuits and risk of counterfeit of products. These risks are rendered in Figure number 2, as follows:

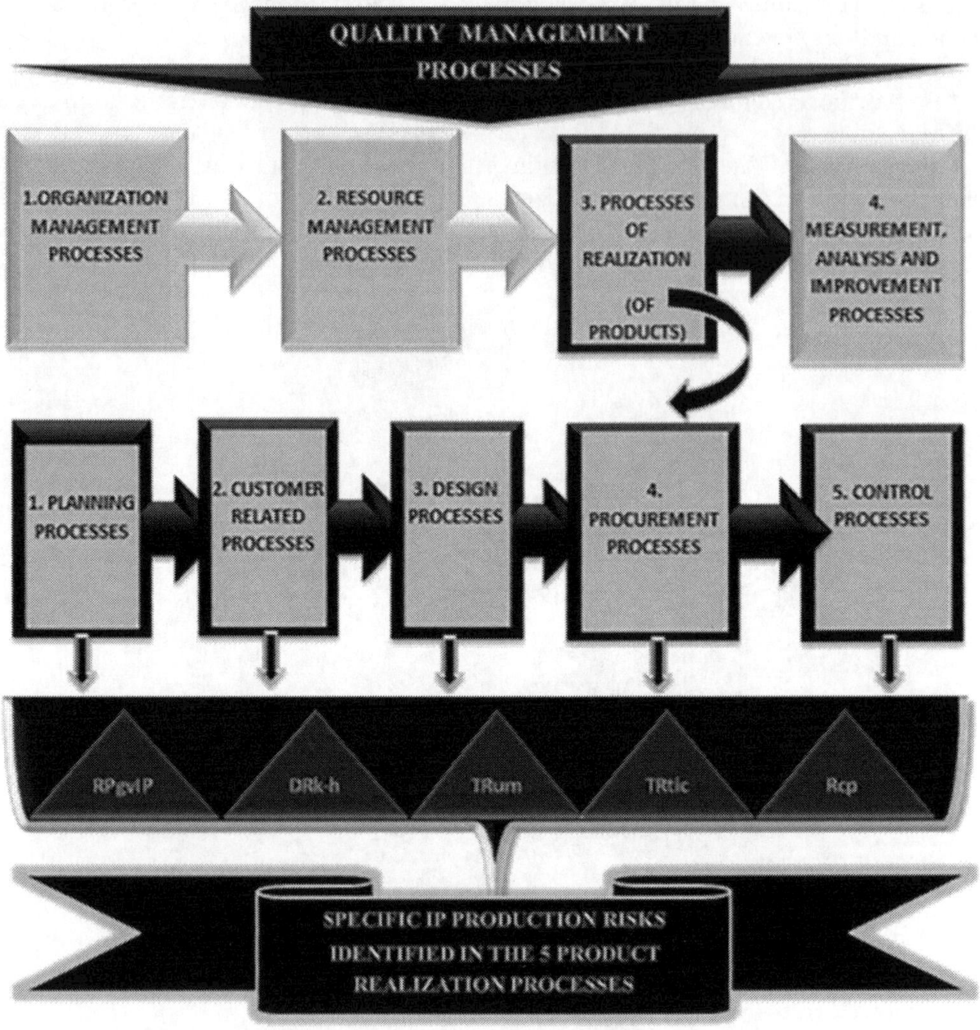

Fig. 2. Specific IP production risks identified in 5 product realization processes [3].

Growing the managerial performance can be achieved by implementing a quality risk and intellectual property management that leads to excellency. This also increases the competition on the targeted market through the increase of the quality of the fabricated products, by reducing the production risks that also include intellectual property risks [11, 12] that can appear in the production process (identified in Figure 2) and through an intelligent management of all intellectual property actives that capitalizes them and leads to an increase of financial revenues.

DOI: 10.1051/matecconf/20179406014

2 Managing specific intellectual property production risks

The management of specific intellectual property production risks can be achieved through (Figure 3):

Fig. 3. Managing specific intellectual property productions risks.

A more complex theoretical evaluation was proposed in a previous research [3], where 3 situations for the detection of specific IP production risks were analyzed and new formulas were proposed for their evaluation.

In order to exemplify we will choose the **Disclosure risk of know-how, knowledge of production, trade secret (noted *DRk-h*)** identified in the product realization process.
The situations of **know-how, production knowledge, trade secret disclosure risk *(DRk-h)*,** based on their apparition time (moment) [3]:
1. detecting DRk-h ***in the anticipation of the production process.***
2. detecting DRk-h ***in the production process.***
3. detecting DRk-h ***in the production's commercialization process.***
Based on the 3 ***DRk-h*** detection moments from the production process, we propose the following evaluation hypothesis in which the following indicators are used:
PA= probability of risk apparition;
FRpi = IP risks frequency;
IRpi = IP risks intensity;
TRpi = IP risks detection time (moment);
FrpiD = IP risks decreasing cumulated frequency;
FrpiC = IP risks increasing cumulated frequency;
FrpiM= IP risks average frequency;
TRpi min = IP risks detection time (moment) in the anticipation stage;
TRpi max = IP risks detection time (moment) during the production process;
TRpi average = IP risks detection time (moment) in the production commercialization process.

In our case, the proposed general formula for the specific ***DRk-h*** production risks is [3]:

$$DR_{k-h} = PA * \frac{FR_{PI} * IR_{PI}}{TR_{PI}} \tag{1}$$

Hypothesis 1: *Detecting* ***DRk-h*** *in the anticipation stage of the production process.*
In this stage [3]:
PA = 50%

$$FR_{PI}D = \sum_{X_{1<X}} Rpi\ 1 + Rpi\ 2 + \cdots RpiN\ (TR_{PI}\ min) \tag{2}$$

TRpi min = 0
IRpi min = 0%
Results that [3]:

$$DR_{k-h} = PA * \frac{FR_{PI}D\ *\ IR_{PI}\ min}{TR_{PI}\ min} \tag{3}$$

In this situation ***DRk-h*** = 0 => detecting ***DRk-h*** in the anticipation stage leads to the possibility of reducing this risk to 0. [3]
Hypothesis 2: *Detecting* ***DRk-h*** *in the production process time.*
In this stage [3]:
PA = 50%

$$FR_{PI}M = \frac{FR_{PI}D + FR_{PI}C}{2} \tag{4}$$

FRpiM= 50
TRpi average = 50
IRpi = 50%

Results that [3]:

$$DR_{k-h} = PA * \frac{FR_{PI}M \ * \ IR_{PI} \ average}{TR_{PI} \ average} \quad (5)$$

In this situation ***DRk-h*** = 50 => detecting ***DRk-h*** within the production process can lead to a great loss of the enterprises capital ≥ 50%. [3]

Hypothesis 3: *Detecting* ***DRk-h*** *in the production's merchandising process.*

In this stage [3]:

PA = 50%

$$FR_{PI}C = \sum_{X_{1},X_{max}} Rpi\ 1 + Rpi\ 2 + \cdots RpiN\ (TR_{PI}\ max) \quad (6)$$

FRpiC = 100

TRpi max = 100

IRpi = 100%

Results that [3]:

$$DR_{k-h} = PA * \frac{FR_{PI}C \ * \ IR_{PI}\, max}{TR_{PI}\, max} \quad (7)$$

In this situation the ***DRk-h's*** impact on the company can be of 100% => detecting ***DRk-h*** in the merchandising stage can lead to losses of up to 100% => the company's bankruptcy risk grows considerably [3].

We must specify the fact that this study is just a general example. The evaluation differs based on the identified production risk and other factors, depending on each case.

In order to exemplify a real situation for the apparition of a specific intellectual property production risks, a case study was chosen based on a specialized expertise [4-8].

3 Analyzing a case study regarding a patent invention's lapse of rights

The subject of this study is represented by the evolution and extension of the engineering work on the trajectory "project-patented invention-technical work" as a result of a patent invention's lapse of right [3, 6, 7].

The case study's subject concerns the patent invention number "x" from 26.04.1982 – Hydraulic order block (Figure 4), owned by ALFA company and offers solutions that can be applied within large companies from the machine constructing industry that were once owned by the state, but avoided from unknown motives. [3, 6, 7] We would like to mention that the case study has been developed in Romania.

On 21.09.1981 company ALFA requires from OSIM (State Office for Inventions and Trademarks in Romania) the urgent patent of an invention that 2 engineers have projected and then cession based on Law number 62 from 1974 regarding inventions and innovations (This Law is applied in Romania). According to the law, the invention patent was assigned but without the right of exploiting the object of the invention patent [3, 6, 7].

The inventors are remunerated, according to the law from that time until the date of 01.04.1983.

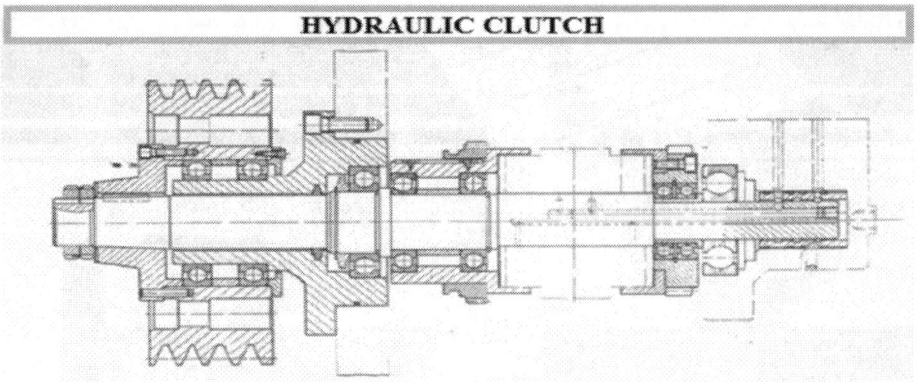

Fig. 4. Hydraulic clutch, subject of Patent– Hydraulic order block. [3, 6, 7].

In the period 01.04.1983 – 2002, the invention is applied by ALFA, either as general ensemble (Figure 5), either as spare parts.

Fig. 5. Three from the ALFA lathes on which the patent was applied– Hydraulic order block. [3, 6, 7].

Starting with 31.03.1987 ALFA which has become company BETA doesn't pay the taxes for OSIM and the invention loses its rights.

Company BETA doesn't announce its inventors that she isn't interested anymore in paying the annual taxes for maintaining the invention in a protection regime. Starting with this date there doesn't exist any agreement regarding a remuneration model [3, 6, 7].

However, company BETA continues to apply the old invention. The inventor's financial rights weren't negotiated between the inventor and the firm that has applied the invention. The company had the right to announce it, based on ART. 41/Law number 64/1991(This Law is applied in Romania), "The holder can renounce completely or in part to the patent, based on a written declaration registered at OSIM. For the inventions from art. 5 line 1, letter a) and line 2, as well as for the inventions that were the object of transfers according to art. 5 line. 1 letter. b), the holder of the patent is obliged to announce the inventor of its intention to renounce; at the inventor's request, the holder is obliged to transmit to him the right over the patent"**.** As a consequence, this is a case of usage without a right [3, 6, 7]. Company BETA has relied upon the stipulation from art.41/L64-91(This Law is applied in Romania): "*The invention or part of it that has been renounced can be freely exploited by third parties*". However, the reliance was made illegally because it can't be separated from the entire context of art.41.

Company BETA continues to apply its previous invention on technical work, without clearing its accounts with the inventors and by falsely applying part 4 from art.41/Law number 64/1991 (This Law is applied in Romania). As a result, the inventors sue company BETA this time for not respecting the CSJ's (The Supreme Court) decision of not producing anymore the object of the invention, as well as for additional damages. For this

DOI: 10.1051/matecconf/20179406014

second period, company BETA hasn't realized any license or cession contracts with its right owners [3, 6, 7].

The lawyers have invoked an extinctive prescription, based on Decree number 167/58 (This Decree is applied in Romania) acknowledging that there might exist a claim right but only because it is prescribed [3, 6, 7].

Based on the analyzed study, the following aspects were concluded:

- in the case in which the owner is no longer interested in applying the invention, he has to announce the inventor so that he and OSIM can assume it. The exploitation of an invention without a contract represents fraud based on art.5, letter b)/Law number 64/1991 (This Law is applied in Romania) [7].
- according to Art.59/Law number 64/1991 (This Law is applied in Romania), the infraction of counterfeit was present, together with the rights of the rightful holder.

4 Conclusions

Managing specific intellectual property production risks can be a challenge, but must be considered more like a necessity for the creation of a powerful, modern and intuitive intellectual property management. This implies a knowledge of the protection legislation, knowing the IP risks preventive forms together with the action laws against intellectual property right breeches.

The importance of a performant management is also emphasized by the increasing number of expertise from this domain [7], which is foreseen as one of the most complex multidisciplinary scientific domains of analysis. This fact proves the necessity of evaluating compatible elements [9, 10] of some principles and sciences considered not long ago as separate: engineering-economy-law.

Besides the important contributions of enriching the specialty literature and based on the realized analysis, this study has however some limitations that can be structured as follows:

- even though the risk analysis is a common theme in the national and international specialty literature, when we bring into discussion the specific intellectual property production risks, the situation changes due to the fact that they are less investigated and acknowledged by industrial companies.
- the impossibility to obtain detailed information regarding the financial situation of the analyzed company hasn't allowed us to apply the proposed formulas for evaluating the specific intellectual property production risks.
- the uncertainty of industrial companies regarding the disclosure of information regarding their activities and results has led to an analysis of the case study only from a legal perspective. Further investigations will have to apply these formulas by using dedicated software.

References

1. D. Săvescu, A. Budală, *Intellectual Property in Romania and in some EU countries* (Lux Libr. Pub. House, Brasov, 2008)
2. C. Greenhalgh, M. Rogers, *Innovation Intellectual Property and Economic Growth* (Prin. Univ. Press, Princeton and Oxford, 2010)
3. R. Pakocs, *Quality and intellectual property risk management in the industrial companies* (Ph.D, Tr. Univ., Brasov, 2015)
4. R. Pakocs, N. B. Lupulescu, *Investigations regarding the management of specific intellectual property risks in companies from Romania* (AFASES, Publ. House of "Henri Coanda" Air Force Acad., Brasov, 2016)

5. R. Pakocs, N. B. Lupulescu, *Need to implement a highly-performing intellectual property management in the technical-engineering field* (AFASES, Publ. House of "Henri Coanda" Air Force Acad., Brasov, 2015)
6. R. Pakocs, N. B. Lupulescu, *Risk management and risk type analysis specific to intellectual property in industrial companies* (AFASES, Publ. House of "Henri Coanda" Air Force Acad., Brasov, 2014)
7. R. S. Fântână, *Examination of goods import-export* (ProUniv. Publ. House, 2010)
8. G. Secară, R. S. Fântână, *General Management* (ProUniv. Publi. House, 2010)
9. R. S. Fântână, N. Minculete, *Extension of Liskov Substitution Principle and Application to Curriculum Management* (Acta Poly. Hun., 2014)
10. R. F. Reilly, R. P. Schweihs, *The Handbook of Business Valuation and Intellectual Property Analysis* (The McGr. – Hill Comp., 2004)
11. D. Hillson, *Project Risk Management: Future Developments* (Available at: www.risk-doctor.com/pdf-files/fut0798.pdf, 2004)
12. M. S. Mcnulty, Applied Risk Analysis of Complex Systems, The Third Conf. on Mathem. Meth. in Reliab., **4** (2002)

 DOI: 10.1051/matecconf/20179406015

Scheduling of an assembly process of a chosen technical mean using the critical chain approach

Iwona Paprocka [1,*], *Aleksander* Gwiazda [1], and *Magdalena* Bączkowicz [1]

[1] Silesian University of Technology, Faculty of Mechanical Engineering, Konarskiego 18A Str., 44-100 Gliwice, Poland

Abstract. In the work is presented the CCPM (Critical Chain Project Management) method that differs from other network approaches. The differences are related with: the scheduling process, with restrictions of the project that are taken into account and with the time buffers allowing avoiding delays in realized tasks. Each project should be such organized to effectively achieve the previously assumed goals. The paper presents all the steps involved with utilization the CCPM method for determining the schedule of an assembly process of a mining shearer machine. It is presented the elaboration of the network of tasks and preparation the process schedule. It is also realized the resource optimization and the safety buffers introduction.

1 Introduction

The Critical Chain Project Management (CCPM) is one of the methods of project management using network planning. This method is called also as the Critical Chain method. It is derived from the Theory of Constraints (ToC) and nowadays it is the main tool used in ToC. CCPM method was developed in 1997 by E. Goldratt and was presented in the book "Critical Chain". It could be described as a method of planning and project management, which is focused on resources needed to perform planned tasks [1, 2]. This method differs from other methods based on network planning and using the idea of a critical path. The major difference is that the CCPM uses an alternative analysis of this critical path, called the critical chain. It is distinguished by the optimization of resources included in the critical chain and no in the critical path. Additionally, during project managing using CCPM, are used new system components that improve the planning procedure. These additional components are: Project Buffer, Feeding Buffer and Resource Buffer [3]. These buffers are used in project planning as safety buffers. The main objective of CCPM is to shorten the duration of a project and simultaneously balancing utilized resources. This in a consequence leads to costs reducing and increasing probability of the project execution on required time.

The analysed method is partly heuristic. It is based largely on the Continuous Path Method (CPM) and certain properties of the distribution of the sum of random variables as

* Corresponding author : iwona.wosik@polsl.pl

well as on some psychological behaviour, to which people are subjected [3]. Project management using CCPM is very advantageous because it leads to reducing the multitasking of resources, to eliminate conflicts related with the allocation of resources, eliminating a common practice starting job at the last possible moment (so called Student's Syndrome). Moreover its application allows preventing artificial increasing the amount of work needed to complete the scheduled time (so called Parkinson's Law). Additionally CCPM allows utilize the so called relay-race effect (systematic use of the time saved on individual tasks), more efficiently controlling projects realization (through monitoring the buffers utilization). Finally utilization of this approach lets to avoid the negative influence of the Murphy right (anything that can go wrong will go wrong). The loss of time and resources is mainly due to waste of these factors. Critical Chain Project Management allows avoiding loss of time and loss of resources. This is why projects are implemented about 20-30% faster, and the risk of delays is significantly reduced. In the end projects are cheaper, and the risk of cost overruns is significantly reduced [4, 5, 6].

Not only the balance of resources and the introduction of buffers are factors that distinguish CCPM method from traditional ones. The factors of CCPM are also times of tasks that are reduced from conservative estimates to realistic ones. The next factor is the moment of beginning the given task. All activities associated with this task begin as late as possible in accordance with the ALAP principle (As Late As Possible) [7, 8, 9].

2 Characteristics of the analysed production process

In this chapter are described all the steps considered with CCPM utilization for determining the schedule of manufacturing process of a mining shearer loader (Figure 1). The analysis begins with creating the network of dependences, through schedule creating, resource optimization to the introduction of safety buffers. The production process of a shearer loader is generally a one piece one. The whole shearer loader is assembled on one stand [10, 11] using specific materials [12, 13, 14].

Fig. 1. Mining shearer loader [9].

All components and subassembly are delivered to this stand. It is a long process. This was the main reason due to which it was decided to use the CCPM approach. It was realized as a special technological project. The project consists of 17 tasks, 9 of which are manufacturing type tasks, and the remaining 8 are the tasks of assembly type (Table 1). To each task is assigned its duration and the person, who is responsible for it.

Table 1. Characterisation of realized tasks.

Task No	**Task code**	**Task type**	**Task name** (in parenthesis number of subassemblies)	**Preceding task No**	**Task duration [days]**	**Worker symbol**
1.	**A**	**MANUFACTURING OF**	the ranging arm	-	15	a
2.	**B**		the loader	-	35	b
3.	**C**		the cutting system	-	35	g
4.	**D**		the cutting head	-	20	h
5.	**E**		the transmission of the feed drive (2)	-	45	e
6.	**F**		the drive system	-	45	c
7.	**G**		the electric drive (2)	-	75	b
8.	**H**		the transforming-converting unit BTP-1	-	70	f
9.	**I**		the block of electric devices BAE-18 (2)	-	15	g
10.	**J**	**ASSEMBLY OF**	the cutting head and the cutting system	C, D	30	a
11.	**K**		the ranging arm with the loader and subassembly obtained in the task **10**	A, B, J	30	h
12.	**L**		the drive system with the electric drive (2)	F, G	20	e
13.	**Ł**		the transmission of the feed drive and the subassembly obtained in the task **12**	E, L	35	d
14.	**M**		the subassembly obtained in the task **13** and the transforming-converting unit BTP-1	Ł, H	25	c
15.	**N**		the subassembly obtained in the task **13** and the block of electric devices BAE-18	Ł, I	15	c
16.	**O**		**14 + 15**	M, N	15	d
17.	**P**		**11 + 16**	K, O	10	e

In order to determine the critical path it was elaborated a network of dependencies between particular tasks that should be realized. This network is presented in Figure 2. On the basis of the elaborated network of dependences it was determined the tasks that are non-critical and the ones that are critical. The sequence of tasks within the critical path is marked in red, while the remaining tasks are marked in blue. According to the established relationships between the tasks and their duration it was also determined the duration of the entire project, which is 175 days. It should be mentioned that numbers of some subassemblies of the shearer loader are odd. This concern particularly to these related with ranging arms. This fact is indicated in the table by the symbol (2). Also some tasks are defined by determining number of task, in which are assembled used subassemblies.

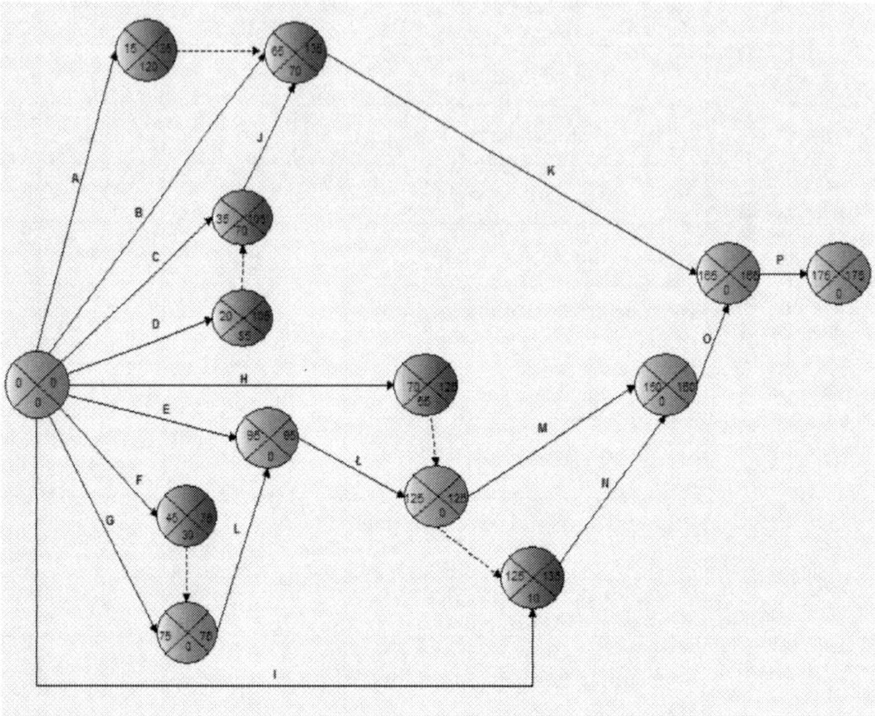

Fig. 2. Network of dependences for the project of a shearer loader manufacturing.

In the diagram above the continuous arrow means a real task and dashed one – an apparent one. The used symbols are presented in Figure 3. In the symbol also are determined three different factors characterising realization of particular tasks. The time t_i means the earliest moment of the task realization, T_i – the latest moment of the task realization and L_i – time reserve for the task realization.

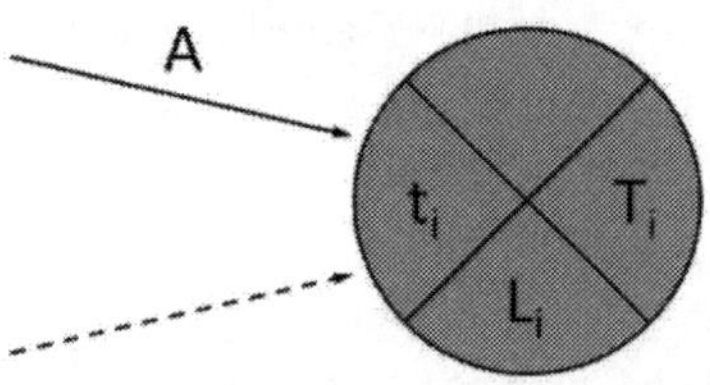

Fig. 3. Symbols used in the network of dependences.

In the next chapter is presented the elaboration of the schedule for the presented project of mining shearer loader manufacturing. The schedule is based on the presented network of dependences but also includes the priority rules.

 DOI: 10.1051/matecconf/20179406015

3 Schedule elaboration

As it was stated the next step is to determine the schedule according to the ALAP principle, that means starting with the last and ending with the initial task. In such elaborated schedule tasks are initiated as late as possible. In order to illustrate the difference between the scheduling from the beginning task and from the last one there have been elaborated two schedules. The first one is elaborated according the ASAP (As Soon As Possible) principle. It is presented in Figure 4. The second schedule is elaborated according to the ALAP principle and is presented in Figure 5.For elaboration the schedules was utilized the program Gantt Project. Below are presented windows of this program. Elaborating the presented schedules it was taken into account the durations of the project tasks described in Table 1. In this schedules also were considered days off, weekends and public holidays. It was done to make more real the elaborated schedules. It was also assumed a hypothetical date of completion of the project. The time unit in the schedule is one week.

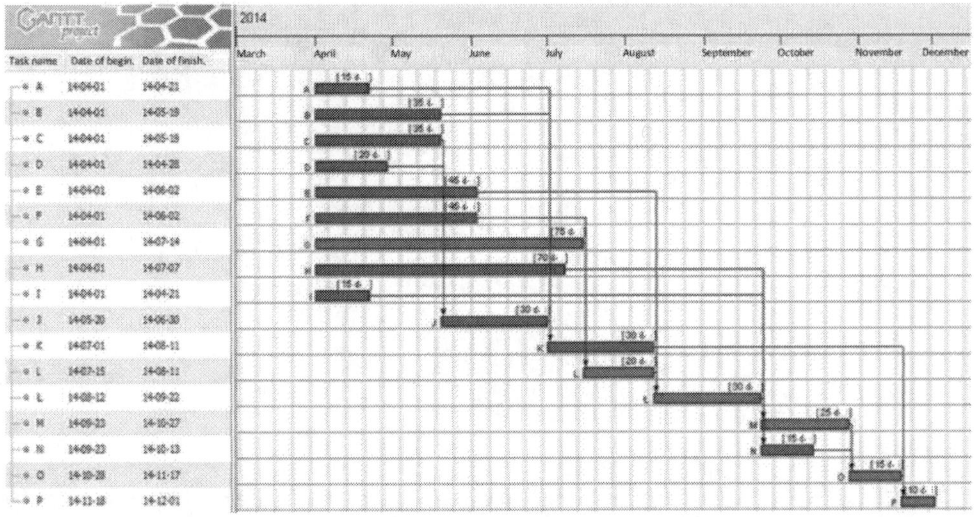

Fig. 4. Schedule elaborated according the ASAP principle (d. – days).

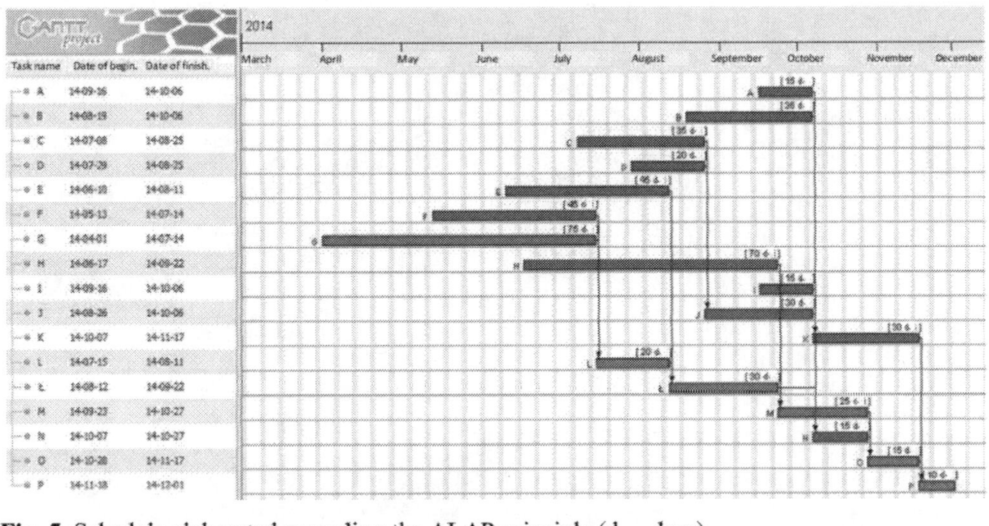

Fig. 5. Schedule elaborated according the ALAP principle (d. – days).

Critical chain is the longest sequence of tasks, together with balanced resources. To determine the critical chain one should take into account both the interdependencies between tasks, and dependences concerning the availability of resources. In the case of this project the most considered problem is the availability of workers performing specific design tasks (as a specific type of scarce resources). For this purpose it was built the schedule with the adopted time unit (one week), presented in Figure 6. The analysis of this schedule allows determining resource conflicts in the elaborated plan (marked in yellow). For example, it was identified the resource conflict between tasks E and L, as at the same time the same employee was assigned to perform work in these two tasks simultaneously. Conflicts were also observed between tasks M and N as well as between A and J.

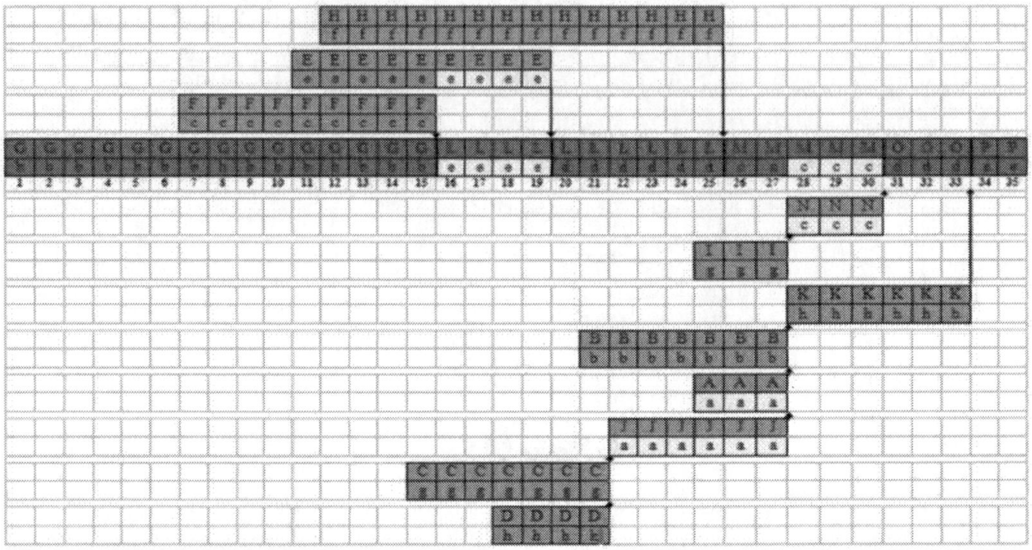

Fig. 6. Illustration of resources conflict.

In order to solve resource conflicts, one should balance the order of the tasks of the project. The schedule after the conflicts elimination, and thus the determination of the critical chain, is shown in Figure 7 (ordered resources are marked in green).

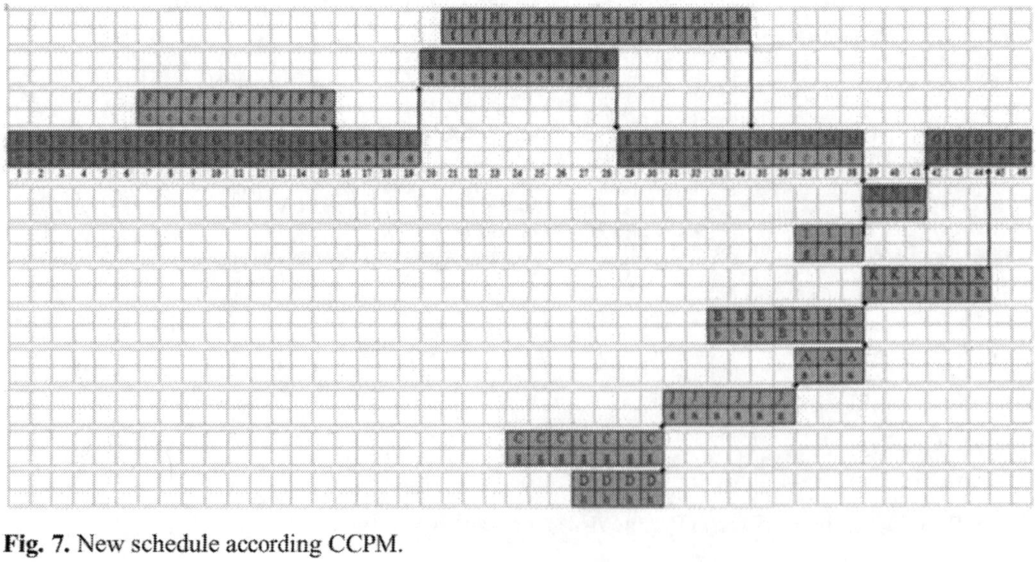

Fig. 7. New schedule according CCPM.

 DOI: 10.1051/matecconf/20179406015

Determination of the order of tasks, which were the source of the conflict, does not affect the length of the duration of the whole project. This is why it was not needed to assign them special, additional priorities. Including these dependences in the schedule elaborated according the ALAP principle, one receives the schedule that is presented in Figure 8. It shows the sequence of re-ordered tasks taking into account the problems considered scares resources. The critical chain is marked in red. Other tasks (non-critical) are marked in blue.

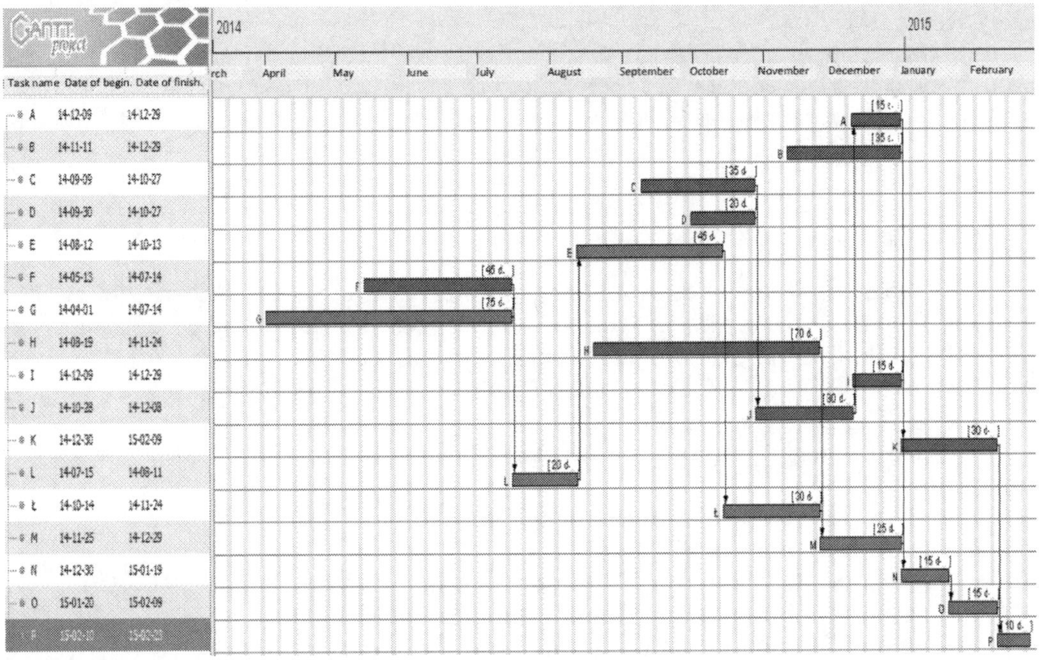

Fig. 8. Schedule with the critical chain (d. – days).

The introduction of new dependence taking into account the availability of resources resulted in extending the duration of the entire project. The duration of the project was extended from initial eight months to eleven. It gives another three months but currently the project is scheduled without hidden resources conflicts.

4 Conclusions

Basing on the conducted investigations one could state that CCPM application is characterised by many advantages and raises a number of problems. Firstly, elaboration of a network of dependences, for project tasks of different types (manufacturing and assembly ones) required modification because of the resource conflicts. It was introduced one initial task for manufacturing tasks having no preceding task. At this stage, the duration of the project was 175 days (approximately 8 moths). Secondly, by elaborating both the schedule according the ASAP principle and the ALAP one, it was possible to illustrate the difference between the CCPM approach, and other methods of network planning like CPM/COST, CPM/PERT etc., as these methods function only according the ASAP principle. So only CCPM allows taking into consideration the problem of resources conflicts. In the presented analysis, three resources conflict related with ordering different tasks to the same staff at the same time were detected. Using the ASAP principle it is probable that this conflicts would be detected during the realization of the project and resulted in unplanned project rescheduling. Thirdly, to avoid conflict one should optimize the primary schedule and to determine the critical chain that allows scheduling both tasks and resources. However such approach results in extending the duration of the project (up to 230 days) but on the other hand the optimized schedule is more realistic.

This work has shown the importance of the selection of the appropriate method of project management [15, 16, 17]. On it depends whether the achievement of the project objective will be realized without disturbances and breaks or it would be implemented with the emergency attempts to resolve conflicts considered scares resource and often convened ad hoc meetings (so called task forces). It is also important to resolve the problem of data analysis [18, 19].

CCPM is the only method that focuses not only on the reduction of time to minimize project costs, but also enables the optimization of resources. In comparison to traditional project management techniques it is a method which overcomes the waste of time and financial resources especially.

References

1. E. M. Goldratt, *Theory of Constraints* (The North River Press, Great Barrington, 1999)
2. E. M. Goldratt, *Critical Chain* (The North River Press, Great Barrington, 2002)
3. L. P. Leach, Critical Chain Project Management (Artech House, London, 2004)
4. J. M. Antill, R. W. Woodhead, *Critical Path Methods in Construction Practice* (Wiley-Interscience, New York, 1990)
5. A. Qazi, J. Quigley, A. Dickson, K. Kirytopoulos, Int. J. of Proj. Man. **34/7**, 1183 (2016)
6. G. Ćwikła, App. Mech. and Mate. **474,** 27 (2014)
7. K. Foit, Adv. Mater. Res. **837**, 577 (2014)
8. H. A. Mesa, K. R. Molenaar, L. F. Alarcón, Int. J. of Pr. Man. **34/7**, 1089 (2016)
9. E. Houghton, V. Portougal, Int. Trans. in Op. Res. **4/5-6**, (1997)
10. http://www.directindustry.com/prod/caterpillar-global-mining/product-55229-916883.html
11. A. Gwiazda, Adv. Mat. Res. **837,** 393 (2014)
12. M. Bączkowicz, A. Gwiazda, IOP Conf. Ser.: Mater. Sci. Eng. **95** (2015)
13. S. Topolska, J. Łabanowski, Mat in Tech. **49/4**, 481 (2015)
14. J. Łabanowski, A. Świerczyńska, S. Topolska, Pol. Mar. Res. **21/4**, 108 (2014)
15. A. Grajcar, A. Płachcińska, S. Topolska, M. Kciuk, Mat in Tech. **49/6**, 889 (2015)
16. I. Paprocka, W. Kempa, K. Kalinowski, C. Grabowik, IOP Conf. Ser.: Mater. Sci. Eng. **95,** (2015)
17. W. Kempa, I. Paprocka, K. Kalinowski, C. Grabowik, Int. J. of Mod. Manuf. Technol. **7/2**, 95 (2015)
18. G. Ćwikła, Adv. Mat. Res. **837,** 618 (2014)
19. K. Foit, A. Gwiazda, W. Banaś, A. Sękala, P. Hryniewicz, IOP Conf. Ser. Mater. Sci. Eng. **95** (2015)

MATEC Web of Conferences **94**, 06016 (2017) DOI: 10.1051/matecconf/20179406016
CoSME'16

Continuous improvement processes using Lean Management tools. A case study

Luminița Pârv[1,*]

[1]Transilvania University of Brașov, Department of Manufacturing Engineering, Mihai Viteazu No.5, Brașov, Romania

Abstract. The paper describes how Lean Management may be applied in the university setting to improve the management processes. The correlation of didactic, educational and research activities with the stakeholders needs is one of the main objectives of the university. In this respect, an indicator used to analyse a university, for the purposes of fulfilling its mission, respectively for the purposes of streamlining its didactic and scientific activity, is related to the number of graduates on the labour market, acting in their area of specialization. This work presents a best practice of Lean Management at Transilvania University of Brasov, Romania.

1 Introduction

Lean practices are the tools and methods commonly associated with the Toyota production system [1]. The main Lean principles are *Continuous Improvement* and *Respect for People*.

A study of use of continuous improvement to achieve instructional improvement and instructional technology improvement with the help of an example of a high school which applied lean framework and resulted at use of screen capture technology was presented by Flumerfelt and Green, in [2].

The conclusion presented the ratio of time allocated to task/relational activities, the ratio of time allocated to passive/active learning, the amount of time allocated to new learning opportunity, the amount of time allocated to individualization, and the amount of time allocated to differentiation.

Lean management is defined as a three levels management system focused on creating value for customers and eliminating waste, unevenness and unreasonableness using the scientific method [3].

The tactical level is based on the following five principles [2]:

- Specify Value –it is necessary to specify what adds value from the end customer's perspective, so that all the non-value activities can be identified and removed; in fact for most production processes only a small part of the total production time and effort adds value for the end customer;
- Identify the Value stream–identify all activities that do not aggregate any value to the final product;

* Corresponding author: luminita.parv@unitbv.ro

- Create Flow - promote continuous flow in the process by eliminating wastes;
- Customer Pull – respond to customer demand; produce what customers' wants at the right time;
- Pursue Perfection – continually remove wastes, striving for a perfect process.

At the operational level, the practices and tools are: Value stream mapping, 5 S, Kaizen, One piece flow, JIT, TQM and Six Sigma.

Fig. 1. Levels of the Lean Management

Lean education is the application of Lean principles and practices to the education field [4].

2 Lean in higher education

The purpose of Lean teaching parallels the purpose of Lean management [1, 2, 5], as used in organizations, which is to improve the value of goods and services (higher education) for end-use customers (students, payers, employers and society) [1] started to implement concepts of Lean teaching, tools, and processes in a college with three functions teaching, research, and service.

Emiliani [6, 7] used kaizen process for ten courses contained in a part-time executive MS degree program in management and the conclusion was that Kaizen is an effective process for improving graduate business school courses and the value proposition for students.

Doman [5] used lean principles and practices to improve university's grade change administrative processes in higher education through learning experience involving undergraduate students working as a team to identify waste and redesign process in a seminar course. The authors have emphasized that team used process mapping and value stream mapping lean tools to predominantly focus on four wastes namely delay, over processing or incorrect processing, correction and knowledge disconnection.

According to [3, 4, 8 - 10], the end-use customers in higher education will be students, employers and society.

The strategic management plan of the Transilvania University formulates the mid- and long-term aims of our university, by considering the actual and future necessities of the working market.

DOI: 10.1051/matecconf/20179406016

Through the perspective of the bounds established with the economic environment, the strategy aims to determine how the university can achieve the best competitive impact for industry-university collaboration.

The team management proposes, in order to improve the educational process, the implementation of the Lean Together strategy in learning-together in practice.

A SIPOC (Suppliers/Input/Process/Output/Customer) diagram is used by the team to identify all relevant elements of the educational process improvement project before work begins.

The SIPOC tool answers at these questions:
- Who supplies inputs to the process?
- What specifications are placed on the inputs?
- Who are the true customers of the process?
- What are the requirements of the customers?

Table 1. The implementation of the strategy is accomplished through projects developed together with the economic environment

Supplier	Input	Process	Output	Customer
Higher education providers	Students	Orientation	Successful graduates	Academic organisations
Employers from industry	Certificated faculties	Practical placement	Unemployed graduates	Government Organizations
		Learning	Projects	Industries

3 Together in learning-together in practice as a PDSA cycle

In order to apply the Together in learning-together in practice the Plan-Do-Study-Act cycle will be used.

Following the conceptual framework's logic of PDSA, the process of improvement application started with the Plan phase, in which a detailed program was designed 100% student-centred (costumer focused).

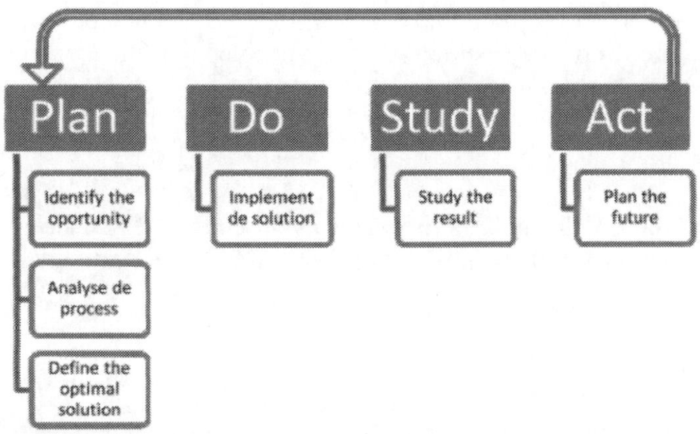

Fig. 2. The PDSA phases

3.1 Plan phase

Plan part suppose opportunity identifier, process analyse, optimal solution development.

Together in learning-together in practice are materialized in two projects:

- Organization of an annual conference, where graduating students of the BSc or MSc programs, who want to participate, have the possibility to present in front of company representatives the diploma or dissertation project. The Graduates in Front of Companies (AFCO) bring face to face our alumni and the representatives of the economic and socio-cultural environment; specifically, the students may enrol for this conference and may present, as poster, significant aspects of diploma paper;
- Furthermore, the university has developed the project Podium of the Companies. This way, in the largest complex of our institution, in the central lobby, it launched a project enabling the main regional companies to present themselves, to discuss with the Transilvania university students and teaching staff, to submit technical and research subjects, to propose diploma-project themes, to offer part-time or full-time jobs. Thereby, by enhanced dialogue, the interests of both parties (university and economic environment) are intertwined and the graduates' chances to succeed on the labour market substantially increase.

The project's objective is to improve the cooperation with the socio-economic environment and the regional integration, by:

- Attracting the socio-economic environment in the process of determining the skills needed for higher education graduates in order to integrate them as quickly as possible within the labor market;
- Enhancing the transfer of innovation from academia to the socio-economic environment;
- Developing the cooperation between the academic and socio-economic environment by developing joint investment projects / programs, research, internship or volunteer programs
- Attracting prestigious specialists from the socio-economic environment in the educational process;
- Promoting the principles of sustainable development in the provision of higher education services;

Proposed activities:

- the promotion of partner companies' activities in which the university's students can be involved;
- presenting to the students the characteristics of the partner companies and the opportunity that the students are recruited by these companies;
- other activities proposed by the companies that respect the objectives of the project in agreement with the university structures responsible for the relationship with the companies.

3.2 Do phase

The Do phase of the continuous improvement cycle comes next. During this phase, the projects were implemented (2012-starting date for AFCO, 2013-starting date for PC).

3.3 Study phase

The third phase of the PDSA cycle is the validation (study) phase. During this phase, university gets to know the evaluation process in a Lean oriented approach. The evaluation of the Together in learning-together in practice strategy was carried out on the platform of the INFO-HE project.

DOI: 10.1051/matecconf/20179406016

The study encompassed quantitative, survey data obtained from questionnaire-based surveys (with graduates and employers) from 2010 and 2014 graduates.

Table 2. Option- Projects proposed by the companies

		2010		2014	
		Absolute frequency	**Relative frequency**	**Absolute frequency**	**Relative frequency**
Projects proposed by the companies	1 – Strongly Disagree	4	10	25	10.04
	2	8	20	37	14.86
	3	8	20	54	21.69
	4	7	17.5	60	24.1
	5 – Strongly Agree	13	32.5	73	29.32
	Total:	**40**		**249**	
	Average	3.42		3.48	
	Standard deviation	1.48		1.49	

In order to assess the effectiveness of the projects implemented by Transilvania University of Brasov, the chapters on personal development and professional career included specific questions aiming at measuring the usefulness of the graduates' skills specific to the field of studies followed.

The first set of questions relates to the teaching methods used in university. As an example: **C9. To what extent have been used the following teaching – learning methods within the license program you graduated in our university?**

The results were present in Table 2. and Figure 3.

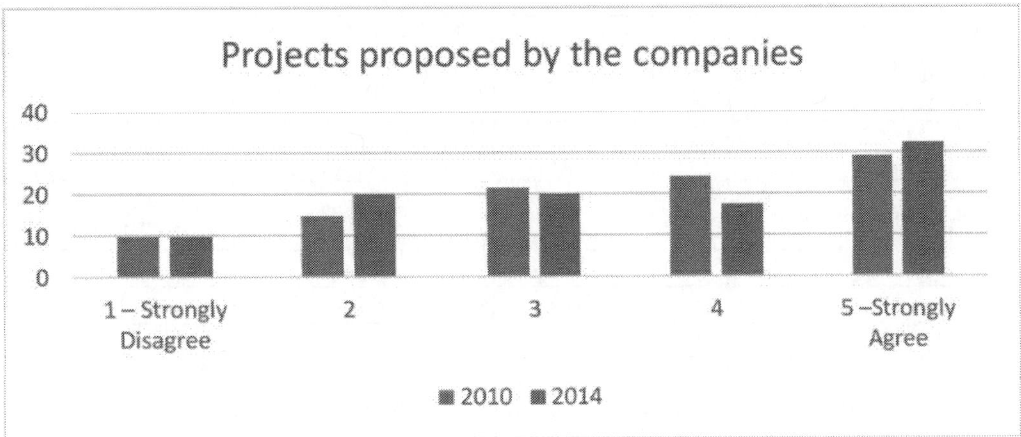

Fig. 3. Evolution of the projects proposed by the companies

3.4 Act phase

Each quarter, the PDSA cycle allows generating the Act phase: the opportunity to improve the educational process.

Starting from the Lean project objective, which is to correlate the academic offer and the learning outcomes with the employment requirements using the Lean methodology.

29% of the respondents (graduated in 2014) are currently working in a position which corresponds to the study program they have graduated from.

The cooperation with the economic and socio-cultural community knew a real improvement lately at Transilvania University of Brasov the correlation between the companies' needs and students didactical and research activities, being one of the main objectives considered by executive management team at Transilvania University. In this respect, a preliminary study on applying Lean Management was carried out for the university.

According to the survey of 2010 graduates, during the first 6 months from graduation, 13.4% of graduates were hired in a corresponding domain.

The question in the questionnaire is: F1. Which of the following would suit best your situation within the first six months after graduation?

The results below refer to promotion of 2014.

Table 3. Survey results at the question F1

	No of people who answered the question	**Frequency (%) of the total no of people (% is over 100% because it is a multiple choice question)**	**Frequency (%) of total responses**
I work in a field that corresponds to the license program I graduated	104	42.62	28.34
I work in a different field	53	21.72	14.44
I continued to work where I have already worked	50	20.49	13.62
I opened my own business	6	2.46	1.63
I performed an internship	15	6.15	4.09
I begun to work as a volunteer	12	4.92	3.27
I enrolled in Master's or other postgraduate university courses	88	36.07	23.98
I enrolled in a new college (undergraduate)	5	2.05	1.36
I looked for a job without success	20	8.2	5.45
Other:	4	1.64	1.09
	367		

The identified improvements were mapped with standard Lean tools from manufacturing.

The improvements and changes that were and are being made are centered on the PDSA improvement cycle and aligned with the students' needs.

4 Conclusion

Following stakeholders' feedback, the university management team standardises the good practices oriented towards the Lean philosophy, in other words, towards continuous improvement.

Table 4. Continuous improvement using Lean tools

Improvement opportunity	Improvement made
PLAN	First, a general frame goal was established, oriented by the Together in learning-together in practice
DO	Accomplishing projects
STUDY	The evaluation have been made with the implication of the main stakeholders
ACT	Given the results of the evaluation, the existing projects will improve and new projects will be implemented

As such, the correction cycle is presented in a continuous way each quarter, in that same order of ideas. Positive results were obtained. This can be achieved with a strong industry-academia relationship grounded on a multitude of partnerships (Project-Based learning, internships, final projects, capstone projects) that provide students and faculty with real life projects, following what many authors have been advocating.

We hereby acknowledge the structural funds project POSDRU 155/1.2/S/141278 for providing the software tool used in this work.

References

1. S. Flumerfelt, G. Green, G., EdTechnSoc, **16 (1)** (2013)
2. J. K. Liker, *The Toyota Way - 14 Management Principles from the World's Greatest Manufacturer* (2004)
3. J. Antony, IntJProdPerfManag, **64(6)** (2015)
4. J. Antony, IntJProdPerfManag, **63(2)** (2014)
5. M.S. Doman, QualAssurEd, **19(3)** (2011)
6. M.L. Emiliani, QualAssurEd, **13(1)** (2005)
7. M.L. Emiliani, QualAssurEd, **12(4)** (2004)
8. R. Stasiak-Betlejewska, M. Kaye, M. Dyason, K. Stachova, ERE, 767 (2014)
9. J. Sujanova, P. Gabris, M. Licko, P. Pavlenda, R. Stasiak-Betlejewska, *13th EuConfKnowManag,* **1-2**, 1135 (2012)
10. R. Ulewicz, R. Kucęba, EcManag, **8(1)** (2016)

MATEC Web of Conferences **94**, 06017 (2017)   DOI: 10.1051/matecconf/20179406017
CoSME'16

The use of IT tools for the analysis and evaluation of psychomotor efficiency of employees

Joanna Sadłowska – Wrzesińska[1], *Izabela* Gabryelewicz[2] , and *Patryk* Krupa[2,*]

[1]Poznan University of Technology, ul. Strzelecka 11, Poznań 60-965, Poland
[2]University of Zielona Góra, ul. Szafrana 4, 65-516 Zielona Góra, Poland

Abstract. The paper presents a computer application ePiórkowski, which allows to carry out psychomotor efficiency tests using a computer. The software enables to measure the response time of mental skills such as: psychomotor response time, resistance to fatigue and stress, concentration and decision making time. The application is a precise mapping of standard measurement equipment called Piorkowski apparatus. The paper describes the rule of operation and the way of operating the application. It also includes a comparison of the application with the standard equipment and a presentation of additional functions, not offered by the traditional apparatus. What is more, the paper describes the construction, interface and technology used for the software development.

1 Introduction

An effective operation of a machine depends of its reliability as well activities of people who operate it. As a result, when analysing the work process, it is essential to consider not only the machine itself but also a man and the environment in which he/she together with the machine realises set goals. In accordance to the systems theory, a system composed of a man, a machine and an environment is called a sociotechnical system.

The following stages can be distinguished in the process of technical object exploitation:
- pre-use processes,
- machines use processes,
- process of providing ability,
- supply processes,
- machines liquidation processes,
- processes supporting operation.

A man, being a deciding party, plays a superior role in a sociotechnical system. A superior role of men is especially visible in case of operators of machines and devices. In situations when working cycles are related with numerous start-ups and stops, there occur changes of tasks, noise, toxic substances, limited visibility, necessity of observing signalling devices and prompt decision making. These states result in fast fatigue and limitations of the capabilities

* Corresponding author: pkrupa@uz.zgora.pl

DOI: 10.1051/matecconf/20179406017

to properly operate machines. Thus, apart from knowledge and skills, the following factors are decisive in evaluating employee's usability for work:
1. Psychomotor performance:
- response pace,
- adequacy of responses,
- visual-motor coordination.
2. Intellectual fitness and cognitive processes:
- observing and caution,
- understanding of situations,
- anticipation.
3. Personality:
- social maturity,
- emotional maturity.

The importance of the human factor is reflected in the Guide to application of the Machinery Directive 2006/42/EC, where, in accordance to norm EN ISO 6385:2004 – Ergonomic principles in the design of work systems, ergonomics is defined as follows: „Ergonomics (or the study of human factors) is the scientific discipline concerned with the understanding of interactions among human and other elements of a system, and the profession that applies theory, principles, data and methods to design in order to optimize human well-being and overall system performance". Requirements concerning machines and devices are presented in Figure 1. Those requirements concern the human factor. That means, the adjustment of a work place to psychomotor capabilities of human [2, 3].

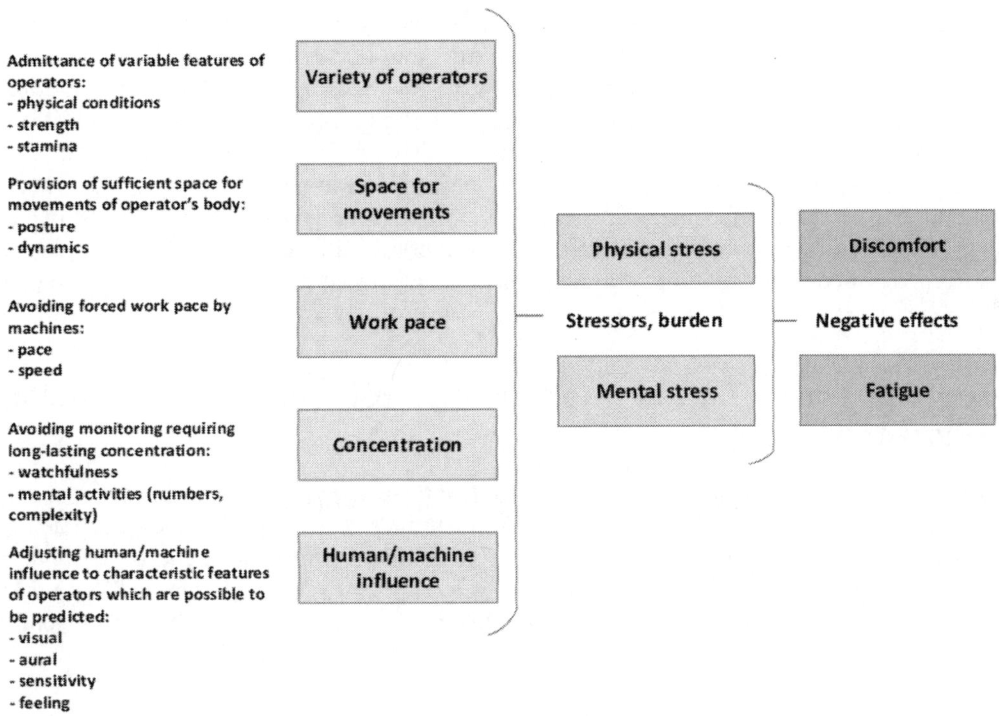

Fig. 1. Ergonomic requirements set for work places.

Source: Guide to application of the Machinery Directive 2006/42/EC, Rev. 2. June 2010, edited by Ian Freser, European Commission – Enterprise and Industry

However, it is not always possible to adjust a work place to psychomotor capabilities of human in an optimal way. Such lack of adjustment can result in professional illnesses of employees. That is why, employees' psychomotor capability is included in provisions of law. The importance of psychomotor capability in work performance is included in the Order of the Minister of Work and Social Policy of 28 May 1996 concerning types of work requiring particular psychomotor efficiency (Journal of Laws no 62, item 287). The list covers 36 types of work that require particular psychomotor efficiency. Also in the Order of the Minister of Health and Social Welfare of 30 May 1996 on carrying out of medical examinations of workers, scope of preventive health care and physician's notices issued in cases prescribed by the Labour Code, which provides six categories of noxious and arduous factors, that can occur at work places [9, 10]. The fifth category includes Other factor, such as:

- Adverse psycho-social factors:
    - Threats resulting from constant significant information inflow and readiness to respond;
    - Threats resulting from working on a decisive and related with responsibility positions;
    - Threats resulting from exposure to death;
    - Threats resulting from work monotony;
- Works requiring full psychomotor efficiency;
- Works at heights;
- Working on forced position;
- Work requiring monotype movements of limbs;
- Work requiring constant and long-lasting effort of voice;

Machines and devices are mainly used for production and services. In the manufacturing process, human work must be considered as well. It constitutes the basic factor: launches and controls any manufacturing factors. The efficiency and productivity of machines depend on the fitness of a worker, his qualifications and work intensity. Natural traits and capabilities of workers are not uniform, but human can adjust to various work conditions.

According to studies, accidents and work disturbances occur when there is a disorder of motor, psychic or mental coordination, caused by various reasons. A lack of concentration causes e.g. a lack of stability of the whole system and often it is a threat to health and life of human, what is a direct cause of accidents, e.g. of drivers. Due to that, it is essential to eliminate fatigue, overworking, poor work conditions, excessive pace of work, and stress. Management over use of machines consists in giving and receiving orders and controlling as well as coordinating such process. Thus, psychomotor efficiency of workers should be regularly examined, what may have a deciding role in allowing one to perform particular work.

2 Testing psychomotor efficiency of employees using ePiórkowski tool

ePiórkowski tool allows conduction of psychological tests aimed at evaluation of visual and motor coordination using a linear arrangement of incentives. Testing psychomotor efficiency is addressed to examining employees working at positions requiring particular psychomotor efficiency. Figure 2 presents the traditional Piórkowski apparatus on the left and ePiórkowski application on the right. A participant under the test is to click the button which is currently highlighted.

 DOI: 10.1051/matecconf/20179406017

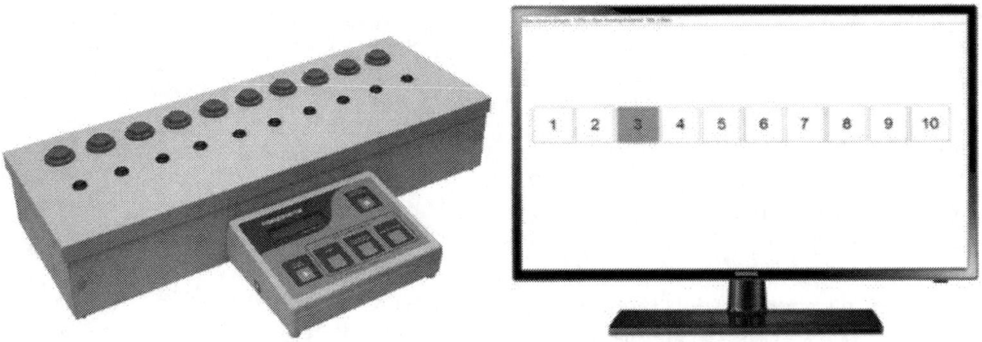

Fig. 2. Left - Piórkowski apparatus [1] in the traditional version; right – software version

Source: own elaboration on the basis of [1]

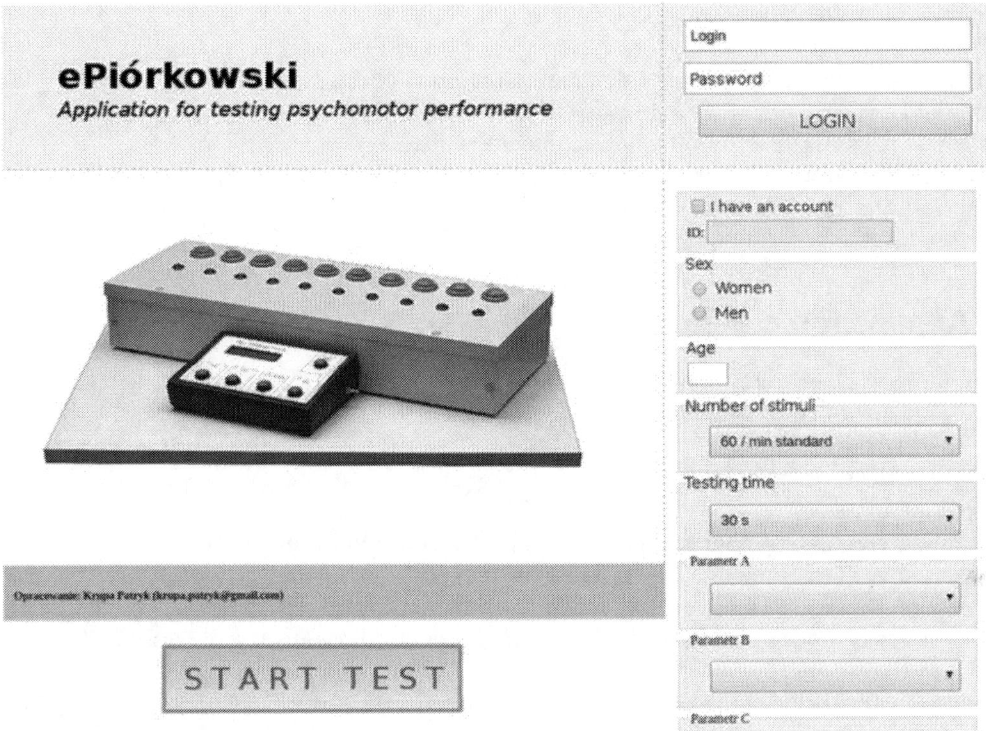

Fig 3. Interface – ePiórkowski [own elaboration]

The software was developed on the basis of the standard Piórkowski apparatus. It is intended to run the same amount of psychomotor efficiency examinations as in the case of the traditional device. The application is developed in order to enrich the testing device by additional functions, which the traditional instrument does not offer. The software version of the testing device is composed of several modules which can be later developed by further functions.

Currently, the software tool has the following inbuilt functions:

- database;
- analyser of results with variables: age, gender, date (exact date or a period), hour (exact hour or range of hours), pace, number of incentives, user ID, analysis ID, additional analysis parameter;
- possibility of users management (analysed persons) and administrators;

- option of automatic generation of charts, including: normal and faulty reactions, obtained times, median of responses, particular response (in the case of selected analysis);
- option of analysis impression with additional parameters;
- adding comments to tests.

Features tested by apparatus / software [1]:
- time of psychomotor response in forced pace;
- visual-motor coordination;
- concentration;
- fitness in noticing elements displayed on the screen *;
- fitness in using pointing tools (computer mouse)*;

* = Extended features (not available in the standard device).

The apparatus allows conducting tests in forced paces: 24*, 30*, 60, 75, 93, 100*, 107, 120*, 125, 150 incentives per minute and the testing time of: 30, 60, 90, or 120 sec.

The apparatus calculates [1]:
- number of presented incentives,
- number of proper responses,
- number of incorrect responses * (including incorrect clicks *, omitted incentives*),
- average response time for all incentives*,
- average response time for only correct incentives*,
- longest time of waiting for response*,
- shortest time of waiting for response*,

* = Extended features (not available in the standard device).

3 Application construction

In order to work properly, the application requires:
- PC with any operating system;
- HTTP server;
- MySQL database server;
- PHP interpreter;
- pointing tool, e.g. computer mouse (if tests are conducted in such form);
- touch screen (recommended for tests by the author).

Figure 4 present the general scheme of the software operation . The application construction is based on interrelated files, including PHP, JavaScript programming languages and HTML markup language. The application design is developed using CSS – cascade style sheets. Files are divided in accordance to their role in the software (each block in the scheme is a separate file).

The application operation way is presented using block diagrams in Figure 4, which illustrates possibilities of moving around the program.

DOI: 10.1051/matecconf/20179406017

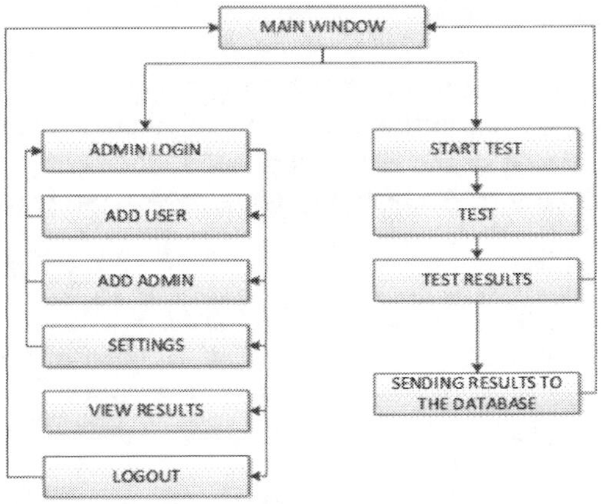

Fig. 4. Scheme of the application functioning [own elaboration]

Figure 5 presents variants of conducting tests. As you can see, the test can be conducted in numerous ways, depending on requirements:

- user has no account – anonymous test; tested person does not provide any personal data (mode used for appointing tested persons with the apparatus);
- used has no account – testing with basic information about tested person (age, gender);
- user has an account, name and ID number - we know his/her age and gender (an account must be created by an administrator);

There are also allowed additional combinations with three optional parameters A, B, C, which can be combined by selecting any configuration, e.g. parameters A+B, A+C or A+B+C. Parameters are generated by the administrator, who can input additional information about the test. The administrator can include places and conditions of conducting tests or factors influencing tested persons, e.g. assembly room, noise of over 85dB, lighting below 500 luxes. Green arrows indicate data which can be input but they are not required. The test can be initiated even if these options are not selected.

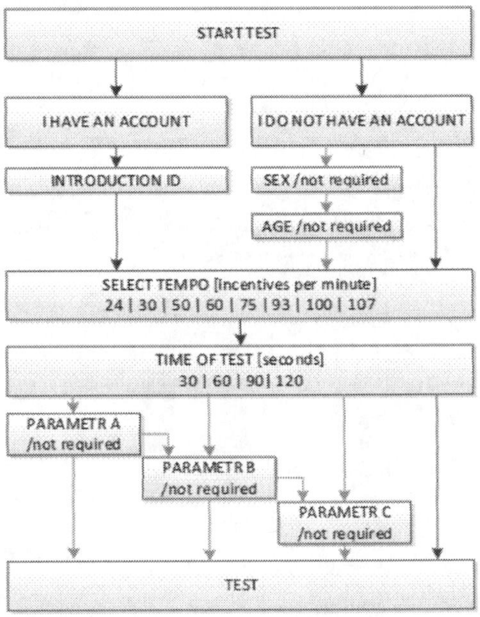

Fig. 5. Scheme of test start [own elaboration]

4 Program interface - Test

Figure 6 presents the test window of ePiórkowski application. Similarly as in the classical apparatus, there are presented ten buttons, only one of which sends stimuli in proper time intervals. After the completed test, the user is directed to the results page (Figure 7).

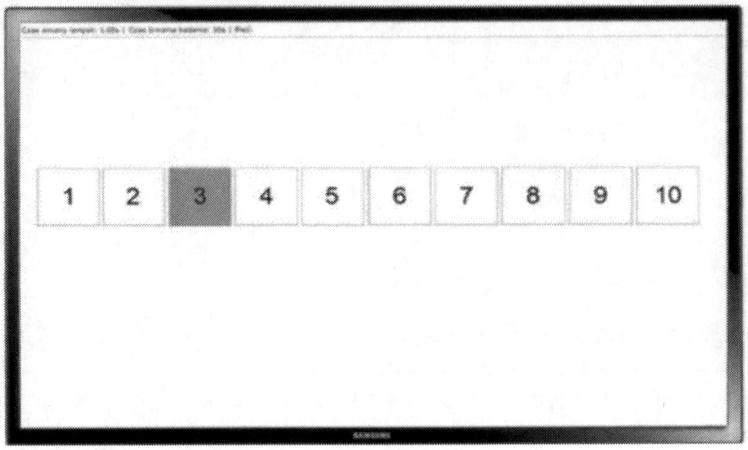

Fig. 6. Interface – test [own elaboration]

The test summary is displayed automatically immediately after its completion. The window of the final result is presented in Figure 7 and it includes results of particular responses and their summaries. Results are divided into three categories:
- 1 (green background) – normal response,
- 0 (red background) – invalid response (clicking a wrong button),
- CM (red background) – invalid response (lack of any response);

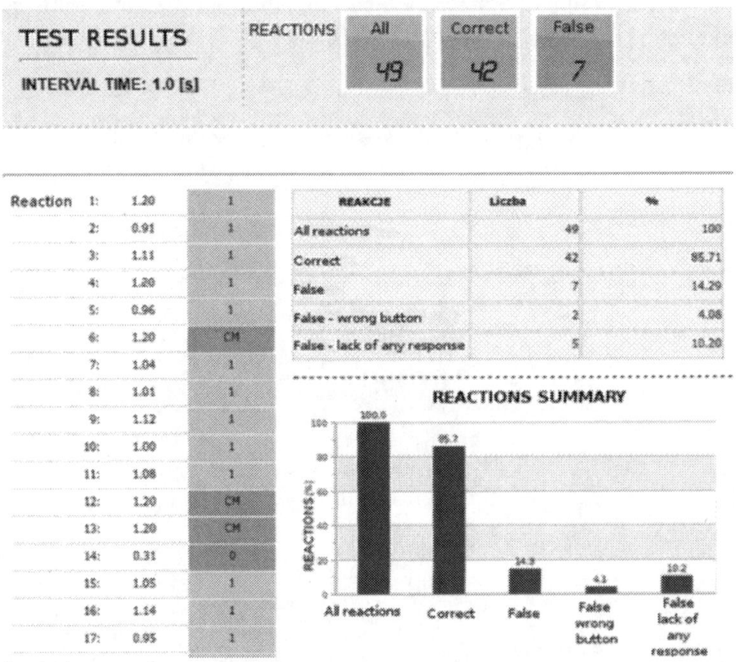

Fig. 7. Interface - test results [own elaboration]

5 Program interface – Administration panel

Persons allowed to log in to the application are those having own login and password. Such persons have an access to the administration panel, which includes:

- test results analyser,
- all test results,
- users and administrators management,
- management of test additional parameters (each admin can add parameters according to own needs).

What is more, the panel allows addition and removal of users and administrators (Figure 8). The access to the panel is password protected in order to keep database safe from an access unauthorised persons. The analysed results are presented in the form of tables and dynamically generated charts. After a selection of tests in accordance to selected parameter, acquired information can be easily copied to word processors or other applications.

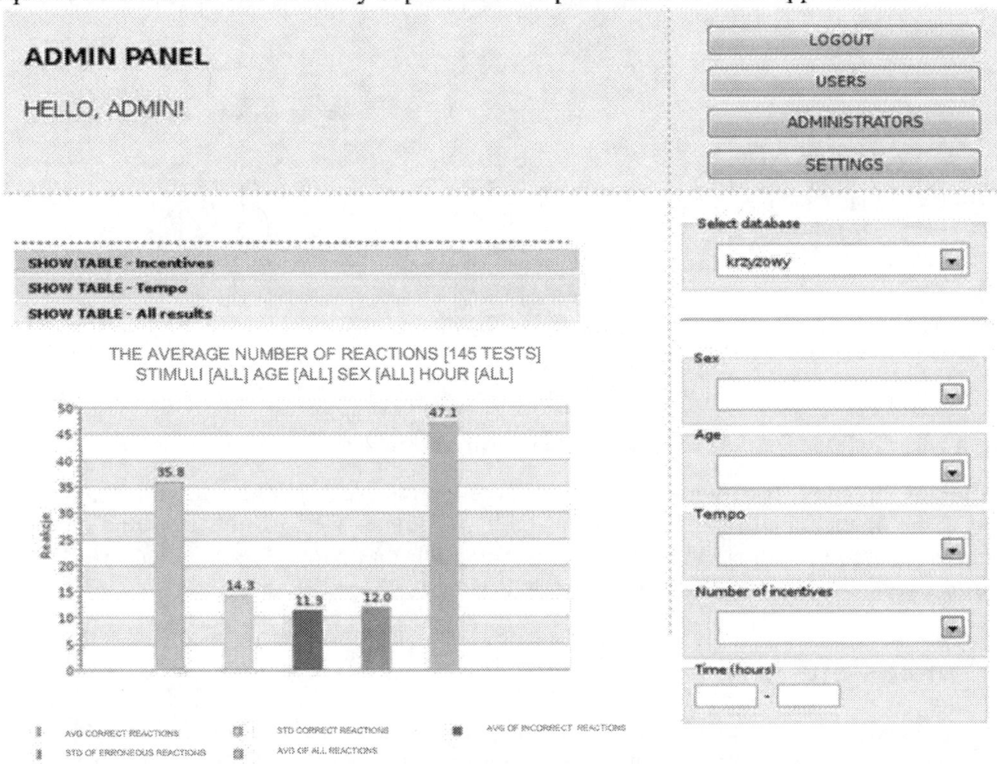

Fig. 8. Interface – administration panel [own elaboration]

6 Used tools and technologies

This part of the work presents all tools and technologies thanks to which this application for testing and analysing psychomotor function was developed. The following generally available free software was used for the application development

- Linux Mint – operating system,
- Apache – www server (application available for MS Windows and Linux),
- Bluefish – editor of HTML, PHP, CSS, SQL codes
- MySQL – database.

The application was developed using various programming languages, including:

- HTML,
- PHP,
- JavaScript,
- SQL,
- CSS.

Thanks to the above mentioned tools and technologies, one can state that it is possible to develop a full value software using solely non-profit technological solutions and applications [5, 6, 8].
The application utilises technologies, which enable the conduction of tests using numerous computers simultaneously. It is possible thanks to a local network and server applications.

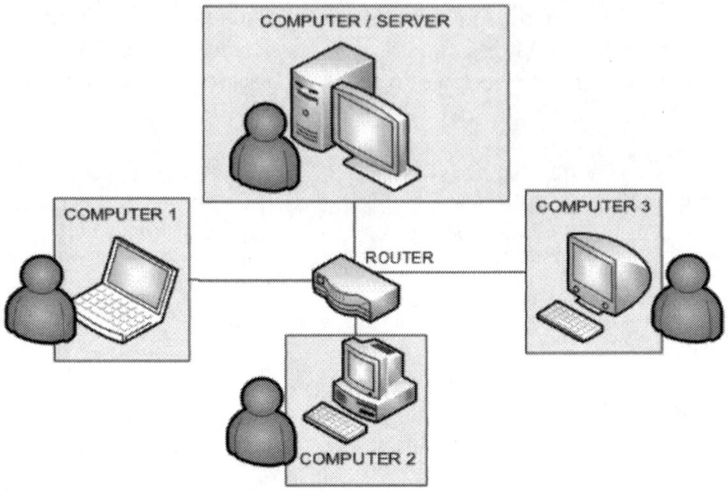

Fig. 9. Diagram – Network use of the application [own elaboration]

In order to conduct network tests, several computers are required. These devices do not need to be of the same type (PC, laptop, tablet, Mac), as well as not the same operating systems (Windows, Linux, MacOS). It is just enough that computers handled operation of local networks and had web browsers installed (Figure 9).

7 Conclusions

Tests of people working as operators of operating panels of multi-functional and multi-task technological devices and at monitoring (works requiring full fitness - control desks, signalling devices, control centres), should take the work position character into consideration. This means that tests should not be conducted using conventional devices but computer software. That can fully allow identify not only response time, stress-resistance, concentration, fatigue, but also worker's adaptation to work with computer screens (including touch screens).

Research results can be significantly influenced by past experience in working on computers (also with computer games). In numerous cases, the author experienced a negative approach to computer tests exhibited by elder people. On the other hand, young persons had a positive attitude to computer assisted studies and regarded them as play [7]. The main aim of the work was to develop an application, enabling testing of psychomotor efficiency. The developed application perfectly represents the operation method of the traditional instrument. What is more, the application is equipped in additional, very helpful tools for analysing the test results. The concept assumed the possibility of further development of programs.

Computer assisted analyses have an advantage of technical possibilities and a further development options. In further directions of studies, one can conduct tests using computer apparatuses, which will be related with the influence of e.g. [7]:
- colour of displayed incentives,
- size (of displayed picture),
- shape of incentives.

References

1. Alfa Electronics - nowoczesne aparaty do badań psychologicznych (Alfa Electronics - modern equipment for psychological research), http://www.alfa-electronics.eu/index.php/aparat-piorkowski (2016)
2. EN ISO 6385:2004, Zasady ergonomiczne w projektowaniu systemów pracy (Ergonomic principles in the design of work systems) (2004)
3. I. Fraser, *Guide to application of the Machinery Directive 2006/42/EC*, 2nd Edition (2010)
4. I. Gabryelewicz, P. Krupa, J. Sadłowska-Wrzesińska, Koncepcja aplikacji wspomagającej ocenę poziomu kultury bezpieczeństwa w przedsiębiorstwie. Część 1. Zastosowane narzędzia i technologie, Logistyka, 5 (Concept of application assisting the assessment of the level of the culture of the safety in the enterprise. Part 1. Applied tools and technologies, Logistics, 5) (2015)
5. L. Welling, L. Thomso, *PHP and MySQL Web Development, Second Edition*, (2003)
6. *HTML & Web Design Tips & Techniques* (McGraw-Hill, 2002)
7. P. Krupa, E. Kowal, G. Dudarski, Ocena sprawności psychomotorycznej z wykorzystaniem aplikacji komputerowej Zeszyty Naukowe Politechniki Poznańskiej. Organizacja i Zarządzanie, 67, (Evaluation of the psychomotor efficiency with using the computer application, Scientific booklets of the Poznań Technical University. The organization and Management, **67**) (2015)
8. R. Sokół., *ABC Linux* (Helion, Gliwice, 2004)
9. Rozporządzenie Ministra Pracy i Polityki Socjalnej z dnia 28 maja 1996 r. w sprawie rodzajów prac wymagających szczególnej sprawności psychofizycznej (Dz. U. Nr 62, poz. 287); (Order of the Minister of Work and Social Policy of 28 May 1996 concerning types of work requiring particular psychomotor efficiency (Journal of Laws, **62**, 287) (1996)
10. Rozporządzenie Ministra Zdrowia i Opieki Społecznej z dnia 30 maja 1996 r. w sprawie przeprowadzania badań lekarskich pracowników, zakresu profilaktyki opieki zdrowotnej nad pracownikami oraz orzeczeń lekarskich wydawanych do celów przewidzianych w Kodeksie pracy (Regulation of the Health Minister and the Social Welfare of 30 May 1996 conducting medical check-ups of employees, scope of the prevention of the health care above employees and medical certificates issued at targets stipulated in the Labour Code) (2015)
11. U. Vazirani, C. Papadimitriou, S. Dasgupta, *Algorytmy (Algorithms)*, (PWN, Warszawa 2012)

Analysis heat transfer in engineering barriers used FEM and experimental method

Dariusz Andrzejewski[1,*], and *Mariusz* Prazmowski[1]

[1]Opole University of Technology, Faculty of Mechanical Engineering, 5 Mikolajczyka St., Opole, Poland

Abstract. Purpose of the calculations is to determine the temperature field at various points in the area in question. Estimate the losses of heat given off by the analysed structures to the environment. Numerical algorithm based on the Finite Element Method for steady flow of heat and areas of heterogeneous. To analyse was used windows with double-glazed unit made PCV with different option gasses and glasses. Verification the simulation which used numerical procedures by experimental method.

1 Introduction

To analysis we have an engineering barrier (windows panel set double glazed resized specially for experimental check in heater chamber with different glasses and used gases) which will be realised algorithm numerical simulation. Besides checking the optimally distances for provide the best insulation between double glazed set has been proven impact on the size panel set for thermal insulation. Specify the temperature field in the distinguished points of the area in question arising from the way the geometrical discretization of the continuum to estimate the losses of heat given off by the design of this analysed. To solve a issues was applicable finite element method (FEM) in variant published in the[1].To finally calculation was used author program to automatic generation net differential where finally equations was develop iterative method Gauss-Seidel [2,3].

We try find optimum distance between two set of glasses in window panel for smallest heat transfer coefficient, use correct glass, gasses and material for construction and design in order to improve save energy used author program and compare to commercial program "Fluent". As well opportunity to use the internal heating plates of glass to eliminate thermal bridge causing condensation of the water at the bottom of the glass. Opportunity to use specific system of cooling the glass to blocked move the heat from the outside to the inside.

2 Material and methods

2.1 Mathematics process

In mathematics problem it's described differential equations [2-4]:

*Corresponding author: d.andrzejewski@po.opole.pl

$$\frac{\partial}{\partial x_j}\left[\lambda_j\left(\frac{\partial T_j}{\partial x_j}\right)\right] = 0e{=}1, 2,\ldots,E \tag{1}$$

λ_e-thermal conductivity

T_e-temperature field in area

For testing calculation checked effective value for numerical calculation. Any convection in "empty"area made effective coefficient instead real thermal conductivity.

Equations for calculation:

$$\frac{\partial}{\partial x_1}\left[\lambda_e\left(\frac{\partial T_e}{\partial x_1}\right)\right] + \frac{\partial}{\partial x_2}\left[\lambda_e\left(\frac{\partial T_e}{\partial x_2}\right)\right] = 0 \tag{2}$$

Adopted boundary conditions Robin:

$$x \in \Gamma_z: -\lambda_e\frac{\partial T_e}{\partial n} = \alpha_z[T_e - T_z] \text{ and } x \in \Gamma_w: -\lambda_e\frac{\partial T_e}{\partial n} = \alpha_w[T_e - T_w] \tag{3}$$

w - area inside, z-area outside, α- heat transfer coefficient $\partial/\partial n$- derivative normal to the edge of the area.

At the contact surface there are continuity conditions [4, 5]:

$$x \in \Gamma_{e,\grave{e}}: -\lambda_e\frac{\partial T_e}{\partial n} = -\lambda_{\grave{e}}\frac{\partial T_{\grave{e}}}{\partial n}, T_e = T_{\grave{e}} \tag{4}$$

e, è- ideal contact surface, for surface no flux conditions

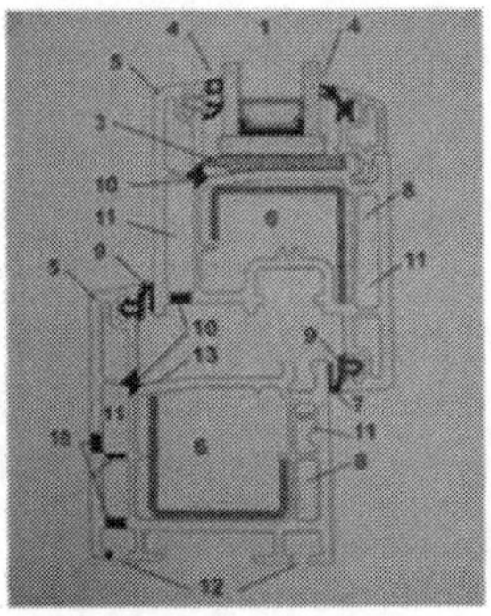

Fig.1. Analysing area: 1-glasses, 3-distance block, 4-gasket, 5-PVC profile, 6- aluminium strengthening, 7-Gasket connector, 8,11- air chamber, 10- drainage hole, 12-wall connector [6].

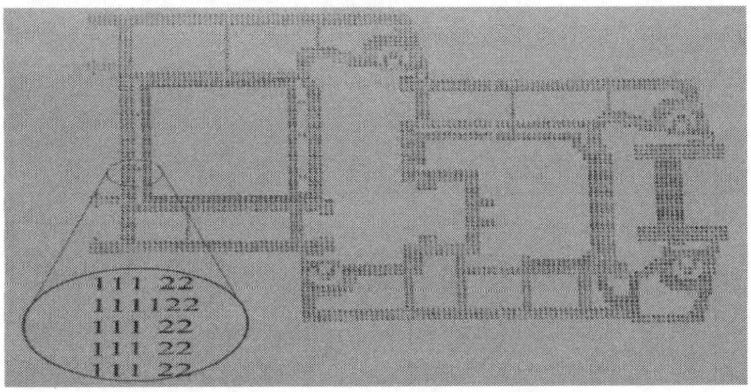

Fig.2. Consideration area and matrix identification [6,13-16].

2.2 Numerical model

Consider the area was covered with a net differential (Fig. 3).

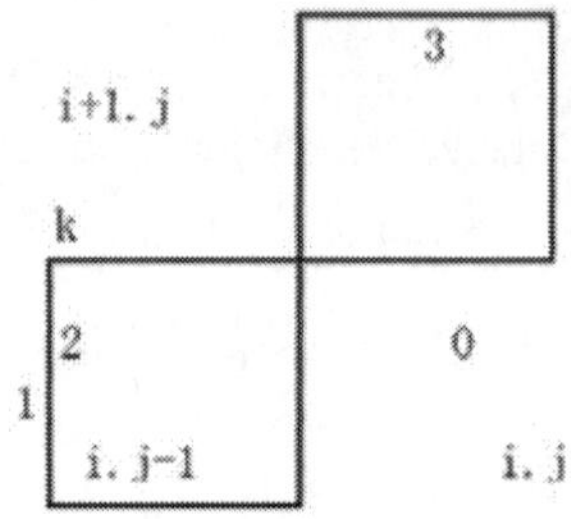

Fig. 3. Five points star net.

Approximation equations shows the relation between central node and neighbor nodes.

$$\frac{T_{i+1,j}-T_{i,j}}{R_{i+1,j}}\Phi_{i+1,j}+\frac{T_{i-1,j}-T_{i,j}}{R_{i-1,j}}\Phi_{i-1,j}+\frac{T_{i,j-1}-T_{i,j}}{R_{i,j-1}}\Phi_{i,j-1}+\frac{T_{i,j+1}-T_{i,j}}{R_{i,j+1}}\Phi_{i,j+1}=0 \quad (5)$$

R- resistance to heat flow, $\Phi=\frac{1}{h}$.

$$A_0T_{i,j}-A_1T_{i+1,j}-A_2T_{i-1,j}-A_3T_{i,j-1}-A_4T_{i,j+1}=0 \quad (6)$$

$$T_{i,j}=\frac{A_1}{A_0}T_{i+1,j}-\frac{A_2}{A_0}T_{i-1,j}-\frac{A_3}{A_0}T_{i,j-1}-\frac{A_4}{A_0}T_{i,j+1} \quad (7)$$

2.3 Numerical realization

The area was covered with net it has steps h=0,001m [13,14].

Area which was considered has got 10576 points. The matrix type integer (Fig.2).

Window size: S 1465mm x H 2195mm, glass 4mm,distance between glazing: 12-18mm,T_{inside}=20^{0}C, $T_{outside}$ = -20, -15, -10, -5,0 (in subsequent calculations),α_w=8.1W/m^{2}K, α_z=23 W/m^{2}K, λ_{sz1}=0.744 W/mK, λ_{sz2}=1.338 W/mK.

Numerical realization was done used special for this prepared author program.

3 Results and discussions

Results simulation show parameters k and q for different set points temperatures in term of –20 ° C to 0° C. When we take into account λ_{ef} results was increase about 57.5% what good correlation to real experimental results.

Table 1. Numerical simulation results.

N°	$T_{z\,(outside)}$ [C^0]	$T_{w\,(inside)}$ [C^0]	q [W/m^2]	k [W/ m^{2}K]
1	- 20	20	48.21	1.205
2	-10	20	36.17	1.206
3	0	20	24.12	1.206
When changed for effective conductivity coefficient ($\lambda_{ef}=\lambda_{air}$)				
4	-20	20	75.46	1.89
5	-10	20	56.60	1.89
6	0	20	37.72	1.89

Graphical presentation simulation which result show in Table 1 (Author program).

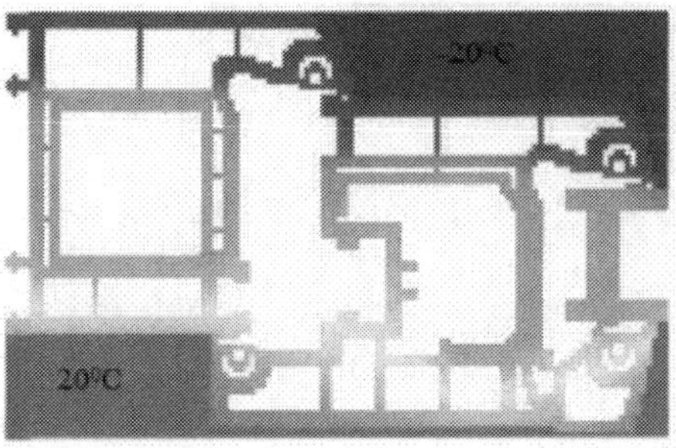

Fig. 4. Window set: -20^{0}C outside 20^{0}C inside.

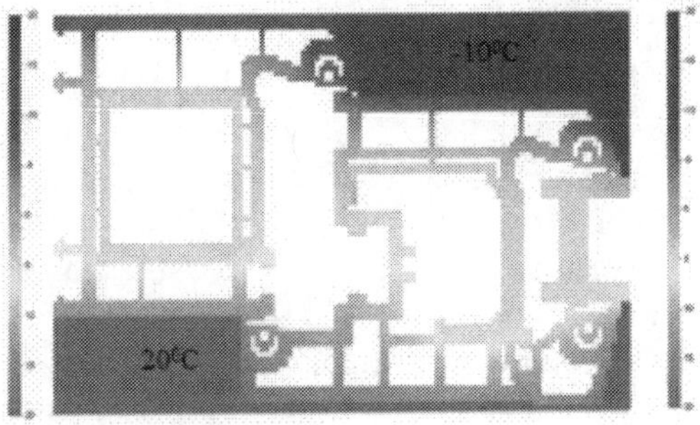

Fig. 5. Window set: -10^{0}C outside 20^{0}C inside.

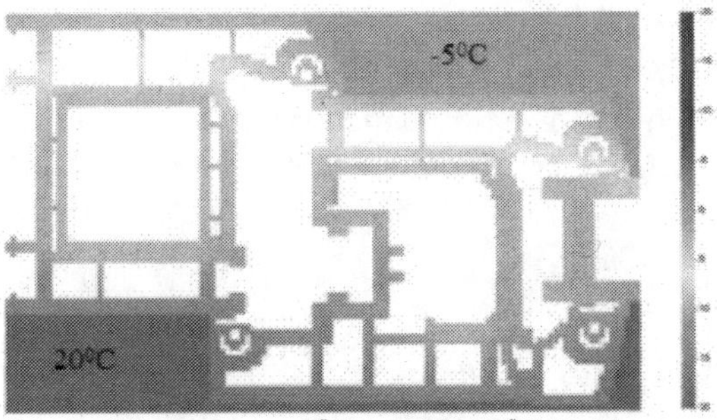

Fig. 6. Window set: -5^{0}C outside 20^{0}C inside.

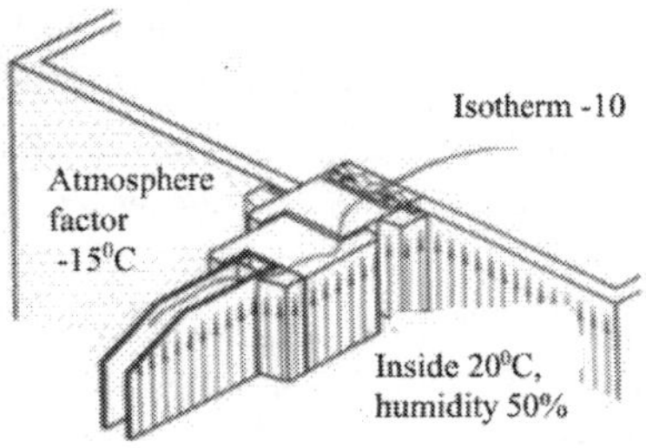

Fig. 7. Isotherm -10^{0}C inside panel [1].

Graphical presentation show areas when different colours are assigned to a specific temperature. Area in specific colour show distribution temperature for interpretation when we have place which must be structural remodel.

Table 2. Show termoisolation optimum distance between glasses, size affected heat transfer coefficient and gases have got changes for result [7].

N°	Distance between glasses [mm]	Panel H[mm]	gas	T outside[°C]	glass	Heat lost[J]	k[W/m²K]
1.	15	1400	Air	-20	quartz	92	1.64
2.	15	1400	Argon	-20	float	67.7	1.21
3.	15	1400	Air	-20	float	83	1.48
4.	15	1400	Air	-10	float	61.4	1.46
5.	15	1400	Air	0	float	37.5	1.34
6.	18	1400	Air	-20	float	83.2	1.49
7.	12	1400	Air	-20	float	90.2	1.61
3'	15	1400	Air	-20	float	83	1.48
8.	15	2600	Air	-20	float	154.1	1.48

In table 2 we have results part of interesting simulations which shows optimal distance between glasses (15 mm) which guarantee maximum isolation. When we increases or decreases the distance between glasses, k changes what means heat transfer have to high level and we lost optimum termoisolation. Similar situation we have when we use quartz glass not float glass. Heat transfer coefficient increase about 11% when we have in windows quartz glass. If we use different gasses not air, example Argon or Krypton heat transfer coefficient ensure better termoisolation about 22%. When we would like the same result when we used Argon or Krypton we must between glasses used "air vacuum".

Interesting situation we have when we used higher window panel and compare to standard panel.

Heat transfer coefficient it is constant. Meaning highest panel set not decrease termoisolation.

We were check different construction as well for find better opportunity. Double glasses window which have between glasses on the middle spectroselective foil (in PVD technology was sputtered metal oxides). Results: k =1 W/m^{2}K. Radiation UV=0. Short IR only 6%. Because this foil stopped radiation longer than 380nm flowers don't have light to life!

As well shadow coefficient increased to 0.66, resulting for this we have better protection against overheating. Visibility 88%.

Graphical presentation simulation which result show in Table 2.

Convection which we see on the Fig. 8. and Fig. 10 was simulation done in commercial program fluent 5 shows Bernard cells. Fig. 8, 9, 11,12 shows graphical presentation results from Table 2.

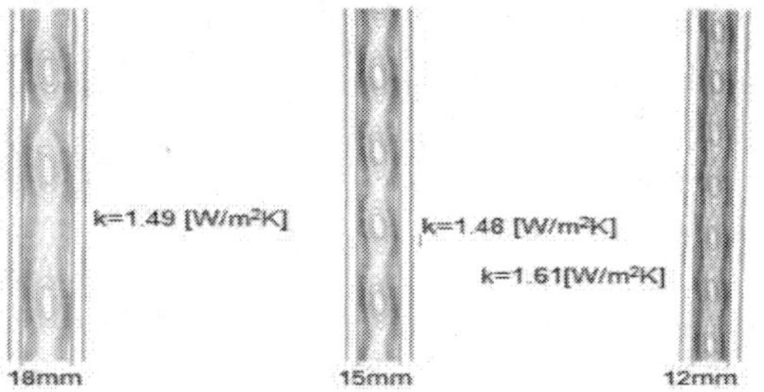

Fig. 8. Set of glass for optimality distance between glazed [8].

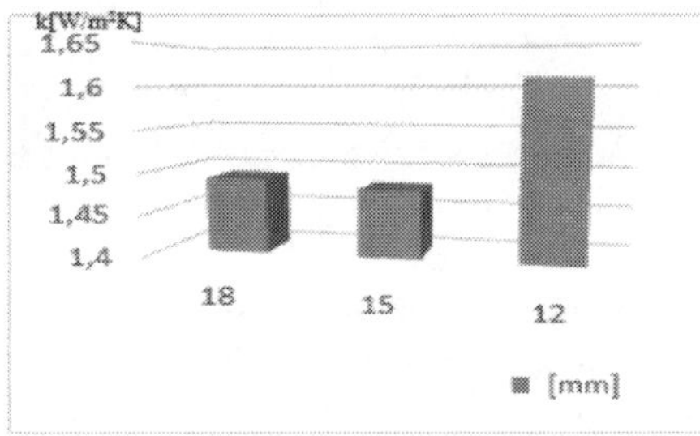

Fig. 9. Optimum distance between glasses.

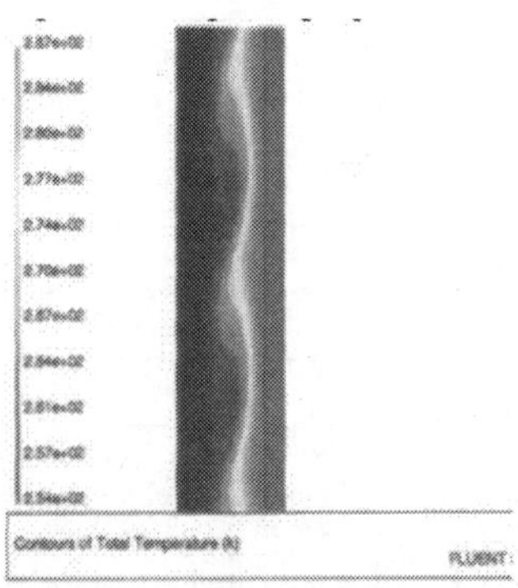

Fig. 10. We have Bernard cells characteristic for free convection in close space between glasses. [7, 9].

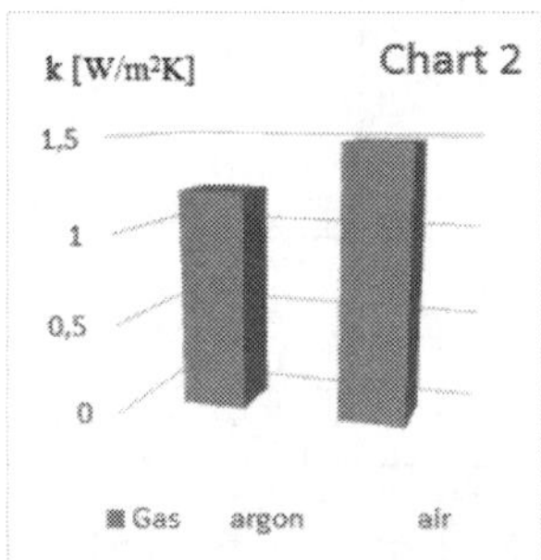

Fig. 11. Affect heat transfer coefficient in glass panel used different gasses [10].

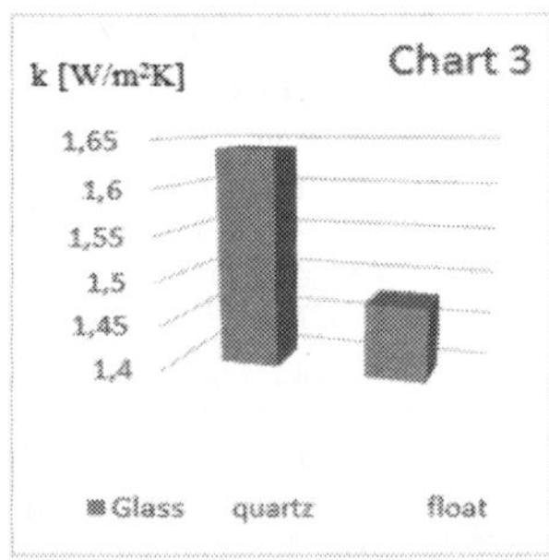

Fig. 12. Affect heat transfer coefficient in glass panel used different glass [10].

4 Experimental study

Results measured in heating chamber have a very good correlation with simulation results. Experimental verification helped confirm the correctness of the tests simulation.

Differences of 3% between the method of calculation and experimental confirms that the application of the calculation method is justified in order to search for the optimal design solution for partitions heat.

Windows samples were tested in a heating chamber [13].

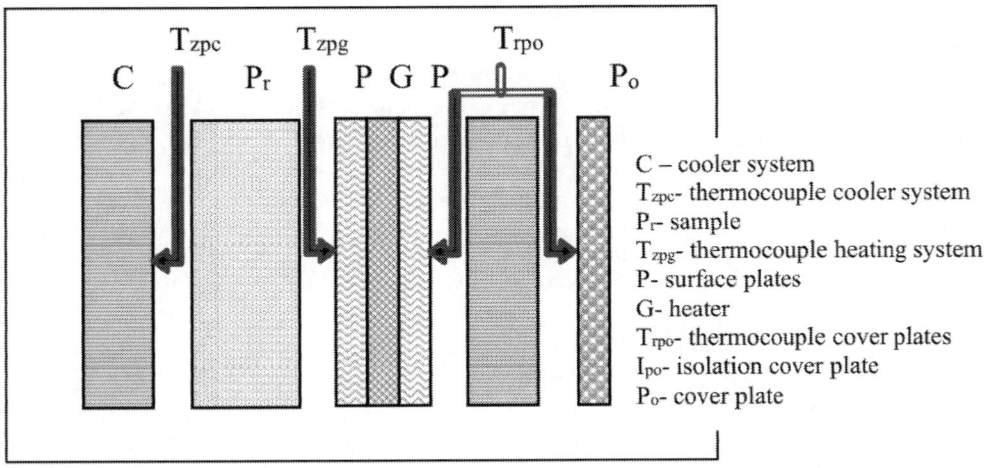

Fig. 13. Heating chamber scheme [11, 12].

Parameters for samples: T_1=20° C, T_2= -20° C, h_e=23 W/m^{2}K, h_i=8 W/m^{2}K

Table 3. Results measurements in heating chamber.

Set of glass in panel	Gas between set of glasses	k [W/m^{2}K]
4-15-4 float	air	1.49
4-15-4 quartz	air	1.67
4-15-4 float	air	1.62
4-15-4 float	air	1.51
4-15-4 float	argon	1.23

5 Conclusions

1. Between both methods we have a good correlation simulation where for set 4-15-4 float was k=1.48 W/m^{2}K, used argon k=1.21 W/m^{2}K and for quartz k=1.64 W/m^{2}K and experimental where was 4-15-4 float was k=1.52 W/m^{2}K, used argon k=1.23W/m^{2}K and for quartz k=1.67 W/m^{2}K means only 3% difference, this allow the use of simulation methods in research application.
2. We found optimum distance between glasses in window panel for the best isolation, it is 15mm between glasses in window panel. If we try increase distance between glasses slowly increases k (from k=1.48 W/m^{2}K for 15mm to k=1.49 W/m^{2}K for 18mm), if we try decrease distance much faster changed k (from k=1.48 W/m^{2}K for 15mm to k=1.61 W/m^{2}K for 12mm)
3. Used different gasses between glasses changed termoisolations in set of window. Argon changed approximately 26% heat transfer coefficient from k=1.48 W/m^{2}K(air) tok=1.21 W/m^{2}K (argon)
4. Different kind of glass affect heat transfer coefficient too. Set 4-15-4 float glass with air

k=1.48 W/m^{2}K but the same set with air for quartz glass k=1.64 W/m^{2}K this is different more than 9%.
5. Possibility used electric wire to heat up inside glass for better isolation and reduce heater bridge on the button for considered size window 160 W-300 W according to temperature outside building.
6. Alternative connection to solar panel on sun water panel to heat up.
7. Possibility connect to geothermal heat pomp for heat up frame and glass or in summer cool down to 12.2°C and will be works like small air-condition.

References

1. B. Mochnacki, J.S. Suchy, *Numerical methods in computations of foundry processes*, (PFTA, Cracow, 1995)
2. D. Andrzejewski, Mechanika **263**, 63 (2000)
3. J. S. Suchy, D. Andrzejewski and others, *State Committee for Scientific Research, Research project no. PC-2511*
4. S. J. Gdula, R. Bialecki, K. Kurpisz, A.J. Nowak , A. Sucheta, *Heat conduction* (PWN, Warszawa,1984)
5. E. Kostowski, series CMP, **97** (1988)
6. Information from KBE Sp .z o.o. Wroclaw
7. *Fluent Manual, Heat Transfer and Numerical Algorithms. Fluent Software* (MTG-96-173, Fluent Europe Ltd.)
8. D.K. Edwards, Heat transfer **12**, 4 (1976)
9. D. Andrzejewski, Mechanika **279**, 72 (2001)
10. Andrzejewski D., *Laboratory tests - Protection against noise and reducing energy needs through the use of modern joinery, III Forum Innovation and Technology Transfer* (Opole, 2001)
11. Polish Committee for Standardization, *Building components and building elements-thermal resistance and heat transfer coefficient PN-EN ISO 6946: 1998*
12. Polish Committee for Standardization, *Determination of the coefficient of heat transfer U PN-EN 674:1999*
13. J. C. Cavendish, D. A. Field, W.H. Frey, *Automatic Mesh Generation: a Finite Element/ Computer aided geometric interface. mathematics of Finite Elements and Applications, Mafelap V* (Academic press, 1984)
14. Y. Depeursinge, *Interactive Generation of Three dimensional Meshes., In: Mathematics of Finite elements and Applications Mafelap IV* (Academic Press, 1981)
15. J. Haüser, C. Taylor, *Numerical grid generation in Computational Fluid dynamics* (Pineridge Press, Swansea, 1986)
16. O.C. Zientkiewicz, *The Finite Element Method* (McGraw, London, 1977)

Research experimental of hipbone arthroplasties using tensometric models

Horatiu Bulea[1,*]

[1]Transilvania University of Brasov, Department of Manufacturing Engineering, Mihai Viteazu No.5, Braşov, Romania

Abstract. Hip prostheses functionality is limited in time, inter alia, by mechanical damage to the femoral components (protrusion, limestone resorption femoral loosening), and by of accurate positioning of the piece femoral. To assess the importance of accurate positioning of the piece femoral hip arthroplasties we performed a tensometric study on experimental models on femurs dehydrated and lubricated, the prostheses were implanted cemented in varum, normal and valgus position. The specific model distortions were determined by applying tensometric stance on the external and internal cortical in maximal stress positions and request using static strength values similar to those that require bone during normal gait. Aim of this experiment is to determine the maximum load areas femur which has been implanted partial prostheses.

1 Introduction

Hip replacement is a surgical procedure in which the hip joint is replaced by a prosthetic implant. Hip replacement surgery can be performed as a total replacement or a hemi (half) replacement. Such joint replacement orthopaedic surgery is generally conducted to relieve arthritis pain or in some hip fractures. A total hip replacement (total hip arthroplasty) consists of replacing both the acetabulum and the femoral head while hemi arthroplasty generally only replaces the femoral head.
Hip replacement is currently the most common orthopaedic operation, though patient satisfaction short and long-term varies widely. In a partial hip replacement also called a hemi arthroplasty, only the femoral head is replaced and the acetabulum is left in place [1]. A partial hip replacement is done mainly to repair fractured hips.

The prosthetic device typically used in partial hip replacements incorporates a one-piece solid metal ball and stem that replaces the femoral head and upper femur, unlike the two-piece stem and ball design used in total hip replacements [2].

The Austin Moore and Thompson prostheses are common examples of the one-piece design used in partial hip replacements. Other prosthetic devices are considered modular and allow for different combinations of stem, neck length, and head size [3].

Improper placement of the prosthesis and the resulting biomechanical disturbances within the hip joint (excessive elongation or shortening of the extremity or improper

* Corresponding author: bulea@unitbv.ro

rotation of the implant) are responsible for failure of hemiarthroplasty [4]. Inadequate p calcar seating, insufficient residual femoral neck length, insufficient metaphyseal fill, and errors in sizing the prosthesis are all associated with early failure of the hemiarthroplasty [5].

Because of technical possibility during the operation, (shape, size and position of the bone) prosthesis can be implanted always in normal position, resulting varum and valgus positions. The objective of the present study is to analyze and evaluate the importance of a correct femoral part positioning in arthroplasties.

2 The implant condition of the prosthesis and experimental methods

2.1 The implant condition of the prosthesis

Hip prosthesis functionality is time-limited, in part by the femoral piece mechanical impairment (protrusion, femoral calcar resorption, de cementation).

In order to evaluate the importance of a correct femoral part positioning in arthroplasties, we performed a tensometric study on experimental models: dehydrated and degreased thighbones in which Moore prostheses were implanted and cemented in varum, normal and valgus position.Fig.1 designed approximately similar to anatomical shape of femoral head size: 45 mm.

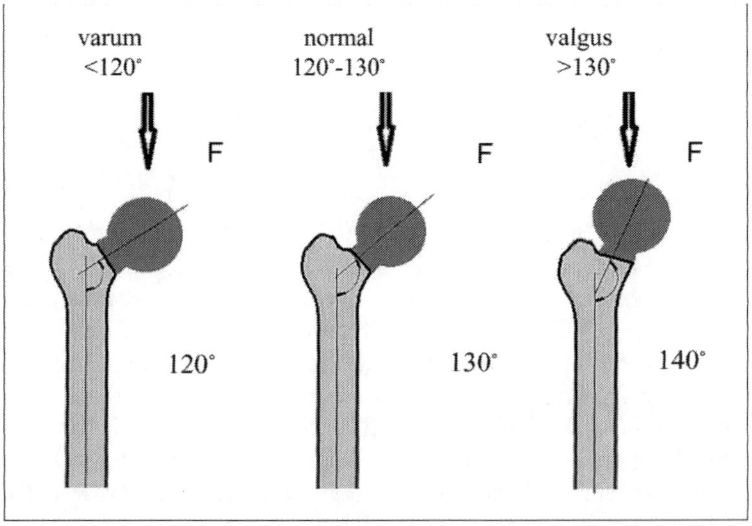

Fig. 1. Forces apply on spherical head of prostheses implanted and cemented in varum, normal and valgus position.

2.2 The experimental methods

The specific model distortions were determined by applying different forces on spherical head of prostheses in conditions similar to real ones they support bone in the human body and measurement with strain gauges deformations that occur on the surface of bone in areas highlighted on the external and internal cortical in maximal stress positions Fig. 2 and the experimental bone Fig. 3.

It is known that Young's modulus in bones has different values according to individual age and measurement area. Specific distortions were therefore determined only in points where tensometric stamps had been applied.

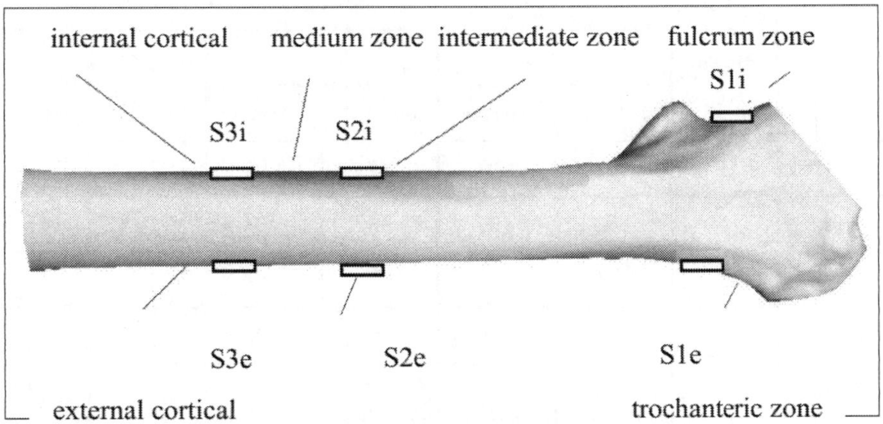

Fig. 2. Tensometric stamps highlighted on the external and internal cortical.

To conduct this experiment we are using the apparatus of Fig. 4.

The specific distortions Ds[mA] was measured for each tensometric stamp ten times for each of the five driving forces F_1 =250 [N], F_2 =300 [N], F_3 =350 [N],F_4 =400[N],F_5 =450[N], for each positions of prostheses: varum, normal and valgus, the values considered are the average of measured values, the results are presented in Table. 1 Forces have been developed by means of a mechanic dynamometer, calibrated in the physics laboratory of the Transylvania University.

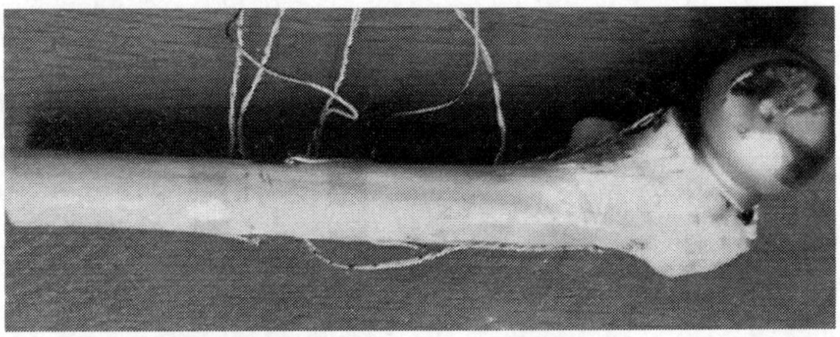

Fig. 3. Experimental bone with prostheses .

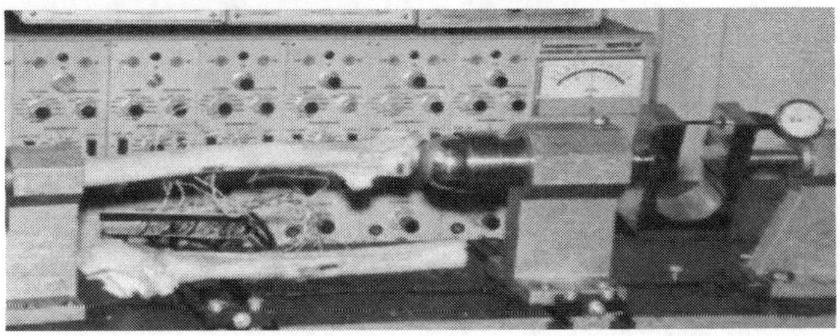

Fig. 4. Experimental installation.

Because the modulus of elasticity of bone material shows different values depending on the age of subject, measuring deformations were determined only specific points of placement of strain gauges. On external cortical they were obtained tensile stresses and compression on the internal cortical.

Table 1. Mean values of the Ds deviation unit in mA.

F [N]	Tensometric stamp	Ds[mA]					
		Varum		Normal		Valgus	
		internal	external	internal	external	internal	external
250	1	101	130	81	120	74	112
300	1	109	134	86	127	79	118
350	1	116	143	91	132	84	123
400	1	125	158	98	139	91	131
450	1	133	166	105	146	97	139
250	2	56	109	53	90	49	82
300	2	67	116	64	97	54	91
350	2	73	121	71	103	61	98
400	2	81	133	79	111	68	106
450	2	89	143	85	126	74	118
250	3	105	193	92	180	97	188
300	3	111	201	101	189	105	196
350	3	121	216	109	198	113	205
400	3	129	228	115	207	119	216
450	3	138	239	123	216	127	224

The variation of Ds [mA] parameter in the conditions presented above, are presented in Fig. 5 for tensometric stamps $S1_i$ on fulcrum zone and in Fig. 6 for tensometric stamps $S1_e$ on trochanteric zone in varum, normal and valgus position of implanted prosthesis.

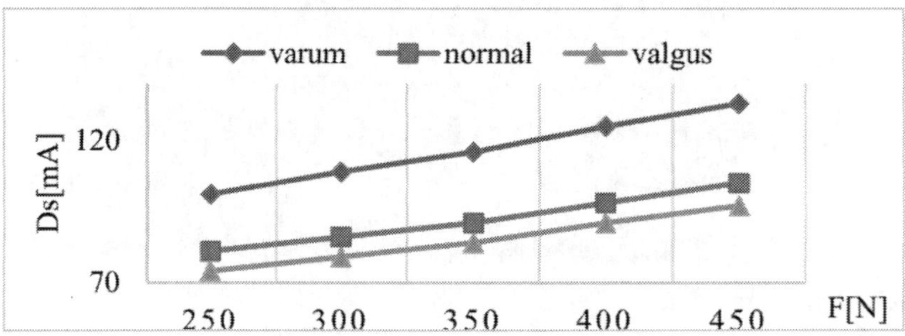

Fig. 5. The dependence of specific distortions Ds[mA] from variation of force request F[N] for tensometric stamps $S1_i$ on fulcrum zone.

DOI: 10.1051/matecconf/20179407002

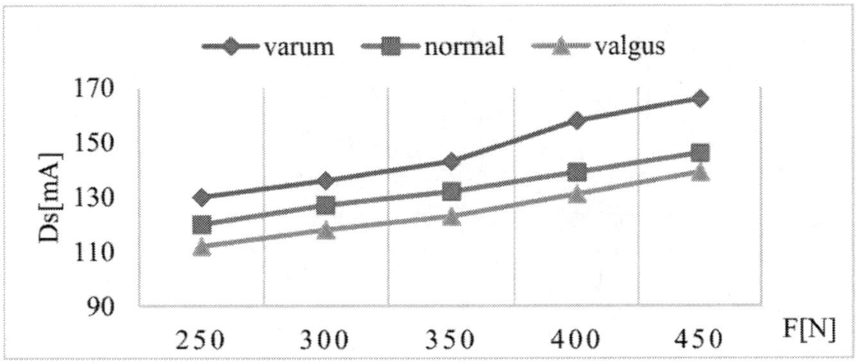

Fig. 6. The dependence of specific distortions Ds[mA] from variation of force request F[N] for tensometric stamps $S1_e$ on trochanteric zone.

The variation of Ds[mA] parameter in the conditions presented above, are presented in Fig. 7 for tensometric stamps $S2_i$ on internal cortical in intermediate zone and in Fig. 8 for tensometric stamps $S2_e$ on external cortical in intermediate zone in varum, normal and valgus position of implanted prosthesis.

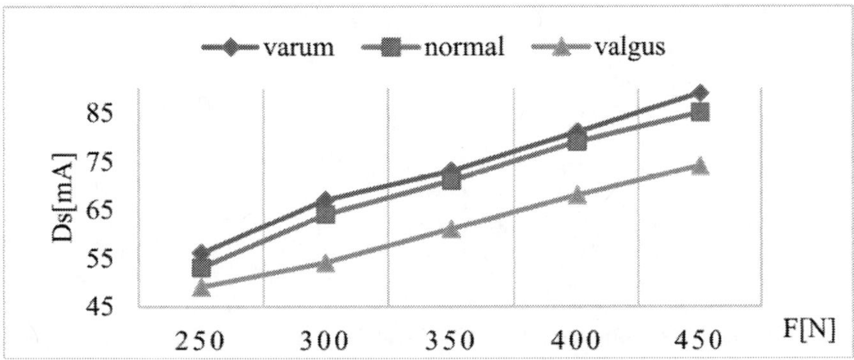

Fig. 7. The dependence of specific distortions Ds[mA] from variation of force request F[N] for tensometric stamps $S2_i$ on internal cortical.

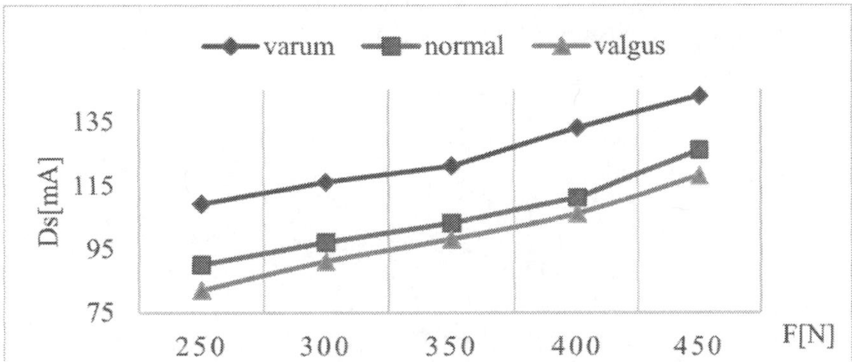

Fig. 8. The dependence of specific distortions Ds[mA] from variation of force request F[N] for tensometric stamps $S2_e$ on external cortical.

The variation of Ds[mA] parameter in the conditions presented above, are presented in Fig. 9 for tensometric stamps $S3_i$ on internal cortical n medium in varum, normal and valgus position of implanted prosthesis.

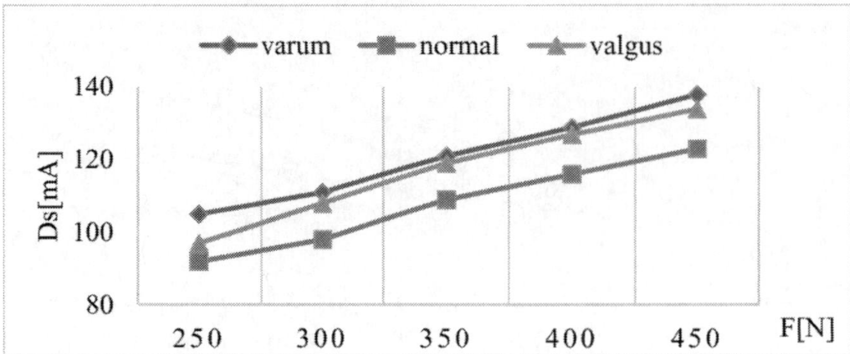

Fig. 9. The dependence of specific distortions Ds[mA] from variation of force request F[N] for tensometric stamps $S3_i$ on internal cortical.

The variation of Ds[mA] parameter in the conditions presented above, are presented in Fig. 10 for tensometric stamps $S3_e$ on external cortical medium zone in varum, normal and valgus position of implanted prosthesis.

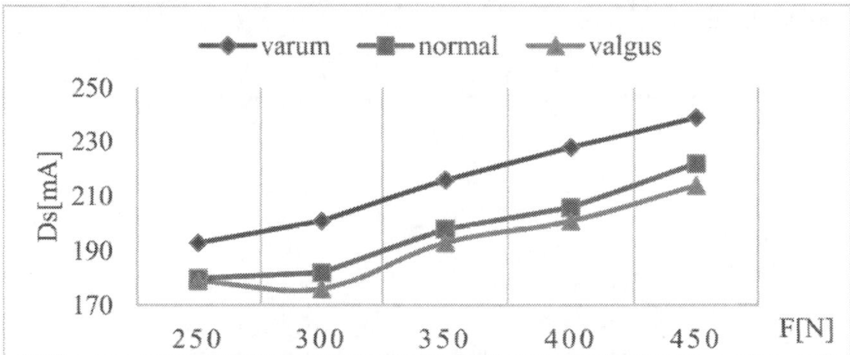

Fig. 10. The dependence of specific distortions Ds[mA] from variation of force request F[N] for tensometric stamps $S3_e$ on external cortical.

The variation of Ds[mA] parameter in the conditions presented above, are presented in Fig. 11 for tensometric stamps $S1_i$ on fulcrum zone and for tensometric stamps $S1_e$ on trochanteric zone in varum, position of implanted prosthesis.

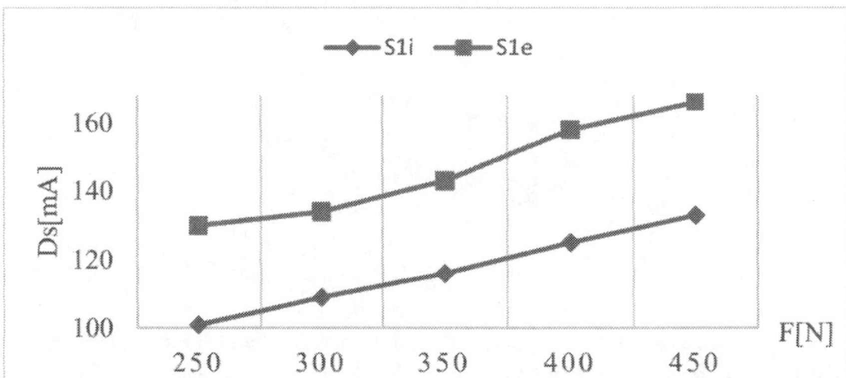

Fig. 11. The dependence of specific distortions Ds[mA] from variation of force request F[N] for tensometric stamps $S1_i$, $S1_e$ in varum, position of implanted prosthesis.

The variation of Ds[mA] parameter in the conditions presented above, are presented in Fig. 12 for tensometric stamps on intermediate zone S2i, for internal cortical and $S2_e$ on external cortical for normal position of implanted prosthesis.

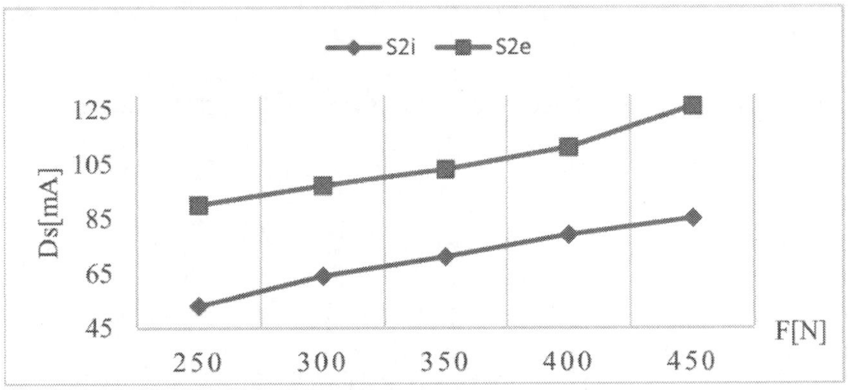

Fig. 12. The dependence of specific distortions Ds[mA] from variation of force request F[N] on intermediate zone for tensometric stamps $S2_i$, $S2_e$ for normal position of implanted prosthesis.

The variation of Ds[mA] parameter in the conditions presented above, are presented in Fig. 13 for tensometric stamps on medium zone S3i, for internal cortical and $S3_e$ on external cortical for valgus position of implanted prosthesis.

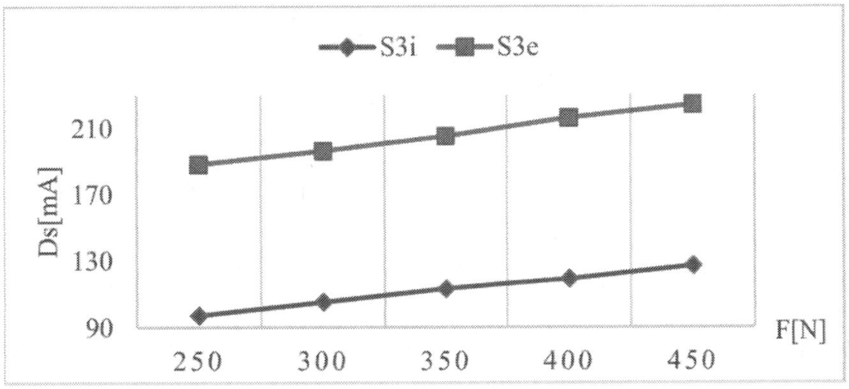

Fig. 13. The dependence of specific distortions Ds[mA] from variation of force request F[N] on medium zone for tensometric stamps $S3_i$, $S3_e$ for valgus position of implanted prosthesis.

3 Conclusions

Ascending the analysis results presented in tables and above graphs the following conclusions:

1. Analyzing the experimental parameters, important influence of deformations of the bone are: first the size of the force apply on spherical head of prostheses, and second the position in which it was implanted prostheses.
2. For same forces apply on spherical head of prostheses deformation being significant higher in varum position of prostheses.
3. For same forces apply on spherical head of prostheses deformation is significant higher in middle zone of bone tail end of prostheses.
4 .The cortical external requests are on cortical tension and on internal compression.
5. For same forces apply on spherical head of prostheses deformation is higher for external cortical that internal cortical in all zone of bone.

 DOI: 10.1051/matecconf/20179407002

6. With increasing strain force deformation increases more in varum position of the prosthesis for all zone.
7. Stress diagrams were set out for the specific distortions they showed an almost linear variation within the limits of the charges applied, thus proving that bones submit to Hooke's law.
8. The most important conclusion resulting from the experiments is that deformations are real, they lead to a request extra bone in the contact areas of the prosthesis.

References

1. V. Meulen, M.C.H. Allen, W.A. Giddings, V.L. Athanasiou, K.A. Poser, R.D. Goodman, S.B. Smith, R.L. Beaupré, Effect of hemi arthroplasty on acetabular cartilage, Project Reports. VA Palo Alto Health Care System's Bone and Joint Rehabilitation Research and Development Center (1996)
2. A. Mecke, I. Lee, J.R. Baker jr., M.M. Banaszak Holl, B.G. Orr, Eur. Phys. J. E **14**, 7 (2004)
3. **Coding for Hip Replacement Surgery, AHIMA, **25/13**, 26 (2013)
4. P. Bhosale, A.V. Suryawanshi, Total hip arthroplasty for failed aseptic Austin Moore prosthesisand, Amber Mittal Indian J Orthop. **46**, 297–303 (2012)
5. P. Weinrauch, Intra-operative error during Austin Moore hemiarthroplasty, J Orthop Surg **14,** 249 (2006)

MATEC Web of Conferences **94**, 07003 (2017) DOI: 10.1051/matecconf/20179407003
CoSME'16

Leading the creativity

Simona Sofia Duicu[1,*]

[1]Transilvania University of Brasov, Department of Manufacturing Engineering, Mihai Viteazu No. 5, Brasov, Romania

Abstract. The paper presents some consideration about creative learning in technical education. Over the last century, different theories were born about human hemispheres behaviour and the source of creativity. As the answer is not simple and complex cognitive function are required in engineering, may be is the best to associate creativity with other important concepts as originality, fast shift between rational and visual approaches, learning system development.

1 Introduction

In 1998 a world organization named "Campaign for Learning" leads a research on the human motivation of developing themselves, stretching their brains and improve the learning capacity. The conclusion is that both adults and young people enjoy learning new things encouraged by their current or possible future jobs rather than to their personal interests [1]. Adults prefer to learn doing practical things or self-study, young people prefer learning in groups.

Learning is an interesting process. In particular, technical education is following the '80 trends becoming more and more preoccupied by virtual reality and standardization. Since the beginning of the industrial revolution process, fragmentation increases productivity which is how left hemisphere of our brain imagines the world. This is an entirely and abstract model that has procedures (laws to do so) and defined area (boxes containing people and/or goods).

A better understanding of our brain cognitive functions can help us to become more productive, efficient and creative.

2 How our brain works

2.1 The map of the brain

Each individual thinks in a different way from another. But there is a common pattern regarding cognitive functions.

Left hemisphere is involved in sequential matters, mathematics and logics, sees details and puts them together to form the bigger picture, handles organization and logic and loves rules and regulations.

[*]Corresponding author: simonaduicu@unitbv.ro

The right part is the root of our emotions, intuition and visualization, sees all the visual references as a whole, works with curiosity and associativity.

Roger Sperry defined in the '60 the specific activities of our hemispheres, and since than almost everybody considered the right part, the creative one (figure 1). But thinking further, we may discover that creativity is an attribute of both hemisphere but in different ways.

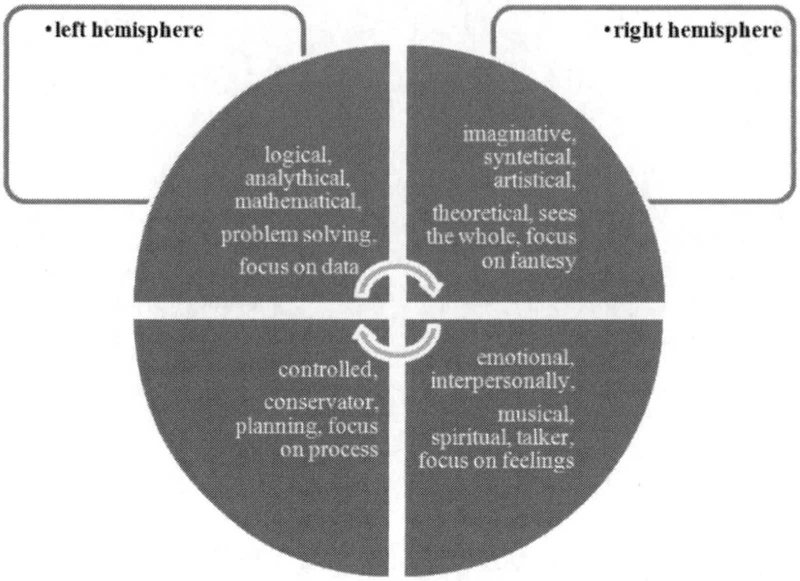

Fig. 1. A brain "map".

All four quadrants contain in a very simplify manner different type of characteristics [2], without ignoring the fact that hemispheres are full time connected by the corpus callosum.

The second important idea is that we are using our intelligent brain to interact with other people or with the surrounding world. When we do so, we are using our left hemisphere by developing the capacity of being able to use words and harmonize ourselves for our benefit. We also use our left hemisphere to imagine gesture, to create language, to nominate tools and to associate experiences with "ideograms" in our brain. And these ideograms are defined with high precision to be used later. This is how our left brain recreates the reality: from small parts to the big puzzle; is like every time we experience something new we add a flag to a map we will use later [3]. The right hemisphere will make all the time the comparison between what we already lived and the current reality. The right part of our brain understands contexts, the meaning of facts, even the body language, emotional expression, metaphors; it understands individuality, makes categories, in working in process.

The right hemisphere doubts, analyses, the left one performs mechanical. Even if on the left part is created the idea of performing an act, in the right one we will find the reason to do it. The profound statement is that reason and imagination don't have their own specific places in different hemisphere.

If we consider the left hemisphere responsible with the creation of the language, than we have to agree about certain mechanisms responsible with this: abstraction, item isolation, static definition, taking out from the context, nomination; the right hemisphere works with all concepts created by the left one, makes connections, understands life, changing and evolution, but creates also other connections between things and processes and relations between living things or humans. And all these lived and analysed experiences creates in our right hemisphere new knowledge and paths linking existing items and abilities to use what we know.

Modern society must understand that creativity is important for a sustainable and majestic development. But this can happen only if the logical left hemisphere works in conjunction with the powerful, visual and introspecting right one. This means to step back and come with new fresh idea by cultivating mental flexibility, confidence and patience in alternative problem solving. Both our hemisphere have the ability to create.

2.2 Where is creativity born?

Creativity can be defined as the use of imagination to generate original ideas. But where is born the creativity, in the left or in the right hemisphere? Left part of our brain takes everything out from its context if something new appears and creates a new image or a word for this. The right part analyses this new created items in a continuously changing context. So which part is more creative? How can we translate this sentence: "I have a feeling that it will work". Feeling is the attribute of the left hemisphere, but the sense of doing, the meaningful is the right's one. Einstein said that intuitive mind is a sacred gift and the rational one is a faithful servant, and all of this because the world surrounding us is changing. Our brain compares new realities with existing models. When is no model there, there are two paths: one to create a new one from the beginning, and the second, to ask a computer to simulate one for us.

Viewing problems from a new perspective, creativity could be the ability to find alternative solutions to problems or a new approach. Means to cultivate mental flexibility and the desire to face challenges or boredom.

2.2.1 Barriers against creativity

There are seven accepted barriers against creativity: school, sarcasm, social pressure, status-quo, standardization, silos and costumes [2]. Table 1 exposes the barriers vs. creativity resources.

Table 1. Barriers against creativity vs. creativity resources exploiting.

Barriers	**Resources to be exploited**
School	Formulate tests with multiple answers; flexibility in testing methods
Sarcasm	Postpone critical judgments during problem solving; brainstorming
Social pressure	Avoid social stratification during problem solving
Status-quo	Broke status-quo, myths, prejudices
Standardization	Rethink the problem from zero level, don't use old thinking patterns
Silos	Create an open culture of the organization, don't keep departments, curricula or people isolated in closed deposits
Costumes	Use the outfit to send messages, to express yourselves

We may train ourselves to be creative if we will be motivated by this desire. Training creativity for students may implicate mixture of disciplines like descriptive geometry and technical drawing, or geometry and design. I met creative students who can't work and studying a schedule, dreamy but not irresponsible, full of quiet energy; they worked with a big passion for their projects without caring about the time; they need to be encouraged, to instill confidence because pride and the sense of achievement is very important to them.

Empathy is also a feature that can be exploited with this kind of people who are, by the way, very objective about their work.

I ask my students at the first seminar to follow my example in solving a problem, than I ask somebody else to take my role by solving and explaining to their colleagues as they were a teacher. Without knowing what they do, they had to play a new role, to break the boundaries between student and teacher, to assume their new status and to rethink the entire process from a completely different position. Students who accepted that challenge had a better understanding of the problem and become more confident in the next sessions. They were able to find new solutions looking from a different angle.

2.2.2 Boost creativity by making liaisons

New ideas are very rare, most of them are creative liaisons between existing or old one. Nicholas Negroponte from MIT said that new ideas come unexpectedly, born from contradiction, from another perspective or from a complex mix of different theories [4]. So, a good exercise is to make unexpected connections between existing ideas, dialog with other people or taking out the problem from the context.

Playing with liaisons is a very simple exercise; here there are some very simple examples about how new idea was born:

- technical drawing + computer = computer aided design CAD;
- mathematics + logics + computer = computer programming;
- long life learning + computer skills = e-learning;
- imaination + new technologies = virtual reality.

A good sketch of this kind of training requires: problem identification, growing the data base of possibilities and serve results. If the first and the last step are easy to settle, data base of possibilities has to provide step by step a clear and obvious solution.

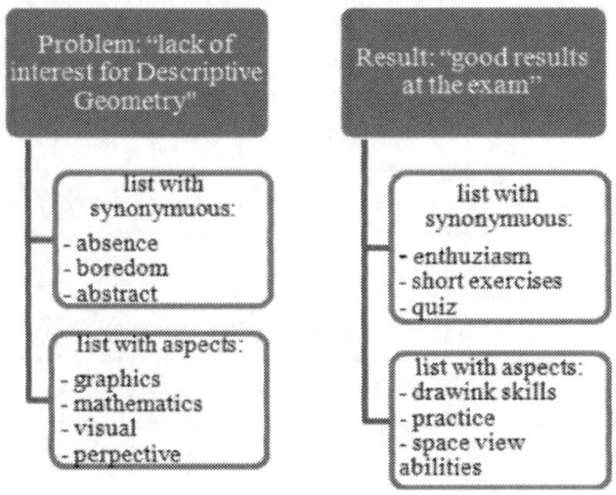

Fig. 2. Methods to enlarge data base.

Some methods to obtain the data base:

- make a list with synonymous for the problem and the result;
- write down as many aspects you find about the problem and the result;
- try to make visual or colour associations between problem and result;
- focus on problem details and result one;
- take short brakes, drink water.

Figure 2 exemplifies the process with only two methods.

2.2.3 *Playing different roles*

The idea of the playing different roles when a creative solving problem has to be done belongs to Edward de Bono and Bill Lucas [2]; the method is called "creative hats" and associates six colourful hats to six different roles played during problem solving:

- white for data, numbers and information input;
- red for sensations and intuitions;
- black for the moments when we think logical and negative;
- yellow for the moments when we think logical and constructive;
- green when we bring creative idea to move forward;
- blue when we act like a leader to manage all the process successfully.

Splitting a creative team by different roles can be an interesting approach in project management. It can be a way to organize students during lab activities or implement the system for each of them for their homework. This will engage flexibility in taking decisions and also develops lateral thinking which is one of the most important approach in creativity development [5].

Some of the "hats", the red and the green one are working with intuition. This brain capacity is not coming from our evaluated hemispheres but from the inner one, the limbic system responsible with feeling analysis, long term memory and processing information provided by the senses. So, it's nothing to do with learning but with our ability to understand our instincts. These creative energies can be focused to change the personal perception over a certain subject.

A good example how to use this method is a new projection about transportation. Being an automotive engineer I'm all the time concerned about this fast consumed and high technology involved in vehicles. An example regarding how to manage a creative concept is figured in caption 3.

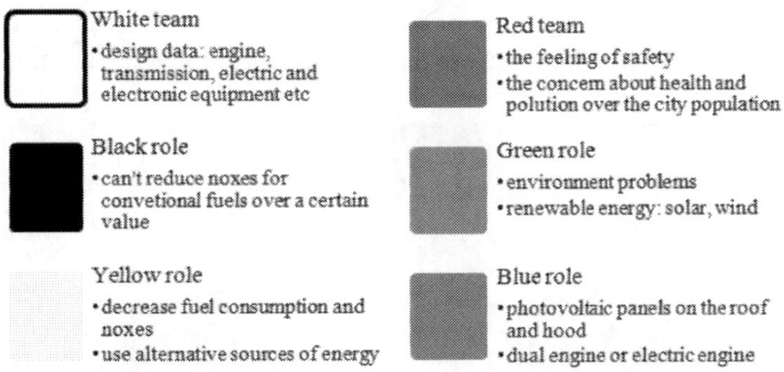

Fig. 3. Managing a creative concept.

The result of this approach can be the creation of a new paint with photovoltaic properties or a photovoltaic network covering the body of the car.

3 Shift for a new approach

Howard Gardner proposes a multiple approach of the knowledge [6] and identifies six access ways depending the way of thinking: the narrative way, the quantitative and numerical, the existential or the fundamental, the aesthetical, the practical and the social one. Going deeper in the subject, we may associate these ways with the creative hats or the colourful roles.

The quantitative and numerical way is the white one. People who access this path are interested developing patterns and procedures type of measurements.

The fundamental and existential way is the black one, suitable for people interested in fundamental, basic questions. They ask themselves about important aspects of life like: who we are and where we came from. They doubt, they are interested in the human capacity, they borrow idea from myths that guide their life.

The practical way is the yellow one, involving experiments, material manipulation, counting, building something, being logical and constructive.

The aesthetical way is green, inspiring equilibrium, harmony, a balance in composition; "green people" have an interior reference.

The social way is blue, connecting people, assuming roles and managing them. People who follow this path are capable to observe other people perspective, to interact with them and to harmonize them.

The red way is the narrative access path suitable for people who learn through stories. Remembering the red role dealing with intuition and sensations, we may understand that this kind of people need stories to imagine conflicts and problems to be solved. They need the dramatic tension to be challenged.

Some attributes of the roles need the left hemisphere, some the right one, but serving the information to each people in a specific mode is a new approach. It is considered that those with dominance in the right brain can be more creative, having a natural creativity; left brained people can be as creative as the others but in a different way [7].

3.1 Training the pragmatism

"Learning is the process in which knowledge are created through experience transformation" said David Kolb in 1984 [8]. What is interesting in this approach is the focus on the pragmatism; the training is combining thinking capacities and action (figure 4). This means that we have to train people according to their past experience. This will engage all the time a link with their past, their different stages of education.

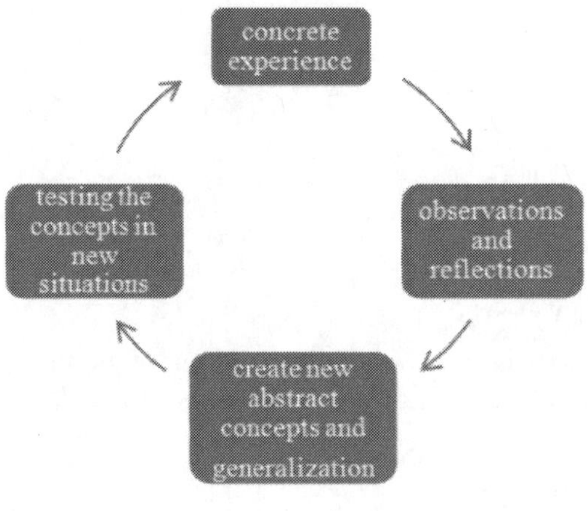

Fig. 4. Learning cycle after Kolb.

This diagram shows how, during a long period of time, education was so focused on individuals that the relation between the subject and the environment is forgotten. Moving the training from the internal mind in the social structure can define a more integrative thinking.

This pragmatic approach is very often choose in the technical education. We train our students to be practical, but is difficult for us and for them to understand the role of the creativity and innovation. It comes a time when we can't understand the concept of recycling their knowledge. This recycling will renew practice, it will develop the critical thinking and all the idea about action would be connected through language and other communication environments (like computer programming languages etc.) with others. And here emotions and conflicts can appear. Who says that we can be practical and original in the same time? If we go deeper in Kolb's diagram we can't say in which stage we are using our left or right brain more.

3.2 Training the creativity – an integrative approach

In figure 5 I adjust the learning cycle after John Heron [9].

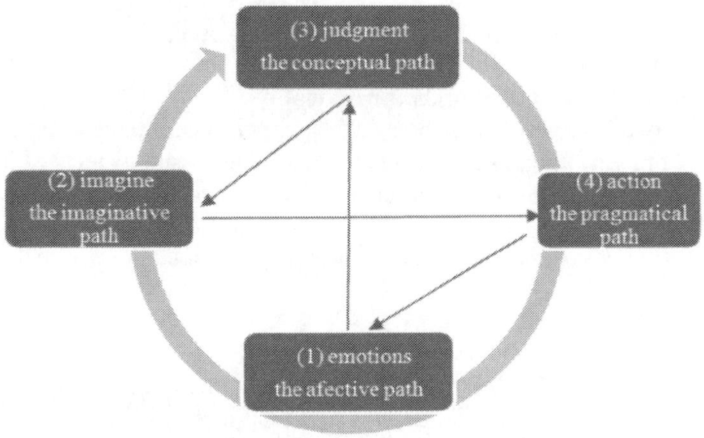

Fig. 5. Reverse learning cycle.

In the first stage, teacher has to establish an emphatic connection with the students, understands his interest and student's needs. This is the world of "presence" of "being". Passing in the second stage, teacher has to ask questions, and talk with students to fill the complete image of their level of education. In this stage teachers or trainers has to explore analogical the situation of the student (he must imagine the training plan comparing the actual picture with one of the existing patterns or curricula). It is more an intuitive perception because there was no previous examination of the student aptitudes. In the third stage teacher has to judge what best educational plan must be applied by making different hypotheses and, in the final, to propose the training plan. This circle is a normal learning circle. If we consider training like a simple already establish knowledge acquisition, than this is only learning. If new bring in this acquisition some new information, this will require research. Than learning plus research means creativity training.

Technical education is using the reverse learning cycle: in the first stage students realize they have no idea or only little information about the subject. It is the emotional stage. If we can exploit this stage by increase enthusiasm but offering in the same time a challenging situation, we may boost the premises of the creativity. This stage is more linked with the third one; students observe and identify themselves their level. In this moment they are using intensively their left hemisphere they work with their existing concepts about the subject trying to imagine (stage three) what is about the subject. This conceptual stage can be experience in a positive state of mind, preparing an open future approach. I think this is the rational stage in which is born the idea of "personal development". Concept evaluation followed by new, alternative idea may compose a complete new model of behaviour.

Student's attention migrates from frustration and ignorance to the image of a future (stage three). This projection is a right hemisphere responsibility even the roots come from judgment, from the right one. If they could apply something they imagine (stage four), the action can modify the initial state of mind and the emotions from the first stage to be shifted.

So, a new cycle is born. The normal training cycle was exemplified for teachers, the reverse one, was for students.

4 Conclusion

The shift for a new creativity education starts with the idea that we are individuals, we can anticipate some of the thinking and acting patterns but we have to provide a multiple faces of the information to develop ourselves. People can improve their originality in every stage of their education:

- train their left and right hemisphere with specific exercises;
- play different roles when a problem has to be solved;
- follow different paths according to their personality;
- modify their way of learning through narrative, experience, images, and practice ways;
- conscious assuming of the learning cycle in every significant aspects of it.

There are no specific tools and resources to access to organize creativity training. There is no specific field in which creativity is less important. Constant improvement of this human attribute can make us more flexible in thinking and more adaptable to change.

References

1. ERIC number ED **422468,** *Attitude to learning '98. MORI State of the Nation Survey: Summary report*, T.Greany editor (*1998),* http://eric.ed.gov/?id=ED422468
2. L. Bill, *Power up your mind. Learn faster, work smarter.* Meteor Publishing, (2016)
3. I. McGilchist, *The divided brain & the making of the western world, RSA Animate,* https://www.ted.com/talks/iain_mcgilchrist_the_divided_brain
4. N. Negroponte, *Where do new ideas come from,*http://www.wired.com/1996/01/negroponte-53
5. S.S. Duicu,*About changing paradigms in technical education,* **10**,13, (2015) http://www.sorging.ro/en/revista/10-special-issue-fascicle-5/
6. H. Gardner *Instructional-Design Theories and Models: A new paradigm of instructional theory,* in *Contemporany Theories of Learning,*(Editura trei, 2014)
7. K. Knight, *Understanding your brain for better design: left vs. right,* http://www.webdesignerdepot.com/2009/11/understanding-your-brain-for-better-design-left-vs-right/
8. D. Kolb, in *Contemporany Theories of Learning* (Editura trei, 2014)
9. J. Heron,*The cycle of life and the cycle of learning* in *Contemporany Theories of Learning (*Editura trei, 2014)

 DOI: 10.1051/matecconf/20179407004

Parametric programming of CNC machine tools

Rafał Gołębski [1,*]
[1] Czestochowa University of Technology, Institute for Mechanical Technologies, Al. Armii Krajowej 21, 42-201 Czestochowa, Poland

Abstract. The article presents the possibilities of parametric programming of CNC machine tools for the SINUMERIK 840D sl control system. The kinds and types of the definition of variables for the control system under discussion described. On the example of the longitudinal cutting cycle, parametric programming possibilities are shown. The program's code and its implementation in the control system is described in detail. The principle of parametric programming in a high-level language is also explained.

1 Introduction

CNC machine tools, including multi-axial multi-purpose machine tools, find increasingly wide application in many branches of industry. With the development of the design of CNC machine tools, a continuous development of their control systems, as well as CAD/CAM type software for their programming, follows. While the machining of geometrically simple items can be programmed manually in the EIA/ISO code or using dialogue programming (even not knowing G-codes), the programming of the machining of a geometrically complex part on a multi-axial CNC machine tool without the automatic generation of the machine tool control code using CAD/CAM software is very difficult, if not impossible. The ease of programming CNC machine tools allows their effective use in small-lot or even unit production.

In practice, many plants manufacture, in larger or smaller lots, different geometrically similar simple parts, for the programming of which parametric programming is the cheapest solution. The parametric programming of CNC machine tools is a type of programming, in which controlled axis positions, feed functions and velocities will be defined in the program as parametric expressions [1]. Introducing variables in a parametric form to the program allows the program being created to be made more flexible. The architecture of the created program needs to be reduced solely to the exchange of data in the declared parameters. It is also possible to perform computation on variables, to make use of conditional instructions, jump commands, branching, loops, or, for example, to use complex array definition instructions in the programming process [2]. The majority of CNC machine tool control systems enable the user to use the parametric programming method; the parametric program can be a complement to the main program, as a subprogram, or can also make a separate cycle for a given specificity of machining [3, 4].

* Corresponding author: rafal@itm.pcz.pl

Using the parametric code in the area of the machining program's iteration and geometric variable parametrization operations may substantially reduce the program's volume and, as a consequence, improve its performance [5, 6].

After building a parametric program, the user can create his own dialogue window to enable the efficient management of the program and, using relevant program structures, to reduce the likelihood of making errors.

2 Parametric programming in the SINUMERIK control system

Parametric programming in the architecture of the SINUMERIK 840D sl control system involves using the machine tool's control system and its available, usually limited memory for storing variables. The control system memory is furnished with a one-dimensional array, to the cells of which reference can be made via the cell addresses. In those cells, different types of data can be used as variables in the control system to perform, with their use, different (mathematical, structural) functions, like in other high-level languages (e.g. C, Pascal). Such a solution can be very convenient in the case of programs for machining parts with a limited number of parameters, which can be readily changed on the control panel, if necessary.

In the SINUMERIK control system, three types of variables can be defined:

– *User defined variables:*

LUD (Local User Data); this variable is applicable in the entire program.

GUD (Global User Data); the range of GUD variable influence refers to the control system's kernel (NCK – Numerical Control Kernel).

PUD (Program User Data); the PUD variable applies to programs and subprograms, in which it is defined; this variable should be declared under a unique name and it may not contain words reserved for control system commands. PG variables are created with the start of the machining program and removed upon its ending or resetting.

These variables can be accessed in the machining program, however, only upon meeting the condition, when the $MN_LUD_EXTENDED_SCOPE machine variable parameter is set to the value "1". Otherwise, user-defined variables defined in a given main program will only be active in that program.

– *Computational variables.* These are real-type variables predefined by the manufacturer, and referred to as so-called R parameters. The R variable is used for computations and transformations in the program. It has an R address reserved to it, with a successive number, e.g.: R512. Computation parameters, and essentially their values, can be assigned in the program to NC control addresses. In the basic system, the computation parameter number R is contained in the range 0-99, which can be extended through the modification of the control system machine data. R parameters were used in older SINUMERIK control systems; presently, however, using R parameters is gradually becoming unjustified. The use of the user's variable ensures better flexibility in programming, while the employed language is slowly beginning to resemble higher-level programming languages.

– *System variables.* These variables have an impact on the control system's configuration and its functioning in relation with the written program. System variables can be stored in, and read out from the control system's area, and the following can be distinguished here: zero point shift variables, tool correction definition variables, current position value variables, and measurement value variables. To be specially distinguished, system variables always start with the sign $. Thus, the relevant machine variable defining the starting position of the tool in a given axis prior to program simulation is defined as $MAS_SIM_START_POSITION and is available in the machine axis data tab and, like other variables, can be edited.

 DOI: 10.1051/matecconf/20179407004

Defining variables in the program involves specifying the allowable data type [7]. For predefined R variables and system variables, their type is predetermined, while in the case of user variables, the variables can be defined – Table 1.

Table 1. The types of variables and the range of their values.

Type of variable	Definition	Allowable value range
INT	Integer numerical values	-2147483646 ... +2147483647
REAL	Fractional numbers with a decimal point	$\pm (2{,}2 * 10^{-308} ... 1{,}8 * 10^{+308})$
BOOL	Logic values: TRUE (1), FALSE (0)	1 or 0
CHAR	ASCII character	Respectively, according to the ASCII code, it is possible to assign Polish characters
STRING	Character chain in brackets [ ]	Up to 200 characters can be defined
AXIS	Axis identifiers (addresses)	Axis identifiers or spindles existing in a given channel
FRAME	Geometrical data of coordinate system transformations	

Defining user variables always starts with the DEF command. The definition syntax is as follows:

DEF <variable type> < variable name> = <variable value>

When defining a variable it should remembered that the definition must always take place in a separate block. Many variables of one type may be defined in a block. A variable name may not exceed 31 characters, and its first two characters must always be letters or underscores. The $ sign and commands characteristic of a given type of NC control may not be used as the name of a variable. A constant or computational value may always be assigned to a variable. In the case, where no value has been assigned to the variable in the variable definition, the system will automatically assign it a default value equal to 0.

The SINUMERIK control system is furnished, as standard, with a definite number of cycles for the most frequent turning, milling or drilling operations. Very often, cycles offered by the manufacturer are limited, and their extension by not advanced user by adding further technological solutions is practically impossible due to their undescribed structure. The SINUMERIK control system offers the capability to define the user's own cycles as subroutines with a parametric structure. They can be stored in the user cycle directory, and after restarting the control system these subroutines will be treated as cycles by the control system. It is also important to make sure that the EXTERN command be removed from the program; in that case, the parametric subroutine will be then treated by the control system as a cycle.

2.1 Example 1

The creation of the user's own cycle is presented on the example of building a longitudinal turning subroutine. At the beginning of the program, all parameters, their names and their entering order must be defined.

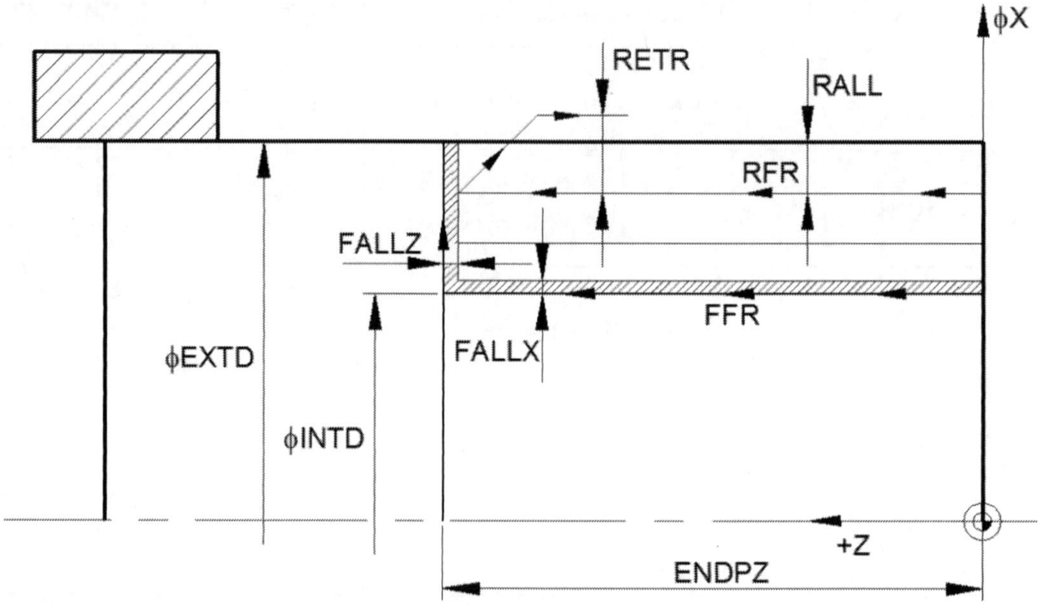

Fig. 1. Longitudinal turning parameters for the TURNING cycle.

For the cycle defined by the name TURNING, the following parameters have been defined in Figure 1:

- **EXTD – external diameter** (type REAL) defines the machining starting location in terms of diameter. For the correct division of the allowance to the first pass, this value should coincide with the blank diameter – the coordinate in the direction of the X axis is given in terms of diameter [mm].
- **INTD – internal diameter** (type REAL) defines the diameter after machining – the X coordinate is given in terms of diameter [mm].
- **ENDPZ - end point Z** type REAL, defines the turning length with reference to the zero point of the workpiece, the coordinate in the direction of the Z axis is given with the positive sign [mm].
- **RALL – roughing allowance** (type REAL) defines the thickness of the machined layer in the roughing pass; an absolute value on the radius [mm] .
- **FALLX – finishing allowance X** (type REAL) defines the amount of the allowance left for roughing in the direction of the X axis; an absolute value on the radius [mm].
- **FALLZ – finishing allowance Z** (type REAL) defines the amount of the allowance left for roughing in the direction of the Z axis; an absolute value in the Z axis direction [mm].
- **RFR – roughing feed rate** (type REAL) defines the magnitude of the synchronous feed for roughing; the value defined in [mm/rev].

- **FFR – finishing feed rate** (type REAL) defines the magnitude of the synchronous feed for finishing; the value defined in [mm/rev].
- **RETR – retract tool** (type REAL) defines the magnitude of tool retraction after performing the roughing pass in the X axis direction; an absolute value on the radius [mm].
- **LAYER_C** (type INT) defines the number of cycle layers; the parameter value calculated based on the parameters declared for the cycle.
- **COUNTER** (type INT) – the counter of the current cycle layer.

After defining the parameters (PUD variables), the structure of the subroutine is shown below. The program code can be transferred to SINUMERIK, where, as a file with the extension *.SPF (Sub Program File), it will perform the function of the subroutine:

```
PROC TURNING_CYCLE(REAL EXTD, REAL INTD, REAL ENDPZ, REAL RALL,
REAL FALLX, REAL FALLZ, REAL RFR, REAL RFR, REAL RETR) SAVE
; definition of the procedure TURNING_CYCLE and individual cycle parameters
DEF INT LAYER_C, COUNTER=0
            ; definition of the parameter of the total number of layers and of the layer counter
LAYER_C=((EXTD/2-INTD/2)-FALLX) DIV ()
    IF (((EXTD/2-INTD/2)-FALLX) MOD ()) <> 0
            ; computing the total number of cycle layers, LAYER_C, verifying if
            ; the thickness of the machined layer divides in a fractional manner, and if so:
    LAYER_C=LAYER_C+1
 ENDIF
 =(EXTD-INTD-ENDPZ)/LAYER_C
            ; otherwise, computing the number of layers without a fraction:
G90 G0 X=EXTD+1 Z=1 DIAMON
  G0 X=EXTD
            ; establishing the absolute dimensioning, activating diametral
            ; dimensioning, running the tool to the point before the workpiece
CYCLE_START:    ; starting the cycle procedure
MSG ("**ROUGHING** CURRENT LAYER:"<< COUNTER+1 << " from " <<
LAYER_C)
            ; displaying messages on the screen
            ; about the machining type and the currently machined layer
G91 G1 X=- F=RFR
  G1 Z=(1+ENDPZ-FALLZ)
  G1 X=RETR
  G1 Z=-(1+ENDPZ-FALLZ)
  G0 X=-RETR
            ; declaration of incremental dimensioning, parametric cycle description
GOTOF CYCLE_END
            ; procedure of forward jumping to procedure CYCLE_END
CYCLE_END:
COUNTER=COUNTER+1
IF COUNTER<LAYER_C GOTOB CYCLE_START
            ; program ending procedure, if the number of layers is greater than
            ; from the current layer, jumping back to procedure CYCLE_START
G90 X=EXTD+50 Z50
            ; absolute dimensioning, retract to a safe distance from material.
MSG ("FINISHING")        ; message "finishing"
```

```
T="FINISHING_T35 A"    ; declaration of the finishing machining tool
G90 G0 X=INTD Z=1
  G1Z=ENDPZ
  G1X=EXTD+1
  G0Z=1
             ; absolute dimensioning, parametric description of the finishing cycle
MSG()        ; message bar clearing
M17          ; end of subroutine
```

In the subprogram, blocks preceded by a semicolon are treated by the control system as comments. In the above code, the procedure of the finishing machining cycle has been incorporated. It has been preceded by the departure and the tool exchange function. This solution is correct. The finishing machining procedure can also be included in a separate subroutine, in that case, this would require the reference to the subprogram to be appropriately addressed in the main program.

The structure of the main program is shown below:

```
EXTERN TURNING_CYCLE(REAL, REAL, REAL, REAL, REAL, REAL, REAL,
REAL, REAL)
T1 D1 S2700 M3
             ; assigning the TURNING_CYCLE subroutine and handing over the parameters
G18 G54 G96 S300 LIMS=3200
             ; definition of the machining plane XZ of the workpiece's coordinate system
             ; switching on the constant turning speed, limiting
             ; spindle rotational speed
G90 G0 X50 Z50
             ; absolute dimensioning, running the tool to the material
             ; to a safe distance
 WORKPIECE(,,,"CYLINDER",192,0,-170,-150,100)
             ; blank declaration, a 100mm-diameter 170mm-long cylinder
             ; allowance for end face planning, 0mm
             ; available safe turning length smaller than 150mm

                        ;EXTD--INTD--ENDPZ-RALL-FALLX-FALLZ-RFR-FFR--RETR
TURNING_CYCLE (100.00, 50.00, -100.00, 2.50, 0.25, 0.25, 0.25, 0.1, 1.0)
             ; declaration of cycle parameter values entered in the order
             ; of defining variables in the subprogram
             ;
M30          ; end of the main program
```

The above code can be copied to the control system and stored with the extension *.MPF (Main Program File) to perform the function of a program. At the next step it was necessary to test the running of the program. For this purpose, the SINUMERIK Operate 4.5.ed.II software was used. It has a graphical dialogue programming overlay, SHOP-Turn, within which simulation was carried out – Figure 2. On the top bar, information about the program's progress is displayed, including the machining type and information on the currently machined layer.

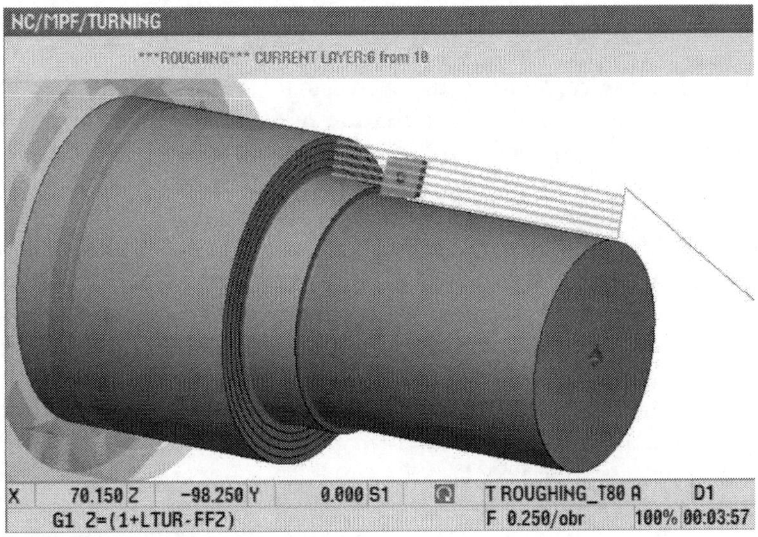

Fig. 2. A view of the cycle simulation window.

The program simulation progressed smoothly, which indicates that the program has been written correctly. The subprogram can be copied to the user cycle directory. Restarting the control system will make the program visible to the NCK. We remove the EXTERN command from the subprogram's code. At the next step, the program was run from the machine tool's control panel level. During the operation of the program, the currently executed machining block is displayed on the machine's screen. In the case of parametric programming, the program code in the parametric system will be displayed. For an operator, who has not created the program, this may pose a great impediment. The solution is to activate the 'BASIC BLOCKS' option, which will allow the operator to observe the functions used by the program and the coordinates after the translation of the program into the ISO code. The lower left window presents the currently executed block in the parametric code, while the lower right window, in the ISO code – Figure 3.

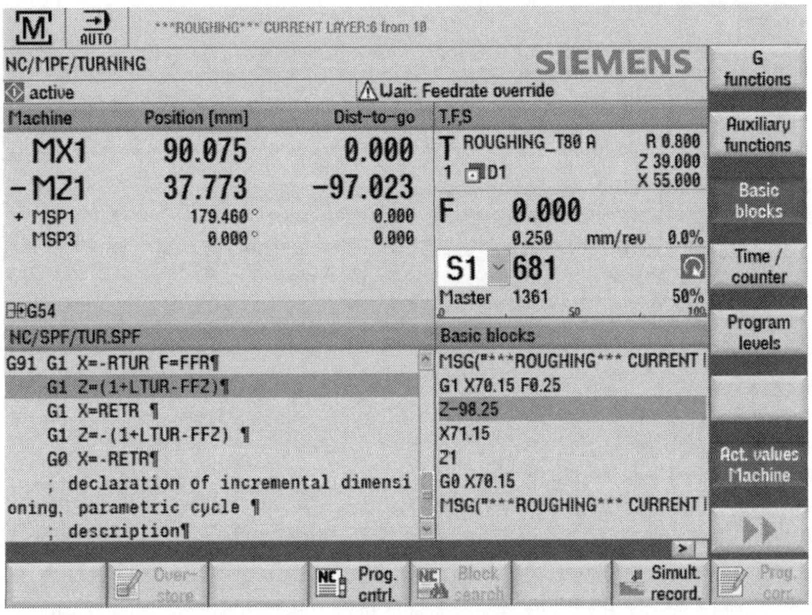

Fig. 3. The window of running the program on a machine tool.

 DOI: 10.1051/matecconf/20179407004

3 Parametric programming in a high-level language using a PC

Considering the increasing capabilities and increasingly wide application of the advanced CAD/CAM software, as well as the centralization of CNC machine tool software, due to the costs and striving to construct Flexible Manufacturing Systems (FMS), CNC machine tool software is more and more often transferred outside the machine tools (also because of the Ethernet operation and remote data transmission capabilities).

In this case, the aim is to utilize the capabilities of the PC for engineering calculations and to generate the CNC machine tool code in the computation program [8]. It is possible to fully utilize the capability of the high-level programming language and to generate, from a single program, control codes for different machine tools in a given plant in the form of a parametric program. Only the indispensable selected parameters and the generated machine tool control code are entered to the final machine tool control program, whereas in the program's computational module, which can be very complex, the number of variables of multi-dimensional arrays may be very large.

3.1 Example 2

A program for machining worms of an arbitrary profile has been developed for VCN-type (multi-purpose machine tool) and QTN-type (turning center) Integrex CNC machine tools manufactured by MAZAK [9]. The program in Delphi includes a module for determining the tool positioning parameters, depending on the cutter diameter, the preset worm axial profile and the required machining accuracy.

The program - Figure 4 generates also the machine tool control code – Figure 5. It is important that the program allows the utilization of the capacity of a PC and enables it to run under the Windows environment, perform mathematical operations and use subprograms in any high-level programming language, and makes it possible to generate the control program for different machine tools, which have been used in the developed machine tool control program - Figure 5. In the example under discussion, the worm is machined by the step-by-step method – Figure 6a.

The machining performance depends on the cutting parameters and the number of tool passes.

```
for i:=number_n downto 1 do
 if bvnl[3,i]<>0.0 then
  begin
   l_mazak:=l_mazak+1;
   Writeln(outf,'G90 G0 X#'+IntToStr(600+l_mazak));
   Writeln(outf,'G95 G01 C0.0 F#122');
   Writeln(outf,'G0 Z[#'+IntTostr(500+l_mazak)+']');
   Writeln(outf,'G91 G95 G1 C2979.383 X[#115+#116+#117] F#105');
   Writeln(outf,'G90 G0 Z[#120+2*#114]');
   Writeln(outf,'G0 X#'+IntToStr(800+l_mazak));
   Writeln(outf,'G95 G01 C0.0 F#122');
   Writeln(outf,'G0 Z[#'+IntTostr(700+l_mazak)+']');
   Writeln(outf,'G91 G95 G01 C2979.383 X[#115+#116+#117] F#105');
   Writeln(outf,'G90 G0 Z[#120+2*#114]');
  end;
```

Fig. 4. A fragment of the machine tool control code generating program.

The tool positioning coordinates for each cutter pass are calculated in the computation program and transferred to the machine tool control system. These values are stored in the memory cells of the control system, and therefore their number is limited by the storage capacity and may not be arbitrary. An advantage of the adopted machining method is the capability to make a worm of any profile and an arbitrary module with a universal tool (a finger ball-end mill) – Figure 6b.

```
G90 G0 Z[#120+2*#114]
G90 G0 X#610
G95 G01 C0.0 F#122
G0 Z[#510]
G91 G95 G1 C2979.383 X[#115+#116+#117] F#105
G90 G0 Z[#120+2*#114]
G0 X#810
G95 G01 C0.0 F#122
G0 Z[#710]
.
.
.
G90 G0 Z[#120+2*#114]
G0 X#813
G95 G01 C0.0 F#122
G0 Z[#713]
G91 G95 G01 C2979.383 X[#115+#116+#117] F#105
```

Fig. 5. A fragment of the machine tool control program code.

The parametric variables are stored in the program's memory under addresses indicated by the sign #. The control program was read in to machine tool control system via a USB (the program can also be transmitted through the Ethernet). The parameters can be displayed and modified on the machine tool control panel. The machine tool control program generated automatically from the universal CAM software program will generally be much longer, and any modifications could be very difficult, if not impossible, to carry out. The CNC machine tool control program written in a high-level language using parametric variables will be a cheaper (there is no need for using the expensive CAM software) and optimal (easy to modify) solution.

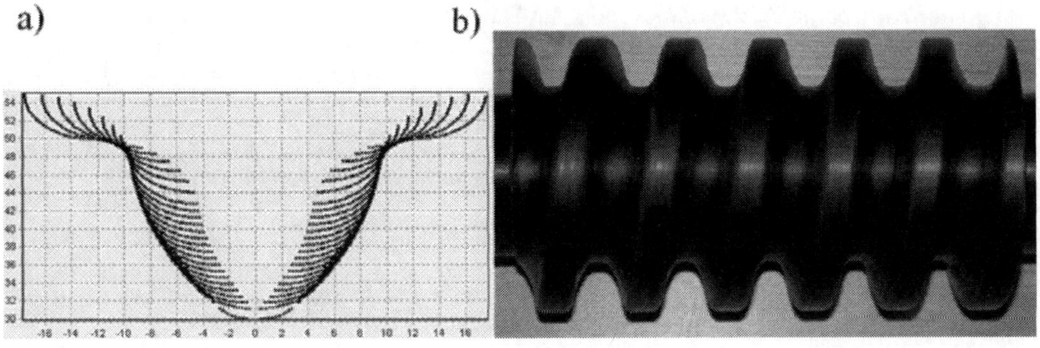

Fig. 6. An example of a worm cut by the step-by-step method (m=10): a) the worm profile and tool profiles in the worm axial section, b) a finished worm.

4 Conclusions

Parametric programming requires the knowledge of G-codes and does not require any additional tools to facilitate programming, often available optionally, (e.g. 3D visualization), and is cheap (is used in the automotive industry).

Parametric programming is convenient and easy to master, and is frequently used in machining geometrically simple parts. However, considering the capacities of the one-dimensional arrays of control systems to store variables, these programs have some limitations.

Parametric programming in a high-level language involves performing engineering calculations, while fully utilizing the capabilities of such calculations, and ultimately generating the parametric CNC machine tool control code. This is usually very difficult (the engineering calculation module may be very complex), but the capabilities of such programming are greater. The machine tool resources, the programming language capability, the ability to generate the code for several machine tools and the possibility of carrying out complicated calculations can all be utilized for machining geometrically more advanced parts.

For the discussed control system, parametric programming provides a perfect alternative to the SHOP-Turn and SHOP-Mill dialogue programming overlays being in use, which are a costly option for the SINUMERIK. Moreover, for a given parametric cycle, there is a possibility of building the user's own graphic window for declaring cycle parameters.

To sum up, with its open architecture, the SINUMERIK environment provides excellent opportunities for using parametric machine tool programming solutions.

References

1. M. Lynch, *Parametric programming for computer numerical control machine tools and touch probes* (Society of Manufacturing Engineers, 1997)
2. R. Stryczek, B. Pytlak, *Flexible programming of machine tools,* PWN (Scientific Publishing House, 2011)
3. D. Manocher, Comp. & Ind. Eng. **35**, 33, (1998)
4. G. Nikiel, Adv. in Man. Sci. and Techn. **33**, 33, (2009)
5. G. Nikiel, *Programming of CNC machine tools on the example of the Sinumerik 810D/840D control system* (ATH University of Bielska-Biala, 2004)
6. J. Szadkowski, R. Stryczek, G. Nikiel G, *The design of technological processes for numerically controlled machine tools* (ATH University of Bielska-Biala, 1995)
7. SINUMERIK 840D, *Job planning, Programming Manual* (2006)
8. T. Nieszporek, A. Piotrowski, *Robotics in Theory and Practice. Applied Mechanics and Materials,* 203, (Trans Tech Publications, Zurich, 2013)
9. Yamazaki, INTEGREX Revolutionary Multi-tasking Machine, https://www.mazak.com.sg/machines/process/multi-tasking/Yamazaki, (2006)

MATEC Web of Conferences **94**, 07005 (2017) DOI: 10.1051/matecconf/20179407005
CoSME'16

Full factorial DOE to determine the influence of the process parameters in cleaning water jets used in sewer cleaning

Nicolae Medan[1,*], *Mircea* Lobontiu[1], and *Mihai* Banica[1]

[1]Technical University of Cluj Napoca, North University Center Baia Mare, 62A, Dr. Victor Babes Street, 430083 Baia Mare, Romania

Abstract. Industrial cleaning is a classic application of water jets technology. In the late 1950s, when reliable high pressure pumps were built, the usage of water jets spread widely in the field of pipes and sewerage cleaning. In our particular case, water jets using pressure between 100 and 200 bars are being used. The operational behaviour of these systems is dependent on the process parameters, which can vary, causing variations in the impact force. The experiment consists in the research method used in the study of the impact forces that occur in the sewerage cleaning head system. To determine the influence process parameters have on impact forces, the full factorial method was used. In order to measure the impact forces between the water jet and a flat and rigid surface, a stand was designed and built for generating pressure water jets, as well as a device to measure the impact forces.

1 Introduction

Phenomena that occur in the cleaning water jets are complex. Adler [1] describes mechanisms occurring at the impact of a jet with a surface. Leach et al [2], Leu et al [3] and Guha et al [4] analysed pressure distribution along the centreline of the water jet. A number of papers have studied the influence of nozzle geometry on water jet [5-7].

The aim of this paper is to determine the influence of process parameters on impact forces produced by water jets from the sewerage system cleaning head, impact forces that depend on the process parameters. In order to measure the impact forces between the water jet and a flat and rigid surface, a stand for generating pressure water jets, as well as a device to measure the impact forces have been designed and built. In the measurements, water jets using pressure between 100 and 200 bars have been used. The values of these pressures correspond to low pressure water jets [8].

To carry out an experimental research, it is recommended to follow a series of steps [9]: 1) recognition of and statement of the problem, 2) selection of the response variable, 3) choice of factors (process parameters), levels, and ranges, 4) choice of experimental design, 5) performing the experiment, 6) statistical analysis of the data, 7) conclusions and recommendations.

* Corresponding author: Nicolae.Medan@cunbm.utcluj.ro

To determine the influence of process parameters on impact forces produced by water jets from the sewerage system cleaning head the research method used is full factorial method.

2 Apparatus used and methodology of the measurements

In order to measure the impact forces between the water jet and a flat and rigid surface, a stand for generating pressure water jets, as well as a device to measure the impact forces were designed and built [10].

2.1 The stand to generate pressure water jet

Schematic diagram of the stand to generate a pressure jet is shown in figure 1.

The component parts of the stand are: 1) electric motor 2) flexible coupling; 3) high pressure pump, 4) pressure regulator, 5) pressure gauge, 6) nozzle, 7) tap water, 8) water tank, 9) chassis.

Water coming out of the high pressure pump (3) goes into the pressure regulator (4). The pressure regulator adjusts the pressure and flow of the water in the path of the high pressure water. This pressure corresponds to the one at the outlet of the nozzle.

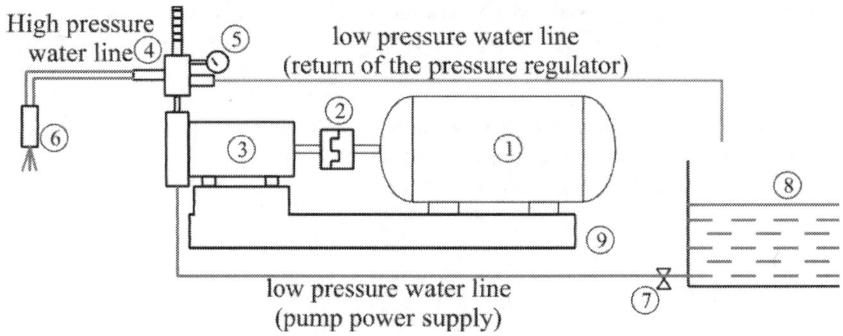

Fig. 1. Schematic diagram of the stand to generate pressure jet.

2.2 The device to measure the impact forces

Figure 2 represents the principle diagram of the device for the measurement of the impact force between the water jet and a flat and rigid surface.

The main component parts of the device are:1) high-pressure water hose, 2) support nozzle, 3) nozzle block, 4) nozzle, 5) water jet, 6) flat and rigid target plate, 7) collection path water, 8) scaled container, 9) piezoelectric sensor mounting, 10) piezoelectric sensor, 11) data acquisition Personal Daq/3000, 12) computer for the processing of data; 13) support plate, 14) acrylic tube, 15) rods for adjusting distance x.

From the high pressure water hose (1) the water comes at a certain pressure p desired. A water jet is generated at the outlet of the nozzle (5) that strikes the target plate (6), which it located at a certain distance x in front of the nozzle. The jet (5) generates an impact force at a time when it meets the target plate (6). This force produces axial movement of the target plate. This movement is converted into an electric signal by the piezoelectric sensor (10). Electrical signals are collected by data acquisition Personal Daq/3000 (11), which forwards data to a computer (12) using DaqView soft processes. Thus, accurate data is obtained.

 DOI: 10.1051/matecconf/20179407005

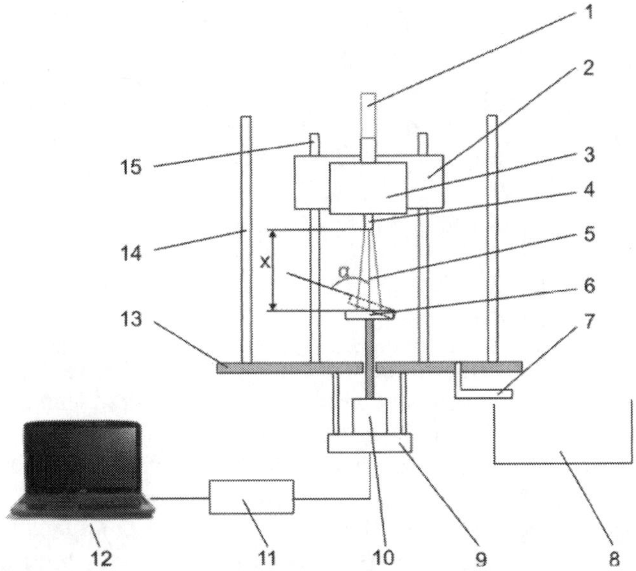

Fig. 2. Diagram of the device for the measurement of the impact force of the water jet.

2.3 Process parameters and geometric configuration of nozzles

In the water jet cleaning process a series of parameters are involved [8]. These parameters can be divided into two major groups, namely: 1) target parameters which shall be defined according to the contact area between the water jet and the surface to be cleaned and 2) process parameters. In the measurement of the impact forces of a stationary water jet and flat and rigid surface the process parameters are involved (figure 3).

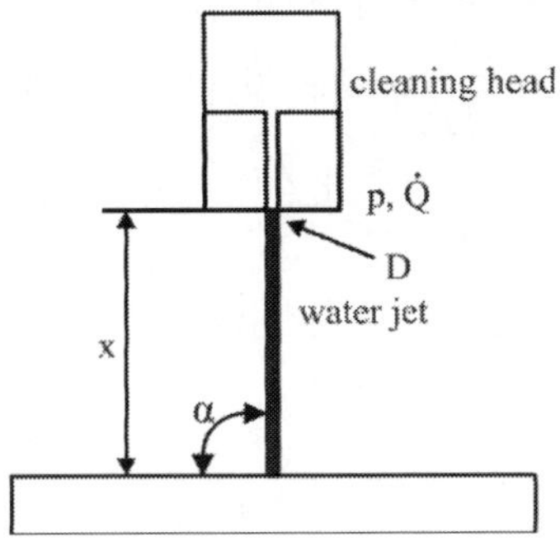

Fig. 3. Process parameters.

They can be divided into two groups, namely hydraulic parameters and the performance parameters. Hydraulic parameters characterized by the system's high pressure pump-nozzle are represented by work pressure (p), volume flow ($\dot{Q}$) and nozzle diameter (D). The performance parameters are more related to the process performance and include the following: cleaning distance (x), impact angle (α) and traverse rate (v_T). For a stationary water jet v_T= 0 m/s.

Setting the values of process parameters, as follows:

1)The diameter D of the nozzle. The values used are D=1mm, 1.5 mm and 2 mm. These are common values used in equipment designed for maintenance and cleaning of sewers. The material used for the nozzles is stainless steel. Figure 4 shows the geometric configuration of the nozzles used.

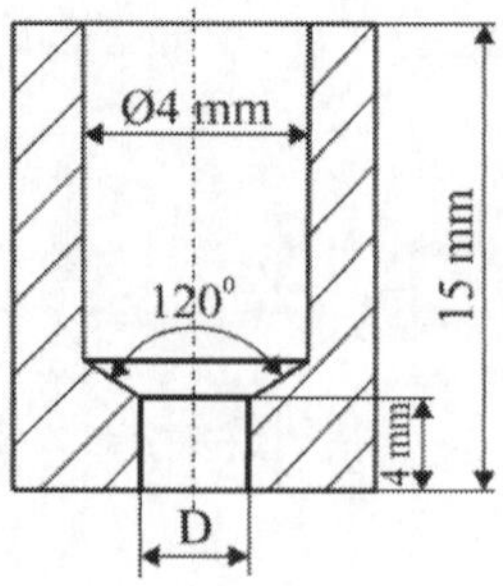

Fig. 4. Geometric configuration of the nozzles.

2)The pressures p used to perform the measurements have the values p=100 bar, p=120 bar, p=140 bar, p=160 bar, p=180 bar and p=200 bar. For the maintenance sewers, maximum pressure used are 200 bars to prevent the damage of the sewer pipe.

3)Cleaning distance x (distance from the nozzle to the target plate). Distance x has been fixed at the values from x=25 mm to x=200 mm, with a step of 25 mm.

4)α – impact angle (the angle formed by the jet with the target plate); in this work α =60^0, 75^0 and 90^0. For cleaning heads the usual value of the angle of impact α is 75^0. If impact angle α decreases below 60^0 it leads to a drop in of the impact forces.

3 Results

For determining the influence of the process parameters the method used shall be full factorial design.

3.1 Choice of levels and values for process parameters

Table 1 presents the values of the parameters used for full factorial method.

Table 1. The values of the parameters.

Parameter	Values
Nozzle diameter D [mm]	1, 1.5, 2
Pressure p [bar]	100, 120,140,160,180,200
Impact angle α [^{0}]	60, 75, 90
Distance x [mm]	25, 50, 75, 100, 125, 150, 175, 200

3.2 Performing the experiment

Taking into account that there are four parameters, the number of values which they each take and using the Full factorial design results in a number of 144 experiments. For each experiment three measures of impact force were performed and still it has worked with F_{med}, representing the arithmetic average of the three forces measured. As a result there are a number of 432 measurements. With this in mind, for the brief of this paper, we dare not present the table with the values obtained.

Based on the values of impact forces obtained using the full factorial method, figures 5, 6 and 7 present the values of impact forces F_{med} which depend on process parameters (values of process parameters presented in table 1).

DOI: 10.1051/matecconf/20179407005

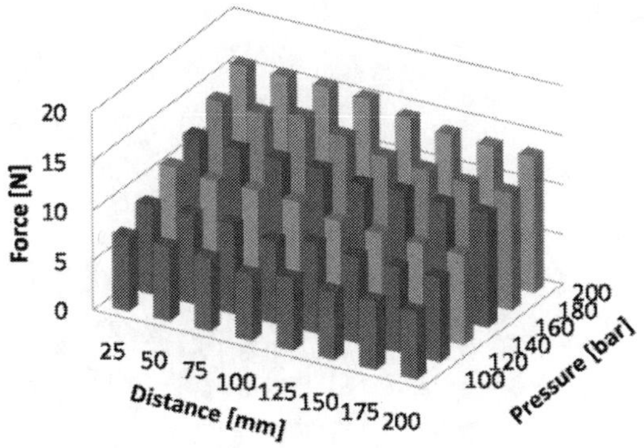

c) α = 60^0

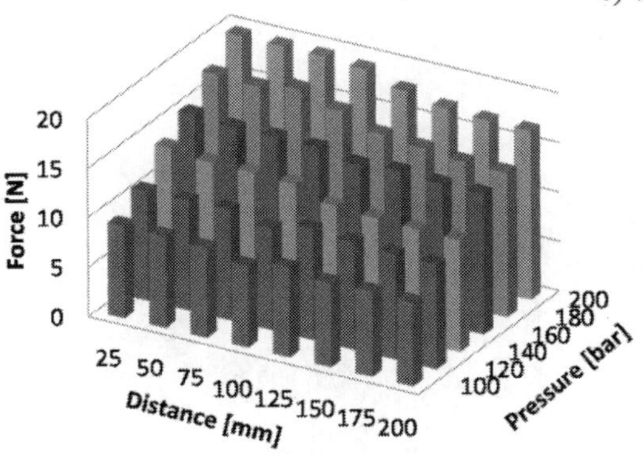

a) α = 90^0

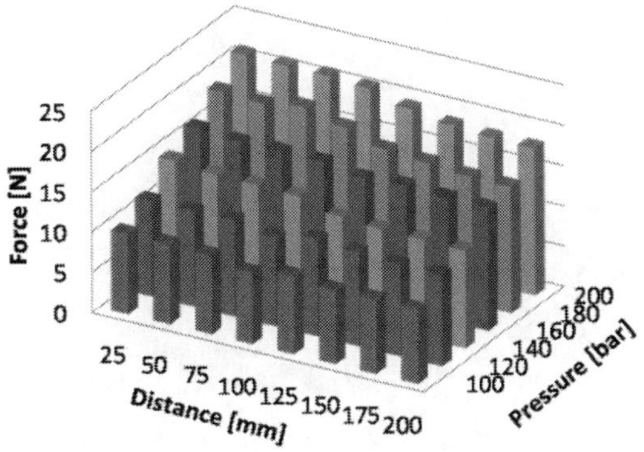

b) α = 75^0

Fig. 5. Values of impact forces for nozzle diameter D=1mm for different values of impact angle α.

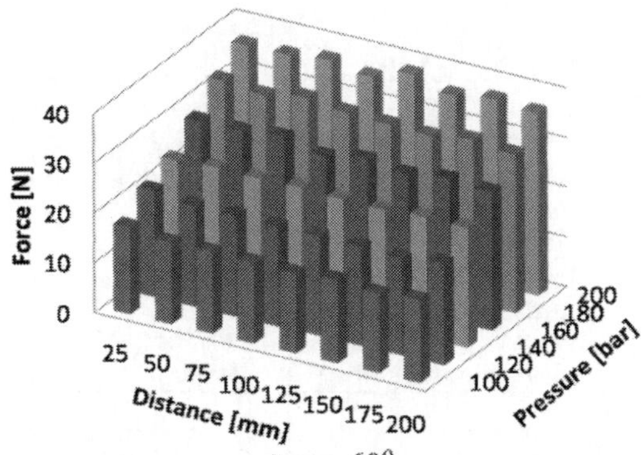

c) α = 60^0

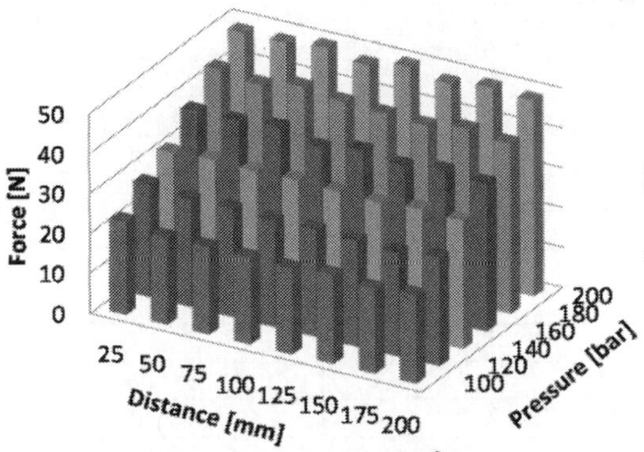

a) α = 90^0

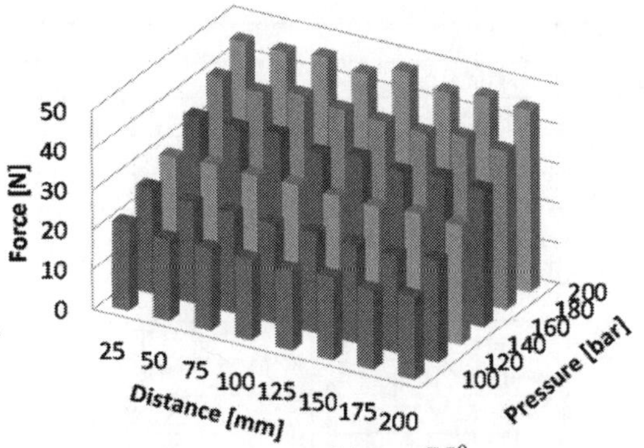

b) α = 75^0

Fig. 6. Values of impact forces for nozzle diameter D=1.5 mm for different values of impact angle α.

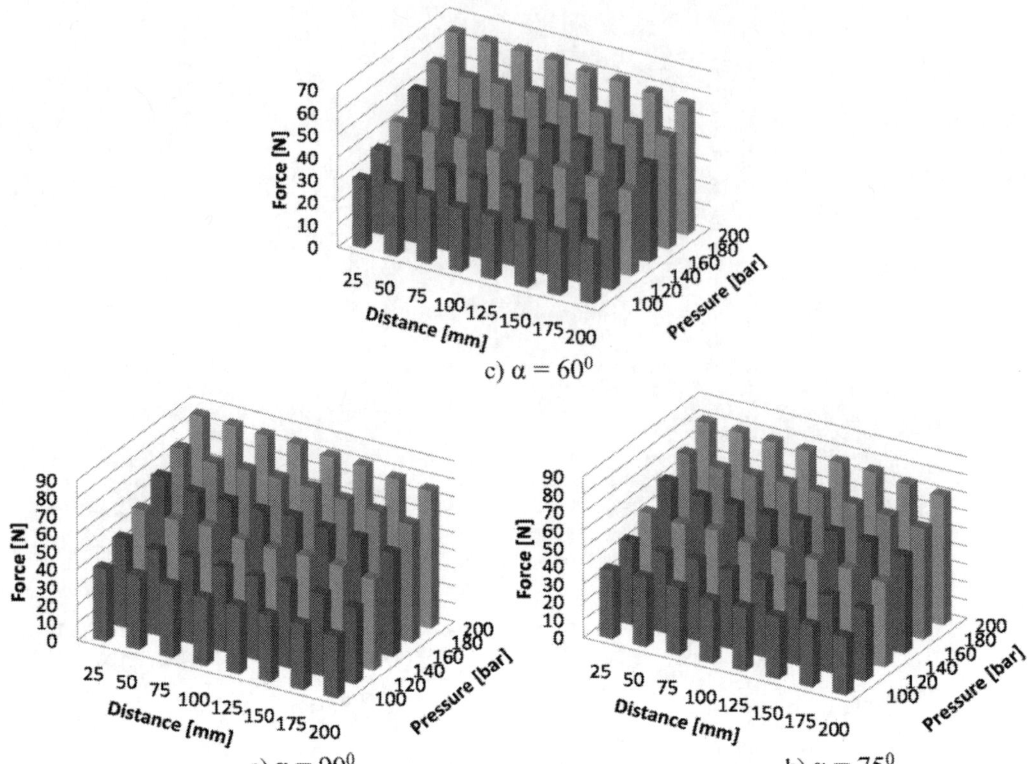

Fig. 7. Values of impact forces for nozzle diameter D=2 mm for different values of impact angle α.

3.3 Determining the Influence of the process parameters

Using the Full factorial method has conducted an analysis of the variance to determine the contribution of each parameter and their interactions on the impact force. Table 2 presents the analysis of the variance, using Minitab 17.

Table 2. Analysis of Variance (Minitab 17).

Source	DF (degree of freedom)	Seq. SS (sum of square)	Contribution
Model	103	163,747	99.87%
Linear	16	153,677	93.73%
Diameter	2	117,239	71.50%
Pressure	5	29,641	18.08%
Distance	7	410	0.25%
Angle	2	6,387	3.90%
2-Way Interactions	87	10,070	6.14%
Diameter*Pressure	10	7,626	4.65%
Diameter*Distance	14	333	0.20%
Diameter*Angle	4	1,636	1.00%
Pressure*Distance	35	55	0.03%
Pressure*Angle	10	414	0.25%
Distance*Angle	14	6	0.00%
Error	328	217	0.13%
Total	431	163,965	100.00%

4 Conclusions

In this paper is presented an experimental research methodology, which uses full factorial method. This allows determining the influence of process parameters to impact forces produced by water jets.

In order to measure the impact forces between the water jet and a flat and rigid surface, a stand was designed and built for generating pressure water jets, as well as a device to measure the impact forces.

Based on the results of experimental work presented in this paper it was determined that the contribution of the process parameters and their interactions are, as follows:
-The most important parameter is the diameter of the nozzle, with a percentage of influence of 71.50%;
-The second parameter is the pressure of water jet, with a percentage of 18.08%;
-The third parameter is the impact angle, with a percentage of 3.90%;
-The distance parameter, with a percentage of 0.25% is virtually insignificant;
-Between interactions only diameter and pressure have an important contribution 4.65%;
-All other interactions have a low importance (from 1% to 0%).

For the experimental domain defined, only three of the four parameters have influence on impact forces. Distance x has an influence only 0.25%, so maybe neglected.

References

1. W. F. Adler, *The Mechanics of Liquid Impact. Treatise on Materials Science and Technology* (Ney York: Academic, 1979)
2. S. J. Leach, G. L. Walker, Some Aspects of Rock cutting by High Speed Water Jets. Philosophical Transactions of the Royal Society of London, **260**, 295-308 (1966)
3. M. Leu, C.P. Meng, E.S. Geskin, L. Tismeneskiy, Mathematical modelling and experimental verification of stationary waterjet cleaning process, ASME Journal of Manufacturing Science and Engineering, **120**, 571-579 (1998)
4. A. Guha, R.M. Barron, R. Balachandar, An Experimental and Numerical Study of Water Jet Cleaning Process, Journal of Materials Processing Technology, 610-618 (2011)
5. M. Annoni, L. Cristaldi, M. Faifer, M. Norgia, Orifice Coefficients Evaluation for Water Jet Application, 16th IMECO TC4 Symposium, 125-130 (2008)
6. E. Ghassemieh, H.K. Versteeg, M. Acar, The effect of nozzle geometry on the flow characteristics of small water jet, Proceedings of the Institution of Mechanical Engineers - Part C, **220** (2006)
7. Y. Liu, F. Ma, H. Xie, Y. Li, Development of Impact Test System for Waterjet Descaling Nozzles with LabVIEW, International Conference in Web Information Systems and Mining, **1** (2010)
8. A. W. Momber, *Hydroblasting and Coating of Stell structures* (Oxford Elsevier Ltd, 2003)
9. D.C. Montgomery, *Design and analysis of experiments* (Wiley, 2013)
10. N. Medan, *Research and Contributions to the Functioning of Cleaning Head-Sewer System* (PhD Thesis, Baia Mare, 2014)

An analysis of gearing

Tadeusz Nieszporek[1], *Piotr* Boral[1] , and *Rafał* Gołębski[1,*]

[1]Czestochowa University of Technology, Institute for Mechanical Technologies, Al. Al. Armii Krajowej 21, 42-201 Czestochowa, Poland

Abstract. The article has presented a numerical method for determining the gear ratio function and the gearing line of the toothed gear. The described algorithm, explained on the example of planar gearing, can be used also for the analysis of spatial gearings. Assembly errors have been allowed for in the gear model.

1 Introduction

The quality of a gear is determined by the analysis of the gearing [1-3], the key element of which is the determination of the gear ratio function [4]. The gearing analysis is used chiefly for determining the kinematic errors that result from toothing cutting errors and gear assembly errors, as well as for optimizing the gear synthesis (especially for the analysis of spatial gearing) [5, 6]. Generally, this is a process involving numerical computation by successive approximation methods.

The method of planar gearing analysis was first formulated by Litvin (1968) [4]. This method can be used not only for toothed wheels and gears, but also for the majority of simple mechanisms (such as cam mechanisms). A similar discussion as for the planar gearing can be conducted for the spatial gearing. The gearing analysis, as supplemented by the analysis of the gear contact trace, is called Tooth Contact Analysis (TCA) [4]. Programs for spatial TCA were first developed by Gleason Works (1960) and by Litvin and Gutman (1981).

It is assumed that the tooth profiles are given, and the distance between the gear wheel axes is also given. The gear ratio function (relationship between gear wheel rotations) and the gearing lines need to be determined.

2 Conditions for tooth profile tangency

Coordinate systems S_1, S_2 and S_f are introduced, which are rigidly tied, respectively, with gear wheels *1* and *2* and with the gear casing. In the case of a planar gearing, the tooth profiles Σ_1 and Σ_2 (of gear wheels *1* and *2*) and the unit vectors of the normals to the profiles are equal [4] - Figure 1.

* Corresponding author: rafal@itm.pcz.pl

DOI: 10.1051/matecconf/20179407006

$$\left\{\theta_1 \in \langle a_1,\ b_1\rangle : \mathbf{r}_1^{(1)} = \mathbf{r}_1^{(1)}(\theta_1) \in C^2\right\} \quad \frac{\partial \mathbf{r}_1^{(1)}}{\partial \theta_1} \neq 0 \tag{1}$$

$$\mathbf{n}_1^{(1)} = \mathbf{n}_1^{(1)}(\theta_1) \tag{2}$$

$$\left\{\theta_2 \in \langle a_2,\ b_2\rangle : \mathbf{r}_2^{(2)} = \mathbf{r}_2^{(2)}(\theta_2) \in C^2\right\} \quad \frac{\partial \mathbf{r}_2^{(2)}}{\partial \theta_2} \neq 0 \tag{3}$$

$$\mathbf{n}_2^{(2)} = \mathbf{n}_2^{(2)}(\theta_2) \tag{4}$$

where: θ_1, θ_2 - parameters of the tooth profiles Σ_1 and Σ_2 (of the gear wheels *1* and *2*).

The superscripts (1) and (2) at the vectors refer to the tooth profiles Σ_1 and Σ_2, while the subscripts indicate the coordinate system, in which these quantities are written.

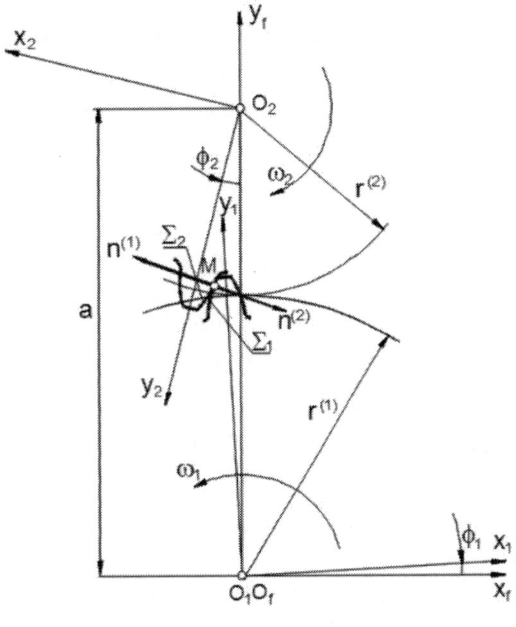

Fig. 1. The coordinate systems and gear wheel tooth profiles.

The transformation of the vector coordinates and the normal unit vector components from the coordinate systems S_1 and S_2 to the stationary reference system S_f can be written as follows [7, 8]

$$\mathbf{r}_f^{(i)}(\theta_i,\ \phi_i) = \left[M_{fi}(\phi_i)\right]\mathbf{r}_i^{(i)}(\theta_i) \qquad i = 1,2 \tag{5ab}$$

$$\mathbf{n}_f^{(i)}(\theta_i,\ \phi_i) = \left[L_{fi}(\phi_i)\right]\ \mathbf{n}_i^{(i)}(\theta_i) \tag{6ab}$$

where: ϕ_1, ϕ_2 - rotation angles of gear wheels *1* and *2*.

The matrix $[M]$ (with a dimension of 4 x 4) considers the mutual displacement and rotation of the coordinate systems, while the matrix $[L]$ (with a dimension of 3 x 3) considers only the rotation.

The tangency of the profiles occurs only when the profiles Σ_1 and Σ_2 contact at the same point and their normal unit vectors are collinear (parallel) at that point. Hence

$$\mathbf{r}_f^{(1)}(\theta_1,\ \phi_1)=\mathbf{r}_f^{(2)}(\theta_2,\ \phi_2) \quad \mathbf{n}_f^{(1)}(\theta_1,\ \phi_1)=\mathbf{n}_f^{(2)}(\theta_2,\ \phi_2) \tag{7ab}$$

3 Gearing analysis

The components of the unit vectors $\mathbf{n}_f^{(1)}$ and $\mathbf{n}_f^{(2)}$ for the planar gearing (with the components $n_{fz}^{(1)}=n_{fz}^{(2)}=0$) meet the condition

$$n_{fx}^{(1)}=n_{fx}^{(2)} \quad \Rightarrow \quad n_{fy}^{(1)}=n_{fy}^{(2)} \tag{8ab}$$

where: $n_{f*}^{(i)}$ - component $*$ of the unit vector i ($i=1,2$) in the system S_f.

Therefore, the projections of the position vectors (7a) and the normal unit vectors (7b) onto the axes of the coordinate system S_f yield four scalar equations, of which only three are independent. Equations (7) i (8) give a system of three equations with four unknowns [4]

$$f_j(\theta_1,\ \phi_1,\ \theta_2,\ \phi_2)=0 \qquad j=1,\ldots,3 \tag{9a-c}$$

where

$$\{f_1,\ f_2,\ f_3\}\in C^1 \tag{10}$$

The analysis of the planar gearing will be defined, if equations (10) can be transformed into three equations as a function of one variable

$$\{\ \theta_1(\phi_1),\ \theta_2(\phi_1),\ \phi_2(\phi_1)\ \}\in C^1 \tag{11}$$

The function $\phi_2(\phi_1)$ represents the relationship between gear wheel rotation angles, or the gear ratio function. The functions $\theta_1(\phi_1)$ and $\theta_2(\phi_1)$ define the points on the profiles Σ_1 and Σ_2, at which the profiles are tangent for a given rotation angle value of gear wheel *1*. The gearing line of the profiles Σ_1 and Σ_2 is described by the function

$$\mathbf{r}_f^{(1)}=\mathbf{r}_f^{(1)}(\theta_1(\phi_1),\ \phi_1) \ \vee \quad \mathbf{r}_f^{(2)}=\mathbf{r}_f^{(2)}(\theta_2(\phi_1),\ \phi_2(\phi_1)) \tag{12ab}$$

4 The analytical algorithm

Litvin assumes [4] that a starting point of the iteration $P^{(0)}=\left(\theta_1^{(0)},\ \phi_1^{(0)},\ \theta_2^{(0)},\ \phi_2^{(0)}\right)$ is given, which satisfies equations (9). In the next step, the successive point is determined, which satisfies equations (9)

$$P^{(1)}=\left(\theta_1^{(1)},\ \phi_1^{(1)},\ \theta_2^{(1)},\ \phi_2^{(1)}\right) \tag{13}$$

DOI: 10.1051/matecconf/20179407006

where the superscript denotes the successive approximation.

A system of two equations is chosen from the system of equations (9). It is assumed that the following equations are chosen:

$$f_2(\theta_1,\ \phi_1,\ \theta_2,\ \phi_2)=0 \qquad f_3(\theta_1,\ \phi_1,\ \theta_2,\ \phi_2)=0 \tag{14ab}$$

Taking $\phi_1 = \phi_1^{(1)}$ as given, it should be assumed that $\theta_1 = \theta_1^{(1)}$. In view of the above, the system of equations (14) can be solved with respect to θ_2 and ϕ_2. Let the solution be $\theta_2^{(1)}$ and $\phi_2^{(1)}$. The new set of parameters (13) is defined, if it satisfies the remaining equation

$$f_1(\theta_1,\ \phi_1,\ \theta_2,\ \phi_2)=0 \tag{15}$$

by the set of parameters $\left(\theta_1^{(1)},\ \phi_1^{(1)},\ \theta_2^{(1)},\ \phi_2^{(1)}\right)$. If equation (15) is not satisfied, then a new value should be taken for $\theta_1^{(1)}$ and the system of equations (14) should be solved again, and then the condition (15) needs to be rechecked.

The above algorithm involves splitting the system (9) of three equations into two subsystems, one of them containing two equations and the other one, and solving them independently.

5 The numerical algorithm

The parameters ϕ_1 and ϕ_2 are the angles of rotations of gear wheels *1* and *2*, while the parameters θ_1 and θ_2 are the parameters of the gear wheel profiles Σ_1 and Σ_2.

In the reference system Σ_f, in the system of coordinates $(x_f\ \ y_f)$, a grid of ordinate lines with fixed values of the abscissae x_{fl} $(l=1,..,m)$, uniformly arranged (with a fixed step), is introduced - Figure 2a (with unrealistic values of the ordinates $y_{f\max}$). For the fixed value of parameter $\phi_1 = \phi_{1r}$ $(r=1,..,t)$, from the profile equation Σ_1

$$\mathbf{r}_f^{(1)} = \mathbf{r}_f^{(1)}(\theta_1,\ \phi_1) \tag{16}$$

for successive values of the coordinates of parameter $\theta_1 = \theta_{1j}$ $(j=1..n)$, the points of the profile Σ_1 are determined in the system of coordinates $(x_f\ \ y_f)$. In doing so, values of parameter θ_1 can be taken uniformly with a fixed step within the assumed interval. The profile points, in a general case, do not lie on the ordinate lines with the abscissae x_{fl} of the system $(x_f\ \ y_f)$ – Figure 2b. As the grid lines should intersect with the profile only once, the grid is taken in the computation program either on the axis of abscissae – Figure 2 or on the axis of ordinates, respectively.

 DOI: 10.1051/matecconf/20179407006

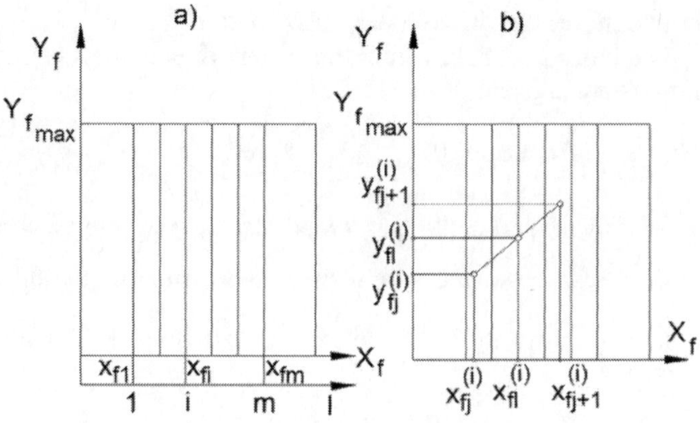

Fig. 2. The grid of abscissae: a) and the gear wheel tooth profile, b) in the system $(x_f \quad y_f)$.

The ordinates of the line segment points lying on the ordinate lines are determined from the following relationship – Figure 2b

$$y_{fl}^{(i)} = y_{fj}^{(i)} + \frac{y_{fj+1}^{(i)} - y_{fj}^{(i)}}{x_{fj+1}^{(i)} - x_{fj}^{(i)}}\left(x_{fl}^{(i)} - x_{fj}^{(i)}\right) \qquad i = 1,2 \tag{17}$$

Substituting for x_f the values of the abscissae of the grid ordinate lines lying within a given segment (or intersecting the segment under consideration), we will obtain the points of the curve on those ordinate lines (the points of grid ordinate line intersection with the analysed segment).

By joining successive profile points with line segments and determining the points of intersection of those segments with the ordinate lines with the abscissae x_{fl} in the system $(x_f \quad y_f)$, the profile Σ_1 is substituted with a set of ordered (normalized) points.

The profile Σ_2 should be dealt with in a similar manner. For the fixed value of parameter $\phi_2 = \phi_{2s}$ $(s = 1,..,v)$, from the profile equation Σ_2

$$\mathbf{r}_f^{(2)} = \mathbf{r}_f^{(2)}(\theta_2, \quad \phi_2) \tag{18}$$

for successive values of the coordinates of parameter $\theta_2 = \theta_{2k}$ $(k = 1..p)$, the points of the profile Σ_2 are determined in the system of coordinates $(x_f \quad y_f)$.

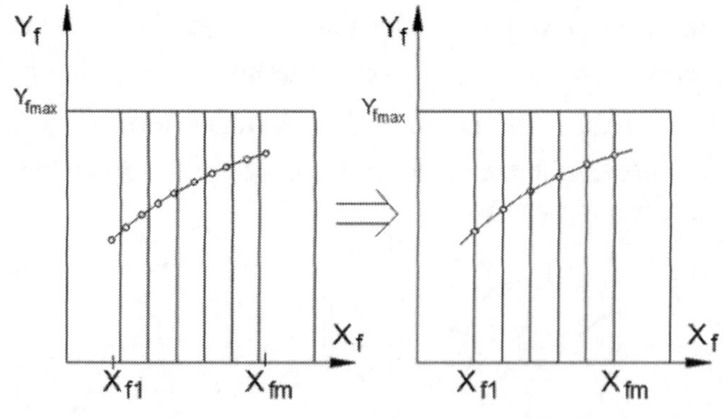

Fig. 3. The arrangement of profile points.

 DOI: 10.1051/matecconf/20179407006

By joining successive profile points with line segments and determining the points of intersection of those segments with the ordinate lines with the abscissac x_{fl} in the system $(x_f \quad y_f)$, the profile Σ_2 is substituted with a set of ordered points – Figure 3.

By comparing the ordinates of the points of the profiles Σ_1 and Σ_2 for the same abscissae x_{fi}, the distances of those points are determined – Figure 4

$$\Delta_{fl} = y_{fl}^{(2)} - y_{fl}^{(1)} \tag{19}$$

where the subscript f identifies the system of coordinates, while the subscript l, the ordinate (with the abscissa x_{fi}). At the point of contact (on condition that the profiles do not interpenetrate at other points) the distance is $\Delta_{fl} = 0.0$ (with the assumed computation accuracy). If this condition is not met, the value of angle $\phi_2 = \phi_{2s}$ should be changed, each time (for successive values of this parameter) repeating the computation process to determine the profile contact point. With a fixed value of ϕ_1, the value of parameter ϕ_2 is determined by the step halving method [8].

The entire computation process should be repeated for successive values of angle $\phi_1 = \phi_{1r}$ to ultimately determine the gear ratio function $\phi_2(\phi_1)$.

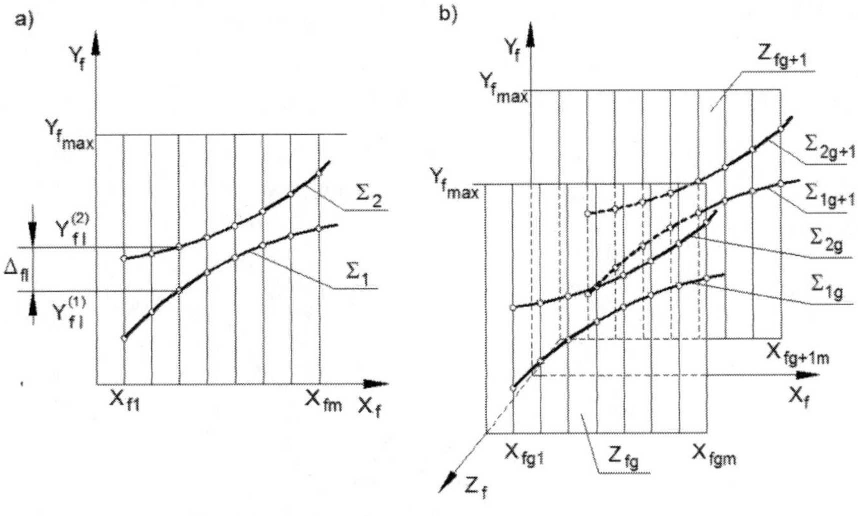

Fig. 4. Determination of the gear wheel tooth contact point.

The algorithm described above is simple, which constitutes its great asset. It can also be used for the analysis of spatial gearing.

In the case of spatial gearing analysis, that is the analysis of toothing surface mating, the surfaces should be determined as sets of profiles in parallel cutting planes – Figure 4b. The above-described algorithm should be repeated for successive cutting planes z_{fg} $(g = 1,..w)$. The surface equations can be written as

$$\mathbf{r}_i^{(i)} = \mathbf{r}_i^{(i)}\left(\theta_i, \ z_{ig}\right) \quad i = 1,2 \tag{20}$$

where z_{ig} defines the position of the cutting plane, while being also a surface parameter.

The surface equations in the reference system Σ_f are as follows

$$\mathbf{r}_f^{(1)} = \mathbf{r}_f^{(i)} \left(\theta_i, \quad \phi_i, \quad z_{ig} \right) \; i = 1,2 \tag{21}$$

By comparing the ordinates of surface points for the same abscissae x_{fgi} (in a given cutting plane), the distances of those points are determined

$$\Delta_{fgl} = y_{fgl}^{(2)} - y_{fgl}^{(1)} \tag{22}$$

where the subscript g identifies the cutting plane, while the subscript l, the ordinate (with the abscissa x_{fgi}) in that plane.

As a result, the surfaces will be determined in the form of sets of normalized points lying on ordinate lines in cutting planes, which enables the examination of their mutual position and determination of the gear ratio function. Whereas, the surfaces may only contact linearly or be tangent in one point only. Therefore, for a fixed value of angle ϕ_{1r}, a value of angle ϕ_{2s} should be determined (by the successive approximation method) such, that in individual cutting planes there will be $\Delta_{fgl} > 0.0$, and at least in one plane and at one point, there will be $\Delta_{fgl} = 0.0$.

6 Example

A spur gear with an involute tooth profile is taken – Figure 5.

The following is assumed: module, $m = 5\,[mm]$; the number of the teeth of gear wheel *1*, $z_1 = 30$; the number of the teeth of gear wheel *2*, $z_2 = 20$; profile angle, $\alpha_0 = 20$ [deg]; axis distance error, $e = -1, 0, 1\,[mm]$; profile modification depth, $t_m = 2\,[mm]$; profile modification length, $g_m = 1\,[mm]$.

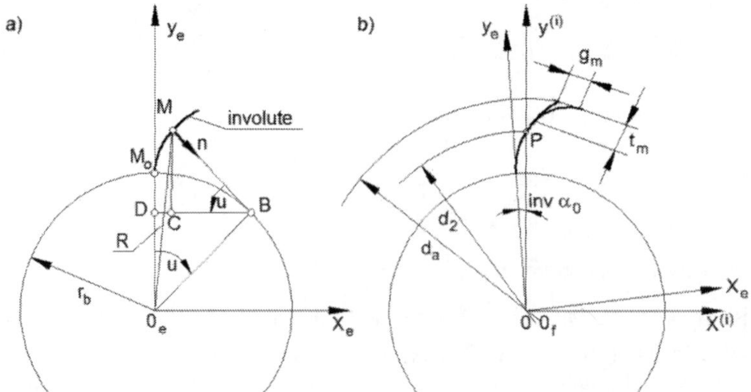

Fig. 5. The profile: a) involute, b) of the gear wheel.

Figure 5a shows that the profile of the involute and the unit vector normal to the profile can be written as

$$\mathbf{r}_e = r_b \left[su - ucu, \quad cu + u\, su \right] \tag{23}$$

$$\mathbf{n}_e = \left[cu, \quad -su \right] \tag{24}$$

DOI: 10.1051/matecconf/20179407006

where the following designations are taken

$$r_b = \frac{z\,m}{2} c\alpha_0 \qquad c* = \cos * \qquad s* = \sin* \tag{25a-c}$$

As follows from Figure 5b, the gear wheel tooth profile and the unit vector of the normal to the profile can be written as

$$\mathbf{r}_i^{(i)} = r_{bi}\left[s(u - inv\alpha_0) - u\, c(u - inv\alpha_0),\;\; c(u - inv\alpha_0) + u\, s(u - inv\alpha_0)\right] \qquad i = 1,2 \tag{26}$$

$$\mathbf{n}_i^{(i)} = \left[c(u - inv\alpha_0),\;\; -s(u - inv\alpha_0)\right] \tag{27}$$

where

$$inv\alpha_0 = tg\alpha_0 - \alpha_0 \tag{28}$$

The matrices of transition from the gear wheel coordinate systems to the stationary coordinate system linked with the gear casing are as follows – Figure 1

$$[M_{f1}] = \begin{bmatrix} c\phi_1 & -s\phi_1 & 0 & 0 \\ s\phi_1 & c\phi_1 & 0 & 0 \\ 0 & 0 & 1 & 0 \\ 0 & 0 & 0 & 1 \end{bmatrix} \qquad [L_{f1}] = \begin{bmatrix} c\phi_1 & -s\phi_1 & 0 \\ s\phi_1 & c\phi_1 & 0 \\ 0 & 0 & 1 \end{bmatrix} \tag{29ab}$$

$$[M_{f2}] = \begin{bmatrix} -c\phi_2 & -s\phi_2 & 0 & 0 \\ s\phi_2 & -c\phi_2 & 0 & a \\ 0 & 0 & 1 & 0 \\ 0 & 0 & 0 & 1 \end{bmatrix} \qquad [L_{f2}] = \begin{bmatrix} -c\phi_2 & -s\phi_2 & 0 \\ s\phi_2 & -c\phi_2 & 0 \\ 0 & 0 & 1 \end{bmatrix} \tag{30ab}$$

where the distance of the axis is equal to

$$a = (z_1 + z_2)\, m / 2.0 + e \tag{31}$$

Based on relationships (5) and (6), we can write:

$$\mathbf{r}_f^{(1)} = r_{b1}\left[s\beta - u\, c\beta,\;\; c\beta + u\, s\beta \right] \qquad \mathbf{n}_f^{(1)} = \left[c\beta,\;\; -s\beta \right] \tag{32ab}$$

$$\mathbf{r}_f^{(2)} = \left[r_{b2}(-s\gamma + u\, c\gamma),\;\; r_{b2}(-c\gamma - u\, s\gamma) + a \right] \qquad \mathbf{n}_f^{(2)} = \left[-c\gamma,\;\; s\gamma \right] \tag{33ab}$$

where the following designation is taken

$$\beta = u - inv\alpha_0 - \phi_1 \qquad \gamma = u - inv\alpha_0 + \phi_2 \tag{34ab}$$

where: r_{b1}, r_{b2} - base circle radii.

Therefore, the equations to be used for further computation are as follows:
1 - analytical method

$$f_2 = r_{b1}(c\beta + u\, s\beta) - r_{b2}(-c\gamma - u\, s\gamma) - a = 0 \tag{35}$$

$$f_3 = c\beta + c\gamma = 0 \tag{36}$$

$$f_1 = r_{b1}(s\beta - u\, c\beta) - r_{b2}(-s\gamma + u\, c\gamma) = 0 \tag{37}$$

2 - numerical method

$$\mathbf{r}_f^{(1)} = r_{b1}[s\beta - u\, c\beta, \quad c\beta + u\, s\beta] \tag{38}$$

$$\mathbf{r}_f^{(2)} = [r_{b2}(-s\gamma + u\, c\gamma), \quad r_{b2}(-c\gamma - u\, s\gamma) + a] \tag{39}$$

The gear ratio function is a linear function, which means that the gear ratio is constant. The error of the gear wheel axis distance has caused a parallel shift in the gear ratio function – Figure 6a. On the other hand, the modification of the profile by either rounding off or truncating the tooth tip – Figure 7, affects the gear ratio function – Figure 6b. The result is a gear ratio error and uneven gear operation.

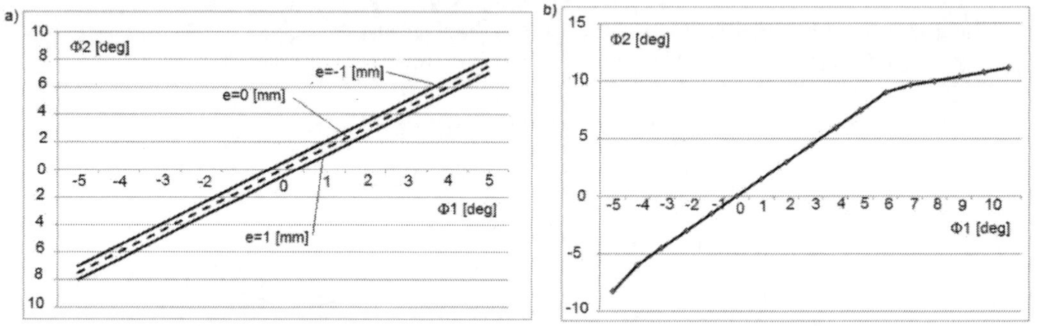

Fig. 6. The gear ratio function: a) with no gear wheel tooth profile modification – with allowance being made for the axis distance error; b) with tooth profile top modification.

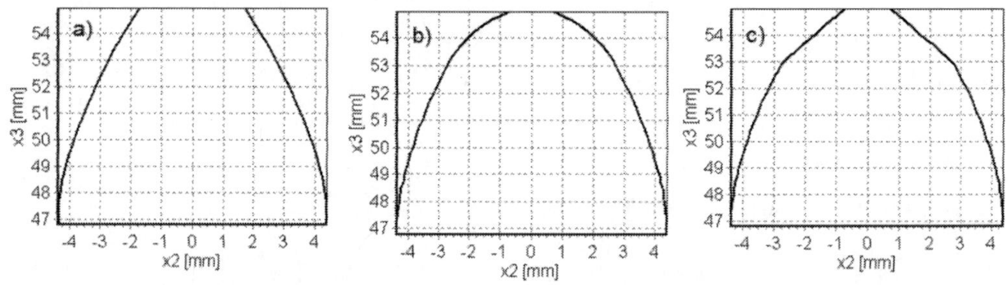

Fig. 7. The tooth profile of gear wheel 1 - a) without modification, b) with modification along the circular arc, c) with modification along the straight line.

7 Conclusions

The presented algorithm does not require the solving of systems of equations, in contrast to the analytical method using the Newtonian method for solving the system of equations. There is no need for allowing for the condition of normal unit vector collinearity, either.

The program consists of a module, in which the user enters profiles to be examined, and a universal module, in which the profile contact points are determined.

The effectiveness of the Newtonian method depends largely on the accuracy of the first approximations of roots, while in the proposed method, root variability ranges can be taken with a greater approximation.

The developed algorithm and computation program can also be used for the analysis of spatial gearing.

The computation accuracy depends on the number of iteration steps, the number of profile points and the number of grid lines, and can be easily changed.

References

1. I. Dudas, *The Theory and Practice of Worm Gear Drives* (Penton Press, London, 2000)
2. F.L. Litvin, A. Fuentes, *Gear Geometry and Applied Theory* (Cambridge University Press, 2004)
3. G. Sheveleva, V.I. Medvedev, A.E. Volkov, Mathematical simulation of spiral bevel gears production and process with contact and bending stressing, Proc. IX World Congress on the Theory of Machines and Mechanisms, **1** (1995)
4. F. L. Litvin, Theory of Gearing, NASA Reference Publication 1212, AVSCOM Technical Report 88-C-035 (1989)
5. F. L. Litvin, Development of Gear Technology and Theory of Gearing, NASA RP-1406 (1998)
6. F. L. Litvin, C. L. Hasiao, Comp. Meth. in Appl. Mech. and Eng. **102**, 337 (1993)
7. T. Nieszporek, *The fundamentals of the theory of synthesis and analysis of approximate gearing* (Czestochowa University of Technology, Czestochowa, 1998)
8. G. A. Korn, T. Korn, *Mathematics for scientific workers and engineers* (PWN, Warsaw, 1983)

Examination of the cylindrical worm profile

Tadeusz Nieszporek[1] , and *Piotr* Boral[1,*]

[1]Czestochowa University of Technology, Institute of Mechanical Technologies, Al. Armii Krajowej 21, 42-201 Czestochowa, Poland

Abstract. The article has presented a special case of machining a cylindrical worm with a fixed and variable pitch. The worm is machined with a cylindrical finger-type mill positioned perpendicularly to the worm axis. The axial profile and axial profile angle of the worm have been calculated and measured. The effect of the offset on the accuracy of the worm axial profile has been examined.

1 Introduction

Fixed-pitch cylindrical worms are used chiefly in worm gears. Fixed and variable-pitch cylindrical worms are used also in extruding presses and feeders in plastic processing and in the food industry. The shaping of fixed-pitch cylindrical helicoids has been extensively covered in the literature [1, 2], while the technology of variable-pitch helical surfaces is not addressed. The helical surfaces of worms are most commonly machined by the hobbing method using rotary tools [1, 3].

Cone-derivative helical surfaces are predominant, which are formed using a tool with the rectilinear axial profile of the tool action surface [4, 5]. The tool profile angle is, in a general case, different from zero, and the cylindrical finger-type mill is a special case of the helical surface forming rotary tool. The equations of the cone-derivative cylindrical helicoid, in a general form, consider disc, ring, finger and cup-type tools [6], but they do not include the cylindrical finger-type mill due to the surd in the envelope condition [7].

2 Worm surface

In a general case, the following worm helical surface geometrical parameters, among others, are given – Figure 1: d_a - outer diameter, d_f - cut bottom diameter, L - length, h_1 - pitch at the beginning, h_2 - pitch at the end, s_1 - offset at the beginning, s_2 - offset at the end. A linear variation in pitch and offset over the worm length is assumed.

In the case of cutting the helical surface using a cylindrical finger mill with a rectilinear tool action surface profile, the profile angle will be equal to pi/2 - Figure 2. In the general case of forming a cone-derivative helical surface with a cylindrical finger mill, a surd will occur in the envelope equation [7]. Therefore, this machining method has been treated as a

* Corresponding author: piotrek@itm.pcz.pl

special case of helical surface forming. This is, however, a worm machining technology that is used in practice.

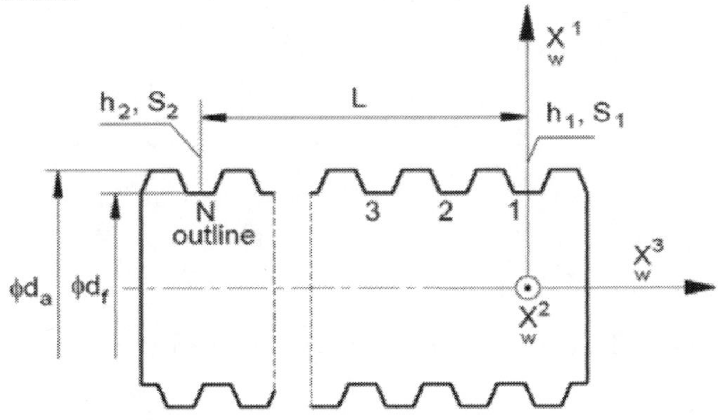

Fig. 1. Worm geometrical parameters.

This helical surface machining method can also be used for initial rough cutting of a worm with an arbitrary profile. This is so because the tool is geometrically simple and cheap – Figure 2.

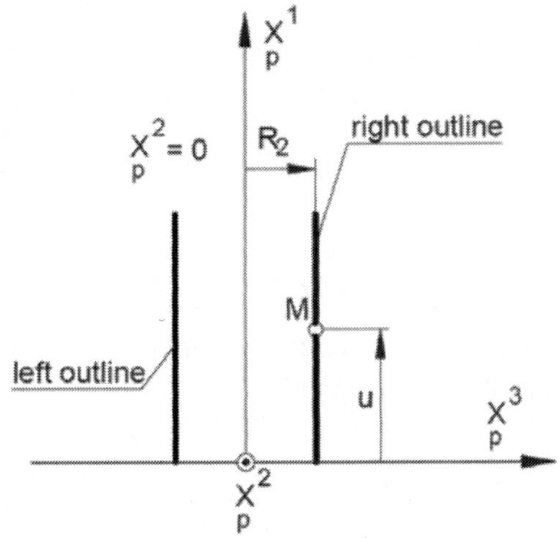

Fig. 2. The axial profile of the cylindrical finger mill action surface.

Comparison of the mill profile in the coordinate system $\{p\}$ is as follows:

$$u \in \langle u_{min}, u_{max} \rangle \rightarrow \underset{p}{\mathbf{x}}(u) = [u,\ \ 0,\ \ R_2]^T \in \mathrm{E} \tag{1}$$

where: u - tool axial profile parameter; R_2 - mill radius; $\underset{*}{\mathbf{x}}$ - subscript $*$ identifies the coordinate system.

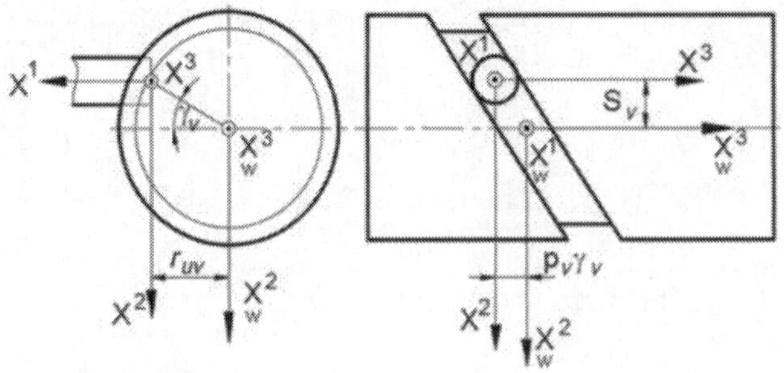

Fig. 3. Tool and worm relative positioning.

The relative positioning of the tool in the worm's system can be described with the equation – Figure 3

$$\mathbf{x} = [1, \quad \varphi]\underset{p}{\mathbf{x}} + [r_{uv}, \quad -s_v, \quad -p_v\gamma_v]^T \tag{2}$$

where: φ - tool action surface parameter; r_{uv} - mill distance from the axial plane X^2X^3 of the helicoid; s_v - mill offset over the worm length; p_v - helical surface parameter; γ_v - angle between the plane X^1X^3 of the helicoid and the radius vector of the point on the tool axis.

The pitch and the offset are variable over the worm length - Figure 1. It is assumed that the worm length is the multiple of the average pitch, or

$$o = \frac{4\pi L}{h_1 + h_2} \tag{3}$$

where: o - the number of relative tool and worm revolutions over the worm length.

So, the helical surface parameter is equal to

$$p_v = \frac{1}{2\pi}(h_1 + a_h v) \qquad a_h = \frac{h_2 - h_1}{o} \tag{4ab}$$

where: a_h - coefficient of helical surface parameter variation over the worm length.

In a similar manner - Figure 3, the parameters of tool positioning in the worm's system are described

$$s_v = \frac{1}{2\pi}(s_1 + a_s v) \qquad a_s = \frac{s_2 - s_1}{o} \tag{5ab}$$

$$r_{uv} = \sqrt{\left(\frac{d_f}{2}\right)^2 - (s_v)^2} \qquad \gamma_v = arctg\frac{s_v}{r_{uv}} \tag{6ab}$$

where: a_s - coefficient of offset variation over the worm length.

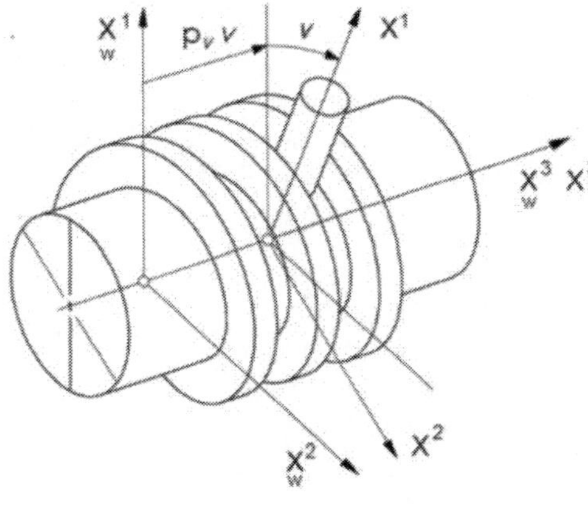

Fig. 4. Relative worm and mill motion.

If we consider the relative motion of the tool and the worm in the machining process – Figure 4, then the family of tool action surfaces can be written with the following equation

$$\underset{w}{\mathbf{x}} = [3, \;\; -v]\,\mathbf{x} + [0, \;\; 0, \;\; \pm p_v v]^T \tag{7}$$

where: $\pm$ - for the right-hand and left-hand worm, respectively.

In order to describe the worm surface, the envelope condition should be included in Equation (7), which can be generally written in the form of the triple product of three vectors [3, 5]

$$f = \frac{\partial \underset{w}{\mathbf{x}}}{\partial u} \frac{\partial \underset{w}{\mathbf{x}}}{\partial \varphi} \frac{\partial \underset{w}{\mathbf{x}}}{\partial v} = 0 \tag{8}$$

After transformations, this equation has the following form

$$\left\{ \begin{array}{l} -[1, \;\; \varphi]^T [3] \left([1, \;\; \varphi] \underset{p}{\mathbf{x}} + [r_{uv}, \;\; -s_v, \;\; -p_v \gamma_v]^T \right) + \\ + [1, \;\; \varphi]^T \left[\frac{-a_s s_v}{r_{uv}}, \;\; -a_s, \;\; \begin{array}{c} -\dfrac{a_h \gamma_v}{2\pi} - \dfrac{p_v (r_{uv})^2}{M} - \\ -\dfrac{a_s r_{uv} + \dfrac{a_s (s_v)^2}{r_{uv}}}{(r_{uv})^2} \end{array} \right]^T \\ + [1, \;\; \varphi]^T [3, \;\; -v]^T \left[0, \;\; 0, \;\; \pm \left(\frac{a_h}{2\pi} v + p_v \right) \right]^T \end{array} \right\} + \cdot \left\{ [1, \;\; 0, \;\; 0]^T \times \left([1] \underset{p}{\mathbf{x}} \right) \right\} = 0 \tag{9}$$

where the following denotation is adopted

$$M = (r_{uv})^2 + (s_v)^2 \tag{10}$$

The profile angle was defined by determining the angle of the straight line passing through the extreme points of the profile.

3 Calculation results

The axial profile of a worm machined with a cylindrical finger mill with a rectilinear tool action surface axial profile is non-rectilinear, which results from the specificity of hobbing. In the case of a worm with a fixed pitch and a fixed offset, the worm profile is constant – Figure 5a. For a worm with a variable pitch and a fixed offset, the worm profile varies over the worm length – Figure 5b.

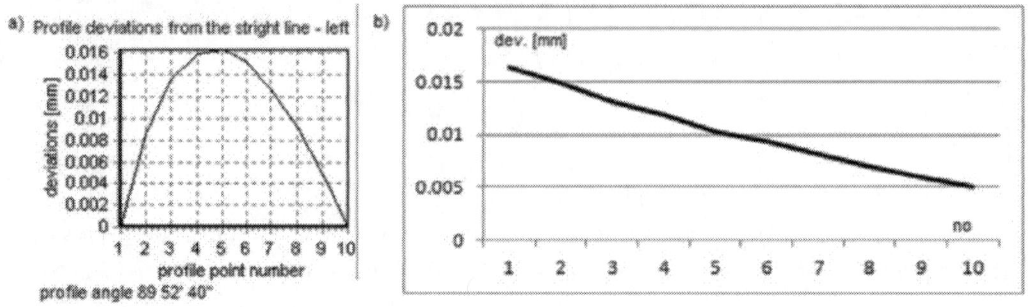

Fig. 5. A worm $L = 150\,mm$, $d_f = 28\ mm$, $s_1 = s_2 = 0\ mm$: a) axial profile ($h_2 = h_1$), b) maximum deviations of successive axial profiles from the straight line ($h_1 = 10\ mm$, $h_2 = 20\ mm$).

A very great effect on the worm profile is exerted by the offset – Figure 6, whereas, for machining the two sides of the worm cut, the offset should be equal in value and opposite in sign for the profiles for the both cut sides be the same. The relationship between the worm axial profile deviation from the straight line and the offset is linear - Figure 6.

So, in the case of variable-pitch worms, the worm profile can be influenced by changing the offset – Table 1. A continuous change in pitch and offset can be accomplished on a CNC machine tool. By changing the offset, worm axial profile errors can be reduced, and also profile angle errors can be diminished.

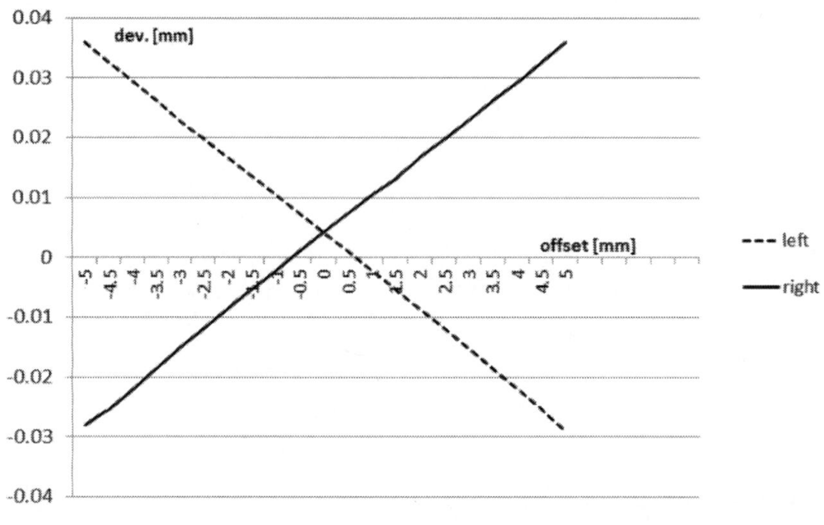

Fig. 6. Deviations of the worm axial profile from the straight line as a function of the offset – $L = 150\,mm$, $d_f = 28\ mm$ ($h_2 = h_1 = 10\ mm$).

 DOI: 10.1051/matecconf/20179407007

Table 1. Offset the impact on the outline of a worm: $L = 150\ mm$, $d_f = 28\ mm$, $h_1 = 10\ mm$, $h_2 = 20\ mm$.

Profile No.	Offset [mm]	Deviation [mm]	Angle	Offset [mm]	Deviation [mm]	Angle
1	0.0	0.0163	89 31'25''	1.25	0.00102	89 48'07''
2		0.0148	89 34'01''	1.15	0.00135	89 50'31''
3		0.0133	89 36'39''	1.00	0.00139	89 54'07''
4		0.0118	89 39'07''	0.95	0.00191	89 54'39''
5		0.0105	89 41'27''	0.86	0.00204	89 56'09''
6		0.0093	89 43'40''	0.82	0.00169	89 56'20''
7		0.0081	89 45'46''	0.75	0.00161	89 57'10''
8		0.0070	89 47'42''	0.70	0.00137	89 57'28''
9		0.0060	89 49'30''	0.65	0.00115	89 57'46''
10		0.0050	89 51'10''	0.62	0.00081	89 57'43''
11		0.0041	89 52'40''	0.60	0.00045	89 57'32''

4 Conclusions

In variable-pitch cylindrical worms, the axial profile varies over the worm length.

The effect of the offset on the worm axial profile is profound. To ensure the constancy of the profile of a variable-pitch worm, the offset needs to be variable over the worm length. With an opposite offset value for either side, the left-hand and right-hand worm cut sides must be machined separately.

The fixed-pitch cylindrical worm can be considered a special case of the variable-pitch cylindrical worm.

References

1. I. Dudas, *The Theory and Practice of Worm Gear Drives* (Penton Press, London, 2000)
2. F. L. Litvin, *Theory of Gearing*, NASA Reference Publication 1212, AVSCOM Technical Report 88-C-035, (1989)
3. F. L. Litvin, A. Fuentes, *Gear Geometry and Applied Theory* (Cambridge University Press, 2004)
4. F. L. Litvin, *Development of Gear Technology and Theory of Gearing*, NASA RP-1406, (1998)
5. F. L. Litvin, C, L. Hasiao, *Computerized simulation of meshing and contact of eneveloping gear tooth surfaces,* Comp. Meth. in Appl. Mech. and Eng., **102**, 337, (1993)
6. T. Nieszporek, *The fundamentals of the theory of synthesis and analysis of approximate gearig,* Series Monografie, **54**, Czestochowa University of Technology, Czestochowa (1998)
7. T. Nieszporek, *Konstrukcja narzędzi skrawających i technologia walcowych uzębień zewnętrznych,* Series Monographs, **265**, Czestochowa University of Technology Czestochowa, (2013)

The preparation for being a researcher in Romania today

Eugen Pămînțaș[1,*], and *Felicia Veronica* Banciu[1]

[1]Politehnica University of Timişoara, Materials and Manufacturing Engineering Department, 300222, Timişoara, Romania

Abstract. Higher technical education in Romania is organized in three cycles: Bachelor, Masters and PhD. Theoretically, the powers provided by the three levels of education should be in correspondence with the requirements of COR (ISCO-88-COM European Community and the UN: ISCO 88) for generic jobs, specified in previous herewith: production engineer, research engineer, and researcher. This paper explores the last two aspects in order to see to what extent the names of study disciplines and especially their content gives the graduate competences and skills of research. We are looking for causes for which young graduates of master and doctoral especially not find appropriate jobs in the labor market preparation. From the position of academics with experience in teaching the final year of Bachelor and Master levels, trying to answer to the questions: What should we teach, How should we teach? Who should we teach? We train for right people? The answers are tasks for interdisciplinary collective of teachers, researchers and employers. Therefore, without claiming an exhaustive treatment of the matter under discussion, the paper is intended as a call a call to quickly find viable solutions. Where answers are not appropriate, will have to find solutions but the decision in their choice and then their implementation in the short time it is a governmental task.

1 About research methodology

The authors teaching experience have found that the generic methodology creates some problems, which is why it was considered useful to make some terminological remarks. This is why hereinafter briefly present some considerations regarding the methodology of scientific research, as is almost universally accepted. The purpose of the presentation is to provide support in analyzing the extent to which these concepts are found in the disciplines of master degrees. Master study cycle is declared as the deepening of a narrower field of study and research initiation into it. The next step is continuing the research activity during doctoral studies, which gives graduate researcher attributes to be confirmed by the results of quantitative and qualitative research work at scientific research institutes or universities, public or private.

* Corresponding author: eugen.pamintas@upt.ro

2 Conceptual delimitations

2.1 Methodology, method, technique, research tool

In a broad sense the methodology means all methods used in scientific research or simply the science of researching, (from Greek: *metodos*- road, path and *logos* - science).

In fact, through research methodology we understand the existence of at least one discipline in which are developed methods for developing the scientific knowledge and all the elements involved in the research. The fundamental aim of the methodology is to assist in understanding - in terms widest possible - not so much the products of science, but especially the functioning of knowledge process itself.

To be able to fulfill its function, scientific research methodology includes both, the proper definition of the studied field, a number of principles and rules for investigation purposes, the instrumentation of work for collecting and interpret data, strategies for theoretical construction or reconstruction as well as results assessment.

Method - the first use of the term may create some difficulties, therefore, in order to better understand, we offer a classification without claiming an exhaustive treatment:

a) on the criterion time:

•Transversal methods - analysis of phenomena or facts at a time (the status quo, operating parameters or currently indicators, tests etc.).

•Longitudinal methods - analysis of facts or phenomena in their evolution (case study, bibliographic analysis, the study panel etc).

b) by type of the investigation:

- Quantitative methods – with orientation type positivist-explanatory (currently system response on experiment, the basic parameters).
- Qualitative methods - phenomenological orientation, comprehensive (participant observation, meeting the permissible limits, case study, etc.).
- Method of intersection - between the approach combining quantitative and qualitative (analysis overview with links, history analysis of relevant information etc.).

c) by the place in the investigative process:

•Data collection methods (analog, digital, sequential, continuous, etc.).

• Data processing methods (statistical and mathematical, comprehensive analysis etc.).

•Data interpretation methods (inductive, deductive, comparative, causal explanation).

Research technique - is subsumed to a method and represents the manner to use various investigative tools, helping to collect or process data (dynamometers, sensors etc.).

Research tool - is the materialization of a method (measurement sheet is a tool of analog data collection method, observation sheet is a tool for participative observation etc.).

2.2 Methodological principles of scientific research

The deployment of scientific research assumes the following general principles:

•*The principle of unity between theoretical and empirical.* Theoretic presence in empirical research, regardless of its weight, it is often the default form of assumptions that must be acknowledged and explained, before starting the research.

•*The principle of unity between understanding (comprehension) and explanation* calls into question the relationship between subject and object of knowledge.

• *The principle of unity between quantity and quality*, require compatible approaches in quantitative and qualitative methods in order to achieve complementarity and interference both in the general epistemological and other particular levels.

•*The principle of unity between the observantand evaluative judgments* requires hiring a researcher moral support generally valid values in any field of human knowledge.

2.3 Assessment of scientific research

The task of evaluation of scientific research lies primarily those who make them. The evaluation of research is intended to constitute a critical attitude in receiving results, preventing improvisation and misrepresentation, through theoretical elaborations sophisticated to support or promote partisan interests.

Although scientific literature designed to evaluate the scientific research is not too rich, nevertheless the criteria were developed aimed mainly:
•the correctness of *of* the theoretical framework research;
•the quality of *the* conceptual analysis and defining the scope of research;
•the *adequacy* of methods, techniques and working tools to the specific characteristics of the studied object and its operating;
•the *validity* and reliability of measurement instruments;
•the *representativeness* of the study object
•the *degree* of favorability of the conduct of investigations to ensure objectivity;
•the *quality* of the processing, analysis and interpretation of information;
•the *manner* of theoretical construction.

Assessment of theoretical and methodological framework is based on the following indicators (top four concerns the theoretical framework and the following, the methodology):
a) appropriate definition of concepts, or scientific validity thereof;
b) the operationalization of concepts to bring to the foreground measurable and relevant indicators;
c) formulation of hypotheses to be a convincing explanatory model and be testable;
d) determining the scale and activity of the research topic;
e) the choice of methods and research techniques in line with the specific objectives pursued and studied field;
f) the tools for working in agreement with the contents topic and defining the operational concepts;
g) the adequacy of research instruments to the characteristics of the subject;
h) evaluating the measurements taken.

Assessing the degree of validity and reliability
a) *validity analysis* - establishes whether the instruments used mode and capture what was proposed. There are the following types of validity:
- Content validity (internal) - designed to measure the degree to which indicators which supposedly is measured;
- Predictive validity (external) - aims extent to which measurements provide an insight into relations with other measurements and forecasting ability of the subjects;
- Construct validity - refers to the explanatory factors of a particular result obtained by measurement. The extent of this type of validity is given by the manner of correlation between the variables studied.

b) *fidelity analysis* - targeting the confidence and stability of tools. The degree of fidelity is given the extent to which through measurements, we can get the true values that characterize the manifestation of a phenomenon or social fact.

3 Research educations

Taking in the consideration the strong emphasis on the research in all countries, as an educational endeavor, we were collectively confusing by the lack of engagement with one of the central concepts within the field of education: curriculum.

Rethinking the preparation for research in these terms directs attention to the forms of knowledge in which it is grounded and how these are articulated in the documentation and assessment of the future job.

But curriculum may be about more than this. The special ad hoc group made up of the Conference "Diaspora and his friends" held at the Polytechnic University and West University of Timişoara between 24 to 28 April 2016 under the auspices of the Romanian Presidential Administration, brought together experienced teachers and researchers in a range of countries, and with varying interpretations of 'the researcher curriculum'. As panel we did not speak with one voice, but rather sought to engage each other and our audience with this rich, powerful, contested, difficult and sometimes slippery notion.

As was noted from the start was that an educational curriculum or books published on education for research is more or less nonexistent in Europe and even more in Romania. The Bologna process has had some impact, since it has forced us to articulate the intended learning outcomes of coursework as competencies; this has been a challenge to many of us. But to articulate the curriculum and pedagogies at work in our daily practice demands some reflection. In our department of Material and Manufacturing Engineering (at the University of Timisoara), which hosts three masters fields, advanced materials, industrial engineering and advanced welding technologies, we have agreed to a common structure for quality assurance through the master's program but it does not provide the necessary skills for research training.

The research results must be disseminated. Following this, there is also pressure for writing in English. This means that master students (and by all means, also senior researchers), who do not have English as their mother language, may always feel a step behind. Many of our master or doctoral students have never written anything in English before entering their research or doctoral education. There is a curricular pedagogical challenge there, how to support those writing skills when you are not yourself a native English speaker.

Research education does not occur in a vacuum, but rather is influenced by, among other factors, policies, procedures, global developments, and significant others support services and resources. What may be broadly termed the environment or context in which master and doctoral studies are undertaken have an important bearing on the curriculum. Furthermore, the examination process is a substantial factor in the curriculum, not only as an assessment strategy, but also as a specific process in which the power difference between the candidate and supervisor is particularly evident. Finally, the outcomes and outputs of the research education experience are critical factors which shape the research curriculum [1].

Practitioner research is a process of discovering and framing questions, collecting data, and analyzing data to answer the questions [2].

It is important that practitioner researchers have more than one method of gathering information so that data can be triangulated [3]. Triangulation, which entails the use of multiple data sources to confirm and illuminate one another, is strength of practitioner research as well as of university research [4-6].

Data is analyzed, with an emphasis on finding patterns. Drawing on their prior knowledge and experience, practitioners can then formulate hypotheses about emerging patterns.

The preparation for research activities "involves disciplined inquiry" [7] which means that "studies are intentional and systematically conducted" [8].

At the center of this action *research plan* was curriculum inquiry, "similar to the ways detectives and journalists do their work—asking probing questions that led to finding truth" [9]. Simms goes on to note that focusing on questions, including emphasizing the use of what, how, who, why, where, and when, served to remind teachers that they need to ask these questions in support of creating curriculum that motivates their students to ask questions and investigate issues and events related to culture.

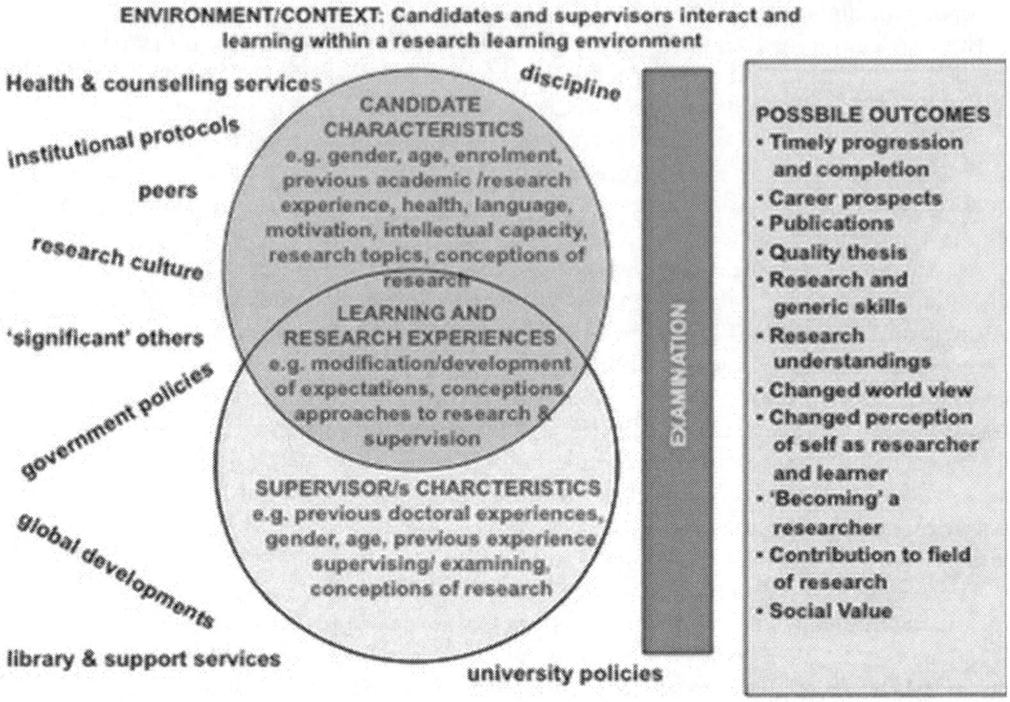

Fig. 1. A model of influences on the doctoral curriculum [1].

It has been well documented that students in science based faculties have greater satisfaction with their candidatures than humanities students, as the science projects often have some laboratory-based component, often more of a team atmosphere and more opportunities for interactions with supervisors, staff and other students and the results here were consistent with previous reports.

Another observation was international students have greater 'agreement' with all scales (with lower agreement with 'infrastructure' and 'research climate') than domestic students. The data provided does not account for this observation and the underlying factors at this stage could only be the subject of speculation.

The quality of the research climate has been shown to be an important part of the research experience and has a role in optimizing output and minimizing student attrition during their candidature. On reflecting on this data the authors make some suggestions. Organizing student review and feedback on the items of the SREQ, including what items they would suggest modifying, omitting or adding and why, and how the changes may have a positive effect on their candidatures.

This exercise could also be utilized to investigate the 'research climate' items and students' perceptions on the importance of these items and why a neutral response to these items is more prevalent than items of other scales [10].

4 Research training curricula

Order of the Ministry of Research and Education of Romania no. 4491/2005 06/07/2005 Published in the Official Gazette, Part I no. 589 of 07/07/2005, regarding the organization of doctoral studies in the academic year 2005/2006, inter alia, provides: "....PhD is a distinct type of learning and research, allowing doctoral students to obtain both peak powers in the specialized field of study and personal and managerial skills generic:
a) develop research projects in order to obtain the resources to finance them;
b) discover innovative solutions and communicate them effectively;
c) ensure good management of time and resources involved in research;
d) continuously monitor the quality and performance in their own work and the work of the team belongs."

Therefore neither this time, nothing about research training curricula, only sequence of "competences" whose assessment methods are mainly subjective, at theirs supervisor discretion.

Because the job of scientists to be attractive for junior researchers, the employment conditions and payroll are some particularly important. Although in Romania of 0.2 percent of GDP for R&D activity is insufficient, recent statistics show that research expenditure structure is a well balanced one both the performance sectors (figure 2) and the types of research (figure 3) [11]. We can say that the will is there, the structure is healthy but content of training researchers is deeply deficient through curricula that do not offer the support required.

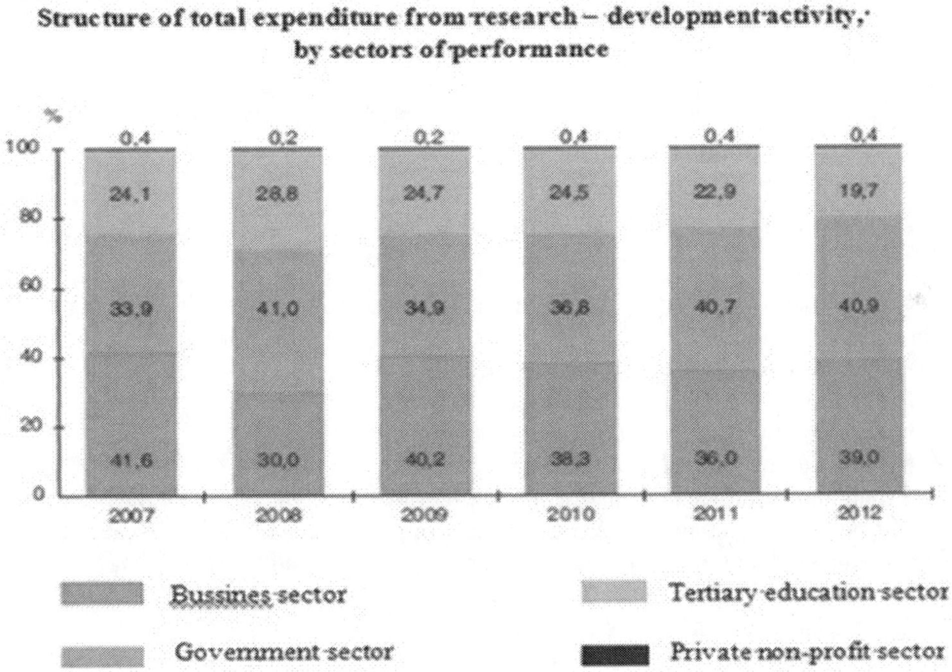

Fig. 2. Structure of total expenditure from R&D activity, by sectors of performances (Romanian Statistical Yearbook 2013).

In support of better understanding the role that it is preparing the master's and doctoral researcher training in the field of engineering, the authors examined in a comprehensive review of both literature and the requirements of employers, the following: the characteristics that must possess researchers from academia and manufacturing, their training and their work environment. Analyzed areas should be developed by faculty and departments dealing with assessment, career development, professional socialization, organizational development, higher education and in research and development enterprises.

The results obtained show that, in addition to terms of knowledge and skills in an area of research, scientists successfully values and attitudes derived from the specific experiences of socialization academic, the significant support given by mentors and colleagues or the organization of activities their research department.

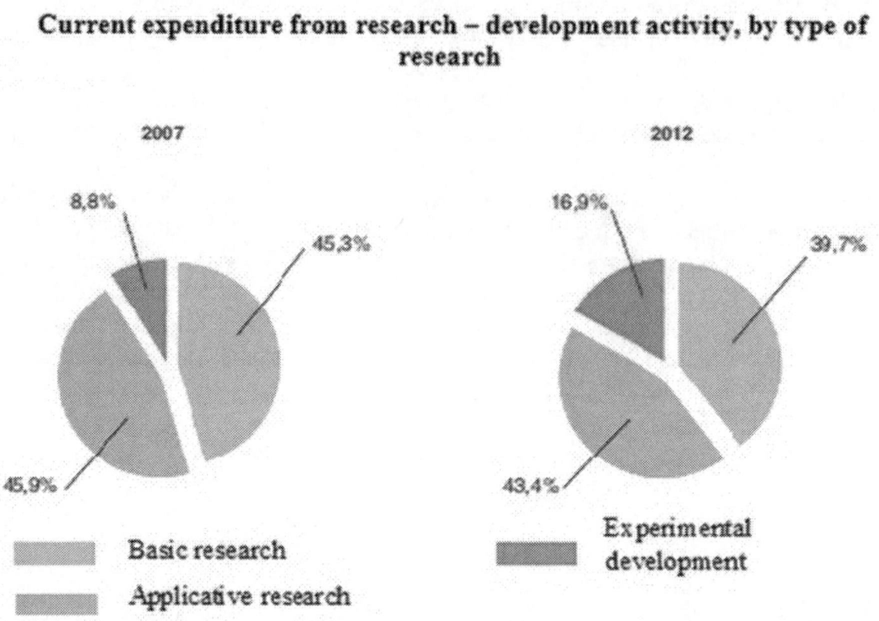

Fig. 3. Current expenditure from R&D activity by type of research (Romanian Statistical Yearbook 2013).

A review of these studies [12] has highlighted substantial variation in academics' views of research, in terms of their:
•research intentions-who is affected by the research;
•research outcomes-the anticipated nature of the effect;
•research questions-the nature of the object of study, and
•research process-how research is undertaken.

As a group, these studies have shown that academics do not necessarily agree on what it means to be university researcher, to conduct research or to be successful in research.

Furthermore, such variation in views of the nature of university research has been shown to occur:
•within disciplines as well as across disciplines;
•within research intensive universities as well as non-research intensive universities;
•amongst senior academics as well as early career academics.

5 Conclusions

We manifest our concern that in Romania today researcher or professor and research worker operates out of the concerns for which was prepared. This divergence is likely to lead to an irrelevant theory and practice unauthorized and invalid. Such results will damage the reputation of universities as trainers of researchers and ultimately will lead to our extinction as caste.

The authors also suggest that both academic unit personnel and students need to be made aware of the importance of the research climate and all must take some collective responsibility to maintain and enhance this climate. Academic unit staff may need regular reminding of the importance in instigating and supporting research climate activities and students may require induction into what a research climate is, why it is important to them and how they can take an active contributory role in its maintenance. Furthermore it would be useful for the institution or academic units to assist master and doctoral candidates to develop strategies to deal with some of the regular issues that arise when working within a competitive, somewhat constrained and bureaucratic environment with regular workshops on these issues with successful established researchers.

References

1. N. Hopwood, D. Boud, A. Lee, M. Abrandt Dahlgren, M. Kiley, Proceedings of 9th QPR conference, Educating Researchers for the 21st Century, **83** (2010)
2. K. H. Campbell, Democracy& Education **20**, 7 (2013)
3. B. Berg, *Qualitative research methods for the social sciences* (MA: Allyn &Bacon, Boston, 2001)
4. M. Cochran-Smith, S. Lytle, *Inside/outside: Teacher research and knowledge* (NY: Teachers College Press, New York, 1993)
5. R. Hubbard, B. Power, B, *Living the questions: A guide for teacher-researchers* (ME: Stenhouse, New York, 1999)
6. R. Shagoury, B.M. Power, *Living the questions: A guide for teacher researchers* (York, ME: Stenhouse, 2012)
7. L. Shulman, L., *Autonomy and obligation.* (NY: Longman, 1983)
8. J. Duffield, S. Townsend, Action in Teacher Education **21**, 13–20 (1999)
9. M. Simms, Democracy&Education, **21(2)**, 3 (2012). Available online at http://democracyeducationjournal.org/home/vol 21/Iss2/3
10. M. Govendir, B. Govendir, Proceedings of 9th QPR conference, Educating Researchers for the 21st Century 13-15, **65** (2010)
11. www.immromania.ro/a-aparut-anuarul-statistic-al-romaniei-2013-7639.htm
12. G. S. Åkerlind, Higher Education **55**, 241 (2008)

MATEC Web of Conferences **94**, 07009 (2017) DOI: 10.1051/matecconf/20179407009
CoSME'16

Impact of employee age on the safe performance of production tasks

Małgorzata Rembiasz[1,*]

[1]Poznań University of Technology, Faculty of Management Engineering, Chair of Entrepreneurship and Business Communication, 11 Strzelecka St., 60-965 Poznań, Poland

Abstract. The number of occupational accidents reported in Polish industry and the extent of the resulting losses show that safety is critical for the effective performance of production tasks. In view of such high accident rates, a pressing need has been recognized for identifying and undertaking improvement measures. The factors which require a response to secure adequate safety levels for the labor force include worker age. Such age affects worker behaviors, influences the speed of reaction to threats and translates into the scope and nature of effective improvement measures. Thus, worker age ranks among the key criteria for ensuring the safe performance of production tasks. The study revealed relationships between worker age and safety at workstations. Links between the two were found to be particularly pronounced among workers aged 50+. The study's outcomes were used to define the impact of worker age on the occurrence of threats that impair the ability to perform production tasks.

1 Introduction

The ageing of societies in highly developed countries caused by longer lifespans and low population growth is bound to make age management central to human resource management. Between 1989 and 2013, the number of elderly people in Poland rose by close to 1.9 million. Their share in the general population climbed from 10% in 1989 to 14.7% in 2013. The percentage of persons aged above 65 is set to continue growing in the next two decades [1].

As is equally important, elderly persons in Poland are becoming increasingly active vocationally. Eurostat data for the 55-64 age bracket show that merely 26.2% were vocationally active in 2004, 36.3% in 2009 and as many as 42.5% in 2014. Despite the increase, Poland continues to lag behind the EU average which has recently reached 51.8%. The highest rates in Europe have been reported by Scandinavian countries: 83.6% in Iceland, 74.0% in Sweden and 72.2% in Norway. Meanwhile, the lowest rates can be found in Greece: 34.0%, Slovenia: 35.4% and Croatia: 36.2%. As for the rest of the world, notable examples include a solid 68.7% of people aged 55 to 64 who continue to work in Japan and 61.3% of workers in the same age group in the USA [2]. These statistics reveal a massive potential of unutilized knowledge and skills of a large segment of the 50+ population that lies fallow in Poland's economy. A number of factors contribute to this state of affairs. One of them, without a doubt, is that working

* Corresponding author: malgorzta.rembiasz@put.poznan.pl

 DOI: 10.1051/matecconf/20179407009

environment factors are poorly suited to the needs and capabilities of older workers, rendering them prematurely unfit for work [3].

It is increasingly critical to employ persons aged 50+. The enterprises that succeed in doing that stand to benefit from the assets of mature workers. In order to employ them, businesses need to adequately prepare their working environments. Working environment modifications to accommodate the needs of mature workers with a view to proactively ensuring their work ability should be seen as an integral part of a comprehensive strategy of age management in an enterprise [4]. Age management is crucial to companies whose labor forces are multi-generational. According to one of the earliest definitions, age management refers to: "*the various dimensions by which human resources are managed within organizations with an explicit focus on agei*ng". [5]. Over the last dozen plus years, both the theory and practice of age management has been advancing in Western Europe, Finland, the United States and Japan. It has become the subject of hosts of scholarly and popular science publications promoting good practices in the field.

The use of an age management strategy by enterprises generates tangible benefits for both mature and younger workers. Jobs designed in compliance with ergonomic principles are likely to contribute to prolonging life expectancy and improving the health of all workers as well as increasing work efficiency in enterprises.

2 Working conditions modification to accommodate 50+ workers

2.1 Effects of age on work ability

In ensuring conditions that enable workers to operate properly in their working environment, one should bear in mind that humans are an organization's most valuable resource. However, to utilize their potential, they need access to an environment that will allow them to operate normally. An enterprise's success or failure in enabling employees to satisfy their needs makes an enormous difference for the effects of their work [6, 7].

By reducing stress be establishing proper working conditions, enterprises can cut their losses, improve productivity and increase their ability to carry out tasks and run processes [8-10].

To ensure the capability to carry out vocational tasks, work must be designed in recognition of the capacities of persons of a specific age and in a specific environment [11, 12]. It is crucial to coordinate psychological and physical effects with the ability to cope with specific situations [13].

A person's work ability declines with age. Both physical fitness and mental function become impaired. Chronic diseases and certain other health issues become exacerbated. The bodily ageing process varies widely from one individual to another depending on the factors that accelerate it or slow it down. It is very difficult to pinpoint the exact moment at which the human body begins to age and at which work ability begins to decline. Improved living conditions not only prolong life spans but also extend the time of efficient operation. Persons of the same calendar age may differ in their biological age and therefore also in their ability to function in society. Thus, to assess a person's work ability, one should define the so called functional age measured not only by the number of years one has lived but also by his/her physical and mental acumen [14].

One of the measures of the state of health that reflects a person's ability to engage in physical exertion in professional and personal life is physical functional capacity. Physical capacity defines a body's ability to perform arduous manual tasks over prolonged time and specifically to engage large muscle groups without triggering a steep increase in fatigue levels and considerable changes within the body. The measure reflects the ability to tolerate effort and

 DOI: 10.1051/matecconf/20179407009

rapidly recover from fatigue once work has ceased. With age, people lose muscular strength and circulatory and pulmonary system efficiency, which in turn hugely compromises the ability to perform strenuous physical labor [15].

The overall work ability can be measured with the work ability index (WAI), which represents a subjective seven-question survey tool. The index measures work ability on a scale from 7 to 49. The higher the score, the greater the ability to perform work. WAI commonly classifies workers into the four ability levels of poor, moderate, good and excellent. Work ability usually declines with age although workers aged 20 to 65 commonly score good or excellent. Some 30% of the workers aged above 45 display a sharp decline in their WAI ratings. This applies to both blue and white-collar workers [15]. Notably, work ability ratios can be modified and improved by enhancing working conditions and preserving the health of workers [16].

2.2 Modifications of working environment to fit the needs of persons aged 50+

The key to creating optimal working conditions is to account for such human characteristics as are vital for ensuring work ability. These are primarily physical and mental fitness, psychological qualities as well as the physical and psychological stress resulting from the nature of the work performed. For instance, mature persons are in less of a need of personal and vocational development but need more social acceptance [17].

Measures taken in keeping with a company's procedures must ensure the achievement of the adopted goals and consequently guarantee safe work performance. To identify adequate and necessary working condition improvement measures, an enterprise needs to be aware of risk levels and profiles as well as worker characteristics. By incorporating the modifications of working conditions into the guidelines [18], companies gain the ability to operate effectively under circumstances in which it is critical to avoid losses associated with failures to ensure proper working conditions to workers operating in a working environment [19, 20].

The desired improvement level can be achieved by properly executing corrective, correction and preventive measures that address specific risks and are sufficiently practicable, effectively ensuring the protection of specified members of the work force. Examples of actions taken in the follow-up to occupational risk assessment accounting for employee age are provided in Table 1.

In accordance with the recommendations of the International Labor Organization, mature workers require measures that [21]:

- remedy those conditions of work and of the working environment that are likely to hasten the ageing process,
- modify the forms of work organization and working time which lead to stress or to an excessive pace of work,
- adapt the job and its content to the worker by recourse to all available technical means and, in particular, to ergonomic principles, so as to preserve health, prevent accidents and maintain working capacity,
- provide for a more systematic supervision of the workers' state of health.

It should also be noted that the key areas of occupational activity in which particular attention needs to be paid to worker age are arduous physical work, work in a hot microclimate and shift work.

The above may be considered in view of ergonomic requirements. The aim of ergonomics is to optimize working environment stresses associated with the performance of work [12, 18, 22]. To preserve or improve the health and skills of ageing workers, care should be taken to improve their working conditions.

Table 1. Actions taken in the follow-up to occupational risk assessment accounting for employee age [20, 22].

Action	Employee age-related factors
Elimination of risks	- ensuring conditions that allow operation on company premises, including safe mobility at workstation, - prevention of direct exposure to existing hazards
Work organization and planning	- work organization in view of workers' capacity to work at a particular age
Risk warnings	- risk notification methods suited to the specific abilities of workers, including their capacity to acquire knowledge, - risk notices designed to correspond to worker experience
Employee training	- training conducted in recognition of reduced ability to acquire knowledge and skills
Use of personal protection equipment	- use of personal protection equipment suited for people of a particular age

Such an effort may be considered in terms of the working environment, which is essential for work ability. The thermal working environment forms a major factor for workers' well-being, health, work efficiency as well as occupational health and safety. Manual labor in a hot environment placed the human body under significant stress. High temperatures may cause fainting, exhaustion and even heat stroke. Numerous studies show that older people are less tolerant to heat in connection with the performance of work. Work under such conditions weakens the cardio-vascular system in 50+ workers compared to younger employees. Other observations have found that mature workers with more years of service in a hot environment behind them tend to retain physical fitness compared to other workers in the same age bracket. Such capacity is the result of heat tolerance linked to the physical fitness of the body [23].

To improve working conditions, enterprises may resort to any of a number of solutions. These include:

- flexible working time,
- job rotation,
- teams mixed in terms of age (allowing more physically demanding tasks to be performed by younger team members),
- opportunities to change jobs (including the securing of relevant training),
- health preservation.

Another category of solutions aimed at improving working conditions for 50+ workers is technical in nature. Such solutions include measures that rely on:

- ergonomic principles in workstation design,
- technical means that eliminate or reduce the amount of arduous physical labor,
- safe technologies,
- proper collective and personal protection equipment suited for a particular job.

3 Worker age and occupational accidents

3.1 Impact of worker age on accident statistics

For a number of years, the relevant literature has depicted worker age ambiguously: on the one hand, such age has been presented as a factor for lowering the risk of occupational accidents while on the other, it has been associated with a greater frequency and higher severity of accidents [24].

According to statistics, the number of accidents at work is the highest among young and middle-aged workers. In 2014, employees aged 20 to 49 accounted for 72% of all occupational accident victims in Poland [25].

Studies have found that as workers gain experience, they also become knowledgeable about occupational safety and more capable of noticing potentially dangerous situations and responding to hazards [25-27].

National statistics provide insights into accident rates and severity in manufacturing companies. This data has been supported with the author's study.

The study has been conducted in three steel-structure manufacturers. Accident rates were examined for workers employed as metal workers, steel structure fitters, welders and welder's hands. The workers operated in a two-shift system.

Outcomes of the study are provided in Table 2.

Table 2. Accident rates in manufacturing companies by worker age.

Manufacturing company	Number of accidents at work between 2014 and 2015			
	Workers aged 18 to 25 years		Workers aged 26 to 39 years	
	Headcount	Number of accidents / accident rate	Headcount	Number of accidents / accident rate
Enterprise A	16	2 / 0.125	18	1 / 0.056
Enterprise B	19	1 / 0.053	19	3 / 0.158
Enterprise C	21	3 / 0.143	22	2 / 0.091

Manufacturing company	Number of accidents at work between 2014 and 2015			
	Workers aged 40 to 50 years		Workers aged above 50 years	
	Headcount	Number of accidents / accident rate	Headcount	Number of accidents / accident rate
Enterprise A	10	1 / 0.100	9	3 / 0.333
Enterprise B	7	2 / 0.286	8	2 / 0.250
Enterprise C	13	2 / 0.154	11	3 / 0.273

The study outcomes (accident rates) are shown in Figure 1.

The above accident statistics show clearly that workers above 50 years of age are the most likely to become involved in occupational accidents. The trend applies regardless of the professional training given workers have received to prepare them for the jobs [3, 20]. However, worker age can also be seen as a mitigator of accident likelihood. To benefit from this influence, employers need to design work properly and make best use of the potential offered by older individuals.

A large proportion of occupational accidents are caused by humans. It seems that the advice and watchful eye of senior colleagues can offset the careless tendencies of younger workers which often result from lack of knowledge and recklessness in the face of hazards.

This said, it should nevertheless be noted that the occupational accidents affecting older workers are more likely to be severe. The ratio of fatal accidents at work per 1000 workers is the lowest among young employees and grows almost linearly with the age of victims to peak in the 60 to 64 age bracket. Such a high frequency of fatal accidents among mature workers may be the result of declines in their physical capacity due to age, among other factors. Furthermore, the time a victim needs to fully recover from an accident also grows with employee age [27-29].

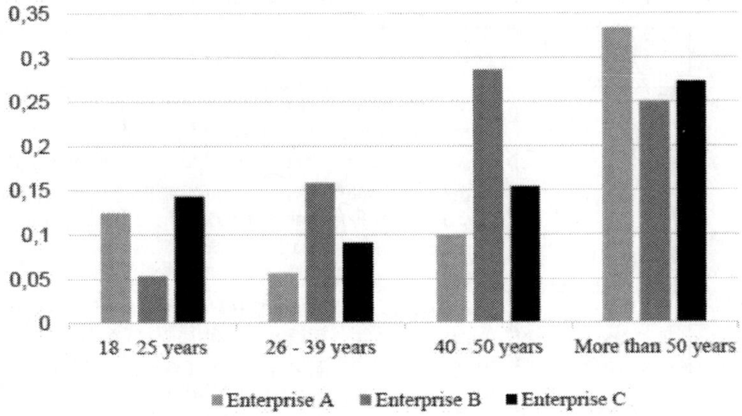

Fig. 1. Accident rates as a function of worker age.

Research suggests that preventive measures should rely on differentiated means and channels to reach potential accident victims and provide them with different types of information depending on their profiles. While electronic information channels work are embraced readily by young people, older workers are best targeted with printed materials as well as campaigns in the traditional news media (the press, radio, television) and provided with information meetings on workplace risks [30].

3.2 Sample age management improvement measures

In many business organizations, job security depends heavily on the management measures taken in specific fields. Such fields include employee age and work experience. An organization's ability to develop a comprehensive age management strategy depends on a number of factors which influence the measures it undertakes. The key factors of this kind include the awareness of the need for such measures.

The most critical age management measures taken by companies are to [31]:

- improve employee competencies by designing individual training schemes for workers to equip them with the knowledge they need in their specific jobs,
- offer a broader spectrum of flexible forms of employment that includes part-time employment opportunities for 50+ workers,
- improve work comfort,
- improve workers' physical fitness and reduce number of tasks causing excessive fatigue.

Such measures may substantially improve working conditions and work organization as well as boost productivity and the quality of job performance.

Good management practices may inspire the design of age management strategies that reflect the specific operating profiles of enterprises. Businesses looking to achieve lasting improvements should therefore adopt these concepts more broadly. Examples of such measures are provided in Table 3.

Table 3. Sample age management measures [32-34].

Business field	Improvement area	Sample measure
Waste treatment company	Ergonomics, improvement of working conditions	To protect older employees from excessive fatigue, the company adopted standard P90, which prohibits assigning employees to jobs which exceed the physical capacities of 90% of the labor force in an average worker group of a specific age. Furthermore, to reduce physical loads, technical solutions were adopted that ease work-related strain.
Manufacturer of would products	Professional improvement of workers	Towards the end of their on-the-job training, interns are employed by the company and then gradually inducted to perform tasks of growing complexity. The company has developed a competence improvement strategy that allows it to benefit from the knowledge of retiring workers (ensure knowledge transfers). To that end, the employer has offered advanced employee training, internal workshops, external improvement opportunities, etc.
Trading company	Ensuring health and well-being	The company offers workshops that increase the awareness of stress and its impact on employee propensity to go on sick leave. By reducing stress, raising creativity levels and increasing loyalty, the enterprise has been able to boost productivity. The company also reminds its workers that each of them is responsible for their own health.

By adopting comprehensive age management strategies, business organizations make better use of their existing human resources. This results mainly from ensuring working conditions that reflect worker age and capabilities and facilitating collaboration among employees of different ages.

4 Conclusions

The high proportion of older people in the overall population results from contemporary socio-economic processes and should be seen as more of a challenge than a threat.

Ergonomically-compliant work organization is likely to prolong worker life spans and improve their health. The design of workstations and physical working environments in keeping with ergonomic principles generates measurable benefits for workers (especially the mature ones) and employers alike. These include reduced work fatigue, fewer cases of work-related diseases and occupational accidents and better work efficiency. Thus, ergonomic requirements address the primary objective of any enterprise (as postulated by economic sciences), which is to maximize profit.

Job satisfaction and work ability are particularly crucial for older workers who face a number of limitations in the performance of work that come with age. Therefore, employers need to take a number of measures that mitigate the impact of such limitations. By employing 50+ workers, an organization benefits from their strengths, thereby adding to its own capacities. One of the side benefits of such employment is greater loyalty of younger workers who grow to perceive their jobs as secure and associate them with career opportunities.

Employees of all ages, including those in preretirement age, can be highly efficient. The only prerequisite for their efficiency is to modify the nature of their work to best fit their capabilities (which include their state of health and predispositions). Such modifications will allow an enterprise to make better use of the worker's potential and increase their motivation to work.

"Good practices" tend to inspire the design of age management strategies that reflect the specific profiles of a company's operations. Organizations that strive to ensure lasting growth should therefore engage in this area of improvement.

In macroeconomic terms, modifying working conditions to the needs of 50+ employees is a way to retain persons aged above 50 in employment. Many mature individuals would like to continue to put their vocational skills and experience to good use. Keeping persons aged above 50 professionally active is certainly a way to ensure better compliance with the principles of knowledge-based economy and to further a country's economic growth. A company's management should be aware of the need to employ older workers, ensure they can function properly in the working environment and make appropriate use of their strengths.

References

1. *Sytuacja demograficzna osób starszych i konsekwencje starzenia się ludności Polski w świetle prognozy na lata 2014-2050* (GUS, Warszawa 2014)
2. EUROSTAT, http://ec.europa.eu/eurostat/statistics-explained/index.php/Employment_statistics/pl (date of access: 01.12.2015)
3. M. Rembiasz, A. Górny, Zeszyty Naukowe Politechniki Poznańskiej, Organizacja i Zarządzanie, **65**, 115-126 (2015)
4. M. Rembiasz, Zeszyty Naukowe Uniwersytetu Szczecińskiego, **849**, 155-167 (2015)
5. A. Walker, *Combating Age Barriers in Employment. European Research Report* (European Foundation for the Improvement of Living and Working Conditions, Dublin 1997)
6. A. Górny, M. Rembiasz, Praca i Zabezpieczenie Społeczne, **8**, 98-110 (2015)
7. A. Kawecka-Endler, B. Mrugalska, Human-Computer Interaction: Applications and Services, III, **8512**, 700-709 (2014)
8. M. Butlewski, Technologies, Materials. Int. Virtual J., **VII (11)**, 24-28 (2013)
9. A. Górny, Foundation of Control and Management Sciences, **11**, 127-138 (2008)
10. B. Mrugalska, Need of safety integration in robust design, In: P. Vink (ed.), *Advances in social and organizational factors*, 260-266 (AHFE 2014)
11. M. Rembiasz, Zeszyty Naukowe Uniwersytetu Szczecińskiego, **809**, 499-508 (2014)
12. A. Górny, Ergonomics Aspects of CSR in System Shaping the Quality of Work Environment, In: P. Vink (Ed.), *Advances in Social and Organizational Factors*, 541-550 (AHFE 2012)
13. D. Efraty, M. J. Sirgy, Social Indicators Research, **22(1)**, 31–47 (1990)
14. J. Bugajska, T. Makowiec-Dąbrowska, E. Wągrowska-Koski, Medycyna Pracy, **60,** 16 (2010)
15. J. Bugajska, E. Łastowiecka, Int. J. of Occupational Safety and Ergonomics, **12**, 231-240 (2006)
16. J. Ilmarinen, *Wspieranie aktywnego starzenia się w miejscu pracy*, https://osha.europa.eu/pl (date of access: 12.07.2014)
17. G. Richenhagen G., Personalführung, **8**, 44-51 (2007)
18. M. Butlewski, E. Tytyk, K. Wróbel, Macroergonomic model of quality of life of elderly employees for design purposes, In: P. Vink (Ed.), *Advances in Social and Organizational Factors*, 252-260 (AHFE 2014)
19. P. F. Drucker, Sk*uteczne zarządzanie: zadania ekonomiczne a decyzje związane z ryzykiem* (PWN, Warszawa 1976)
20. A. Górny, *Zarządzanie ryzykiem zawodowym* (Wydawnictwo Polietchniki Poznańskiej, Poznań 2011)

21. *Recommendation concerning Older Workers (No. 162,* 66th ILC session (23 Jun 1980) (International Labour Organization (ILO), Geneva 1980)
22. A. Górny, The Elements of Work Environment in the Improvement Process of Quality Management System Structure, In: W. Karwowski, G. Salvendy (Eds.), *Advances in Human Factors, Ergonomics, and Safety in Manufacturing and Service Industries*, 599-606 (AHFE 2011)
23. T. Makowiec-Dąbrowska i in., *Zawodowe i pozazawodowe determinanty zdolności do pracy starszych pracowników*, Raport końcowy PCZ 21-21, Zadanie 8 (CIOP-PIB, Warszawa 2004)
24. L. Laflamme, E. Menckel, Safety Science, **21**, 145-161 (1995)
25. *Wypadki przy pracy w 2014 roku* (GUS, Warszawa 2015)
26. S. A. Gyekye, S. Salminen, JOSE, **16(4)**, 441-443 (2010)
27. *Analiza przyczyn i skutków wypadków przy pracy w latach 2008 - 2011 ze szczególnym uwzględnieniem obszarów, w których wypadki przy pracy powodują największe straty ekonomiczne i społeczne oraz określenie rekomendacji dotyczących działań prewencyjnych* (RESOURCE 2012)
28. S. Ordysiński, Bezpieczeństwo Pracy, **7**, 22-26 (2013)
29. A. Górny, M. Rembiasz, Zeszyty Naukowe Politechniki Poznańskiej, Organizacja i Zarządzanie, **64**, 49-64 (2015)
30. A. Kwiatkiewicz, *Analiza dobrych praktyk dotyczących zarządzania wiekiem w polskich przedsiębiorstwach – studium przypadku* (PARP, Warszawa 2010)
31. M. Rembiasz, Zeszyty Naukowe Uniwersytetu Szczecińskiego, **848**, 139-148 (2015)
32. J. Litwiński, *Opis dobrych praktyk dotyczących zarządzania wiekiem w przedsiębiorstwach polskich oraz innych krajów UE* (PARP, Warszawa 2010)
33. European Observatory of Working Life, http://www.eurofound.europa.eu/areas/ populationandsociety /cases/si004.htm (date of access: 01.05.2015)
34. M. Rembiasz, Aktywność zawodowa osób w wieku emerytalnym – wybrane problemy, In: M. Szczepański (Ed.), *Pension reforms - comparison and evaluation*, 329-339 (Poznan University of Technology, Pozanń 2013)

DOI: 10.1051/matecconf/20179407010

Elbow joint rehabilitation equipment actuated by pneumatic muscles

Georgiana Vetrice[1,*], and *Andrea* Deaconescu[1]

[1]Transilvania University of Braşov, Department of Industrial Engineering and Management, 500036 Braşov, 29 Eroilor Bd., Romania

Abstract. The mobility of the limbs is an essential prerequisite for the individual's physical autonomy. For persons suffering from post-traumatic affections of the elbow joint such limited mobility results in barriers in fulfilling personal or professional tasks. Passive motion has certain specific advantages and beneficial effects, thus being highly recommended for the recovery of injured joints. The paper presents a model of rehabilitation equipment that induces continuous passive motion of the elbow, as part of a recovery programme. The equipment is actuated by pneumatic muscles, using compressed air as the source of energy that generates force and motion. The main benefits of the pneumatic actuation system compared to other driving systems are its low cost, simple and robust construction and swift response to commands. Its constructive simplicity and reduced cost adds to the system's eligibility for deployment in medical units.

1 Introduction

The movements of the upper limbs are essential for the individual. Lack of mobility can have a major impact on social life. In the case of patients suffering from post-traumatic affections of the elbow joint (Fig. 1.), natural recovery of mobility can take up to several months. In order to ensure swift rehabilitation continuous passive motion is recommended, a postoperative treatment protocol which consists in applying a range of slow, rhythmical and continuous movements to the joint, without contracting the patient's muscles. In order to obtain fast results, this procedure can be used from the next day after surgery.

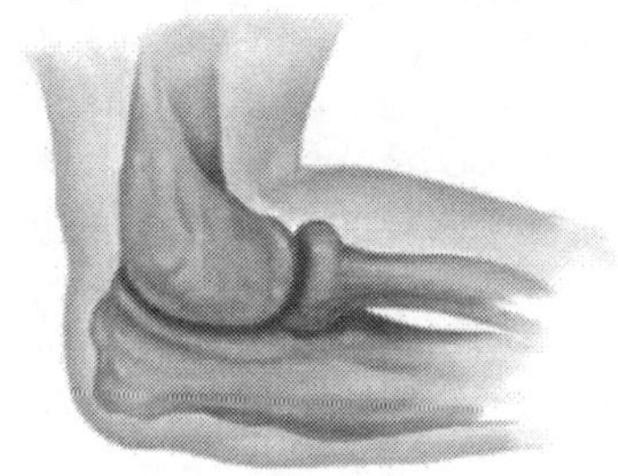

Fig. 1. The elbow joint.

* Corresponding author: vetrice.georgiana@unitbv.ro

The elbow joint is responsible for two movements: flexion-extension and pronation-supination. Flexion is obtained by moving the forearm close to the arm, and extension by moving it away from the arm. Pronation is the rotation of the forearm by its longitudinal axis, so that the thumb is turned towards the body, while supination is the rotation of the forearm by its longitudinal axis, when thumb is turned away from the body. In the wake of surgery, the individual's capacity of conducting these movements may be limited, thus calling for specially designed equipment for the speedy rehabilitation of the elbow joint.

2 Elbow rehabilitation equipment

Using rehabilitation equipment considerably reduces the recovery period and the quantity of drugs consumed. Thus, the total costs entailed by recovery are diminished [1]. Most rehabilitation equipment which is nowadays available on the market is driven by electric motors and has a rigid structure. Several such examples are:

- E3 Elbow CPM Device, (Fig. 2a), is a product of OttoBock company, which allows both flexion-extension and pronation-supination [2];
- Fisiotek LT-G Device, (Fig. 2b) also allows both movements, flexion-extension and pronation-supination [3];
- Kinetec 6080 Elbow CPM Device, (Fig. 2c), performs two types of movements: flexion-extension or flexion-extension with synchronized pronation-supination [4].

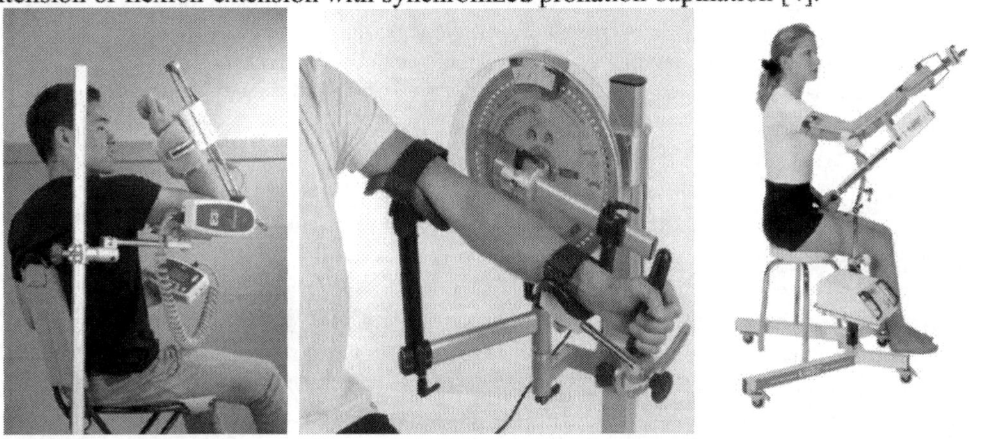

a. E3 Elbow CPM Device **b.** Fisiotek LT-G Device **c.** Kinetec 6080 Elbow CPM Device

Fig. 2. Electrically driven devices.

Another category is that of rehabilitation equipment endowed with pneumatic actuation systems, characterized by the low weight of the components, favourable response to commands, compliance due to air compressibility, and low maintenance. The following pneumatic muscle actuated rehabilitation devices were patented.

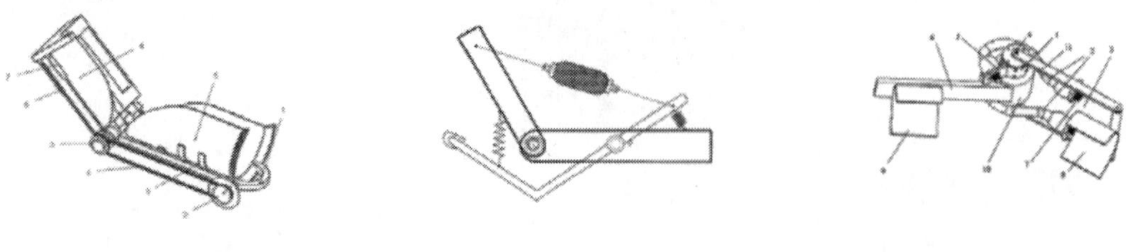

Fig. 3. Pneumatically actuated devices.

Chinese Xi'an Jiaotong University developed a rehabilitation equipment (Fig. 3a) which, by contracting the pneumatic muscle causes the movement of the forearm and thus, of the joint [5].

Jiliang University of China developed a pneumatic muscle actuated joint (Fig. 3b). The pneumatic muscle contraction/relaxation generates the rotation of the left bar around the joint [6].

The patent granted to Zhejiang University of China refers to a device actuated by two pneumatic muscles (Fig. 3c), which rotates the elbow joint and the upper-arm by controlling the air pressure [7].

This paper presents a constructive solution for elbow rehabilitation equipment using continuous passive motion. The absolute novelty consists in its actuation (by means of pneumatic muscles) and in making the joint movement possible by two axes.

3 Working principle of the pneumatic muscles

The pneumatic muscle is an elastic element which, under air pressure, enlarges its diameter and reduces its length. Thus, it carries out a stroke directly proportional with the feed pressure. Fig. 4 first shows the muscle in relaxed state, for zero pressure, and then the contracted muscle for a positive value of the pressure.

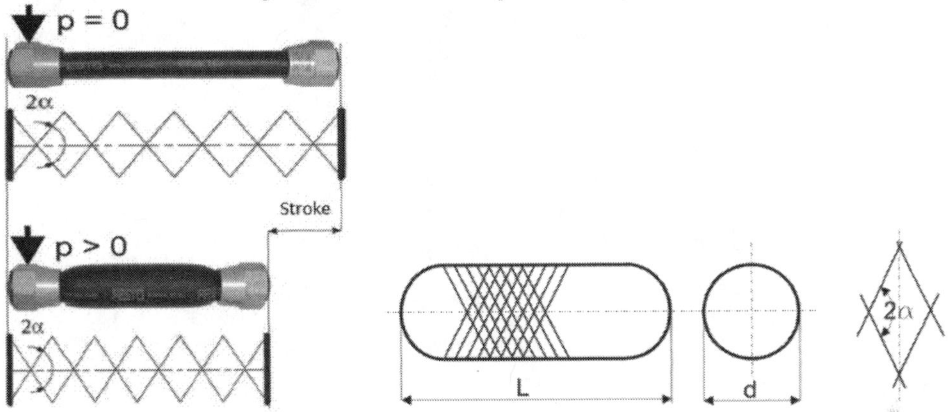

Fig. 4. Working principle of the pneumatic muscle.

An artificial pneumatic muscle includes an interior tube made from an elastic material, usually neoprene. The tube is wrapped in a multilayer tissue made from nylon, in order to give it strength and to protect it from the influences of the working environment. Figure 4 also presents the geometry of the tissue that the interior tube is wrapped in. It can be observed that, in order to perform a stroke, the behaviour of the tissue is similar to the one of a pair of scissors. The behaviour of the pneumatic muscle is similar to that of a spring, meaning that at the end of the stroke the developed force is zero.

If one muscle is not able to achieve the desired movements, various solutions for series or parallel mounting of several muscles can be found.

The main advantages of pneumatic muscles are low price, high power to weight ratio, the possibility of deployment in wet environments, silent operation etc. [8].

In the case of the rehabilitation equipment proposed in this study two pairs of muscles were used, one for each movement performed by the elbow joint. Fig. 5 illustrates the operation of the muscle causing pronation-supination. It can be noticed that the rotation of the joint by means of pneumatic muscles is achieved similarly to the operation of human muscles, based on the agonist-antagonist principle. While a muscle is contracted, the other one is relaxed.

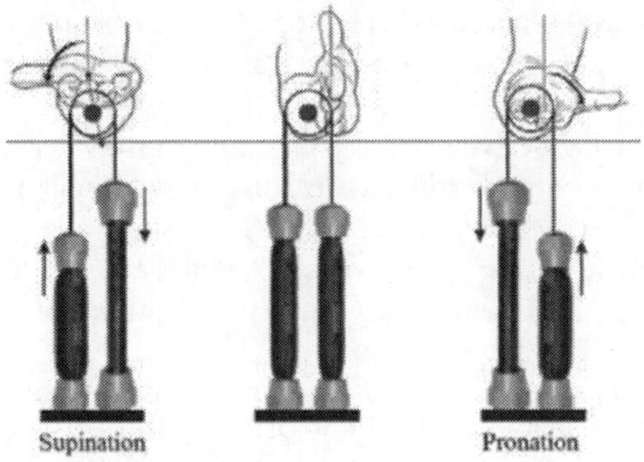

Fig. 5. The operation of the muscle causing pronation-supination.

The same principle is used in the case of flexion-extension (Fig. 6). At zero position, the feed pressure in both muscles is equal; for the movement to take place one muscle is fed compressed air, while the other one relaxes.

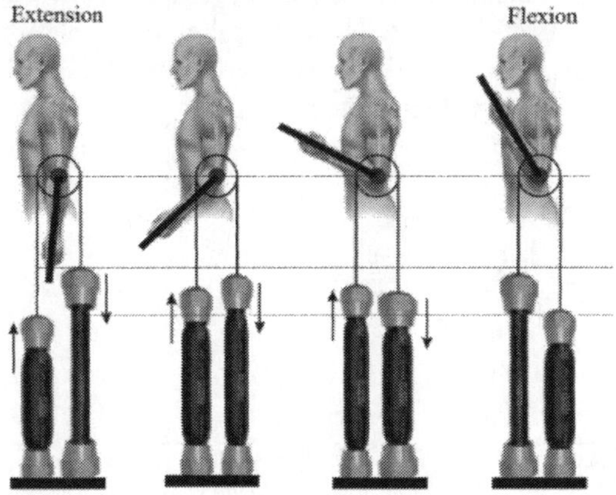

Fig. 6. The operation of the muscle causing flexion-extension.

4 Proposed rehabilitation equipment

Fig. 7 shows the kinematic diagram of the proposed equipment, and also the positioning the pneumatic muscles in relation to the human arm in order to perform the movements necessary for recovery. The first pair of muscles causes flexion-extension, while the second one, placed perpendicularly to the first pair, causes pronation-supination movement. These movements are conducted one at a time, separately and not simultaneously. For the correct deployment of the equipment, the patient needs to be in a comfortable position.

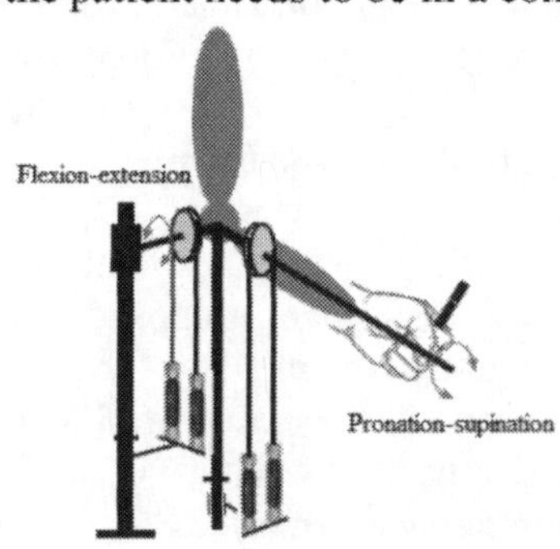

Fig. 7. Position of the patient upper limb in the equipment.

The geometrical model of the elbow rehabilitation equipment (Fig. 8) was developed based on the proposed kinematic diagram.

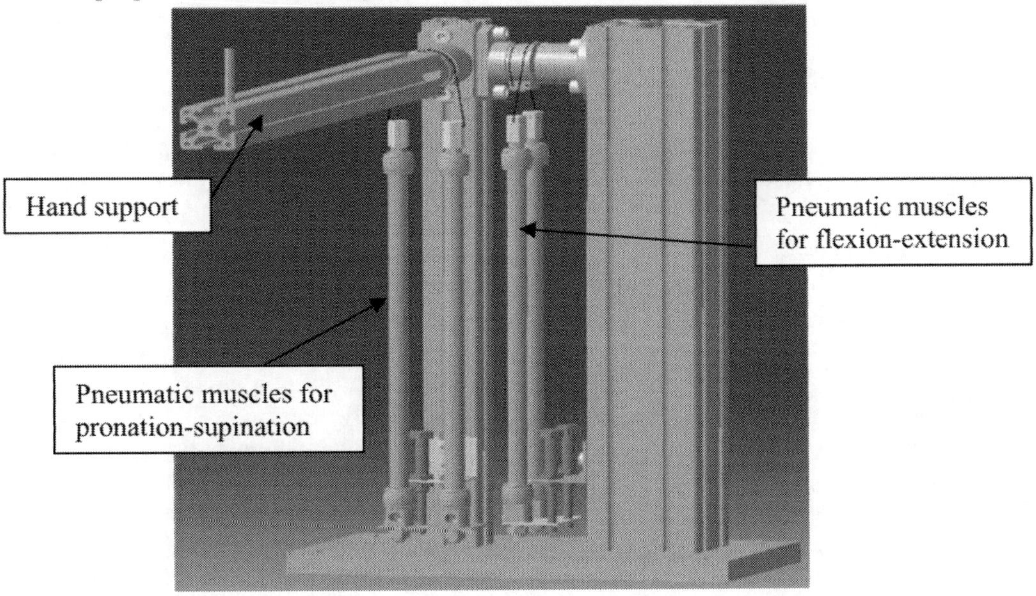

Fig. 8. Construction of the rehabilitation equipment.

The movements made by the hand support are the elbow movements that need to be recovered. For flexion-extension, the rotation of the hand support is generated by using a pair of pneumatic muscles, working in agonist-antagonist mode. For pronation-supination, the movement of the hand support is obtained by rotating it by its longitudinal axis, by means of the second pair of muscles, also working in agonist-antagonist mode.

Fig. 9 presents the construction of the two rotation modules of the rehabilitation equipment.

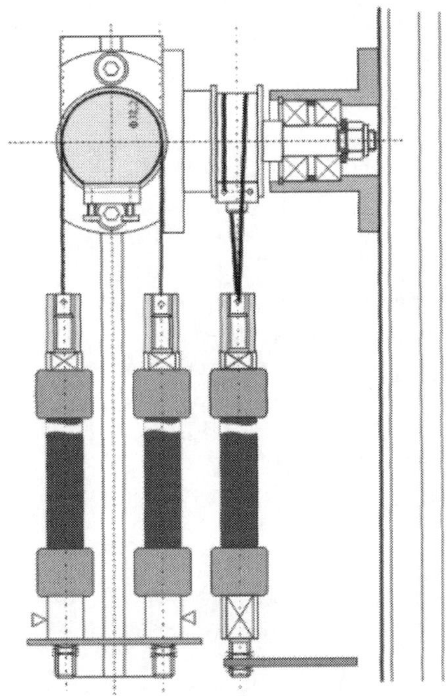

Fig. 9. Rotation modules of the rehabilitation equipment.

As can be noticed, the free ends of the pneumatic muscles are connected by a wire, drawn over a pulley and fixed in a specially designed structure. Thus, by feeding of compressed air, the movement of the pair muscles one against the other causes the rotation of the shaft by its axis and also the movement of the hand support. The working principle is suitable for both flexion-extension and pronation-supination.

The two pairs of pneumatic muscles used in the construction of the equipment are of 10 mm interior diameter and initial length of 300 mm. The maximum possible stroke of the free end of the muscle is approximately 20% of its length in relaxed state (that is 60 mm).

5 Conclusion

Regaining the mobility of the limbs is very important in the case of patients with post-traumatic disorders, such as to enable them to satisfy their needs and achieve their goals. The deployment of rehabilitation equipment aimed at the recovery of the functions responsible for movement is bound to considerably improve patient personal and professional life. Rehabilitation equipment based on continuous passive motion has the advantage of being usable in both medical units and in patients' homes. This is important for them because in this way they have the chance to use it in the comfort of their own homes. It is well-known that most of the devices developed so far for the rehabilitation of patients with post-traumatic disabilities are electrically driven and thus lacking user-friendliness.

This paper presents and discusses the mechanical structure and the actuating system of a piece of innovative elbow rehabilitation equipment for patients with post-traumatic afflictions. The purpose of designing such equipment was explained together with its concept.

In summary, the pneumatic actuating system endows the equipment the advantages of compliance, low cost and operational simplicity. Therefore, the authors feel the pneumatic equipment is eligible to replace the electrical devices. Further on, this new type of equipment will be computer simulated, built and tested, in order to verify the compliant behaviour of these actuators.

References

1. A. Deaconescu, T. Deaconescu, IMECS, **2**, 1823 (2009)
2. Otto Bock, http://www.ottobock.com/ Accessed: 20.04.2016
3. Rimec, http://www.rimec.it/ Accessed: 20.04.2016
4. Kinetec, http://kinetec.fr/en/ Accessed: 20.04.2016
5. J. Liu, Z. Yifang, S. Yingshuang, Z. Yongqiang, Z. Wei; Y. Qingtai, *Elbow medical rehabilitation device based on pneumatic artificial muscle drive*. CN203263743 (2013)
6. Y. Dongming, W. Binrui, J. Yinglian, *Pneumatic muscle flexible elbow joint device with buffer spring and flexible shaft sleeve*. CN202071080 (2011)
7. Y. Canjun, Z. Jiafan, Z. Jie, C. Ying, *Flexible exoskeleton elbow joint based on pneumatic power*. CN200984250 (2007)
8. A. Deaconescu, T. Deaconescu, WCE, **2**, 1516 (2009)

AUTHOR INDEX

AUTHOR INDEX